Grundlehren der mathematischen Wissenschaften 345

A Series of Comprehensive Studies in Mathematics

For further volumes:
www.springer.com/series/138

Tomasz Komorowski • Claudio Landim •
Stefano Olla

Fluctuations in Markov Processes

Time Symmetry and Martingale Approximation

Springer

Tomasz Komorowski
Institute of Mathematics
Maria Curie-Skłodowska University
Lublin, Poland
and
Institute of Mathematics
Polish Academy of Sciences
Warsaw, Poland

Claudio Landim
Instituto de Matemática (IMPA)
Rio de Janeiro, Brazil
and
CNRS UMR 6085
Université de Rouen
Saint-Étienne-du-Rouvray, France

Stefano Olla
CEREMADE, CNRS UMR 7534
Université Paris-Dauphine
Paris, France

ISSN 0072-7830 Grundlehren der mathematischen Wissenschaften
ISBN 978-3-642-29879-0 ISBN 978-3-642-29880-6 (eBook)
DOI 10.1007/978-3-642-29880-6
Springer Heidelberg New York Dordrecht London

Library of Congress Control Number: 2012943039

Mathematics Subject Classification: 60F05, 60J25, 60K35, 60K37, 60H25, 60G60, 35B27, 76M50

Printed on acid-free paper

Springer is part of Springer Science+Business Media (www.springer.com)

To Raghu Varadhan, a continuous source of inspiration

Preface

In statistical mechanics as in many other fields, diffusion phenomena may arise from Markovian stochastic modeling. This has motivated, since the early investigations of Kolmogorov and Doeblin's seminal paper (Doeblin, 1938), an increasing interest in central limit theorems for Markov processes. Doeblin made a close connection between the validity of such theorems and the time mixing properties of the Markov process. Such mixing conditions, nowadays called Doeblin's conditions, are sometimes difficult to verify. In particular many problems in statistical mechanics are intrinsically infinite dimensional, and in that case the Doeblin mixing assumption typically does not hold.

In the 1960s Gordin (1969) introduced another more analytical approach, based on martingale approximation of additive functionals of Markov processes. Consider a Markov process $\{X_t,\, t \geq 0\}$ with an ergodic stationary probability measure $\pi(dx)$, and let $V(x)$ be a function on the state space of the process with mean zero with respect to π. Let L be the generator of the corresponding Markovian semigroup $P_t u(x) = \mathbb{E}_x[u(X_t)]$. Then if u is a function in the domain of L, the process $u(X_t) - u(X_0) - \int_0^t Lu(X_s)ds$ is a martingale. Solving the Poisson equation $Lu = V$ with a *nice* function u in the domain of L, one can express $\int_0^t V(X_s)ds$ as a martingale plus a boundary term, and reduce the problem to proving a central limit theorem for martingales. As we will explain in Chaps. 1 and 2, central limit theorems for martingales requires only ergodicity of the process and a global control of the corresponding quadratic variation.

Of course the possibility of inverting the generator L is related to the mixing properties of the process, but it permits to tailor the conditions to the particular function V we are studying. If L possesses a spectral gap, then the corresponding Poisson equation can be solved for any function V with mean zero.

However in many applications in statistical mechanics, the generator L does not have a spectral gap. The typical infinite dimensional situation deals with a dynamics that has conservation laws and there is an entire family of stationary ergodic measures. The classical example is given by the problem of the macroscopic diffusive behavior of a *tagged particle* in a system of interacting particles. The dynamics of the interacting particles may conserve the density and eventually some other quantities.

The problem of the tagged particle in exclusion processes (studied in detail in Part II of this book) motivated Kipnis and Varadhan to develop a general central limit theorem that exploits the time symmetries of the process. The idea in the article (Kipnis and Varadhan, 1986) can be summarized as follows: the asymptotic variance for $t^{-1/2}\int_0^t V(X_s)ds$ is given by

$$\sigma^2(V) = 2\int_0^\infty \mathbb{E}_\pi\big[V(X_t)V(X_0)\big]\,dt,$$

where $\mathbb{E}_\pi$ is the expectation with respect to the path measure of the process starting from the stationary measure π. If the measure is time-reversible, i.e. the generator L is self-adjoint in $L^2(\pi)$, then $\mathbb{E}_\pi[V(X_t)V(X_0)] \geq 0$ for all $t \geq 0$ and $\sigma^2(V)$ will be finite if this time correlation decays sufficiently fast. The condition $\sigma^2(V) < +\infty$ is much weaker than asking that V belongs to the range of L, and it translates to an integrability condition on the corresponding spectral measure (supported on the real line) around zero that is relatively easy to verify. A simple argument using this spectral measure shows that the martingale approximation can be done using the resolvent solution $u_\lambda = (\lambda - L)^{-1}V$, even when V is outside the range of L. In the introductory Chap. 1 we will explain this approach in detail for a simple example of a reversible (discrete time) Markov chain.

Some of the ideas discussed above were already present in the literature dealing with the problem of homogenization of diffusions in stationary ergodic random environments (cf. Kozlov 1979; Papanicolaou and Varadhan 1981), but it is in the aforementioned work of Kipnis and Varadhan (1986) that the connection with reversibility has been exploited fully and the finiteness of the variance $\sigma^2(V)$ has been formulated as a sufficient condition for the central limit theorems for reversible Markov processes.

Later on the theory has been extended to certain classes of *non-reversible* Markov processes, i.e. Markov processes with a stationary ergodic (but non-reversible) probability measure π. When this measure is explicitly known, the generator can be decomposed as the sum of a symmetric and an anti-symmetric operator in $L^2(\pi)$, $L = S + A$. If for a sufficiently large set C, that is a common core of L and S, there exists a finite constant K such that

$$\left(\int f L g\,d\pi\right)^2 \leq K \int f(-S)f\,d\pi \int g(-S)g\,d\pi, \quad f, g \in C$$

then we say that L satisfies a *sector condition*. The name illustrates the fact that the spectrum of L, in general a subset of the complex plane, is contained now in a cone around the negative reals touching the imaginary axis only at 0, where the constant K is the tangent of the corresponding semiangle of the cone. There are many interesting examples of Markov processes that satisfy the sector condition: asymmetric exclusion processes with null average jump rate (studied in Sect. 5.3), *cyclic* random walks in random environment (Sect. 3.3), doubly stochastic random walks in one dimension (Sect. 3.6), diffusions with the random generator in a divergence form (Chap. 9).

Another class of processes where the theory can be extended is provided by Markov processes with *normal* generators L and their bounded perturbations (Sect. 2.7.5). This condition allows to deal with examples like random walks, or diffusions in time dependent random environment (Sect. 9.9).

A further extension concerns processes where the generator L satisfies a *graded sector condition*: this is the case where the space $L^2(\pi)$ can be decomposed into a direct sum of orthogonal subspaces, $L^2(\pi) = \bigoplus_{n\geq 0} \mathscr{A}_n$, and L satisfies a sector condition on each subspace with a constant K_n eventually growing to infinity not too fast (see Sect. 2.7.4 for the precise formulation of this condition). Examples include asymmetric exclusion processes in dimension $d \geq 3$ (Sect. 5.5, this example motivated the extension), and diffusions with Gaussian drifts (Chap. 12).

We also present some examples that go beyond sector type conditions:

- *diffusions in divergence free fields* with the stream matrix that is square integrable (i.e. finite Péclet number), can be dealt with by an approximation procedure (Chap. 11);
- *doubly stochastic random walks* in space-mixing environment in dimension 3 or higher, where an approximation procedure can be implemented (Sect. 3.5);
- *Ornstein–Uhlenbeck process in a random potential* (position-velocity Langevin diffusion). This is a very degenerate diffusion (noise acting only on the velocity), but time symmetry of the Gibbs measure can be exploited by changing sign of the velocity in the reversed process (see Chap. 13).

All these examples of the central limit theorem for processes with the generator not satisfying any sector condition exploit particular features of the dynamics. What is missing is a general theorem for non-reversible Markov processes. We expected the validity of the central limit theorem for any Markov process with stationary, ergodic measure π and a function V such that $(-S)^{-1/2}V$ belongs to $L^2(\pi)$. We are not aware of any counterexamples to this statement.

Description of the Content of This Book Part I concerns the general theory, and could be used as the base for a graduate course. Only some basic probability is required, in particular ergodic theorems and martingales. In the first chapter we expose the central limit theorem for a countable state space, discrete time, reversible Markov chains. This is the most elementary, non-trivial set-up where we can illustrate the basic ideas without spending much time on technicalities. In Chap. 2 we develop the theory for general continuous time Markov processes, and introduce the various sector conditions. In Chap. 3 we apply the theory to random walks in random environment on $\mathbb{Z}^d$, that are nice and relatively simple, although at the same time quite non-trivial examples of the general theory. We conclude this part of the book with Chap. 4 that contains estimates and variational formulas for the asymptotic variance $\sigma^2(V)$, which turn out to be quite useful in applications. In fact in some examples the central limit theorem follows by the application of the general theory, while proving strict positivity of the variance requires some extra effort.

Part II is completely dedicated to central limit theorems for exclusion processes, the problem which motivated the development of the theory presented in this book.

In Chap. 5, we introduce the simple exclusion process, prove the main properties of its generator, and examine central limit theorems for additive functionals. In the case where the jump rates of the particles are symmetric, this result is a straightforward consequence of the first part of the book since the generator is self-adjoint. If the jump rates have mean zero, the generator of the exclusion process satisfies a sector condition and we may still apply the general theory.

In the asymmetric case, the picture is different. The duality introduces an orthogonal decomposition of the L^2 space which permits to consider the central limit theorem from the perspective of the graded sector condition. However, one needs to assume that the dimension is larger than or equal to three to prove that the asymmetric part of the generator which changes the degree of a function satisfies hypothesis (2.45) assumed in the general theory. Moreover, the asymmetric part of the generator which keeps the degrees of the functions does not satisfy condition (2.50). To overcome this obstacle a method, known as the removal of the hard core interaction, has been developed and is presented in Theorem 5.19.

To proceed step by step, increasing progressively the level of difficulty, we first present a central limit theorem for additive functionals of asymmetric exclusion processes in dimension $d \geq 3$ when the density of particles is equal to $1/2$, in which case the asymmetric part of the generator which keeps the degree of the functions does not appear. In the following section we examine the full asymmetric case in $d \geq 3$. In the last section of this chapter we present some results on transient Markov chains needed in the chapter and which have intrinsic interest.

In the following two chapters, we extend the central limit theorem to the case of a tagged particle, deriving the self-diffusion coefficient of exclusion processes, and to the case of a second class particle which is connected to the equilibrium fluctuations from the hydrodynamic limit of the empirical measure, Kipnis and Landim (1999).

In the last chapter of this second part of the book we prove that the asymptotic variance depends smoothly on the density of particles. In particular, we show that the self-diffusion and the bulk diffusion coefficients are smooth functions, a property which has important consequences in the theory of hydrodynamic limits.

Part III deals entirely with diffusions in random environments. Chapter 9 contains the main applications of the general theory. It starts with the periodic environments, where a spectral gap is present in the dynamics. In the quasi-periodic case we illustrate the loss of compactness and of the spectral gap property. This motivates the study of general ergodic random environments. Chapter 10 contains variational principles for the *homogenized diffusion matrix*, while the following chapters contain some other applications and extensions: divergence free drifts (Chap. 11, where the homogenized diffusion is enhanced by the microscopic convection), Gaussian drifts (see Chap. 12), where we have included also the discussion on superdiffusion effect due to large convection. Chapter 13 deals with the Ornstein–Uhlenbeck process in a random potential. The final Chap. 14 is dedicated to the relation of this probabilistic approach with the classical analytic homogenization theory and the notions of G-convergence of operators and Γ-convergence of quadratic forms.

There are many problems and results related to this theory that we have not included:

- The theory is about Markov processes in a stationary, ergodic state, and is tailor made for situations where there can be many ergodic measures. So the central limit theorem obtained refers to the particular ergodic measure chosen, the diffusion coefficient may depend on it, and the convergence of the laws are *in probability* with respect to the initial chosen ergodic measure. There are many *almost sure* results for diffusions in random environments, and it remains an open problem in the case of interacting particles systems, like for the self-diffusion of the tagged particles (Part II), even in the reversible case.
- For the same reason we do not deal with non-stationary problems, or locally ergodic environments etc. Some limited results in these directions do exist in the literature.

Tomasz Komorowski expresses his gratitude for hospitality to CIRM (Luminy), Université Paris-Dauphine and Université de Nice-Sophia Antipolis. During the course of writing the book he has been supported by Polish Ministry of Higher Education grants NN 201 419139, NN 201 045 31/3732 and Nr. 2 PO3A 031 23, as well as EC FP6 Marie Curie ToK programme SPADE2.

Claudio Landim wishes to thank the Scuola Normale de Pisa, the Polish Academy of Sciences at Warsaw, the University of Rome, La Sapienza, and the University of Tokyo where parts of this book were written. He has been supported by a fellowship from the John S. Guggenheim Memorial Foundation, by the grants Cientistas do Nosso Estado, FAPERJ, E-26 102.774/2008, Bolsa de Produtividade CNPq, and by the French Ministry of Education through the grants ANR-2007-BLAN-2-184264 (LHMSHE) and ANR-2010-BLAN-0108-03 (SHEPI).

Stefano Olla has been supported by the European Advanced Grant *Macroscopic Laws and Dynamical Systems* (MALADY) (ERC AdG 246953), and by the French Ministry of Education through the grant ANR-07-BLAN-2-184264 (LHMSHE).

We thank the CIRM (Luminy), the Polish Academy of Sciences (Warsaw), the Centre Borel of the IHP (Paris), for their kind hospitality and support while we worked on this book.

Many colleagues and friends encouraged us and helped with suggestions and corrections during the long years we have been working on this project. It is impossible to mention all of them here, but we thank them warmly. A special thank to Kenkichi Tsunoda and Lu Xu of Tokyo University, for the careful checking of the first two parts of the manuscript.

Lublin, Poland — Tomasz Komorowski
Rio de Janeiro, Brazil — Claudio Landim
Paris, France — Stefano Olla

Contents

Part I
General Theory

Chapter 1
A Warming-Up Example

The purpose of this chapter is to present, in the simplest possible context, some of the ideas that will appear recurrently in this book. We assume that the reader is familiar with the basic theory of Markov chains (e.g. Chap. 7 of Breiman 1968 or Chap. 5 of Durrett 1996) and with the spectral theory of bounded symmetric operators (Sect. 107 in Riesz and Sz.-Nagy 1990, Sect. XI.6 in Yosida 1995).

Consider a Markov chain $\{X_j : j \geq 0\}$ on a countable state space E, stationary and ergodic with respect to a probability measure π. The problem is to find necessary and sufficient conditions on a function $V: E \to \mathbb{R}$ to guarantee a central limit theorem for

$$\frac{1}{\sqrt{N}} \sum_{j=0}^{N-1} V(X_j). \tag{1.1}$$

We assume that $E_\pi[V] = 0$, where E_π stands for the expectation with respect to the probability measure π. The idea is to relate this question to the well-known martingale central limit theorems.

Denote by P the transition probability of the Markov chain and fix a function V in $L^2(\pi)$, the space of functions $f : E \to \mathbb{R}$ square integrable with respect to π. Assume the existence of a solution of the Poisson equation

$$V = (I - P)f \tag{1.2}$$

for some function f in $L^2(\pi)$, where I stands for the identity. For $j \geq 1$, let

$$Z_j = f(X_j) - (Pf)(X_{j-1}).$$

It is easy to check that $M_0 = 0$, $M_N = \sum_{1 \leq j \leq N} Z_j$, $N \geq 1$, is a martingale with respect to the filtration $\{\mathscr{F}_j : j \geq 0\}$, $\mathscr{F}_j = \sigma(X_0, \ldots, X_j)$, and that

$$\sum_{j=0}^{N-1} V(X_j) = M_N - f(X_N) + f(X_0). \tag{1.3}$$

T. Komorowski et al., *Fluctuations in Markov Processes*,
Grundlehren der mathematischen Wissenschaften 345,
DOI 10.1007/978-3-642-29880-6_1,

Since f is square integrable, the last two terms divided by $N^{1/2}$ vanish as $N \uparrow \infty$ and the central limit theorem for $N^{-1/2} \sum_{0 \le j < N} V(X_j)$ follows from the central limit theorem for the martingale M_N.

If the Markov chain has good *mixing* properties which guarantee the convergence of the series $\sum_{j \ge 0} P^j V$ in $L^2(\pi)$, the Poisson equation (1.2) has a solution given by $f = \sum_{j \ge 0} P^j V$. Unfortunately, as will be seen in this book, this is not the typical situation. Nevertheless, the central limit theorem can be established with weaker conditions, approximating V, in a proper norm, by functions in the range of $I - P$.

The optimal situation is achieved in the special case where the invariant state π is *reversible*. It is shown in Theorem 1.10, the main result of this chapter, that in this case the finiteness of the limit variance

$$\sigma^2(V) = \lim_{N \to \infty} \frac{1}{N} \mathbb{E}\Bigg[\bigg(\sum_{j=0}^{N-1} V(X_j)\bigg)^2\Bigg] \tag{1.4}$$

is a necessary and sufficient condition for a central limit theorem for (1.1). *Non-reversible* chains require a deeper analysis, presented in the next chapter.

The material is organized as follows. In Sect. 1.1 we introduce some terminology and prove a few elementary facts on Markov chains on countable state spaces. In the second section, we prove a central limit theorem for the sequence $N^{-1/2} \sum_{0 \le j < N} V(X_j)$ assuming that the solution of the Poisson equation (1.2) belongs to $L^2(\pi)$. In Sect. 1.3 we prove a central limit theorem for a stationary and ergodic sequence of random variables whose partial sums form a square integrable martingale. In the fourth section we obtain necessary and sufficient conditions for the limit variance (1.4) to be finite. This computation leads us to introduce some Hilbert spaces associated to the transition probability of the Markov chain which are examined in detail in Sect. 1.6. In Sect. 1.5 we prove a central limit theorem for the sequence $N^{-1/2} \sum_{0 \le j < N} V(X_j)$ showing that this sum can be approximated by a martingale.

1.1 Ergodic Markov Chains

In this section, we present some elementary results on Markov chains. Fix a countable state space E and a transition probability function $P\colon E \times E \to \mathbb{R}$:

$$P(x,y) \ge 0, \quad x, y \in E, \qquad \sum_{y \in E} P(x,y) = 1, \quad x \in E.$$

A sequence of random variables $\{X_j : j \ge 0\}$ defined on some probability space $(\Omega, \mathscr{F}, \mathbb{P})$ and taking values in E is a time-homogeneous Markov chain on E if

$$\mathbb{P}[X_{j+1} = y | X_j, \ldots, X_0] = P(X_j, y) \tag{1.5}$$

for all $j \geq 0$, y in E. $P(x, y)$ is called the probability of jump from x to y in one step. Notice that it does not depend on time, which explains the terminology of a time-homogeneous chain. The law of X_0 is called the initial state of the chain.

Assume furthermore that on $(\Omega, \mathscr{F})$ we are given a family of measures $\mathbb{P}_z$, $z \in E$, each satisfying (1.5) and such that $\mathbb{P}_x[X_0 = x] = 1$. We call it a Markov family that corresponds to the transition probabilities $P(\cdot, \cdot)$. For a given probability measure μ on E, let $\mathbb{P}_\mu = \sum_{x \in E} \mu(x)\mathbb{P}_x$. Observe that μ is the initial state of the chain under $\mathbb{P}_\mu$. We shall denote by $\mathbb{E}_\mu$ the expectation with respect to that measure and by $\mathbb{E}_x$ the expectation with respect to $\mathbb{P}_x$.

The transition probability P can be considered as an operator on $C_b(E)$, the space of (continuous) bounded functions on E. In this case, for f in $C_b(E)$, $Pf: E \to E$ is defined by

$$(Pf)(x) = \sum_{y \in E} P(x, y) f(y) = \mathbb{E}\big[f(X_1) | X_0 = x\big]. \tag{1.6}$$

We use the same notation P for the transition probability and for the operator on $C_b(E)$. In the countable case, we can think of Pf as the product of the square matrix P with the column vector f.

Let μP be the state of the process at time 1 if the Markov chain starts from μ, i.e., the distribution at time 1 of the process starting at time 0 from μ:

$$(\mu P)(x) = \mathbb{P}_\mu[X_1 = x] = \sum_{y \in E} \mu(y) P(y, x).$$

In the countable case, we can think of μP as the product of the line vector μ with the square matrix P.

For $n \geq 1$, we denote by P^n the n-fold composition of P with itself so that

$$P^n(x, y) = \sum_{x_1, \ldots, x_{n-1} \in E} P(x, x_1) P(x_1, x_2) \cdots P(x_{n-2}, x_{n-1}) P(x_{n-1}, y)$$

for all x, y in E. In particular,

$$\big(P^n f\big)(x) = \mathbb{E}_x\big[f(X_n)\big], \qquad \big(\mu P^n\big)(x) = \sum_{y \in E} \mu(y) \mathbb{P}_y[X_n = x]$$

for all bounded functions f and all probability measures μ. Hence, μP^n stands for the state of the process at time n if it starts from μ. By convention we let $P^0 = I$—the identity operator.

A probability measure π is said to be stationary or invariant for the chain if $\pi P = \pi$. This happens if and only if the sequence of random variables $\{X_j : j \geq 0\}$ is stationary under $\mathbb{P}_\pi$. We do not assume the chain to be indecomposable. There might exist, in particular, more than one invariant measure. Denote by E_π the expectation with respect to π, not to be confused with $\mathbb{E}_\pi$—the expectation with respect to $\mathbb{P}_\pi$.

We say that an invariant measure π is ergodic if any bounded function f satisfying $(I - P)f = 0$ is π-almost everywhere constant. It can be shown, see e.g.

the proof of Theorem 7.16 in Breiman (1968), that π is ergodic if and only if the sequence of random variables $\{X_j : j \geq 0\}$ is ergodic when considered over $(\Omega, \mathscr{F}, \mathbb{P}_\pi)$.

We may extend the domain of definition of the operator P given in (1.6) to $L^2(\pi)$, the space of π-square integrable functions. It is indeed clear, by Schwarz inequality, that Pf defined by (1.6) belongs to $L^2(\pi)$ if f does since

$$\begin{aligned}\sum_{x\in E} \pi(x)\big[(Pf)(x)\big]^2 &= \sum_{x\in E} \pi(x)\Big\{\sum_{y\in E} P(x,y)f(y)\Big\}^2 \\ &\leq \sum_{x\in E} \pi(x) \sum_{y\in E} P(x,y)f(y)^2 = \sum_{y\in E} (\pi P)(y) f(y)^2 \\ &= \sum_{y\in E} \pi(y) f(y)^2\end{aligned}$$

because π is invariant. We have thus proved that P is a contraction in $L^2(\pi)$:

$$\langle Pf, Pf\rangle_\pi \leq \langle f, f\rangle_\pi, \tag{1.7}$$

where $\langle\cdot,\cdot\rangle_\pi$ stands for the scalar product in $L^2(\pi)$. Let $\|\cdot\|$ be the norm associated to the scalar product $\langle\cdot,\cdot\rangle_\pi$.

1.2 Almost Sure Central Limit Theorem for Ergodic Markov Chains

Consider a time-homogeneous irreducible (or indecomposable in the terminology of Breiman 1968) Markov chain $\{X_j : j \geq 0\}$ on a countable state space E with transition probability function $P : E \times E \to \mathbb{R}_+$. Assume that there exists a stationary probability measure, denoted by π. By (Breiman, 1968, Theorem 7.16), π is unique and ergodic. In particular, for any bounded function $g : E \to \mathbb{R}$ and any x in E,

$$\lim_{N\to\infty} \frac{1}{N} \sum_{j=0}^{N-1} \big(P^j g\big)(x) = E_\pi[g].$$

Fix a function $V : E \to \mathbb{R}$ in $L^2(\pi)$ which has mean zero with respect to π. In this section, we prove a central limit theorem for the sequence $N^{-1/2}\sum_{j=0}^{N-1} V(X_j)$ assuming that the solution of the Poisson equation (1.2) belongs to $L^2(\pi)$. Under this hypothesis we obtain a central limit theorem which holds π-a.s. with respect to the initial state.

Theorem 1.1 *Fix a function $V : E \to \mathbb{R}$ in $L^2(\pi)$ which has mean zero with respect to π. Assume that there exists a solution f in $L^2(\pi)$ of the Poisson equation* (1.2).

Then, for all x in E, as $N \uparrow \infty$,

$$\frac{1}{\sqrt{N}} \sum_{j=0}^{N-1} V(X_j)$$

converges in $\mathbb{P}_x$ distribution to a mean zero Gaussian random variable with variance $\sigma^2(V) = E_\pi[f^2] - E_\pi[(Pf)^2]$.

Proof Fix a mean zero function V in $L^2(\pi)$ and an initial state x in E. By assumption, there exists a solution f in $L^2(\pi)$ of the Poisson equation (1.2). Consider the sequence $\{Z_j : j \ge 1\}$ of random variables defined by

$$Z_j = f(X_j) - Pf(X_{j-1}). \tag{1.8}$$

The sequence $\{Z_j : j \ge 1\}$ is adapted to the natural filtration of the Markov chain $\mathscr{F}_j = \sigma(X_0, \dots, X_j)$. Let $M_0 = 0$, $M_j = \sum_{1\le k\le j} Z_k$, $j \ge 1$. A simple computation shows that $\{M_j : j \ge 0\}$ is a martingale adapted to the filtration $\{\mathscr{F}_j : j \ge 0\}$.

Assume first that the solution f of the Poisson equation (1.2) is bounded. In this case, the random variables $\{Z_j : j \ge 1\}$ are bounded. Thus, for $|\theta|$ small enough, we may define

$$A_j(\theta) := \log \mathbb{E}_x\big[\exp\{i\theta Z_j\} | \mathscr{F}_{(j-1)}\big]. \tag{1.9}$$

Fix θ in $\mathbb{R}$. An elementary computation shows that for all N large enough,

$$\mathbb{E}_x\left[\exp\left\{(i\theta/\sqrt{N})M_N - \sum_{j=1}^{N} A_j(\theta/\sqrt{N})\right\}\right] = 1.$$

It follows from a second order Taylor expansion that

$$\sum_{j=1}^{N} A_j(\theta/\sqrt{N}) = -\frac{\theta^2}{2N}\sum_{j=1}^{N} \mathbb{E}_x\big[Z_j^2 | \mathscr{F}_{(j-1)}\big] + \frac{1}{\sqrt{N}} R_N,$$

for some random variable R_N bounded above by a constant. Since

$$\mathbb{E}_x\big[Z_j^2 | \mathscr{F}_{(j-1)}\big] = (Pf^2)(X_{j-1}) - (Pf)^2(X_{j-1}),$$

by the ergodic theorem, as $N \uparrow \infty$, $\sum_{1\le j\le N} A_j(\theta/\sqrt{N})$ converges $\mathbb{P}_x$-a.s. to $-(\theta^2/2)\, E_\pi[(Pf^2) - (Pf)^2] = -(\theta^2/2) E_\pi[f^2 - (Pf)^2]$. In particular,

$$\lim_{N\to\infty} \mathbb{E}_x\big[\exp\big\{(i\theta/\sqrt{N})M_N\big\}\big] = e^{-\theta^2\sigma^2/2},$$

where $\sigma^2 = E_\pi[f^2 - (Pf)^2]$. The central limit theorem for $(1/\sqrt{N}) \times \sum_{0\le j<N} V(X_j)$ follows from this result and identity (1.3).

Assume now that f belongs to $L^2(\pi)$. Let $\{f_n : n \geq 1\}$ be a sequence of bounded functions which converge to f in $L^2(\pi)$. For a fixed $n \geq 1$, let $Z_j^{(n)} = f_n(X_j) - Pf_n(X_{j-1})$, $j \geq 1$, and let $M_0^{(n)} = 0$, $M_j^{(n)} = \sum_{1\leq k\leq j} Z_k^{(n)}$ be the martingale associated to the sequence $\{Z_j^{(n)} : j \geq 1\}$. By the first part of the proof, for every $n \geq 1$,

$$\lim_{N\to\infty} \mathbb{E}_x\big[\exp\big\{(i\theta/\sqrt{N})M_N^{(n)}\big\}\big] = e^{-\theta^2\sigma_n^2/2},$$

where $\sigma_n^2 = E_\pi[f_n^2 - (Pf_n)^2]$.

Since $E_\pi[f_n^2 - (Pf_n)^2]$ converges to $E_\pi[f^2 - (Pf)^2]$ as $n \uparrow \infty$, to conclude the proof of the theorem we need to show that

$$\lim_{n\to\infty} \limsup_{N\to\infty} \big|\mathbb{E}_x\big[\exp\big\{(i\theta/\sqrt{N})M_N^{(n)}\big\}\big] - \mathbb{E}_x\big[\exp\big\{(i\theta/\sqrt{N})M_N\big\}\big]\big| = 0.$$

Since $|\exp\{ix\} - \exp\{iy\}| \leq |x-y|$, the previous difference is absolutely bounded

$$\frac{\theta}{\sqrt{N}}\mathbb{E}_x\big[\big|M_N^{(n)} - M_N\big|\big] \leq \left\{\frac{\theta^2}{N}\mathbb{E}_x\big[\big(M_N^{(n)} - M_N\big)^2\big]\right\}^{1/2},$$

where we used Schwarz inequality in the last step. Recalling the representation of the martingales M_N, $M_N^{(n)}$ in terms of the sequences $\{Z_j : j \geq 1\}$, $\{Z_j^{(n)} : j \geq 1\}$, by orthogonality of these variables in $L^2(\mathbb{P}_x)$, the expression inside braces is equal to

$$\frac{\theta^2}{N}\sum_{j=1}^{N} \mathbb{E}_x\big[\big(Z_j^{(n)} - Z_j\big)^2\big].$$

Let $F_n = f_n - f$ so that $Z_j^{(n)} - Z_j = F_n(X_j) - PF_n(X_{j-1})$. With this notation the previous sum becomes

$$\frac{\theta^2}{N}\sum_{j=1}^{N} \big\{\big(P^j F_n^2\big)(x) - \big[P^{j-1}(PF_n)^2\big](x)\big\}.$$

By the ergodic theorem this average converges to $\theta^2 E_\pi[F_n^2 - (PF_n)^2]$. This expression vanishes as $n \uparrow \infty$ because F_n converges to 0 in $L^2(\pi)$. This proves the central limit theorem for the martingale M_N and the theorem in view of identity (1.3). □

In general there is no solution of the Poisson equation in $L^2(\pi)$, but one still expects a central limit theorem for $N^{-1/2}\sum_{j=0}^{N-1} V(X_j)$ if its variance remains finite. We prove such a result in the following sections under the assumption of reversibility of the stationary measure π. The approach relies on a central limit theorem for martingales presented below.

1.3 Central Limit Theorem for Martingales

Fix a probability space $(\Omega, \mathscr{F}, \mathbb{P})$ and an increasing filtration $\{\mathscr{F}_j : j \geq 0\}$. Denote by $\mathbb{E}$ the expectation with respect to the probability measure $\mathbb{P}$. Let $\{Z_j : j \geq 1\}$ be a stationary and ergodic sequence of random variables adapted to the filtration $\{\mathscr{F}_j\}$ and such that

$$\mathbb{E}\big[Z_1^2\big] < \infty, \qquad \mathbb{E}[Z_{j+1}|\mathscr{F}_j] = 0, \quad j \geq 0. \tag{1.10}$$

The variables $\{Z_j : j \geq 1\}$ are usually called martingale differences because the process $\{M_j : j \geq 0\}$ defined as $M_0 := 0$, $M_j := \sum_{1\leq k\leq j} Z_k$, $j \geq 1$, is a zero-mean, square integrable martingale with respect to the filtration $\{\mathscr{F}_j : j \geq 0\}$.

Theorem 1.2 *Let $\{Z_j : j \geq 1\}$ be a sequence of stationary, ergodic random variables satisfying* (1.10). *Then, $N^{-1/2}\sum_{1\leq j\leq N} Z_j$ converges in distribution, as $N \uparrow \infty$, to a Gaussian law with zero mean and variance $\sigma^2 = \mathbb{E}[Z_1^2]$.*

Proof If one assumes that the martingale differences $\{Z_j\}$ are bounded, the proof is elementary and follows from the ergodic assumption. Suppose therefore that $|Z_1| \leq C_0$, $\mathbb{P}$-a.s. for some finite constant C_0.

We first build exponential martingales. Since $\{Z_j\}$ are martingale differences, $\mathbb{E}[\sum_{j+1\leq k\leq j+K} Z_k|\mathscr{F}_j] = 0$ for all $j \geq 0$, $K \geq 1$. Therefore, since $|e^{ix} - 1 - ix| \leq x^2/2$, $x \in \mathbb{R}$, subtracting $\mathbb{E}[i\theta\sum_{j+1\leq k\leq j+K} Z_k|\mathscr{F}_j]$ from the expression on the left-hand side in the next formula we obtain that

$$\left|\mathbb{E}\left[\exp\left\{i\theta\sum_{k=j+1}^{j+K} Z_k\right\}\Big|\mathscr{F}_j\right] - 1\right| \leq \frac{\theta^2}{2}\mathbb{E}\left[\left(\sum_{k=j+1}^{j+K} Z_k\right)^2\Big|\mathscr{F}_j\right].$$

Since the variables $\{Z_j\}$ are martingale differences, we may replace $(\sum_k Z_k)^2$ by $\sum_k Z_k^2$ in the above conditional expectation to obtain that this expression is bounded above by $(\theta C_0)^2 K/2$. The left-hand side of the previous displayed equation is thus bounded by $1/2$ if $|\theta| \leq 1/\sqrt{K}C_0$. We can therefore define for θ in this range and for $j \geq 1$ the compensator

$$A_j(\theta) := \log\mathbb{E}\left[\exp\left\{i\theta\sum_{k=(j-1)K+1}^{jK} Z_k\right\}\Big|\mathscr{F}_{(j-1)K}\right],$$

with the usual definition of logarithm $\log(1+z) := z - z^2/2 + z^3/3 - \cdots$ valid for $|z| < 1$.

Fix $1 \ll K \ll N$. This means that K increases to infinity after N. Let $m_j := M_{jK}$ for $j \geq 0$. Clearly, for every admissible θ, $\exp\{i\theta m_j - \sum_{1\leq k\leq j} A_k(\theta)\}$ is a mean one exponential martingale with respect to the filtration $\{\mathscr{F}_{jK} : j \geq 0\}$.

Assume without loss of generality that $N = \ell K$ for some integer ℓ and fix $\theta \in \mathbb{R}$. An elementary third order Taylor expansion shows that for N sufficiently large

$$\sum_{j=1}^{\ell} A_j(\theta/\sqrt{N}) = -\frac{\theta^2}{2N}\sum_{j=1}^{\ell}\mathbb{E}\Bigg[\sum_{k=(j-1)K+1}^{jK} Z_k^2\Big|\mathscr{F}_{(j-1)K}\Bigg] + \frac{K^2}{\sqrt{N}}R_{N,K}$$

for some random variables $R_{N,K}$ that can be deterministically bounded independently of N and K. Since $\{\exp\{i\theta m_j - \sum_{1\le k\le j} A_k(\theta)\} : j \ge 0\}$ is a mean one exponential martingale, for any θ in $\mathbb{R}$ and N sufficiently large we obtain

$$\begin{aligned}
1 &= \mathbb{E}\Bigg[\exp\Bigg\{i(\theta/\sqrt{N})m_\ell - \sum_{k=1}^{\ell} A_k(\theta/\sqrt{N})\Bigg\}\Bigg] \\
&= \mathbb{E}\Bigg[\exp\Bigg\{\frac{i\theta}{\sqrt{N}}M_N + \frac{\theta^2}{2N}\sum_{j=0}^{\ell-1}\mathbb{E}\Bigg[\sum_{k=jK+1}^{(j+1)K} Z_k^2\Big|\mathscr{F}_{jK}\Bigg] - \frac{K^2}{\sqrt{N}}R_{N,K}\Bigg\}\Bigg].
\end{aligned}$$

We prove below that

$$\lim_{K\to\infty}\sup_{\ell\ge 1}\mathbb{E}\Bigg[\Bigg|\exp\Bigg\{\frac{\theta^2}{2N}\sum_{j=0}^{\ell-1}\mathbb{E}\Bigg[\sum_{k=jK+1}^{(j+1)K}\big(Z_k^2-\sigma^2\big)\Big|\mathscr{F}_{jK}\Bigg]\Bigg\} - 1\Bigg|\Bigg] = 0. \tag{1.11}$$

Therefore, since $R_{N,K}$ are uniformly bounded random variables, for every θ in $\mathbb{R}$,

$$\lim_{N\to\infty}\mathbb{E}\Bigg[\exp\Bigg\{\frac{i\theta}{\sqrt{N}}M_N + \frac{\theta^2\sigma^2}{2}\Bigg\}\Bigg] = 1,$$

which proves the central limit theorem in the case of bounded martingale differences.

It remains to show that (1.11) is in force. The expression inside braces is bounded by a finite constant which depends on C_0 and θ. Since $|e^x - 1| \le |x|e^{|x|}$, $x \in \mathbb{R}$, the expectation in (1.11) is less than or equal to

$$\frac{C_1\theta^2}{\ell}\sum_{j=1}^{\ell}\mathbb{E}\Bigg[\Bigg|\frac{1}{K}\sum_{k=jK+1}^{(j+1)K}\big(Z_k^2-\sigma^2\big)\Bigg|\Bigg]$$

for some finite constant C_1. Since $\{Z_k\}$ is a stationary sequence, this expression does not depend on ℓ and is equal to

$$C_1\theta^2\mathbb{E}\Bigg[\Bigg|\frac{1}{K}\sum_{k=1}^{K}\big(Z_k^2-\sigma^2\big)\Bigg|\Bigg]. \tag{1.12}$$

By the ergodic theorem (1.12) vanishes as $K \uparrow \infty$.

The general case can be deduced from the previous one by approximating the martingale differences $\{Z_j : j \geq 0\}$ by bounded martingales differences. The most natural way consists in fixing a cut-off level $\kappa \geq 1$ and to define

$$Z_j^{(\kappa)} := \phi_\kappa(Z_j) - \mathbb{E}\big[\phi_\kappa(Z_j)\big|\mathscr{F}_{j-1}\big]$$

for $j \geq 1$. Here $\phi_\kappa : \mathbb{R} \to [-\kappa, \kappa]$ stands for a cut-off function which can be taken as $\phi_\kappa(x) = x\mathbf{1}\{|x| \leq \kappa\}$ for instance.

Note that $\{Z_j^{(\kappa)} : j \geq 1\}$ forms a sequence of bounded martingale differences. Although $\{\phi_\kappa(Z_j)\}$ inherits stationarity and ergodicity from the original sequence, the random variables $\{Z_j^{(\kappa)} : j \geq 1\}$ may lose these properties due to the presence of the conditional expectation.

Let $\sigma_\kappa^2 := \mathbb{E}[\phi_\kappa(Z_1)^2]$. By the dominated convergence theorem, σ_κ^2 converges to σ^2 as $\kappa \uparrow \infty$. Moreover, $Z_1^{(\kappa)}$ converges to Z_1 in $L^2(\mathbb{P})$, as $\kappa \uparrow \infty$, since

$$\mathbb{E}\big[\big(Z_1 - Z_1^{(\kappa)}\big)^2\big] \leq \mathbb{E}\big[Z_1^2\mathbf{1}\{|Z_1| > \kappa\}\big].$$

To derive this inequality we expanded the square and used the identity $\mathbb{E}[Z_1|\mathscr{F}_0] = 0$ which follows from the fact that $\{Z_j\}$ are martingale differences.

Let $\{M_j^{(\kappa)} : j \geq 0\}$ be the martingale associated to the sequence $\{Z_j^{(\kappa)} : j \geq 1\}$: $M_0^{(\kappa)} := 0$, $M_N^{(\kappa)} := \sum_{1\leq j\leq N} Z_j^{(\kappa)}$, $N \geq 1$. We show below that

$$\lim_{\kappa\to\infty}\lim_{N\to\infty}\big|\mathbb{E}\big[\exp\big\{(i\theta/\sqrt{N})M_N^{(\kappa)}\big\}\big] - e^{-\theta^2\sigma_\kappa^2/2}\big| = 0. \tag{1.13}$$

This claim does not follow from the first part of the proof because the sequence $\{Z_j^{(\kappa)} : j \geq 1\}$ may be neither stationary nor ergodic.

On the other hand, we have seen above that $\exp\{-\theta^2\sigma_\kappa^2/2\}$ converges, as $\kappa \uparrow \infty$, to $\exp\{-\theta^2\sigma^2/2\}$. Therefore, to conclude the proof of the theorem we need to show that the difference

$$\mathbb{E}\big[\exp\big\{(i\theta/\sqrt{N})M_N\big\}\big] - \mathbb{E}\big[\exp\big\{(i\theta/\sqrt{N})M_N^{(\kappa)}\big\}\big]$$

vanishes as $N \uparrow \infty$ and then $\kappa \uparrow \infty$.

Since $|e^{ix} - e^{iy}| \leq |x - y|$ for $x, y \in \mathbb{R}$, by Schwarz inequality, the previous expression is absolutely bounded by

$$\theta\bigg\{\frac{1}{N}\mathbb{E}\big[\big(M_N^{(\kappa)} - M_N\big)^2\big]\bigg\}^{1/2}.$$

Since the random variables $\{Z_j - Z_j^{(\kappa)} : j \geq 1\}$ are orthogonal, the expression inside braces is equal to the average of $\mathbb{E}[(Z_j^{(\kappa)} - Z_j)^2]$. We estimated above a similar term showing that it is less than or equal to $\mathbb{E}[Z_1^2\mathbf{1}\{|Z_1| \geq \kappa\}]$. Analogous arguments apply here. This expectation vanishes as $\kappa \uparrow \infty$, concluding the proof.

We now turn to the claim (1.13). Recall the proof of the central limit theorem for bounded martingale differences and observe that up to (1.11) we did not use the stationarity or the ergodicity of the sequence. The term we need to estimate is the exponential of

$$\frac{\theta^2}{2N}\sum_{j=0}^{\ell-1}\mathbb{E}\Bigg[\sum_{k=jK+1}^{(j+1)K}\big(Z_k^{(\kappa)}\big)^2\Big|\mathscr{F}_{jK}\Bigg]$$

$$=\frac{\theta^2}{2N}\sum_{j=0}^{\ell-1}\mathbb{E}\Bigg[\sum_{k=jK+1}^{(j+1)K}\big\{\phi_\kappa(Z_k)^2-\mathbb{E}\big[\phi_\kappa(Z_k)|\mathscr{F}_{k-1}\big]^2\big\}\Big|\mathscr{F}_{jK}\Bigg],$$

where the identity follows from the definition of the variables $Z_k^{(\kappa)}$. Notice that the positive term $\mathbb{E}[\phi_\kappa(Z_k)|\mathscr{F}_{k-1}]^2$ has a negative sign in front. In particular, its exponential is bounded by one. On the other hand, since for each fixed κ, $\{\phi_\kappa(Z_k) : j\ge 1\}$ is a stationary ergodic sequence, by (1.11) we may replace $\phi_\kappa(Z_k)^2$ by σ_κ^2 in the exponential.

It remains to estimate

$$\mathbb{E}\Bigg[\Bigg|\exp\Bigg\{\frac{-\theta^2}{2N}\sum_{j=0}^{\ell-1}\mathbb{E}\Bigg[\sum_{k=jK+1}^{(j+1)K}\mathbb{E}\big[\phi_\kappa(Z_k)|\mathscr{F}_{k-1}\big]^2\big|\mathscr{F}_{jK}\Bigg]\Bigg\}-1\Bigg|\Bigg].$$

Since the expression inside the exponential is negative and since $1-e^{-x}\le x$ for $x\ge 0$, the previous expression is bounded above by

$$\frac{\theta^2}{2N}\sum_{k=1}^{N}\mathbb{E}\big[\mathbb{E}[\phi_\kappa(Z_k)|\mathscr{F}_{k-1}]^2\big].$$

Since $\{Z_k : k\ge 1\}$ is a martingale difference, we may subtract Z_k in the conditional expectation without affecting it. Applying Schwarz inequality we bound the previous expression by $(\theta^2/2)\mathbb{E}[Z_1^2\mathbf{1}\{|Z_1|>\kappa\}]$ which vanishes as $\kappa\uparrow\infty$. This proves (1.13) and concludes the proof of the theorem. □

In the proof of Theorem 1.2 all estimates were carried through in $L^1(\mathbb{P})$. A straightforward adaptation of the arguments gives a conditional central limit theorem:

Theorem 1.3 *Under the assumptions of Theorem* 1.2, *for every* θ *in* $\mathbb{R}$,

$$\lim_{N\to\infty}\mathbb{E}\big[\big|\mathbb{E}\big[\exp\{i\theta(M_N/\sqrt{N})\}\big|\mathscr{F}_0\big]-e^{-\theta^2\sigma^2/2}\big|\big]=0,$$

where $\sigma^2=\mathbb{E}[Z_1^2]$.

Remark 1.4 Deeper analysis permits to prove convergence of the partial sums to a Brownian motion with diffusion coefficient $\sigma^2 = \mathbb{E}[Z_1^2]$: For every $T > 0$,

$$Z^N(t) = \frac{1}{\sqrt{N}} \sum_{j=0}^{[Nt]} Z_j + \frac{Nt - [Nt]}{\sqrt{N}} Z_{[Nt]+1}$$

converges to a Brownian motion in $C([0, T])$, the space of continuous functions in $[0, T]$. In this formula, $[a]$ stands for the integer part of $a \in \mathbb{R}$: $[a] = \sup\{n \in \mathbb{Z} : n \le a\}$. We refer to Theorem 2.29.

1.4 Time-Variance in Reversible Markov Chains

In this section, we examine the asymptotic behavior of the variance of

$$\frac{1}{\sqrt{N}} \sum_{j=0}^{N-1} V(X_j)$$

for square integrable functions V in the context of reversible Markov chains. Reversibility with respect to π means that P is a symmetric operator in $L^2(\pi)$:

$$\langle Pf, g\rangle_\pi = \langle f, Pg\rangle_\pi$$

for all f, g in $L^2(\pi)$. It is easy to check that a probability measure π is reversible if and only if it satisfies the *detailed balance condition:*

$$\pi(x)P(x, y) = \pi(y)P(y, x)$$

for all x, y in E, which means that

$$\mathbb{P}_\pi[X_n = x, X_{n+1} = y] = \mathbb{P}_\pi[X_n = y, X_{n+1} = x].$$

A reversible measure is necessarily invariant since

$$(\pi P)(x) = \sum_{y\in E} \pi(y)P(y, x) = \sum_{y\in E} \pi(x)P(x, y) = \pi(x).$$

In this section, we prove that the following limit exists:

$$\sigma^2(V) = \lim_{N\to\infty} \mathbb{E}_\pi\left[\left(\frac{1}{\sqrt{N}} \sum_{j=0}^{N-1} V(X_j)\right)^2\right],$$

where we admit $+\infty$ as a possible value, and we find necessary and sufficient conditions for $\sigma^2(V)$ to be finite. We also introduce Hilbert spaces associated to the transition operator P which will play a central role in the following chapters.

Fix an invariant probability measure π and a function V in $L^2(\pi)$. An elementary computation gives that

$$\mathbb{E}_\pi\Big[\Big(\frac{1}{\sqrt{N}}\sum_{j=0}^{N-1} V(X_j)\Big)^2\Big] = \frac{1}{N}\sum_{j,k=0}^{N-1} \mathbb{E}_\pi\big[V(X_j)V(X_k)\big]$$

$$= \frac{1}{N}\sum_{j=0}^{N-1} \mathbb{E}_\pi\big[V(X_j)^2\big] + \frac{2}{N}\sum_{j<k} \mathbb{E}_\pi\big[V(X_j)V(X_k)\big].$$

Since π is a stationary measure, $\mathbb{E}_\pi[V(X_j)^2] = \langle V, V\rangle_\pi$ and, for $j < k$, $\mathbb{E}_\pi[V(X_j)V(X_k)] = \langle V, P^{k-j}V\rangle_\pi$ so that the second term is equal to

$$\frac{2}{N}\sum_{j<k}\big\langle V, P^{k-j}V\big\rangle_\pi = 2\sum_{i=1}^{N-1}\big[1-(i/N)\big]\big\langle V, P^i V\big\rangle_\pi.$$

In conclusion,

$$\mathbb{E}_\pi\Big[\Big(\frac{1}{\sqrt{N}}\sum_{j=0}^{N-1} V(X_j)\Big)^2\Big] = \langle V, V\rangle_\pi + 2\sum_{i=1}^{N-1}\big[1-(i/N)\big]\big\langle V, P^i V\big\rangle_\pi.$$

To estimate the second expression we rely on the spectral decomposition of the operator P. Since P is symmetric in $L^2(\pi)$, all its eigenvalues are real and P admits a spectral decomposition:

$$P = \int_{\mathbb{R}} \varphi \, dE_\varphi.$$

By (1.7), P is a contraction, its spectrum is contained in $[-1, 1]$ so that

$$P = \int_{-1}^{1} \varphi \, dE_\varphi.$$

Notice that 1 is an eigenvalue associated to the constants because $P\mathbf{1} = \mathbf{1}$ if $\mathbf{1}$ is the constant function equal to 1.

The spectral decomposition of P permits to represent the scalar product $\langle P^k V, V\rangle_\pi$ in terms of the spectral measure of V:

$$\big\langle V, P^k V\big\rangle_\pi = \Big\langle V, \int_{-1}^{1} \varphi^k \, dE_\varphi V\Big\rangle_\pi = \int_{-1}^{1} \varphi^k \, d\langle V, E_\varphi V\rangle_\pi.$$

Denote the spectral measure $d\langle V, E_\varphi V\rangle_\pi$ of V by $\mu_V(d\varphi)$ and notice that μ_V is a finite measure on $[-1, 1]$ with total mass equal to $\langle V, V\rangle_\pi$. With this notation,

$$\mathbb{E}_\pi\Big[\Big(\frac{1}{\sqrt{N}}\sum_{j=0}^{N-1} V(X_j)\Big)^2\Big] = \langle V, V\rangle_\pi + 2\int_{-1}^{1}\sum_{i=1}^{N-1}\big[1-(i/N)\big]\varphi^i \, \mu_V(d\varphi).$$

The second term on the right-hand side can be rewritten as

$$2\int_{-1}^{1}\sum_{i\geq 1}[1-(i/N)]^{+}\varphi^{i}\mu_{V}(d\varphi),$$

where a^{+} stands for the positive part of a. For $-1\leq\varphi\leq 0$, an elementary computation shows that $\sum_{i\geq 1}[1-(i/N)]^{+}\varphi^{i}$ is absolutely bounded by a finite constant and that it converges to $\varphi/(1-\varphi)$, as $N\uparrow\infty$. On the other hand, for $0\leq\varphi\leq 1$, $\sum_{i\geq 1}[1-(i/N)]^{+}\varphi^{i}$ increases to $\varphi/(1-\varphi)$. Therefore, by the monotone and by the dominated convergence theorem, the previous integral converges to

$$2\int_{-1}^{1}\frac{\varphi}{1-\varphi}\mu_{V}(d\varphi).$$

We have thus proved the following result.

Lemma 1.5 *For any function V in $L^2(\pi)$,*

$$\sigma^2(V)<\infty \quad \textit{if and only if} \quad \int_{-1}^{1}\frac{1}{1-\varphi}\mu_{V}(d\varphi)<\infty. \tag{1.14}$$

In this case,

$$\sigma^2(V)=\int_{-1}^{1}\frac{1+\varphi}{1-\varphi}\mu_{V}(d\varphi). \tag{1.15}$$

Note that in this computation we did not use the fact that V has mean zero. However, $\sigma^2(V)=\infty$ if $E_\pi[V]\neq 0$ since for such functions the spectral measure gives a positive weight to 1: $\mu_V(\{1\})>0$.

The previous computation leads to the following Hilbert space. Since P is an operator bounded by 1, $I-P$ is non-negative so that

$$\langle f,g\rangle_1=\langle f,(I-P)g\rangle_\pi$$

defines a semi-definite scalar product in $L^2(\pi)$. It is semi-definite because $\langle \mathbf{1},\mathbf{1}\rangle_1=0$. Denote by $\mathcal{H}_1$ the Hilbert space induced by $L^2(\pi)$ endowed with the scalar product $\langle\cdot,\cdot\rangle_1$. Let $\|\cdot\|_1$ be the norm associated to this scalar product. The Hilbert space $\mathcal{H}_1$ is examined in detail in Sect. 1.6.

For $\lambda>0$, consider the resolvent equation

$$\lambda f_\lambda+(I-P)f_\lambda=V. \tag{1.16}$$

In contrast with the Poisson equation, the resolvent equation always has a solution f_λ in $L^2(\pi)$ because $(1+\lambda)I-P$ is invertible for all $\lambda>0$. It is given explicitly by $f_\lambda=(1+\lambda)^{-1}\sum_{j\geq 0}(1+\lambda)^{-j}P^jV$.

The proof of the central limit theorem for $N^{-1/2}\sum_{0\leq j<N}V(X_j)$ relies on the following two estimates on the solution of the resolvent equation.

Lemma 1.6 *Let f_λ be the solution of the resolvent equation* (1.16) *for some zero-mean function V with finite time-variance $\sigma^2(V)$. Then,*

$$\lim_{\lambda\to 0} \lambda\langle f_\lambda, f_\lambda\rangle_\pi = 0.$$

Proof Since f_λ is the solution of the resolvent equation (1.16),

$$\lambda\langle f_\lambda, f_\lambda\rangle_\pi = \lambda E_\pi\big[\big\{\big[(1+\lambda)I - P\big]^{-1}V\big\}^2\big] = \int_{-1}^{1} \frac{\lambda}{(1+\lambda-\varphi)^2}\,\mu_V(d\varphi).$$

Since $\lambda/(1+\lambda-\varphi)^2 \le (1-\varphi)^{-1}$ and since $\mu_V(d\varphi)$ integrates $(1-\varphi)^{-1}$, by the dominated convergence theorem, the previous integral vanishes as $\lambda \downarrow 0$. □

Lemma 1.7 *Let f_λ be the solution of the resolvent equation* (1.16) *for some zero-mean function V with finite time-variance $\sigma^2(V)$. The sequence f_λ is a Cauchy sequence in $\mathcal{H}_1$: for every $\varepsilon > 0$, there exists $\lambda_0 > 0$ such that for any $\lambda_1, \lambda_2 < \lambda_0$*

$$\big\langle f_{\lambda_1} - f_{\lambda_2}, (I-P)(f_{\lambda_1} - f_{\lambda_2})\big\rangle_\pi < \varepsilon.$$

Proof Since f_λ is the solution of the resolvent equation,

$$\begin{aligned}
&\big\langle f_{\lambda_1} - f_{\lambda_2}, (I-P)(f_{\lambda_1} - f_{\lambda_2})\big\rangle_\pi \\
&\quad = \int_{-1}^{1} (1-\varphi)\left\{\frac{1}{1+\lambda_2-\varphi} - \frac{1}{1+\lambda_1-\varphi}\right\}^2 \mu_V(d\varphi) \\
&\quad = \int_{-1}^{1} \frac{(1-\varphi)(\lambda_2-\lambda_1)^2}{[1+\lambda_1-\varphi]^2[1+\lambda_2-\varphi]^2}\,\mu_V(d\varphi).
\end{aligned}$$

Since the integrand is bounded above by $(1-\varphi)^{-1}$ and since the spectral measure of V integrates $(1-\varphi)^{-1}$, the integral converges to 0 as $\lambda_1, \lambda_2 \downarrow 0$. □

A similar computation to the one presented in the previous proof shows that

$$\sigma^2(V) = \lim_{\lambda\to 0}\big\langle f_\lambda, \big(I-P^2\big)f_\lambda\big\rangle_\pi. \tag{1.17}$$

Indeed, since f_λ solves the resolvent equation, by the spectral representation of the operator P,

$$\big\langle f_\lambda, \big(I-P^2\big)f_\lambda\big\rangle_\pi = \int_{-1}^{1} \frac{(1-\varphi^2)}{(1+\lambda-\varphi)^2}\,\mu_V(d\varphi).$$

Since the integrand is bounded above by $(1-\varphi)^{-1}$, by the dominated convergence theorem, as $\lambda \downarrow 0$, the previous expression converges to

$$\int_{-1}^{1} \frac{1+\varphi}{1-\varphi}\,\mu_V(d\varphi),$$

which is equal to $\sigma^2(V)$ in view of (1.15).

1.5 Central Limit Theorem for Reversible Markov Chains

In this section, we prove a central limit theorem for additive functionals of reversible Markov chains. Fix a zero-mean function V in $L^2(\pi)$. We have seen in the beginning of this chapter that a central limit theorem for the additive functional $N^{-1/2}\sum_{0\le j<N} V(X_j)$ follows easily from a central limit theorem for martingales if V belongs to the range of $I-P$, i.e., if there is a solution in $L^2(\pi)$ of the Poisson equation $(I-P)f=V$. This assumption is too strong and should be relaxed. A natural condition to impose on V is to require that its time-variance $\sigma^2(V)$ is finite. In this case we may try to repeat the approach presented in the beginning of the chapter replacing the solution of the Poisson equation $(I-P)f=V$, which may not exist, by the solution f_λ of the resolvent equation $\lambda f_\lambda+(I-P)f_\lambda=V$ which always exists.

Fix therefore a zero-mean function V and assume that its variance $\sigma^2(V)$ is finite. Let f_λ be the solution of the resolvent equation (1.16). For $N\ge 1$,

$$\begin{aligned}\sum_{j=0}^{N-1} V(X_j) &= \lambda\sum_{j=0}^{N-1} f_\lambda(X_j)+\sum_{j=0}^{N-1}\{f_\lambda(X_j)-(Pf_\lambda)(X_j)\}\\ &= M_N^\lambda+f_\lambda(X_0)-f_\lambda(X_N)+\lambda\sum_{j=0}^{N-1} f_\lambda(X_j),\end{aligned} \tag{1.18}$$

where $\{M_N^\lambda : N\ge 0\}$ is the martingale with respect to the filtration $\{\mathscr{F}_j : j\ge 0\}$, $\mathscr{F}_j=\sigma(X_0,\dots,X_j)$, defined by $M_0^\lambda:=0$,

$$M_N^\lambda:=\sum_{j=1}^{N} Z_j^\lambda,$$

for $Z_j^\lambda=f_\lambda(X_j)-(Pf_\lambda)(X_{j-1})$ for $j\ge 1$.

Lemma 1.8 *For each $N\ge 1$, as $\lambda\downarrow 0$, M_N^λ converges in $L^2(\mathbb{P}_\pi)$ to some variable M_N. The limit process $\{M_j, j\ge 0\}$ is a martingale with respect to the filtration $\{\mathscr{F}_j\}$ and*

$$\sum_{j=0}^{N-1} V(X_j)=M_N+R_N$$

for some process R_N in $L^2(\mathbb{P}_\pi)$.

Proof Since N is fixed, to prove the convergence of the martingale M_N^λ, we just need to show that each term $Z_j^\lambda=f_\lambda(X_j)-(Pf_\lambda)(X_{j-1})$ converges in $L^2(\mathbb{P}_\pi)$. We will prove that this sequence is Cauchy. For $0 < \lambda_1 < \lambda_2$, since P is a symmetric operator and π is a stationary measure,

$$\begin{aligned}
&\mathbb{E}_\pi\big[\big(f_{\lambda_2}(X_{j+1}) - f_{\lambda_1}(X_{j+1}) - (Pf_{\lambda_2})(X_j) + (Pf_{\lambda_1})(X_j)\big)^2\big] \\
&\quad = E_\pi\big[(f_{\lambda_2} - f_{\lambda_1})^2\big] - E_\pi\big[(Pf_{\lambda_2} - Pf_{\lambda_1})^2\big] \\
&\quad = \big\langle f_{\lambda_2} - f_{\lambda_1}, \big(I - P^2\big) f_{\lambda_2} - f_{\lambda_1}\big\rangle_\pi .
\end{aligned}$$

Since $I + P \le 2I$ and $I - P \ge 0$, we have $I - P^2 = (I - P)(I + P) \le 2(I - P)$. Therefore, the above quantity is bounded by

$$2\big\langle f_{\lambda_2} - f_{\lambda_1}, (I - P)(f_{\lambda_2} - f_{\lambda_1})\big\rangle_\pi ,$$

which vanishes as $\lambda_1, \lambda_2 \downarrow 0$, by virtue of Lemma 1.7. We have thus proved that the martingale M_N^λ is a Cauchy sequence in $L^2(\mathbb{P}_\pi)$ as $\lambda \downarrow 0$. It converges, in particular, to some process M_N, which is a square-integrable martingale because the convergence takes place in $L^2(\mathbb{P}_\pi)$.

To conclude the proof of the lemma it remains to recall the decomposition of $\sum_{0\le j<N} V(X_j)$ given in (1.18). Since the left-hand side does not depend on λ and since we just proved the convergence of the first term on the right-hand side, the second must also converge to some limit that we denote by R_N. □

Lemma 1.9 *Recall the decomposition of $\sum_{0\le j<N} V(X_j)$ obtained in the previous lemma. $N^{-1/2} R_N$ converges to 0 in $L^2(\mathbb{P}_\pi)$ as $N \uparrow \infty$.*

Proof It follows from the decomposition given in (1.18) and the one presented in Lemma 1.8 that

$$R_N = M_N^\lambda - M_N + f_\lambda(X_0) - f_\lambda(X_N) + \lambda \sum_{j=0}^{N-1} f_\lambda(X_j)$$

for all $\lambda > 0$. Choose $\lambda = N^{-1}$. We claim that each term on the right-hand side divided by $\sqrt{N}$ vanishes in $L^2(\mathbb{P}_\pi)$. By Schwarz inequality and by the invariance of π,

$$\frac{1}{N}\mathbb{E}_\pi\left[\left(\lambda \sum_{j=0}^{N-1} f_\lambda(X_j)\right)^2\right] \le \frac{1}{N} E_\pi\big[f_\lambda^2\big] = \lambda\langle f_\lambda, f_\lambda\rangle_\pi$$

because $\lambda = N^{-1}$. We proved in Lemma 1.6 that this expression vanishes as $\lambda \downarrow 0$. By similar reasons, $N^{-1/2} f_\lambda(X_N)$ and $N^{-1/2} f_\lambda(X_0)$ vanish in $L^2(\mathbb{P}_\pi)$ as $N \uparrow \infty$.

It remains to consider the martingale part. By orthogonality of the increments of the martingales and by stationarity

$$N^{-1}\mathbb{E}_\pi\big[\big(M_N^\lambda - M_N\big)^2\big] = \frac{1}{N}\sum_{j=0}^{N-1} \mathbb{E}_\pi\big[\big(Z_j^\lambda - Z_j\big)^2\big] = \mathbb{E}_\pi\big[\big(Z_1^\lambda - Z_1\big)^2\big].$$

We proved in Lemma 1.8 that this expression vanishes as $N \uparrow \infty$, which concludes the proof of the lemma. □

We are now in a position to state the main result of this chapter.

Theorem 1.10 *Consider a Markov chain $\{X_j, j \geq 0\}$ on a countable state space E, ergodic and reversible with respect to some invariant state π. Let $V: E \to \mathbb{R}$ be a zero-mean function in $L^2(\pi)$ with finite time-variance $\sigma^2(V) < \infty$. Then, under $\mathbb{P}_\pi$,*

$$\frac{1}{\sqrt{N}} \sum_{j=0}^{N-1} V(X_j)$$

converges in distribution to a zero-mean Gaussian law with variance $\sigma^2(V)$.

Proof By Lemma 1.8, each variable Z_j^λ converges in $L^2(\mathbb{P}_\pi)$, as $\lambda \downarrow 0$. Denote by Z_j its limit. We claim that the sequence $\{Z_j : j \geq 1\}$ satisfies the assumptions of Theorem 1.2 with respect to the filtration $\mathscr{F}_j = \sigma(X_0, \ldots, X_j)$, $j \geq 0$, and the probability measure $\mathbb{P}_\pi$.

The sequence $\{Z_j\}$ inherits the stationary from $\{Z_j^\lambda\}$ as well as the measurability with respect to the filtration $\{\mathscr{F}_j\}$. Since Z_j^λ converges to Z_j in $L^2(\mathbb{P}_\pi)$ and since $\{Z_j^\lambda\}$ are martingale differences with respect to the filtration $\{\mathscr{F}_j\}$,

$$\mathbb{E}_\pi\big[Z_1^2\big] < \infty \quad \text{and} \quad \mathbb{E}_\pi[Z_{j+1}|\mathscr{F}_j] = 0, \; j \geq 0.$$

To show that the sequence $\{Z_j\}$ is ergodic, let ν be the probability measure on $E \times E$ defined by $\nu(x, y) := \pi(x)P(x, y)$. For $\lambda > 0$, let $\Psi_\lambda : E \times E \to \mathbb{R}$ be defined by $\Psi_\lambda(x, y) := f_\lambda(y) - P f_\lambda(x)$. Note that $Z_k^\lambda = \Psi_\lambda(X_k, X_{k-1})$. Moreover,

$$\|\Psi_\lambda\|_\nu^2 = \frac{1}{2}\langle f_\lambda, \big(I - P^2\big) f_\lambda\rangle_\pi \leq \|f_\lambda\|_1^2.$$

In particular, by the proof of Lemma 1.8, Ψ_λ is a Cauchy sequence in $L^2(\nu)$. Denote by Ψ its limit. Since $Z_k^\lambda = \Psi_\lambda(X_k, X_{k-1})$, $Z_k = \Psi(X_k, X_{k-1})$ for $k \geq 1$. This shows that the sequence $\{Z_k\}$ is ergodic.

We just showed that all assumptions of Theorem 1.2 are in force. Thus, $N^{-1/2} M_N$ converges in distribution to a zero-mean Gaussian variable with variance

$$\sigma^2 = \mathbb{E}_\pi\big[Z_1^2\big] = \lim_{\lambda \to 0} \mathbb{E}_\pi\big[\big(Z_1^\lambda\big)^2\big] = \lim_{\lambda \to 0}\langle f_\lambda, \big(I - P^2\big) f_\lambda\rangle_\pi.$$

By (1.17), this expression is equal to $\sigma^2(V)$.

By Lemma 1.9, $N^{-1/2} R_N$ converges to 0 in $L^2(\mathbb{P}_\pi)$. Therefore, in view of the decomposition presented in Lemma 1.8, $N^{-1/2} \sum_{0 \leq j < N} V(X_j)$ converges in distribution to a zero-mean Gaussian variable with variance $\sigma^2(V)$. □

The computations performed in Sect. 1.4 show that not only does the sequence $N^{-1/2}\sum_{j=0}^{N-1} V(X_j)$ converge in distribution to a Gaussian law, but also the variance of this sequence converges to the variance of the limiting distribution.

Remark 1.11 Under the assumptions of Theorem 1.10, the variance of the sequence $N^{-1/2}\sum_{j=0}^{N-1} V(X_j)$ converges to $\sigma^2(V)$.

There are several concepts of convergence. In Sect. 1.2, the central limit theorem holds with respect to all measures $\mathbb{P}_x$. In other words, with respect to π-almost all starting points. In Theorem 1.3 we presented a central limit theorem for martingales which holds in probability with respect to the initial data. This result provides the following statement for Markov chains.

Remark 1.12 Under the assumptions of Theorem 1.10, for every bounded continuous function $f : \mathbb{R} \to \mathbb{R}$,

$$\lim_{N\to\infty} \sum_{x\in E} \pi(x) \left| \mathbb{E}_x \left[f\left(\frac{1}{\sqrt{N}} \sum_{j=0}^{N-1} V(X_j) \right) \right] - \int f(u) \Phi_{\sigma^2(V)}(u)\, du \right| = 0,$$

where Φ_{σ^2} is the density of the mean zero Gaussian distribution with variance σ^2.

A central limit theorem, almost sure with respect to the initial state, follows from this result in the countable case. This is a special feature of the discrete setting. In general we will have to content ourselves with a central limit theorem in probability as stated above. This result is weaker than the almost sure result, known in the literature as a quenched statement, but stronger than the annealed result stated in Theorem 1.10.

Remark 1.13 The Markov chain may have different ergodic measures π. In this case the central limit theorem is valid for each ergodic measure, but in general the asymptotic variance will depend on the particular ergodic measure: $\sigma^2(V) = \sigma^2(V, \pi)$. This is the case of the exclusion processes which will be examined in detail in the following chapters.

Remark 1.14 A slightly deeper analysis permits to prove convergence of the partial sums to a Brownian motion with diffusion coefficient given by (1.15): For every $T > 0$,

$$\frac{1}{\sqrt{N}} \sum_{j=0}^{[Nt]} V(X_j) + \frac{Nt - [Nt]}{\sqrt{N}} V(X_{[Nt]+1})$$

converges to a Brownian motion in $C([0, T])$ (cf. Theorem 2.32).

Remark 1.15 Existence of a solution of the Poisson equation (1.2) in $L^2(\mathbb{P}_\pi)$ corresponds to the condition $(I - P)^{-1}V \in L^2(\pi)$. In terms of the spectral measure of

V, this is equivalent to requiring that

$$\int_{-1}^{1} \frac{1}{(1-\varphi)^2} \mu_V(d\varphi) < \infty,$$

a stronger assumption than the hypothesis that V has a finite time-variance since in this case we only require (1.14).

1.6 The Space of Finite Time-Variance Functions

In this section, we examine the subspace of $L^2(\pi)$ of functions with finite time-variance. Assume without loss of generality that the set E is connected in the sense that for any x, y in E, there exists a path from x to y. Here and below, a path from x to y is a finite sequence $\{x_0, \dots, x_n\}$ such that $x_0 = x$, $x_n = y$, $P(x_i, x_{i+1}) > 0$ for $0 \le i < n$. It follows from the connectivity and the reversibility that $\pi(x) > 0$ for all x in E.

1.6.1 The Space $\mathscr{H}_1$

A. An Explicit Formula for $\|\cdot\|_1$ Recall from Sect. 1.4 the definition of the semi-scalar product $\langle \cdot, \cdot \rangle_1$. Fix a function f in $L^2(\pi)$. A simple computation shows that

$$\langle f, (I-P)f \rangle_\pi = \sum_{x,y \in E} \pi(x) P(x,y) f(x) \big[f(x) - f(y) \big].$$

We have seen that $\pi(x)P(x,y) = \pi(y)P(y,x)$ because the process is reversible. We may therefore rewrite the previous expression as

$$(1/2) \sum_{x,y \in E} \pi(x) P(x,y) f(x) \big[f(x) - f(y) \big]$$
$$+ (1/2) \sum_{x,y \in E} \pi(y) P(y,x) f(x) \big[f(x) - f(y) \big].$$

Renaming x and y in the second sum and adding the two sums, we conclude that

$$\langle f, (I-P)f \rangle_\pi = (1/2) \sum_{x,y \in E} \pi(x) P(x,y) \big[f(x) - f(y) \big]^2. \tag{1.19}$$

B. The Space $\mathbb{D}$ Denote the right-hand side of (1.19) by $\|f\|_1^2$, which is well defined for any function $f : E \to \mathbb{R}$. Let $\tilde{\mathbb{D}}$ be the space of functions $f : E \to \mathbb{R}$ such that $\|f\|_1 < \infty$:

$$\tilde{\mathbb{D}} = \big\{ f : E \to \mathbb{R} : \|f\|_1 < \infty \big\}.$$

It follows from (1.19) that $\|\cdot\|_1$ is a semi-norm. Its kernel corresponds to the constant functions: Suppose that $\|f\|_1 = 0$ for some function $f : E \to \mathbb{R}$. By (1.19), $f(y) = f(x)$ if $P(x, y) > 0$. By the connectivity of E, f is constant.

Define the equivalence relation $\sim$ in $\tilde{\mathbb{D}}$ by stating that $f \sim g$ if $\|f - g\|_1 = 0$, i.e., if $f - g$ is constant. Denote by $\mathbb{D}$ the equivalence classes of $\tilde{\mathbb{D}}$: $\mathbb{D} = \tilde{\mathbb{D}}|_{\sim}$.

The space $\mathbb{D}$ is complete. Consider a Cauchy sequence $\{f_n : n \geq 1\}$ in $\mathbb{D}$. For any x, y such that $P(x, y) > 0$, $f_n(y) - f_n(x)$ is a Cauchy sequence and thus converges. Since E is connected, $f_n(y) - f_n(x)$ converges for all x, y in E. Fix a site x^* in E and consider the function $f : E \to \mathbb{R}$ defined by $f(x^*) = 0$,

$$f(y) = \lim_{n\to\infty} f_n(y) - f_n(x^*).$$

It is not difficult to show that f belongs to $\mathbb{D}$ and that f_n converges to f in $\mathbb{D}$.

Since the norm $\|\cdot\|_1$ satisfies the parallelogram identity, by Sect. 87 in Riesz and Sz.-Nagy (1990), $\mathbb{D}$ is a Hilbert space with the scalar product $\langle\cdot,\cdot\rangle_1$ defined by $\langle f, g\rangle_1 = (1/4)\{\|f + g\|_1^2 - \|f - g\|_1^2\}$. From (1.19) we have that

$$\langle g, f\rangle_1 = (1/2) \sum_{x,y\in E} \pi(x)P(x, y)\big[f(x) - f(y)\big]\big[g(x) - g(y)\big] \tag{1.20}$$

for any function f, g in $\mathbb{D}$.

C. The Space $\mathscr{H}_1$ Since $(a-b)^2 \leq 2a^2+2b^2$, it follows from (1.19) that $\|f\|_1^2 \leq 2\|f\|^2$ for any function f in $L^2(\pi)$. Thus $L^2(\pi)$ is contained in $\mathbb{D}$. Denote by $\mathscr{H}_1$ the subspace of $\mathbb{D}$ generated by the functions in $L^2(\pi)$ so that

$$L^2(\pi) \subset \mathscr{H}_1 \subset \mathbb{D}.$$

D. Exact Forms Let $\mathbb{B}$ (for bonds) be the set of pairs (x, y) in $E \times E$ such that $\pi(x)P(x, y) > 0$. Let ν be the measure on $\mathbb{B}$ defined by $\nu(x, y) = (1/2)\pi(x)P(x, y)$ and denote by $\|\cdot\|_\nu$ the norm of $L^2(\nu)$. A function D in $L^2(\nu)$ is called an exact form if for all x in E and all finite paths $\{x_0, x_1, \ldots, x_n\}$ from x to x,

$$\sum_{i=0}^{n-1} D(x_i, x_{i+1}) = 0. \tag{1.21}$$

Let $\mathfrak{F}$ be the subspace of exact forms in $L^2(\nu)$. Note that $\mathfrak{F}$ is closed.

Fix an exact form D, a vertex x^* in E and a constant c in $\mathbb{R}$, since we assumed E to be connected, we may define a function $f_{D,c,x^*} = f_D : E \to \mathbb{R}$ by

$$f_D(y) = c + \sum_{i=0}^{n-1} D(x_i, x_{i+1}),$$

where $\{x_0, \ldots, x_n\}$ is a path from x^* to y. The value of f_D at y does not depend on the specific path because D is an exact form. It is, of course, not difficult to conceive examples where f_D does not belong to $L^2(\pi)$.

Reciprocally, given a function f in $L^2(\pi)$, define $D_f : \mathbb{B} \to \mathbb{R}$ by

$$D_f(x, y) = f(y) - f(x).$$

It is easy to check that D_f fulfills (1.21) and that D_f belongs to $L^2(\nu)$, by (1.19). Let $\mathfrak{F}_0$ be the subspace of $L^2(\nu)$ generated by exact forms associated to functions in $L^2(\pi)$: $\mathfrak{F}_0 = \{D_f \in L^2(\nu) : f \in L^2(\pi)\}$. We have just shown that $\mathfrak{F}_0 \subset \mathfrak{F}$.

The space $\mathfrak{F}$ represents the set of functions $f : E \to \mathbb{R}$ with derivative D_f in $L^2(\nu)$ and $\mathfrak{F}_0$ the subspace of functions f in $L^2(\pi)$ with derivative D_f in $L^2(\nu)$. The space $\mathscr{H}_1$ may be identified with $\overline{\mathfrak{F}_0}$, the closure of $\mathfrak{F}_0$ in $L^2(\nu)$, since, by (1.19), the application which maps f to D_f is an isometry: $\|f\|_1 = \|D_f\|_\nu$.

E. Liouville D-property A Markov chain is said to have the Liouville D-property if all solutions $f : E \to \mathbb{R}$ of the Liouville problem

$$\begin{cases} (I - P)f = 0, \\ \|f\|_1 < \infty, \end{cases} \tag{1.22}$$

are constant. Notice that $(I - P)f$ is well defined because $\|f\|_1$ is finite.

A Markov chain has the Liouville D-property if and only if $\overline{\mathfrak{F}_0}$ coincides with $\mathfrak{F}$. To prove this statement, assume that $\overline{\mathfrak{F}_0} \neq \mathfrak{F}$. In this case, there exists a non-zero D in $\mathfrak{F}$ which is orthogonal to $\mathfrak{F}_0$. Let $f : E \to \mathbb{R}$ be such that $D = D_f$. To show that f is a solution of the Liouville problem (1.22), fix x_0 in E and let $\delta_{x_0} \in L^2(\pi)$ be the indicator function of the singleton $\{x_0\}$. By orthogonality, by the explicit form of the scalar product in $L^2(\nu)$ and by the computations presented at the beginning of this section,

$$0 = \langle D_f, D_{\delta_{x_0}} \rangle_\nu = \pi(x_0) \sum_{y \in E} P(x_0, y)\big\{f(x_0) - f(y)\big\},$$

where $\langle \cdot, \cdot \rangle_\nu$ stands for the inner product in $L^2(\nu)$. The right-hand side is equal to $\pi(x_0)[(I - P)f](x_0)$. The connectivity of the set E gives that $\pi(x_0) > 0$. Since the identity holds for every x_0, $(I - P)f = 0$. Thus, f is a non-constant solution of the Liouville problem (1.22).

The same arguments show that any non-constant solution f of the Liouville problem (1.22) provides a non-zero derivative D_f in $\mathfrak{F}$ orthogonal to $\mathfrak{F}_0$.

A Markov chain reversible with respect to a probability measure π has the Liouville D-property. Assume that f is a solution of (1.22) so that $(I - P)f = 0$. For $A > 0$ denote by $\phi_A : \mathbb{R} \to [-A, A]$ the cut-off function $\phi_A(x) = \max\{-A, \min\{x, A\}\}$. Since $(I - P)f = 0$, $\phi_A(f)$ is bounded and D_f belongs to $L^2(\nu)$,

$$\begin{aligned}0 &= \langle \phi_A(f), (I-P)f \rangle_\pi \\ &= (1/2) \sum_{x,y \in E} \pi(x)P(x,y)\big[f(x)-f(y)\big]\big[\phi_A(f)(x)-\phi_A(f)(y)\big].\end{aligned}$$

Each term on the right-hand side is positive. Thus for any pair (x, y) such that $P(x, y) > 0$, $[f(x) - f(y)][\phi_A(f)(x) - \phi_A(f)(y)] = 0$ for all $A > 0$. Letting $A \uparrow \infty$ we conclude that $f(y) = f(x)$. The connectivity of E guarantees that f is constant, proving the claim.

The next example shows that the Liouville D-property may hold even when π is not a probability measure or when the process is transient.

F. Symmetric Random Walks on $\mathbb{Z}^d$ Finite range symmetric random walks on $\mathbb{Z}^d$ have the Liouville D-property.

Denote by π the counting measure on $\mathbb{Z}^d$ and consider a symmetric random walk on $\mathbb{Z}^d$. Denote by $\{e_j : 1 \le j \le d\}$ the canonical basis of $\mathbb{R}^d$ and assume, without loss of generality, that $P(0, \pm e_j) = 1/2d$.

Assume that $f : \mathbb{Z}^d \to \mathbb{R}$ is a solution of the Liouville problem (1.22). Hence, $\sum_{1\le j\le d} \sum_{x\in\mathbb{Z}^d} [f(x+e_j) - f(x)]^2 < \infty$ and $0 = (I-P)f = -(1/2d)\Delta f$, where Δ stands for the discrete Laplacian. Let D be the exact form associated to f and let $D_j(x) = D(x, x+e_j)$, $1 \le j \le d$. D_j belongs to $\ell^2(\mathbb{Z}^d)$, the space of square summable series.

Denote by $\hat{D}_j : (-\pi, \pi]^d \to \mathbb{R}$ the Fourier transform of D_j:

$$\hat{D}_j(\theta) = \sum_{x\in\mathbb{Z}^d} e^{ix\cdot\theta} D_j(x),$$

where $\cdot$ stands for the usual scalar product in $\mathbb{R}^d$. Since D is an exact form, by (1.21), $D_j(x) + D_k(x+e_j) = D_k(x) + D_j(x+e_k)$. Thus, $\hat{D}_j(\theta)(1 - e^{-i\theta_k}) = \hat{D}_k(\theta)(1 - e^{-i\theta_j})$. In particular, $\hat{D}_j(\theta) = (1 - e^{-i\theta_j})b(\theta)$ for some function $b : (-\pi, \pi]^d \to \mathbb{R}$.

It follows from the identity $\Delta f = 0$ that $\sum_{1\le j\le d} \hat{D}_j(\theta)(1 - e^{i\theta_j}) = 0$. Therefore, $0 = b(\theta) \sum_{1\le j\le d} (1 - \cos\theta_j)$. Since the latter function is strictly positive for $\theta \ne 0$, $b(\theta) = 0$. This proves that D_j vanishes and therefore the claim.

1.6.2 The Space $\mathscr{H}_{-1}$

For a function f in $L^2(\pi)$, let

$$\|f\|_{-1}^2 = \sup_{g\in L^2(\pi)} \big\{ 2\langle f, g\rangle_\pi - \langle g, g\rangle_1 \big\}. \tag{1.23}$$

It is easy to check that this variational formula defines a semi-norm.

A. $\mathscr{H}_{-1} \subset L^2(\pi)$ Fix a function f in $L^2(\pi)$. Since $\|g\|_1^2 \le 2\|g\|^2$, it follows from the variational formula (1.23) that $2\|f\|_{-1}^2 \ge \|f\|^2$. In particular, $\|\cdot\|_{-1}$ is a norm in $L^2(\pi)$. Observe also that $\|\mathbf{1}\|_{-1} = \infty$. Indeed, if we take $g = a\mathbf{1}$, $a > 0$, in the definition of the norm $\|\cdot\|_{-1}$, we get that $\|\mathbf{1}\|_{-1} \ge a$ because $\langle \mathbf{1}, \mathbf{1}\rangle_1 = 0$. Letting $a \uparrow \infty$, we conclude.

B. The $\mathscr{H}_{-1}$ Space Denote by $\mathscr{H}_{-1}$ the set of functions in $L^2(\pi)$ with finite $\|\cdot\|_{-1}$ norm: $\mathscr{H}_{-1} = \{h \in L^2(\pi), \|h\|_{-1} < \infty\}$. The set $\mathscr{H}_{-1}$ endowed with the norm $\|\cdot\|_{-1}$ is complete.

Consider a Cauchy sequence $\{f_n : n \ge 1\}$ in $\mathscr{H}_{-1}$. By **A** above, $\{f_n : n \ge 1\}$ is a Cauchy sequence in $L^2(\pi)$. Denote by f its limit in $L^2(\pi)$.

The function f belongs to $\mathscr{H}_{-1}$. Let $C_0 := \sup_{n\ge1} \|f_n\|_{-1}^2$, which is finite because $\{f_n : n \ge 1\}$ is a Cauchy sequence. Assume by contradiction that $\|f\|_{-1} = \infty$. Then, by definition of the $\mathscr{H}_{-1}$ norm, there exists g in $L^2(\pi)$ such that $2\langle f, g\rangle_\pi - \|g\|_1^2 \ge C_0 + 2$. Since f_n converges to f in $L^2(\pi)$, $2\langle f - f_n, g\rangle_\pi$ is absolutely bounded by 1 for n large enough. Therefore, $\|f_n\|_{-1}^2 \ge 2\langle f_n, g\rangle_\pi - \|g\|_1^2 \ge C_0 + 1$, contradicting the definition of C_0.

The sequence $\{f_n\}$ converges to f in $\mathscr{H}_{-1}$. Fix $\epsilon > 0$. Since $\{f_n\}$ is a Cauchy sequence in $\mathscr{H}_{-1}$, there exists $n_0 \ge 1$ such that $\|f_m - f_n\|_{-1}^2 \le \epsilon$ for $n, m \ge n_0$. Fix $n \ge n_0$. There exists g in $L^2(\pi)$ such that $\|f_n - f\|_{-1}^2 \le 2\langle f_n - f, g\rangle_\pi - \|g\|_1^2 + \epsilon$. Introducing f_m in the linear term, the right-hand side of the previous inequality becomes

$$2\langle f_n - f_m, g\rangle_\pi - \|g\|_1^2 + 2\langle f_m - f, g\rangle_\pi + \epsilon.$$

The first two terms are bounded by $\|f_n - f_m\|_{-1}^2$, which is smaller than ϵ for $m \ge n_0$. Thus, $\|f_n - f\|_{-1}^2 \le 2\epsilon + 2\langle f_m - f, g\rangle_\pi$ for m large enough. Since g depends on f_n but not on f_m, the last term vanishes as $m \uparrow \infty$ because f_m converges to f in $L^2(\pi)$. This concludes the proof.

C. The Parallelogram Identity The norm $\|\cdot\|_{-1}$ satisfies the parallelogram identity:

$$\|f + g\|_{-1}^2 + \|f - g\|_{-1}^2 = 2\|f\|_{-1}^2 + 2\|g\|_{-1}^2$$

for all f, g in $\mathscr{H}_{-1}$. This follows from the parallelogram identity for the $\mathscr{H}_1$ norm. By Sect. 87 in Riesz and Sz.-Nagy (1990), $\mathscr{H}_{-1}$ is thus a Hilbert space with scalar product $\langle\cdot,\cdot\rangle_{-1}$ given by polarization:

$$\langle f, g\rangle_{-1} = \frac{1}{4}\{\|f + g\|_{-1}^2 - \|f - g\|_{-1}^2\}.$$

Hence, $\mathscr{H}_{-1}$ is a Hilbert space contained in $L^2(\pi)$ which itself is contained in $\mathscr{H}_1$:

$$\mathscr{H}_{-1} \subset L^2(\pi) \subset \mathscr{H}_1.$$

D. Representation of Elements of $\mathscr{H}_{-1}$ Any function g in $\mathscr{H}_{-1}$ can be represented as $(I-P)G$ for some function G in $\mathscr{H}_1$ and $\|g\|_{-1} = \|G\|_1$.

Fix a function f in $\mathbb{D}$ so that $(I-P)f$ is well defined. Schwarz inequality shows that $(I-P)f$ belongs to $L^2(\pi)$. In particular, $(I-P)f$ belongs to $L^2(\pi)$ for any f in $\mathscr{H}_1$.

Fix a function g in $\mathscr{H}_{-1} \subset L^2(\pi)$. Since g belongs to $\mathscr{H}_{-1}$, the linear functional $\mathscr{L}_g : L^2(\pi) \to \mathbb{R}$ defined by $\mathscr{L}_g(f) = \langle g, f\rangle_\pi$ is bounded in $\mathscr{H}_1$. By Riesz representation theorem, there exists G in $\mathscr{H}_1$ such that $\langle g, f\rangle_\pi = \langle G, f\rangle_1$ for all f in $L^2(\pi)$ and $\|g\|_{-1} = \|G\|_1$.

It follows from the explicit formula (1.20) for the scalar product in $\mathscr{H}_1$ and an elementary computation, similar to the one performed at the beginning of this section, that

$$\langle G, f\rangle_1 = -\sum_{x,y\in E} \pi(x)P(x,y)f(x)\big\{G(y) - G(x)\big\} = \big\langle f, (I-P)G\big\rangle_\pi$$

for any function f in $L^2(\pi)$ and any G in $\mathscr{H}_1$. Therefore, the last identity shows that $\langle f, (I-P)G\rangle_\pi = \langle f, g\rangle_\pi$ for any f in $L^2(\pi)$. Recall that we proved in the beginning of section **D** that $(I-P)G$ belongs to $L^2(\pi)$. It thus follows from the previous identity that $g = (I-P)G$.

The above computation shows that the map $(I-P) : \mathscr{H}_1 \to \mathscr{H}_{-1}$ is an isometry: $\|(I-P)f\|_{-1} = \|f\|_1$. We showed in the first part of the proof that this map is surjective. Thus, formally $(I-P)\mathscr{H}_1 = \mathscr{H}_{-1}$.

E. Schwarz Inequality For every f in $\mathscr{H}_{-1}$ and g in $L^2(\pi)$,

$$\langle f, g\rangle_\pi^2 \le \langle f, f\rangle_{-1}\langle g, g\rangle_1. \tag{1.24}$$

Indeed, by (1.23), for every a in $\mathbb{R}$, $2a\langle f, g\rangle_\pi - a^2\langle g, g\rangle_1 \le \langle f, f\rangle_{-1}$. Maximizing over a we obtain the result.

F. Alternative Formula for the $\mathscr{H}_{-1}$ Norm

Lemma 1.16 *A function V in $L^2(\pi)$ belongs to $\mathscr{H}_{-1}$ if and only if there exists a finite constant C such that $\langle V, g\rangle_\pi^2 \le C\langle g, g\rangle_1$ for every g in $L^2(\pi)$. Moreover,*

$$\|V\|_{-1}^2 = \inf\left\{C > 0 : \frac{\langle V, g\rangle_\pi^2}{\langle g, g\rangle_1} \le C \textit{ for all } g \in L^2(\pi);\ \langle g, g\rangle_1 \ne 0\right\}. \tag{1.25}$$

Proof Assume the existence of a finite constant C such that

$$\langle V, g\rangle_\pi^2 \le C\langle g, g\rangle_1$$

for every g in $L^2(\pi)$. By (1.23), $\langle V, V\rangle_{-1}$ is bounded above by $\sup_{a>0}\{2\sqrt{C}a - a^2\} = C$. This proves that V belongs to $\mathscr{H}_{-1}$ and that $\langle V, V\rangle_{-1} \le A$, if A stands for the infimum on the right-hand side of (1.25). Assume now that V belongs to $\mathscr{H}_{-1}$. By (1.24), for any function f in $L^2(\pi)$, $\langle f, V\rangle_\pi^2 \le \langle V, V\rangle_{-1}\langle f, f\rangle_1$, proving the reverse assertion and that $A \le \langle V, V\rangle_{-1}$. □

G. Alternative Characterization of $\mathscr{H}_{-1}$ Since $I-P$ is a positive, bounded, self-adjoint operator, we may define the operator $\sqrt{I-P}$. The next lemma provides a characterization of the space $\mathscr{H}_{-1}$ in terms of this operator.

Lemma 1.17 *A function V in $L^2(\pi)$ belongs to $\mathscr{H}_{-1}$ if and only if there exists f in $L^2(\pi)$ such that $V=\sqrt{I-P}f$. In this case, $\langle V,V\rangle_{-1}=\langle f,f\rangle_\pi$.*

Proof Assume first that $V=\sqrt{I-P}f$ for some f in $L^2(\pi)$. Fix g in $L^2(\pi)$. By the symmetry of P and by Schwarz inequality, $\langle V,g\rangle_\pi^2=\langle\sqrt{I-P}f,g\rangle_\pi^2=\langle f,\sqrt{I-P}g\rangle_\pi^2\le\langle f,f\rangle_\pi\langle g,(I-P)g\rangle_\pi=\langle f,f\rangle_\pi\langle g,g\rangle_1$. In particular, by Lemma 1.16, V belongs to $\mathscr{H}_{-1}$ and $\langle V,V\rangle_{-1}\le\langle f,f\rangle_\pi$.

Assume now that V belongs to $\mathscr{H}_{-1}$. We want to show the existence of f in $L^2(\pi)$ such that $V=\sqrt{I-P}f$. The natural candidate is of course

$$f=\{I-P\}^{-1/2}V,$$

but $\sqrt{I-P}$ is not invertible and $\{I-P\}^{-1/2}V$ is not defined. To circumvent this problem, we introduce a sequence of approximating functions by setting $f_\lambda=\{(1+\lambda)I-P\}^{-1/2}V$ for $\lambda\downarrow 0$. The idea is to prove that f_λ converges, as $\lambda\downarrow 0$, to some function f which solves the equation $V=\sqrt{I-P}f$.

We first prove that the sequence f_λ is bounded in $L^2(\pi)$ and admits therefore a weakly converging subsequence. Fix g in $L^2(\pi)$. By definition of f_λ,

$$\langle g,f_\lambda\rangle_\pi^2=\big\langle g,\big\{(1+\lambda)I-P\big\}^{-1/2}V\big\rangle_\pi^2=\big\langle\big\{(1+\lambda)I-P\big\}^{-1/2}g,V\big\rangle_\pi^2.$$

Since V belongs to $\mathscr{H}_{-1}$, by (1.24), the previous expression is bounded above by

$$\langle V,V\rangle_{-1}\big\langle\big\{(1+\lambda)I-P\big\}^{-1/2}g,(I-P)\big\{(1+\lambda)I-P\big\}^{-1/2}g\big\rangle_\pi.$$

Since $I-P\le(1+\lambda)I-P$, the last term is less than or equal to $\langle\{(1+\lambda)I-P\}^{-1/2}g,[(1+\lambda)I-P]\{(1+\lambda)I-P\}^{-1/2}g\rangle_\pi=\langle g,g\rangle_\pi$. We have thus obtained that for all functions g in $L^2(\pi)$,

$$\langle g,f_\lambda\rangle_\pi^2\le\langle V,V\rangle_{-1}\langle g,g\rangle_\pi,$$

which proves that $\{f_\lambda,0<\lambda\le 1\}$ is a bounded sequence in $L^2(\pi)$ such that $\langle f_\lambda,f_\lambda\rangle_\pi\le\langle V,V\rangle_{-1}$.

Denote by f a weak limit point and assume, without loss of generality, that f_λ converges weakly to f in $L^2(\pi)$. First of all, by the weak lower semi-continuity of norms,

$$\langle f,f\rangle_\pi\le\langle V,V\rangle_{-1}\tag{1.26}$$

because mass can only be lost. We claim also that $V=\sqrt{I-P}f$. To prove this identity, we only need to show that $\langle V,g\rangle_\pi=\langle\sqrt{I-P}f,g\rangle_\pi$ for all g in $L^2(\pi)$ or, equivalently, that $\langle V,g\rangle_\pi=\langle f,\sqrt{I-P}g\rangle_\pi$. Since f is the weak limit of f_λ,

$$\langle f,\sqrt{I-P}g\rangle_\pi=\lim_{\lambda\to 0}\langle f_\lambda,\sqrt{I-P}g\rangle_\pi$$

$$= \lim_{\lambda \to 0} \big\langle \{(1+\lambda)I - P\}^{-1/2} V, \sqrt{I-P}g \big\rangle_\pi$$
$$= \lim_{\lambda \to 0} \big\langle V, \{(1+\lambda)I - P\}^{-1/2} \sqrt{I-P}g \big\rangle_\pi .$$

The last scalar product is equal to

$$\langle V, g\rangle_\pi + \big\langle V, \{(1+\lambda)I - P\}^{-1/2} \sqrt{I-P}g - g \big\rangle_\pi .$$

To conclude the proof we need to show that the second term vanishes as $\lambda \downarrow 0$. By Schwarz inequality, the square of the second term is bounded above by $\langle V, V\rangle_\pi$ times

$$\big\langle \big[\{(1+\lambda)I - P\}^{-1/2} \sqrt{I-P} - I\big]g, \big[\{(1+\lambda)I - P\}^{-1/2} \sqrt{I-P} - I\big]g \big\rangle_\pi .$$

By the spectral decomposition of P, this scalar product is equal to

$$\int_{-1}^{1} \left(\frac{\sqrt{1-\varphi}}{\sqrt{1+\lambda-\varphi}} - 1 \right)^2 \mu_g(d\varphi).$$

The integrand is bounded and converges to 0 as $\lambda \downarrow 0$. In particular, by the dominated convergence theorem, the previous integral vanishes in the limit, which proves that $V = \sqrt{I-P}f$.

Up to this point, we have shown that a function V in $L^2(\pi)$ belongs to $\mathcal{H}_{-1}$ if and only if there exists f in $L^2(\pi)$ such that $V = \sqrt{I-P}f$. We have also seen at the beginning of the proof that in the case of such a representation, $\langle V, V\rangle_{-1} \le \langle f, f\rangle_\pi$. It remains to recall (1.26) to conclude the proof of the lemma. □

It follows from the previous lemma that

$$\langle V, V\rangle_{-1} = \big\langle \{I-P\}^{-1/2} V, \{I-P\}^{-1/2} V \big\rangle_\pi$$

so that

$$\langle V, V\rangle_{-1} = \int_{-1}^{1} \frac{1}{1-\varphi} \mu_V(d\varphi).$$

In particular, for the variance $\sigma^2(V)$ defined in (1.15) to be finite it is necessary and sufficient that V belongs to $\mathcal{H}_{-1}$.

1.7 Comments and References

Central Limit Theorems for martingales can be found in many textbooks, Billingsley (1995); Durrett (1996); Ethier and Kurtz (1986); Varadhan (2001), for instance. We refer to Whitt (2007) for a recent account.

To our knowledge, the first central limit theorem for Markov chains goes back to Doeblin (1938) who reduced the problem to the case of independent identically distributed random variables. We refer to Nagaev (1957) for a proof along the line of Doeblin's idea. Gordin (1969) and Gordin and Lifšic (1978) showed that

$$\frac{1}{\sqrt{N}}\sum_{j=0}^{N-1} V(X_j) \tag{1.27}$$

converges to a mean zero Gaussian random variable if V belongs to the range of the operator $I - P$ in $L^2(\pi)$. Lawler (1982) proved an invariance principle for a Markov chain in random environment.

Kozlov (1985) and Kipnis and Varadhan (1986) proposed independently a general method to prove central limit theorems for additive functionals of Markov chains from martingale central limit theorems. The approach presented here follows Kipnis and Varadhan (1986). This seminal paper has been the starting point of much research on asymptotic normality of additive functionals of ergodic Markov chains which is reviewed in the following chapters. De Masi et al. (1989) and Goldstein (1995) considered anti-symmetric additive functionals of reversible Markov chains.

Maxwell and Woodroofe (2000) proved that the sequence (1.27) is asymptotically normal for stationary ergodic Markov chains $\{X_j : j \geq 0\}$ provided V has mean zero with respect to the stationary measure π and

$$\sum_{n\geq 1} n^{-3/2}\left\| \sum_{j=0}^{n-1} P^j V \right\| < \infty .$$

Quenched Invariance Principles Derriennic and Lin (2001a, 2003) presented an alternative proof of the previous result which requires slightly stronger assumptions but which holds for Markov chains starting from a fixed point. An almost sure invariance principle is also proved in Rassoul-Agha and Seppäläinen (2008) based on a bound on the moment of the resolvent.

Conditional Invariance Principles Wu and Woodroofe (2004) and Zhao and Woodroofe (2008b) obtained necessary and sufficient conditions for conditional asymptotic normality. Ouchti and Volný (2008) presented an example of a Markov chain which satisfies central limit theorem in probability, as stated in Remark 1.12, but which does not satisfy a quenched central limit theorem. Relying on a maximal inequality for stationary sequences of random variables, Peligrad and Utev (2005) proved a conditional invariance principle for (1.27). Let us mention also that the law of the iterated logarithm is deduced in Zhao and Woodroofe (2008a).

Positively and Negatively Dependent Random Variables A function $f : \mathbb{R}^d \to \mathbb{R}$ is said to be increasing if $f(x_1, \dots, x_d) \leq f(y_1, \dots, y_d)$ whenever $x_i \leq y_i$, $1 \leq i \leq d$. A random vector $\{X_1, \dots, X_d\}$ is said to satisfy the FKG inequalities if for any increasing functions f, g such that $E[f(X_1, \dots, X_d)^2] < \infty$,

$E[g(X_1, \ldots, X_d)^2] < \infty$, $\mathrm{Cov}(f(X_1, \ldots, X_d), g(X_1, \ldots, X_d)) \geq 0$. A sequence of random variables is said to satisfy the FKG inequalities if every finite subset satisfies them.

Newman (1980) and Newman and Wright (1981) proved an invariance principle for a sequence of stationary random variables with finite second moments, satisfying the FKG inequalities, and such that $\sum_k \mathrm{Cov}(X_0, X_k)$ is finite. The central limit theorem is generalized in Newman (1983) to sequences of functions, not necessarily monotone, of variables satisfying the FKG inequalities. Newman (1984) presents a review of these results as well as a survey of results which give independence as a consequence of zero covariance for associated random variables.

We mention also Newman and Wright (1982) where some well-known martingales inequalities are extended to dependent random variables satisfying FKG inequalities.

Time-Variance We obtained in (1.15) an explicit formula for the asymptotic variance and observed in Remark 1.11 that the time-variance converges. Häggström and Rosenthal (2007) compare three different expressions for the asymptotic variance in the central limit theorem for Markov chains in a more general context than ours.

Exact Forms Exact forms on graphs and their connection to Markov chains and to the D-Liouville property are discussed in Soardi (1994).

Poisson Equation We have seen in this chapter the close connexion between solutions of the Poisson equation and the central limit theorem. We proved, in fact, a central limit theorem under the condition that $(I - P)^{1/2} f = V$ has a solution f in $L^2(\pi)$. Derriennic and Lin (2001b) examines the fractional Poisson equation

$$(I - P)^{\alpha} f = V,$$

for $0 < \alpha < 1$.

References

Billingsley P (1995) Probability and measure. Wiley series in probability and mathematical statistics: probability and mathematical statistics, 3rd edn. Wiley, New York. A Wiley-Interscience Publication

Breiman L (1968) Probability. Addison-Wesley, Reading

De Masi A, Ferrari PA, Goldstein S, Wick WD (1989) An invariance principle for reversible Markov processes. Applications to random motions in random environments. J Stat Phys 55(3–4):787–855

Derriennic Y, Lin M (2001a) The central limit theorem for Markov chains with normal transition operators, started at a point. Probab Theory Relat Fields 119(4):508–528

Derriennic Y, Lin M (2001b) Fractional Poisson equations and ergodic theorems for fractional coboundaries. Isr J Math 123:93–130

Derriennic Y, Lin M (2003) The central limit theorem for Markov chains started at a point. Probab Theory Relat Fields 125(1):73–76

Doeblin W (1938) Sur deux problèmes de M Kolmogoroff concernant les chaînes dénombrables. Bull Soc Math Fr 66:210–220

Durrett R (1996) Probability: theory and examples, 2nd edn. Duxbury Press, Belmont

Ethier SN, Kurtz TG (1986) Markov processes: characterization and convergence. Wiley series in probability and mathematical statistics: probability and mathematical statistics. Wiley, New York

Goldstein S (1995) Antisymmetric functionals of reversible Markov processes. Ann Inst Henri Poincaré Probab Stat 31(1):177–190

Gordin MI (1969) The central limit theorem for stationary processes. Dokl Akad Nauk SSSR 188:739–741

Gordin MI, Lifšic BA (1978) Central limit theorem for stationary Markov processes. Dokl Akad Nauk SSSR 239(4):766–767

Häggström O, Rosenthal JS (2007) On variance conditions for Markov chain CLTs. Electron Commun Probab 12:454–464 (electronic)

Kipnis C, Varadhan SRS (1986) Central limit theorem for additive functionals of reversible Markov processes and applications to simple exclusions. Commun Math Phys 104(1):1–19

Kozlov SM (1985) The averaging method and walks in inhomogeneous environments. Usp Mat Nauk 40(2(242)):61–120, 238

Lawler GF (1982) Weak convergence of a random walk in a random environment. Commun Math Phys 87(1):81–87

Maxwell M, Woodroofe M (2000) Central limit theorems for additive functionals of Markov chains. Ann Probab 28(2):713–724

Nagaev SV (1957) Some limit theorems for stationary Markov chains. Teor Veroâtn Ee Primen 2:389–416

Newman CM (1980) Normal fluctuations and the FKG inequalities. Commun Math Phys 74(2):119–128

Newman CM (1983) A general central limit theorem for FKG systems. Commun Math Phys 91(1):75–80

Newman CM (1984) Asymptotic independence and limit theorems for positively and negatively dependent random variables. In: Inequalities in statistics and probability, Lincoln, NE, 1982. IMS lect notes monogr ser, vol 5. Inst Math Stat, Hayward, pp 127–140

Newman CM, Wright AL (1981) An invariance principle for certain dependent sequences. Ann Probab 9(4):671–675

Newman CM, Wright AL (1982) Associated random variables and martingale inequalities. Z Wahrscheinlichkeitstheor Verw Geb 59(3):361–371

Ouchti L, Volný D (2008) A conditional CLT which fails for ergodic components. J Theor Probab 21(3):687–703

Peligrad M, Utev S (2005) A new maximal inequality and invariance principle for stationary sequences. Ann Probab 33(2):798–815

Rassoul-Agha F, Seppäläinen T (2008) An almost sure invariance principle for additive functionals of Markov chains. Stat Probab Lett 78(7):854–860

Riesz F, Sz.-Nagy B (1990) Functional analysis. Dover books on advanced mathematics. Dover, New York. Translated from the second French edition by Leo F Boron, reprint of the 1955 original

Soardi PM (1994) Potential theory on infinite networks. Lecture notes in math, vol 1590. Springer, Berlin

Varadhan SRS (2001) Probability theory. Courant lecture notes in mathematics, vol 7. New York University, Courant Institute of Mathematical Sciences, New York

Whitt W (2007) Proofs of the martingale FCLT. Probab Surv 4:268–302

Wu WB, Woodroofe M (2004) Martingale approximations for sums of stationary processes. Ann Probab 32(2):1674–1690

Yosida K (1995) Functional analysis. Classics in mathematics. Springer, Berlin. Reprint of the sixth 1980 edition

Zhao O, Woodroofe M (2008a) Law of the iterated logarithm for stationary processes. Ann Probab 36(1):127–142

Zhao O, Woodroofe M (2008b) On martingale approximations. Ann Appl Probab 18(5):1831–1847

Chapter 2
Central Limit Theorems

In this chapter, we extend the ideas introduced in Chap. 1 to continuous-time Markov processes on general state spaces with a stationary ergodic measure allowed to be non-reversible. We refer to Chap. 1 of Blumenthal and Getoor (1968), Chaps. 1 and 2 of Chung and Walsh (2005), Chap. 1 of Ethier and Kurtz (1986) and Chap. 3 of Rogers and Williams (2000) for the terminology and the basic properties of Markov processes appearing in this chapter.

Let E be a complete and separable metric space endowed with its Borel σ-algebra $\mathscr{E}$. Denote by $B(E)$ the set of bounded measurable functions on E and by $C_0(E)$ the space of continuous functions on E which vanish at infinity, regarded as a Banach space with the norm $\|f\| = \sup_x |f(x)|$.

Let $D([0,\infty), E)$ be the set of all functions $X : [0,\infty) \to E$ which are right continuous and have left limits (r.c.l.l.). Denote by $\Pi_s : D([0,\infty), E) \to E$, $s \geq 0$, the canonical projections defined by $\Pi_s(X) = X_s$. Let $\mathscr{F}^o$ be the smallest σ-algebra on $D([0,\infty), E)$ which turns the projections Π_s, $s \geq 0$, measurable; and let $\mathscr{F}_t^o$ be the natural filtration, i.e. the smallest σ-algebra relative to which all the mappings Π_s, $0 \leq s \leq t$, are measurable.

Let $\{P_t : t \geq 0\}$ be a strictly Markovian, Feller semigroup of linear operators on $C_0(E)$ (Rogers and Williams, 2000, Definition III.6.5), (Liggett, 1985, Definition I.1.4). Strictly Markovian means that $P_t\mathbf{1} = \mathbf{1}$ for all $t \geq 0$, where $\mathbf{1}$ is the function constant equal to 1, which does not belong to $C_0(E)$ if E is not compact, but the semigroup $\{P_t : t \geq 0\}$ can clearly be extended to $B(E)$.

Consider a normal Markov process on E associated to the semigroup $\{P_t : t \geq 0\}$. This is a family $\{\mathbb{P}_x : x \in E\}$ of probability measures defined on $(D([0,\infty), E), \mathscr{F}^o)$ such that

(1) $\mathbb{P}_x[X_0 = x] = 1$ for all $x \in E$ (normality);
(2) For every $A \in \mathscr{F}^o$, the mapping $x \to \mathbb{P}_x[A]$ is measurable;
(3) For all $x \in E$, $f \in C_0(E)$,

$$\mathbb{E}_x\big[f(X_{t+s})|\mathscr{F}_s^o\big] = (P_t f)(X_s), \quad \mathbb{P}_x \text{ a.s.},$$

where $\mathbb{E}_x$ stands for the expectation with respect to $\mathbb{P}_x$.

T. Komorowski et al., *Fluctuations in Markov Processes*,
Grundlehren der mathematischen Wissenschaften 345,
DOI 10.1007/978-3-642-29880-6_2, © Springer-Verlag Berlin Heidelberg 2012

The reader will find in Chap. 1 of Blumenthal and Getoor (1968) and in Sect. III.7 of Rogers and Williams (2000) a proof of the existence and uniqueness of a normal Markov process associated to a Feller semigroup.

For a probability measure μ in $(E, \mathscr{E})$, denote by $\mathbb{P}_\mu$ the measure on $(D([0,\infty), E), \mathscr{F}^o)$ given by $\int \mathbb{P}_x \, \mu(dx)$. Expectation with respect to $\mathbb{P}_\mu$ is denoted by $\mathbb{E}_\mu$. Assume that a probability measure π on E is stationary for the semigroup: $\langle P_t f\rangle_\pi = \langle f\rangle_\pi$ for all $f \in C_0(E)$, where $\langle\cdot\rangle_\pi$ stands for the expectation with respect to π.

Denote by $(D([0,\infty), E), \mathscr{F}, \mathbb{P}_\pi, \{\mathscr{F}_t : t \ge 0\})$ the usual augmentation of the filtered space $(D([0,\infty), E), \mathscr{F}^o, \mathbb{P}_\pi, \{\mathscr{F}_t^o : t \ge 0\})$ which, by definition, satisfies the usual conditions (Rogers and Williams, 2000, Sect. II.67). By Theorem 8.11 and Proposition 8.12 of Chap. 1 of Blumenthal and Getoor (1968) the right continuous Feller process $\{X_t : t \ge 0\}$ is strong Markov with respect to the augmented filtration.

The property that $\{X_t : t \ge 0\}$ is a strong Markov process with respect to a filtration which satisfies the usual condition is needed to be able to define later the predictable quadratic variation of the Dynkin martingales, which requires the use of the Doob–Meyer decomposition theorem.

Let $L^2(\pi)$ be the Hilbert space of π-square integrable functions and denote by $\langle\cdot,\cdot\rangle_\pi$ the scalar product in $L^2(\pi)$ and by $\|\cdot\|_\pi$ the norm associated to this scalar product. Denote by $L^p(\pi)$, $p \ge 1$, the space of measurable functions $f : E \to \mathbb{R}$ such that $\langle |f|^p\rangle_\pi < \infty$. The semigroup $\{P_t : t \ge 0\}$ extends then to a semigroup of positive contractions on any $L^p(\pi)$, $p \ge 1$. We assume that this extension is strongly continuous for any $p \in [1, +\infty)$. We suppose furthermore that the measure π is ergodic: any $f \in L^1(\pi)$ such that $P_t f = f$ for all $t \ge 0$ is constant π-a.e. Let L be the generator of the semigroup in $L^2(\pi)$ with $\mathscr{D}(L)$ denoting its domain. Let $\mathscr{C} \subset \mathscr{D}(L)$ be a core for the operator L (Liggett, 1985, Definition I.2.11). Denote by L^* the adjoint of L in $L^2(\pi)$ and assume that $\mathscr{C} \subset \mathscr{D}(L^*)$. Since π is stationary, L^* is itself the generator of a Markov process. On $\mathscr{C}$ we can define $S = (1/2)(L + L^*)$ and $A := (1/2)(L - L^*)$ the symmetric and anti-symmetric parts of the generator. We shall suppose that S is essentially self-adjoint.

In this chapter, we will use repeatedly the following property. If $\mathscr{C}$ is a core for an operator G, for any function f in the domain of G, there exists a sequence $\{f_k : k \ge 1\}$ in $\mathscr{C}$ such that f_k, Gf_k converge to f, Gf, respectively.

Recall that we denote by ω a trajectory of $D(\mathbb{R}_+, E)$. Let $\{\theta_t : t \ge 0\}$ be the semigroup of shift operators, $\theta_t : D(\mathbb{R}_+, E) \to D(\mathbb{R}_+, E)$, $(\theta_t\omega)(s) = \omega(t+s)$. Since π is a stationary ergodic measure, $\mathbb{P}_\pi$ is invariant and ergodic under the flow of transformations $\{\theta_t : t \ge 0\}$.

Fix a function $V : E \to \mathbb{R}$ in $L^2(\pi)$ such that $\langle V\rangle_\pi = 0$. The object of this section is to find conditions on V which guarantee a central limit theorem for

$$\frac{1}{\sqrt{t}} \int_0^t V(X_s)\, ds. \tag{2.1}$$

Assume first that there exists a solution f in $\mathscr{D}(L)$ of the Poisson equation

$$-Lf = V. \tag{2.2}$$

In this case a central limit theorem follows from the central limit theorem for martingales stated in Lemma 2.1 below. Indeed, since f belongs to the domain of the generator,

$$M_t = f(X_t) - f(X_0) - \int_0^t (Lf)(X_s)\,ds$$

is a martingale with respect to $\{\mathscr{F}_s : s \geq 0\}$. With the additional assumption that also f^2 belongs to $\mathscr{D}(L)$, the predictable quadratic variation of M_t is given by

$$\langle M, M\rangle_t = \int_0^t \big[(Lf^2)(X_s) - 2f(X_s)(Lf)(X_s)\big]\,ds$$

so that $\mathbb{E}_\pi[\langle M, M\rangle_t] = 2t\langle f, (-L)f\rangle_\pi$. If only f belongs to $\mathscr{D}(L)$, then the previous formula for the quadratic variation of M may not be valid, but $\langle M, M\rangle_t$ is still an increasing additive functional of the process with expectation given by $2t\langle f, -Lf\rangle_\pi$. This can be proven by standard approximation arguments.

Since f is the solution of the Poisson equation (2.2), we may write the additive functional in terms of the martingale M_t:

$$\frac{1}{\sqrt{t}}\int_0^t V(X_s)\,ds = \frac{M_t}{\sqrt{t}} + \frac{f(X_0) - f(X_t)}{\sqrt{t}}.$$

Since f belongs to $L^2(\pi)$ and the measure is stationary, $[f(X_0) - f(X_t)]/\sqrt{t}$ vanishes in $L^2(\mathbb{P}_\pi)$ as $t \uparrow \infty$. It remains to check that the martingale M_t satisfies the assumptions of Lemma 2.1.

The increments of the martingale M_t are stationary because X_t under $\mathbb{P}_\pi$ is stationary. On the other hand, since π is ergodic, in view of the formula for the quadratic variation of the martingale M_t, by the ergodic theorem, $t^{-1}\langle M, M\rangle_t$ converges in $L^1(\mathbb{P}_\pi)$, as $t \uparrow \infty$, to $2\langle f, (-L)f\rangle_\pi$. Since the martingale M_t has stationary increments, by Theorem 2.1, $t^{-1/2}M_t$, and therefore the law of $t^{-1/2}\int_0^t V(X_s)\,ds$, converges in distribution to a Gaussian law with mean zero and variance $\sigma^2(V) = 2\langle f, (-L)f\rangle_\pi$.

However, similarly to the discrete case, the existence of a solution f in $L^2(\pi)$ of the Poisson equation (2.2) is too strong a condition and can be weakened. The purpose of this chapter is to give sufficient conditions for the validity of the central limit theorem for an additive functional of the Markov process $\{X_t : t \geq 0\}$.

This chapter is conceived as follows. In Sect. 2.1, we prove a central limit theorem for continuous time right-continuous martingales. In Sect. 2.2, we introduce the Hilbert spaces $\mathscr{H}_1$ and $\mathscr{H}_{-1}$, already encountered in the previous chapter, which are related to the space of finite time-variance functions. In Sect. 2.3, the resolvent equation (2.13) is introduced and, in Sect. 2.5, the variance of $\int_0^t V(X_s)ds$ is estimated in terms of the $\mathscr{H}_{-1}$ norm of V. In Sect. 2.6, a central limit theorem for the additive function $\int_0^t V(X_s)ds$ is proved, in probability with respect to the initial condition, provided V has mean zero and belongs to $L^2(\pi) \cap \mathscr{H}_{-1}$. The proof relies solely on a uniform estimate (2.28) of the $\mathscr{H}_{-1}$ norm of Lf_λ, $0 < \lambda \leq 1$, where f_λ is the solution of the resolvent equation (2.13). In Sect. 2.7 we show that this condition (2.28) is fulfilled by Markov processes whose generator satisfy a sector, or a graded sector, condition and by Markov processes whose generator is a perturbation of a normal operator. In Sect. 2.8 we refine the previous arguments to the multidimensional case and we derive invariance principles. We conclude in Sect. 2.9 with comments on the literature.

2.1 Central Limit Theorem for Continuous Time Martingales

On a probability space $(\Omega, \mathbb{P}, \mathscr{F})$ consider a right continuous, square-integrable martingale $\{M_t : t \geq 0\}$ with respect to a given filtration $\{\mathscr{F}_t : t \geq 0\}$ satisfying the usual conditions. We refer to Jacod and Shiryaev (1987) for the terminology adopted and some elementary properties of martingales used without further comments. Assume that $M_0 = 0$ and denote by $\langle M, M\rangle_t$ its predictable quadratic variation. Denote by $\mathbb{E}$ expectation with respect to $\mathbb{P}$.

Theorem 2.1 *Assume that the increments of the martingale M_t are stationary: for every $t \geq 0$, $n \geq 1$ and $0 \leq s_0 < \cdots < s_n$, the random vectors $(M_{s_1} - M_{s_0}, \dots, M_{s_n} - M_{s_{n-1}})$, $(M_{t+s_1} - M_{t+s_0}, \dots, M_{t+s_n} - M_{t+s_{n-1}})$ have the same distribution. Assume also that the predictable quadratic variation converges in $L^1(\mathbb{P})$ to $\sigma^2 = \mathbb{E}M_1^2$:*

$$\lim_{n\to\infty} \mathbb{E}\left|\frac{\langle M, M\rangle_n}{n} - \sigma^2\right| = 0.$$

Then, the distribution of $M_t/\sqrt{t}$ conditioned on $\mathscr{F}_0$ converges in probability, as $t \uparrow \infty$, to a mean zero Gaussian law with variance σ^2:

$$\lim_{t\to\infty} \mathbb{E}\big[\big|\mathbb{E}\big[e^{i\theta M_t/\sqrt{t}}|\mathscr{F}_0\big] - e^{-\sigma^2\theta^2/2}\big|\big] = 0$$

for all θ in $\mathbb{R}$.

The proof of this theorem relies on the next lemma which reduces the problem to proving the central limit theorem for integer times.

Lemma 2.2 *Under the assumptions of Theorem* 2.1,

$$\lim_{n\to\infty} \mathbb{E}\Big[\sup_{n\leq t\leq n+1} \big|\mathbb{E}\big[e^{i\theta M_t/\sqrt{t}}|\mathscr{F}_0\big] - \mathbb{E}\big[e^{i\theta M_n/\sqrt{n}}|\mathscr{F}_0\big]\big|\Big] = 0.$$

Proof The difference of conditional expectations appearing in the statement of the lemma equals

$$\mathbb{E}\big[\big(\exp\big\{i\theta[M_t/\sqrt{t} - M_n/\sqrt{n}]\big\} - 1\big)e^{i\theta M_n/\sqrt{n}}|\mathscr{F}_0\big].$$

Since M_t is a martingale, we may subtract inside the parenthesis the expression $\theta(M_t/\sqrt{t} - M_n/\sqrt{n})$. Using the elementary estimate $|e^{ix} - 1 - ix| \le x^2/2$, valid for all $x \in \mathbb{R}$, we conclude that the norm of the previous expression is bounded by

$$\frac{\theta^2}{2}\mathbb{E}\left[\left(\frac{M_t}{\sqrt{t}} - \frac{M_n}{\sqrt{n}}\right)^2 \Big| \mathscr{F}_0\right].$$

Adding and subtracting $M_n/\sqrt{t}$, since $(a+b)^2 \le 2(a^2+b^2)$ and since $\langle M, M\rangle_t$ is the predictable quadratic variation of the martingale M_t, we obtain that the previous conditional expectation is less than or equal to

$$\frac{\theta^2}{t}\mathbb{E}\big[\langle M, M\rangle_t - \langle M, M\rangle_n | \mathscr{F}_0\big] + \theta^2\left(1 - \frac{\sqrt{n}}{\sqrt{t}}\right)^2 \mathbb{E}\left[\frac{\langle M, M\rangle_n}{n} \Big| \mathscr{F}_0\right].$$

We may still estimate this expression replacing t either by n or by $n+1$.

Up to this point we proved that the expectation appearing in the statement of the lemma is bounded by

$$\frac{\theta^2}{n}\mathbb{E}\big[\langle M, M\rangle_{n+1} - \langle M, M\rangle_n\big] + \theta^2\left(1 - \frac{\sqrt{n}}{\sqrt{n+1}}\right)^2 \mathbb{E}\left[\frac{\langle M, M\rangle_n}{n}\right].$$

It remains to recall that $\langle M, M\rangle_n/n$ converges in $L^1(\mathbb{P})$ to σ^2 as $n \uparrow \infty$. □

Proof of Theorem 2.1 We want to show that the conditional expectation $\mathbb{E}[\exp\{i\theta M_N/\sqrt{N}\}|\mathscr{F}_0]$ converges in $L^1(\mathbb{P})$ to $\exp\{-\theta^2\sigma^2/2\}$, as $N \uparrow \infty$, for all θ in $\mathbb{R}$. Since M_j is a martingale, the natural idea is to condition on $\mathscr{F}_j$ to obtain a recursive formula for $\mathbb{E}[\exp\{i\theta M_{j+1}/\sqrt{N}\}|\mathscr{F}_0]$.

Define the stationary sequence $Z_j := M_j - M_{j-1}$, $j \ge 1$. It is obviously adapted to the filtration $\{\mathscr{F}_j : j \ge 0\}$ and $M_0 = 0$, $M_j = \sum_{1 \le k \le j} Z_k$. Recall the expansion

$$e^{ia} = 1 + ia - a^2/2 - R(a)a^2,$$

where

$$R(a) := a^{-2}\int_0^a da_1 \int_0^{a_1} \big(e^{ix} - 1\big)\, dx.$$

Note that $|R(a)| \le 1$, $a \in \mathbb{R}$.

Denote by $\mathbb{E}_0[\cdot]$ the conditional expectation $\mathbb{E}[\cdot|\mathscr{F}_0]$. Since $\{M_j : j \ge 0\}$ is adapted and since $\mathbb{E}_0[Z_{j+1}|\mathscr{F}_j] = 0$,

$$\begin{aligned}
&\mathbb{E}_0\big[e^{i(\theta/\sqrt{N})M_{j+1}}\big] \\
&\quad = \mathbb{E}_0\big[e^{i(\theta/\sqrt{N})M_j}\,\mathbb{E}_0\big[e^{i(\theta/\sqrt{N})Z_{j+1}}|\mathscr{F}_j\big]\big] \\
&\quad = \mathbb{E}_0\left[e^{i(\theta/\sqrt{N})M_j}\left\{1 + \mathbb{E}_0\left[e^{i(\theta/\sqrt{N})Z_{j+1}} - 1 - \frac{i\theta}{\sqrt{N}}Z_{j+1}\Big|\mathscr{F}_j\right]\right\}\right].
\end{aligned}$$

It follows from the expansion of e^{ia} that

$$\mathbb{E}_0\big[e^{i(\theta/\sqrt{N})M_{j+1}}\big] = \mathbb{E}_0\big[e^{i(\theta/\sqrt{N})M_j}\big] - \frac{\theta^2}{2N}\mathbb{E}_0\big[e^{i(\theta/\sqrt{N})M_j}Z_{j+1}^2\big] - \frac{\theta^2}{N}\mathbb{E}_0\big[e^{i(\theta/\sqrt{N})M_j}Z_{j+1}^2 R(\theta Z_{j+1}/\sqrt{N})\big].$$

To derive a recursive formula for $\exp\{\theta^2\sigma^2 j/(2N)\}\mathbb{E}_0[\exp\{i(\theta/\sqrt{N})M_j\}]$, write

$$\begin{aligned} &e^{\theta^2\sigma^2(j+1)/(2N)}\mathbb{E}_0\big[e^{i(\theta/\sqrt{N})M_{j+1}}\big] - e^{\theta^2\sigma^2 j/(2N)}\mathbb{E}_0\big[e^{i(\theta/\sqrt{N})M_j}\big] \\ &\quad= e^{\theta^2\sigma^2(j+1)/(2N)}\big\{\mathbb{E}_0\big[e^{i(\theta/\sqrt{N})M_{j+1}}\big] - \mathbb{E}_0\big[e^{i(\theta/\sqrt{N})M_j}\big]\big\} \\ &\qquad+ e^{\theta^2\sigma^2(j+1)/(2N)}\big(1 - e^{-\theta^2\sigma^2/(2N)}\big)\mathbb{E}_0\big[e^{i(\theta/\sqrt{N})M_j}\big]. \end{aligned}$$

It follows from the previous identity that this expression equals

$$\begin{aligned} &-\frac{\theta^2}{2N}e^{\theta^2\sigma^2(j+1)/(2N)}\mathbb{E}_0\big[e^{i(\theta/\sqrt{N})M_j}\big(Z_{j+1}^2 - \sigma^2\big)\big] \\ &\quad-\frac{\theta^2}{N}e^{\theta^2\sigma^2(j+1)/(2N)}\mathbb{E}_0\big[e^{i(\theta/\sqrt{N})M_j}Z_{j+1}^2 R(\theta Z_{j+1}/\sqrt{N})\big] \\ &\quad+ e^{\theta^2\sigma^2(j+1)/(2N)}\left\{1 - \frac{(\theta\sigma)^2}{2N} - e^{-\theta^2\sigma^2/(2N)}\right\}\mathbb{E}_0\big[e^{i(\theta/\sqrt{N})M_j}\big]. \end{aligned}$$

Since $M_0 = 0$, adding from $j = 0$ up to $N - 1$ we obtain that $\exp\{\theta^2\sigma^2/2\}$ $\mathbb{E}_0[\exp\{i(\theta/\sqrt{N})M_N\}] - 1$ is equal to

$$\begin{aligned} &-\frac{\theta^2}{2N}\sum_{j=0}^{N-1} e^{\theta^2\sigma^2(j+1)/(2N)}\mathbb{E}_0\big[e^{i(\theta/\sqrt{N})M_j}\big(Z_{j+1}^2 - \sigma^2\big)\big] \\ &\quad-\frac{\theta^2}{N}\sum_{j=0}^{N-1} e^{\theta^2\sigma^2(j+1)/(2N)}\mathbb{E}_0\big[e^{i(\theta/\sqrt{N})M_j}Z_{j+1}^2 R(\theta Z_{j+1}/\sqrt{N})\big] \\ &\quad+\sum_{j=0}^{N-1} e^{\theta^2\sigma^2(j+1)/(2N)}\left\{1 - \frac{(\theta\sigma)^2}{2N} - e^{-\theta^2\sigma^2/(2N)}\right\}\mathbb{E}_0\big[e^{i(\theta/\sqrt{N})M_j}\big]. \end{aligned} \tag{2.3}$$

Observe that to conclude the proof of the theorem, it is enough to show that this expression vanishes in $L^1(\mathbb{P})$ as $N \uparrow \infty$. We examine each term separately. The third one is absolutely bounded by C_0/N for some finite deterministic constant C_0.

To estimate the second term, choose an arbitrary $\varepsilon > 0$ and bound the expectation of its absolute value by

$$\frac{C_0}{N}\sum_{j=0}^{N-1}\mathbb{E}\big[Z_{j+1}^2\big|R(\theta Z_{j+1}/\sqrt{N})\big|\mathbf{1}\big\{|\theta Z_{j+1}|\geq\varepsilon\sqrt{N}\big\}\big]$$

$$+\frac{C_0}{N}\sum_{j=0}^{N-1}\mathbb{E}\big[Z_{j+1}^2\big|R(\theta Z_{j+1}/\sqrt{N})\big|\mathbf{1}\big\{|\theta Z_{j+1}|<\varepsilon\sqrt{N}\big\}\big]$$

for some finite constant C_0 which depends only on θ and σ^2. From the stationarity of $\{Z_j : j \geq 1\}$, which follows from the stationarity of the increments of the martingale, and the boundedness of R, we have that the first term is less than or equal to

$$C_0\mathbb{E}\big[Z_1^2\mathbf{1}\big\{|\theta Z_1|\geq\varepsilon\sqrt{N}\big\}\big]$$

for some finite constant C_0. Since $Z_1 = M_1$ belongs to $L^2(\mathbb{P})$ this expression vanishes as $N\uparrow\infty$. On the other hand, the second term is bounded above by

$$C_0\sup_{|h|\leq\varepsilon}\big|R(h)\big|\mathbb{E}\big[Z_1^2\big].$$

This expression vanishes as $\varepsilon\downarrow 0$ in view of the explicit formula for the remainder R.

It remains to estimate the first term in (2.3). Since $M_t^2 - \langle M, M\rangle_t$ is martingale, taking conditional expectations, we replace Z_{j+1}^2 by $Y_{j+1} = \langle M, M\rangle_{j+1} - \langle M, M\rangle_j$. Note that the random variables $\{Y_j : j\geq 1\}$ are positive and that, by assumption, $N^{-1}\sum_{j=1}^N Y_j = \langle M, M\rangle_N/N$ converges in $L^1(\mathbb{P})$ to σ^2.

Fix $K\geq 1$ and divide the set $\Lambda_N = \{0,\dots,N-1\}$ in $\ell = [N/K]$ contiguous subintervals of size K or $K+1$:

$$\Lambda_N = \bigcup_{m=1}^{\ell} I_m,\qquad I_m\cap I_n = \varnothing\quad\text{for } m\neq n,$$

$$I_m = \{j_m,\dots,j_m+K-1\}\quad\text{or}\quad I_m = \{j_m,\dots,j_m+K\}$$

for some integer j_m. Here $[a]$ stands for the integer part of $a\in\mathbb{R}$.

We need to estimate

$$\frac{1}{N}\sum_{j=0}^{N-1}e^{j\beta/N}\mathbb{E}_0\big[e^{i(\theta/\sqrt{N})M_j}\big\{\sigma^2 - Y_{j+1}\big\}\big]$$

$$=\frac{1}{N}\sum_{k=1}^{\ell}\sum_{j\in I_k}e^{j\beta/N}\mathbb{E}_0\big[e^{i(\theta/\sqrt{N})M_j}\big\{\sigma^2 - Y_{j+1}\big\}\big]$$

for some fixed β in $\mathbb{R}$. To take advantage of the law of large numbers for the variables Y_j, we need to replace Y_{j+1} by the sum $\sum_{j\in I_k} Y_{j+1}$. With this in mind, we estimate the expectation of the absolute value of the previous expression by

$$\frac{CK}{N} + \frac{1}{N}\sum_{k=1}^{\ell} e^{j_k\beta/N}\mathbb{E}\Big[\Big|\sum_{j\in I_k} e^{i(\theta/\sqrt{N})M_j}\{\sigma^2 - Y_{j+1}\}\Big|\Big]$$

for some finite constant $C = C(\beta,\sigma)$. The second term of this sum is less than or equal to

$$\frac{C}{N}\sum_{k=1}^{\ell} |I_k|\mathbb{E}\Big[\Big|\frac{1}{|I_k|}\sum_{j\in I_k}\{\sigma^2 - Y_{j+1}\}\Big|\Big]$$
$$+\frac{C}{N}\sum_{k=1}^{\ell}\mathbb{E}\Big[\Big|\sum_{j\in I_k}\{e^{i(\theta/\sqrt{N})M_j} - e^{i(\theta/\sqrt{N})M_{j_k}}\}\{\sigma^2 - Y_{j+1}\}\Big|\Big]$$

for some finite constant $C = C(\beta)$. To keep notation simple, assume that all intervals I_m have the same length K. Since $\{Y_j,\, j\geq 1\}$ is a stationary sequence in $L^1(\mathbb{P})$, the first term is equal to

$$C\mathbb{E}\Big[\Big|\frac{1}{K}\sum_{j=1}^{K}\{\sigma^2 - Y_{j+1}\}\Big|\Big] = C\mathbb{E}\Big[\Big|\frac{\langle M,M\rangle_K}{K} - \sigma^2\Big|\Big],$$

which converges to 0 as $K\uparrow\infty$ by assumption. By the same reasons, the second term is bounded above by

$$\frac{C}{K}\sum_{j=1}^{K}\mathbb{E}\big[\big|e^{i(\theta/\sqrt{N})M_j} - 1\big|\{\sigma^2 + Y_{j+1}\}\big].$$

Since Y_j belongs to $L^1(\mathbb{P})$, by the dominated convergence theorem, for any fixed $K\geq 1$, this expression vanishes as $N\uparrow\infty$. This concludes the proof of the theorem. □

2.2 The Spaces $\mathscr{H}_1$ and $\mathscr{H}_{-1}$

We assume throughout this section that $\mathscr{C}$ is a core for the operators L and L^*. Consider the semi-norm $\|\cdot\|_1$ defined on the common core $\mathscr{C}$ by

$$\|f\|_1^2 = \langle f, (-L)f\rangle_\pi. \tag{2.4}$$

Let $\sim_1$ be the equivalence relation in $\mathscr{C}$ defined by $f\sim_1 g$ if $\|f-g\|_1 = 0$ and denote by $\mathscr{G}_1$ the normed space $(\mathscr{C}|_{\sim_1}, \|\cdot\|_1)$. It is easy to see from definition (2.4) that the norm $\|\cdot\|_1$ satisfies the parallelogram identity so that $\mathscr{H}_1$, the completion of $\mathscr{G}_1$ with respect to the norm $\|\cdot\|_1$, is a Hilbert space with inner product $\langle\cdot,\cdot\rangle_1$ given by polarization:

$$\langle f, g\rangle_1 = \frac{1}{4}\big\{\|f+g\|_1^2 - \|f-g\|_1^2\big\}.$$

Notice that in this definition only the symmetric part of the generator, $S = (1/2)(L+L^*)$, plays a role because

$$\|f\|_1^2 = \langle f, (-L)f\rangle_\pi = \langle f, (-S)f\rangle_\pi .$$

It is also easy to check that for any f, g in $\mathscr{C}$

$$\langle f, g\rangle_1 = \langle f, (-S)g\rangle_\pi \tag{2.5}$$

and that $\|c\|_1 = 0$ for any constant c. Note, finally, that in general neither $\mathscr{H}_1$ nor $L^2(\pi)$ are subspaces of each other, but in the case when L is bounded we have $L^2(\pi) \subset \mathscr{H}_1$.

By definition, $\mathscr{H}_1$ consists of sequences $\{f_n : n \ge 1\}$ of functions in the core $\mathscr{C}$ which are Cauchy in $\mathscr{H}_1$. We adopt the following convention. If such a sequence converges in $L^2(\pi)$ to some function f, we identify the sequence $\{f_n : n \ge 1\}$ with f and say that f belongs to $\mathscr{H}_1$.

There is no ambiguity in this postulate. If $\{f_n : n \ge 1\}$, $\{g_n : n \ge 1\}$ are two sequences of functions in the core $\mathscr{C}$ which are Cauchy in $\mathscr{H}_1$ and which converge in $L^2(\pi)$ to some function f, $f_n - g_n$ vanishes in $\mathscr{H}_1$ as $n \uparrow \infty$. Indeed, denote by F the limit in $\mathscr{H}_1$ of $f_n - g_n$ and fix some function h in the core $\mathscr{C}$. By (2.5) and since $\{f_n : n \ge 1\}$ and $\{g_n : n \ge 1\}$ converge in $L^2(\pi)$ to f,

$$\langle F, h\rangle_1 = \lim_{n\to\infty} \langle f_n - g_n, h\rangle_1 = \lim_{n\to\infty} \langle f_n - g_n, (-S)h\rangle_\pi = 0.$$

Since $\mathscr{C}$ is dense in $\mathscr{H}_1$, $F = 0$, which proves the claim.

This convention leads us to introduce the notion of approximating sequences. Consider a Hilbert space $\mathscr{H}$ constructed, as above, as the completion of a subspace $\mathscr{C}_0$ of $L^2(\pi)$ endowed with some scalar product $\langle\cdot,\cdot\rangle_{\mathscr{H}}$. A sequence of functions $\{f_n : n \ge 1\}$ in $\mathscr{C}_0$ such that f_n converges to f in $L^2(\pi)$ and f_n is Cauchy in $\mathscr{H}$ is called a $(\mathscr{C}_0, \mathscr{H})$-approximating sequence of f. When the context specifies the space $\mathscr{C}_0$, this sequence is called an $\mathscr{H}$-approximating sequence of f.

Hence, we declared a function f in $L^2(\pi)$ to belong to $\mathscr{H}_1$ if there exists a $(\mathscr{C}, \mathscr{H}_1)$-approximating sequence of f.

The domain $\mathscr{D}(S)$ is contained in $\mathscr{H}_1$. Indeed, fix a function f in $\mathscr{D}(S)$. Since, by assumption, $\mathscr{C}$ is a core for $\mathscr{D}(S)$, there exists a sequence $\{f_n : n \ge 1\}$ of functions in $\mathscr{C}$ such that f_n, Sf_n converge in $L^2(\pi)$ as $n \uparrow \infty$ to f, Sf, respectively. It follows from (2.5) that the sequence $\{f_n : n \ge 1\}$ is Cauchy in $\mathscr{H}_1$. Thus, $\{f_n : n \ge 1\}$ is a $(\mathscr{C}, \mathscr{H}_1)$-approximating sequence of f and $\mathscr{D}(S) \subset \mathscr{H}_1$. The same argument applies to $\mathscr{D}(L)$, $\mathscr{D}(L^*)$.

Associated to the Hilbert space $\mathscr{H}_1$, is the dual space $\mathscr{H}_{-1}$ defined as follows. For f in $L^2(\pi)$, let

$$\|f\|_{-1}^2 = \sup_{g\in\mathscr{C}}\big\{2\langle f, g\rangle_\pi - \|g\|_1^2\big\}. \tag{2.6}$$

Denote by $\mathscr{G}_{-1}^*$ the subspace of $L^2(\pi)$ of all functions with finite $\|\cdot\|_{-1}$ norm. Introduce in $\mathscr{G}_{-1}^*$ the equivalence relation $\sim_{-1}$ by stating that $f\sim_{-1} g$ if $\|f - g\|_{-1} = 0$ and denote by $\mathscr{G}_{-1}$ the normed space $(\mathscr{G}_{-1}^*|_{\sim_{-1}}, \|\cdot\|_{-1})$. The completion of $\mathscr{G}_{-1}$ with respect to the norm $\|\cdot\|_{-1}$, denoted by $\mathscr{H}_{-1}$, is again a Hilbert space with inner product defined through polarization.

We conclude this section with some remarks concerning the spaces $\mathscr{H}_1$, $\mathscr{H}_{-1}$ which will be important later in the chapter.

Claim A (S can be extended as a bounded operator from $\mathscr{H}_1$ to $\mathscr{H}_{-1}$) Note that Sf belongs to $\mathscr{H}_{-1}$ for any f in $\mathscr{C}$. Indeed, for any g in $\mathscr{C}$, since $-S$ is a non-negative operator, by Schwarz inequality,

$$\langle Sf, g\rangle_\pi^2 \le \big\langle(-S)f, f\big\rangle_\pi\big\langle(-S)g, g\big\rangle_\pi = \|f\|_1^2\|g\|_1^2.$$

In particular, by the definition (2.6) of the $\mathscr{H}_{-1}$ norm, $\|Sf\|_{-1} \le \|f\|_1$. This proves Claim A.

Claim B ($\mathscr{H}_{-1}$ is the closure of $\{Sf : f\in\mathscr{C}\}$) Fix g in $L^2(\pi)\cap\mathscr{H}_{-1}$ and f in $\mathscr{C}$. We have just proved that Sf belongs to $\mathscr{H}_{-1}$. We claim that

$$\big\langle g, (-S)f\big\rangle_{-1} = \langle g, f\rangle_\pi. \tag{2.7}$$

Indeed, since the scalar product in $\mathscr{H}_{-1}$ is defined through polarization, we have that $\langle g, (-S)f\rangle_{-1} = (1/4)\{\|g - Sf\|_{-1}^2 - \|g + Sf\|_{-1}^2\}$. By (2.6),

$$\|g - Sf\|_{-1}^2 = \sup_{h\in\mathscr{C}}\big\{2\langle g - Sf, h\rangle_\pi - \|h\|_1^2\big\}.$$

Since f and h belong to $\mathscr{C}$, by (2.5), $\langle(-S)f, h\rangle_\pi = \langle f, h\rangle_1$. Thus,

$$2\big\langle(-S)f, h\big\rangle_\pi - \|h\|_1^2 = \|f\|_1^2 - \|h - f\|_1^2$$

so that

$$\|g - Sf\|_{-1}^2 = \|f\|_1^2 + \sup_{h\in\mathscr{C}}\big\{2\langle g, h\rangle_\pi - \|h - f\|_1^2\big\}.$$

As f belongs to $\mathscr{C}$, replacing h by $h' = f - h$, we obtain that the variational term is equal to $2\langle g, f\rangle_\pi + \|g\|_{-1}^2$ so that

$$\|g - Sf\|_{-1}^2 = \|f\|_1^2 + 2\langle g, f\rangle_\pi + \|g\|_{-1}^2.$$

Replacing f by $-f$ we get,

$$\|g + Sf\|_{-1}^2 = \|f\|_1^2 - 2\langle g, f\rangle_\pi + \|g\|_{-1}^2,$$

which proves (2.7).

It follows from (2.7) that $\mathscr{H}_{-1} \cap L^2(\pi)$ is contained in the $\mathscr{H}_{-1}$-closure of $\{Sf : f \in \mathscr{C}\}$: Fix g in $\mathscr{H}_{-1} \cap L^2(\pi)$ and assume that $\langle g, Sf\rangle_{-1} = 0$ for all f in $\mathscr{C}$. By (2.7), $\langle g, f\rangle_\pi = 0$ for all f in $\mathscr{C}$. This implies that $g = 0$ in $L^2(\pi)$, because the core $\mathscr{C}$ is dense in $L^2(\pi)$, and, by (2.6), that $g = 0$ in $\mathscr{H}_{-1}$.

Claim B follows from the previous observation since $\mathscr{H}_{-1}$ is the $\mathscr{H}_{-1}$-closure of $\mathscr{H}_{-1} \cap L^2(\pi)$.

Claim C (Extension of the scalar product $\langle\cdot,\cdot\rangle_\pi$ to $\mathscr{H}_1 \times \mathscr{H}_{-1}$) It is easy to check from the variational formula (2.6) that for every function f in $\mathscr{C}$ and every function g in $L^2(\pi) \cap \mathscr{H}_{-1}$

$$\big|\langle f, g\rangle_\pi\big| \le \|f\|_1 \|g\|_{-1}. \tag{2.8}$$

Indeed, fix such functions f, g and fix a in $\mathbb{R}$. Since af belongs to $\mathscr{C}$, by the variational formula (2.6),

$$2a\langle f, g\rangle_\pi \le a^2\|f\|_1^2 + \|g\|_{-1}^2.$$

Dividing by a and minimizing over a we prove (2.8).

The inner product $\langle\cdot,\cdot\rangle_\pi$ can be extended to $\mathscr{H}_1 \times \mathscr{H}_{-1}$: Fix g in $\mathscr{H}_{-1}$ and f in $\mathscr{H}_1$. Consider a sequence $\{g_n : n \ge 1\}$ in $L^2(\pi)$ (resp. $\{f_n : n \ge 1\}$ in $\mathscr{C}$) converging to g in $\mathscr{H}_{-1}$ (resp. f in $\mathscr{H}_1$). Define

$$\langle f, g\rangle_\pi := \lim_{n\to\infty} \langle f_n, g_n\rangle_\pi.$$

In view of (2.8), the definition does not depend on the sequence chosen. Moreover, Schwarz inequality holds in this more general context: By (2.8) once again,

$$\big|\langle f, g\rangle_\pi\big| = \lim_{n\to\infty} \big|\langle f_n, g_n\rangle_\pi\big| \le \lim_{n\to\infty} \|f_n\|_1 \|g_n\|_{-1} = \|f\|_1 \|g\|_{-1}. \tag{2.9}$$

The extension of the scalar product $\langle\cdot,\cdot\rangle_\pi$ to $\mathscr{H}_1 \times \mathscr{H}_{-1}$ permits to generalize relation (2.5). For f, g in $\mathscr{H}_1$,

$$\big\langle f, (-S)g\big\rangle_\pi = \langle f, g\rangle_1. \tag{2.10}$$

By Claim A, Sg belongs to $\mathscr{H}_{-1}$. The left-hand side of this expression has to be understood as the extended scalar product between a function f in $\mathscr{H}_1$ with a function Sg in $\mathscr{H}_{-1}$, while the right-hand side is the usual scalar product of $\mathscr{H}_1$. To prove (2.10), consider sequences $\{f_n : n \ge 1\}$, $\{g_n : n \ge 1\}$ of functions in $\mathscr{C}$ converging to f and g in $\mathscr{H}_1$, respectively. By Claim A, Sg_n converges to Sg in $\mathscr{H}_{-1}$. Since Sg_n belong to $L^2(\pi)$, by definition of the extension of the scalar product and by (2.5),

$$\big\langle f, (-S)g\big\rangle_\pi = \lim_{n\to\infty} \big\langle f_n, (-S)g_n\big\rangle_\pi = \lim_{n\to\infty} \langle f_n, g_n\rangle_1 = \langle f, g\rangle_1,$$

where the last identity follows because f_n, g_n converge to f, g in $\mathscr{H}_1$, respectively.

Likewise, one can extend (2.7) and obtain that

$$\big\langle g, (-S)f\big\rangle_{-1} = \langle f, g\rangle_\pi \tag{2.11}$$

for all $f \in \mathscr{H}_1$ and $g \in \mathscr{H}_{-1}$. Consider a sequence $\{f_n : n \geq 1\}$ (resp. $\{g_n : n \geq 1\}$) in $\mathscr{C}$ (resp. $L^2(\pi) \cap \mathscr{H}_{-1}$) converging to f (resp. g) in $\mathscr{H}_1$ (resp. $\mathscr{H}_{-1}$). By Claim A, Sf_n converges to Sf in $\mathscr{H}_{-1}$. Therefore, by Claim C and (2.7),

$$\langle f, g\rangle_\pi = \lim_{n\to\infty} \langle f_n, g_n\rangle_\pi = \lim_{n\to\infty} \big\langle (-S) f_n, g_n\big\rangle_{-1} = \big\langle (-S) f, g\big\rangle_{-1},$$

proving (2.11).

Claim D ($-S$ is an isometric equivalence between $\mathscr{H}_1$ and $\mathscr{H}_{-1}$ and $\mathscr{H}_{-1} = S\mathscr{H}_1$) Indeed, letting $g = (-S)f$ in (2.7) we obtain that $\|Sf\|_{-1} = \|f\|_1$ for f in $\mathscr{C}$.

In Claim A we proved that $S\mathscr{H}_1 \subset \mathscr{H}_{-1}$. Conversely, fix h in $\mathscr{H}_{-1}$. By Claim B, there exists a sequence $\{f_n : n \geq 1\}$ in $\mathscr{C}$ such that Sf_n converges to h in $\mathscr{H}_{-1}$. By the first part of this claim, $\{f_n : n \geq 1\}$ is Cauchy in $\mathscr{H}_1$ and therefore converges to some f in $\mathscr{H}_1$. By Claim A, Sf_n converges to Sf in $\mathscr{H}_{-1}$. Thus, $h = \lim_n Sf_n = Sf$, where the limits are understood in $\mathscr{H}_{-1}$.

Note that we just proved that for any h in $\mathscr{H}_{-1}$, there exists f in $\mathscr{H}_1$ such that

$$Sf = h.$$

Claim E (Weak convergence in $\mathscr{H}_1$) Assume that a sequence $\{f_n : n \geq 1\}$ converges weakly to f in $\mathscr{H}_1$. Then, for all g in $\mathscr{H}_{-1}$,

$$\lim_{n\to\infty} \langle f_n, g\rangle_\pi = \langle f, g\rangle_\pi .$$

Indeed, fix g in $\mathscr{H}_{-1}$. By Claim D, there exists h in $\mathscr{H}_1$ such that $g = (-S)h$. Thus, by (2.10) and by the weak convergence of f_n to f in $\mathscr{H}_1$,

$$\begin{aligned} \lim_{n\to\infty} \langle f_n, g\rangle_\pi &= \lim_{n\to\infty} \big\langle f_n, (-S)h\big\rangle_\pi = \lim_{n\to\infty} \langle f_n, h\rangle_1 \\ &= \langle f, h\rangle_1 = \big\langle f, (-S)h\big\rangle_\pi = \langle f, g\rangle_\pi . \end{aligned}$$

Claim F (The space $\mathscr{H}_{-1}$ as the dual of $\mathscr{H}_1$) The space $\mathscr{H}_{-1}$ can be identified with the space of bounded linear functionals on $\mathscr{H}_1$.

Fix a bounded linear functional $\ell : \mathscr{H}_1 \to \mathbb{R}$. There exists g in $\mathscr{H}_{-1}$ such that

$$\ell(f) = \langle f, g\rangle_\pi$$

for all f in $\mathscr{H}_1$. Here, the scalar product $\langle f, g\rangle_\pi$ has to be understood in the generalized sense defined in Claim C above.

Since $\mathscr{H}_1$ is a Hilbert space, by Riesz representation theorem, there exists h in $\mathscr{H}_1$ such that

$$\ell(f) = \langle f, h\rangle_1 = \big\langle f, (-S)h\big\rangle_\pi$$

for all f in $\mathscr{H}_1$. The last equality follows from (2.10). By Claim A, Sh belongs to $\mathscr{H}_{-1}$, which proves the first part of the claim. Conversely, suppose that g belongs to $\mathscr{H}_{-1}$. Then, $\ell(f) = \langle f, g\rangle_\pi$ defines a bounded functional on $\mathscr{H}_1$ thanks to (2.9).

Claim G (Sufficient conditions for belonging to $\mathscr{H}_{-1}$) By (2.6), a function f in $L^2(\pi)$ belongs to $\mathscr{H}_{-1}$ if and only if there exists a finite constant C_0 such that

$$\langle f, g\rangle_\pi \le C_0 \|g\|_1 \tag{2.12}$$

for every g in $\mathscr{C}$. In this case, $\|f\|_{-1} \le C_0$.

Claim H (Replacing the core $\mathscr{C}$ by the domain $\mathscr{D}(S)$) We have seen just below (2.5) that the domains $\mathscr{D}(L)$ and $\mathscr{D}(S)$ are contained in $\mathscr{H}_1$. For any f in $\mathscr{D}(L)$, g in $\mathscr{D}(S)$,

$$\|f\|_1^2 = \langle f, (-L)f\rangle_\pi, \qquad \|g\|_1^2 = \langle g, (-S)g\rangle_\pi, \qquad \langle f, g\rangle_1 = \langle f, (-S)g\rangle_\pi .$$

Indeed, consider sequences $\{f_n : n \ge 1\}$, $\{g_n : n \ge 1\}$ in $\mathscr{C}$ such that f_n, g_n, Lf_n, Sg_n converge in $L^2(\pi)$ as $n \uparrow \infty$ to f, g, Lg, Sg, respectively. Since, by convention, f represents in $\mathscr{H}_1$ the Cauchy sequence $\{f_n : n \ge 1\}$, by (2.5),

$$\|f\|_1^2 = \lim_{n\to\infty} \|f_n\|_1^2 = \lim_{n\to\infty} \langle f_n, (-S)f_n\rangle_\pi = \lim_{n\to\infty} \langle f_n, (-L)f_n\rangle_\pi = \langle f, (-L)f\rangle_\pi .$$

The third identity follows from the fact that f_n belongs to $\mathscr{C}$ which is a core for both S and L. By similar reasons, $\|g\|_1^2 = \langle g, (-S)g\rangle_\pi$ and $\langle f, g\rangle_1 = \langle f, (-S)g\rangle_\pi$.

The arguments presented in Claim A and the statements of the previous paragraph show that Sf belongs to $\mathscr{H}_{-1}$ for any f in $\mathscr{D}(S)$ and that $\|Sf\|_{-1} \le \|f\|_1$.

We may replace in the variational formula (2.6) the core $\mathscr{C}$ by the domain $\mathscr{D}(L)$ because for any function g in $\mathscr{D}(L)$ there exists a sequence $\{g_n : n \ge 1\}$ in $\mathscr{C}$ such that g_n, Lg_n converge in $L^2(\pi)$ to g, Lg, respectively. The same is true for $\mathscr{D}(S)$ in place of $\mathscr{D}(L)$.

Finally, we claim that for any g in $L^2(\pi) \cap \mathscr{H}_{-1}$ and f in $\mathscr{D}(S)$,

$$\langle g, (-S)f\rangle_{-1} = \langle g, f\rangle_\pi .$$

This claim follows from the arguments presented in the proof of Claim B, replacing in the definition (2.6) of the $\mathscr{H}_{-1}$ norm the core $\mathscr{C}$ by the domain $\mathscr{D}(S)$.

2.3 The Resolvent Equation

Fix a function V in $L^2(\pi) \cap \mathscr{H}_{-1}$, $\lambda > 0$ and consider the resolvent equation

$$\lambda f_\lambda - L f_\lambda = V. \tag{2.13}$$

Note that $f_\lambda = (\lambda - L)^{-1} V$ belongs to the domain of the generator L. Taking the scalar product with respect to f_λ on both sides of this equation we get that

$$\lambda \langle f_\lambda, f_\lambda\rangle_\pi + \|f_\lambda\|_1^2 = \langle V, f_\lambda\rangle_\pi . \tag{2.14}$$

Hence, by Schwarz inequality (2.9),

$$\lambda\langle f_\lambda, f_\lambda\rangle_\pi + \|f_\lambda\|_1^2 \le \|f_\lambda\|_1 \|V\|_{-1}$$

so that $\|f_\lambda\|_1 \le \|V\|_{-1}$. Combining the two previous bounds we easily obtain the stronger estimate

$$\lambda\langle f_\lambda, f_\lambda\rangle_\pi + \|f_\lambda\|_1^2 \le \|V\|_{-1}^2. \tag{2.15}$$

From the above estimate we conclude that λf_λ vanishes in $L^2(\pi)$ as $\lambda \downarrow 0$ and that $\{f_\lambda : 0 < \lambda \le 1\}$ forms a bounded sequence in $\mathscr{H}_1$ and is therefore weakly precompact.

Another simple consequence of (2.15) is that $(\lambda - L)^{-1}$ extends to a bounded mapping from $\mathscr{H}_{-1}$ to $\mathscr{H}_1$:

Lemma 2.3 *The operator $(\lambda - L)^{-1}$ extends from $L^2(\pi)$ to a bounded mapping from $\mathscr{H}_{-1}$ to $\mathscr{H}_1$. Moreover, for any $V \in \mathscr{H}_{-1}$ we have*

$$\left\|(\lambda - L)^{-1}V\right\|_1 \le \|V\|_{-1}.$$

We wish to formulate sufficient conditions for the central limit theorem of $t^{-1/2}\int_0^t V(X_s)ds$ in terms of the asymptotic behavior, as $\lambda \downarrow 0$, of the solutions f_λ of the resolvent equation (2.13). We first observe in Sect. 2.5 that the condition $V \in \mathscr{H}_{-1}$ guarantees that the $L^2(\mathbb{P}_\pi)$ norm of $t^{-1/2}\int_0^t V(X_s)ds$ remains bounded for large t. Next, in Theorem 2.7, we show that a central limit theorem is valid, provided the following two conditions are satisfied:

$$\lim_{\lambda\to 0} \lambda\|f_\lambda\|_\pi^2 = 0 \quad \text{and} \quad \lim_{\lambda\to 0} \|f_\lambda - f\|_1 = 0$$

for some f in $\mathscr{H}_1$. In Theorem 2.14, we prove that the bound $\sup_{0<\lambda\le 1}\|Lf_\lambda\|_{-1} < \infty$ implies the previous two conditions. Therefore, a central limit theorem holds if $\sup_{0<\lambda\le 1}\|Lf_\lambda\|_{-1} < +\infty$.

2.4 Dynkin's Martingales

Fix a function f in the domain of the generator L. Consider the so-called Dynkin martingale $M_t = f(X_t) - f(X_0) - \int_0^t (Lf)(X_s)\,ds$, $t \ge 0$. These martingales will appear repeatedly in the following sections with the solution f_λ of the resolvent equation (2.13) in place of f.

Note that M_t may not be right continuous because functions in the domain of the generator may not be continuous. We need, however, the right continuity of the martingale for several reasons, the main one being the existence of the predictable quadratic variation of the martingale, which relies on the Doob–Meyer decomposition (cf. Sect. 1.4 in Karatzas and Shreve, 1991) which requires the martingale to be right continuous and the filtration to satisfy the usual conditions.

There are two possible ways to circumvent the lack of right continuity of the Dynkin martingales. The first one is to assume that the core $\mathscr{C}$ consists of continuous functions and to approximate the solution of the resolvent equation by functions in the core. Such approximation is discussed in detail in Sect. 2.7.3 after formula (2.37), and in Sect. 2.7.4 at the end of the proof of Lemma 2.21.

We may also consider a r.c.l.l. modification of the martingale M_t, which exists in view of Theorem 1.3.13 in Karatzas and Shreve (1991).

Let M_t be the Dynkin martingale associated to a function f in the domain of the generator. We claim that

$$\mathbb{E}_\pi\big[\langle M, M\rangle_t\big] = 2t\,\|f\|_1^2. \tag{2.16}$$

Indeed, since M_t is a martingale vanishing at $t = 0$, expanding the square, we get that

$$\mathbb{E}_\pi\big[(M_t)^2\big] = \mathbb{E}_\pi\bigg[\bigg\{\int_0^t (Lf)(X_s)\,ds\bigg\}^2 - 2f(X_t)\int_0^t (Lf)(X_s)\,ds\bigg].$$

We may rewrite the previous difference as

$$-2\int_0^t (Lf)(X_s)\bigg\{f(X_t) - \int_s^t (Lf)(X_r)\,dr\bigg\}\,ds.$$

To conclude the proof of (2.16), it remains to add and subtract $f(X_s)$ in the term inside braces and to observe that one of them becomes $M_t - M_s$.

2.5 $\mathscr{H}_{-1}$ Estimates of the Time-Variance

In this section, we estimate the limiting variance of the integral $\int_0^t V(X_s)ds$ for a mean zero function V in $L^2(\pi)$. Assume that $\mathscr{C}$ is a core for the operators L and L^*. Let

$$\sigma^2(V) = \limsup_{t\to\infty} \mathbb{E}_\pi\bigg[\bigg(\frac{1}{\sqrt{t}}\int_0^t V(X_s)\,ds\bigg)^2\bigg].$$

Since π is invariant, a change of variables shows that for each fixed t the expectation on the right-hand side of the previous formula is equal to

$$\begin{aligned}\frac{1}{t}\int_0^t ds\int_0^t \mathbb{E}_\pi\big[V(X_{|s-r|})V(X_0)\big]\,dr &= \frac{1}{t}\int_0^t ds\int_0^t \langle P_{|s-r|}V, V\rangle_\pi\,dr\\ &= 2\int_0^t \big[1-(s/t)\big]\langle P_s V, V\rangle_\pi\,ds.\end{aligned}$$

Denote by a^+ the positive part of a: $a^+ = \max\{0, a\}$, so that

$$\sigma^2(V) = 2\limsup_{t\to\infty}\int_0^\infty \big[1-(s/t)\big]^+\langle P_s V, V\rangle_\pi\,ds. \tag{2.17}$$

In general, it is not clear whether this upper limit is in fact a limit, or whether it is finite without some restrictions on V. However, in the reversible case, the integral is an increasing function of t because $\langle P_s V, V\rangle_\pi = \langle P_{s/2}V, P_{s/2}V\rangle_\pi$ is positive. Hence, in this particular case, by the monotone convergence theorem,

$$\sigma^2(V) = 2\lim_{t\to\infty}\int_0^\infty \left[1-(s/t)\right]^+\langle P_s V, V\rangle_\pi\, ds = 2\int_0^\infty \langle P_s V, V\rangle_\pi\, ds.$$

In the general case, one can show that $\sigma^2(V)$ defined in (2.17) is finite provided the function V belongs to the space $\mathscr{H}_{-1}$. Namely, there exists a universal constant C_0 such that

$$\sigma^2(V) \le C_0 \|V\|_{-1}^2 \tag{2.18}$$

for all functions $V \in \mathscr{H}_{-1}\cap L^2(\pi)$. The main difference between $\sigma^2(V)$ and $\|V\|_{-1}^2$ is that while in the definition of the latter only the symmetric part of the generator is involved, in the first term the full generator appears. Formally,

$$\sigma^2(V) = 2\int_0^\infty \langle P_s V, V\rangle_\pi = 2\big\langle V, (-L)^{-1}V\big\rangle_\pi = 2\big\langle V, \big[(-L)^{-1}\big]^s V\big\rangle_\pi,$$

while $\|V\|_{-1}^2 = \langle V, (-S)^{-1}V\rangle_\pi$. Here and in what follows, B^s represents the symmetric part of the operator B. Recalling that S, A stand for the respective symmetric and anti-symmetric parts of the generator L, we can write, at least formally, that

$$\big\{\big[(-L)^{-1}\big]^s\big\}^{-1} = -S + A^*(-S)^{-1}A \ge -S,$$

where A^* stands for the adjoint of A. We have therefore that $[(-L)^{-1}]^s \le (-S)^{-1}$, from which it follows that $\sigma^2(V) \le 2\|V\|_{-1}^2$. We now present a rigorous version of this informal argument.

Lemma 2.4 *Fix $T > 0$ and a function V in $L^2(\pi)\cap\mathscr{H}_{-1}$. Then,*

$$\mathbb{E}_\pi\left[\sup_{0\le t\le T}\left(\int_0^t V(X_s)\, ds\right)^2\right] \le 24T\|V\|_{-1}^2.$$

Proof Fix $\varepsilon > 0$. Assume first that there exists a function h in the core $\mathscr{C}$ such that

$$\|Sh - V\|_\pi \le \varepsilon, \qquad \|h\|_1 \le \|V\|_{-1} + \varepsilon. \tag{2.19}$$

We shall remove this assumption at the end of the proof.

Recall that $\{\mathscr{F}_s : s \ge 0\}$ is the augmented filtration corresponding to $\{X_r : 0 \le r \le s\}$. Since h belongs to the domain of the generator we can define a zero-mean martingale M_t with respect to $\{\mathscr{F}_s : s \ge 0\}$ by

$$M_t = h(X_t) - h(X_0) - \int_0^t (Lh)(X_s)\, ds. \tag{2.20}$$

In the same way, denote by $\{\mathscr{F}_t^- : t \in [0, T]\}$ the augmented backward filtration generated by $\{X_{T-t} : t \in [0, T]\}$. Recall that L^* stands for the adjoint of the generator L with respect to the invariant measure π. Since we assumed $\mathscr{C}$ to be a core for the adjoint operator L^*, h belongs to the domain of L^*. It is easy to check that the process $\{M_t^- : 0 \le t \le T\}$ defined by

$$M_t^- = h(X_{T-t}) - h(X_T) - \int_0^t (L^*h)(X_{T-s})\,ds \tag{2.21}$$

is a $(\mathbb{P}_\pi, \{\mathscr{F}_t^-, t \in [0, T]\})$ martingale, called the backward martingale.

A change of variables gives that

$$M_T^- - M_{T-t}^- = h(X_0) - h(X_t) - \int_0^t (L^*h)(X_s)\,ds.$$

In particular,

$$M_t + M_T^- - M_{T-t}^- = -2\int_0^t (Sh)(X_s)\,ds = -2\int_0^t \{V(X_s) - R(X_s)\}\,ds,$$

where $R := V - Sh$. Therefore,

$$\begin{aligned}&\mathbb{E}_\pi\Big[\sup_{0\le t\le T}\Big(\int_0^t V(X_s)\,ds\Big)^2\Big]\\&\quad = (1/4)\mathbb{E}_\pi\Big[\sup_{0\le t\le T}\Big(M_t + M_T^- - M_{T-t}^- - 2\int_0^t R(X_s)\,ds\Big)^2\Big].\end{aligned}$$

Since, according to (2.19), $\|R\|_\pi \le \varepsilon$, by Schwarz' and Doob's inequalities, the previous expression is bounded above by

$$4\big\{2\mathbb{E}_\pi\big[\big(M_T^-\big)^2\big] + \mathbb{E}_\pi\big[(M_T)^2\big] + T^2\varepsilon^2\big\}.$$

By (2.16), the variances of the forward and backward martingales are both equal to $2T\|h\|_1^2 \le 2T(\|V\|_{-1} + \varepsilon)^2$. The previous expression is thus bounded above by $24T(\|V\|_{-1}+\varepsilon)^2 + 4T^2\varepsilon^2$. After sending $\varepsilon \downarrow 0$ we conclude the proof of the lemma.

We now return to approximation (2.19). Fix $\varepsilon > 0$. For $\lambda > 0$ denote by h_λ the solution of the resolvent equation (2.13) with S in place of L and $-V$ instead of V. By (2.15), there exists λ sufficiently small for which $\|Sh_\lambda - V\|_\pi \le \varepsilon$ and $\|h_\lambda\|_1 \le \|V\|_{-1}$. Since h_λ belongs to the domain of S, there exists h in the core $\mathscr{C}$ such that $\|h - h_\lambda\|_\pi \le \varepsilon$, $\|Sh - Sh_\lambda\|_\pi \le \varepsilon$. Thus $\|Sh - V\|_\pi \le 2\varepsilon$ and $\|h\|_1 \le \|h_\lambda\|_1 + \|h - h_\lambda\|_1 \le \|V\|_{-1} + \varepsilon$ because $\|g\|_1^2 \le \|g\|_\pi \|Sg\|_\pi$ for every g in the domain of S. This concludes the proof of (2.19) and completes the demonstration of the lemma. □

Remark 2.5 We showed that $\sigma^2(V) \le 24\|V\|_{-1}^2$. We present in the last section of Chap. 5 examples of functions V **not** in $\mathscr{H}_{-1}$, but with finite variance.

2.6 Central Limit Theorem for Markov Processes

It follows from (2.15) that

$$\sup_{0<\lambda\le 1} \lambda \| f_\lambda \|_\pi^2 < \infty \quad \text{and} \quad \sup_{0<\lambda\le 1} \| f_\lambda \|_1 < \infty. \tag{2.22}$$

Our first aim in this section is to show that a central limit theorem for the additive functional $t^{-1/2}\int_0^t V(X_s)ds$ holds provided we can prove the following stronger statements:

$$\lim_{\lambda\to 0} \lambda \| f_\lambda \|_\pi^2 = 0 \quad \text{and} \quad \lim_{\lambda\to 0} \| f_\lambda - f \|_1 = 0 \tag{2.23}$$

for some f in $\mathscr{H}_1$.

Before formulating the first main result of this chapter, we introduce some terminology to be used recurrently. On a probability space $(\Omega, \mathscr{F}, \mathbb{P})$, consider a σ-algebra $\mathscr{G} \subset \mathscr{F}$ and a stochastic process $\{Y_t : t \ge 0\}$.

Definition 2.6 The laws of a stochastic process $\{Y_t : t \ge 0\}$ conditioned on the σ-algebra $\mathscr{G}$ converge in probability to a mean zero Gaussian distribution with variance σ^2 if

$$\lim_{t\to\infty} \mathbb{E}\big[\big|\mathbb{E}\big[e^{i\theta Y_t}\big|\mathscr{G}\big] - e^{-\sigma^2\theta^2/2}\big|\big] = 0$$

for all θ in $\mathbb{R}$.

Note that this concept is stronger than the annealed convergence and the stable convergence, but weaker than the quenched version. It stands for the typical central limit theorem we prove in this book. The next result is an example.

Theorem 2.7 *Fix a function V in $L^2(\pi) \cap \mathscr{H}_{-1}$ and assume that* (2.23) *holds. Then, the law of $t^{-1/2}\int_0^t V(X_s)\,ds$ conditioned on $\mathscr{F}_0$ converges in probability to a mean zero Gaussian distribution with variance*

$$\sigma^2(V) = 2 \lim_{\lambda\to 0} \| f_\lambda \|_1^2.$$

It follows from (2.23) and (2.14) that the *asymptotic variance* $\sigma^2(V)$ is given by

$$\sigma^2(V) = 2 \lim_{\lambda\to 0} \| f_\lambda \|_1^2 = 2 \lim_{\lambda\to 0} \langle V, f_\lambda \rangle_\pi . \tag{2.24}$$

The idea of the proof of Theorem 2.7 is again to express $\int_0^t V(X_s)\,ds$ as the sum of a martingale and a term which vanishes in the limit. This is proved in (2.26) and Lemma 2.10 below. We start taking advantage of the resolvent equation (2.13) to build up a martingale which approximates $\int_0^t V(X_s)\,ds$ up to the order of magnitude $o(\sqrt{t})$. For each fixed $\lambda > 0$, let M_t^λ be the martingale defined by

$$M_t^\lambda = f_\lambda(X_t) - f_\lambda(X_0) - \int_0^t (L f_\lambda)(X_s)\,ds$$

so that

$$\int_0^t V(X_s)\,ds = M_t^\lambda + f_\lambda(X_0) - f_\lambda(X_t) + \lambda \int_0^t f_\lambda(X_s)\,ds. \tag{2.25}$$

Denote by R_t^λ the remainder:

$$R_t^\lambda := f_\lambda(X_0) - f_\lambda(X_t) + \lambda \int_0^t f_\lambda(X_s)\,ds.$$

Proposition 2.8 *Fix a function V in $L^2(\pi) \cap \mathcal{H}_{-1}$ and assume that* (2.23) *holds. Then, $t^{-1/2} R_t^\lambda$ vanishes in $L^2(\mathbb{P}_\pi)$ as $\lambda \downarrow 0$ and then $t \uparrow \infty$:*

$$\lim_{t\to\infty} \lim_{\lambda\to 0} \left\| t^{-1/2} R_t^\lambda \right\|_\pi = 0.$$

Proposition 2.8 follows from Lemmas 2.9 and 2.10 below.

Lemma 2.9 *The martingales $\{M_t^\lambda : t \geq 0\}$ converge in $L^2(\mathbb{P}_\pi)$, as $\lambda \downarrow 0$, to a certain square integrable martingale $\{M_t : t \geq 0\}$: for each $T > 0$,*

$$\lim_{\lambda\to 0} \mathbb{E}_\pi \Big[\sup_{t\in[0,T]} \big(M_t^\lambda - M_t\big)^2 \Big] = 0.$$

Proof Fix $T > 0$. For λ, $\lambda' > 0$, since π is an invariant state, by (2.16) the expectation of the quadratic variation of the martingale $M_T^\lambda - M_T^{\lambda'}$ is equal to

$$2T\big\langle (f_\lambda - f_{\lambda'}), (-L)(f_\lambda - f_{\lambda'}) \big\rangle_\pi = 2T \| f_\lambda - f_{\lambda'} \|_1^2.$$

By assumption (2.23), f_λ converges in $\mathcal{H}_1$. In particular, M_T^λ is a Cauchy sequence in $L^2(\mathbb{P}_\pi)$ and converges to a certain M_T. Using a standard argument one can choose a version of $\{M_t : t \geq 0\}$ that is a right continuous, square integrable martingale. This proves the lemma. □

It follows from this result and from identity (2.25) that R_t^λ also converges in $L^2(\mathbb{P}_\pi)$ as $\lambda \downarrow 0$. Denote this limit by $\{R_t : t \geq 0\}$ so that

$$\int_0^t V(X_s)\,ds = M_t + R_t. \tag{2.26}$$

Lemma 2.10 *$t^{-1/2} R_t$ vanishes in $L^2(\mathbb{P}_\pi)$ as $t \uparrow \infty$.*

Proof Putting together Eq. (2.25) with (2.26), we get that

$$\frac{R_t}{\sqrt{t}} = \frac{1}{\sqrt{t}} \left\{ M_t^\lambda - M_t + f_\lambda(X_0) - f_\lambda(X_t) + \lambda \int_0^t f_\lambda(X_s)\,ds \right\}. \tag{2.27}$$

We consider separately each term on the right-hand side of the above expression. Since M_t^λ converges to M_t in $L^2(\mathbb{P}_\pi)$,

$$\frac{1}{t}\mathbb{E}_\pi\big[\big(M_t^\lambda - M_t\big)^2\big] = \frac{1}{t}\lim_{\lambda'\to 0}\mathbb{E}_\pi\big[\big(M_t^\lambda - M_t^{\lambda'}\big)^2\big].$$

In the previous lemma, we computed the expectation of the quadratic variation of the martingale $M_t^\lambda - M_t^{\lambda'}$. This calculation shows that the previous expression is equal to

$$2\lim_{\lambda'\to 0}\|f_\lambda - f_{\lambda'}\|_1^2 = 2\|f_\lambda - f\|_1^2.$$

In the last step we used assumption (2.23) which states that f_λ converges in $\mathscr{H}_1$ to some f.

We now estimate the term R_t^λ appearing in (2.27). We can write that

$$\mathbb{E}_\pi\big[\big(t^{-1/2}R_t^\lambda\big)^2\big] \le 3t^{-1}\mathbb{E}_\pi\big[f_\lambda(X_t)^2\big] + 3t^{-1}\mathbb{E}_\pi\big[f_\lambda(X_0)^2\big] + 3t^{-1}\lambda^2\mathbb{E}_\pi\left[\left\{\int_0^t f_\lambda(X_s)\,ds\right\}^2\right].$$

Since $\{X_t, t \ge 0\}$ is stationary with the initial distribution π we obtain that the first two expressions on the right-hand side taken together yield $6t^{-1}\|f_\lambda\|_\pi^2$. On the other hand, by Schwarz inequality, the third term is bounded by $3t\lambda^2\|f_\lambda\|_\pi^2$. Putting together all the previous estimates, we obtain that

$$\frac{1}{t}\mathbb{E}_\pi\big[R_t^2\big] \le 2\|f_\lambda - f\|_1^2 + 2\big(6t^{-1} + 3t\lambda^2\big)\|f_\lambda\|_\pi^2$$

for all $\lambda > 0$. Setting $\lambda = t^{-1}$ we conclude the proof of the lemma in view of hypotheses (2.23). □

Proof of Theorem 2.7 We claim that the martingale $\{M_t : t \ge 0\}$ of the decomposition (2.26) satisfies the assumptions of Theorem 2.1. Obviously, the martingale differences are stationary. Recall from the beginning of the chapter, the definition of the shift operator $\{\theta_t : t \ge 0\}$. Since π is stationary, $\{\theta_t : t \ge 0\}$ is a measure preserving semiflow on the space $(D(\mathbb{R}_+, E), \mathscr{D}, \mathbb{P}_\pi)$, where $\mathscr{D}$ is the Borel σ-field of $D(\mathbb{R}_+, E)$.

To apply Kingman's ergodic theorem (Krengel, 1985, Theorem 1.5.6), for $0 \le s < t$, let $F_{s,t} = \langle M, M\rangle_t - \langle M, M\rangle_s$. It is easy to check that $F_{s,t} \circ \theta_u = F_{s+u,t+u}$, $0 \le s < t$, $u \ge 0$, $F_{s,t} = F_{s,u} + F_{u,t}$, $0 \le s < u < t$, $F_{0,t} \ge 0$, $t > 0$. Hence, $\{F_{s,t} : 0 \le s < t\}$ is a subadditive process. It is also clear that $F_{s,t}$ is separable because the predictable quadratic variation is a monotone, right continuous function.

Since $\langle M, M\rangle_1$ has finite expected value and since $\mathbb{P}_\pi$ is ergodic with respect to the shifts, by Krengel (1985, Theorem 1.5.6), $t^{-1}\langle M, M\rangle_t$ converges a.s. and in $L^1(\mathbb{P}_\pi)$ to $\mathbb{E}_\pi[\langle M, M\rangle_1]$. Thus, according to Theorem 2.1, the law of $M_t/\sqrt{t}$

conditioned on $\mathscr{F}_0$ converges in probability, as $t \uparrow \infty$, to a mean zero Gaussian law with variance

$$\begin{aligned}\mathbb{E}_\pi\big[\langle M, M\rangle_1\big] &= \mathbb{E}_\pi\big[M_1^2\big] = \lim_{\lambda\to 0}\mathbb{E}_\pi\big[\big(M_1^\lambda\big)^2\big] \\ &= \lim_{\lambda\to 0}\mathbb{E}_\pi\big[\big\langle M^\lambda, M^\lambda\big\rangle_1\big] = 2\lim_{\lambda\to 0}\|f_\lambda\|_1^2.\end{aligned}$$

The last identity follows from (2.16).

To finish the proof recall the decomposition (2.26) and observe that for any $\theta \in \mathbb{R}$

$$\big|\mathbb{E}_\pi\big[e^{i\theta\int_0^t V(X_s)ds/\sqrt{t}}|\mathscr{F}_0\big] - \mathbb{E}_\pi\big[e^{i\theta M_t/\sqrt{t}}|\mathscr{F}_0\big]\big| \le \mathbb{E}_\pi\big[|\theta R_t|/\sqrt{t}|\mathscr{F}_0\big].$$

By Lemma 2.10, the right-hand side vanishes in $L^2(\mathbb{P}_\pi)$ as $t \uparrow \infty$. □

Corollary 2.11 *Fix a function V in $L^2(\pi) \cap \mathscr{H}_{-1}$. Under the assumptions of Theorem 2.7, the variance of $t^{-1/2}\int_0^t V(X_s)\,ds$ converges to the asymptotic variance $\sigma^2(V)$:*

$$\lim_{t\to\infty}\mathbb{E}_\pi\bigg[\bigg(\frac{1}{\sqrt{t}}\int_0^t V(X_s)\,ds\bigg)^2\bigg] = \sigma^2(V).$$

Proof Recall the decomposition (2.26). By Lemma 2.10, by the proof of Theorem 2.7 and by (2.24),

$$\begin{aligned}\lim_{t\to\infty}\mathbb{E}_\pi\bigg[\bigg(\frac{1}{\sqrt{t}}\int_0^t V(X_s)\,ds\bigg)^2\bigg] &= \lim_{t\to\infty}\frac{1}{t}\mathbb{E}_\pi\big[M_t^2\big] \\ &= \lim_{t\to\infty}\frac{1}{t}\mathbb{E}_\pi\big[\langle M, M\rangle_t\big] = 2\lim_{\lambda\to 0}\|f_\lambda\|_1^2 = \sigma^2(V).\end{aligned}$$

□

Remark 2.12 In the proof of Theorem 2.7, we showed that the quadratic variation of the martingale M of the decomposition (2.26) converges a.s. to the asymptotic variance $\sigma^2(V)$:

$$\lim_{t\to\infty}\frac{1}{t}\langle M, M\rangle_t = \sigma^2(V) \quad \mathbb{P}_\pi \text{ a.s.}$$

Remark 2.13 In the statement of Theorem 2.7 we did not assume the existence of a common core for L and L^*.

Our next task is to show the central limit theorem under the condition of boundedness of $\mathscr{H}_{-1}$ norm for $\{Lf_\lambda : 0 < \lambda \le 1\}$.

Theorem 2.14 *Suppose that Lf_λ belongs to $\mathscr{H}_{-1}$, $0 < \lambda \le 1$, and that*

$$\sup_{0<\lambda\le 1}\|Lf_\lambda\|_{-1} < \infty. \tag{2.28}$$

Then, the conclusion of Theorem 2.7 holds.

Remark 2.15 Notice that

$$\sup_{0<\lambda\le 1} \|Lf_\lambda\|_{-1} < \infty \quad \text{if and only if} \quad \sup_{0<\lambda\le 1} \|\lambda f_\lambda\|_{-1} < \infty \tag{2.29}$$

because V belongs to $\mathscr{H}_{-1}$.

The conclusion of Theorem 2.14 follows directly from Theorem 2.7 and the following

Lemma 2.16 *Fix a function V in $\mathscr{H}_{-1} \cap L^2(\pi)$ and denote by $\{f_\lambda : \lambda > 0\}$ the solution of the resolvent equation* (2.13). *Assume that* $\sup_{\lambda>0} \|Lf_\lambda\|_{-1} \le C_0$ *for some finite constant C_0. Then, there exists f in $\mathscr{H}_1$ such that* (2.23) *holds.*

Proof We have already proved in (2.15) that

$$\sup_{0<\lambda\le 1} \|f_\lambda\|_1 \le \|V\|_{-1} \quad \text{and} \quad \sup_{0<\lambda\le 1} \lambda\|f_\lambda\|_\pi^2 \le \|V\|_{-1}^2.$$

In particular, λf_λ converges to 0 in $L^2(\pi)$, as $\lambda \downarrow 0$. The proof is divided into several claims.

Claim 1 (Lf_λ converges weakly in $\mathscr{H}_{-1}$, as $\lambda \downarrow 0$, to $-V$) For that purpose it suffices to prove that $\lim_{\lambda\to 0} \lambda f_\lambda = 0$ weakly in $\mathscr{H}_{-1}$. Since $\sup_{\lambda>0} \|Lf_\lambda\|_{-1}$ is bounded, so is $\sup_{\lambda>0} \|\lambda f_\lambda\|_{-1}$. As a result, any sequence $\{\lambda_n f_{\lambda_n} : n \ge 1\}$ is weakly precompact in $\mathscr{H}_{-1}$ when $\lambda_n \downarrow 0$. We show that 0 is its only weak limiting point. Suppose, therefore, that $\lambda_n f_{\lambda_n}$ converges to g weakly in $\mathscr{H}_{-1}$. From (2.11) and since λf_λ converges strongly to 0 in $L^2(\pi)$, for any h in $\mathscr{C}$,

$$\langle g, (-S)h\rangle_{-1} = \lim_{n\to\infty} \langle \lambda_n f_{\lambda_n}, (-S)h\rangle_{-1} = \lim_{n\to\infty} \langle \lambda_n f_{\lambda_n}, h\rangle_\pi = 0.$$

By Claim B of Sect. 2.2, $\{Sf : f \in \mathscr{C}\}$ is dense in $\mathscr{H}_{-1}$ so that $g = 0$. This concludes the proof of the claim.

In the same way, since $\sup_{\lambda>0} \|f_\lambda\|_1$ is bounded, each sequence $\lambda_n \downarrow 0$ has a subsequence still denoted by λ_n, for which f_{λ_n} converges weakly in $\mathscr{H}_1$ to some function, denoted by W.

Claim 2 (W satisfies the relation $\|W\|_1^2 = \langle W, V\rangle_\pi$) To check this identity, apply Mazur's theorem (Yosida, 1995, Section V.1) to the sequences f_{λ_n}, Lf_{λ_n} to obtain sequences g_n, Lg_n which converge strongly in $\mathscr{H}_1$ to W, strongly in $\mathscr{H}_{-1}$ to $-V$, respectively. Keep in mind that each g_n is obtained as a finite convex combination of functions f_{λ_k} and therefore belongs to the domain $\mathscr{D}(L)$. On the one hand, since g_n (resp. Lg_n) converges strongly in $\mathscr{H}_1$ (resp. $\mathscr{H}_{-1}$) to W (resp. $-V$), $\langle g_n, Lg_n\rangle_\pi$ converges to $-\langle W, V\rangle_\pi$. Here the scalar product $\langle\cdot,\cdot\rangle_\pi$ has to be understood in the generalized sense explained in Claim C, Sect. 2.2. On the other hand, since $-\langle g_n, Lg_n\rangle_\pi = \|g_n\|_1^2$, it converges to $\|W\|_1^2$. Therefore, $\|W\|_1^2 = \langle W, V\rangle_\pi$.

Claim 3 ($\lim_{\lambda\to 0}\lambda\|f_\lambda\|_\pi^2=0$) Suppose by contradiction that $\lambda\|f_\lambda\|_\pi^2$ does not converge to 0 as $\lambda\downarrow 0$. In this case there exists $\varepsilon>0$ and a subsequence $\lambda_n\downarrow 0$ such that $\lambda_n\|f_{\lambda_n}\|_\pi^2\ge\varepsilon$ for all n. We have just shown the existence of a sub-subsequence $\lambda_{n'}$ for which $f_{\lambda_{n'}}$ converges weakly in $\mathscr{H}_1$ to some W satisfying the relation $\langle W,V\rangle_\pi=\|W\|_1^2$. Since f_λ is the solution of the resolvent equation,

$$\begin{aligned}\limsup_{n'\to\infty}\|f_{\lambda_{n'}}\|_1^2&\le\limsup_{n'\to\infty}\big(\lambda_{n'}\|f_{\lambda_{n'}}\|_\pi^2+\|f_{\lambda_{n'}}\|_1^2\big)\\&=\limsup_{n'\to\infty}\langle f_{\lambda_{n'}},V\rangle_\pi=\langle W,V\rangle_\pi=\|W\|_1^2\le\limsup_{n'\to\infty}\|f_{\lambda_{n'}}\|_1^2\cdot\end{aligned}\tag{2.30}$$

The second identity follows from Claim E in Sect. 2.2. These inequalities contradict the fact that $\lambda_n\|f_{\lambda_n}\|_\pi^2\ge\varepsilon$ for all n, so that $\lim_{\lambda\to 0}\lambda\|f_\lambda\|_\pi^2=0$.

Claim 4 (f_λ converges strongly in $\mathscr{H}_1$ as $\lambda\downarrow 0$) It follows also from the previous argument that $f_{\lambda_{n'}}$ converges to W strongly in $\mathscr{H}_1$. In particular, all sequences λ_n have subsequences $\lambda_{n'}$ for which $f_{\lambda_{n'}}$ converges strongly in $\mathscr{H}_1$. To show that f_λ converges strongly, it remains to check uniqueness of the limit. Consider two decreasing sequences λ_n, μ_n, vanishing as $n\uparrow\infty$. Denote by W_1, W_2 the strong limit in $\mathscr{H}_1$ of f_{λ_n}, f_{μ_n}, respectively. Since f_λ is the solution of the resolvent equation,

$$\langle\lambda_n f_{\lambda_n}-\mu_n f_{\mu_n},f_{\lambda_n}-f_{\mu_n}\rangle_\pi+\|f_{\lambda_n}-f_{\mu_n}\|_1^2=0\tag{2.31}$$

for all n. Since f_{λ_n}, f_{μ_n} converges strongly to W_1, W_2 in $\mathscr{H}_1$,

$$\lim_{n\to\infty}\|f_{\lambda_n}-f_{\mu_n}\|_1^2=\|W_1-W_2\|_1^2.\tag{2.32}$$

On the other hand, since $\lambda\|f_\lambda\|_\pi^2$ vanishes as $\lambda\downarrow 0$,

$$\begin{aligned}&\lim_{n\to\infty}\langle\lambda_n f_{\lambda_n}-\mu_n f_{\mu_n},f_{\lambda_n}-f_{\mu_n}\rangle_\pi\\&\quad=-\lim_{n\to\infty}\big(\langle\lambda_n f_{\lambda_n},f_{\mu_n}\rangle_\pi+\langle\mu_n f_{\mu_n},f_{\lambda_n}\rangle_\pi\big).\end{aligned}$$

Each of these terms vanishes as $n\uparrow\infty$. Indeed,

$$\lambda_n\langle f_{\lambda_n},f_{\mu_n}\rangle_\pi=\lambda_n\langle f_{\lambda_n},f_{\mu_n}-W_2\rangle_\pi+\lambda_n\langle f_{\lambda_n},W_2\rangle_\pi.$$

These scalar products have to be understood as the scalar product of a function $\lambda_n f_{\lambda_n}$ in $\mathscr{H}_{-1}$ with functions $f_{\mu_n}-W_2$, W_2 in $\mathscr{H}_1$. By Schwarz inequality (2.9), the first term on the right-hand side is bounded above by $\|\lambda_n f_{\lambda_n}\|_{-1}\|f_{\mu_n}-W_2\|_1$, which vanishes because λf_λ is bounded in $\mathscr{H}_{-1}$ and f_{μ_n} converges to W_2 in $\mathscr{H}_1$. To show that the second term of the previous formula also vanishes, fix $\varepsilon>0$. Since $\mathscr{H}_1$ is obtained as the completion of $\mathscr{C}$, there exists g in $\mathscr{C}$ such that $\|W_2-g\|_1\le\varepsilon$. By Schwarz inequality (2.9), the second term is then absolutely bounded by $\sup_{0<\lambda\le 1}\|\lambda f_\lambda\|_{-1}\varepsilon + |\langle \lambda_n f_{\lambda_n}, g\rangle_\pi|$. Since λf_λ vanishes in $L^2(\pi)$ as $\lambda \downarrow 0$, the second term on the right-hand side of the previous displayed formula converges to 0 as $n \uparrow \infty$. This concludes the proof of the lemma. □

We conclude this section by observing that one can still prove the central limit theorem if we replace the bound $\sup_{0<\lambda\le 1}\|Lf_\lambda\|_{-1} < \infty$ by an energy identity. We have already shown that the family $\{f_\lambda : 0 < \lambda \le 1\}$ is weakly precompact in $\mathscr{H}_1$. Assume that f_λ converges weakly in $\mathscr{H}_1$, as $\lambda \downarrow 0$, to some W and that

$$\|W\|_1^2 = \langle W, V\rangle_\pi. \tag{2.33}$$

This relation is sometimes called an *energy identity*.

Theorem 2.17 *Fix a function V in $L^2(\pi)\cap\mathscr{H}_{-1}$ and assume that f_λ converges weakly in $\mathscr{H}_1$, as $\lambda \downarrow 0$, to some W which satisfies the energy identity* (2.33). *Then, both conditions* (2.23) *hold. In particular, the conditional law of $t^{-1/2}\int_0^t V(X_s)\,ds$ with respect to $\mathscr{F}_0$ converges in probability, as $t \uparrow \infty$, to a mean zero Gaussian distribution with variance given by* (2.24).

Proof The argument made in Claim 3 above shows that $\lim_{\lambda\to 0}\lambda\|f_\lambda\|_\pi^2 = 0$. In addition, by (2.30), f_λ converges strongly to W in $\mathscr{H}_1$, as $\lambda \downarrow 0$. Hence, both conditions (2.23) hold. As a result, by Theorem 2.7, the central limit theorem follows. □

2.7 Some Examples

Fix a function V in $L^2(\pi)\cap\mathscr{H}_{-1}$. In this section we present four different set of conditions which guarantee that the solution f_λ of the resolvent equation (2.13) satisfies the bound (2.28). Assume that $\mathscr{C}$ is a core for the operators L and L^*.

2.7.1 Reversibility

Assume that the generator L is self-adjoint in $L^2(\pi)$. In this case, as we have seen in Claim H of Sect. 2.2, Lf belongs to $\mathscr{H}_{-1}$ for any f in $\mathscr{D}(L)$, and

$$\|Lf\|_{-1} \le \|f\|_1.$$

In fact the equality holds. In particular, in the reversible case (2.28) follows from the elementary estimate (2.22).

In the reversible case the asymptotic variance $\sigma^2(V) = 2\|V\|_{-1}^2$. Indeed, in the resolvent equation, take the inner product in $\mathscr{H}_{-1}$ with respect to V on both sides of the equation. The right-hand side is equal to $\|V\|_{-1}^2$, while the left-hand side is equal to

$$\langle \lambda f_\lambda, V\rangle_{-1} + \langle -Lf_\lambda, V\rangle_{-1}.$$

By Claim 1 in the Proof of Lemma 2.16, λf_λ converges weakly to 0 in $\mathscr{H}_{-1}$ as $\lambda \downarrow 0$. In particular, the first term vanishes in the limit. On the other hand, by Claim H in Sect. 2.2, $\langle -Lg, h\rangle_{-1} = \langle g, h\rangle_\pi$ if g belongs to the domain $\mathscr{D}(L)$ and h to $L^2(\pi) \cap \mathscr{H}_{-1}$. In particular, $\langle -Lf_\lambda, V\rangle_{-1} = \langle f_\lambda, V\rangle_\pi$. Hence, by (2.24),

$$\sigma^2(V) = 2 \lim_{\lambda\to 0} \langle f_\lambda, V\rangle_\pi = 2\|V\|_{-1}^2.$$

2.7.2 Spectral Gap

Assume that the generator L satisfies a spectral gap condition: there exists a finite constant γ such that

$$\|f\|_\pi^2 - \langle f\rangle_\pi^2 \le \gamma \big\langle (-L)f, f\big\rangle_\pi \tag{2.34}$$

for every function f in the domain $\mathscr{D}(L)$.

Theorem 2.18 *Fix $V \in L^2(\pi)$ such that $\langle V\rangle_\pi = 0$ and suppose that the generator L satisfies the spectral gap condition. Then, the conclusions of Theorem 2.7 hold.*

Proof Observe that (2.34) implies that $L_0^2(\pi)$, the space of all mean zero, square integrable functions, is embedded into $\mathscr{H}_{-1}$. Indeed, for any f orthogonal to the constants in $L^2(\pi)$ and for any g belonging to the common core of L and L^* we have

$$\langle f, g\rangle_\pi^2 \le \|f\|_\pi^2 \big\{\|g\|_\pi^2 - \langle g\rangle_\pi^2\big\} \le \gamma \|f\|_\pi^2 \|g\|_1^2.$$

It follows from this estimate that

$$\|f\|_{-1}^2 \le \gamma \|f\|_\pi^2 \tag{2.35}$$

for all $f \in L_0^2(\pi)$. This of course implies that $V \in \mathscr{H}_{-1}$ and that $\lim_{\lambda\to 0} \lambda \|f_\lambda\|_{-1} = 0$, in view of (2.15), which in turn yields (2.28). The conclusion of the theorem follows directly from Theorem 2.14. □

2.7.3 Sector Condition

Assume now that the generator L satisfies the sector condition

$$\langle f, Lg\rangle_\pi^2 \le C_0 \big\langle f, (-L)f\big\rangle_\pi \big\langle g, (-L)g\big\rangle_\pi \tag{2.36}$$

for some finite constant C_0 and every function f, g in the domain $\mathscr{D}(L)$ of the generator. In view of (2.12), for any function g in $\mathscr{D}(L)$,

$$\|Lg\|_{-1}^2 \le C_0 \|g\|_1^2$$

and condition (2.28) follows from estimate (2.22).

Of course, L satisfies a sector condition if and only if A, the asymmetric part of the generator, satisfies it:

$$\langle f, Ag\rangle_\pi^2 \le C_0 \langle f, (-L)f\rangle_\pi \langle g, (-L)g\rangle_\pi$$

for some finite constant C_0 and every function f, g in the core. In this case, A can be extended as a bounded operator from $\mathscr{H}_1$ to $\mathscr{H}_{-1}$:

Lemma 2.19 *Fix an operator B. Assume that $\mathscr{C}$ is a core for B and that B satisfies the sector condition*

$$\langle f, Bg\rangle_\pi^2 \le C_0 \langle f, (-L)f\rangle_\pi \langle g, (-L)g\rangle_\pi$$

for all f, g in $\mathscr{C}$ and a finite constant C_0. Then, the operator B extends from $\mathscr{C}$ to a bounded, linear mapping from $\mathscr{H}_1$ to $\mathscr{H}_{-1}$.

Proof Suppose that f belongs to $\mathscr{C}$. By definition of the $\mathscr{H}_{-1}$ norm and the sector condition,

$$\|Bf\|_{-1}^2 = \sup_{g\in\mathscr{C}} \bigl\{2\langle Bf, g\rangle_\pi - \|g\|_1^2\bigr\} \le C_0 \|f\|_1^2.$$

This proves the lemma. □

Formally, we just obtained that

$$\langle g, A^*(-S)^{-1}Ag\rangle_\pi = \|Ag\|_{-1}^2 \le C_0\|g\|_1^2 = C_0\langle g, (-S)g\rangle_\pi$$

for all functions g in $\mathscr{C}$. Hence, the sector condition requires that

$$A^*(-S)^{-1}A \le C_0(-S)$$

for some finite constant C_0. This inequality states that the asymmetric part of the generator can be estimated by the symmetric part. Furthermore, in this case, in view of the computations performed just after (2.18),

$$(-S) \le (-S) + A^*(-S)^{-1}A \le (1+C_0)(-S)$$

so that

$$C_1^{-1}\sigma^2(V) \le \|V\|_{-1}^2 \le C_1\sigma^2(V)$$

for some finite constant C_1. This means that under the sector condition, the limiting variance is finite if and only if the function belongs to $\mathscr{H}_{-1}$.

This formal argument can be made rigorous. On the one hand, we have seen in Lemma 2.4 that $\sigma^2(V) \le C_0\|V\|_{-1}^2$ for some finite constant C_0. On the other hand, we have seen in the previous subsection that

$$\|V\|_{-1}^2 = \lim_{\lambda\to 0}\langle -Lf_\lambda, V\rangle_{-1},$$

where the inner product in $\mathscr{H}_{-1}$ appears on the right-hand side.

We claim that

$$\langle -Lf_\lambda, V\rangle_{-1} \le \langle f_\lambda, V\rangle_\pi + (C_0/2)\|f_\lambda\|_1^2 + (1/2)\|V\|_{-1}^2, \tag{2.37}$$

where C_0 is the finite constant appearing in (2.36). To prove this estimate write

$$\langle -Lf_\lambda, V\rangle_{-1} = \langle -Sf_\lambda, V\rangle_{-1} + \langle -Af_\lambda, V\rangle_{-1}.$$

Note that Sf_λ and Af_λ may not be defined. However, we assumed $\mathscr{C}$ to be a core for these two operators. We may therefore approximate f_λ by a function in the core, still denoted by f_λ. This step is explained in detail at the end of the argument. As before $\langle -Sf_\lambda, V\rangle_{-1} = \langle f_\lambda, V\rangle_\pi$. On the other hand, by Schwarz inequality and Lemma 2.19,

$$\left|\langle -Af_\lambda, V\rangle_{-1}\right| \le \|Af_\lambda\|_{-1}\|V\|_{-1} \le \sqrt{C_0}\|f_\lambda\|_1\|V\|_{-1}.$$

Since $2ab \le Aa^2 + A^{-1}b^2$ for every $A > 0$, the last expression is bounded above by $(C_0/2)\|f_\lambda\|_1^2 + (1/2)\|V\|_{-1}^2$, proving the claim.

Since $\langle f_\lambda, V\rangle_\pi$ and $\|f_\lambda\|_1^2$ converge to $(1/2)\sigma^2(V)$ as $\lambda \downarrow 0$, we conclude from the previous estimates that

$$\|V\|_{-1}^2 \le (1/2)\{1 + C_0\}\sigma^2(V) + (1/2)\|V\|_{-1}^2,$$

so that $\|V\|_{-1}^2 \le \{1 + C_0\}\sigma^2(V)$, as claimed.

We conclude this subsection by justifying the replacement of the solution of the resolvent equation by a function in the core in the proof of (2.37). By definition of a core, for each $\lambda > 0$, there exists a sequence of functions $\{g_k : k \ge 1\}$ in the core $\mathscr{C}$ such that g_k, Lg_k converge to f_λ, Lf_λ in $L^2(\pi)$, respectively. Therefore, g_k converges to f_λ in $\mathscr{H}_1$ and, by Lemma 2.19, Lg_k converge to Lf_λ in $\mathscr{H}_{-1}$.

Rewrite $\langle -Lf_\lambda, V\rangle_{-1}$ as $\langle -Lg_k, V\rangle_{-1} - \langle L(f_\lambda - g_k), V\rangle_{-1}$. By the proof presented above,

$$\langle -Lg_k, V\rangle_{-1} \le \langle g_k, V\rangle_\pi + (C_0/2)\|g_k\|_1^2 + (1/2)\|V\|_{-1}^2.$$

On the other hand, by Schwarz inequality, $\langle -L(f_\lambda - g_k), V\rangle_{-1}$ is absolutely bounded by $\|L(f_\lambda - g_k)\|_{-1}\|V\|_{-1}$. Therefore,

$$\langle -Lf_\lambda, V\rangle_{-1} \le \langle g_k, V\rangle_\pi + (C_0/2)\|g_k\|_1^2 + (1/2)\|V\|_{-1}^2 + \left\|L(f_\lambda - g_k)\right\|_{-1}\|V\|_{-1}$$

for all $k \ge 1$. To deduce (2.37), it remains to let $k \uparrow \infty$ and to recall the properties of the sequence $\{g_k : k \ge 1\}$.

2.7.4 Graded Sector Condition

Instead of assuming that the generator satisfies a sector condition on the whole space, we decompose $L^2(\pi)$ as a direct sum of orthogonal spaces $\mathscr{A}_n$ and assume that on each subspace $\mathscr{A}_n$ the generator satisfies a sector condition with a constant which may be different on each $\mathscr{A}_n$.

Assume that $L^2(\pi)$ can be decomposed as a direct sum $\bigoplus_{n\geq 0}\mathscr{A}_n$ of orthogonal spaces. Let $\mathscr{G}_n = \bigoplus_{0\leq k\leq n}\mathscr{A}_k$. Functions in $\mathscr{G}_n\setminus\mathscr{G}_{n-1}$, $\mathscr{G} = \bigcup_{n\geq 0}\mathscr{G}_n$ are said to have degree n, finite degree, respectively, while functions in $\mathscr{A}_n$ are said to be monomials of degree n. For $n\geq 0$, denote by Π_n the orthogonal projection on $\mathscr{A}_n$ so that

$$f = \sum_{n\geq 0}\Pi_n f, \quad \Pi_n f\in\mathscr{A}_n, n\geq 0, \quad \text{and} \quad \|f\|_\pi^2 = \sum_{n\geq 0}\|\Pi_n f\|_\pi^2 \tag{2.38}$$

for all f in $L^2(\pi)$. We assume that the core $\mathscr{C}$ is closed under $\{\Pi_n : n\geq 0\}$ and that all functions in the core $\mathscr{C}$ have finite degree:

$$\Pi_n f \text{ belongs to } \mathscr{C} \text{ for all } f \text{ in } \mathscr{C},\ n\geq 0, \text{ and } \mathscr{C}\subset\mathscr{G}. \tag{2.39}$$

Supposethat the generator L keeps the degree of a monomial or changes it by one: $L\colon \mathscr{C}\cap\mathscr{A}_n\to\mathscr{A}_{n-1}\oplus\mathscr{A}_n\oplus\mathscr{A}_{n+1}$. Denote by L_- (resp. L_+ and L_0) the piece of the generator L which decreases (resp. increases and keeps) the degree of a monomial. Since all functions in the core have finite degree, the operators L_+, L_- and L_0 which in principle are only defined on $\bigcup_{n\geq 0}\{\mathscr{C}\cap\mathscr{G}_n\} = \mathscr{C}\cap\mathscr{G}$, are in fact defined on $\mathscr{C}$. Hence, we may take $\mathscr{C}$ as the common domain of the operators L_+, L_- and L_0.

Recall that we denote by L^* the adjoint of L in $L^2(\pi)$ and that we assumed $\mathscr{C}$ to be a core for L^*. Since $\langle Lf, g\rangle_\pi = 0$ for all $f\in\mathscr{C}\cap\mathscr{A}_n$, $g\in\mathscr{C}\cap\mathscr{A}_m$, $|m-n|>1$, L^* also keeps the degree of a monomial or changes it by one: $L^*\colon\mathscr{C}\cap\mathscr{A}_n\to\mathscr{A}_{n-1}\oplus\mathscr{A}_n\oplus\mathscr{A}_{n+1}$. Denote by $(L^*)_-$, $(L^*)_+$, $(L^*)_0$ the piece of the generator L^* which decreases, increases, keeps, respectively, the degree of a monomial. As above, we take $\mathscr{C}$ as the common domain of the operators $(L^*)_-$, $(L^*)_+$, $(L^*)_0$.

An elementary computation shows that for any $f, g\in\mathscr{C}$,

$$\langle L_+ f, g\rangle_\pi = \big\langle f, (L^*)_- g\big\rangle_\pi .$$

Therefore, $\mathscr{C}$ is contained in the domain of $(L_+)^*$, the adjoint of L_+, and $(L_+)^*g = (L^*)_- g$, $g\in\mathscr{C}$. For the same reasons, $(L_-)^*g = (L^*)_+ g$, $(L_0)^*g = (L^*)_0 g$, $g\in\mathscr{C}$.

Assume that the operator L_0 can be decomposed as a sum $S_0 + B_0$, where $-S_0$ is a non-negative symmetric operator which preserve the degrees of functions and which is bounded by $-C_0 L$ for some positive constant C_0:

$$0\leq\big\langle f, (-S_0)f\big\rangle_\pi \leq C_0\big\langle f, (-L)f\big\rangle_\pi \tag{2.40}$$

for all functions f in $\mathscr{C}$.

Since $-S_0$ is a positive operator, repeating the steps of Sect. 2.2 with S_0 in place of L, we define the Sobolev spaces $\mathscr{H}_{0,1}$, $\mathscr{H}_{0,-1}$ and the norms $\|\cdot\|_{0,1}$, $\|\cdot\|_{0,-1}$ associated to S_0. As S_0 keeps the degree of a monomial, for every f in $\mathscr{C}$,

$$\|f\|_{0,1}^2 = \big\langle f, (-S_0)f\big\rangle_\pi = \Big\langle\sum_{n\geq 0}\Pi_n f, (-S_0)\sum_{n\geq 0}\Pi_n f\Big\rangle_\pi = \sum_{n\geq 0}\big\langle\Pi_n f, (-S_0)\Pi_n f\big\rangle_\pi$$

so that

$$\| f \|_{0,1}^2 = \sum_{n \geq 0} \| \Pi_n f \|_{0,1}^2 \tag{2.41}$$

for all functions f in the core $\mathscr{C}$. Moreover, in terms of the new norm $\| \cdot \|_{0,1}$, (2.40) translates to

$$\| f \|_{0,1} \leq \sqrt{C_0} \| f \|_1 \tag{2.42}$$

for all functions f in $\mathscr{C}$ and some finite constant C_0.

In the definition of the space $\mathscr{H}_1$, recall the convention adopted just after (2.5). We claim that

Lemma 2.20 *The domain $\mathscr{D}(L)$ is contained in $\mathscr{H}_{0,1}$ and $\Pi_j f$ belongs to $\mathscr{H}_{0,1}$ for every $f \in \mathscr{D}(L)$, $j \geq 0$. Moreover, for every f in $\mathscr{D}(L)$, (2.41) holds.*

Proof We have already seen that $\mathscr{D}(L)$ is contained in $\mathscr{H}_1$. On the other hand, by (2.42), every $(\mathscr{C}, \mathscr{H}_1)$-approximating sequence $\{f_n : n \geq 1\}$ of a function f in $\mathscr{D}(L)$ is also a $(\mathscr{C}, \mathscr{H}_{0,1})$-approximating sequence. This proves the first assertion of the lemma.

To prove the second one, fix $f \in \mathscr{D}(L)$ and $j \geq 0$. Since $\mathscr{D}(L) \subset \mathscr{H}_1$, there exists a $(\mathscr{C}, \mathscr{H}_1)$-approximating sequence $\{f_n : n \geq 1\}$ of f, which, by (2.42), is also a $(\mathscr{C}, \mathscr{H}_{1,0})$-approximating sequence. Since $\mathscr{C}$ is closed under the projection Π_j, $\{\Pi_j f_n : n \geq 1\}$ is a sequence of functions in $\mathscr{C}$ which converges to $\Pi_j f$ in $L^2(\pi)$. By (2.41), $\{\Pi_j f_n : n \geq 1\}$ is a Cauchy sequence in $\mathscr{H}_{1,0}$, proving that $\Pi_j f$ belongs to $\mathscr{H}_{1,0}$. This proves the second claim.

To prove the third statement, fix a function f in $\mathscr{D}(L)$ and consider a $(\mathscr{C}, \mathscr{H}_{0,1})$-approximating sequence $\{f_n : n \geq 1\}$ of f. On the one hand, by (2.41),

$$\| f \|_{0,1}^2 = \lim_{n \to \infty} \| f_n \|_{0,1}^2 = \lim_{n \to \infty} \sum_{j \geq 0} \| \Pi_j f_n \|_{0,1}^2 .$$

For every $K \geq 1$, the previous expression is bounded below by

$$\lim_{n \to \infty} \sum_{j=0}^{K} \| \Pi_j f_n \|_{0,1}^2 = \sum_{j=0}^{K} \| \Pi_j f \|_{0,1}^2$$

because $\{\Pi_j f_n : n \geq 1\}$ is a $(\mathscr{C}, \mathscr{H}_{0,1})$-approximating sequence of $\Pi_j f$ for every $j \geq 0$. Therefore $\| f \|_{0,1}^2 \geq \sum_{j \geq 0} \| \Pi_j f \|_{0,1}^2$.

On the other hand, since $\{f_n : n \geq 1\}$ is a Cauchy sequence in $\mathscr{H}_{0,1}$, for every $\varepsilon > 0$, there exists $n_0 = n_0(\varepsilon) \geq 1$ such that $\| f_n - f_m \|_{0,1}^2 \leq \varepsilon$ for n, $m \geq n_0$. For this n_0 fixed, there exists $k = k(\varepsilon)$ such that $\sum_{j \geq k} \| \Pi_j f_{n_0} \|_{0,1}^2 \leq \varepsilon$. Thus, for all $n \geq n_0$,

$$\sum_{j \geq k} \| \Pi_j f_n \|_{0,1}^2 \leq 2 \sum_{j \geq 0} \big\| \Pi_j (f_n - f_{n_0}) \big\|_{0,1}^2 + 2 \sum_{j \geq k} \| \Pi_j f_{n_0} \|_{0,1}^2 \leq 4\varepsilon.$$

Therefore, in view of (2.41),

$$\|f\|_{0,1}^2 = \lim_{n\to\infty} \|f_n\|_{0,1}^2 \le \lim_{n\to\infty} \sum_{j<k} \|\Pi_j f_n\|_{0,1}^2 + 4\varepsilon.$$

Since $\{\Pi_j f_n : n \ge 1\}$ is a $(\mathscr{C}, \mathscr{H}_{0,1})$-approximating sequence of $\Pi_j f$ for every $j \ge 0$,

$$\|f\|_{0,1}^2 \le \sum_{j\ge 0} \|\Pi_j f\|_{0,1}^2 + 4\varepsilon,$$

which proves the lemma. □

By (2.41), for a function f in $L^2(\pi) \cap \mathscr{H}_{0,-1}$,

$$\|f\|_{0,-1}^2 = \sup_{g\in\mathscr{C}} \big\{2\langle f, g\rangle_\pi - \|g\|_{0,1}^2\big\} = \sum_{n\ge 0} \|\Pi_n f\|_{0,-1}^2. \tag{2.43}$$

Furthermore, it follows from the last inequality and from the variational formula for the $\mathscr{H}_{-1}$, $\mathscr{H}_{0,-1}$ norms that

$$\|f\|_{-1} \le \sqrt{C_0}\,\|f\|_{0,-1} \tag{2.44}$$

for all functions f in $L^2(\pi)$ and the same finite constant C_0.

Suppose now that a sector condition holds on each subspace $\mathscr{A}_n$ with a constant which depends on n. More precisely, assume that there exist $\beta < 1$ and a finite constant C_0 such that for all $n \ge 0$,

$$\begin{aligned} \big\langle f, (-L_+)g\big\rangle_\pi^2 &\le C_0(n+1)^{2\beta}\big\langle f, (-S_0)f\big\rangle_\pi \big\langle g, (-S_0)g\big\rangle_\pi, \\ \big\langle g, (-L_-)f\big\rangle_\pi^2 &\le C_0(n+1)^{2\beta}\big\langle f, (-S_0)f\big\rangle_\pi \big\langle g, (-S_0)g\big\rangle_\pi \end{aligned} \tag{2.45}$$

for all g in $\mathscr{C} \cap \mathscr{A}_n$ and f in $\mathscr{C} \cap \mathscr{A}_{n+1}$.

The first estimate extends to functions f in the domain $\mathscr{D}(L)$. More precisely, fix a function f in $\mathscr{D}(L)$ and g in $\mathscr{C} \cap \mathscr{A}_n$. We claim that

$$\big\langle f, (-L_+)g\big\rangle_\pi^2 \le C_0(n+1)^{2\beta} \|\Pi_{n+1} f\|_{0,1}^2 \|g\|_{0,1}^2.$$

Consider a $(\mathscr{C}, \mathscr{H}_{0,1})$-approximating sequence $\{f_k : k \ge 1\}$ of f. By assumption (2.39) and by (2.45), the previous inequality holds with $\Pi_{n+1} f_k$ in place of f. It remains to let $k \uparrow \infty$ and to observe that $\langle \Pi_{n+1} f_k, (-L_+)g\rangle_\pi$, $\|\Pi_{n+1} f_k\|_{0,1}$ converge to $\langle \Pi_{n+1} f, (-L_+)g\rangle_\pi = \langle f, (-L_+)g\rangle_\pi$, $\|\Pi_{n+1} f\|_{0,1}$, respectively. Similarly, we can show that the second estimate in (2.45) holds for functions g in $\mathscr{D}(L)$ if we replace on the right-hand side $\langle g, (-S_0)g\rangle_\pi$ by $\|\Pi_n g\|_{0,1}^2$.

Furthermore, since $(L_+)^*h = (L^*)_-h$, $(L_-)^*h = (L^*)_+h$, $h \in \mathscr{C}$, it follows from (2.45) and the previous argument that there exists a finite constant C_0, which may by different from the previous one, such that for all $n \geq 0$,

$$\begin{aligned} \langle f, (-L^*)_+ g\rangle_\pi^2 &\leq C_0(n+1)^{2\beta} \|\Pi_{n+1} f\|_{0,1}^2 \|g\|_{0,1}^2, \\ \langle G, (-L^*)_- F\rangle_\pi^2 &\leq C_0(n+1)^{2\beta} \|F\|_{0,1}^2 \|\Pi_n G\|_{0,1}^2 \end{aligned} \tag{2.46}$$

for all f, G in $\mathscr{D}(L)$, g in $\mathscr{C} \cap \mathscr{A}_n$, F in $\mathscr{C} \cap \mathscr{A}_{n+1}$.

Finally, it follows from (2.45) and from the variational formula for the $\|\cdot\|_{-1,0}$ norm that

$$\begin{aligned} &\|L_+ g\|_{0,-1} \leq \sqrt{C_0}(n+1)^\beta \|g\|_{0,1}, && \|L_- f\|_{0,-1} \leq \sqrt{C_0}(n+1)^\beta \|f\|_{0,1}, \\ &\|(L^*)_+ g\|_{0,-1} \leq \sqrt{C_0}(n+1)^\beta \|g\|_{0,1}, && \|(L^*)_- f\|_{0,-1} \leq \sqrt{C_0}(n+1)^\beta \|f\|_{0,1} \end{aligned} \tag{2.47}$$

for all $n \geq 0$ and all g in $\mathscr{C} \cap \mathscr{A}_n$ and f in $\mathscr{C} \cap \mathscr{A}_{n+1}$. The proof of Lemma 2.21 below shows that the restriction $\beta < 1$ is crucial.

Fix $k \geq 0$ and define the triple norms $|||\cdot|||_{k,\pi}$, $|||\cdot|||_{k,1}$ and $|||\cdot|||_{k,-1}$ by

$$\begin{aligned} &|||f|||_{k,\pi}^2 = \sum_{n\geq 0} (n+1)^{2k} \|\Pi_n f\|_\pi^2, \qquad |||f|||_{k,1}^2 = \sum_{n\geq 0} (n+1)^{2k} \|\Pi_n f\|_{0,1}^2, \\ &|||f|||_{k,-1}^2 = \sum_{n\geq 0} (n+1)^{2k} \|\Pi_n f\|_{0,-1}^2. \end{aligned} \tag{2.48}$$

On several occasions we omit the dependence of the norms on k.

Lemma 2.21 *Assume that the generator L satisfies conditions* (2.39), (2.40), (2.45), *and that $\beta < 1$. Let V be a function in $L^2(\pi)$ such that*

$$|||V|||_{k,-1} < \infty$$

for some $k \geq 1$. Denote by f_λ the solution of the resolvent equation (2.13). *There exists a finite constant C_1 depending only on β, k and C_0 such that*

$$\lambda |||f_\lambda|||_{k,\pi}^2 + |||f_\lambda|||_{k,1}^2 \leq C_1 |||V|||_{k,-1}^2.$$

Proof Consider an increasing sequence $\{t_n : n \geq 0\}$ eventually constant, to be fixed later, and denote by $T\colon L^2(\pi) \to L^2(\pi)$ the operator which is a multiple of the identity on each subspace $\mathscr{A}_n$:

$$Tf = \sum_{n\geq 0} t_n \Pi_n f.$$

Assume that f_λ belongs to the core. We show at the end of the proof how to proceed if this is not the case. Apply T to both sides of the resolvent equation, and take the inner product with respect to Tf_λ on both sides of the identity to obtain that

$$\lambda \langle Tf_\lambda, Tf_\lambda\rangle_\pi - \langle Tf_\lambda, LTf_\lambda\rangle_\pi = \langle Tf_\lambda, TV\rangle_\pi - \langle Tf_\lambda, [L,T]f_\lambda\rangle_\pi.$$

In this formula, $[L, T]$ stands for the commutator of L and T and is given by $LT - TL$. Since the sequence $\{t_k : k \geq 0\}$ is eventually constant, $Tf_\lambda = t_n f_\lambda + \sum_{0\leq j\leq n}(t_j - t_n)\Pi_j f_\lambda$ for some $n \geq 0$, so that Tf_λ belongs to $\mathscr{C}$ by hypothesis (2.39).

By assumption (2.40), the second term on the left-hand side is bounded below by

$$C_0^{-1}\langle Tf_\lambda, (-S_0)Tf_\lambda\rangle_\pi = C_0^{-1}\|Tf_\lambda\|_{0,1}^2 = C_0^{-1}\sum_{n\geq 0} t_n^2\|\Pi_n f_\lambda\|_{0,1}^2 .$$

Let $\delta > 0$. We now estimate the scalar product $\langle Tf_\lambda, [L, T]f_\lambda\rangle_\pi$ in terms of $\|Tf_\lambda\|_{0,1}^2$. Since T commutes with any operator which keeps the degree, $[L, T] = [L_+ + L_-, T]$. To fix ideas, consider the operator $[L_+, T]$, the other expression being estimated in a similar way. Since L_+ increases the degree by one, by definition of the commutator,

$$\Pi_n[L_+, T]f = L_+T\Pi_{n-1}f - TL_+\Pi_{n-1}f = (t_{n-1} - t_n)L_+\Pi_{n-1}f$$

for all functions f in $\mathscr{C}$, $n \geq 1$. Note that $\Pi_0[L_+, T]f = 0$ for all functions f in the core. Therefore,

$$\begin{aligned}\langle Tf_\lambda, [L_+, T]f_\lambda\rangle_\pi &= \sum_{n\geq 1}\langle \Pi_n Tf_\lambda, \Pi_n[L_+, T]f_\lambda\rangle_\pi \\ &= \sum_{n\geq 1}(t_{n-1} - t_n)t_n\langle \Pi_n f_\lambda, L_+\Pi_{n-1}f_\lambda\rangle_\pi .\end{aligned}$$

By (2.45) and since the sequence t_n is increasing, the previous expression is bounded above by

$$\begin{aligned}&\sqrt{C_0}\sum_{n\geq 1}(t_n - t_{n-1})t_n n^\beta\|\Pi_n f_\lambda\|_{0,1}\|\Pi_{n-1}f_\lambda\|_{0,1} \\ &\leq \frac{\sqrt{C_0}}{2}\sum_{n\geq 1}(t_n - t_{n-1})t_n n^\beta\|\Pi_n f_\lambda\|_{0,1}^2 \\ &\quad + \frac{\sqrt{C_0}}{2}\sum_{n\geq 1}(t_n - t_{n-1})t_n n^\beta\|\Pi_{n-1}f_\lambda\|_{0,1}^2 .\end{aligned}$$

Since $\beta < 1$, there exists $n_1 = n_1(C_0, \beta, \delta, k)$ such that

$$\sqrt{C_0}n^\beta\Big\{1 - \frac{(n-1)^k}{n^k}\Big\} \leq \delta, \qquad \sqrt{C_0}n^\beta\Big\{\frac{n^k}{(n-1)^k} - 1\Big\}\frac{n^k}{(n-1)^k} \leq \delta$$

for all $n \geq n_1$. Fix $n_2 > n_1$ and set $t_n = n_1^k\mathbf{1}\{n < n_1\} + n^k\mathbf{1}\{n_1 \leq n \leq n_2\} + n_2^k\mathbf{1}\{n > n_2\}$. With this definition, we obtain that the previous expression is bounded by

$$\delta\sum_{n\geq 0} t_n^2\|\Pi_n f_\lambda\|_{0,1}^2 = \delta\|Tf_\lambda\|_{0,1}^2 .$$

It remains to estimate $\langle Tf_\lambda, TV\rangle_\pi$. By (2.9), and since $2ab \le A^{-1}a^2 + Ab^2$ for every $A > 0$,

$$\begin{aligned}\langle Tf_\lambda, TV\rangle_\pi &= \sum_{n\ge 0} t_n^2 \langle \Pi_n f_\lambda, \Pi_n V\rangle_\pi \le \sum_{n\ge 0} t_n^2 \|\Pi_n f_\lambda\|_{0,1} \|\Pi_n V\|_{0,-1} \\ &\le \delta \sum_{n\ge 0} t_n^2 \|\Pi_n f_\lambda\|_{0,1}^2 + \delta^{-1} \sum_{n\ge 0} t_n^2 \|\Pi_n V\|_{0,-1}^2 \\ &= \delta \|Tf_\lambda\|_{0,1}^2 + \delta^{-1}\|TV\|_{0,-1}^2.\end{aligned}$$

Putting together the previous three estimates, we obtain that

$$\lambda \|Tf_\lambda\|_\pi^2 + C_0^{-1}\|Tf_\lambda\|_{0,1}^2 \le 3\delta \|Tf_\lambda\|_{0,1}^2 + \delta^{-1}\|TV\|_{0,-1}^2.$$

There is a factor 3, instead of 2, multiplying δ due to the estimate of the commutator $[L_-, T]f_\lambda$ which we omitted. Therefore,

$$\lambda \|Tf_\lambda\|_\pi^2 + \delta \|Tf_\lambda\|_{0,1}^2 \le \delta^{-1}\|TV\|_{0,-1}^2 \tag{2.49}$$

if we choose $\delta = 1/4C_0$.

Recall the definition of the sequence t_n and the identities (2.38), (2.41), (2.43). Since, the previous estimate holds uniformly in n_2, let $n_2 \uparrow \infty$ and define T' as the operator associated to the sequence t_n', where $t_n' = n^k \mathbf{1}\{n \ge n_1\} + n_1^k \mathbf{1}\{n < n_1\}$, to deduce that

$$|||f_\lambda|||_{k,1}^2 \le \left\|T' f_\lambda\right\|_{0,1}^2 \le \delta^{-2}\left\|T'V\right\|_{0,-1}^2 \le \delta^{-2} n_1^{2k} |||V|||_{k,-1}^2.$$

A similar argument shows that

$$\lambda |||f_\lambda|||_{k,\pi}^2 \le \delta^{-1} n_1^{2k} |||V|||_{k,-1}^2.$$

To conclude the proof of the lemma recall that we fixed $\delta = 1/4C_0$ and that $n_1 = n_1(C_0, k, \beta, \delta)$.

It remains to justify the replacement in the resolvent equation of its solution f_λ by a function g in the core. Rewrite the resolvent equation as

$$\lambda g - Lg = V + \lambda(g - f_\lambda) - L(g - f_\lambda),$$

where g is a function in the core. Apply the operator T and take the inner product on both sides of the identity with respect to Tg. At this point we follow the proof presented above with g replacing f_λ. Instead of (2.49), we obtain the estimate

$$\lambda \|Tg\|_\pi^2 + \delta \|Tg\|_{0,1}^2 \le \delta^{-1}\|TV\|_{0,-1}^2 + \lambda \big\langle Tg, T(g - f_\lambda)\big\rangle_\pi - \big\langle Tg, TL(g - f_\lambda)\big\rangle_\pi$$

for any function g in the core. Choose a sequence $\{g_k : k \ge 1\}$ in $\mathscr{C}$ such that g_k, Lg_k converge in $L^2(\pi)$, as $k \uparrow \infty$, to f_λ, Lf_λ, respectively. Clearly, in view of (2.42), $\|Tg_k\|_\pi$, $\|Tg_k\|_{0,1}$ converge to $\|Tf_\lambda\|_\pi$, $\|Tf_\lambda\|_{0,1}$. On the other hand, since

$\|Th\|_\pi \le n_2^k \|h\|_\pi$, $h \in L^2(\pi)$, $\langle Tg_k, T(g_k - f_\lambda)\rangle_\pi$ and $\langle Tg_k, TL(g_k - f_\lambda)\rangle_\pi$ vanish as $k \uparrow \infty$ because g_k and Lg_k converge in $L^2(\pi)$ to f_λ, Lf_λ, respectively. This proves (2.49). □

Assume now that $L_0 = S_0 + B_0$ satisfies a sector condition on each subset $\mathscr{A}_n$, $n \ge 0$:

$$\langle g, (-L_0)f\rangle_\pi^2 \le C_0(n+1)^{2\gamma}\langle f, (-S_0)f\rangle_\pi \langle g, (-S_0)g\rangle_\pi \tag{2.50}$$

for some $\gamma > 0$ and all functions f, g in $\mathscr{C} \cap \mathscr{A}_n$. Notice that we do not impose any condition on γ. As before, we may extend this estimate to all functions g in the domain $\mathscr{D}(L)$ and to the adjoint $(-L^*)_0$:

$$\begin{aligned} \langle g, (-L_0)f\rangle_\pi^2 &\le C_0(n+1)^{2\gamma}\,\|f\|_{0,1}^2\,\|\Pi_n g\|_{0,1}^2, \\ \langle g, (-L^*)_0 f\rangle_\pi^2 &\le C_0(n+1)^{2\gamma}\,\|f\|_{0,1}^2\,\|\Pi_n g\|_{0,1}^2 \end{aligned} \tag{2.51}$$

for all $n \ge 0$ and all functions g in $\mathscr{D}(L)$, f in $\mathscr{C} \cap \mathscr{A}_n$. Moreover, by (2.50) and by the variational formula for the norm $\|\cdot\|_{0,-1}$,

$$\|L_0 f\|_{0,-1} \le \sqrt{C_0}(n+1)^{\gamma}\,\|f\|_{0,1} \tag{2.52}$$

for all $n \ge 0$ and all functions f in $\mathscr{C} \cap \mathscr{A}_n$.

Assumptions (2.45), (2.50) permit to estimate the $\mathscr{H}_1$ norm of a function in terms of a triple norm:

Lemma 2.22 *Let $\alpha = \beta \vee \gamma := \max\{\beta, \gamma\}$. There exists a finite constant C_0 such that for every function f in $L^2(\pi)$,*

$$\|f\|_1^2 \le C_0 \sum_{n \ge 0} (n+1)^{2\alpha}\,\|\Pi_n f\|_{0,1}^2.$$

Proof We first prove the estimate for functions in the core $\mathscr{C}$. Fix such a function g. Since L may be decomposed as $L_0 + L_+ + L_-$, $\|g\|_1^2 = \langle g, (-L)g\rangle_\pi$ can be written as

$$\sum_{n\ge 0}\langle \Pi_n g, (-L_0)\Pi_n g\rangle_\pi + \sum_{n \ge 0}\langle \Pi_{n+1} g, (-L_+)\Pi_n g\rangle_\pi + \sum_{n\ge 1}\langle \Pi_{n-1} g, (-L_-)\Pi_n g\rangle_\pi.$$

Note that all sums are finite since we assumed g to belong to $\mathscr{C}$. By assumptions (2.45), (2.50) the previous sum is bounded above by

$$C_0 \sum_{n \ge 0} (n+1)^{2\alpha}\,\|\Pi_n g\|_{0,1}^2$$

for some finite constant C_0. This proves the lemma for functions in the core $\mathscr{C}$.

Fix now a function f in $L^2(\pi)$ and assume that $\sum_{n\geq 0}(n+1)^{2\alpha}\|\Pi_n f\|_{0,1}^2$ is finite. Since $\Pi_n f$ belongs to $\mathscr{H}_{0,1}$ there is a sequence $\{g_{n,k} : k \geq 1\}$ of functions in $\mathscr{C} \cap \mathscr{A}_n$ which converges to $\Pi_n f$ in $L^2(\pi)$ and in $\mathscr{H}_{0,1}$. By adding these functions, one obtains a sequence $\{g_k : k \geq 1\}$ of functions in the core $\mathscr{C}$ which converges to f in $L^2(\pi)$ and such that

$$\lim_{k\to\infty} \sum_{n\geq 0}(n+1)^{2\alpha} \big\| \Pi_n\{f - g_k\}\big\|_{0,1}^2 = 0.$$

It follows from this fact and from the statement of the lemma applied to functions in $\mathscr{C}$ that $\{g_k : k \geq 1\}$ is a Cauchy sequence in $\mathscr{H}_1$. Since g_k converges to f in $L^2(\pi)$, the sequence $\{g_k : k \geq 1\}$ is a $(\mathscr{C}, \mathscr{H}_1)$-approximating sequence of f. Therefore, f belongs to $\mathscr{H}_1$ and

$$\begin{aligned}\|f\|_1^2 &= \lim_{k\to\infty} \|g_k\|_1^2 \leq C_0 \lim_{k\to\infty} \sum_{n\geq 0}(n+1)^{2\alpha}\|\Pi_n g_k\|_{0,1}^2 \\ &= C_0 \sum_{n\geq 0}(n+1)^{2\alpha}\|\Pi_n f\|_{0,1}^2,\end{aligned}$$

which proves the lemma. □

Theorem 2.23 *Suppose that the generator L satisfies hypotheses* (2.39), (2.40), (2.45) *and* (2.50). *Fix a function V such that*

$$|||V|||_{k,-1} < \infty$$

for some $k \geq (\beta \vee \gamma)$. Let f_λ be the solution of the resolvent equation (2.13). *Then,*

$$\sup_{0<\lambda\leq 1} \|Lf_\lambda\|_{0,-1} < \infty.$$

Proof It follows from (2.43) that

$$\|Lf_\lambda\|_{0,-1}^2 = \sum_{n\geq 0} \|\Pi_n L f_\lambda\|_{0,-1}^2. \tag{2.53}$$

Fix $n \geq 0$ and recall the variational formula for the $\mathscr{H}_{0,-1}$ norm of a function in $L^2(\pi)$. Since $\Pi_n L f_\lambda$ belongs to $\mathscr{A}_n$, by (2.41) we may restrict the supremum to functions h in $\mathscr{C} \cap \mathscr{A}_n$ so that

$$\|\Pi_n L f_\lambda\|_{0,-1}^2 = \sup_{h\in\mathscr{C}\cap\mathscr{A}_n} \big\{2\langle f_\lambda, L^*h\rangle_\pi - \|h\|_{0,1}^2\big\}.$$

Since $L^* = (L^*)_+ + (L^*)_- + (L^*)_0$, by (2.46) and (2.51), $\langle f_\lambda, L^*h\rangle_\pi$ is absolutely bounded by

$$C_0\big\{(n+1)^\beta\|\Pi_{n-1}f_\lambda\|_{0,1} + (n+1)^\gamma\|\Pi_n f_\lambda\|_{0,1} + (n+1)^\beta\|\Pi_{n+1}f_\lambda\|_{0,1}\big\}\|h\|_{0,1}$$

for some finite constant C_0. By Lemma 2.21 and since $k \geq (\beta \vee \gamma)$, the right-hand side of (2.53) is bounded above by

$$C_1 \sum_{n\geq 0} (n+1)^{2k} \|\Pi_n f_\lambda\|_{0,1}^2 \leq C_1 \sum_{n\geq 0} (n+1)^{2k} \|\Pi_n V\|_{0,-1}^2$$

for some finite constant C_1 depending only on C_0, β and γ. □

In fact, we have proved the result below.

Corollary 2.24 *Fix $f \in \mathscr{D}(L)$. Then, for every $k \geq 0$ there exists a finite constant C_k such that*

$$|||Lf|||_{k,-1} \leq C_k |||f|||_{k+m,1},$$

where $m = \max\{\beta, \gamma\}$. In particular, the function Lf belongs to $\mathscr{H}_{-1,0}$, and therefore to $\mathscr{H}_{-1}$, if $|||f|||_{m,1} < \infty$.

2.7.5 Perturbations of Normal Operators

In this section, we suppose that L can be decomposed on the core $\mathscr{C}$ as a normal operator plus a perturbation which satisfies a sector condition with respect to the normal operator. More specifically, assume that the operator L can be written as $L = L_0 + B$ and that $\mathscr{C}$ is a common core for each of the operators L_0, L_0^*, B and B^*. We impose three conditions on the operators L_0, B.

Assume that $L_0 : \mathscr{C} \to L^2(\pi)$ is *normal*, i.e. the operators $S_0 := (1/2)(L_0 + L_0^*)$ and $iA_0 := (i/2)(L_0 - L_0^*)$ are essentially self-adjoint and their spectral resolutions commute. Observe that the latter condition follows e.g. for some $\lambda > 0$ the resolvents $R_\lambda(S_0)$, $R_\lambda(iA)$ corresponding to S_0 and iA_0 commute. Indeed, these operators are bounded and normal and since they commute this also holds for their respective spectral resolutions, say $E_1(\cdot)$ and $E_2(\cdot)$. Since the spectral resolutions of the operators in question can be obtained by appropriate change of variables in $E_1(\cdot)$ and $E_2(\cdot)$ this proves that they also have to commute.

Suppose that the Dirichlet form associated to L_0 is equivalent to the Dirichlet form associated to L: there exist constants $0 < c_* < C_* < +\infty$ such that

$$c_* \langle f, (-L)f \rangle_\pi \leq \langle f, (-L_0)f \rangle_\pi \leq C_* \langle f, (-L)f \rangle_\pi, \tag{2.54}$$

for every function f in the core $\mathscr{C}$. Denote by $\|\cdot\|_{0,\pm 1}$ the respective $\mathscr{H}_{\pm 1}$ norms for L_0. It follows from this condition that there exist constants $0 < c_1 < C_1 < +\infty$ such that

$$c_1 \|f\|_{0,\pm 1} \leq \|f\|_{\pm 1} \leq C_1 \|f\|_{0,\pm 1} \tag{2.55}$$

for all f in $\mathscr{C}$.

Assume that B satisfies a sector condition relative to L_0:

$$\langle f, Bg\rangle_\pi^2 \le C_0\langle f, (-L_0)f\rangle_\pi\langle g, (-L_0)g\rangle_\pi \tag{2.56}$$

for some finite constant C_0 and every function $f, g \in \mathscr{C}$. In view of (2.55), a sector condition for B relative to L_0 holds if and only if it holds relative to L. It follows from this sector condition that B is a bounded mapping from $\mathscr{H}_1$ to $\mathscr{H}_{-1}$: There exists a finite constant $C_B > 0$ such that

$$\|Bf\|_{-1} \le C_B\|f\|_1 \tag{2.57}$$

for all f in $\mathscr{C}$. Indeed, by definition of the $\mathscr{H}_{-1}$ norm, the sector condition (2.56) and the bound (2.55),

$$\|Bf\|_{-1} = \sup_{\|\phi\|_1=1}\langle Bf, \phi\rangle_\pi \le C_0\|f\|_{0,1} \le \frac{C_0}{c_1}\|f\|_1.$$

Proposition 2.25 *Assume that L_0 is normal and hypotheses* (2.54), (2.56). *Fix a function V in $\mathscr{H}_{-1}\cap L^2(\pi)$ and denote by f_λ the solution of the resolvent equation* (2.13). *Then,*

$$\sup_{0<\lambda\le 1}\|Lf_\lambda\|_{-1} < +\infty.$$

Proof Fix a function V in $\mathscr{H}_{-1}\cap L^2(\pi)$. For $\lambda > 0$, denote by $f_\lambda^{(0)}$ the solution of the resolvent equation

$$\lambda f_\lambda^{(0)} - L_0 f_\lambda^{(0)} = V.$$

By the spectral decomposition of the normal operator L_0,

$$f_\lambda^{(0)} = \int_0^{+\infty}\int_{-\infty}^{+\infty}\frac{1}{\lambda+\varphi+i\tau}E(d\varphi, d\tau)V,$$

where $E(d\varphi, d\tau)$ is the spectral resolution of identity which corresponds to the normal operator L_0. Since V belongs to $\mathscr{H}_{-1}\cap L^2(\pi)$,

$$\|V\|_{0,-1}^2 = \int_0^{+\infty}\int_{-\infty}^{+\infty}\frac{1}{\varphi}\mu_V(d\varphi, d\tau) < \infty,$$

where $\mu_V(d\varphi, d\tau)$ stands for the spectral measure of V: $\mu_V(d\varphi, d\tau) = \langle E(d\varphi, d\tau)V, V\rangle_\pi$. Hence,

$$\big\|L_0 f_\lambda^{(0)}\big\|_{0,-1}^2 = \int_0^{+\infty}\int_{-\infty}^{+\infty}\left|\frac{\varphi+i\tau}{\lambda+\varphi+i\tau}\right|^2\frac{\mu_V(d\varphi, d\tau)}{\varphi} \le \|V\|_{0,-1}^2.$$

In particular, by (2.55),

$$\big\|L_0 f_\lambda^{(0)}\big\|_{-1} \le \frac{C_1}{c_1}\|V\|_{-1}. \tag{2.58}$$

Denote by f_λ the solution of the full resolvent equation $(\lambda - L)f_\lambda = V$. Since $L = L_0 + B$, we may rewrite the resolvent equation as

$$(\lambda - L_0)f_\lambda = V + Bf_\lambda.$$

Since $L = L_0 + B$, by (2.58) with $V + Bf_\lambda$ in place of V,

$$\begin{aligned}\sup_{0<\lambda\le 1} \|Lf_\lambda\|_{-1} &\le \sup_{0<\lambda\le 1} \|L_0 f_\lambda\|_{-1} + \sup_{0<\lambda\le 1} \|Bf_\lambda\|_{-1} \\ &\le \frac{C_1}{c_1} \sup_{0<\lambda\le 1} \|V + Bf_\lambda\|_{-1} + \sup_{0<\lambda\le 1} \|Bf_\lambda\|_{-1}.\end{aligned}$$

By (2.57), the previous expression is bounded above by

$$\begin{aligned}&\frac{C_1}{c_1}\|V\|_{-1} + \Big(1 + \frac{C_1}{c_1}\Big) \sup_{0<\lambda\le 1} \|Bf_\lambda\|_{-1} \\ &\quad \le \frac{C_1}{c_1}\|V\|_{-1} + C_B\Big(1 + \frac{C_1}{c_1}\Big) \sup_{0<\lambda\le 1} \|f_\lambda\|_1.\end{aligned}$$

By (2.15), this sum is less than or equal to $C_2\|V\|_{-1}$ for some finite constant C_2. Therefore, $\sup_{0<\lambda\le 1} \|Lf_\lambda\|_{-1} \le C_2\|V\|_{-1} < \infty$, which concludes the proof of the proposition. □

2.8 Invariance Principles in the Multidimensional Case

Suppose that $\{X_t : t \ge 0\}$ is an $\mathbb{R}^d$-valued stochastic processes whose realizations are a.s. right continuous and possess left limits. Sometimes it is useful to know not only that the random variables $X_t/\sqrt{t}$ satisfy the central limit theorem but also that the laws of the scaled processes $\{\varepsilon X_{t\varepsilon^{-2}} : t \ge 0\}$, considered over $D(\mathbb{R}_+, \mathbb{R}^d)$, converge weakly, as probability measures, as $\varepsilon \downarrow 0$, to a Wiener measure. This kind of statement is called *an invariance principle*. It can be verified by establishing *tightness* of the laws of the scaled processes, say for $\varepsilon \in (0, 1]$, in $D(\mathbb{R}_+, \mathbb{R}^d)$ and proving for arbitrary $0 \le t_1 \le \cdots \le t_N$ the weak convergence of the finite dimensional distributions $(\varepsilon X_{t_1\varepsilon^{-2}}, \dots, \varepsilon X_{t_N\varepsilon^{-2}})$, as $\varepsilon \downarrow 0$, to the law of $(W_{t_1}, \dots, W_{t_N})$, where $\{W_t : t \ge 0\}$ is a Wiener process. In fact, the proofs of all the central limit theorems shown in this chapter, namely Theorem 2.1 for martingales and Theorem 2.7 for Markov processes, can be easily modified to demonstrate also the convergence of finite dimensional distributions. The only remaining question is tightness of the laws of the respective scaled processes.

2.8.1 *Martingales with Stationary Increments*

Before proving an invariance principle for martingales, we extend Theorem 2.1 to the multidimensional case. Let $\{M_t = (M_{t,1}, \dots, M_{t,d}) : t \ge 0\}$ be a r.c.l.l., $\mathbb{R}^d$-

valued, square integrable martingale with respect to a filtration $\{\mathscr{F}_t : t \geq 0\}$ satisfying the usual conditions and such that $M_0 = 0$. Denote by $\{\langle M_i, M_j\rangle_t : t \geq 0\}$, $i, j = 1, \dots, d$, its predictable quadratic covariation processes.

Denote by $\langle\cdot,\cdot\rangle_{\mathbb{R}^d}$ the scalar product of $\mathbb{R}^d$. Using Theorem 2.1 for the scalar valued martingales $\{\langle M_t, \theta\rangle_{\mathbb{R}^d} : t \geq 0\}$, $\theta \in \mathbb{R}^d$, one can establish the following multidimensional generalization of the central limit theorem for martingales with stationary increments.

Theorem 2.26 *Let $\{M_t = (M_{t,1}, \dots, M_{t,d}) : t \geq 0\}$ be a r.c.l.l., $\mathbb{R}^d$-valued, square integrable martingale with respect to a filtration $\{\mathscr{F}_t : t \geq 0\}$ satisfying the usual conditions. Assume that the increments are stationary, as defined in Theorem 2.1, and that $M_0 = 0$. Suppose also that*

$$\lim_{n\to\infty} \mathbb{E}\left| \frac{\langle M_i, M_j\rangle_n}{n} - \Sigma_{ij} \right| = 0,$$

where $\Sigma_{ij} = \mathbb{E}[M_{1,i} M_{1,j}]$, $i, j = 1, \dots, d$. Then, the conditional laws of $M_t/\sqrt{t}$ on $\mathscr{F}_0$ converge in probability, as $t \uparrow \infty$, to a mean zero Gaussian law with covariance matrix $\Sigma = \{\Sigma_{ij} : 1 \leq i, j \leq d\}$:

$$\lim_{t\to\infty} \mathbb{E}\big[\big|\mathbb{E}\big[\exp\big\{i\big\langle\theta, t^{-1/2} M_t\big\rangle_{\mathbb{R}^d}\big\} |\mathscr{F}_0\big] - \exp\big\{-(1/2)\langle\Sigma\theta, \theta\rangle_{\mathbb{R}^d}\big\}\big|\big] = 0$$

for all θ in $\mathbb{R}^d$.

To prove an invariance principle for multidimensional martingales, we rely on Theorem 6.4.13 of Jacod and Shiryaev (1987), to examine the tightness issue.

Theorem 2.27 *Suppose that $\{M_t^{(n)} = (M_{t,1}^{(n)}, \dots, M_{t,d}^{(n)}) : t \geq 0\}$ is a family of r.c.l.l., square integrable, $\mathbb{R}^d$-valued martingales with respect to a filtration $\{\mathscr{F}_t : t \geq 0\}$ satisfying the usual conditions, indexed by $n \geq 1$. Let also $\langle M^{(n)}\rangle_t$ be the sum of the predictable quadratic variations: $\langle M^{(n)}\rangle_t := \sum_{i=1}^d \langle M_i^{(n)}, M_i^{(n)}\rangle_t$. Then, for the family to be tight it suffices that:*

(i) *The sequence $\{M_0^{(n)} : n \geq 1\}$ is tight in $\mathbb{R}^d$;*
(ii) *The laws of $\{\langle M^{(n)}\rangle_t : t \geq 0\}$, $n \geq 1$ are tight in $D(\mathbb{R}_+, \mathbb{R}^d)$ and any of its weak limiting points is supported in $C(\mathbb{R}_+, \mathbb{R}^d)$.*

Remark 2.28 If in addition we know that the martingales of the family $\{M_t^{(n)} = (M_{t,1}^{(n)}, \dots, M_{t,d}^{(n)}) : t \geq 0\}$, $n \geq 1$, have continuous trajectories, tightness of their laws can be claimed in $C(\mathbb{R}_+, \mathbb{R}^d)$.

The invariance principle for martingales with stationary increments follows from the above theorem.

Theorem 2.29 *Suppose that the martingale $\{M_t : t \geq 0\}$ satisfies the assumptions of Theorem 2.26. In addition to the L^1 convergence of the predictable quadratic covariations assumed there, we suppose also that*

$$\lim_{n\to\infty} \frac{\langle M_i, M_j\rangle_n}{n} = \Sigma_{ij} \quad \textit{almost surely.}$$

Then, the laws of $\{\varepsilon M_{t\varepsilon^{-2}} : t \geq 0\}$ converge weakly in $D(\mathbb{R}_+, \mathbb{R}^d)$, as $\varepsilon \downarrow 0$, to a Wiener measure with mean zero and covariance matrix Σ.

Proof Since $M_0 = 0$, according to Theorem 2.27 it suffices to prove that for an arbitrary sequence $\varepsilon_n \downarrow 0$, as $n \uparrow \infty$, the family $\{\varepsilon_n^2 \langle M\rangle_{t\varepsilon_n^{-2}} : t \geq 0\}$ is tight in $D(\mathbb{R}_+, \mathbb{R}^d)$ and that any of its weak limiting points is supported in the space of continuous functions. We show that for an arbitrary $T > 0$ we have

$$\lim_{n\to\infty} \mathbb{E}\Big[\sup_{t\in[0,T]} \big|\varepsilon_n^2 \langle M\rangle_{t\varepsilon_n^{-2}} - \sigma^2 t\big|\Big] = 0, \tag{2.59}$$

where σ^2 is the trace of Σ.

Let $\delta > 0$ be arbitrarily fixed. Then, $|\varepsilon_n^2 \langle M\rangle_{t\varepsilon_n^{-2}} - \sigma^2 t|$ is less than or equal to

$$\varepsilon_n^2 \big\{\langle M\rangle_{[t\varepsilon_n^{-2}]+1} - \langle M\rangle_{[t\varepsilon_n^{-2}]}\big\} + \varepsilon_n^2\big[t\varepsilon_n^{-2}\big]\left|\frac{\langle M\rangle_{[t\varepsilon_n^{-2}]}}{[t\varepsilon_n^{-2}]} - \sigma^2\right|$$
$$+ \sigma^2\big\{t - \varepsilon_n^2\big[t\varepsilon_n^{-2}\big]\big\}.$$

The last term is bounded by $(\varepsilon_n\sigma)^2$ for all $t \geq 0$. Denote the suprema on $[\delta, T]$ of the remaining first and second term by A_N and B_N, respectively. Let also $m_n := [\delta\varepsilon_n^{-2}]$ and $M_n := [T\varepsilon_n^{-2}] + 1$.

On the one hand, A_N is bounded by $\varepsilon_n^2 \max\{X_m : m \leq M_n\}$, where $X_k := \langle M\rangle_{k+1} - \langle M\rangle_k$. The sequence $\{X_k : k \geq 1\}$ is stationary and $\mathbb{E}X_0 = \mathbb{E}[\langle M\rangle_1]$ is finite. By Lemma 2.30 below, $\lim_{N\to\infty} A_N = 0$ a.s. and in $L^1(\mathbb{P})$. On the other hand, by assumption,

$$B_N \leq T \sup_{m\in[m_n, M_n]} \left|\frac{\langle M\rangle_m}{m} - \sigma^2\right| \to 0, \quad \text{a.s.}$$

as $n \to \infty$. In fact, the above convergence also takes place in $L^1(\mathbb{P})$ since we can bound the expression under the supremum by $m_n^{-1}\langle M\rangle_{M_n} + \sigma^2$, which converges in $L^1(\mathbb{P})$, as assumed in Theorem 2.26.

Therefore,

$$\lim_{n\to\infty} \mathbb{E}\Big[\sup_{t\in[0,T]} \big|\varepsilon_n^2 \langle M\rangle_{t\varepsilon_n^{-2}} - \sigma^2 t\big|\Big] \leq \lim_{n\to\infty} \mathbb{E}\Big[\sup_{t\in[0,\delta]} \big|\varepsilon_n^2 \langle M\rangle_{t\varepsilon_n^{-2}} - \sigma^2 t\big|\Big].$$

The expectation on the right-hand side is bounded above by $\varepsilon_n^2 \mathbb{E}[\langle M\rangle_{\delta\varepsilon_n^{-2}}] + \sigma^2\delta$, which converges to $2\sigma^2\delta$ by the L^1 convergence of the predictable quadratic covariation. Since $\delta > 0$ was arbitrarily chosen the theorem is proved. □

Note that the a.s. convergence assumed in the previous theorem can be replaced by the weaker assumption (2.59).

Lemma 2.30 *Suppose that a stationary sequence $\{X_k : k \geq 0\}$ is such that $X_0 \geq 0$ a.s. and $\mathbb{E}[X_0] < +\infty$. Then,*

$$\frac{1}{N} \max\{X_0, \dots, X_N\} \to 0,$$

as $N \uparrow \infty$, both a.s. and in $L^1(\mathbb{P})$.

Proof Let $S_0 := X_0$ and $S_N := \sum_{n=0}^{N} X_n$. By virtue of the ergodic theorem we know that $N^{-1}S_N \to Y$, as $N \uparrow \infty$, both a.s. and in $L^1(\mathbb{P})$. Choosing an arbitrary $\varepsilon > 0$, for $\mathbb{P}$-a.s. ω we can find $N_0(\omega)$ so that $|N^{-1}S_N(\omega) - Y(\omega)| < \varepsilon$ for all $N \geq N_0(\omega)$. We can estimate therefore

$$\begin{aligned}\frac{1}{N} \max\bigl\{X_0(\omega), \dots, X_N(\omega)\bigr\} &\leq \frac{1}{N} \max\bigl\{X_0(\omega), \dots, X_{N_0(\omega)-1}(\omega)\bigr\} \\ &\quad + \frac{1}{N} \max\bigl\{X_{N_0(\omega)}(\omega), \dots, X_N(\omega)\bigr\}.\end{aligned}$$

The first term on the right-hand side clearly tends to 0 as $N \uparrow \infty$ while any of the terms under the second maximum can be written as

$$\begin{aligned}\frac{S_{n+1}(\omega)}{n+1}\frac{n+1}{N} - \frac{S_n(\omega)}{n}\frac{n}{N} &\leq \bigl(Y(\omega)+\varepsilon\bigr)\frac{n+1}{N} - \bigl(Y(\omega)-\varepsilon\bigr)\frac{n}{N} \\ &< 2\varepsilon + \frac{Y(\omega)}{N}\end{aligned}$$

for all $N_0(\omega) \leq n < N$. From here we obtain the a.s. statement of the lemma. To conclude the convergence in $L^1(\mathbb{P})$, it suffices to observe that $N^{-1}\max\{X_0, \dots, X_N\}$ is bounded by the $L^1(\mathbb{P})$ convergent sequence $N^{-1}S_N$. □

2.8.2 Additive Functionals of Markov Processes

In this section, we prove an invariance principle for additive functionals of Markov processes. We start with a multidimensional version of the central limit theorem and then examine the tightness issue.

Let $V = (V_1, \dots, V_d) : E \to \mathbb{R}^d$ be a vector valued function whose components belong to $L^2(\pi) \cap \mathscr{H}_{-1}$. Let $f_{k,\lambda}$ be the solution of the resolvent equation (2.13) with V_k on the right-hand side instead of V. Define $f_\lambda := (f_{1,\lambda}, \dots, f_{d,\lambda})$. One can easily adapt to the multidimensional case the argument made in Sect. 2.6 and prove multidimensional versions of Theorems 2.7, 2.14 and 2.17. For example, the multidimensional correspondent of the first of these results can be stated as follows.

Theorem 2.31 *Suppose that the components of* $V = (V_1, \dots, V_d)$ *belong to* $L^2(\pi) \cap \mathscr{H}_{-1}$ *and*

$$\lim_{\lambda\to 0} \lambda \| f_{k,\lambda} \|_\pi^2 = 0 \quad \text{and} \quad \lim_{\lambda\to 0} \| f_{k,\lambda} - f_k \|_1 = 0$$

for some f_k *in* $\mathscr{H}_1$, $k = 1, \dots, d$. *Then, the conditional laws of* $t^{-1/2} \int_0^t V(X_s)\, ds$ *on* $\mathscr{F}_0$ *converge in probability, as* $t \uparrow \infty$, *to a mean zero,* d*-dimensional Gaussian law whose covariance matrix* $\sigma^2(V) = \{\sigma_{k,l}^2(V) : 1 \le k, l \le d\}$ *is given by*

$$\sigma_{k,l}^2(V) = 2 \lim_{\lambda\to 0} \langle f_{k,\lambda}, f_{l,\lambda} \rangle_1 = 2 \langle f_k, f_l \rangle_1$$

for $1 \le k, l \le d$.

In the statement of the invariance principle, we first assume V to be a real function. In the proof we need the following simple fact. For δ, $\eta > 0$, let $A_{\delta,\eta}$ be the subset of $D([0, T], \mathbb{R})$ defined by

$$A_{\delta,\eta} = \Big\{ \omega : \sup_{\substack{0\le s,t\le T \\ |t-s|\le\delta}} \big|\omega(t) - \omega(s)\big| \ge \eta \Big\}.$$

One can check that $A_{\delta,\eta}$ is not closed for the Skorohod topology. However, if we denote by $C_T = C([0, T], \mathbb{R})$ the (closed) subspace of continuous trajectories of $D([0, T], \mathbb{R})$ and if $\overline{A_{\delta,\eta}}$ represents the closure of $A_{\delta,\eta}$ with respect to the Skorohod topology,

$$\overline{A_{\delta,\eta}} \cap C_T = A_{\delta,\eta} \cap C_T .$$

In particular, if a sequence of probability measures $\{\mathbb{Q}_N : N \ge 1\}$ on $D([0, T], \mathbb{R})$ converges to a measure $\mathbb{Q}$ concentrated on continuous trajectories, $\mathbb{Q}[C_T] = 1$, then

$$\begin{aligned} \limsup_{N\to\infty} \mathbb{Q}_N[A_{\delta,\eta}] &\le \limsup_{N\to\infty} \mathbb{Q}_N[\overline{A_{\delta,\eta}}] \le \mathbb{Q}[\overline{A_{\delta,\eta}}] \\ &= \mathbb{Q}[C_T \cap \overline{A_{\delta,\eta}}] = \mathbb{Q}[C_T \cap A_{\delta,\eta}] = \mathbb{Q}[A_{\delta,\eta}]. \end{aligned}$$

Letting $\delta \downarrow 0$, since $\mathbb{Q}$ is concentrated on continuous paths, we obtain that for all $\eta > 0$,

$$\lim_{\delta\to 0} \limsup_{N\to\infty} \mathbb{Q}_N[A_{\delta,\eta}] = 0. \tag{2.60}$$

Theorem 2.32 *Besides the hypotheses of Theorem* 2.7, *assume that* π *is ergodic under the backwards dynamics given by the adjoint semigroup* $\{P_t^* : t \ge 0\}$. *Then, the laws of the processes* $\{\varepsilon \int_0^{t\varepsilon^{-2}} V(X_s) ds : t \ge 0\}$ *converge weakly in* $C(\mathbb{R}_+, \mathbb{R})$, *as* $\varepsilon \downarrow 0$, *to the Wiener measure with zero mean and the variance given by* (2.24).

Proof The proof of Theorem 2.7 can be easily modified to demonstrate the convergence of the finite dimensional distributions of the process $\{\varepsilon \int_0^{t\varepsilon^{-2}} V(X_s)ds : t \geq 0\}$. The only remaining question is tightness.

Fix $T > 0$ and h belonging to $\mathscr{C}$—the common core of L and L^*. Let $Y_\varepsilon(t) := \varepsilon \int_0^{t\varepsilon^{-2}} V(X_s)\,ds$. Using the backward-forward martingale decomposition, presented in the proof of Lemma 2.4, one can write

$$Y_\varepsilon(t) = (1/2)\varepsilon M_{t\varepsilon^{-2}} + (1/2)\varepsilon M^-_{T\varepsilon^{-2}} - (1/2)\varepsilon M^-_{(T-t)\varepsilon^{-2}} + \varepsilon \int_0^{t\varepsilon^{-2}} R(X_s)ds$$

for $0 \leq t \leq T$, where $R := V - Sh$, $\{M_t : t \geq 0\}$ is the square integrable martingale given by (2.20) and $\{M_t^- : t \geq 0\}$ is the backward martingales given by (2.21), with T replaced there by $T\varepsilon^{-2}$. Observe that the forward and backward martingales depend on the function h. One can show, see Karatzas and Shreve (1991, Theorem 1.3.13), that both martingales have equivalent r.c.l.l. versions.

By Theorem 2.29 and Remark 2.12, the laws of $\{M_\varepsilon^+(t) := \varepsilon M_{t\varepsilon^{-2}} : t \in [0,T]\}$ converge to the Wiener measure with mean zero and variance $\sigma_h^2 := 2\langle h, (-L)h\rangle_\pi$.

Consider the backward martingales $\{\varepsilon M^-_{t\varepsilon^{-2}} : t \in [0,T]\}$. We claim that

$$\lim_{n\to\infty} \mathbb{E}\Big[\sup_{t\in[0,T]} \big|\varepsilon_n^2 \langle M^-, M^-\rangle_{t\varepsilon_n^{-2}} - \sigma_h^2 t\big|\Big] = 0$$

for any arbitrary sequence $\varepsilon_n \downarrow 0$. Indeed, since π is ergodic under the backwards dynamics, we show exactly as in the proof of Theorem 2.7 and Remark 2.12 that

$$\lim_{n\to\infty} \frac{1}{n} \langle M^-, M^-\rangle_n = \sigma_h^2$$

both a.s. and in $L^1(\mathbb{P})$. To strengthen the result for the supremum, we repeat the argument made in the proof of Theorem 2.29. In particular, by the same theorem, the laws of $\{M_\varepsilon^-(t) := \varepsilon(M^-_{T\varepsilon^{-2}} - M^-_{(T-t)\varepsilon^{-2}}) : t \in [0,T]\}$ also converge to the Wiener measure with mean zero and variance $\sigma_h^2 := 2\langle h, (-L)h\rangle_\pi$.

Since the laws of M_ε^+, M_ε^- converge over $D([0,T],\mathbb{R})$ to a Wiener measure concentrated on continuous paths, by (2.60), for all $\eta > 0$,

$$\lim_{\delta\to 0} \limsup_{\varepsilon\to 0} \mathbb{P}_\pi\Big[\sup_{\substack{0\leq s,t\leq T\\ |t-s|\leq\delta}} \big|M_\varepsilon^\pm(t) - M_\varepsilon^\pm(s)\big| \geq \eta\Big] = 0. \tag{2.61}$$

On the other hand, according to Lemma 2.4,

$$\mathbb{E}_\pi\Big[\sup_{t\in[0,T]} \Big(\varepsilon \int_0^{t\varepsilon^{-2}} R(X_s)\,ds\Big)^2\Big] \leq 24T\|V - Sh\|_{-1}^2. \tag{2.62}$$

We have now all elements to prove the theorem. Fix $\alpha > 0$. From the decomposition of $Y_\varepsilon(t)$, we obtain that

$$\mathbb{P}_\pi\Big[\sup_{\substack{0\le s,t\le T\\ |t-s|\le\delta}} \big|Y_\varepsilon(t) - Y_\varepsilon(s)\big| \ge 3\eta\Big] \le \mathbb{P}_\pi\Big[\sup_{\substack{0\le s,t\le T\\ |t-s|\le\delta}} \big|M_\varepsilon^+(t) - M_\varepsilon^+(s)\big| \ge 2\eta\Big]$$
$$+ \mathbb{P}_\pi\Big[\sup_{\substack{0\le s,t\le T\\ |t-s|\le\delta}} \big|M_\varepsilon^-(t) - M_\varepsilon^-(s)\big| \ge 2\eta\Big]$$
$$+ \mathbb{P}_\pi\left[\sup_{\substack{0\le s,t\le T\\ |t-s|\le\delta}} \left|\varepsilon\int_{s\varepsilon^{-2}}^{t\varepsilon^{-2}} R(X_r)\,dr\right| \ge \eta\right].$$

In view of (2.62), the third term on the right-hand side is bounded above by

$$\mathbb{P}_\pi\left[\sup_{0\le t\le T} \left|\varepsilon\int_0^{t\varepsilon^{-2}} R(X_r)\,dr\right| \ge (\eta/2)\right] \le C_0 T\eta^{-2}\|Sh - V\|_{-1}^2$$

for some finite constant C_0. Note that this estimate is uniform in $\varepsilon > 0$. Choose h in $\mathscr{C}$ so that the right-hand side is bounded by α. Hence, in view of (2.61),

$$\lim_{\delta\to 0}\limsup_{\varepsilon\to 0} \mathbb{P}_\pi\Big[\sup_{\substack{0\le s,t\le T\\ |t-s|\le\delta}} \big|Y_\varepsilon(t) - Y_\varepsilon(s)\big| \ge \eta\Big] \le \alpha.$$

Since this holds for all $\alpha > 0$, the sequence $\{Y_\varepsilon(t) : 0 \le t \le T\}$ is tight, which proves the theorem. □

As in the martingale context, this theorem can be easily generalized to the case where $V : E \to \mathbb{R}^d$ is a vector valued function.

Theorem 2.33 *Under the hypotheses of Theorem* 2.32 *the laws of the processes* $\{\varepsilon\int_0^{t\varepsilon^{-2}} V(X_s)ds : t \ge 0\}$ *converge weakly in* $C(\mathbb{R}_+, \mathbb{R}^d)$, *as* $\varepsilon \downarrow 0$, *to the Wiener measure with zero mean and covariance matrix* $\sigma^2(V)$ *defined in Theorem* 2.31.

2.9 Comments and References

Central limit theorems for continuous-time martingales have been proven in many versions with different conditions (Ethier and Kurtz, 1986; Helland, 1982; Jacod and Shiryaev, 1987; Rebolledo, 1980). We refer to Whitt (2007) for a recent account. As for the discrete time case, we present here a version that is tailored for our use, and that is a direct consequence of the discrete version. A simple direct proof of the central limit theorem for martingales with continuous paths can be found in Olla (1994a).

In this chapter, we presented the ideas developed in Kipnis and Varadhan (1986) to prove an invariance principle for reversible Markov processes. This method was extended to non-reversible processes satisfying a sector condition in Osada (1998a); Osada and Saitoh (1995); Varadhan (1995); and to non-reversible Markov processes satisfying a graded sector condition in Komorowski and Olla (2003b); Landim and Yau (1997); Sethuraman et al. (2000). Horváth et al. (2010) introduced a weaker version of the graded sector condition. The extension to normal operator is due to Derriennic and Lin (1996, 2001a).

Tóth (1986) proved that in the non-reversible case a central limit theorem for additive functionals follows from equations similar to the ones presented in (2.23), with resolvent estimates in place of spectral estimates. He deduced from this result a central limit theorem for a non-reversible random walk in random environment. De Masi et al. (1989) and Goldstein (1995) proved an invariance principle for additive functionals of reversible Markov processes anti-symmetric by time reversal. Note that the additive functionals (2.1) is symmetric by time reversal.

The central limit theorem for Markov processes holds in probability, with respect to the initial configuration. It should be distinguished from the weaker statement where averaging of the initial state is also performed. Sometimes these last type of results are called *annealed* or *averaged*. One can have situations where the averaged central limit theorem has a different limit variance than the central limit theorem in probability. This is due to the fluctuations of the initial distribution, contained in the averaged central limit theorem, but not in the central limit theorem in probability. An example is given by the tagged particle in the asymmetric one-dimensional nearest neighbor exclusion process, studied by Kipnis (1986) (cf. Chap. 6).

Lemma 2.4 is taken from Wu (1999) and Sethuraman et al. (2000). This general estimate for additive functionals of Markov processes in the stationary state is very useful for proving tightness in invariance principles.

Edgeworth Expansions Bolthausen (1980, 1982) obtained a Berry–Esseen bound for additive functionals of strongly mixing Harris recurrent Markov chains. This result is extended in Lezaud (2001) to ergodic, reversible Markov processes admitting a spectral gap and in Kontoyiannis and Meyn (2003) to geometrically ergodic Markov chains.

Stein's Method Stein's method has been recently adapted to the context of Markov chains to obtain rates of convergence to the stationary state. We refer to the book (Diaconis and Holmes, 2004) and the recent review (Chatterjee et al., 2005) for references on the subject.

Spectral Gap We have seen in this chapter that estimates on the spectral gap plays an important role in the proof of the central limit theorem. Cheeger (1970) proved a lower bound on the smallest strictly positive eigenvalue of the Laplacian on a compact Riemannian manifold in terms of an isoperimetric constant for the manifold. Lawler and Sokal (1988) proved a general version of Cheeger's inequality for positive recurrent discrete-time Markov chains and continuous-time jump Markov processes, both reversible and non-reversible, with general state spaces. The upper and lower bounds are expressed in terms of the probability flow, the rate at which the Markov chain jumps in the stationary state from a set A to its complement normalized by the invariant probabilities of A and A^c. We refer to Kipnis and Landim (1999) and Levin et al. (2009) for results on spectral gaps of Markov processes and interacting particle systems.

References

Blumenthal RM, Getoor RK (1968) Markov processes and potential theory. Pure and applied mathematics, vol 29. Academic Press, New York

Bolthausen E (1980) The Berry-Esseen theorem for functionals of discrete Markov chains. Z Wahrscheinlichkeitstheor Verw Geb 54(1):59–73

Bolthausen E (1982) The Berry-Esseén theorem for strongly mixing Harris recurrent Markov chains. Z Wahrscheinlichkeitstheor Verw Geb 60(3):283–289

Chatterjee S, Diaconis P, Meckes E (2005) Exchangeable pairs and Poisson approximation. Probab Surv 2:64–106 (electronic)

Cheeger J (1970) A lower bound for the smallest eigenvalue of the Laplacian. In: Problems in analysis (Papers dedicated to Salomon Bochner, 1969). Princeton University Press, Princeton, pp 195–199

Chung KL, Walsh JB (2005) Markov processes, Brownian motion, and time symmetry. Grundlehren der Mathematischen Wissenschaften [Fundamental principles of mathematical sciences], vol 249, 2nd edn. Springer, New York

De Masi A, Ferrari PA, Goldstein S, Wick WD (1989) An invariance principle for reversible Markov processes. Applications to random motions in random environments. J Stat Phys 55(3–4):787–855

Derriennic Y, Lin M (1996) Sur le théorème limite central de Kipnis et Varadhan pour les chaînes réversibles ou normales. C R Acad Sci Paris Sér I Math 323(9):1053–1057

Derriennic Y, Lin M (2001a) The central limit theorem for Markov chains with normal transition operators, started at a point. Probab Theory Relat Fields 119(4):508–528

Diaconis P, Holmes S (eds) (2004) Stein's method: expository lectures and applications. IMS lect notes monogr ser, vol 46. Inst Math Stat, Beachwood. Papers from the workshop on Stein's method held at Stanford University, Stanford, CA, 1998

Ethier SN, Kurtz TG (1986) Markov processes: characterization and convergence. Wiley series in probability and mathematical statistics: probability and mathematical statistics. Wiley, New York

Goldstein S (1995) Antisymmetric functionals of reversible Markov processes. Ann Inst Henri Poincaré Probab Stat 31(1):177–190

Helland IS (1982) Central limit theorems for martingales with discrete or continuous time. Scand J Stat 9(2):79–94

Horváth I, Tóth B, Vető B (2010) Diffusive limits for "true" (or myopic) self-avoiding random walks and self-repellent Brownian polymers in $d \geq 3$. arXiv:1009.0401

Jacod J, Shiryaev AN (1987) Limit theorems for stochastic processes. Grundlehren der Mathematischen Wissenschaften [Fundamental principles of mathematical sciences], vol 288. Springer, Berlin

Karatzas I, Shreve SE (1991) Brownian motion and stochastic calculus. Graduate texts in mathematics, vol 113, 2nd edn. Springer, New York

Kipnis C (1986) Central limit theorems for infinite series of queues and applications to simple exclusion. Ann Probab 14(2):397–408

Kipnis C, Landim C (1999) Scaling limits of interacting particle systems. Grundlehren der Mathematischen Wissenschaften [Fundamental principles of mathematical sciences], vol 320. Springer, Berlin

Kipnis C, Varadhan SRS (1986) Central limit theorem for additive functionals of reversible Markov processes and applications to simple exclusions. Commun Math Phys 104(1):1–19

Komorowski T, Olla S (2003b) On the sector condition and homogenization of diffusions with a Gaussian drift. J Funct Anal 197(1):179–211

Kontoyiannis I, Meyn SP (2003) Spectral theory and limit theorems for geometrically ergodic Markov processes. Ann Appl Probab 13(1):304–362

Krengel U (1985) Ergodic theorems: with a supplement by Antoine Brunel. de Gruyter studies in mathematics, vol 6. de Gruyter, Berlin

Landim C, Yau HT (1997) Fluctuation-dissipation equation of asymmetric simple exclusion processes. Probab Theory Relat Fields 108(3):321–356

Lawler GF, Sokal AD (1988) Bounds on the L^2 spectrum for Markov chains and Markov processes: a generalization of Cheeger's inequality. Trans Am Math Soc 309(2):557–580

Levin DA, Peres Y, Wilmer EL (2009) Markov chains and mixing times: with a chapter by James G Propp and David B Wilson. Am Math Soc, Providence

Lezaud P (2001) Chernoff and Berry-Esséen inequalities for Markov processes. ESAIM Probab Stat 5:183–201 (electronic)

Liggett TM (1985) Interacting particle systems. Grundlehren der Mathematischen Wissenschaften [Fundamental principles of mathematical sciences], vol 276. Springer, New York

Olla S (1994a) Homogenization of diffusion processes in random fields. Centre de mathématiques appliquées, ecole polytechnique. Available at http://www.ceremade.dauphine.fr/~olla/lho.ps

Osada H (1998a) An invariance principle for Markov processes and Brownian particles with singular interaction. Ann Inst Henri Poincaré Probab Stat 34(2):217–248

Osada H, Saitoh T (1995) An invariance principle for non-symmetric Markov processes and reflecting diffusions in random domains. Probab Theory Relat Fields 101(1):45–63

Rebolledo R (1980) Central limit theorems for local martingales. Z Wahrscheinlichkeitstheor Verw Geb 51(3):269–286

Rogers LCG, Williams D (2000) Diffusions, Markov processes, and martingales, vol 1: foundations. Cambridge mathematical library. Cambridge University Press, Cambridge. Reprint of the second 1994 edition

Sethuraman S, Varadhan SRS, Yau HT (2000) Diffusive limit of a tagged particle in asymmetric simple exclusion processes. Commun Pure Appl Math 53(8):972–1006

Tóth B (1986) Persistent random walks in random environment. Probab Theory Relat Fields 71(4):615–625

Varadhan SRS (1995) Self-diffusion of a tagged particle in equilibrium for asymmetric mean zero random walk with simple exclusion. Ann Inst Henri Poincaré Probab Stat 31(1):273–285

Whitt W (2007) Proofs of the martingale FCLT. Probab Surv 4:268–302

Wu L (1999) Forward-backward martingale decomposition and compactness results for additive functionals of stationary ergodic Markov processes. Ann Inst Henri Poincaré Probab Stat 35(2):121–141

Yosida K (1995) Functional analysis. Classics in mathematics. Springer, Berlin. Reprint of the sixth 1980 edition

Chapter 3
Random Walks in Random Environment

To illustrate the theory presented in the two previous chapters, we prove here central limit theorems for random walks in random environment which hold in L^1 with respect to the environment.

We start in Sect. 3.1 with an elementary example, random walks with random conductances, whose dynamics can be informally described as follows. Consider a sequence of strictly positive i.i.d. random variables $\{\xi(x,y)\}$ indexed by the bonds of $\mathbb{Z}^d$. For each fixed environment ξ, let $\{X_t^\xi : t \geq 0\}$ be the random walk on $\mathbb{Z}^d$ which jumps from x (resp. y) to y (resp. x) at rate $\xi(x,y)$. We show in Lemma 3.1 below that the random environment as seen from the random walk evolves according to a Markov process, ergodic and reversible with respect to the distribution $\mathbb{Q}$ of the random variables $\{\xi(x,y)\}$. The position of the random walk X_t^ξ can be expressed as the sum of a martingale $\mathfrak{M}(t)$ with an additive functional $\int_0^t V(\eta(s))\,ds$ of the environment process $\eta(t)$. Elementary computations show that the local drift V belongs to $\mathscr{H}_{-1}$. Hence, by the theory developed in the previous chapter, the additive functional $\int_0^t V(\eta(s))\,ds$ can be represented as the sum of a martingale $\mathfrak{m}(t)$ and a remainder $R(t)$ such that $t^{-1/2}R(t)$ vanishes in L^2 as $t \uparrow \infty$. In particular, $X_t^\xi = \mathfrak{M}(t) + \mathfrak{m}(t) + o(t^{1/2})$ and a central limit theorem for X_t^ξ follows from the central limit theorem for martingales.

In Sect. 3.2, we introduce a class of random walks in random environment, the so-called doubly stochastic random walks. We are given a family of random variables $p(x,y) : \Omega \to \mathbb{R}_+$, $x, y \in \mathbb{Z}^d$, defined on some probability space $(\Omega, \mathscr{F}, \mathbb{Q})$, which represent the rate at which a random walk X_t^ω jumps from x to y in the environment ω. The main assumption requires the rate at which the random walk leaves a site x to be equal to the rate at which it jumps to x: $\sum_{y\in\mathbb{Z}^d} p(x,y;\omega) = \sum_{y\in\mathbb{Z}^d} p(y,x;\omega)$ for all x in $\mathbb{Z}^d$ and $\mathbb{Q}$-a.a. ω's. As in the previous section, the evolution of the environment process $\eta(t)$ is Markovian and $\mathbb{Q}$ is seen to be an ergodic, stationary state for the dynamics. Moreover, the position of the random walk X_t^ω can be represented as the sum of a martingale $\mathfrak{M}(t)$ and an additive functional $\int_0^t V(\eta(s))\,ds$ of the environment process $\eta(t)$. The main result of this section states that a central limit theorem for the random walk holds, in L^1 with respect to the

T. Komorowski et al., *Fluctuations in Markov Processes*,
Grundlehren der mathematischen Wissenschaften 345,
DOI 10.1007/978-3-642-29880-6_3,

environment, if the generator L of the environment process $\eta(t)$ satisfies a sector condition and if the local drift V has $\mathbb{Q}$-mean zero and belongs to $\mathscr{H}_{-1}$.

We examine in Sect. 3.3 a special class of doubly stochastic random walks, the cyclic random walks, in which one can prove a sector condition. The dynamics can be described as follows. Fix a finite sequence $C = (0 = y_0, y_1, \dots, y_{n-1}, 0)$ such that $y_i \neq y_j$, $0 \le i \neq j \le n-1$. We may define a random walk on C which jumps from y_j to y_{j+1} at rate 1. Translating this sequence by z, we obtain a new cycle $C + z = (z, z + y_1, \dots, z + y_{n-1}, z)$ in which we may define a similar random walk which jumps from $z + y_j$ to $z + y_{j+1}$ at rate 1. Superposing all these dynamics we derive a random walk on $\mathbb{Z}^d$ which jumps from x to y at a rate equal to the number of sites z such that $(x, y) = (z + y_j, z + y_{j+1})$ for some $0 \le j \le n-1$. This picture can be slightly generalized if we speed up the jump rates by a function of z. This means that we allow the jump rates over the cycle $z + C$ to be equal to $W(z)$ instead of 1. To conclude, we choose the rates $W(z)$ according to a stationary ergodic random field $W(z) = W(z, \omega)$. It turns out that these random walks satisfy a strong sector condition.

In Sect. 3.4, we demonstrate that a sector condition for the generator of the environment process is not needed to prove the central limit theorem for the random walk provided the assumptions on the local drift are strengthened. In this section, we derive a central limit theorem assuming that the rates satisfy the ellipticity condition (3.20) and that the local drift is the divergence of a matrix in L^d, where $d \ge 3$ is the dimension.

In Sect. 3.5 we show that the local drift satisfies the hypotheses of the previous section if the random field obtained by translating the jump rates of the random walk is sufficiently strongly mixing.

In Sect. 3.6, we restrict our investigation to one-dimensional doubly stochastic random walks under an ellipticity condition. The main result of that section states that the generator of the environment process satisfy a sector condition provided the local drift V has zero mean with respect to the measure $\mathbb{Q}$ and the random rates satisfy the ellipticity condition (3.20). This statement follows from the fact, presented in Theorem 3.17, that in dimension 1 the generator of the environment process associated to a doubly stochastic random walk can be written as the sum of its symmetric part with an operator in divergence form with bounded coefficients if the expectation of the local drift vanishes. Using the theory of bounded perturbations of normal operators, the central limit theorem is extended to the case where the local drift does not have mean zero. Finally, in Sect. 3.7, we prove some estimates on the transition probability of a simple random walk that we use in this chapter.

3.1 Random Walks with Random Conductances

Denote by $\{e_j : 1 \le j \le d\}$ the canonical basis of $\mathbb{R}^d$ and denote by $\xi = \{\xi(x, x + e_j) : x \in \mathbb{Z}^d, 1 \le j \le d\}$ a collection of strictly positive conductances indexed by the bonds of $\mathbb{Z}^d$. For each ξ, let $\{X_t^\xi : t \ge 0\}$ be the random walk on $\mathbb{Z}^d$ which jumps from x (resp. $x + e_j$) to $x + e_j$ (resp. x) at rate $\xi(x, x + e_j)$. In this section, we prove a central limit theorem for X_t^ξ in the case where the conductances $\{\xi(x, x+e_j)\}$ are picked from a set of i.i.d. random variables bounded below by a strictly positive constant and bounded above by a finite constant.

To state the result we need to introduce some notation. We first define the set of conductances. Fix $0 < a < b < \infty$ and denote by ν a probability measure defined on the Borel sets of $[a, b]$. Let Ω be the product space $[a, b]^{\mathbb{B}_d}$, where $\mathbb{B}_d$ stands for the set of bonds of $\mathbb{Z}^d$: $\mathbb{B}_d = \{(x, x+e_j) : x \in \mathbb{Z}^d, 1 \le j \le d\}$. Denote by $\mathscr{F}$ the Borel σ-algebra of Ω generated by the cylinder, or finite dimensional, Borel sets of Ω. Finally, let $\mathbb{Q}$ be the product probability measure on $\mathscr{F}$ whose one-dimensional marginals are equal to ν.

Let $B(\mathbb{Z}^d)$ be the set of bounded functions on $\mathbb{Z}^d$ endowed with the sup norm $\|\cdot\|_\infty$. For each random conductance ξ in Ω, let $L_\xi : B(\mathbb{Z}^d) \to B(\mathbb{Z}^d)$ be the operator defined by

$$(L_\xi f)(x) = \sum_{j=1}^{2d} \xi(x, x+e_j)\big\{f(x+e_j) - f(x)\big\}. \tag{3.1}$$

Here, to keep notation simple, $e_{d+j} = -e_j$, $1 \le j \le d$, and $\xi(x+e_j, x) = \xi(x, x+e_j)$, $x \in \mathbb{Z}^d$, $1 \le j \le d$. It is easy to check that L_ξ is the generator of a Markov process on $\mathbb{Z}^d$ which jumps from x (resp. $x+e_j$) to $x+e_j$ (resp. x) at rate $\xi(x, x+e_j)$. Denote by $\{X_t^\xi : t \ge 0\}$ this random walk, defined on a probability space $(\Sigma, \mathscr{A}, \mathbb{P})$.

For each ξ in Ω, let $\mathbb{P}_x^\xi$ be the probability measure on $(\Sigma, \mathscr{A})$ under which $\{X_t^\xi : t \ge 0\}$ is a Markov process with generator L_ξ starting from x: $\mathbb{P}_x^\xi[X_0^\xi = x] = 1$. Expectation with respect to $\mathbb{P}_x^\xi$ is denoted by $\mathbb{E}_x^\xi$. The superscript ξ of X_t^ξ is often omitted and we sometimes denote X_t^ξ by $X^\xi(t)$.

The counting measure on $\mathbb{Z}^d$ is reversible. Indeed, if $f, g : \mathbb{Z}^d \to \mathbb{R}$ are finitely supported functions,

$$\sum_{x \in \mathbb{Z}^d} (L_\xi f)(x)\, g(x) = \sum_{x \in \mathbb{Z}^d} (L_\xi g)(x)\, f(x).$$

Denote by $p_t^\xi : \mathbb{Z}^d \times \mathbb{Z}^d \to [0, 1]$, $t \ge 0$, the transition probability functions of the Markov process $\{X_t^\xi : t \ge 0\}$. Thus, $p_t^\xi(x, y)$ represents the probability to be at y at time t if the initial position is x.

Denote by $\{\tau_x : x \in \mathbb{Z}^d\}$ the group of translations on Ω: $(\tau_x \xi)(y, y+e_j) = \xi(x+y, x+y+e_j)$, x, y in $\mathbb{Z}^d$, $1 \le j \le d$. Of course, if we translate the environment, the transition probability functions are translated by the same amount:

$$p_t^{\tau_z \xi}(x, y) = p_t^\xi(x+z, y+z) \tag{3.2}$$

for x, y, z in $\mathbb{Z}^d$, $t \ge 0$.

The environment process, next defined, plays a central role in the book. Let $\{\eta(t) : t \ge 0\}$ be the state of the environment as seen from the position of the random walk:

$$\eta(t) = \tau_{X_t^\xi}\xi.$$

Denote by $B(\Omega)$ the space of bounded measurable functions $f : \Omega \to \mathbb{R}$ endowed with the sup norm.

Lemma 3.1 *$\{\eta(t) : t \geq 0\}$ is a Markov process with values on Ω. Its generator L acts on $B(\Omega)$ as*

$$(Lf)(\xi) = \sum_{j=1}^{2d} \xi(0, e_j)\big\{f(\tau_{e_j}\xi) - f(\xi)\big\}.$$

The measure $\mathbb{Q}$ is reversible and ergodic under the transition probability semigroup associated to the generator L.

Proof Fix an environment ξ, $n \geq 1$, $0 \leq t_1 \leq \cdots \leq t_n \leq t$, $h \geq 0$ and bounded measurable functions $g_1, \dots, g_n, f : \Omega \to \mathbb{R}$. From the definition of $\eta(t)$ we obtain

$$\mathbb{E}_x^\xi\left[\prod_{j=1}^n g_j\big(\eta(t_j)\big) f\big(\eta(t+h)\big)\right] = \mathbb{E}_x^\xi\left[\prod_{j=1}^n g_j(\tau_{X(t_j)}\xi) f(\tau_{X(t+h)}\xi)\right],$$

where we omitted the superscript ξ of X_t^ξ. By the Markov property of the random walk, for a fixed environment ξ we can rewrite the right-hand side of the previous identity as

$$\mathbb{E}_x^\xi\left[\prod_{j=1}^n g_j(\tau_{X(t_j)}\xi)\left(\sum_{y\in\mathbb{Z}^d} p_h^\xi\big(X^\xi(t), y\big) f(\tau_y\xi)\right)\right]$$
$$= \mathbb{E}_x^\xi\left[\prod_{j=1}^n g_j(\tau_{X(t_j)}\xi)\left(\sum_{y\in\mathbb{Z}^d} p_h^{\eta(t)}\big(0, y - X^\xi(t)\big) f(\tau_y\xi)\right)\right],$$

where we have used (3.2) in the last step. Substituting $y := y - X^\xi(t)$ in the summation above and using again the definition of the environment process we obtain that the right-hand side equals

$$\mathbb{E}_x^\xi\left[\prod_{j=1}^n g_j(\tau_{X^\xi(t_j)}\xi) P_h f\big(\eta(t)\big)\right],$$

where $\{P_t : t \geq 0\}$ is the transition probability semigroup of $\{\eta(t) : t \geq 0\}$ defined on $B(\Omega)$ by

$$P_t f(\xi) := \sum_{y\in\mathbb{Z}^d} p_t^\xi(0, y) f(\tau_y\xi).$$

This proves the Markov property for the environment process and also identifies the transition probability semigroup. The formula for the generator presented in the statement of the lemma is obtained by differentiating the previous expression with respect to h at $h=0$.

It remains to prove that the measure $\mathbb{Q}$ is reversible and ergodic for $\{P_t : t \ge 0\}$. Denote by $\langle \cdot, \cdot \rangle_{\mathbb{Q}}$ the inner product of $L^2(\mathbb{Q})$. On the one hand, by the definition of L and since the measure $\mathbb{Q}$ is a product, the change of variables $\xi' = \tau_{e_j}\xi$ gives that

$$\langle g, Lf \rangle_{\mathbb{Q}} = \sum_{j=1}^{2d} E_{\mathbb{Q}}\big[\xi(0,e_j) g(\xi) \big\{ f(\tau_{e_j}\xi) - f(\xi) \big\} \big] = \langle Lg, f \rangle_{\mathbb{Q}}$$

for any two functions f, g in $B(\Omega)$. Since the generator L is bounded in $L^2(\mathbb{Q})$, it can be extended to the entire space $L^2(\mathbb{Q})$ and the previous identity shows that it is self-adjoint.

To show ergodicity we calculate the Dirichlet form of the process. The same change of variables performed in the previous paragraph and our convention that $e_{d+j} = -e_j$, gives that

$$\langle -Lf, f \rangle_{\mathbb{Q}} = \sum_{j=1}^{d} E_{\mathbb{Q}}\big[\xi(0,e_j) \big\{ f(\tau_{e_j}\xi) - f(\xi) \big\}^2 \big].$$

Fix $f \in L^2(\mathbb{Q})$ such that $Lf = 0$. Thus $f(\tau_{e_j}\xi) = f(\xi)$ $\mathbb{Q}$-a.s. Since $\mathbb{Q}$ is a product measure, $\mathbb{Q}$ is ergodic for the shift and f is $\mathbb{Q}$-a.s. constant. □

Recall from Sect. 2.2 the definition of the spaces $\mathscr{H}_1$, $\mathscr{H}_{-1}$. Note that all functions f in $L^2(\mathbb{Q})$ belong to $\mathscr{H}_1$ because the generator is a bounded operator. We already computed in the proof of the previous lemma the $\mathscr{H}_1$ norm of a function f in $L^2(\mathbb{Q})$:

$$\|f\|_1^2 = \sum_{j=1}^{d} E_{\mathbb{Q}}\big[\xi(0,e_j) \big\{ f(\tau_{e_j}\xi) - f(\xi) \big\}^2 \big].$$

Our goal is to prove a central limit theorem for X_t^ξ. The position of the random walk can be recovered from the process $\eta(t)$ by counting its number of jumps. More precisely, denote by $N_j(t)$, $1 \le j \le 2d$, the number of times that the process $\eta(t)$ jumped from a state η to the state $\tau_{e_j}\eta$ in the time interval $[0,t]$. Clearly,

$$X_t^\xi = \sum_{j=1}^{d} e_j \big\{ N_j(t) - N_{d+j}(t) \big\}.$$

This identity permits to express X_t^ξ as the sum of a martingale with an additive functional of the process $\eta(t)$.

Denote by $\mathfrak{m}_j(t)$, $1 \le j \le 2d$, the elementary martingales defined by $\mathfrak{m}_j(t) = N_j(t) - \int_0^t \eta_s(0, e_j)\, ds$, where η_s stands for $\eta(s)$. Since these martingales are associated to counting processes and since they do not have common jumps, by Theorem 6.2.17 in Dacunha-Castelle and Duflo (1986) they are orthogonal in $L^2(\mathbb{Q})$. Moreover, their predictable quadratic variations are given by $\int_0^t \eta_s(0, e_j)\, ds$.

With this notation we have that

$$X_t^\xi = \mathfrak{M}(t) + \int_0^t V\big(\eta(s)\big)\, ds,$$

where

$$\mathfrak{M}(t) = \sum_{j=1}^d e_j \big\{\mathfrak{m}_j(t) - \mathfrak{m}_{d+j}(t)\big\} \quad \text{and} \quad V(\eta) = \sum_{j=1}^d e_j \big\{\eta(0, e_j) - \eta(-e_j, 0)\big\}$$

is the *local drift* of the random walk X_t^ξ. Denote by $\mathfrak{M}_j(t)$, V_j the j-th component of $\mathfrak{M}(t)$, V, respectively: $\mathfrak{M}_j(t) = \mathfrak{m}_j(t) - \mathfrak{m}_{d+j}(t)$ and $V_j(\eta) = \eta(0, e_j) - \eta(-e_j, 0)$, $1 \le j \le d$.

Each function V_j has mean zero with respect to $\mathbb{Q}$ and belongs to $L^2(\mathbb{Q})$. We claim that V_j also belongs to $\mathcal{H}_{-1}$. In view of (2.12), to prove this statement it is enough to show that there exists a finite constant C_0 such that $\langle V_j, g\rangle_{\mathbb{Q}}$ is absolutely bounded by $C_0 \|g\|_1$ for all functions g in $L^2(\mathbb{Q})$. Fix such a functions g. A change of variables $\xi' = \tau_{-e_j}\xi$ shows that

$$\langle V_j, g\rangle_{\mathbb{Q}} = \int \big\{\xi(0, e_j) - \xi(-e_j, 0)\big\} g(\xi)\, d\mathbb{Q} = \int \xi(0, e_j) \big\{g(\xi) - g(\tau_{e_j}\xi)\big\}\, d\mathbb{Q}.$$

Thus, by Schwarz inequality and the explicit form for the $\mathcal{H}_1$ norm of g,

$$\langle V_j, g\rangle_{\mathbb{Q}}^2 \le E_{\mathbb{Q}}\big[\xi(0, e_j)\big] \|g\|_1^2,$$

which proves that V_j belongs to $\mathcal{H}_{-1}$ and that $\|V_j\|_{-1}^2 \le E_{\mathbb{Q}}[\xi(0, e_j)]$.

Following the strategy presented in the previous chapter, to prove a central limit theorem for X_t^ξ, we express the additive functional $\int_0^t V_j(\eta(s))\, ds$ as the sum of a martingale with a negligible remainder through the resolvent equation.

For $\lambda > 0$, $1 \le j \le d$, let $f_{j,\lambda}$ be the solution of the resolvent equation

$$\lambda f_{j,\lambda} - L f_{j,\lambda} = V_j \tag{3.3}$$

and let $m_{j,\lambda}(t)$, $t \ge 0$, be the martingale given by

$$m_{j,\lambda}(t) := f_{j,\lambda}\big(\eta(t)\big) - f_{j,\lambda}(\xi) - \int_0^t L f_{j,\lambda}\big(\eta(s)\big)\, ds$$

so that

$$X_t^\xi = \sum_{j=1}^d e_j \big\{\mathfrak{M}_j(t) + m_{j,\lambda}(t)\big\} + \sum_{j=1}^d e_j R_{j,\lambda}(t), \tag{3.4}$$

where the remainder term $R_{j,\lambda}(t)$ is given by

$$R_{j,\lambda}(t) = f_{j,\lambda}(\xi) - f_{j,\lambda}\big(\eta(t)\big) + \lambda \int_0^t f_{j,\lambda}\big(\eta(s)\big)\, ds.$$

We analyze separately the martingale terms and the remainder terms. We start with the latter ones. It follows from Proposition 2.8 that the remainders $t^{-1/2} R_{j,\lambda}(t)$ vanish in $L^2(\mathbb{Q})$ as $\lambda \downarrow 0$ and then $t \uparrow \infty$ if

$$\lim_{\lambda \to 0} \lambda \| f_{j,\lambda} \|_{\mathbb{Q}}^2 = 0, \qquad \lim_{\lambda \to 0} \| f_{j,\lambda} - f_j \|_1 = 0 \tag{3.5}$$

for some f_j in $\mathscr{H}_1$. By Lemma 2.16, conditions (3.5) are in force as soon as

$$\sup_{0<\lambda\le 1} \| L f_{j,\lambda} \|_{-1} < \infty. \tag{3.6}$$

Since the generator L is self-adjoint, by Sect. 2.7.1, (3.6) follows from the fact that V_j belongs to $\mathscr{H}_{-1}$.

We turn now to the martingales $\mathfrak{M}_j(t) + m_{j,\lambda}(t)$. To compute their predictable quadratic covariations, we first express the martingales $m_{j,\lambda}(t)$ in terms of the elementary martingales $\mathfrak{m}_k(t)$.

Fix a function f in $L^2(\mathbb{Q})$ and denote by $m^f(t)$ the Dynkin martingale associated to f:

$$m^f(t) = f\big(\eta(t)\big) - f\big(\eta(0)\big) - \int_0^t (Lf)\big(\eta(s)\big)\, ds.$$

Recall the definition of the counting processes $N_k(t)$, $1 \le k \le 2d$, defined above. We may express the difference $f(\eta(t)) - f(\eta(0))$ as

$$\begin{aligned} f\big(\eta(t)\big) - f\big(\eta(0)\big) &= \sum_{k=1}^{2d} \int_0^t \big\{ f\big(\eta(s)\big) - f\big(\eta(s-)\big) \big\}\, dN_k(s) \\ &= \sum_{k=1}^{2d} \int_0^t \big\{ f\big(\tau_{e_k}\eta(s-)\big) - f\big(\eta(s-)\big) \big\}\, dN_k(s). \end{aligned}$$

Since $\mathfrak{m}_k(t) = N_k(t) - \int_0^t \eta_s(0, e_k)\, ds$ is a martingale, the previous sum can be rewritten as

$$\begin{aligned} &\sum_{k=1}^{2d} \int_0^t \big\{ f\big(\tau_{e_k}\eta(s-)\big) - f\big(\eta(s-)\big) \big\}\, d\mathfrak{m}_k(s) \\ &\quad + \sum_{k=1}^{2d} \int_0^t \eta_s(0, e_k) \big\{ f\big(\tau_{e_k}\eta(s)\big) - f\big(\eta(s)\big) \big\}\, ds. \end{aligned}$$

Here, we replaced $\eta(s-)$ by $\eta(s)$ in the second integral because $\eta(s-) = \eta(s)$ for Lebesgue almost all $0 \le s \le t$. Note that the second sum is exactly $\int_0^t (Lf)(\eta(s))\, ds$.

Hence, the martingale $m^f(t)$ can be expressed in terms of the elementary martingales $\mathfrak{m}_k(t)$, $1 \le k \le 2d$, as

$$m^f(t) = \sum_{k=1}^{2d} \int_0^t \big\{ f\big(\tau_{e_k}\eta(s-)\big) - f\big(\eta(s-)\big)\big\}\, d\mathfrak{m}_k(s).$$

It is now easy to compute the predictable quadratic covariations of the martingales $\mathfrak{M}_j(t) + m_{j,\lambda}(t)$ because the elementary martingales $\mathfrak{m}_k(t)$ are orthogonal and have quadratic variations equal to $\int_0^t \eta_s(0, e_k)\, ds$. For two martingales M_t, N_t, $t \ge 0$, denote by $\langle M, N\rangle_t$ their predictable quadratic covariations. A straightforward calculation gives that

$$\begin{aligned}
&\langle \mathfrak{M}_i + m_{i,\lambda}, \mathfrak{M}_j + m_{j,\lambda}\rangle_t \\
&\quad = \sum_{k=1}^{2d} \int_0^t \eta_s(0, e_k)\big\{e_i \cdot e_k + (D_{e_k} f_{i,\lambda})\big(\eta(s)\big)\big\}\big\{e_j \cdot e_k + (D_{e_k} f_{j,\lambda})\big(\eta(s)\big)\big\}\, ds,
\end{aligned}$$

where $(D_{e_k} f)(\eta) = f(\tau_{e_k}\eta) - f(\eta)$ and $u \cdot v$ stands for the inner product in u, $v \in \mathbb{R}^d$.

We may now state the main result of this section.

Theorem 3.2 *As $t \uparrow \infty$, under $\mathbb{Q}$, the random walk $t^{-1/2} X_t^\xi$ converges in distribution to a mean zero Gaussian random variable with covariance matrix $\sigma^2 = \{\sigma_{i,j}^2 : 1 \le i, j \le d\}$ given by*

$$\sigma_{i,j}^2 = \lim_{\lambda \to 0} \sum_{k=1}^{2d} E_{\mathbb{Q}}\big[\eta(0, e_k)\big\{e_i \cdot e_k + (D_{e_k} f_{i,\lambda})(\eta)\big\}\big\{e_j \cdot e_k + (D_{e_k} f_{j,\lambda})(\eta)\big\}\big],$$

where $f_{j,\lambda}$ are the solutions of the resolvent equation (3.3). *The convergence takes place in $L^1(\mathbb{Q})$ with respect to the environment*:

$$\lim_{t \to \infty} E_{\mathbb{Q}}\big[\big|\mathbb{E}_0^\xi\big[\exp\{i\theta \cdot t^{-1/2} X_t^\xi\}\big] - \exp\{-(1/2)\theta^* \sigma^2 \theta\}\big|\big] = 0$$

for all θ in $\mathbb{R}^d$, where θ^* represents the transposition of the vector θ.

Proof Recall the decomposition (3.4). We have already shown that the remainders $t^{-1/2} R_{j,\lambda}(t)$ vanish in $L^2(\mathbb{Q})$ as $\lambda \downarrow 0$ and then $t \uparrow \infty$. By Lemma 2.9, the martingales $m_{j,\lambda}(t)$ converge in $L^2(\mathbb{Q})$, as $\lambda \downarrow 0$, to certain martingales denoted by $m_j(t)$.

We claim that the vector-valued martingale $m(t) + \mathfrak{M}(t)$, whose j-th component is $m_j(t) + \mathfrak{M}_j(t)$, satisfy the assumptions of Theorem 2.26. Since $\mathbb{Q}$ is a stationary state for the Markov process $\eta(t)$, the increments of $m(t) + \mathfrak{M}(t)$ are stationary. By the ergodic theorem for the process $\eta(t)$, the predictable quadratic covariations divided by t converge in $L^1(\mathbb{Q})$ to σ^2. Theorem 3.2 follows therefore from Theorem 2.26. □

In dimension 1, the asymptotic variance σ^2 is proportional to the harmonic mean of the rates $\xi(j, j+1)$:

$$\sigma^2 = \frac{2}{E_{\mathbb{Q}}[\xi(0,1)^{-1}]}.$$

The proof is similar to the one presented in Theorem 9.22 for one-dimensional diffusions.

3.2 Doubly Stochastic Random Walks

In the previous section we proved a central limit theorem, in L^1 with respect to the environment, for random walks with random conductances. The approach, derived from the theory developed in Chap. 2, applies to a large class of processes. In this section, we present a general framework, which may appear abstract at a first reading, in which similar arguments can be used. The reader is invited to keep in mind the example of the first section as a particular case of the class of processes introduced below.

On a probability space $(\Omega, \mathscr{F}, \mathbb{Q})$ we are given a group of transformations $\tau_x : \Omega \to \Omega$, $x \in \mathbb{Z}^d$, which preserves the measure $\mathbb{Q}$: τ_0 is the identity, $\tau_x \circ \tau_y = \tau_{x+y}$ and $\mathbb{Q}[\tau_x A] = \mathbb{Q}[A]$ for all $x, y \in \mathbb{Z}^d$, $A \in \mathscr{F}$. We assume that the action of the group is *ergodic*: if A is such that $\mathbb{Q}[(\tau_x A) \bigtriangleup A] = 0$ for all $x \in \mathbb{Z}^d$ then $\mathbb{Q}[A]$ is either 0 or 1. Here $\bigtriangleup$ stands for the symmetric difference: $A \bigtriangleup B = (A \setminus B) \cup (B \setminus A)$.

Suppose that we are given a family of non-negative functions $p_x : \Omega \to [0, \infty)$, $x \in \mathbb{Z}^d$, which represent the rate at which a random walk in random environment jumps from the origin to x if the environment is ω. In the example of the previous section, $\Omega = [a, b]^{\mathbb{B}_d}$ and $p_x(\xi) = \xi(0, e_j)$ if $x = e_j$ for some $1 \le j \le 2d$, $p_x(\xi) = 0$ otherwise. We shall assume that the functions $\{p_x : x \in \mathbb{Z}^d\}$ satisfy the following conditions.

(H1) Boundedness: p_z belongs to $B(\Omega)$ for all $z \in \mathbb{Z}^d$;
(H2) Finite range: There exists a deterministic $R > 0$ such that $p_z = 0$ if $\|z\| \ge R$, where $\|\cdot\|$ stands for the Euclidean norm of $\mathbb{R}^d$;
(H3) Irreducibility: For $\mathbb{Q}$-a.e. ω there exists a random set $\Lambda(\omega)$ generating $\mathbb{Z}^d$ such that $p_z(\omega) > 0$, $z \in \Lambda(\omega)$;
(H4) Double stochasticity: $\sum_{z \in \mathbb{Z}^d} p_{-z}(\tau_z \omega) = \sum_{z \in \mathbb{Z}^d} p_z(\omega)$ for $\mathbb{Q}$-a.e. ω.

As before, $B(\Omega)$ stands for the space of bounded measurable functions $f : \Omega \to \mathbb{R}$ endowed with the sup norm. Moreover, a subset Λ of $\mathbb{Z}^d$ is said to generate $\mathbb{Z}^d$ if for any x in $\mathbb{Z}^d$ one can find $n \ge 1$ and $z_1, \dots, z_n \in \Lambda$ for which $x = \sum_{i=1}^n z_i$. One can easily check that the random walks with random conductances examined in the previous section satisfy all the above conditions.

For each fixed ω, we define a random walk $\{X_t^\omega : t \ge 0\}$ over a probability space $(\Sigma, \mathscr{A}, \mathbb{P})$ whose random jump rate from y to z, denoted by $p(y, z; \omega)$, is given by

$$p(y, z; \omega) := p_{z-y}(\tau_y \omega) \tag{3.7}$$

for y, z in $\mathbb{Z}^d$. This means that the rate at which the random walk jumps from y to z if the environment is ω is equal to the rate at which it jumps from the origin to $z-y$ if the environment is $\tau_y\omega$. We sometimes omit the superscript ω of X_t^ω and denote X_t^ω by $X^\omega(t)$.

Formally, the random walk in random environment is defined as a stochastic process on the product probability space $(\Omega \times \Sigma, \mathscr{F} \times \mathscr{A}, \mathbb{Q} \otimes \mathbb{P})$ by the formula $X(t;\omega,\sigma) = X_t^\omega(\sigma)$.

The generator $\mathscr{L}_\omega$ of this random walk acts on functions f in $B(\mathbb{Z}^d)$ as

$$(\mathscr{L}_\omega f)(x) = \sum_{z\in\mathbb{Z}^d} p(x, x+z;\omega)\big\{f(x+z) - f(x)\big\}.$$

Denote by $\mathbb{P}_x^\omega$, $x \in \mathbb{Z}^d$, a probability measure on $(\Sigma, \mathscr{A})$ under which $\{X_t^\omega : t \geq 0\}$ is a Markov process with generator $\mathscr{L}_\omega$ starting from x: $\mathbb{P}_x^\omega[X_0^\omega = x] = 1$. Expectation with respect to $\mathbb{P}_x^\xi$ is denoted by $\mathbb{E}_x^\omega$.

Condition (**H4**) implies that $\mathbb{Q}$-a.s.,

$$\sum_{y\in\mathbb{Z}^d} p(y, x;\omega) = \sum_{y\in\mathbb{Z}^d} p(x, y;\omega)$$

for all x in $\mathbb{Z}^d$, which means that the random walk is doubly stochastic. In particular, the counting measure in $\mathbb{Z}^d$ is invariant for the random walk: For every finitely supported function $f : \mathbb{Z}^d \to \mathbb{R}$,

$$\sum_{x\in\mathbb{Z}^d} (\mathscr{L}_\omega f)(x) = \sum_{x\in\mathbb{Z}^d}\sum_{y\in\mathbb{Z}^d} p(x, y;\omega) f(y) - \sum_{x\in\mathbb{Z}^d}\sum_{y\in\mathbb{Z}^d} p(x, y;\omega) f(x) = 0.$$

Denote by $p_t^\omega : \mathbb{Z}^d \times \mathbb{Z}^d \to [0,1]$, $t \geq 0$, the transition probability functions of the Markov process $\{X_t^\omega : t \geq 0\}$. Thus, $p_t^\omega(x,y)$ represents the probability to be at y at time t if the initial position is x. As a consequence of (3.7),

$$p(x+z, y+z;\omega) = p(x, y;\tau_z\omega) \tag{3.8}$$

for all x, y, z in $\mathbb{Z}^d$, so the transition probability functions of the process satisfy

$$p_t^\omega(x+z, y+z) = p_t^{\tau_z\omega}(x, y)$$

for all $x, y, z \in \mathbb{Z}^d$, $t \geq 0$ and $\mathbb{Q}$-a.a. $\omega \in \Omega$.

As in the previous section, the *environment process* $\{\eta(t) : t \geq 0\}$ is defined by

$$\eta(t) := \tau_{X^\omega(t)}\omega.$$

Lemma 3.3 *$\{\eta(t) : t \geq 0\}$ is a Markov process whose generator L acts on functions f in $B(\Omega)$ as*

$$(Lf)(\omega) = \sum_{z\in\mathbb{Z}^d} p_z(\omega)\big\{f(\tau_z\omega) - f(\omega)\big\}.$$

The measure $\mathbb{Q}$ is invariant and ergodic under the transition probability semigroup corresponding to the process.

Proof The proof that the process $\{\eta(t) : t \ge 0\}$ is Markovian and has generator L is similar to the one of Lemma 3.1. It relies on the identity stated just after (3.8). Due to condition (**H2**), the sum appearing in the definition of the generator is carried over a finite set. In particular, by (**H1**), the generator L is a bounded operator in $L^2(\mathbb{Q})$.

We now prove that the measure $\mathbb{Q}$ is stationary for the semigroup $\{P_t : t \ge 0\}$. Denote by $\langle\cdot\rangle_{\mathbb{Q}}$ the expectation with respect to the measure $\mathbb{Q}$. Performing the change of variables $\omega' = \tau_z\omega$, since the transformations $\{\tau_x\}$ preserve the measure $\mathbb{Q}$, by assumption (**H4**), for any function f in $B(\Omega)$,

$$\begin{aligned}\langle Lf\rangle_{\mathbb{Q}} &= \sum_{z\in\mathbb{Z}^d} \big\langle p_z(\omega) f(\tau_z\omega)\big\rangle_{\mathbb{Q}} - \sum_{z\in\mathbb{Z}^d} \langle p_z f\rangle_{\mathbb{Q}} \\ &= \sum_{z\in\mathbb{Z}^d} \big\langle p_z(\tau_{-z}\omega) f(\omega)\big\rangle_{\mathbb{Q}} - \sum_{z\in\mathbb{Z}^d} \langle p_z f\rangle_{\mathbb{Q}} = 0.\end{aligned}$$

This proves the stationarity of $\mathbb{Q}$.

To check ergodicity we calculate the Dirichlet form of the process. We first compute the adjoint L^* of L. Fix two functions f, g in $L^2(\mathbb{Q})$. By the same arguments used above to show the stationarity of $\mathbb{Q}$,

$$\begin{aligned}\langle g, Lf\rangle_{\mathbb{Q}} &= \sum_{z\in\mathbb{Z}^d} \big\langle p_z(\omega) f(\tau_z\omega) g(\omega)\big\rangle_{\mathbb{Q}} - \sum_{z\in\mathbb{Z}^d} \big\langle p_z(\omega) f(\omega) g(\omega)\big\rangle_{\mathbb{Q}} \\ &= \sum_{z\in\mathbb{Z}^d} \big\langle p_z(\tau_{-z}\omega) f(\omega) g(\tau_{-z}\omega)\big\rangle_{\mathbb{Q}} - \sum_{z\in\mathbb{Z}^d} \big\langle p_z(\omega) f(\omega) g(\omega)\big\rangle_{\mathbb{Q}}.\end{aligned}$$

Changing z to $-z$ in the first sum and recalling the double stochasticity condition (**H4**), we obtain that the previous sum is equal to

$$\sum_{z\in\mathbb{Z}^d} \big\langle p_{-z}(\tau_z\omega) f(\omega)\big\{g(\tau_z\omega) - g(\omega)\big\}\big\rangle_{\mathbb{Q}},$$

so that

$$\big(L^* f\big)(\omega) = \sum_{z\in\mathbb{Z}^d} p_{-z}(\tau_z\omega)\big\{f(\tau_z\omega) - f(\omega)\big\}.$$

The Dirichlet form is now easy to compute: for any function f in $L^2(\mathbb{Q})$,

$$\begin{aligned}\big\langle f, (-L)f\big\rangle_{\mathbb{Q}} &= (1/2)\big\langle f, -\big(L + L^*\big) f\big\rangle_{\mathbb{Q}} \\ &= (1/4) \sum_{z\in\mathbb{Z}^d} \big\langle \big\{p_z(\omega) + p_{-z}(\tau_z\omega)\big\}\big\{f(\tau_z\omega) - f(\omega)\big\}^2\big\rangle_{\mathbb{Q}}.\end{aligned}$$

Performing the change of variables $\omega' = \tau_z \omega$, $z' = -z$ in the term which is multiplied by $p_{-z}(\tau_z \omega)$, the previous expression becomes

$$(1/2) \sum_{z \in \mathbb{Z}^d} \big\langle p_z(\omega) \big\{ f(\tau_z \omega) - f(\omega) \big\}^2 \big\rangle_{\mathbb{Q}}.$$

Therefore, if some f in $L^2(\mathbb{Q})$ is such that $Lf(\omega) = 0$, then, by the previous formula for the Dirichlet form, $p_z(\omega)\{f(\tau_z\omega) - f(\omega)\}^2$ vanishes $\mathbb{Q}$-a.s. for all z. By the irreducibility assumption (**H3**), f is invariant under the action of the group $\{\tau_x\}$: $f(\tau_x\omega) = f(\omega)$ for all x in $\mathbb{Z}^d$, $\mathbb{Q}$-a.s. Since the action of the group is ergodic, it follows that f is constant $\mathbb{Q}$-a.s. □

The formulas for the generator L and its adjoint L^* permit to compute the symmetric part S and the anti-symmetric part A of the generator L. Denote by D_z, $D_z^* : L^2(\mathbb{Q}) \to L^2(\mathbb{Q})$, $z \in \mathbb{Z}^d$, the operators defined by

$$(D_z f)(\omega) = f(\tau_z\omega) - f(\omega), \qquad \big(D_z^* f\big)(\omega) = f(\tau_{-z}\omega) - f(\omega) \tag{3.9}$$

and note that D_z^* is the adjoint of D_z. With this notation and in view of the formulas obtained for L and L^* in the proof of the previous lemma, we may write the generator and its adjoint as

$$Lf = \sum_{z \in \mathbb{Z}^d} p_z D_z f, \qquad L^* f = \sum_{z \in \mathbb{Z}^d} (p_{-z} \circ \tau_z) D_z f,$$

where $\circ$ stands for the composition of two functions so that $(p_{-z} \circ \tau_z)(\omega) = p_{-z}(\tau_z\omega)$. Expanding the formula for L^*, performing change of variables $z' = -z$ in the term $p_{-z}(\tau_z\omega) f(\tau_z\omega)$ and applying the double stochasticity assumption (**H4**), we may rewrite the adjoint L^* as

$$L^* f = \sum_{z \in \mathbb{Z}^d} D_z^*(p_z f),$$

which is the formula one obtains by computing the scalar product $\langle Lf, g \rangle_{\mathbb{Q}}$, expressing the generator L in terms of the operators D_z and using the fact that D_z^* is the adjoint of D_z.

Changing variables $z' = -z$ in the first formula for the adjoint L^*, we have that

$$2Sf = \big(L + L^*\big) f = \sum_{z \in \mathbb{Z}^d} \big\{ p_z D_z f - (p_z \circ \tau_{-z}) D_z[f \circ \tau_{-z}] \big\}$$

because $D_{-z} f = -D_z[f \circ \tau_{-z}]$. From these considerations we finally obtain a nice divergence-type formula for the symmetric part S of the generator:

$$Sf = -(1/2) \sum_{z \in \mathbb{Z}^d} D_z^*[p_z D_z f]. \tag{3.10}$$

Similar arguments show that the anti-symmetric part A of the generator L is given by

$$Af = (1/2) \sum_{z \in \mathbb{Z}^d} q_z D_z f = -(1/2) \sum_{z \in \mathbb{Z}^d} D_z^*(q_z f), \tag{3.11}$$

where $q_z(\omega) := p_z(\omega) - p_{-z}(\tau_z \omega)$.

Recall from Sect. 2.2 the definition of the spaces $\mathscr{H}_1$, $\mathscr{H}_{-1}$. Note that all functions f in $L^2(\mathbb{Q})$ belong to $\mathscr{H}_1$ because the generator L is a bounded operator. We already computed in the proof of the previous lemma the $\mathscr{H}_1$ norm of a function f in $L^2(\mathbb{Q})$:

$$\|f\|_1^2 = (1/2) \sum_{z \in \mathbb{Z}^d} \langle p_z(\omega) \{ f(\tau_z \omega) - f(\omega) \}^2 \rangle_{\mathbb{Q}}. \tag{3.12}$$

Our goal is to prove a central limit theorem for X_t^ω. The position of the random walk can be recovered from the process $\eta(t)$ by counting its number of jumps. More precisely, denote by $N_z(t)$, $z \in \mathbb{Z}^d$, the number of times that the process $\eta(t)$ jumped from a state η to the state $\tau_z \eta$ in the time interval $[0, t]$. Clearly,

$$X_t^\omega = \sum_{z \in \mathbb{Z}^d} z N_z(t).$$

This identity permits to express X_t^ω as the sum of a martingale with an additive functional of the process $\eta(t)$.

Denote by $\mathfrak{m}_z(t)$, $z \in \mathbb{Z}^d$, the elementary martingales defined by $\mathfrak{m}_z(t) = N_z(t) - \int_0^t p_z(\eta(s))\, ds$. Since these martingales are associated to counting processes and since they do not have common jumps, by Theorem 6.2.17 in Dacunha-Castelle and Duflo (1986) they are orthogonal in $L^2(\mathbb{Q})$. Moreover, their predictable quadratic variations are given by $\int_0^t p_z(\eta(s))\, ds$.

With this notation we have that

$$X_t^\omega = \mathfrak{M}(t) + \int_0^t V(\eta(s))\, ds, \tag{3.13}$$

where $\mathfrak{M}(t)$ is the vector-valued martingale $\sum_{z \in \mathbb{Z}^d} z \mathfrak{m}_z(t)$ and $V(\omega) = \sum_{z \in \mathbb{Z}^d} z p_z(\omega)$ is the *local drift* of the random walk X_t^ω. We denote by $\mathfrak{M}_j(t)$ the components of $\mathfrak{M}(t)$

Denote by V_j, $1 \le j \le d$, the components of the vector-valued function V:

$$V_j = V \cdot e_j = \sum_{z \in \mathbb{Z}^d} (z \cdot e_j) p_z.$$

In contrast with the previous section, there is no reason for V_j to have mean zero, which we need to assume: for $1 \le j \le d$

$$E_{\mathbb{Q}} \Big[\sum_{z \in \mathbb{Z}^d} (z \cdot e_j) p_z \Big] = 0. \tag{3.14}$$

We shall also assume that each

$$V_j \text{ belongs to } \mathcal{H}_{-1}. \tag{3.15}$$

To verify the previous two assumptions, we shall impose further conditions on the rate functions $\{p_z\}$.

Assuming (3.14) and (3.15), we may express the functional $\int_0^t V_j(\eta(s))\,ds$ as the sum of a martingale with a negligible remainder through the resolvent equation. For $\lambda > 0$, $1 \le j \le d$, let $f_{j,\lambda}$ be the solution of the resolvent equation

$$\lambda f_{j,\lambda} - L f_{j,\lambda} = V_j \tag{3.16}$$

and let $m_{j,\lambda}(t)$ be the martingale given by

$$m_{j,\lambda}(t) := f_{j,\lambda}\big(\eta(t)\big) - f_{j,\lambda}(\omega) - \int_0^t L f_{j,\lambda}\big(\eta(s)\big)\,ds$$

so that

$$X_t^{\omega} = \mathfrak{M}(t) + m_\lambda(t) + R_\lambda(t),$$

where $m_\lambda(t)$ is the vector-valued martingale with components $m_{j,\lambda}(t)$ and $R_\lambda(t)$ is the vector-valued remainder with components $R_{j,\lambda}(t)$ given by

$$R_{j,\lambda}(t) = f_{j,\lambda}(\omega) - f_{j,\lambda}\big(\eta(t)\big) + \lambda \int_0^t f_{j,\lambda}\big(\eta(s)\big)\,ds.$$

It follows from Proposition 2.8 that the remainders $t^{-1/2} R_{j,\lambda}(t)$ vanish in $L^2(\mathbb{Q})$ as $\lambda \downarrow 0$ and then $t \uparrow \infty$ if

$$\lim_{\lambda\to 0} \lambda \|f_{j,\lambda}\|_{\mathbb{Q}}^2 = 0, \qquad \lim_{\lambda\to 0} \|f_{j,\lambda} - f\|_1 = 0$$

for some f in $\mathcal{H}_1$. By Lemma 2.16, these statements hold as soon as

$$\sup_{0<\lambda\le 1} \|L f_{j,\lambda}\|_{-1} < \infty.$$

By Sect. 2.7, the latter condition follows from the fact that V_j belongs to $\mathcal{H}_{-1}$ and from a sector condition which we shall assume:

$$\text{The generator } L \text{ satisfies a sector condition.} \tag{3.17}$$

We turn now to the vector-valued martingales $\mathfrak{M}(t) + m_\lambda(t)$. The computations presented at the end of the previous section show that the martingales $m_{j,\lambda}(t)$ can be written as

$$m_{j,\lambda}(t) = \sum_{z\in\mathbb{Z}^d} \int_0^t \big\{ f_{j,\lambda}\big(\tau_z \eta(s-)\big) - f_{j,\lambda}\big(\eta(s-)\big)\big\}\,d\mathfrak{m}_z(s).$$

The predictable quadratic covariations of $\mathfrak{M}(t) + m_\lambda(t)$ are therefore given by

$$
\begin{aligned}
&\langle \mathfrak{M}_j + m_{j,\lambda}, \mathfrak{M}_k + m_{k,\lambda} \rangle_t \\
&\quad = \sum_{z \in \mathbb{Z}^d} \int_0^t \bigl\{ z \cdot e_j + (D_z f_{j,\lambda})\bigl(\eta(s)\bigr)\bigr\}\bigl\{ z \cdot e_k + (D_z f_{k,\lambda})\bigl(\eta(s)\bigr)\bigr\} p_z\bigl(\eta(s)\bigr)\, ds.
\end{aligned}
$$

Theorem 3.4 *Assume conditions* **(H1)**–**(H4)**, (3.14), (3.15), (3.17). *As $t \uparrow \infty$, under $\mathbb{Q}$, the random walk $t^{-1/2} X_t^\omega$ converges in distribution to a mean zero Gaussian random variable with covariance matrix $\sigma^2 = \{\sigma_{i,j}^2 : 1 \le i, j \le d\}$ given by*

$$
\sigma_{j,k}^2 = \lim_{\lambda \to 0} \sum_{z \in \mathbb{Z}^d} E_{\mathbb{Q}}\bigl[\{ z \cdot e_j + (D_z f_{j,\lambda}) \} \{ z \cdot e_k + (D_z f_{k,\lambda}) \} p_z \bigr],
$$

where $f_{j,\lambda}$ are solutions of the resolvent equation (3.16). *The convergence takes place in $L^1(\mathbb{Q})$ with respect to the environment:*

$$
\lim_{t \to \infty} E_{\mathbb{Q}}\bigl[\bigl| \mathbb{E}_0^\omega \bigl[\exp\bigl\{ i\theta \cdot t^{-1/2} X_t^\omega \bigr\} \bigr] - \exp\bigl\{ -(1/2)\theta^* \sigma^2 \theta \bigr\} \bigr| \bigr] = 0
$$

for all θ in $\mathbb{R}^d$, where θ^* represents the transposition of the vector θ.

The proof of this theorem is similar to the one of Theorem 3.2 and therefore omitted. In the next section we present an example where we can prove conditions **(H1)**–**(H4)**, (3.14), (3.15), (3.17).

3.3 Cyclic Random Walks

In this section we give an example of a non-reversible random walk satisfying hypotheses **(H1)**–**(H4)**, (3.14), (3.15), (3.17). We start with the definition of a cycle.

A cycle C of length n is a sequence of n sites of $\mathbb{Z}^d$ starting and ending at the same point: $(y_0, y_1, \dots, y_{n-1}, y_n = y_0)$, $y_i \ne y_{i+1}$, $0 \le i \le n-1$. To a cycle C of length n, we associate a zero-mean probability measure p_C on $\mathbb{Z}^d$ which does not charge the origin defined by

$$
p_C(x) = \frac{1}{n} \sum_{j=0}^{n-1} \mathbf{1}\{x = y_{j+1} - y_j\}. \tag{3.18}
$$

p_C has mean zero since

$$
\sum_{x \in \mathbb{Z}^d} x p_C(x) = \frac{1}{n} \sum_{x \in \mathbb{Z}^d} \sum_{j=0}^{n-1} x \mathbf{1}\{x = y_{j+1} - y_j\}
$$

$$= \frac{1}{n}\sum_{j=0}^{n-1}\{y_{j+1} - y_j\} = \frac{y_n - y_0}{n} = 0.$$

A probability measure associated to a cycle C is called a cyclic probability measure.

The measure associated to a cycle $C = (y_0, y_1, \dots, y_{n-1}, y_0)$ translated by x coincides with the one associated to C: If $C + x = (y_0 + x, y_1 + x, \dots, y_{n-1} + x, y_0 + x)$, $p_{C+x}(\cdot) = p_C(\cdot)$. The same observation holds for the probability measure associated to the cycle $C' = (y_1, y_2, \dots, y_{n-1}, y_0, y_1)$ obtained by shifting the cycle C. We may, in particular, assume that $y_0 = 0$.

A cycle $C = (y_0, y_1, \dots, y_{n-1}, y_0)$ is said to be *irreducible* if $y_i \neq y_j$ for $0 \le i \neq j \le n - 1$. Clearly, a cycle C can always be decomposed in a finite number of irreducible cycles. Moreover, if C is decomposed in irreducible cycles $C_1, \dots, C_k$, p_C is a rational convex combination of $p_{C_1}, \dots, p_{C_k}$: $p_C(\cdot) = \sum_{1\le j\le k} w_j p_{C_j}(\cdot)$ for some strictly positive rationals w_j such that $\sum_{1\le j\le k} w_j = 1$.

The cyclic probability measures are the finite-range zero-mean probability measures on $\mathbb{Z}^d$ taking rational values which do not charge the origin. Indeed, fix such a probability measure p and denote its support by $S = \{x_1, \dots, x_n\}$. There exists a sufficiently large positive integer M for which $p(x) = m(x)/M$, where $m(x)$ are positive integers. Consider the cycle $C = (0, x_1, 2x_1, \dots, m(x_1)x_1, m(x_1)x_1 + x_2, \dots, m(x_1)x_1 + m(x_2)x_2, \dots, m(x_1)x_1 + \cdots + m(x_{n-1})x_{n-1} + (m(x_n) - 1)x_n, 0)$. It is easy to check that the probability measure associated to this cycle is p.

For a positive integer m, finite cycles $\mathbf{C} = \{C_1, \dots, C_m\}$ and a probability measure $\mathbf{w} = \{w_1, \dots, w_m\}$, let $p_{\mathbf{C},\mathbf{w}}(\cdot)$ be the probability measure on $\mathbb{Z}^d$ defined by

$$p_{\mathbf{C},\mathbf{w}}(\cdot) = \sum_{k=1}^{m} w_k p_{C_k}(\cdot).$$

We prove in Lemma 5.6 that all finite-range, zero-mean probability measures on $\mathbb{Z}^d$ which do not charge the origin may be written as a convex combination of cyclic probability measures.

We introduce in the next lines the random walk on $\mathbb{Z}^d$ associated to an *irreducible* cycle $C = (0, y_1, \dots, y_{n-1}, 0)$. For $z \in \mathbb{Z}^d$, define a random walk on $C + z = (z, y_1 + z, \dots, y_{n-1} + z, z)$ which jumps from $y_j + z$ to $y_{j+1} + z$ at rate 1. Superposing the dynamics associated to each cycle $C + z$ we obtain a random walk on $\mathbb{Z}^d$ which jumps from x to y at a rate equal to the number of sites z such that $(x, y) = (z + y_j, z + y_{j+1})$ for some $0 \le j \le n - 1$. This picture can be slightly generalized if we speed up the jump rates by a function of z. This means that we associate a parameter $W(z) > 0$ to each z which represents the rate at which the random walk jumps on the cycle $C + z$. The random walk now jumps from x to y at a rate $r(x, y)$ equal to

$$r(x, y) = \sum_{z\in\mathbb{Z}^d} W(z)\mathbf{1}\{(x, y) \in z + C\},$$

where we understand that a pair (x, y) belongs to a cycle $C = (y_0, \dots, y_{n-1}, y_0)$ if $(x, y) = (y_j, y_{j+1})$ for some $0 \le j \le n - 1$. To complete the definition of the model, it remains to choose the rates according to a stationary ergodic random field $W(z) = W(z,\omega)$.

To include this model in the frame of the previous section, fix a strictly positive function $W : \Omega \to \mathbb{R}$ in $B(\Omega)$ and an irreducible cycle $C = \{0, y_1, \dots, y_{n-1}, 0\}$. For z in $\mathbb{Z}^d$, let

$$p_z(\omega) = \sum_{y\in\mathbb{Z}^d} W(\tau_y\omega)\mathbf{1}\big\{(0,z)\in y+C\big\}.$$

Note that p_z vanishes unless z belongs to the support of the probability measure p_C, which is equal to $\{y_1, y_2 - y_1, \dots, y_{n-1} - y_{n-2}, -y_{n-1}\}$.

Of course, for the sake of generality, one could consider a finite number of cycles $C_1, \dots, C_m$ rooted at the origin and a finite number of strictly positive functions $W_j : \Omega \to \mathbb{R}$ in $B(\Omega)$ and define the jump rates p_z by

$$p_z(\omega) = \sum_{k=1}^{m}\sum_{y\in\mathbb{Z}^d} W_k(\tau_y\omega)\mathbf{1}\big\{(0,z)\in y+C_k\big\}.$$

We assume below that $m = 1$, but all arguments apply to this more general situation.

A simple argument gives an alternative formula for the rates $\{p_z\}$ which simplifies the computations. To require the pair $(0,z)$ to belong to the cycle $y + C$ is the same as to require the origin to be equal to $y + y_j$ and z to be equal to $y + y_{j+1}$ for some $0 \le j \le n-1$. In other words,

$$\mathbf{1}\big\{(0,z)\in y+C\big\} = \sum_{j=0}^{n-1}\mathbf{1}\{0 = y + y_j\}\mathbf{1}\{z = y + y_{j+1}\}.$$

Thus, changing the order of summations, a legal operation because all terms are positive, we obtain that

$$\begin{aligned} p_z(\omega) &= \sum_{j=0}^{n-1}\sum_{y\in\mathbb{Z}^d} W(\tau_y\omega)\mathbf{1}\{0 = y + y_j\}\mathbf{1}\{z = y + y_{j+1}\} \\ &= \sum_{j=0}^{n-1} W(\tau_{-y_j}\omega)\mathbf{1}\{z = y_{j+1} - y_j\}. \end{aligned} \tag{3.19}$$

It is now easy to see that the rates $\{p_z : z \in \mathbb{Z}^d\}$ satisfy the conditions (**H1**) and (**H2**). Since we assumed W to be strictly positive, the irreducibility assumption (**H3**) follows from the irreducibility of the cyclic probability p_C, which we suppose from now on. Finally, to show that condition (**H4**) holds, observe that

$$\sum_{z\in\mathbb{Z}^d} p_z(\omega) = \sum_{j=0}^{n-1} W(\tau_{-y_j}\omega).$$

On the other hand, the sum $\sum_{z\in\mathbb{Z}^d} p_{-z}(\tau_z\omega)$ is equal to

$$\sum_{z\in\mathbb{Z}^d}\sum_{j=0}^{n-1} W(\tau_{z-y_j}\omega)\mathbf{1}\{-z=y_{j+1}-y_j\}=\sum_{j=0}^{n-1} W(\tau_{-y_{j+1}}\omega),$$

which proves the double stochasticity assumption (**H4**) because C is a cycle.

Recall the definition of the local drift V given just after (3.13). In the present context, by (3.19), V takes the form

$$V(\omega)=\sum_{z\in\mathbb{Z}^d} zp_z(\omega)=\sum_{j=0}^{n-1}(y_{j+1}-y_j)W(\tau_{-y_j}\omega).$$

Hypothesis (3.14), which requires V to have mean zero, is straightforward. Since the measure $\mathbb{Q}$ is shift invariant,

$$E_{\mathbb{Q}}[V]=E_{\mathbb{Q}}[W]\sum_{j=0}^{n-1}(y_{j+1}-y_j)=0.$$

To prove condition (3.15), we first derive a formula for the generator L of the environment process $\eta(t)$ introduced in Lemma 3.3 and one for the associated Dirichlet form. By (3.19), the generator L acts on functions f in $L^2(\mathbb{Q})$ as

$$(Lf)(\omega)=\sum_{j=0}^{n-1} W(\tau_{-y_j}\omega)\big\{f(\tau_{y_{j+1}-y_j}\omega)-f(\omega)\big\},$$

while, by (3.12), the Dirichlet form of a function f in $L^2(\mathbb{Q})$ is given by

$$\|f\|_1^2=(1/2)\sum_{j=0}^{n-1} E_{\mathbb{Q}}\big[W(\tau_{-y_j}\omega)\big\{f(\tau_{y_{j+1}-y_j}\omega)-f(\omega)\big\}^2\big].$$

Performing the change of variables $\omega'=\tau_{-y_j}\omega$, we can rewrite the Dirichlet form as

$$\|f\|_1^2=(1/2)\sum_{j=0}^{n-1} E_{\mathbb{Q}}\big[W(\omega)\big\{f(\tau_{y_{j+1}}\omega)-f(\tau_{y_j}\omega)\big\}^2\big].$$

We claim that each component of V belongs to $\mathcal{H}_{-1}$. In view of (2.12), to prove this statement it is enough to show that there exists a finite constant C_0 such that $\|\langle V,f\rangle_{\mathbb{Q}}\|\le C_0\|f\|_1$ for all function f in $L^2(\mathbb{Q})$, where $\|\cdot\|$ stands for the Euclidean norm of $\mathbb{R}^d$. Fix a function f in $L^2(\mathbb{Q})$ and notice that V can be rewritten as

$$V(\omega)=\sum_{j=0}^{n-1} y_{j+1}\big\{W(\tau_{-y_j}\omega)-W(\tau_{-y_{j+1}}\omega)\big\}.$$

Therefore, a change of variables shows that

$$\begin{aligned} E_{\mathbb{Q}}[Vf] &= \sum_{j=0}^{n-1} y_{j+1} E_{\mathbb{Q}}\big[\big\{W(\tau_{-y_j}\omega) - W(\tau_{-y_{j+1}}\omega)\big\} f(\omega)\big] \\ &= \sum_{j=0}^{n-1} y_{j+1} E_{\mathbb{Q}}\big[W(\omega)\big\{f(\tau_{y_j}\omega) - f(\tau_{y_{j+1}}\omega)\big\}\big]. \end{aligned}$$

In particular, by Schwarz inequality,

$$\big\| E_{\mathbb{Q}}[Vf]\big\|^2 \le 2\|f\|_1^2 E_{\mathbb{Q}}[W] \sum_{j=0}^{n-1} \|y_j\|^2,$$

proving that V belongs to $\mathscr{H}_{-1}$.

It remains to check the sector condition (3.17). Fix two functions f, g in $L^2(\mathbb{Q})$. Here we use strongly the cyclic property. By definition of the generator L and the change of variables $\omega' = \tau_{-y_j}\omega$,

$$\langle g, Lf\rangle_{\mathbb{Q}} = \sum_{j=0}^{n-1} E_{\mathbb{Q}}\big[W(\omega) g(\tau_{y_j}\omega)\big\{f(\tau_{y_{j+1}}\omega) - f(\tau_{y_j}\omega)\big\}\big].$$

Since C is a cycle, we may rewrite the right-hand side as

$$\sum_{j=0}^{n-1} E_{\mathbb{Q}}\big[W(\omega)\big\{g(\tau_{y_j}\omega) - g(\omega)\big\}\big\{f(\tau_{y_{j+1}}\omega) - f(\tau_{y_j}\omega)\big\}\big].$$

Since $2ab \le Aa^2 + A^{-1}b^2$, $A > 0$, the previous expression is less than or equal to

$$\begin{aligned} &\frac{1}{2A} \sum_{j=0}^{n-1} E_{\mathbb{Q}}\big[W(\omega)\big\{g(\tau_{y_j}\omega) - g(\omega)\big\}^2\big] \\ &\quad + \frac{A}{2} \sum_{j=0}^{n-1} E_{\mathbb{Q}}\big[W(\omega)\big\{f(\tau_{y_{j+1}}\omega) - f(\tau_{y_j}\omega)\big\}^2\big] \end{aligned}$$

for every $A > 0$. The second term is just $A\|f\|_1^2$. Interpolating the difference $g(\tau_{y_j}\omega) - g(\omega)$ as $\sum_{0\le k\le j-1} g(\tau_{y_{k+1}}\omega) - g(\tau_{y_k}\omega)$, applying Schwarz inequality and inverting the order of summations, we may bound the first term by

$$\frac{n^2}{2A} \sum_{j=0}^{n-1} E_{\mathbb{Q}}\big[W(\omega)\big\{g(\tau_{y_{j+1}}\omega) - g(\tau_{y_j}\omega)\big\}^2\big] = \frac{n^2}{A}\|g\|_1^2.$$

Hence, minimizing over A we get that

$$\big|\langle g, Lf\rangle_{\mathbb{Q}}\big| \le 2n\|f\|_1\|g\|_1,$$

where n is the length of the cycle C. Therefore, the sector conditions holds with a constant proportional to the length of the cycle.

We have just checked all conditions of the previous section. Theorem 3.4 is therefore in force for the cyclic random walks in random environment presented in this section.

3.4 Random Walks with Drift in $\mathscr{H}_{-1}$

Consider a doubly stochastic random walk $\{X^\omega(t) : t \ge 0\}$ as defined in Sect. 3.2. In this section, we show that the sector condition (3.17) is not needed to prove the central limit theorem provided the assumptions on the drift are strengthen.

We shall suppose that the random rates satisfy an *ellipticity condition*, i.e., that there exists a deterministic constant $\kappa > 0$ and a finite *deterministic* subset $\Lambda \subset \mathbb{Z}^d$, *generating the lattice*, such that $\mathbb{Q}$ almost surely

$$\min_{z\in\Lambda} p_z(\omega) \ge \kappa. \tag{3.20}$$

To simplify the exposition, although not needed, we shall also assume that $p_z(\omega) = 0$ for $z \notin \Lambda$. Denote by R a positive integer such that $p_z = 0$ for $\|z\|_\infty > R$, where $\|x\|_\infty$ stands for the max norm, $\|(x_1, \dots, x_d)\|_\infty = \max\{|x_1|, \dots, |x_d|\}$.

To present the extra assumptions needed on the local drift, we first show in Lemma 3.5 below that any function in the space $L^2(\mathbb{Q}) \cap \mathscr{H}_{-1}$ can be represented as $\sum_{z\in\Lambda} D_z^*(p_z\Psi_z)$ for some $\Psi_z \in L^2(\mathbb{Q})$, $z \in \Lambda$.

Let $\mathscr{H}$ be the Hilbert space consisting of all random vectors $F := \{F_z : z \in \Lambda\}$, $F_z : \Omega \to \mathbb{R}$, that satisfy

$$\|F\|^2_{\mathscr{H}} := \frac{1}{2}\sum_{z\in\Lambda}\langle p_z F_z, F_z\rangle_{\mathbb{Q}} < \infty.$$

Denote by $\|\cdot\|_{\mathscr{H}}$ the respective norm.

Recall from (3.9) the definition of the operator $D_z : L^2(\mathbb{Q}) \to L^2(\mathbb{Q})$, $z \in \mathbb{Z}^d$. Let $L^2_0(\mathbb{Q})$ represent the subspace of $L^2(\mathbb{Q})$ consisting of all zero mean elements and let $\mathscr{D} : L^2_0(\mathbb{Q}) \to \mathscr{H}$ be given by $\mathscr{D}g := \{D_z g, z \in \Lambda\}$. The closure of $\mathscr{D}(L^2_0(\mathbb{Q}))$ shall be denoted by $\mathscr{H}_\nabla \subset \mathscr{H}$. It represents the space of gradients.

It follows from the explicit expression of the Dirichlet form that $\mathscr{H}_\nabla$ is isomorphic to $\mathscr{H}_1$, where the isomorphism $\mathscr{D} : \mathscr{H}_1 \to \mathscr{H}_\nabla$ is given by the continuous extension of the operator $\mathscr{D}$.

Lemma 3.5 *Suppose that* $\{p_z, z \in \mathbb{Z}^d\}$ *satisfy conditions* **(H1)**, **(H2)**, **(H4)** *and* (3.20). *A function* f *in* $L^2(\mathbb{Q})$ *belongs to* $\mathscr{H}_{-1}$ *if and only if there exist* $\{\Psi_z : z \in \Lambda\}$

in $\mathscr{H}$ *such that*

$$f = \sum_{z\in\Lambda} D_z^*\{p_z\Psi_z\}.$$

Proof Assume there exist $\{\Psi_z : z \in \Lambda\}$ in $\mathscr{H}$ for which the identity holds. Since D_z^* is the adjoint of D_z in $L^2(\mathbb{Q})$, for every function g in $L^2(\mathbb{Q})$, by Schwarz inequality,

$$\langle f, g\rangle_{\mathbb{Q}}^2 \le 4\|\Psi\|_{\mathscr{H}}^2\|g\|_1^2.$$

This proves that f belongs to $\mathscr{H}_{-1}$ in view of (2.12).

To prove the converse statement, fix a function f which belongs to $L^2(\mathbb{Q}) \cap \mathscr{H}_{-1}$. Since f belongs to $\mathscr{H}_{-1}$, $\langle f\rangle_{\mathbb{Q}} = 0$. Let $L_f : \mathscr{D}(L_0^2(\mathbb{Q})) \to \mathbb{R}$ be the linear functional defined as $L_f(\mathscr{D}g) = \langle f, g\rangle_{\mathbb{Q}}$. This functional is bounded because f belongs to $\mathscr{H}_{-1}$. In particular, it can be extended to a bounded linear functional on $\mathscr{H}_\nabla$. By Riesz representation theorem, there exists a unique $\Psi := \{\Psi_z : z \in \Lambda\}$ in $\mathscr{H}_\nabla$ such that $L_f(\mathscr{D}g) = (1/2)\sum_{z\in\Lambda}\langle p_z\Psi_z, D_z g\rangle_{\mathbb{Q}}$ for all g in $L_0^2(\mathbb{Q})$. Therefore,

$$\langle f, g\rangle_{\mathbb{Q}} = \frac{1}{2}\Big\langle \sum_{z\in\Lambda} D_z^*\{p_z\Psi_z\}, g\Big\rangle_{\mathbb{Q}}$$

for all g in $L_0^2(\mathbb{Q})$. Hence, $f = (1/2)\sum_{z\in\Lambda} D_z^*\{p_z\Psi_z\} + C$ for some finite constant C. Since f and $(1/2)\sum_{z\in\Lambda} D_z^*\{p_z\Psi_z\}$ have mean zero, $C = 0$ which proves the lemma. □

Informally, this lemma states that any function in $L^2(\mathbb{Q}) \cap \mathscr{H}_{-1}$ is the divergence of a matrix in $L^2(\mathbb{Q})$. We shall require the local drift to be the divergence of a matrix in $L^d(\mathbb{Q})$. More precisely, Lemma 3.5 states that if the local drift V belongs to $\mathscr{H}_{-1}$ there exist $H_{k,z} \in L^2(\mathbb{Q})$, $1 \le k \le d$, $z \in \Lambda$, such that

$$V_k := \sum_{z\in\Lambda} D_z^*(p_z H_{k,z}). \tag{3.21}$$

We shall assume furthermore that there exists such a decomposition with the extra property that

$$\begin{cases} H_{k,z} \text{ belongs to } L^d(\mathbb{Q}) \text{ in dimension } d \ge 3, \\ H_{k,z} \text{ belongs to } L^{2+\delta}(\mathbb{Q}) \text{ for some } \delta > 0 \text{ in dimension } d = 2. \end{cases} \tag{3.22}$$

Theorem 3.6 *Suppose that* $\{p_z : z \in \mathbb{Z}^d\}$ *satisfy conditions* (**H1**), (**H2**), (**H4**), (3.20), (3.21) *and* (3.22). *Then, the random walk* $\{X^\omega(t) : t \ge 0\}$ *satisfies the central limit theorem as stated in Theorem* 3.4.

In Sect. 3.5 we show that assumptions (3.21) and (3.22) hold if the random rates $\{p_z : z \in \mathbb{Z}^d\}$ satisfy some mixing conditions. In Sect. 3.6 we examine doubly stochastic random walks in dimension 1.

The central limit theorem for the random walk $\{X_t^\omega : t \geq 0\}$ stated above is a consequence of Theorem 2.17 and Proposition 3.7 below, which establishes an energy identity for weak limiting points in $\mathscr{H}_1$ of the solutions of the resolvent equation.

3.4.1 The Corrector Field

In this subsection, we introduce the corrector field and establish an energy identity. Recall the resolvent equation (3.16) for the local drift V. Multiplying both sides of the resolvent equation by $f_{j,\lambda}$ and integrating over Ω, in view of the explicit formula (3.12) for the $\mathscr{H}_1$ norm and of relation (3.21),

$$\lambda \| f_{j,\lambda} \|_{\mathbb{Q}}^2 + \frac{1}{2} \sum_{z \in \Lambda} \langle p_z D_z f_{j,\lambda}, D_z f_{j,\lambda} \rangle_{\mathbb{Q}} = \sum_{z \in \Lambda} \langle p_z H_{j,z}, D_z f_{j,\lambda} \rangle_{\mathbb{Q}}$$

for $1 \leq j \leq d$. Since $H_{j,z}$ belongs to $L^2(\mathbb{Q})$ and since p_z is strictly elliptic,

$$\sup_{0<\lambda\leq 1} \lambda \| f_{j,\lambda} \|_{\mathbb{Q}}^2 < \infty \quad \text{and} \quad \sup_{0<\lambda\leq 1} \| D_z f_{j,\lambda} \|_{\mathbb{Q}}^2 < \infty$$

for all z in Λ, $1 \leq j \leq d$.

Proposition 3.7 *Suppose that the hypotheses of Theorem* 3.6 *hold and that F_j is the weak limit in $\mathscr{H}_1$ of $\{f_{j,\lambda_n} : n \geq 1\}$, where $\lambda_n \downarrow 0$, as $n \uparrow \infty$. Then, for $1 \leq j \leq d$,*

$$\frac{1}{2} \sum_{z \in \Lambda} \langle p_z D_z F_j, D_z F_j \rangle_{\mathbb{Q}} = \sum_{z \in \Lambda} \langle p_z H_{j,z}, D_z F_j \rangle_{\mathbb{Q}}.$$

This proposition is proved in three steps. In this subsection, we define the corrector field and we obtain two estimates on its asymptotic behavior. In the next subsection, we derive an elliptic difference equation satisfied by the corrector field from which we deduce, in the last subsection, the energy estimate stated above.

Fix a sequence $\{\lambda_n : n \geq 1\}$ which vanishes as $n \uparrow \infty$ and such that f_{j,λ_n} converges weakly to F_j in $\mathscr{H}_1$ for $1 \leq j \leq d$. Since $\{D_z f_{j,\lambda} : 0 < \lambda < 1\}$ is a bounded sequence in $L^2(\mathbb{Q})$, taking a further subsequence if necessary, we may assume that $D_z f_{j,\lambda_n}$ converges weakly in $L^2(\mathbb{Q})$ for all z in Λ, $1 \leq j \leq d$. Let

$$\mathfrak{f}_j(z) = \lim_{n\to\infty} D_z f_{j,\lambda_n}.$$

Clearly, $\mathscr{D} F_j = \{\mathfrak{f}_j(z) : z \in \Lambda\}$ in $L^2(\mathbb{Q})$.

We claim that

$$D_x f_{j,\lambda_n} \text{ converges weakly in } L^2(\mathbb{Q}) \text{ for all } x \text{ in } \mathbb{Z}^d. \tag{3.23}$$

To prove this statement, call a sequence $x = x_0, x_1, \dots, x_n = y$ of vertices in $\mathbb{Z}^d$ such that $x_p - x_{p-1} \in \Lambda$, $1 \le p \le n$, a *proper* path connecting x to y. Fix $x \in \mathbb{Z}^d$, a proper path $0 = x_0, x_1, \dots, x_n = x$ connecting the origin to x and observe that

$$D_x g = g \circ \tau_x - g = \sum_{p=1}^{n} \{g \circ \tau_{x_p} - g \circ \tau_{x_{p-1}}\} = \sum_{p=1}^{n} (D_{x_p - x_{p-1}} g) \circ \tau_{x_{p-1}}$$

for all $g \in L^2(\mathbb{Q})$. This identity together with the weak convergence in $L^2(\mathbb{Q})$ of $D_z f_{j,\lambda_n}$ proves claim (3.23). Denote by $f_j(x)$ the weak limit of $D_x f_{j,\lambda_n}$, $x \in \mathbb{Z}^d$,

$$f_j(x) = \lim_{n \to \infty} D_x f_{j,\lambda_n}, \tag{3.24}$$

and observe that $f_j(0) = 0$ and that

$$f_j(x) = \sum_{p=1}^{n} f_j(x_p - x_{p-1}) \circ \tau_{x_{p-1}}. \tag{3.25}$$

The field $\{f_j(x) : x \in \mathbb{Z}^d\}$ is called *a corrector field*. More precisely,

Definition 3.8 A random field $\{f_j(x;\omega) : x \in \mathbb{Z}^d\}$, $1 \le j \le d$, is a corrector field if there exists a sequence $\{\lambda_n : n \ge 1\}$, vanishing as $n \uparrow \infty$, for which (3.24) holds weakly in $L^2(\mathbb{Q})$ for all $x \in \mathbb{Z}^d$, $1 \le j \le d$.

A corrector field $\{f_j(x) : x \in \mathbb{Z}^d\}$, though not stationary since $f_j(0) \equiv 0$, has stationary increments. Indeed, for $z \in \mathbb{Z}^d$ and a random field $\{h(x;\omega) : x \in \mathbb{Z}^d\}$, $h(x) \in L^2(\mathbb{Q})$, let $(\partial_z h)(x;\omega) = h(x+z;\omega) - h(x,\omega)$. It follows from (3.24) that

$$\partial_z f_j(x) = f_j(z) \circ \tau_x = D_z F_j \circ \tau_x \tag{3.26}$$

for all x in $\mathbb{Z}^d$, $z \in \Lambda$. By (3.24), the first equality also holds for $z \in \mathbb{Z}^d$. To extend the second one, note that it follows from (3.25) and from the identities $f_j(z) = D_z F_j$, $z \in \Lambda$, that $f_j(x) = D_x F_j$ in $L^2(\mathbb{Q})$, $x \in \mathbb{Z}^d$.

We conclude this subsection presenting two estimates of the corrector field.

Lemma 3.9 *Let $\{f_j(x) : x \in \mathbb{Z}^d\}$, $1 \le j \le d$, be a corrector field. For any $K > 0$ and $1 \le j \le d$,*

$$\lim_{\varepsilon \to 0} \sup_{\|x\|_\infty \le K/\varepsilon} \varepsilon \big\| f_j(x) \big\|_{\mathbb{Q}} = 0.$$

Proof Fix $1 \le j \le d$. For a given $x = (x_1, \dots, x_d)$ let $\Gamma(x)$ be the path connecting 0 to x given by the vertices: $z_0, z_1, \dots, z_d$, where $z_0 := 0$, $z_p := \sum_{1 \le k \le p} x_k e_k$, $1 \le p \le d$. Recall that we denote by $\mathfrak{s}(x)$, $x \in \mathbb{Z}$, the sign of x. In view of (3.24),

$$f_j(x) = \sum_{p=1}^{d} \sum_{k=0}^{|x_p|-1} f_j\big(\mathfrak{s}(x_p) e_p\big) \circ \tau_{z_{p-1} + k\mathfrak{s}(x_p) e_p}. \tag{3.27}$$

By convention we omit the summation if its upper limit is smaller than the lower one.

Let $\{U_x : x \in \mathbb{Z}^d\}$, $U_x : L^2(\mathbb{Q}) \to L^2(\mathbb{Q})$, be the unitary maps associated to the group $\{\tau_x : x \in \mathbb{Z}^d\}$ of $\mathbb{Q}$-measure preserving transformations ($U_x f := f \circ \tau_x$). By the spectral resolution of unitary maps (Yosida, 1995, Sect. XI.4), there exists a system of orthogonal projection operators $\{E_\theta : \theta \in \mathbb{T}^d\}$ such that

$$U_x = \int_{\mathbb{T}^d} e^{2\pi i x\cdot\theta}\, dE_\theta,$$

where the integral is understood as a spectral integral.

Let $\hat{F}_p(d\theta)$ be the spectral projection of $f_j(\mathfrak{s}(x_p)e_p)$: $\hat{F}_p(d\theta) = dE_\theta f_j(\mathfrak{s}(x_p)e_p)$. In view of the spectral resolution of the operators U_x,

$$f_j(x) = \sum_{p=1}^{d} \int_{\mathbb{T}^d} e^{2\pi i\theta\cdot z_{p-1}} \left(\frac{e^{2\pi i x_p\theta_p} - 1}{e^{2\pi i \mathfrak{s}(x_p)\theta_p} - 1} \right) \hat{F}_p(d\theta).$$

Hence, by Schwarz inequality,

$$\| f_j(x) \|_{\mathbb{Q}}^2 \le C_0 \sum_{p=1}^{d} \int_{\mathbb{T}^d} \left\| \frac{e^{2\pi i x_p\theta_p} - 1}{e^{2\pi i \mathfrak{s}(x_p)\theta_p} - 1} \right\|^2 \mu_{j,p}(d\theta),$$

for some finite constant C_0 which may change from line to line. In this formula, $\mu_{j,p}(d\theta)$ stands for the spectral measure of $f_j(\mathfrak{s}(x_p)e_p)$: $\mu_{j,p}(d\theta) = \langle dE_\theta f_j(\mathfrak{s}(x_p)e_p), f_j(\mathfrak{s}(x_p)e_p)\rangle_{\mathbb{Q}}$. Fix $\eta > 0$ and $1 \le p \le d$. The expression inside the integral is bounded by $C_0 x_p^2$ on $\mathbb{T}^d$ and by $C(\eta)$ on the set $\{\theta \in \mathbb{T}^d : |\theta_p| \ge \eta\}$. Therefore, for every $\eta > 0$,

$$\begin{aligned} \sup_{\|x\|_\infty \le K/\varepsilon} \varepsilon \| f_j(x) \|_{\mathbb{Q}} &\le \varepsilon C(\eta) \sum_{p=1}^{d} \mu_{j,p}(\mathbb{T}^d)^{1/2} \\ &\quad + C_0 K \sum_{p=1}^{d} \mu_{j,p}\big(\{\theta \in \mathbb{T}^d : |\theta_p| \le \eta\}\big)^{1/2}. \end{aligned}$$

Since $\mu_{j,p}(\mathbb{T}^d) = \| f_j(\mathfrak{s}(x_p)e_p)\|_{\mathbb{Q}}^2$ is finite, letting $\varepsilon \downarrow 0$ and then $\eta \downarrow 0$ we conclude that

$$\lim_{\varepsilon\to 0} \sup_{\|x\|_\infty \le K/\varepsilon} \varepsilon \| f_j(x) \|_{\mathbb{Q}} \le C_0 K \sum_{p=1}^{d} \sqrt{\mu_{j,p}(I_p)},$$

where $I_p = \{\theta \in \mathbb{T}^d : \theta_p = 0\}$.

It remains to show that $\mu_{j,p}(I_p) = 0$. Recall that we represented by $\hat{F}_p(d\theta)$ the spectral projection of $f_j(\mathfrak{s}(x_p)e_p)$. Let $h_{j,p} \in L^2(\mathbb{Q})$, $1 \le p \le d$, be the function defined by

$$h_{j,p} = \int_{\mathbb{T}^d} \mathbf{1}\{\theta \in I_p\} \hat{F}_p(d\theta).$$

With this notation, $\mu_{j,p}(I_p) = \langle h_{j,p}, f_j(\mathfrak{s}(x_p)e_p)\rangle_{\mathbb{Q}}$. Since $f_j(\mathfrak{s}(x_p)e_p)$ has mean zero with respect to $\mathbb{Q}$, to conclude the proof it is enough to show that $h_{j,p}$ is constant.

Since $D_{\mathfrak{s}(x_q)e_q} f_j(\mathfrak{s}(x_p)e_p) = D_{\mathfrak{s}(x_p)e_p} f_j(\mathfrak{s}(x_q)e_q)$, $1 \le p, q \le d$,

$$D_{\mathfrak{s}(x_q)e_q} \int_{\mathbb{T}^d} g(\theta)\hat{F}_p(d\theta) = D_{\mathfrak{s}(x_p)e_p} \int_{\mathbb{T}^d} g(\theta)\hat{F}_q(d\theta)$$

for every bounded function g. Applying this identity to the indicator function of the set I_p and since $\{E_\theta : \theta \in \mathbb{T}^d\}$ is the spectral resolution of the unitary operators $\{U_x\}$, we obtain that

$$\begin{aligned} D_{\mathfrak{s}(x_q)e_q} h_{j,p} &= D_{\mathfrak{s}(x_q)e_q} \int_{\mathbb{T}^d} \mathbf{1}\{\theta \in I_p\}\hat{F}_p(d\theta) = D_{\mathfrak{s}(x_p)e_p} \int_{\mathbb{T}^d} \mathbf{1}\{\theta \in I_p\}\hat{F}_q(d\theta) \\ &= \int_{\mathbb{T}^d} \mathbf{1}\{\theta \in I_p\}\big(e^{2\pi i \mathfrak{s}(x_p)\theta_p} - 1\big)\hat{F}_q(d\theta) = 0. \end{aligned}$$

The previous identity shows that $h_{j,p}$ is invariant under the group of spatial shifts $\{\tau_x : x \in \mathbb{Z}^d\}$. Thus, by the ergodic theorem, $h_{j,p}$ is constant, which concludes the proof of the lemma. □

Fix $\varepsilon > 0$ and denote by $f_{j,\varepsilon} : \varepsilon\mathbb{Z}^d \to \mathbb{R}$ the function given by $f_{j,\varepsilon}(x) := \varepsilon f_j(x\varepsilon^{-1})$. For any $g : \varepsilon\mathbb{Z}^d \to \mathbb{R}$, let

$$\sum_{\varepsilon,K} g(x) := \varepsilon^d \sum_{x\in\varepsilon\mathbb{Z}^d;\, \|x\|_\infty \le K} g(x),$$

where $\nabla^\varepsilon g(x) := (\partial_1^\varepsilon g(x), \dots, \partial_d^\varepsilon g(x))$, and $\partial_j^\varepsilon g(x) := \varepsilon^{-1}[g(x+\varepsilon e_j) - g(x)]$.

Let $\alpha = 2d/(d-2)$ in dimension $d \ge 3$ and let α be arbitrary in the set $[1,\infty)$ in dimension $d = 2$. The Sobolev inequality (Leoni, 2009, Corollary 11.9, Theorem 11.23, Exercise 11.26 and Theorem 12.15) establishes that for every $K > 0$, there exists a constant C_1 depending only on K and d such that for any $g : \varepsilon\mathbb{Z}^d \to \mathbb{R}$

$$\Big\{\sum_{\varepsilon,K} |g(x)|^\alpha\Big\}^{1/\alpha} \le C_1 \Big\{\sum_{\varepsilon,K} |g(x)|^2 + \sum_{\varepsilon,K} \|\nabla^\varepsilon g(x)\|^2\Big\}^{1/2}. \tag{3.28}$$

In dimension 2, the result is proved in Leoni (2009) for $\alpha \ge 2$ and follows for $1 \le \alpha \le 2$ from the elementary estimate

$$\Big\{\sum_{\varepsilon,K} |g(x)|^\alpha\Big\}^{1/\alpha} \le C_1 \Big\{\sum_{\varepsilon,K} |g(x)|^2\Big\}^{1/2}$$

which holds for some finite constant C_1 depending on α. In dimension $d = 1$ one can replace the left-hand side of the previous estimate by the maximum over the range:

$$\max_{x\in\varepsilon\mathbb{Z};|x|\le K} |g(x)| \le C_1\Big\{\sum_{\varepsilon,K} |g(x)|^2 + \sum_{\varepsilon,K} |\nabla^\varepsilon g(x)|^2\Big\}^{1/2}.$$

Lemma 3.10 *Let* $\{f_j(x) : x \in \mathbb{Z}^d\}$, $1 \le j \le d$, *be a corrector field. Fix* $K > 0$ *and* α *as in* (3.28). *Then*, $\mathbb{Q}$-*a.s.*

$$L_*(\omega) := \limsup_{\varepsilon\to 0}\Big\{\sum_{\varepsilon,K} |f_{j,\varepsilon}(x)|^\alpha\Big\}^{1/\alpha} < \infty.$$

Proof By the Sobolev inequality (3.28), the expression appearing in the statement of the lemma is less than or equal to

$$C_1\Big\{\sum_{\varepsilon,K} |f_{j,\varepsilon}(x)|^2 + \sum_{\varepsilon,K} \|\nabla^\varepsilon f_{j,\varepsilon}(x)\|^2\Big\}^{1/2}$$

for some finite constant C_1 depending only on K and d. We examine these two terms separately.

On the one hand, since $\nabla^\varepsilon f_{j,\varepsilon}(x) = Df_j \circ \tau_{\varepsilon^{-1}x}$, where $Df_j = (f_j(e_1), \dots, f_j(e_d))$,

$$\sum_{\varepsilon,K} \|\nabla^\varepsilon f_{j,\varepsilon}(x)\|^2 = \varepsilon^d \sum_{y\in\mathbb{Z}^d;\|y\|_\infty\le\varepsilon^{-1}K} \|Df_j\|^2 \circ \tau_y.$$

By the ergodic theorem, this expression converges $\mathbb{Q}$-a.s. to $C_1\langle\|Df_j\|^2\rangle_{\mathbb{Q}}$ for some deterministic constant C_1 depending only on K and d.

On the other hand, by the definition of $f_{j,\varepsilon}$ and a change of variables,

$$\sum_{\varepsilon,K} |f_{j,\varepsilon}(x)|^2 = \varepsilon^{d+2} \sum_{y\in\mathbb{Z}^d,\|y\|_\infty\le\varepsilon^{-1}K} |f_j(y)|^2.$$

For each $y = (y_1, \dots, y_d)$ in $\mathbb{Z}^d$, consider the path $\Gamma(y) = (0 = z_0, \dots, z_n = y)$ from the origin to y such that $\|z_{i+1} - z_i\| = 1$, $0 \le i < n$, and which first moves on the e_1 direction, than on the e_2 direction and so on. Note that its length n is equal to $\sum_{1\le i\le d} |y_i|$ and that $z_{|y_1|} = y_1e_1$, $z_{|y_1|+|y_2|} = y_1e_1 + y_2e_2$. By (3.27) and by Schwarz inequality, the previous expression is bounded above by

$$\varepsilon^{d+1}C_1(K) \sum_{\substack{y\in\mathbb{Z}^d\\ \|y\|_\infty\le\varepsilon^{-1}K}} \sum_{k=0}^{n-1} |f_j(z_{k+1} - z_k)|^2 \circ \tau_{z_k} \tag{3.29}$$

for some finite constant $C_1(K)$ depending only on K. Note that $z_{k+1} - z_k = \pm e_p$ for some $1 \le p \le d$. Changing the order of summation, we may rewrite this expression as

$$\varepsilon^{d+1} C_1(K) \sum_{m=1}^{2d} \sum_{\substack{z \in \mathbb{Z}^d \\ \|z\|_\infty \le \varepsilon^{-1} K}} \left| f_j(e_m) \right|^2 \circ \tau_z \sum_{\substack{y \in \mathbb{Z}^d \\ \|y\|_\infty \le \varepsilon^{-1} K}} \mathbf{1}\big\{(z, z + e_m) \in \Gamma(y)\big\},$$

where we used the convention that $e_{d+m} = -e_m$, $1 \le m \le d$. In this formula, a pair (z, z') belongs to the path $\Gamma(y) = (0 = z_0, \dots, z_n = y)$ if $z = z_\ell$, $z' = z_{\ell+1}$ for some $0 \le \ell < n$.

Fix a function g in $L^1(\mathbb{Q})$. It is well known that

$$\frac{1}{(2N+1)^\ell} \sum_{\substack{\|z\|_\infty \le N \\ z_{\ell+1}, \dots, z_d = 0}} g \circ \tau_z$$

converges $\mathbb{Q}$-a.s., as $N \uparrow \infty$, to a random variable g_ℓ equal to the conditional expectation of g with respect to the σ-algebra of events which are invariant under the shifts in the directions $e_1, \dots, e_\ell$.

It follows from this observation that (3.29) converges $\mathbb{Q}$-a.s., as $\varepsilon \downarrow 0$, to

$$\sum_{m=1}^{2d} C_1(m, K) Y_{j,m}.$$

In this formula, $C_1(m, K)$, $1 \le m \le d$, are finite constants depending only on K and $Y_{j,m}$, $Y_{j,d+m}$, $1 \le m \le d$, are the conditional expectations of $|f_j(e_m)|^2$, $|f_j(-e_m)|^2$ with respect to the σ-algebra of events which are invariant under the shifts in the directions $e_1, \dots, e_m$. This proves the lemma. □

3.4.2 An Elliptic Equation for the Corrector Field

In this subsection we demonstrate that the corrector field satisfies in a weak sense a linear elliptic difference equation $\mathbb{Q}$-a.s. Let $C_0(\mathbb{Z}^d)$ be the space of all compactly supported functions on $\mathbb{Z}^d$. Recall the definition of the random variables $\{H_{j,z} : z \in \Lambda\}$, $1 \le j \le d$, introduced in (3.21) and recall that $q_z = p_z - p_{-z} \circ \tau_z$.

Proposition 3.11 *Let $\{f_j(x) : x \in \mathbb{Z}^d\}$, $1 \le j \le d$, be a corrector field. For each $1 \le j \le d$ and for every x in $\mathbb{Z}^d$, $\mathbb{Q}$-a.s.,*

$$\begin{aligned}
&\frac{1}{2} \sum_{z \in \Lambda} \partial_z^* \big\{ p(x, x+z) \partial_z f_j(x) \big\} - \frac{1}{2} \sum_{z \in \Lambda^s} q(x, x+z) \partial_z f_j(x) \\
&\quad = \sum_{z \in \Lambda} \partial_z^* \big\{ p(x, x+z) H_{j,z}(x) \big\},
\end{aligned}$$

where $q(x, x+z) := q_z \circ \tau_x$, $\Lambda^s = \{z \in \mathbb{Z}^d : z \text{ or } -z \in \Lambda\}$ *and* ∂_z^* *represents the adjoint of* ∂_z: $(\partial_z^* g)(x) = [g(x-z) - g(x)]$.

Proof Fix a corrector field $\{f_j(x) : x \in \mathbb{Z}^d\}$, $1 \le j \le d$, and consider a sequence $\{\lambda_n : n \ge 1\}$, vanishing as $n \uparrow \infty$, for which (3.24) holds weakly in $L^2(\mathbb{Q})$ for all $x \in \mathbb{Z}^d$. Multiply both sides of the resolvent equation (3.3) by $g \in L^2(\mathbb{Q})$ and integrate with respect to $\mathbb{Q}$. By (3.21), the right-hand side is equal to $\sum_{z\in\Lambda} \langle p_z H_{j,z}, D_z g\rangle_{\mathbb{Q}}$. On the left-hand side, $\langle \lambda_n f_{\lambda_n,j}, g\rangle_{\mathbb{Q}}$ vanishes as $n \uparrow \infty$ because $\lambda_n f_{\lambda_n,j}$ vanishes in $L^2(\mathbb{Q})$. Rewriting the generator L as $S + A$, and recalling the explicit formulas (3.10), (3.11) for S and A as well as the definition of $f_j(z)$, we obtain that

$$\frac{1}{2}\sum_{z\in\Lambda} \langle p_z f_j(z), D_z g\rangle_{\mathbb{Q}} - \frac{1}{2}\sum_{z\in\Lambda^s} \langle q_z f_j(z), g\rangle_{\mathbb{Q}} = \sum_{z\in\Lambda} \langle p_z H_{j,z}, D_z g\rangle_{\mathbb{Q}}.$$

Recall from (3.26) that $f_j(z) \circ \tau_x = \partial_z f_j(x)$. For g in $L^2(\mathbb{Q})$, define the random field $\{g_x : x \in \mathbb{Z}^d\}$, by $g_x = g \circ \tau_x$ so that $H_{j,z}(x) = H_{j,z} \circ \tau_x$, $x \in \mathbb{Z}^d$. Since $\mathbb{Q}$ is shift invariant, for every x in $\mathbb{Z}^d$, we may rewrite the previous identity as

$$\begin{aligned}
&\frac{1}{2}\sum_{z\in\Lambda} \big\langle [p_z \circ \tau_x]\partial_z f_j(x), D_z g_x\big\rangle_{\mathbb{Q}} - \frac{1}{2}\sum_{z\in\Lambda^s} \big\langle [q_z \circ \tau_x]\partial_z f_j(x), g_x\big\rangle_{\mathbb{Q}} \\
&\quad = \sum_{z\in\Lambda} \big\langle [p_z \circ \tau_x] H_{j,z}(x), D_z g_x\big\rangle_{\mathbb{Q}}.
\end{aligned}$$

Fix a function Φ in $C_0(\mathbb{Z}^d)$, multiply both sides of the equation by $\Phi(x)$ and sum over $x \in \mathbb{Z}^d$ to get that

$$\begin{aligned}
&\frac{1}{2}\sum_{\substack{z\in\Lambda \\ x\in\mathbb{Z}^d}} \big\langle [p_z \circ \tau_x]\partial_z f_j(x), D_z g_x\big\rangle_{\mathbb{Q}} \Phi(x) - \frac{1}{2}\sum_{\substack{z\in\Lambda^s \\ x\in\mathbb{Z}^d}} \big\langle [q_z \circ \tau_x]\partial_z f_j(x), g_x \Phi(x)\big\rangle_{\mathbb{Q}} \\
&\quad = \sum_{z\in\Lambda}\sum_{x\in\mathbb{Z}^d} \big\langle [p_z \circ \tau_x] H_{j,z}(x), D_z g_x\big\rangle_{\mathbb{Q}} \Phi(x).
\end{aligned} \tag{3.30}$$

In the first term on the left-hand side, rewrite $D_z g_x$ as $g \circ \tau_{x+z} - g \circ \tau_x$. Consider the expressions associated to each of these terms separately. In the first one, use the shift invariance of $\mathbb{Q}$ to translate all terms by $-z$. With this change of variables this sum becomes

$$\frac{1}{2}\sum_{z\in\Lambda}\sum_{x\in\mathbb{Z}^d} \big\langle [p_z \circ \tau_{x-z}]\partial_z f_j(x-z), g_x\big\rangle_{\mathbb{Q}} \Phi(x)$$

because, by (3.26), $\partial_z f_j(x) \circ \tau_{-z} = f_j(z) \circ \tau_{x-z} = \partial_z f_j(x-z)$. Hence, the change of variables $x' = x - z$ permits to rewrite the first term on the left-hand side of (3.30) as

$$\frac{1}{2}\sum_{z\in\Lambda}\sum_{x\in\mathbb{Z}^d}\big\langle p(x,x+z)\partial_z f_j(x),\,\partial_z\big[g_x\Phi(x)\big]\big\rangle_{\mathbb{Q}}$$

because, by (3.7), $p_z\circ\tau_x = p(x,x+z)$. By similar reasons, the term on the right-hand side of (3.30) can be written as

$$\sum_{z\in\Lambda}\sum_{x\in\mathbb{Z}^d}\big\langle p(x,x+z)H_{j,z}(x),\,\partial_z\big[g_x\Phi(x)\big]\big\rangle_{\mathbb{Q}}.$$

Up to this point we proved that for every g in $L^2(\mathbb{Q})$ and for every Φ in $C_0(\mathbb{Z}^d)$,

$$\begin{aligned}\frac{1}{2}\sum_{\substack{z\in\Lambda\\x\in\mathbb{Z}^d}}&\big\langle p(x,x+z)\partial_z f_j(x),\,\partial_z\big[g_x\Phi(x)\big]\big\rangle_{\mathbb{Q}}\\&\quad-\frac{1}{2}\sum_{\substack{z\in\Lambda^s\\x\in\mathbb{Z}^d}}\big\langle q(x,x+z)\partial_z f_j(x),\,g_x\Phi(x)\big\rangle_{\mathbb{Q}}\\&=\sum_{z\in\Lambda}\sum_{x\in\mathbb{Z}^d}\big\langle p(x,x+z)H_{j,z}(x),\,\partial_z\big[g_x\Phi(x)\big]\big\rangle_{\mathbb{Q}}.\end{aligned}$$

To conclude the proof of the lemma, sum by parts and choose Φ to be the indicator function of a site x_0 in $\mathbb{Z}^d$. Since the identity thus obtained holds for all $g_{x_0} = g\circ\tau_{x_0}$ in $L^2(\mathbb{Q})$, the proposition is proved □

Consider a random field $\{h_x : x\in\mathbb{Z}^d\}$, $h_x\in L^2(\mathbb{Q})$, and a function Φ in $C_0(\mathbb{Z}^d)$. Multiply both sides of the equation appearing in the statement of Proposition 3.11 by $h_x\Phi(x)$, sum over x and sum by parts the first term on the left-hand side and the term on the right-hand side to get that $\mathbb{Q}$-a.s.,

$$\begin{aligned}\frac{1}{2}\sum_{\substack{z\in\Lambda\\x\in\mathbb{Z}^d}}p(x,x+z)\partial_z f_j(x)\partial_z\big[h_x\Phi(x)\big]-\frac{1}{2}\sum_{\substack{z\in\Lambda^s\\x\in\mathbb{Z}^d}}q(x,x+z)\partial_z f_j(x)h_x\Phi(x)\\=\sum_{z\in\Lambda}\sum_{x\in\mathbb{Z}^d}p(x,x+z)H_{j,z}(x)\partial_z\big[h_x\Phi(x)\big].\end{aligned}\tag{3.31}$$

We shall deduce the energy identity from this equation in the next subsection.

3.4.3 The Energy Identity

In this subsection, we prove the energy identity stated in Proposition 3.7. Let $\varphi:\mathbb{R}^d\to[0,\infty)$ be any smooth, non-negative function, supported in the unit cube $(-1,1)^d$ and satisfying $\int_{\mathbb{R}^d}\varphi(x)dx = 1$. For $\varepsilon>0$, let $\varphi_\varepsilon(x) = \varepsilon^d\varphi(\varepsilon x)$.

Replace in (3.31) h, Φ by f_j, φ_ε, respectively. We claim that the first term on the left-hand side of (3.31) converges in probability, as $\varepsilon \downarrow 0$, to

$$\frac{1}{2}\sum_{z\in\Lambda}\langle p_z D_z F_j, D_z F_j\rangle_{\mathbb{Q}}. \tag{3.32}$$

In this subsection, convergence in probability refers to the measure $\mathbb{Q}$.

To prove (3.32), note that the first term on the left-hand side of (3.31), with f_j, φ_ε in place of h, Φ, is equal to

$$\frac{1}{2}\sum_{z\in\Lambda}\sum_{x\in\mathbb{Z}^d} p(x,x+z)\partial_z f_j(x)\partial_z\big[f_j(x)\varphi_\varepsilon(x)\big].$$

In view of the formula for the discrete gradient of a product

$$\partial_z[fg](x) = f(x+z)(\partial_z g)(x) + g(x)(\partial_z f)(x),$$

the previous expression is equal to

$$\frac{1}{2}\sum_{z\in\Lambda}\sum_{x\in\mathbb{Z}^d} \varphi_\varepsilon(x+z)p(x,x+z)\partial_z f_j(x)^2 \\ + \frac{1}{2}\sum_{z\in\Lambda}\sum_{x\in\mathbb{Z}^d} \partial_z\varphi_\varepsilon(x)p(x,x+z)\partial_z f_j(x) f_j(x).$$

Denote the first, second expression by $E_1(\varepsilon)$, $E_2(\varepsilon)$, correspondingly, and recall identity (3.26). Since $p(x,x+z) = p_z\circ\tau_x$, $\partial_z f_j(x) = f_j(z)\circ\tau_x = D_zF_j\circ\tau_x$ and $p_z[D_zF_j]^2$ belongs to $L^1(\mathbb{Q})$, by the ergodic theorem, as $\varepsilon\downarrow 0$, $E_1(\varepsilon)$ converges $\mathbb{Q}$-a.s. to (3.32).

We claim that $E_2(\varepsilon)$ vanishes in $L^1(\mathbb{Q})$ as $\varepsilon\downarrow 0$. To estimate $E_2(\varepsilon)$, recall that the set Λ is finite and observe that

$$(\partial_z\varphi_\varepsilon)(x) = \varepsilon^{d+1}(\nabla\varphi)(\varepsilon x)\cdot z + \varepsilon^{d+2}r_\varepsilon(z,x)\mathbf{1}\big\{\|x\|_\infty \le \varepsilon^{-1}\big\}, \tag{3.33}$$

where the term $r_\varepsilon(z,x)$ is absolutely bounded by a finite constant C_1 uniformly in z, x and $\varepsilon\in(0,1)$. In this formula, $\nabla\varphi$ stands for the gradient of $\varphi : \nabla\varphi = (\partial_{x_1}\varphi,\dots,\partial_{x_d}\varphi)$. Note that we use the same symbol ∂ to represent the discrete derivative ∂_z and the continuous one ∂_{x_i}. We may rewrite $E_2(\varepsilon)$ as the sum $E_{2,1}(\varepsilon)+E_{2,2}(\varepsilon)$, where

$$E_{2,1}(\varepsilon) = \frac{\varepsilon^{d+1}}{2}\sum_{z\in\Lambda}\sum_{x\in\mathbb{Z}^d}(\nabla\varphi)(\varepsilon x)\cdot z p(x,x+z)\partial_z f_j(x) f_j(x).$$

Since $\partial_z f_j(x) = f_j(z)\circ\tau_x$ and $\mathbb{Q}$ is shift invariant, by Schwarz inequality

$$\big\langle\big|E_{2,1}(\varepsilon)\big|\big\rangle_{\mathbb{Q}} \le C_1 \max_{\|x\|_\infty\le\varepsilon^{-1}}\big\|\varepsilon f_j(x)\big\|_{\mathbb{Q}} \max_{z\in\Lambda}\big\|f_j(z)\big\|_{\mathbb{Q}}$$

for some finite constant C_1 which depends only on φ and p_z through its support Λ and the bound $\max_{z\in\Lambda}\|p_z\|_\infty$. By Lemma 3.9, this expression vanishes as $\varepsilon\downarrow 0$. By similar reasons, $E_{2,2}(\varepsilon)$ vanishes in $L^1(\mathbb{Q})$ as $\varepsilon\downarrow 0$. This proves claim (3.32).

The same type of arguments shows that the right-hand side of (3.31) with $f_j(x)\varphi_\varepsilon(x)$ in place of $h_x\Phi(x)$ converges in probability, as $\varepsilon\downarrow 0$, to

$$\sum_{z\in\Lambda}\big\langle p_z H_{j,z}, f_j(z)\big\rangle_{\mathbb{Q}} = \sum_{z\in\Lambda}\langle p_z H_{j,z}, D_z F_j\rangle_{\mathbb{Q}} \tag{3.34}$$

because $H_{j,z}$ belongs to $L^2(\mathbb{Q})$ by assumption. The last identity follows from (3.26).

In view of (3.32) and (3.34), to conclude the proof of the energy identity stated in Proposition 3.7 it remains to show that

$$\sum_{z\in\Lambda^s}\sum_{x\in\mathbb{Z}^d}\varphi_\varepsilon(x)q(x,x+z)\partial_z f_j(x)f_j(x) \tag{3.35}$$

converges in probability to 0 as $\varepsilon\downarrow 0$.

We first claim that

$$\begin{aligned} 2\sum_{\substack{z\in\Lambda^s\\x\in\mathbb{Z}^d}}&\varphi_\varepsilon(x)q(x,x+z)\partial_z f_j(x)f_j(x)\\ &=-\sum_{\substack{z\in\Lambda^s\\x\in\mathbb{Z}^d}}(\partial_z\varphi_\varepsilon)(x)q(x,x+z)f_j(x)^2\\ &\quad-\sum_{z\in\Lambda^s}\sum_{x\in\mathbb{Z}^d}(\partial_z\varphi_\varepsilon)(x)q(x,x+z)\partial_z f_j(x)f_j(x). \end{aligned} \tag{3.36}$$

Indeed, since $q_z=p_z-p_{-z}\circ\tau_z$, by assumption (**H4**), $\sum_z q_z=0$. In particular, since $q(x,x+z)=q_z\circ\tau_x$, one half of the left-hand side of (3.36) is equal to

$$\sum_{z\in\Lambda^s}\sum_{x\in\mathbb{Z}^d}\varphi_\varepsilon(x)[q_z\circ\tau_x]f_j(x+z)f_j(x).$$

Changing variable $x':=x+z$ and using the identity $q_z\circ\tau_{x-z}=-q_{-z}\circ\tau_x$ we obtain that the previous sum equals

$$-\sum_{z\in\Lambda^s}\sum_{x\in\mathbb{Z}^d}\varphi_\varepsilon(x-z)[q_{-z}\circ\tau_x]f_j(x)f_j(x-z).$$

Change variables $z'=-z$ once more, replace $\varphi_\varepsilon(x+z)$, $f_j(x+z)$ by $\varphi_\varepsilon(x)+\partial_z\varphi_\varepsilon(x)$, $f_j(x)+\partial_z f_j(x)$, respectively, and recall that $\sum_z q_z$ vanishes to rewrite the previous expression as

$$-\sum_{z\in\Lambda^s}\sum_{x\in\mathbb{Z}^d}(\partial_z\varphi_\varepsilon)(x)[q_z\circ\tau_x]f_j(x)(\partial_z f_j)(x)-\sum_{z\in\Lambda^s}\sum_{x\in\mathbb{Z}^d}(\partial_z\varphi_\varepsilon)(x)[q_z\circ\tau_x]f_j(x)^2$$
$$-\sum_{z\in\Lambda^s}\sum_{x\in\mathbb{Z}^d}\varphi_\varepsilon(x)[q_z\circ\tau_x]f_j(x)(\partial_z f_j)(x).$$

Since the last sum is equal to one half of the left-hand side of (3.36), the claim is proved.

By (3.36), the proof of (3.35), and therefore the proof of Proposition 3.7, is reduced to the proof that

$$\begin{gathered}\sum_{z\in\Lambda^s}\sum_{x\in\mathbb{Z}^d}(\partial_z\varphi_\varepsilon)(x)[q_z\circ\tau_x]f_j(x)(\partial_z f_j)(x)\\ \text{and}\quad\sum_{z\in\Lambda^s}\sum_{x\in\mathbb{Z}^d}(\partial_z\varphi_\varepsilon)(x)[q_z\circ\tau_x]f_j(x)^2\end{gathered}\tag{3.37}$$

vanish in probability as $\varepsilon\downarrow 0$.

The arguments which permitted to estimate $E_2(\varepsilon)$ show that the first sum vanishes in $L^1(\mathbb{Q})$ as $\varepsilon\downarrow 0$. To estimate the second term we use the expansion (3.33) of $\partial_z\varphi_\varepsilon(x)$ and express the sum as $B_1(\varepsilon)+B_2(\varepsilon)$, where

$$\begin{aligned}B_1(\varepsilon)&=\varepsilon^{d+1}\sum_{z\in\Lambda^s}\sum_{x\in\mathbb{Z}^d}(\nabla\varphi)(\varepsilon x)\cdot z[q_z\circ\tau_x]f_j(x)^2,\\ B_2(\varepsilon)&=\varepsilon^{d+2}\sum_{z\in\Lambda^s}\sum_{\|x\|_\infty\le\varepsilon^{-1}}r_\varepsilon(z,x)[q_z\circ\tau_x]f_j(x)^2.\end{aligned}\tag{3.38}$$

Since q_x is bounded and r_ε is uniformly bounded, $B_2(\varepsilon)$ vanishes in $L^1(\mathbb{Q})$, as $\varepsilon\downarrow 0$, in view of Lemma 3.9. We now turn our attention to $B_1(\varepsilon)$. An elementary computation shows that for each g in $L^2(\mathbb{Q})$,

$$\sum_{z\in\Lambda^s}z\langle q_z,g\rangle_{\mathbb{Q}}=2\sum_{z\in\Lambda}z\langle p_z,g\rangle_{\mathbb{Q}}+\sum_{z\in\Lambda}z\langle p_z,D_zg\rangle_{\mathbb{Q}}.$$

Since this identity holds for all g in $L^2(\mathbb{Q})$, $\sum_z zq_z=2\sum_z zp_z+\sum_z zD_z^*p_z=2V+\sum_z zD_z^*p_z$ by definition of the local drift. Hence, by assumption (3.21), for $1\le k\le d$,

$$\sum_{z\in\Lambda^s}z_kq_z=\sum_{z\in\Lambda}D_z^*G_{k,z},$$

where $G_{k,z}=2p_zH_{k,z}+z_kp_z$. By hypothesis (3.22), $G_{k,z}$ belongs to $L^d(\mathbb{Q})$ in dimension $d\ge 3$ and belongs to $L^{2+\delta}(\mathbb{Q})$ in dimension $d=2$ for some $\delta>0$.

Replacing in the first term of (3.38) $\sum_z z_k q_z$ by $\sum_z D_z^* G_{k,z}$, we obtain that

$$B_1(\varepsilon) = \varepsilon^{d+1} \sum_{k=1}^{d} \sum_{z\in\Lambda} \sum_{x\in\mathbb{Z}^d} (\partial_{x_k}\varphi)(\varepsilon x)\big[D_z^* G_{k,z} \circ \tau_x\big] f_j(x)^2.$$

Since $D_z^* G_{k,z} = G_{k,z} \circ \tau_{-z} - G_{k,z}$, performing the change of variables $x' = x - z$, the previous identity becomes $B_1(\varepsilon) = A_1(\varepsilon) + A_2(\varepsilon)$, where

$$\begin{aligned} A_1(\varepsilon) &= \varepsilon^{d+1} \sum_{k=1}^{d} \sum_{z\in\Lambda} \sum_{x\in\mathbb{Z}^d} (\partial_{x_k}\varphi)\big(\varepsilon[x+z]\big)[G_{k,z}\circ\tau_x]\partial_z f_j(x)\big\{f_j(x+z) \\ &\quad + f_j(x)\big\}, \\ A_2(\varepsilon) &= \varepsilon^{d+1} \sum_{k=1}^{d} \sum_{z\in\Lambda} \sum_{x\in\mathbb{Z}^d} \partial_z(\partial_{x_k}\varphi)(\varepsilon x)[G_{k,z}\circ\tau_x] f_j(x)^2. \end{aligned}$$

Note that ∂_z represents a discrete derivative, while ∂_{x_k} represents a continuous one. In view of (3.38) and the observation following that equation, to conclude the proof of the energy identity we have to show that $A_1(\varepsilon)$ and $A_2(\varepsilon)$ vanish in probability as $\varepsilon \downarrow 0$.

Changing variables $x' := \varepsilon x$ and remembering that the support of φ is contained in $(-1, 1)^d$, we obtain that

$$\begin{aligned} &\big|A_1(\varepsilon)\big| \\ &\quad \le C_1 \varepsilon^{d+1} \sum_{k=1}^{d} \sum_{z\in\Lambda} \sum_{x\in\varepsilon\mathbb{Z}^d} |G_{k,z}\circ\tau_{x\varepsilon^{-1}}|\big|\partial_z f_j\big(x\varepsilon^{-1}\big)\big|\big|f_j\big(x\varepsilon^{-1}+z\big) + f_j\big(x\varepsilon^{-1}\big)\big| \end{aligned}$$

for some finite constant C_1 depending only on φ. The summation over x extends to those $x \in \varepsilon\mathbb{Z}^d$ for which $\|x\|_\infty \le 1$. Recall that $G_{k,z}$ vanishes for $\|z\|_\infty \ge R$.

We consider only the case $d \ge 3$. The argument in dimension $d = 2$ requires only minor modifications. For any $K > 0$ we can estimate the previous sum by

$$\begin{aligned} &C_1 \varepsilon^{d+1} K \sum_{k=1}^{d} \sum_{z\in\Lambda} \sum_{x\in\varepsilon\mathbb{Z}^d} |G_{k,z}\circ\tau_{x\varepsilon^{-1}}| I_{j,x} \\ &\quad + C_1 \varepsilon^{d+1} \sum_{k=1}^{d} \sum_{z\in\Lambda} \sum_{x\in\varepsilon\mathbb{Z}^d} |G_{k,z}\circ\tau_{x\varepsilon^{-1}}|\big|\partial_z f_j\big(x\varepsilon^{-1}\big)\big| \mathbf{1}\big\{\big|\partial_z f_j\big(x\varepsilon^{-1}\big)\big| > K\big\} I_{j,x}, \end{aligned}$$

where $I_{j,x} = |f_j(x\varepsilon^{-1}+z)| + |f_j(x\varepsilon^{-1})|$. Denote the first term by $A_{1,1}(\varepsilon, K)$ and the second one by $A_{1,2}(\varepsilon, K)$. By Schwarz inequality,

$$\big\langle A_{1,1}(\varepsilon, K)\big\rangle_{\mathbb{Q}} \le C_1 K \max_{\|x\|_\infty \le \varepsilon^{-1}} \big\|\varepsilon f_j(x)\big\|_{\mathbb{Q}} \sum_{k=1}^{d} \sum_{z\in\Lambda} \|G_{k,z}\|_{\mathbb{Q}}.$$

By Lemma 3.9 this expression vanishes as $\varepsilon \downarrow 0$. Hence, $A_{1,1}(\varepsilon, K)$ vanishes in $L^1(\mathbb{Q})$ as $\varepsilon \downarrow 0$ for every $K > 0$.

By Hölder inequality, applied to $G_{k,z}$, $\partial_z f_j 1\{|\partial_z f_j| > K\}$ and f_j with exponents d, 2 and $2d/(d-2)$, respectively, $A_{1,2}(\varepsilon, K)$ is bounded above by the sum of two similar terms, the first one being

$$C_1\varepsilon^d \sum_{k=1}^{d}\sum_{z\in\Lambda}\Big\{\sum_{x\in\varepsilon\mathbb{Z}^d} |G_{k,z}\circ\tau_{x\varepsilon^{-1}}|^d\Big\}^{1/d}\Big\{\sum_{x\in\varepsilon\mathbb{Z}^d} |\varepsilon f_j(x\varepsilon^{-1})|^{2d/(d-2)}\Big\}^{(d-2)/2d}$$
$$\times\Big\{\sum_{x\in\varepsilon\mathbb{Z}^d} \{\partial_z f_j(x\varepsilon^{-1})\}^2\mathbf{1}\{|\partial_z f_j(x\varepsilon^{-1})| > K\}\Big\}^{1/2}.$$

The second term in the decomposition of $A_{1,2}(\varepsilon, K)$ is obtained from the previous one by replacing in the second expression $f_j(x\varepsilon^{-1})$ by $f_j(x\varepsilon^{-1}+z)$. Since both terms are estimated in the same way, we just present the details for the first one. Note that assumption (3.21) on the components $H_{k,z}$ of the local drift V appears now clearly.

Since $G_{k,z}$ belongs to $L^d(\mathbb{Q})$ and $\partial_z f_j$ to $L^2(\mathbb{Q})$, by Lemma 3.10 and the ergodic theorem, $\mathbb{Q}$-a.s.,

$$\limsup_{\varepsilon\to 0}\big|A_{1,2}(\varepsilon, K)\big| \le C_1 L_* \sum_{k=1}^{d}\sum_{z\in\Lambda}\langle |G_{k,z}|^d\rangle_{\mathbb{Q}}^{1/d}\langle [f_j(z)]^2\mathbf{1}\{|f_j(z)| > K\}\rangle_{\mathbb{Q}}^{1/2}.$$

Since $\langle [f_j(z)]^2\mathbf{1}\{|f_j(z)| > K\}\rangle_{\mathbb{Q}}$ vanishes as $K \uparrow \infty$, $A_{1,2}(\varepsilon, K)$ converges in probability to 0 as $\varepsilon \downarrow 0$ and then $K \uparrow \infty$. This fact together with the convergence of $A_{1,1}(\varepsilon, K)$ to 0 in $L^1(\mathbb{Q})$ as $\varepsilon \downarrow 0$ for every $K > 0$ shows that $A_1(\varepsilon)$ vanishes in probability as $\varepsilon \downarrow 0$.

We finally examine the term $A_2(\varepsilon)$. Since $\partial_z(\partial_{x_k}\varphi)(\varepsilon x)$ is absolutely bounded by $C_1\varepsilon\mathbf{1}\{\|x\|_\infty \le \varepsilon^{-1}\}$ for some finite constant C_1 which depends only on φ and Λ,

$$\big|A_2(\varepsilon)\big| \le C_1\varepsilon^{d+2}\sum_{k=1}^{d}\sum_{z\in\Lambda}\sum_{\|x\|_\infty\le\varepsilon^{-1}} |G_{k,z}\circ\tau_x| f_j(x)^2.$$

For every $K > 0$, the previous sum is bounded by $A_{2,1}(\varepsilon, K) + A_{2,2}(\varepsilon, K)$, where

$$A_{2,1}(\varepsilon, K) = C_1 K\varepsilon^{d+2}\sum_{k=1}^{d}\sum_{z\in\Lambda}\sum_{\|x\|_\infty\le\varepsilon^{-1}} f_j(x)^2,$$

$$A_{2,2}(\varepsilon, K) = C_1\varepsilon^{d+2}\sum_{k=1}^{d}\sum_{z\in\Lambda}\sum_{\|x\|_\infty\le\varepsilon^{-1}} |G_{k,z}\circ\tau_x|\mathbf{1}\big\{|G_{k,z}\circ\tau_x| > K\big\} f_j(x)^2.$$

In view of Lemma 3.9, for every $K > 0$, the expectation of $A_{2,1}(\varepsilon, K)$ vanishes as $\varepsilon \downarrow 0$. By Hölder inequality, $A_{2,2}(\varepsilon, K)$ is bounded above by

$$C_1 \sum_{k=1}^{d} \sum_{z \in \Lambda} \Big\{ \varepsilon^d \sum_{\|x\|_\infty \le \varepsilon^{-1}} |G_{k,z} \circ \tau_x|^{d/2} \mathbf{1}\big\{ |G_{k,z} \circ \tau_x| > K \big\} \Big\}^{2/d} \\ \times \Big\{ \varepsilon^d \sum_{\|x\|_\infty \le \varepsilon^{-1}} \big[\varepsilon f_j(x)\big]^{2d/(d-2)} \Big\}^{(d-2)/d}.$$

By Lemma 3.10 and by the ergodic theorem, $\mathbb{Q}$-a.s., the limit sup of the previous sum as $\varepsilon \downarrow 0$ is bounded by

$$C_1 L_* \sum_{k=1}^{d} \sum_{z \in \Lambda} \big\langle |G_{k,z}|^{d/2} \mathbf{1}\big\{ |G_{k,z}| > K \big\} \big\rangle^{2/d}.$$

By assumption (3.22), this expression vanishes as $K \uparrow \infty$. This proves that $A_{2,2}(\varepsilon, K)$ converges in probability to 0 as $\varepsilon \downarrow 0$ and then $K \uparrow \infty$. Since $A_{2,1}(\varepsilon, K)$ converges to 0 in $L^1(\mathbb{Q})$ as $\varepsilon \downarrow 0$ for every $K > 0$, we deduce that $A_2(\varepsilon)$ converges in probability to 0 as $\varepsilon \downarrow 0$. This concludes the proof of the energy identity.

3.5 Random Walks in Mixing Environments

In this section, we show that the local drift V satisfies the assumptions (3.21), (3.22) if the random field $\{p_z \circ \tau_x : x \in \mathbb{Z}^d\}$ is sufficiently strongly mixing for each $z \in \Lambda$.

Denote by d the distance on $\mathbb{Z}^d$ defined by $d(x, y) = \sum_{1 \le i \le d} |x_i - y_i|$, x, $y \in \mathbb{Z}^d$. We extend this notion to sets in the natural way. For $x \in \mathbb{Z}^d$ and subsets Γ, Γ_1, Γ_2 of $\mathbb{Z}^d$, $d(x, \Gamma)$ stands for the distance from x to Γ defined by $d(x, \Gamma) = \min_{y \in \Gamma} d(x, y)$, and $d(\Gamma_1, \Gamma_2)$ for the distance between Γ_1 and Γ_2, defined by $d(\Gamma_1, \Gamma_2) = \min_{x \in \Gamma_1, y \in \Gamma_2} d(x, y)$.

For an arbitrary finite subset Γ of $\mathbb{Z}^d$ and a positive integer m, denote by Γ_m^c the set of sites which are a distance at least m from Γ: $\Gamma_m^c = \{x \in \mathbb{Z}^d : d(x, \Gamma) \ge m\}$. For a subset A of $\mathbb{Z}^d$ and $z \in \Lambda$, denote by $\mathscr{F}_z(A)$ the σ-algebra generated by $p_z \circ \tau_x$ for $x \in A$. For $m \ge 1$, a finite subset Γ of $\mathbb{Z}^d$ and $z \in \Lambda$, let

$$\alpha_m(\Gamma; p_z) := \sup\big\{ \big|\mathbb{Q}(A \cap B) - \mathbb{Q}(A)\mathbb{Q}(B)\big| : A \in \mathscr{F}_z(\Gamma), B \in \mathscr{F}_z\big(\Gamma_m^c\big) \big\}.$$

The α-mixing coefficients of the random field $\{p_z \circ \tau_x : x \in \mathbb{Z}^d\}$ are defined as $\alpha_m(p_z) := \sup_\Gamma \alpha_m(\Gamma; p_z)$, where the supremum is carried over all finite subsets Γ of $\mathbb{Z}^d$. The following result which allows to control the covariance of random variables in terms of the mixing coefficient is Theorem 17.2.1 of Ibragimov and Linnik (1971), p. 306 and Lemma 3 in p. 10 of Doukhan (1994).

Lemma 3.12 *Suppose that Γ_1, Γ_2 are finite subsets of $\mathbb{Z}^d$ such that $d(\Gamma_1, \Gamma_2) \geq m$ for some integer $m \geq 1$. Let X_1, X_2 be bounded random variables which are $\mathscr{F}_z(\Gamma_1), \mathscr{F}_z(\Gamma_2)$-measurable, respectively. Then,*

$$\left|\langle X_1 X_2\rangle_{\mathbb{Q}} - \langle X_1\rangle_{\mathbb{Q}}\langle X_2\rangle_{\mathbb{Q}}\right| \leq 4\|X_1\|_\infty \|X_2\|_\infty \alpha_m(p_z).$$

One can easily generalize this result to expectations of an arbitrary number of random variables.

Lemma 3.13 *Suppose that Γ_i, $1 \leq i \leq n$, are finite subsets of $\mathbb{Z}^d$ such that* $\min_{i\neq j}$ *$d(\Gamma_i, \Gamma_j) \geq m$ for some integer $m \geq 1$. Let X_i, $1 \leq i \leq n$, be bounded random variables and assume that X_i is $\mathscr{F}_z(\Gamma_i)$-measurable, $1 \leq i \leq n$. Then,*

$$\left|\left\langle \prod_{i=1}^n X_i \right\rangle_{\mathbb{Q}} - \prod_{i=1}^n \langle X_i\rangle_{\mathbb{Q}}\right| \leq 8n\alpha_m(p_z)\prod_{i=1}^n \|X_i\|_\infty.$$

Proof We prove this result by induction on n. For $n = 2$ it is Lemma 3.12. Suppose that we have shown it for a certain $n \geq 2$. Fix an integer $m \geq 1$ and consider a family Γ_i, $1 \leq i \leq n+1$, of finite subsets of $\mathbb{Z}^d$ such that $\min_{i\neq j} d(\Gamma_i, \Gamma_j) \geq m$. Let X_i, $1 \leq i \leq n+1$, be bounded random variables, X_i being $\mathscr{F}_z(\Gamma_i)$-measurable, $1 \leq i \leq n+1$. Let $\tilde{X}_{n+1} := X_{n+1} - \langle X_{n+1}\rangle_{\mathbb{Q}}$. According to Lemma 3.12,

$$\left|\left\langle \tilde{X}_{n+1} \prod_{i=1}^n X_i \right\rangle_{\mathbb{Q}}\right| \leq 4\alpha_m(p_z)\|\tilde{X}_{n+1}\|_\infty \prod_{i=1}^n \|X_i\|_\infty.$$

The assertion now holds for $n+1$ since

$$\begin{aligned}
&\left|\left\langle \prod_{i=1}^{n+1} X_i \right\rangle_{\mathbb{Q}} - \prod_{i=1}^{n+1} \langle X_i\rangle_{\mathbb{Q}}\right| \\
&\quad \leq \left|\left\langle \tilde{X}_{n+1} \prod_{i=1}^n X_i \right\rangle_{\mathbb{Q}}\right| + \|X_{n+1}\|_\infty \left|\left\langle \prod_{i=1}^n X_i \right\rangle_{\mathbb{Q}} - \prod_{i=1}^n \langle X_i\rangle_{\mathbb{Q}}\right| \\
&\quad \leq 4\alpha_m(p_z)\|\tilde{X}_{n+1}\|_\infty \prod_{i=1}^n \|X_i\|_\infty + 8n\alpha_m(p_z)\prod_{i=1}^{n+1} \|X_i\|_\infty
\end{aligned}$$

in virtue of the previous estimate and the induction hypothesis. This proves the lemma since $\|\tilde{X}_{n+1}\|_\infty \leq 2\|X_{n+1}\|_\infty$. □

Theorem 3.14 *Assume that $d \geq 3$ and fix an integer $n \geq 1$. Assume that the random field $\{p_z \circ \tau_x : x \in \mathbb{Z}^d\}$ is such that*

$$\limsup_{m\to+\infty} m^N \alpha_m(p_z) < \infty$$

for $N \geq [2dn/(d-2)]+1$, where $[a]$ represents the integer part of a. Then, p_z can be represented as

$$p_z - \langle p_z \rangle_{\mathbb{Q}} = \sum_{j=1}^{d} D^*_{e_j} H_j,$$

where H_i belongs to $L^{2n}(\mathbb{Q})$, $1 \leq i \leq d$.

This theorem will be proved at the end of this section. It is easy from this result to rewrite $p_z - \langle p_z \rangle_{\mathbb{Q}}$ as $\sum_{y \in \Lambda} D^*_y(p_y G_y)$ for functions G_y in $L^{2n}(\mathbb{Q})$, $y \in \Lambda$, as required in assumption (3.22). An elementary calculation, using proper paths from the origin to $-e_j$ and similar to the one which led to formula (3.25), shows that

$$p_z - \langle p_z \rangle_{\mathbb{Q}} = \sum_{y \in \Lambda} D^*_y(p_y G_y),$$

where

$$G_y = -\frac{1}{p_y} \sum_{j=1}^{d} \sum_{k=0}^{n_j - 1} \mathbf{1}\{x^j_{k+1} - x^j_k = y\}\{H_j \circ \tau_{x^j_{k+1}}\},$$

where $x^j_0, \dots, x^j_{n_j}$ are proper paths connecting the origin to $-e_j$, $1 \leq j \leq d$. Note, furthermore, that we need $n \geq d/2$ to fulfill assumption (3.22).

Fix a dimension $d \geq 3$ and let $\{X_t : t \geq 0\}$ be a symmetric, simple random walk on $\mathbb{Z}^d$. Denote by $q_t(x, y) = q_t(y - x) = q(t, y - x)$ its transition probability and by $G(x, y) = G(y - x)$ its Green function:

$$G(x) = \int_0^\infty q(t, x)\, dt.$$

Fix $z \in \Lambda$, $m \geq 1$, $1 \leq j \leq d$ and let

$$G^{(m)}_j(\omega) := \sum_{\|x\|_\infty \leq m} (\partial_{e_j} G)(x)[\tilde{p}_z \circ \tau_x],$$

where $\tilde{p}_z = p_z - \langle p_z \rangle_{\mathbb{Q}}$. The proof of Theorem 3.14 relies on the next lemma.

Lemma 3.15 *Suppose that a random field $\{p_z \circ \tau_x : x \in \mathbb{Z}^d\}$ satisfies the assumptions of Theorem 3.14. Then the sequence $\{G^{(m)}_j : m \geq 1\}$ converges in $L^{2n}(\mathbb{Q})$.*

Proof Consider a random field $\{p_z \circ \tau_x : x \in \mathbb{Z}^d\}$ which satisfies the assumptions of Theorem 3.14. Let $\tilde{p}(x, x+z) = p_z \circ \tau_x - \langle p_z \rangle_{\mathbb{Q}}$. To prove the lemma, it suffices to show that for any $\varepsilon > 0$ there exists M such that

$$\int_0^\infty \Big\| \sum_{m_1 \leq \|x\|_\infty < m_2} (\partial_{e_j} q)(t, x) \tilde{p}(x, x+z) \Big\|_{L^{2n}(\mathbb{Q})} dt < \varepsilon$$

for all $m_2 > m_1 \geq M$.

An elementary computation shows that

$$\Big\| \sum_{m_1 \le \|x\|_\infty < m_2} (\partial_{e_j} q)(t,x)\tilde{p}(x,x+z) \Big\|_{L^{2n}(\mathbb{Q})}^{2n}$$
$$= \sum_{x_1,\dots,x_{2n}} \prod_{i=1}^{2n} (\partial_{e_j} q)(t,x_i) \Big\langle \prod_{i=1}^{2n} \tilde{p}(x_i,x_i+z) \Big\rangle_{\mathbb{Q}},$$

where the sum is carried over all $x_1,\dots,x_{2n}$ such that $m_1 \le \|x_i\|_\infty < m_2$, $1 \le i \le 2n$.

Fix ρ in the interval $(n/N, [d-2]/2d)$. This is possible since $2dn/(d-2) < N$. To apply Lemma 3.13, we partition the elements of a sequence $(x_1,\dots,x_{2n})$ into subclasses which are at distance t^ρ from each other. To describe such a partition we introduce some terminology. Consider the equivalence relation which identifies sequences which differ only by a permutation. For a given sequence $(y_1,\dots,y_{2n})$ denote by $\{y_1,\dots,y_{2n}\}$ the corresponding equivalence class and call it *a set of elements with repetitions*. One can easily define the union of any two sets with repetitions. By *a maximal t^ρ-partition* $\pi\{x_1,\dots,x_{2n}\}$ of a given set with repetitions $\{x_1,\dots,x_{2n}\}$ we understand a family of subsets with repetitions $\Gamma_1,\dots,\Gamma_k$ such that

(i) $\bigcup_{1\le i\le k} \Gamma_i = \{x_1,\dots,x_{2n}\}$,
(ii) $d(\Gamma_i,\Gamma_j) \ge t^\rho$ for $1 \le i \ne j \le k$,
(iii) for any $1 \le i \le k$, Γ_i cannot be decomposed as the union of two non-empty sets which are at distances greater than or equal to t^ρ. In other words, if for some $1 \le i \le k$, $\Gamma_i = A \cup B$ where $d(A,B) \ge t^\rho$, then either $A = \varnothing$ or $B = \varnothing$.

With this notation, we may rewrite the right-hand side of the previous displayed formula as

$$\sum_{\Gamma_1,\dots,\Gamma_k} A(\Gamma_1,\dots,\Gamma_k) \prod_{i=1}^{k} \prod_{x\in\Gamma_i} (\partial_{e_j} q)(t,x) \Big\langle \prod_{i=1}^{k} \prod_{x\in\Gamma_i} \tilde{p}(x,x+z) \Big\rangle_{\mathbb{Q}},$$

where the summation $\sum_{\Gamma_1,\dots,\Gamma_k}$ extends over of all maximal t^ρ-partitions of $2n$ sites taken from the set $\{x \in \mathbb{Z}^d : m_1 \le \|x\|_\infty < m_2\}$ and where $A(\Gamma_1,\dots,\Gamma_k)$ stands for the number of vectors $(x_1,\dots,x_{2n})$ whose maximal t^ρ-partition is $\Gamma_1,\dots,\Gamma_k$. By Lemmas 3.13 and 3.22, the absolute value of the previous expression is less than or equal to

$$\frac{C_*^{2n}}{(t+1)^n} \sum_{\Gamma_1,\dots,\Gamma_k} A(\Gamma_1,\dots,\Gamma_k) \prod_{i=1}^{k} \prod_{x\in\Gamma_i} q(c_* t,x) \Big| \prod_{i=1}^{k} \Big\langle \prod_{x\in\Gamma_i} \tilde{p}(x,x+z) \Big\rangle_{\mathbb{Q}} \Big|$$
$$+ \frac{16n\alpha_{t^\rho}(p_z)(C_*\|p_z\|_\infty)^{2n}}{(t+1)^n} \sum_{\Gamma_1,\dots,\Gamma_k} A(\Gamma_1,\dots,\Gamma_k) \prod_{i=1}^{k} \prod_{x\in\Gamma_i} q(c_* t,x)$$

for some finite constants C_*, $c_* > 0$ whose value may change from line to line.

Denote the first and the second term by $E_1(t)$ and $E_2(t)$, respectively. We can exclude from the sum appearing in $E_1(t)$ those partitions where at least one Γ_i is a singleton. Hence,

$$E_1(t) \le \frac{(C_* \|p_z\|_\infty)^{2n}}{(t+1)^n} \sum_{x_1,\dots,x_{2n}} \prod_{i=1}^{2n} q(c_*t, x_i),$$

where the summation extends over all sites $x_1, \dots, x_{2n}$ such that $m_1 \le \|x_i\|_\infty < m_2$, $1 \le i \le 2n$, and such that $\pi\{x_1, \dots, x_{2n}\}$ contains no singleton.

Suppose that the maximal t^ρ-partition of $\{x_1, \dots, x_{2n}\}$ has k, $1 \le k \le n$, sets. In this case, we may bound the previous sum by

$$\sum_{x_1,\dots,x_k} \sum_{y_1,\dots,y_k} \sum_{z_1,\dots,z_{2(n-k)}} \prod_{i=1}^{k} q(c_*t, x_i) \prod_{i=1}^{k} q(c_*t, y_i) \prod_{i=1}^{2(n-k)} q(c_*t, z_i),$$

where the first sum is carried over all sites $x_1, \dots, x_k$ such that $m_1 \le \|x_i\|_\infty < m_2$, the second one extends over all sites $y_1, \dots, y_k$ such that $d(y_i, x_i) \le t^\rho$ and the third one is performed over all sites $z_1, \dots, z_{2(n-k)}$ such that $d(z_i, \{x_1, \dots, x_k\}) \le 2nt^\rho$, $m_1 \le \|z_i\|_\infty < m_2$. Recall from Lemma 3.21 that $q(t,x) \le C_0(1+t)^{-d/2}$ for some finite constant C_0. Apply this estimate to all terms $q(c_*t, y_i)$ and to half of the terms $q(c_*t, z_i)$ to bound the previous expression by

$$\frac{C_0^n (nt^\rho)^{dn}}{(1+t)^{nd/2}} \sum_{x_1,\dots,x_k} \sum_{z_1,\dots,z_{(n-k)}} \prod_{i=1}^{k} q(c_*t, x_i) \prod_{i=1}^{(n-k)} q(c_*t, z_i).$$

If we just keep the restrictions $\|x_i\|_\infty \ge m_1$, $\|z_j\|_\infty \ge m_1$, the previous expression is bounded by

$$\frac{C_0^n (nt^\rho)^{dn}}{(1+t)^{nd/2}} P\big[\|X_{c_*t}\|_\infty \ge m_1\big]^n.$$

Summing over all possible sets in the maximal t^ρ-partition we conclude that

$$E_1(t) \le \frac{(C_* \|p_z\|_\infty n^{d/2} t^{d\rho/2})^{2n}}{(t+1)^{n+nd/2}} P\big[\|X_{c_*t}\|_\infty \ge m_1\big]^n$$

so that for every $a > 0$,

$$\begin{aligned}
&\int_0^\infty E_1(t)^{1/(2n)}\, dt \\
&\quad \le C_0 n^{d/2} \|p_z\|_\infty \int_a^\infty \frac{t^{d\rho/2}}{(t+1)^{1/2+d/4}}\, dt \\
&\qquad + C_0 n^{d/2} \|p_z\|_\infty \sup_{t\in[0,a]} P\big[\|X_{c_*t}\|_\infty \ge m_1\big]^{1/2} \int_0^a \frac{t^{d\rho/2}}{(t+1)^{1/2+d/4}}\, dt,
\end{aligned}$$

for some finite constant C_0 which does not depend on t. Since $\rho < (d-2)/2d$ we have that $(1/2)+(d/4)-(d\rho)/2 > 1$. Choosing a sufficiently large $a > 0$ and then $M > 0$ we get

$$\int_0^\infty E_1(t)^{1/(2n)}\,dt < (\varepsilon/2)$$

for all $m_1 \geq M$.

On the other hand, since we assumed the random field $p_z \circ \tau_x$ to satisfy the assumptions of Theorem 3.14, $\alpha_{t^\rho}(p_z) \leq C_0(t+1)^{-N\rho}$ for some finite constant C_0. We can therefore estimate $E_2(t)$ by

$$\begin{aligned}
&\frac{C_0 n (C_*\|p_z\|_\infty)^{2n}}{(t+1)^{n+N\rho}} \sum_{\Gamma_1,\dots,\Gamma_k} A(\Gamma_1,\dots,\Gamma_k)\prod_{i=1}^{k}\prod_{x\in\Gamma_i} q(c_*t,x)\\
&\quad\leq \frac{C_0 n (C_*\|p_z\|_\infty)^{2n}}{(t+1)^{n+N\rho}} P\big[\|X_{c_*t}\|_\infty \geq m_1\big]^{2n}
\end{aligned}$$

for some finite constant C_0 independent of t, n, ρ and N. Therefore, for every $a > 0$,

$$\begin{aligned}
&\int_0^\infty E_2(t)^{1/(2n)}\,dt\\
&\quad\leq C_0\|p_z\|_\infty \int_a^\infty \frac{1}{(t+1)^{1/2+N\rho/(2n)}}\,dt\\
&\qquad + C_0\|p_z\|_\infty \sup_{t\in[0,a]} P\big[\|X_{c_*t}\|_\infty \geq m_1\big]\int_0^a \frac{1}{(t+1)^{1/2+N\rho/(2n)}}\,dt.
\end{aligned}$$

Since $N\rho > n$, $(1/2)+N\rho/(2n) > 1$. Choose a large and then M large so that

$$\int_0^\infty E_2(t)^{1/(2n)}\,dt < (\varepsilon/2)$$

for all $m_1 \geq M$. This proves the lemma. □

We turn now to the

Proof of Theorem 3.14 Recall the definition of $G_j^{(m)}$ introduced just after the statement of the theorem and let

$$G_j := \lim_{m\to\infty} G_j^{(m)} = \lim_{m\to\infty} \sum_{\|x\|_\infty \leq m} (\partial_{e_j} G)(x)\tilde{p}(x,x+z),$$

which exists in $L^2(\mathbb{Q})$ in view of the previous lemma. A simple calculation shows that

$$D_{e_j} G_j = \lim_{m\to\infty} D_{e_j} G_j^{(m)} = \lim_{m\to\infty} \sum_{\|x\|_\infty \leq m} \partial_{e_j} G(x)\partial_{e_j}\tilde{p}(x,x+z).$$

The limit exists because $G_j^{(m)}$ converges to G_j in $L^2(\mathbb{Q})$. Perform a summation by parts and observe that the boundary terms vanish in the limit by the arguments presented in the Proof of Lemma 3.15. Therefore,

$$D_{e_j} G_j = \lim_{m\to\infty} \sum_{\|x\|_\infty \le m} \big(\partial^*_{e_j} \partial_{e_j} G\big)(x) \tilde{p}(x, x+z).$$

Since $\sum_{j=1}^d \partial^*_{e_j} \partial_{e_j} g(x) = 2d\delta_{x,0}$ we obtain that

$$\sum_{j=1}^d D_{e_j} G_j = 2d\big\{p_z - \langle p_z\rangle_{\mathbb{Q}}\big\}$$

which proves the theorem with $H_j := -(1/2d)(G_j \circ \tau_{e_j})$. □

Corollary 3.16 *Suppose that $d \ge 3$ and that the random field $\{p_z : z \in \mathbb{Z}^d\}$ satisfies conditions* (**H1**), (**H2**), (**H4**), (3.20) *and the assumption of Theorem* 3.14 *with $n \ge d/2$. Then, the corresponding random walk $\{X^\omega(t) : t \ge 0\}$ satisfies the central limit theorem as stated in Theorem* 3.4.

3.6 Doubly Stochastic Random Walks in Dimension $d = 1$

As the title suggests, in this section, we examine, doubly stochastic random walks in dimension $d = 1$ satisfying assumptions (**H1**)–(**H4**). The main result states that the generator of the environment process associated to such randoms walks satisfies a sector condition provided the local drift has zero mean with respect to the ergodic measure $\mathbb{Q}$ and the random rates satisfy an elliptic condition (3.20). This statement follows from the fact, presented in Theorem 3.17 below, that in dimension 1 the generator of the environment process associated to a doubly stochastic random walk can be written as the sum of three pieces: its symmetric part; an operator in divergence form with bounded coefficients; and an operator which vanishes if the expectation of the local drift vanishes.

We start with the decomposition of the generator of the environment process.

Theorem 3.17 *Suppose that the field $\{p_z : z \in \mathbb{Z}^d\}$ satisfies conditions* (**H1**)–(**H4**). *Then, there exist a non-random finite set Λ_0 and a random field $\{a_{y,z} : y, z \in \Lambda_0\}$, $a_{y,z}$ in $B(\Omega)$, such that the generator L of the process $\{\eta(t) : t \ge 0\}$ satisfies*

$$Lf = Sf + \langle V\rangle_{\mathbb{Q}} D_1 f + \sum_{y,z\in\Lambda_0} D^*_y(a_{y,z} D_z f)$$

for all f in $L^2(\mathbb{Q})$, where S is the symmetric part of the generator L given by (3.10) *and D_z, D^*_y are the operators defined by* (3.9).

Proof Since the generator L can be written as $S + A$, we need to show that the anti-symmetric part A is given by the sum of the second and third terms on the right-hand side of the identity appearing in the statement of the theorem. Recall the explicit form (3.11) of the anti-symmetric part of the generator L. By interpolating the differences $f(\tau_z\omega) - f(\omega)$, we may write Af as

$$Af = (1/2)\sum_{z\in\mathbb{Z}} q_z \sum_{j=0}^{|z|-1} [D_{\mathfrak{s}(z)} f] \circ \tau_{\mathfrak{s}(z)j}, \tag{3.39}$$

where $\mathfrak{s}(z)$ stands for the sign of z.

Fix f, g in $L^2(\mathbb{Q})$. In view of the previous formula for A, since the transformations $\{\tau_x\}$ preserve the measure $\mathbb{Q}$,

$$\langle Af, g\rangle_{\mathbb{Q}} = (1/2)\sum_{z\in\mathbb{Z}} \sum_{j=0}^{|z|-1} \big\langle (q_z g) \circ \tau_{-\mathfrak{s}(z)j}, D_{\mathfrak{s}(z)} f\big\rangle_{\mathbb{Q}}.$$

Rewriting $(q_z g) \circ \tau_{-\mathfrak{s}(z)j}$ as $[D_{-\mathfrak{s}(z)j} g][q_z \circ \tau_{-\mathfrak{s}(z)j}] + [q_z \circ \tau_{-\mathfrak{s}(z)j}]g$, we obtain that

$$\langle Af, g\rangle_{\mathbb{Q}} = (1/2)\sum_{z\in\mathbb{Z}} \sum_{j=0}^{|z|-1} \big\langle [q_z \circ \tau_{-\mathfrak{s}(z)j}] D_{\mathfrak{s}(z)} f, g\big\rangle_{\mathbb{Q}} + \mathscr{A}_0(\mathscr{D} f, \mathscr{D} g),$$

where $\mathscr{A}_0$ is the bilinear form of the gradients $\mathscr{D} f = \{D_z f : z \in \mathbb{Z}\}$ and $\mathscr{D} g := \{D_z g : z \in \mathbb{Z}\}$ given by

$$\mathscr{A}_0(\mathscr{D} f, \mathscr{D} g) := (1/2)\sum_{z\in\mathbb{Z}} \sum_{j=0}^{|z|-1} \big\langle [q_z \circ \tau_{-\mathfrak{s}(z)j}] D_{\mathfrak{s}(z)} f, D_{-\mathfrak{s}(z)j} g\big\rangle_{\mathbb{Q}}.$$

To symmetrize the expression for $\langle Af, g\rangle_{\mathbb{Q}}$, we change variables in (3.39) setting $z' = -z$ and obtain that

$$\begin{aligned}\langle Af, g\rangle_{\mathbb{Q}} &= (1/4)\sum_{z\in\mathbb{Z}} \sum_{j=0}^{|z|-1} \big\langle [q_z \circ \tau_{-\mathfrak{s}(z)j}] D_{\mathfrak{s}(z)} f, g\big\rangle_{\mathbb{Q}} \\ &\quad + (1/4)\sum_{z\in\mathbb{Z}} \sum_{j=0}^{|z|-1} \big\langle [q_{-z} \circ \tau_{\mathfrak{s}(z)j}] D_{-\mathfrak{s}(z)} f, g\big\rangle_{\mathbb{Q}} + \mathscr{A}_1(\mathscr{D} f, \mathscr{D} g),\end{aligned}$$

where $\mathscr{A}_1$ is the analogous symmetrization of $\mathscr{A}_0$.

Since $D_{-\mathfrak{s}(z)} = -D_{\mathfrak{s}(z)} \circ \tau_{-\mathfrak{s}(z)}$ and since the operators $\{\tau_y\}$ preserve the measure $\mathbb{Q}$, we can rewrite the second expression on the right-hand side as

$$-(1/4)\sum_{z\in\mathbb{Z}} \sum_{j=0}^{|z|-1} \big\langle [q_{-z} \circ \tau_{\mathfrak{s}(z)(j+1)}] D_{\mathfrak{s}(z)} f, g\big\rangle_{\mathbb{Q}}$$

$$- (1/4) \sum_{z\in\mathbb{Z}} \sum_{j=0}^{|z|-1} \big\langle [q_{-z} \circ \tau_{\mathfrak{s}(z)(j+1)}] D_{\mathfrak{s}(z)} f, D_{\mathfrak{s}(z)} g \big\rangle_{\mathbb{Q}} .$$

We denote the second term by $\mathscr{A}_2(\mathscr{D} f, \mathscr{D} g)$.

Recall that $q_z = p_z - p_{-z} \circ \tau_z$ so that $q_z \circ \tau_{-\mathfrak{s}(z)j} - q_{-z} \circ \tau_{\mathfrak{s}(z)(j+1)} = r_+(z) - r_-(z)$, where

$$r_+(z) := p_z \circ \tau_{-\mathfrak{s}(z)j} + p_z \circ \tau_{-\mathfrak{s}(z)(|z|-j-1)},$$
$$r_-(z) := p_{-z} \circ \tau_{\mathfrak{s}(z)(j+1)} + p_{-z} \circ \tau_{\mathfrak{s}(z)(|z|-j)} .$$

With this notation and in view of the previous formulas, we have that

$$\langle Af, g\rangle_{\mathbb{Q}} = (1/4) \sum_{z\in\mathbb{Z}} \sum_{j=0}^{|z|-1} \big\langle \big[r_+(z) - r_-(z)\big] D_{\mathfrak{s}(z)} f, g \big\rangle_{\mathbb{Q}} + \mathscr{A}_3(\mathscr{D} f, \mathscr{D} g),$$

where $\mathscr{A}_3 = \mathscr{A}_1 + \mathscr{A}_2$.

Performing the change of variables $z' = -z$, since $D_{-\mathfrak{s}(z)} f = -(D_{\mathfrak{s}(z)} f) \circ \tau_{-\mathfrak{s}(z)}$, we obtain that

$$\sum_{z\in\mathbb{Z}} \sum_{j=0}^{|z|-1} \big\langle r_-(z) D_{\mathfrak{s}(z)} f, g \big\rangle_{\mathbb{Q}} = - \sum_{z\in\mathbb{Z}} \sum_{j=0}^{|z|-1} \big\langle \big[r_-(-z) \circ \tau_{\mathfrak{s}(z)}\big] D_{\mathfrak{s}(z)} f, g \big\rangle_{\mathbb{Q}}$$
$$- \sum_{z\in\mathbb{Z}} \sum_{j=0}^{|z|-1} \big\langle \big[r_-(-z) \circ \tau_{\mathfrak{s}(z)}\big] D_{\mathfrak{s}(z)} f, D_{\mathfrak{s}(z)} g \big\rangle_{\mathbb{Q}} .$$

Denote the second term on the right-hand side by $\mathscr{A}_4(\mathscr{D} f, \mathscr{D} g)$ and remark that $r_-(-z) \circ \tau_{\mathfrak{s}(z)} = r_+(z)$. Therefore,

$$\langle Af, g\rangle_{\mathbb{Q}} = (1/2) \sum_{z\in\mathbb{Z}} \sum_{j=0}^{|z|-1} \big\langle r_+(z) D_{\mathfrak{s}(z)} f, g \big\rangle_{\mathbb{Q}} + \mathscr{A}_5(\mathscr{D} f, \mathscr{D} g),$$

where $\mathscr{A}_5 = \mathscr{A}_3 + \mathscr{A}_4$.

It remains to replace $D_{-1} f$ by $D_1 f$ for $z < 0$. Using once again that $D_{-1} f = -[D_1 f] \circ \tau_{-1}$, we obtain that

$$\sum_{z<0} \sum_{j=0}^{|z|-1} \big\langle r_+(z) D_{-1} f, g \big\rangle_{\mathbb{Q}} = - \sum_{z<0} \sum_{j=0}^{|z|-1} \big\langle \big[r_+(z) \circ \tau_1\big] D_1 f, g \big\rangle_{\mathbb{Q}}$$
$$- \sum_{z<0} \sum_{j=0}^{|z|-1} \big\langle \big[r_+(z) \circ \tau_1\big] D_1 f, D_1 g \big\rangle_{\mathbb{Q}} .$$

Denote the second term on the right-hand side by $\mathscr{A}_6(\mathscr{D}f, \mathscr{D}g)$ so that

$$\langle Af, g\rangle_{\mathbb{Q}} = \langle RD_1 f, g\rangle_{\mathbb{Q}} + \mathscr{A}_7(\mathscr{D}f, \mathscr{D}g),$$

where $\mathscr{A}_7 = \mathscr{A}_5 + \mathscr{A}_6$ and

$$R = (1/2)\sum_{z>0}\sum_{j=0}^{z-1} r_+(z) - (1/2)\sum_{z<0}\sum_{j=0}^{-z-1} r_+(z)\circ\tau_1 .$$

We claim that R is constant. Indeed,

$$\begin{aligned} D_1^* R &= (1/2)\sum_{z>0}\sum_{j=0}^{z-1} D_1^* r_+(z) - (1/2)\sum_{z<0}\sum_{j=0}^{-z-1} D_1^*\big[r_+(z)\circ\tau_1\big] \\ &= \sum_{z>0}\{p_z\circ\tau_{-z} - p_z\} - \sum_{z<0}\{p_z - p_z\circ\tau_{-z}\}. \end{aligned}$$

By the double stochasticity hypothesis, this expression vanishes. Thus, $R\circ\tau_{-1} = R$, $\mathbb{Q}$-a.s. By the ergodicity of $\mathbb{Q}$, R is constant so that $R = \langle R\rangle_{\mathbb{Q}}$. Since $\langle r_+(z)\rangle_{\mathbb{Q}} = 2\langle p_z\rangle_{\mathbb{Q}}$, the expectation of R is seen to be equal to $\sum_z z\langle p_z\rangle_{\mathbb{Q}}$, which is the expected value of the local drift V. In conclusion,

$$\langle Af, g\rangle_{\mathbb{Q}} = \langle R\rangle_{\mathbb{Q}}\langle D_1 f, g\rangle_{\mathbb{Q}} + \mathscr{A}_7(\mathscr{D}f, \mathscr{D}g).$$

The bilinear form $\mathscr{A}_7$ can be computed explicitly, but its exact expression does not play any role. In any case, there is a finite non-random set Λ_0 and bounded functions $\{a_{y,z} : y, z\in\Lambda_0\}$, $a_{y,z} : \Omega\to\mathbb{R}$, such that

$$\mathscr{A}_7(\mathscr{D}f, \mathscr{D}g) = \sum_{y,z\in\Lambda_0} \langle a_{y,z} D_z f, D_y g\rangle_{\mathbb{Q}}.$$

This proves that $Af = \langle R\rangle_{\mathbb{Q}} D_1 f + \sum_{y,z\in\Lambda_0} D_y^*[a_{y,z} D_z f]$, as claimed. □

Suppose that the random rates satisfy the *ellipticity condition* (3.20). A sector condition for the generator L follows from this ellipticity condition if the local drift V has mean zero with respect to $\mathbb{Q}$.

Corollary 3.18 *Under the hypotheses of Theorem* 3.17, *assume the ellipticity condition* (3.20) *and that the local drift V has mean zero with respect to $\mathbb{Q}$: $\langle V\rangle_{\mathbb{Q}} = 0$. Then the generator satisfies a sector condition* (2.36).

Proof The proof is straightforward. Use the ellipticity assumption (3.20), the boundedness of the coefficients $\{a_{y,z}\}$ and the elementary inequality $\langle (D_{x+y} f)^2\rangle_{\mathbb{Q}} \le 2\langle (D_x f)^2\rangle_{\mathbb{Q}} + 2\langle (D_y f)^2\rangle_{\mathbb{Q}}$ to estimate $\sum_{y,z\in\Lambda_0}\langle a_{y,z} D_z f, D_y g\rangle_{\mathbb{Q}}$ in terms of the Dirichlet forms of f and g. □

Up to this point we have proved that if the local drift has mean zero, conditions (3.14) and (3.17) are in force. It remains to check that V belongs to $\mathscr{H}_{-1}$ to be in a position to apply Theorem 3.4.

We derive an alternative formula for the local drift V to show that it belongs to $\mathscr{H}_{-1}$. We claim that

$$V(\omega) = -(1/2)\sum_{z\in\mathbb{Z}} z D_z^* p_z(\omega) + \sum_{y,z\in\Lambda_0} z D_y^* a_{y,z}(\omega). \tag{3.40}$$

Formula (3.40) follows from the proof of Theorem 3.17. Rewrite V as the convex combination of its symmetric and anti-symmetric parts:

$$V = (1/2)\sum_{z\in\mathbb{Z}} z\{p_z + p_{-z}\circ\tau_z\} + (1/2)\sum_{z\in\mathbb{Z}} z\{p_z - p_{-z}\circ\tau_z\}.$$

The first piece corresponds to the first term on the right-hand side of (3.40): For a function g in $L^2(\mathbb{Q})$, performing the obvious change of variables we obtain that

$$\Big\langle (1/2)\sum_{z\in\mathbb{Z}} z\{p_z + p_{-z}\circ\tau_z\}, g\Big\rangle_{\mathbb{Q}} = -(1/2)\sum_{z\in\mathbb{Z}} z\langle p_z, D_z g\rangle_{\mathbb{Q}}.$$

To show that the sum of $z\{p_z - p_{-z}\circ\tau_z\}$ corresponds to the second term on the right-hand side of (3.40), we have to proceed as in the proof of Theorem 3.17. Details are left to the reader.

The representation of the local drift as (3.40) permits to prove without difficulty that it belongs to $\mathscr{H}_{-1}$: For any function g in $L^2(\mathbb{Q})$,

$$\langle V, g\rangle_{\mathbb{Q}} = -(1/2)\sum_{z\in\mathbb{Z}} z\langle p_z, D_z g\rangle_{\mathbb{Q}} + \sum_{y,z\in\Lambda_0} z\langle a_{y,z}, D_y g\rangle_{\mathbb{Q}}.$$

It remains to recall the ellipticity assumption (3.20) and apply Schwarz inequality to obtain that $\langle V, g\rangle_{\mathbb{Q}}^2$ is bounded above by $C\|g\|_1^2$, for some finite constant C. This proves that V belongs to $\mathscr{H}_{-1}$.

We have just shown that all assumptions of Sect. 3.2 are in force so that

Theorem 3.19 *Under the hypotheses of Theorem* 3.17, *assume the ellipticity condition* (3.20) *and that the local drift V has mean zero with respect to $\mathbb{Q}$: $\langle V\rangle_{\mathbb{Q}} = 0$. Then, a central limit theorem for the random walk X_t^ω holds in $L^1(\mathbb{Q})$ with respect to the environment.*

We conclude this chapter by showing that the assumption that the local drift V has mean zero is not necessary if one uses the theory of bounded perturbations of normal operators developed in Sect. 2.7.5.

Let $V_0 = V - \langle V\rangle_{\mathbb{Q}}$. In view of (3.13), we have that

$$X^\omega(t) - \langle V\rangle_{\mathbb{Q}} t = \mathfrak{M}(t) + \int_0^t V_0\big(\eta(s)\big)\,ds.$$

Assume that $\langle V\rangle_{\mathbb{Q}} > 0$ and define the operator L_0 in $L^2(\mathbb{Q})$ by

$$L_0 f := -(1/2)D_1^* D_1 f + \langle V\rangle_{\mathbb{Q}} D_1 f$$

for f in $L^2(\mathbb{Q})$. Note that L_0 is a generator because $\langle V\rangle_{\mathbb{Q}} > 0$. Let B be the operator $L - L_0$ so that $L = L_0 + B$.

One can easily check that L_0 is a normal operator. Since $\langle V\rangle_{\mathbb{Q}} D_1$ is a generator, $\langle -L_0 f, f\rangle_{\mathbb{Q}} \geq (1/2)\langle [D_1 f]^2\rangle_{\mathbb{Q}}$. Conditions **(H1)**, **(H2)** and the fact that measure $\mathbb{Q}$ is shift invariant yield that $\langle -Lf, f\rangle_{\mathbb{Q}} = \langle -Sf, f\rangle_{\mathbb{Q}} \leq C_0 \langle [D_1 f]^2\rangle_{\mathbb{Q}}$ for some finite constant C_0. Therefore, $\langle -L_0 f, f\rangle_{\mathbb{Q}} \geq (1/2C_0)\langle -Lf, f\rangle_{\mathbb{Q}}$. On the other hand, by the ellipticity condition (3.20), $\langle -L_0 f, f\rangle_{\mathbb{Q}}$ is bounded above by $C_1\langle -Lf, f\rangle_{\mathbb{Q}}$ for some finite constant C_1. Condition (2.54) is thus in force. Theorem 3.17 provides an explicit formula for the operator B from which condition (2.56) follows. All hypotheses of Sect. 2.7.5 are therefore matched.

It remains to check that V_0 belongs to $\mathscr{H}_{-1}$. This follows from the representation of V_0 as

$$V_0(\omega) = -(1/2)\sum_{z\in\mathbb{Z}} z D_z^* p_z(\omega) + \sum_{y,z\in\Lambda_0} z D_y^* a_{y,z}(\omega),$$

which can be proved in the same way as (3.40) was derived. In conclusion, by Sect. 2.7.5.

Theorem 3.20 *Under the hypotheses of Theorem* 3.17, *assume the ellipticity condition* (3.20). *Then, a central limit theorem for the centered random walk* $\{X_t^\omega - \langle V\rangle_{\mathbb{Q}} t\}$ *holds in* $L^1(\mathbb{Q})$ *with respect to the environment.*

If $\langle V\rangle_{\mathbb{Q}} < 0$, since $D_1 = -D_{-1} - D_{-1}^* D_{-1}$, we may write L as $L_0' + B'$, where

$$L_0' f := -(1/2)D_1^* D_1 f - \langle V\rangle_{\mathbb{Q}} D_{-1} f$$

and $B' = B - \langle V\rangle_{\mathbb{Q}} D_{-1}^* D_{-1}$. In this case, L_0' is a generator and we may repeat the previous arguments.

3.7 Symmetric Random Walks

In this section, we present two results on the transition probability of symmetric random walks used in the chapter. By a symmetric, simple random walk on $\mathbb{Z}^d$ starting at x we understand a random sequence $\{X_n, n \geq 0\}$, defined over a probability space $(\Sigma, \mathscr{W}, \mathbb{Q})$, taking values on a d-dimensional integer lattice and such that $\mathbb{Q}[X_0 = x] = 1$, and

$$\mathbb{Q}[X_{n+1} = x_{n+1} | X_n = x_n, \ldots, X_0 = x_0] = \begin{cases} (2d)^{-1} & \text{if } \|x_n - x_{n+1}\|_\infty = 1, \\ 0 & \text{otherwise.} \end{cases}$$

In the particular case $d = 1$ we have

$$\mathbb{Q}[X_{2n} = 2y] = \frac{1}{2^{2n}}\binom{2n}{n-y}, \qquad \mathbb{Q}[X_{2n+1} = 2y+1] = \frac{1}{2^{2n+1}}\binom{2n+1}{n-y}, \tag{3.41}$$

for $y \in \mathbb{Z}$, where by convention $\binom{n}{m} = 0$ whenever $m > n$, or $m < 0$.

Suppose that $\{a_n, n \geq 1\}$, $\{b_n, n \geq 1\}$ are two sequences of positive numbers. We write $a_n \asymp b_n$ if

$$0 < \liminf_{n\to+\infty} a_n/b_n \leq \limsup_{n\to+\infty} a_n/b_n < +\infty.$$

By Stirling's formula for each $n \geq 1$ there exists $\theta \in (0, 1)$ such that $n! = \sqrt{2\pi n}$ $(ne^{-1})^n\, e^{\theta/(12n)}$. Therefore for any y such that $\|y\|_\infty < n$

$$\begin{aligned}
&\mathbb{Q}[X_{2n} = 2y] \\
&\asymp n^{-1/2}\left[1 - \left(\frac{y}{n}\right)^2\right]^{-1/2} \exp\left\{-n \log\left[1 - \left(\frac{y}{n}\right)^2\right] - y \log\frac{1 + y/n}{1 - y/n}\right\} \\
&= n^{-1/2}\left[1 - \left(\frac{y}{n}\right)^2\right]^{-1/2} \exp\left\{-n \sum_{p=1}^{+\infty} \frac{1}{p(2p-1)}\left(\frac{y}{n}\right)^{2p}\right\}.
\end{aligned}$$

A similar asymptotic formula holds for $\mathbb{Q}[X_{2n+1} = 2y + 1]$. As a result for any $\delta \in (0, 1)$ there exist $0 < c_U < C_U$, depending on δ, such that

$$\frac{c_U}{n^{1/2}} \exp\left\{-\frac{C_U x^2}{n}\right\} \leq \mathbb{Q}[X_n = x] \leq \frac{C_U}{n^{1/2}} \exp\left\{-\frac{c_U x^2}{n}\right\}, \tag{3.42}$$

provided that $\|x\|_\infty \leq (1 - \delta)n$ and $x - n \equiv 0 \bmod 2$. In addition, we can choose constants c_U, C_U in such a way that

$$c_U \exp\{-C_U n\} \leq \mathbb{Q}[X_n = x] \leq C_U \exp\{-c_U n\} \tag{3.43}$$

for $n > \|x\|_\infty > n(1 - \delta)$ and $x - n \equiv 0 \bmod 2$. From the upper bounds (3.42) and (3.43) we conclude that

$$\mathbb{Q}[X_n = x] \leq \frac{C_U}{n^{1/2}}, \quad \forall n \geq 1. \tag{3.44}$$

Suppose that $\{N_t, t \geq 0\}$ is a Poisson process with intensity 1, starting at 0, independent of the random walk $\{X_n, n \geq 0\}$. A symmetric, simple random walk in continuous time starting at x is defined as $X_t := X_{N_t}$, $t \geq 0$. Let $q(t, x) := \mathbb{Q}[X_t = x | X_0 = 0]$.

Lemma 3.21 *There exist $C_*, c_* > 0$ such that*

$$\frac{c_* \mathscr{E}(t, C_* \|x\|_\infty)}{(t+1)^{d/2}} \leq q(t, x) \leq \frac{C_* \mathscr{E}(t, c_* \|x\|_\infty)}{(t+1)^{d/2}}, \quad \forall t > 0,\ x \in \mathbb{Z}^d, \tag{3.45}$$

where

$$\mathscr{E}(t,r) := \exp\left\{-r\,\operatorname{arsinh}\left(\frac{r}{t}\right) - t\left(\sqrt{1+(r/t)^2}-1\right)\right\}, \qquad t,r>0.$$

Proof The upper bound. When $d=1$ we can write

$$q(t,x) = \sum_{k=|x|}^{+\infty} \frac{t^k e^{-t}}{k!} \mathbb{Q}[X_k = x], \quad \forall t \geq 0,\ x \in \mathbb{Z}^d. \tag{3.46}$$

Suppose that $|x| \leq 2t$. From (3.42) and (3.43) we obtain

$$q(t,x) \leq C_U \sum_{k=|x|}^{2|x|} \frac{e^{-c_U k} t^k e^{-t}}{k!} + C_U \sum_{k=2|x|}^{+\infty} \frac{t^k e^{-t}}{k! k^{1/2}} \exp\left\{-\frac{c_U x^2}{k}\right\}. \tag{3.47}$$

Denote the terms on the right-hand side by I and II respectively. For any $\delta \in (0,1)$ we can estimate

$$I \leq C_U e^{-\delta t - c_U|x|}(1-\delta)^{|x|} \sum_{k=|x|}^{2|x|} \frac{[(1-\delta)t]^k e^{-(1-\delta)t}}{k!} \leq C_U e^{-ct - c'_U|x|} \tag{3.48}$$

for some $c'_U > 0$, provided that $c > 0$ is sufficiently small. However, $e^{-ct} \leq C(t+1)^{-1/2}$, for some $C>0$ and all $t>0$. Also $e^{-c'_U|x|} \leq e^{-c'_U x^2/t}$. We obtain therefore

$$I \leq \frac{C_U}{(t+1)^{1/2}} e^{-c_U x^2/t} \tag{3.49}$$

for some $c_U, C_U > 0$ and $|x| \leq 2t$. As for the second term II we write it as $II_1 + II_2$ depending on whether the summation is over $k \leq [t/2]$, or otherwise. We use the convention that $II_1 = 0$ in case $[t/2] < 2|x|$. Sequence $t^k/(k! k^{1/2})$, $k=1,\ldots,[t/2]$ is increasing, therefore

$$\begin{aligned} II_1 &\leq C e^{-c'_U x^2/t} \frac{t^{[t/2]+1} e^{-t}}{[t/2]! [t/2]^{1/2}} \\ &\overset{\text{Stirling's form}}{\leq} C e^{-c'_U x^2/t} \exp\left\{[t/2]\log t - [t/2]\log\left([t/2]/e\right) - t\right\} \\ &\leq C e^{-c'_U x^2/t} e^{-ct} \end{aligned} \tag{3.50}$$

for some $c'_U, c > 0$ and $t > 1$. For $k \geq [t/2]$ we have

$$\begin{aligned} II_2 &\leq \frac{C}{(t+1)^{1/2}} e^{-c_U x^2/(2t)} \sum_{2t \geq k > [t/2]} \frac{t^k e^{-t}}{k!} + \frac{C}{(t+1)^{1/2}} \sum_{k \geq 2t} \frac{t^k e^{-t}}{k!} \\ &\leq \frac{C}{(t+1)^{1/2}} \left\{ e^{-c_U x^2/t} \mathbb{Q}\left[N_t > [t/2]\right] + \frac{t^{[2t]} e^{-t}}{[2t]!} \right\} \end{aligned}$$

$$\asymp \frac{C}{(t+1)^{1/2}} e^{-c_U x^2/t}, \tag{3.51}$$

as $t \to +\infty$. The last asymptotic equality follows from the central limit theorem for $t^{-1/2}(N_t - t)$ and the fact that $t^{[2t]}e^{-t}/[2t]!$ is clearly of order $o(e^{-Ct})$ for any $C > 0$.

Summarizing, from (3.49), (3.50) and (3.51) we obtain

$$q(t,x) \le \frac{C}{(t+1)^{1/2}} e^{-cx^2/t} \quad \text{for } |x| \le 2t.$$

Since $\operatorname{arsinh} r \sim r$ for $r \ll 1$ we have also

$$e^{-cx^2/t} \le \exp\left\{-\frac{cx^2/t}{\sqrt{1+(x/t)^2}+1} - c|x| \operatorname{arsinh}\big(|x|/t\big)\right\} \le C_* \mathscr{E}\big(t, c_*|x|\big)$$

for some $c_*, C_* > 0$ and $|x| \le 2t$. We conclude therefore the upper bound in (3.45) in this case.

When $|x| > 2t$ the sequence $\{t^k/k!, k \ge |x|\}$ is majorized by a geometric progression with ratio $1/2$. Therefore,

$$q(t,x) \le C_U \sum_{k=|x|}^{+\infty} \frac{t^k e^{-t}}{k! k^{1/2}} \le \frac{C_U}{(t+1)^{1/2}} \times \frac{t^{|x|}e^{-t}}{|x|!}.$$

We obtain therefore

$$q(t,x) \le \frac{C}{(t+1)^{1/2}} \exp\left\{-|x| \log\left(\frac{|x|}{et}\right) - t\right\}, \quad \text{for } |x| > 2t. \tag{3.52}$$

Since $r \log[r/(et)] \asymp r \operatorname{arsinh}[r/(et)]$ and $1 + r \asymp \sqrt{1+r^2} - 1$ for $r \gg 1$ we can estimate the expression in (3.52) by $C_*(t+1)^{-1/2}\mathscr{E}(c_* t, |x|)$ also in this case.

The upper bound in the multidimensional case can be concluded from one-dimensional estimates. Indeed, in an arbitrary spatial dimension d, the corresponding transition probability function $q_d(t,x)$ satisfies Kolmogorov's equation

$$\partial_t q_d(t,x) = \frac{1}{2d} \sum_{j=1}^{d} \partial_j^* \partial_j q_d(t,x).$$

Its Fourier transform $\hat{q}_d(t,k) = \sum_x e^{ikx} q_d(t,x)$ therefore satisfies

$$\partial_t \hat{q}_d(t,k) = -\frac{2}{d} \sum_{j=1}^{d} \sin^2(k_j/2) \hat{q}_d(t,k), \quad \hat{q}(0,k) = 1.$$

Hence,

$$q_d(t,x) = \int_{\mathbb{T}^d} e^{ik\cdot x} \exp\left\{-\frac{2t}{d} \sum_{j=1}^{d} \sin^2(k_j/2)\right\} dk = \prod_{j=1}^{d} q_1(t/d, x_j). \tag{3.53}$$

An elementary calculation shows that

$$\mathscr{E}(t,r) = \exp\Big\{-\max_{\lambda}\big[\lambda r - t(\cosh\lambda - 1)\big]\Big\}, \tag{3.54}$$

where the maximum is attained at $\lambda = \operatorname{arsinh}(r/t)$. For any $\lambda_0, r_1, \ldots, r_d > 0$ we have therefore

$$\sum_{j=1}^{d} \max_{\lambda}\big[\lambda r_j - t(\cosh\lambda - 1)\big] \geq \sum_{j=1}^{d}\big[\lambda_0 r_j - t(\cosh\lambda_0 - 1)\big].$$

Since the previous expression is equal to $\lambda_0 \sum_{j=1}^{d} r_j - dt(\cosh\lambda_0 - 1)$, we conclude that

$$\prod_{j=1}^{d} \mathscr{E}(t, r_j) \leq \mathscr{E}\left(dt, \sum_{j=1}^{d} r_j\right). \tag{3.55}$$

The lower bound. Again, we obtain first the lower bound when $d = 1$. Suppose that $|x| \leq t$ and let $\Delta := [t, 2t]$. We have

$$q(t,x) \geq \frac{c_U e^{-C_U x^2/t}}{(t+1)^{1/2}} \sum_{k\in\Delta} \frac{t^k e^{-t}}{k!} = \frac{c_U e^{-C_U x^2/t}}{(t+1)^{1/2}} \mathbb{Q}\big[t^{-1/2}(N_t - t) \in (0, 2\sqrt{t})\big]. \tag{3.56}$$

This sequence is asymptotically equivalent to $c_U(t+1)^{-1/2} e^{-C_U x^2/t}$ as $t \to +\infty$, by virtue of the central limit theorem for $t^{-1/2}(N_t - t)$. On the other hand, when $|x| > t$ the sequence $\{t^k/k!, k \geq |x|\}$ is decreasing therefore, from (3.43)

$$q(t,x) \geq \frac{t^{|x|} e^{-t}}{|x|!} \mathbb{Q}[X_{|x|} = x] \geq c_U \exp\left\{-|x|\log\left(\frac{|x|}{t}\right) - C_U|x|\right\}. \tag{3.57}$$

The argument used after (3.52) can also be applied in this case. The desired estimate therefore follows for $d = 1$. In the multidimensional case we use (3.53) and (3.56) to conclude the lower bound when $\|x\|_\infty \leq t$. In case $\|x\|_\infty > t$ we have that $q_d(t,x) = \prod_{j=1}^{d} q_1(t/d, x_j)$ is bounded below by

$$\prod_{|x_j|<t/d} \left(c_U t^{-1/2} e^{-C_U x_j^2/t}\right) \prod_{|x_j|>t/d} \left(c_U \exp\left\{-|x_j|\log\left(\frac{|x_j|}{t}\right) - C_U|x_j|\right\}\right).$$

The desired lower bound can be obtained by replacing each $|x_j|$ by $\|x\|_\infty$. □

Using the above result we obtain the following.

Lemma 3.22 *There exist $\hat{C}_*, \hat{c}_* > 0$ such that*

$$\big|\nabla_x q_d(t,x)\big| \leq \hat{C}_*(t+1)^{-1/2} q_d(\hat{c}_* t, x), \quad \forall t > 0,\ x \in \mathbb{Z}^d. \tag{3.58}$$

Proof We start with the case $d=1$. Recall $\nabla_x q(t,x)=q(t,x+1)-q(t,x)$. Only the case $|x|\le t$ needs to be considered. Otherwise, using the upper bound in (3.45) (slightly adjusting the constants appearing there), we can write

$$\big|\nabla_x q(t,x)\big|\le q(t,x+1)+q(t,x)\le \frac{C_* e^{-\delta t}}{(t+1)^{1/2}}\mathscr{E}\big(t,c_*|x|\big)$$

for some $C_*, c_*>0$ and $\delta>0$. This of course leads to the desired estimate for $e^{-\delta t}\ll (t+1)^{-k}$ for any $k>0$ when $t\gg 1$.

On the other hand, we have

$$\nabla_x q(t,x)=\frac{1}{2}\sum_{k=0}^{+\infty}\big[(-1)^{k+x}+1\big]\frac{t^k e^{-t}}{k!}\left(\frac{t}{k+1}\mathbb{Q}[X_{k+1}=x+1]-\mathbb{Q}[X_k=x]\right). \tag{3.59}$$

A simple calculation, using (3.41), shows that

$$\big|\mathbb{Q}[X_{k+1}=x+1]-\mathbb{Q}[X_k=x]\big|\le \min\big[1,(k+1)^{-1}\big(|x|+1\big)\big]\mathbb{Q}[X_k=x], \tag{3.60}$$

when $k,x\ge 0$ and $x-k\equiv 0 \bmod 2$. On the other hand,

$$\big|\mathbb{Q}[X_{k-1}=x+1]-\mathbb{Q}[X_k=x]\big|\le \min\big[1,|x|(k+1)^{-1}\big]\mathbb{Q}[X_k=x], \tag{3.61}$$

when $k\ge 0$, $x<0$ and $x-k\equiv 0 \bmod 2$. This estimate also holds when k and $x\ge 0$ are odd, or k and $x<0$ are even. We obtain therefore for $x=2y\ge 0$

$$\begin{aligned}\nabla_x q(t,x)&=\frac{1}{2}\sum_{k=0}^{+\infty}\big[(-1)^{k+x}+1\big]\frac{t^k e^{-t}}{k!}\big(\mathbb{Q}[X_{k+1}=x+1]-\mathbb{Q}[X_k=x]\big)\\&\quad+\frac{1}{2}\sum_{k=0}^{+\infty}\big[(-1)^{k+x}+1\big]\frac{t^k e^{-t}}{k!}\left(\frac{t}{k+1}-1\right)\mathbb{Q}[X_{k+1}=x+1].\end{aligned} \tag{3.62}$$

Denote the first and the second terms on the right-hand side of this equality by I and II, respectively.

By virtue of (3.60) we obtain

$$|I|\le\sum_{k\ge|x|}^{+\infty}\frac{t^k e^{-t}}{k!}\times\min\big[(x+1)(k+1)^{-1/2},1\big]\mathbb{Q}[X_k=x].$$

The argument used to obtain the upper bound in Lemma 3.21 can be repeated for I and it leads to

$$|I|\le\frac{C_*}{t+1}\mathscr{E}(t,c_*x),\qquad \forall t\ge 0,\ x\in\mathbb{Z} \tag{3.63}$$

for some $C_*, c_* > 0$. To estimate of II we write it as $II_1 + II_2 + II_3$, where II_i, $i = 1, 2, 3$ correspond to summation over ranges $k \le [t/2]$, $[2t] > k \ge [t/2]$ and $k \ge [2t]$. Term II_1 can be estimated as in (3.48) and (3.50) leading to

$$|II_1| \le C_* e^{-\delta t} \mathscr{E}\big(t, c_*|x|\big), \quad \forall |x| \le t, t > 0$$

for some $C_*, c_*, \delta > 0$ and we conclude eventually

$$|II_1| \le \frac{C_*}{t+1} \mathscr{E}\big(t, c_*|x|\big), \quad \forall |x| \le t,\ t > 0. \tag{3.64}$$

Using upper bound (3.42) we obtain

$$\begin{aligned} |II_2| \le{}& \frac{C}{(t+1)^{1/2}} e^{-c_U x^2/(2t)} \sum_{[2t] \ge k > [t/2]} \left| \frac{t}{k+1} - 1 \right| \frac{t^k e^{-t}}{k!} \\ & \times \frac{C}{t+1} e^{-c_U x^2/(2t)} \mathbb{E}_{\mathbb{Q}} \left| \frac{N_t - t}{t^{1/2}} \right| \asymp \frac{C}{t+1} \mathscr{E}\big(t, c_*|x|\big), \end{aligned} \tag{3.65}$$

as $t \to +\infty$, for all $|x| \le t$ and some constants $c_*, C > 0$. Finally, since $\{t^k/k!, k \ge [2t]\}$ is dominated by a geometric progression with ratio $1/2$ we obtain that for any $M > 0$

$$|II_3| \le C \sum_{k > [2t]} \left| \frac{t}{k+1} - 1 \right| \frac{t^k e^{-t}}{k!} \le C \frac{t^{[2t]} e^{-t}}{[2t]!} \le C e^{-Mt}$$

for some $C > 0$. This allows us to conclude

$$|II_3| \le \frac{C_*}{t+1} \mathscr{E}\big(t, c_*|x|\big) \tag{3.66}$$

for $|x| \le t$ and some $C_*, c_* > 0$. Summarizing, from (3.63), (3.64), (3.65) and (3.66) it follows that

$$\big|\nabla_x q(t,x)\big| \le \frac{C_*}{t+1} \mathscr{E}\big(t, c_*|x|\big), \quad \forall t \ge 0, \tag{3.67}$$

and $x \ge 0$ even. Estimate (3.58) can be concluded using the lower bound from (3.45). The other cases can be done similarly.

In the multidimensional case we can write (see (3.53))

$$\partial_j q_d(t,x) := q_d(t, x + e_j) - q_d(t,x) = \partial_j q(t/d, x_j) \prod_{i \ne j}^{d} q_1(t/d, x_i),$$

and we obtain

$$\big|\partial_j q_d(t,x)\big| \le \frac{C_*}{(t+1)^{(d+1)/2}} \prod_{i=1}^{d} \mathscr{E}\big(t/d, c_*|x_i|\big)$$

for some $C_*, c_* > 0$. Using estimates (3.54)–(3.55) we conclude from the above that

$$\left|\partial_j q_d(t,x)\right| \le \frac{C_*}{(t+1)^{(d+1)/2}} \mathscr{E}\big(t, c_* \|x\|_\infty\big)$$

for some $C_*, c_* > 0$. The desired estimate follows from the lower bound in (3.45). □

3.8 Comments and References

Central limit theorems for random walks in random environment is a vast subject. We refer to the monographs (Zeitouni, 2004) and (Sznitman, 2004) for a list of references. We review here only some results whose proofs are connected to the ideas presented in this chapter.

Random Walks with Random Conductances and Random Walks on the Infinite Cluster of the Supercritical bond Percolation Anshelevich et al. (1982) proved a central limit theorem for a random walk with symmetric transition probabilities on the d-dimensional lattice and obtained an explicit expression for the diffusion matrix. Künnemann (1983) proved the convergence of random walks with random conductances to a Brownian motion with diffusion matrix given by the effective conductivity. The proof is based on the convergence of the resolvents. De Masi et al. (1989) and Goldstein (1995) examined the asymptotic behavior of a large class of random walks in random environment which includes a random walk evolving in the infinite cluster of the two-dimensional supercritical bond percolation and a random walk with random conductivity. They also proved that the diffusion coefficient of the random walk evolving in the infinite cluster is strictly positive. Sidoravicius and Sznitman (2004) generalized these results by proving an almost sure invariance principle for the random walk on the infinite supercritical cluster of the bond percolation in dimension $d \ge 4$ and for a random walk on a network of i.i.d. strictly elliptic random conductances in any dimension. Berger and Biskup (2007) and Mathieu and Piatnitski (2007), independently, extended the previous result on the cluster of the bond percolation to dimension $d \ge 2$. Mathieu (2008) and Biskup and Prescott (2007), also independently, proved an a.s. invariance principle for a random walk on $\mathbb{Z}^d$, $d \ge 2$, among i.i.d. random conductances. Barlow (2004) obtained Gaussian upper and lower bounds for the transition density of the continuous time simple random walk on a supercritical bond percolation cluster in the lattice $\mathbb{Z}^d$. This result was extended by Barlow and Deuschel (2010) to random walks among random conductances.

Random Walks Among Obstacles Tanemura (1993) considers a random walk among random spherical obstacles in $\mathbb{R}^d$, $d \ge 2$. The proof of the central limit theorem is obtained by writing the position of the walk as an additive functional of the environment and by applying the central limit theorem for additive functionals, exactly as done in this chapter. One of the main difficulties is to prove the ergodicity of the environment process. This result was extended by Osada and Saitoh (1995) for a class of non-symmetric diffusions evolving among random obstacles in which the generator satisfies a sector condition.

Cyclic Random Walks Komorowski and Olla (2003a) proved that cyclic random walks satisfy a sector condition. Mathieu (2006) proved a Carne–Varopoulos bound for cyclic randoms walks. Deuschel and Kösters (2008) proved an almost sure invariance principle for the cyclic random walks presented in this chapter.

Random Walks with Drift in $\mathscr{H}_{-1}$ The mixing arguments presented in Sects. 3.4 and 3.5 are taken from Komorowski and Olla (2003a). The result formulated there claims a central limit theorem when $L^d(\mathbb{Q})$ is replaced by $L^2(\mathbb{Q})$ in assumption (3.22) of Theorem 3.6. The proof is however incomplete. The energy identity stated in Proposition 3.7 has been obtained by Oelschläger (1988) for diffusion processes in divergence free random fields. The central limit theorem for a random walk in mixing environment in dimension 2 is an open question.

Other Models Bezuidenhout and Grimmett (1999) examined random walks on $\mathbb{Z}^d$ evolving among random mirrors. Rassoul-Agha and Seppäläinen (2005) proved an invariance principle for a discrete time random walk on a space time i.i.d. random environment by the method of the process as seen from the tagged particle presented in this chapter. In the same spirit, (Dolgopyat et al., 2008) deduced a quenched central limit theorem for random walks with bounded increments in an environment evolving according to a Markovian dynamics.

References

Anshelevich VV, Khanin KM, Sinaĭ YG (1982) Symmetric random walks in random environments. Commun Math Phys 85(3):449–470

Barlow MT (2004) Random walks on supercritical percolation clusters. Ann Probab 32(4):3024–3084

Barlow MT, Deuschel JD (2010) Invariance principle for the random conductance model with unbounded conductances. Ann Probab 38(1):234–276

Berger N, Biskup M (2007) Quenched invariance principle for simple random walk on percolation clusters. Probab Theory Relat Fields 137(1–2):83–120

Bezuidenhout C, Grimmett G (1999) A central limit theorem for random walks in random labyrinths. Ann Inst Henri Poincaré Probab Stat 35(5):631–683

Biskup M, Prescott TM (2007) Functional CLT for random walk among bounded random conductances. Electron J Probab 12(49):1323–1348 (electronic)

Dacunha-Castelle D, Duflo M (1986) Probability and statistics, vol II. Springer, New York. Translated from the French by David McHale

De Masi A, Ferrari PA, Goldstein S, Wick WD (1989) An invariance principle for reversible Markov processes. Applications to random motions in random environments. J Stat Phys 55(3–4):787–855

Deuschel JD, Kösters H (2008) The quenched invariance principle for random walks in random environments admitting a bounded cycle representation. Ann Inst Henri Poincaré Probab Stat 44(3):574–591

Dolgopyat D, Keller G, Liverani C (2008) Random walk in Markovian environment. Ann Probab 36(5):1676–1710

Doukhan P (1994) Mixing: properties and examples. Lecture notes in statistics, vol 85. Springer, New York

Goldstein S (1995) Antisymmetric functionals of reversible Markov processes. Ann Inst Henri Poincaré Probab Stat 31(1):177–190

Ibragimov IA, Linnik YV (1971) Independent and stationary sequences of random variables: with a supplementary chapter by IA Ibragimov and VV Petrov. Wolters-Noordhoff Publishing, Groningen. Translation from the Russian edited by JFC Kingman

Komorowski T, Olla S (2003a) A note on the central limit theorem for two-fold stochastic random walks in a random environment. Bull Pol Acad Sci, Math 51(2):217–232

Künnemann R (1983) The diffusion limit for reversible jump processes on $\mathbf{Z}^d$ with ergodic random bond conductivities. Commun Math Phys 90(1):27–68

Leoni G (2009) A first course in Sobolev spaces. Graduate studies in mathematics, vol 105. Am Math Soc, Providence

Mathieu P (2006) Carne-Varopoulos bounds for centered random walks. Ann Probab 34(3):987–1011

Mathieu P (2008) Quenched invariance principles for random walks with random conductances. J Stat Phys 130(5):1025–1046

Mathieu P, Piatnitski A (2007) Quenched invariance principles for random walks on percolation clusters. Proc R Soc Lond, Ser A, Math Phys Eng Sci 463(2085):2287–2307

Oelschläger K (1988) Homogenization of a diffusion process in a divergence-free random field. Ann Probab 16(3):1084–1126

Osada H, Saitoh T (1995) An invariance principle for non-symmetric Markov processes and reflecting diffusions in random domains. Probab Theory Relat Fields 101(1):45–63

Rassoul-Agha F, Seppäläinen T (2005) An almost sure invariance principle for random walks in a space-time random environment. Probab Theory Relat Fields 133(3):299–314

Sidoravicius V, Sznitman AS (2004) Quenched invariance principles for walks on clusters of percolation or among random conductances. Probab Theory Relat Fields 129(2):219–244

Sznitman AS (2004) Topics in random walks in random environment. In: School and conference on probability theory. ICTP lect notes, vol XVII. Abdus Salam Int Cent Theoret Phys, Trieste, pp 203–266 (electronic)

Tanemura H (1993) Central limit theorem for a random walk with random obstacles in $\mathbb{R}^d$. Ann Probab 21(2):936–960

Yosida K (1995) Functional analysis. Classics in mathematics. Springer, Berlin. Reprint of the sixth 1980 edition

Zeitouni O (2004) Random walks in random environment. In: Lectures on probability theory and statistics. Lecture notes in math, vol 1837. Springer, Berlin, pp 189–312

Chapter 4
Bounds and Variational Principles for the Asymptotic Variance

In Chap. 2, we presented conditions on the generator of a Markov process $\{X_t : t \geq 0\}$ which guarantee a central limit theorem for the additive functional $\int_0^t V(X_s)\,ds$ for functions V in $L^2 \cap \mathscr{H}_{-1}$. We proved in Theorem 2.7 and in (2.24) that the asymptotic variance $\sigma^2(V)$ is given by

$$\sigma^2(V) = 2 \lim_{\lambda \to 0} \langle (\lambda - L) f_\lambda, f_\lambda \rangle_\pi = 2 \lim_{\lambda \to 0} \langle (\lambda - L)^{-1} V, V \rangle_\pi, \tag{4.1}$$

where f_λ is the solution of the resolvent equation, and we showed in Corollary 2.11 that the variance of $t^{-1/2} \int_0^t V(X_s)\,ds$ converges to $\sigma^2(V)$. In this chapter, we present variational formulas, upper and lower bounds for the asymptotic variance $\sigma^2(V)$.

In Sect. 4.1, we provide two variational formulas for $\langle (\lambda - L)^{-1} V, V \rangle_\pi$, $\lambda > 0$. In Sect. 4.2, we show that an upper and a lower bound for the variance $\sigma^2(V)$ which hold in great generality follow from these variational formulas. Assuming further conditions on V we are also able to derive two variational formulas for the variance $\sigma^2(V)$. In Sect. 4.3, we prove that the conditions imposed on V to derive a variational formula for the variance $\sigma^2(V)$ hold if the generator satisfies the graded sector conditions introduced in Sect. 2.7.4. Finally, in Sect. 4.4 we present some estimates of the variance of an additive functional of a Markov process which will be useful in proving the superdiffusive behavior of a tracer particle in some turbulent diffusion models.

4.1 Quadratic Functional of the Resolvent

Recall the set-up and the notation introduced in the beginning of Chap. 2 and fix a function V in $L^2(\pi)$. Note that we do not assume in this section that V belongs to $\mathscr{H}_{-1}$. Recall that S stands for the closure of the essentially self-adjoint symmetric part of L and $\mathscr{C}$ for a core of the generator L and its adjoint L^*. Assume that

$$(\lambda - S)^{-1}(\mathscr{C}) \subset \mathscr{D}(L^*) \tag{4.2}$$

for each $\lambda > 0$.

T. Komorowski et al., *Fluctuations in Markov Processes*, Grundlehren der mathematischen Wissenschaften 345,
DOI 10.1007/978-3-642-29880-6_4,

For $\lambda > 0$, let $\mathscr{H}_{1,\lambda}$ be the completion of $\mathscr{C}$ under the pre-Hilbert norm defined by $\|f\|_{1,\lambda}^2 := \langle(\lambda - L)f, f\rangle_\pi$. Since the operator $-L$ is non-negative definite, it is clear that $\|\cdot\|_{1,\lambda}$ is a norm. Also, since $\|f\|_{1,\lambda}^2 \geq \lambda\|f\|_\pi^2$, each Cauchy sequence in this norm is also Cauchy in $L^2(\pi)$. This fact allows for an obvious identification of $\mathscr{H}_{1,\lambda}$ with a dense subset of $L^2(\pi)$ since it contains $\mathscr{C}$. For any $f \in L^2(\pi)$ define

$$\|f\|_{-1,\lambda}^2 := \sup_{g\in\mathscr{C}}\{2\langle f, g\rangle_\pi - \|g\|_{1,\lambda}^2\}$$

being finite, or not. Note that the expression inside braces on the right-hand side of the previous formula is smaller than $2\|f\|_\pi\|g\|_\pi - \lambda\|g\|_\pi^2$. In particular,

$$\|f\|_{-1,\lambda}^2 \leq \lambda^{-1}\|f\|_\pi^2 \tag{4.3}$$

for any $f \in L^2(\pi)$. Let $\mathscr{H}_{-1,\lambda}$ be the completion of $L^2(\pi)$ under the norm $\|\cdot\|_{-1,\lambda}$.

Theorem 4.1 *Fix V in $L^2(\pi)$ and assume that* (4.2) *holds. Then, for every $\lambda > 0$,*

$$\begin{aligned}\langle V, (\lambda - L)^{-1}V\rangle_\pi &= \sup_{g\in\mathscr{C}}\{2\langle V, g\rangle_\pi - \|g\|_{1,\lambda}^2 - \|Ag\|_{-1,\lambda}^2\}\\ &= \inf_{g\in\mathscr{C}}\{\|g\|_{1,\lambda}^2 + \|V + Ag\|_{-1,\lambda}^2\}.\end{aligned}$$

Proof Fix $\lambda > 0$ and V in $L^2(\pi)$. Denote by $G_\lambda : L^2(\pi) \to L^2(\pi)$ the resolvent operator $(\lambda - L)^{-1}$, by G_λ^s the symmetric part of G_λ and by $\mathscr{R}_\lambda$ the range of G_λ^s: $\mathscr{R}_\lambda = \{G_\lambda^s f : f \in L^2(\pi)\}$. Since G_λ^s is a bounded positive definite, symmetric operator, by Theorem 3.1.2 of Kesavan (1989),

$$\langle V, (\lambda - L)^{-1}V\rangle_\pi = \langle V, G_\lambda^s V\rangle_\pi = \sup_{g\in\mathscr{R}_\lambda}\{2\langle V, g\rangle_\pi - \langle g, [G_\lambda^s]^{-1}g\rangle_\pi\}.$$

Claim A Denote by $\mathscr{R}_\lambda^*$ the range of $G_\lambda^s(\lambda - L^*)$ over the core $\mathscr{C}$: $\mathscr{R}_\lambda^* = \{G_\lambda^s(\lambda - L^*)g : g \in \mathscr{C}\}$. We claim that we can replace the set $\mathscr{R}_\lambda$ by $\mathscr{R}_\lambda^*$ in the previous variational formula.

Since $\mathscr{C}$ is a core for L^*, $\mathscr{R}_\lambda^* \subset \mathscr{R}_\lambda$ so that

$$\sup_{g\in\mathscr{R}_\lambda^*}\{2\langle V, g\rangle_\pi - \langle g, [G_\lambda^s]^{-1}g\rangle_\pi\} \leq \sup_{g\in\mathscr{R}_\lambda}\{2\langle V, g\rangle_\pi - \langle g, [G_\lambda^s]^{-1}g\rangle_\pi\}.$$

To prove the reverse inequality, it is enough to show that for each g in $\mathscr{R}_\lambda$, there exists a sequence $\{g_n : n \geq 1\}$ of functions in $\mathscr{R}_\lambda^*$ such that $\langle V, g_n\rangle_\pi$, $\langle g_n, [G_\lambda^s]^{-1}g_n\rangle_\pi$ converge, as $n \uparrow \infty$, to $\langle V, g\rangle_\pi$, $\langle g, [G_\lambda^s]^{-1}g\rangle_\pi$, respectively.

Fix a function g in $\mathscr{R}_\lambda$. By definition, $g = G_\lambda^s h$ for some h in $L^2(\pi)$. g may be rewritten as $G_\lambda^s(\lambda - L^*)(\lambda - L^*)^{-1}h$. If $\tilde{h} = (\lambda - L^*)^{-1}h$ belonged to the core $\mathscr{C}$, g would be an element of $\mathscr{R}_\lambda^*$ and there would be nothing to prove. This may not be the case, but since $\tilde{h}$ belongs to the domain of L^*, $\tilde{h}$ may be approximated by functions in the core and one just needs to check that the approximating sequence has the required properties. The rigorous argument is as follows.

Since $(\lambda - L^*)^{-1}$ is a bounded operator in $L^2(\pi)$, $\tilde{h} = (\lambda - L^*)^{-1}h$ is well defined and belongs to $\mathscr{D}(L^*)$, the domain of L^*. Since $\mathscr{C}$ is a core of L^*, there exists a sequence $\{h_n : n \ge 1\}$ in $\mathscr{C}$ such that h_n, $(\lambda - L^*)h_n$ converge in $L^2(\pi)$, as $n \uparrow \infty$, to $\tilde{h}$, $(\lambda - L^*)\tilde{h} = h$, respectively. Since G_λ^s is a bounded operator, $g_n = G_\lambda^s(\lambda - L^*)h_n$ converges in $L^2(\pi)$ to $G_\lambda^s h = g$. For each $n \ge 1$, g_n belongs to $\mathscr{R}_\lambda^*$ because each h_n is in the core $\mathscr{C}$. Since g_n converges to g in $L^2(\pi)$, $\langle V, g_n\rangle_\pi$ converges to $\langle V, g\rangle_\pi$. On the other hand, by definition of g_n, h_n, h, and since g_n, $(\lambda - L^*)h_n$ converge to g, h, respectively,

$$\lim_{n\to\infty}\langle g_n, [G_\lambda^s]^{-1} g_n\rangle_\pi = \lim_{n\to\infty}\langle g_n, (\lambda - L^*)h_n\rangle_\pi = \langle g, h\rangle_\pi = \langle g, [G_\lambda^s]^{-1} g\rangle_\pi,$$

which proves Claim A.

Claim B $\mathscr{R}_\lambda^* = \{(\lambda - L)^{-1}(\lambda - S)h : h \in \mathscr{C}\}$.

To prove this claim, it is enough to show that

$$G_\lambda^s(\lambda - L^*)h = (\lambda - L)^{-1}(\lambda - S)h \tag{4.4}$$

for every h in $\mathscr{C}$. Since $2S = L + L^*$ and $2G_\lambda^s = (\lambda - L)^{-1} + [(\lambda - L)^{-1}]^*$, we only have to show that

$$[(\lambda - L)^{-1}]^*(\lambda - L^*)h = h$$

for every $h \in \mathscr{C}$. Fix a function g in $L^2(\pi)$ and let $f = G_\lambda g \in L^2(\pi)$ so that $(\lambda - L)f = g$. By definition,

$$\begin{aligned}\langle [(\lambda - L)^{-1}]^*(\lambda - L^*)h, g\rangle_\pi &= \langle (\lambda - L^*)h, (\lambda - L)^{-1}g\rangle_\pi = \langle (\lambda - L^*)h, f\rangle_\pi \\ &= \langle h, (\lambda - L)f\rangle_\pi = \langle h, g\rangle_\pi,\end{aligned}$$

which proves Claim B. It follows from (4.4) that $\mathscr{R}_\lambda^* \subset \mathscr{D}(L)$.

Claim C For any V in $L^2(\pi)$,

$$\langle V, (\lambda - L)^{-1}V\rangle_\pi = \sup_{g\in\mathscr{R}_\lambda^*}\{2\langle V, g\rangle_\pi - \|(\lambda - L)g\|_{-1,\lambda}^2\}.$$

Indeed, fix g in $\mathscr{R}_\lambda^*$ and keep in mind that $\mathscr{R}_\lambda^* \subset \mathscr{D}(L)$. By definition, $g = G_\lambda^s(\lambda - L^*)h$ for some h in $\mathscr{C}$. Hence,

$$\langle g, [G_\lambda^s]^{-1} g\rangle_\pi = \langle g, (\lambda - L^*)h\rangle_\pi = \langle (\lambda - L)g, h\rangle_\pi$$

since g belongs to $\mathscr{D}(L)$. In view of (4.4), $g = (\lambda - L)^{-1}(\lambda - S)h$ so that $h = (\lambda - S)^{-1}(\lambda - L)g$. The expression on the right-hand side in the previous displayed formula is thus equal to $\langle(\lambda - L)g, (\lambda - S)^{-1}(\lambda - L)g\rangle_\pi = \|(\lambda - L)g\|_{-1,\lambda}^2$. Claim C follows from this identity and Claim A.

Claim D We may replace the set $\mathscr{R}_\lambda^*$ in the variational formula of Claim C by $\mathscr{C}$.

Fix a function g in $\mathscr{R}_\lambda^*$. By (4.4), g belongs to the domain of L. Since $\mathscr{C}$ is a core for L, there exists a sequence $\{h_n : n \geq 1\}$ of functions in $\mathscr{C}$ such that h_n, $(\lambda - L)h_n$ converge in $L^2(\pi)$ as $n \uparrow \infty$ to g, $(\lambda - L)g$, respectively. By (4.3), the convergence takes place also in $\mathscr{H}_{-1,\lambda}$. In particular

$$\lim_{n\to\infty} \|h_n - g\|_\pi = 0, \qquad \lim_{n\to\infty} \big\|(\lambda - L)h_n\big\|_{-1,\lambda} = \big\|(\lambda - L)g\big\|_{-1,\lambda},$$

so that

$$\sup_{g\in\mathscr{R}_\lambda^*} \big\{2\langle V, g\rangle_\pi - \big\|(\lambda - L)g\big\|_{-1,\lambda}^2\big\} \leq \sup_{g\in\mathscr{C}} \big\{2\langle V, g\rangle_\pi - \big\|(\lambda - L)g\big\|_{-1,\lambda}^2\big\}.$$

To prove the reverse inequality, fix a function g in the core $\mathscr{C}$. $(\lambda - S)^{-1}(\lambda - L)g$ belongs to the domain of S. Since $\mathscr{C}$ is a core for S, there exists a sequence $\{h_n : n \geq 1\}$ in $\mathscr{C}$ such that h_n, $(\lambda - S)h_n$ converge in $L^2(\pi)$ as $n \uparrow \infty$ to $(\lambda - S)^{-1}(\lambda - L)g$, $(\lambda - L)g$, respectively.

Let $g_n = (\lambda - L)^{-1}(\lambda - S)h_n$. By Claim B, g_n belongs to $\mathscr{R}_\lambda^*$. On the other hand, $(\lambda - L)g_n = (\lambda - S)h_n$ converges to $(\lambda - L)g$ in $L^2(\pi)$ and therefore in $\mathscr{H}_{-1,\lambda}$. Moreover, since $(\lambda - L)^{-1}$ is a bounded operator and since $(\lambda - S)h_n$ converges to $(\lambda - L)g$, $g_n = (\lambda - L)^{-1}(\lambda - S)h_n$ converges to $(\lambda - L)^{-1}(\lambda - L)g = g$ in $L^2(\pi)$.

Hence, for a function g in $\mathscr{C}$, we obtained a sequence $\{g_n : n \geq 1\}$ in $\mathscr{R}_\lambda^*$ such that g_n converges to g in $L^2(\pi)$ and $(\lambda - L)g_n$ converges to $(\lambda - L)g$ in $\mathscr{H}_{-1,\lambda}$. This shows that

$$\sup_{g\in\mathscr{C}} \big\{2\langle V, g\rangle_\pi - \big\|(\lambda - L)g\big\|_{-1,\lambda}^2\big\} \leq \sup_{g\in\mathscr{R}_\lambda^*} \big\{2\langle V, g\rangle_\pi - \big\|(\lambda - L)g\big\|_{-1,\lambda}^2\big\}$$

and concludes the proof of the claim.

Claim E Proof of the first statement of the theorem. Since A is an anti-symmetric operator, for $g \in \mathscr{C}$,

$$\begin{aligned}\big\|(\lambda - L)g\big\|_{-1,\lambda}^2 &= \big\langle(\lambda - S)g, g\big\rangle_\pi + \big\langle(\lambda - S)^{-1}Ag, Ag\big\rangle_\pi \\ &= \|g\|_{1,\lambda}^2 + \|Ag\|_{-1,\lambda}^2.\end{aligned}$$

In view of Claim D, this concludes the proof the first variational formula.

Claim F Lower bound for the second variational formula.

To prove the second statement of the theorem, note that for each $g \in \mathscr{C}$,

$$\|Ag\|_{-1,\lambda}^2 = \sup_{h\in\mathscr{C}} \bigl\{2\langle Ag, h\rangle_\pi - \|h\|_{1,\lambda}^2\bigr\}.$$

In particular, by the first part of the theorem and since A is anti-symmetric,

$$\begin{aligned}\bigl\langle V, (\lambda - L)^{-1} V\bigr\rangle_\pi &= \sup_{g\in\mathscr{C}} \inf_{h\in\mathscr{C}} \bigl\{2\langle V + Ah, g\rangle_\pi - \|g\|_{1,\lambda}^2 + \|h\|_{1,\lambda}^2\bigr\} \\ &\le \inf_{h\in\mathscr{C}} \sup_{g\in\mathscr{C}} \bigl\{2\langle V + Ah, g\rangle_\pi - \|g\|_{1,\lambda}^2 + \|h\|_{1,\lambda}^2\bigr\} \\ &= \inf_{h\in\mathscr{C}} \bigl\{\|V + Ah\|_{-1,\lambda}^2 + \|h\|_{1,\lambda}^2\bigr\}.\end{aligned}$$

Claim G Upper bound for the second variational formula.

To prove the reverse inequality, for any $\varepsilon > 0$, it is enough to exhibit a function h in $\mathscr{C}$ such that

$$\bigl\langle V, (\lambda - L)^{-1} V\bigr\rangle_\pi \ge \|V + Ah\|_{-1,\lambda}^2 + \|h\|_{1,\lambda}^2 - \varepsilon. \tag{4.5}$$

Assume first that there exist functions h_0, g_0 in $\mathscr{C}$ such that

$$\begin{cases}(\lambda - L)h_0 = V, \\ (\lambda - L^*)g_0 = V.\end{cases} \tag{4.6}$$

We present below the modifications needed in the case where (4.6) has no solution in $\mathscr{C}$.

Let

$$g = (1/2)\{h_0 + g_0\}, \qquad h = (1/2)\{h_0 - g_0\}.$$

It follows from the identities relating h_0, g_0 to V that

$$\begin{cases}(\lambda - S)g - Ah = V, \\ (\lambda - S)h - Ag = 0.\end{cases}$$

Since $(\lambda - L)h_0 = V$, since $\langle(\lambda - L)h_0, h_0\rangle_\pi = \langle(\lambda - S)h_0, h_0\rangle_\pi$ and since $h_0 = g + h$,

$$\begin{aligned}\bigl\langle V, (\lambda - L)^{-1} V\bigr\rangle_\pi &= \bigl\langle(\lambda - S)h_0, h_0\bigr\rangle_\pi = \bigl\langle(\lambda - S)(g + h), (g + h)\bigr\rangle_\pi \\ &= \bigl\langle(\lambda - S)g, g\bigr\rangle_\pi + \bigl\langle(\lambda - S)h, h\bigr\rangle_\pi + 2\bigl\langle(\lambda - S)h, g\bigr\rangle_\pi.\end{aligned}$$

Since $(\lambda - S)h = Ag$, the last term vanishes. On the other hand, the first term can be rewritten as

$$\bigl\|(\lambda - S)g\bigr\|_{-1,\lambda}^2 = \|Ah + V\|_{-1,\lambda}^2.$$

Hence, the penultimate formula becomes

$$\langle V, (\lambda - L)^{-1} V\rangle_\pi = \|h\|_{1,\lambda}^2 + \|Ah + V\|_{-1,\lambda}^2,$$

which proves the claim.

We now adapt the previous arguments to the case in which solutions to (4.6) do not exist in $\mathscr{C}$. Fix $\varepsilon > 0$. Since $\mathscr{C}$ is a common core for both L and L^* and since $(\lambda - L)^{-1}V$, $(\lambda - L^*)^{-1}V$ belongs to $\mathscr{D}(L)$, $\mathscr{D}(L^*)$, respectively, there exists h_ε, g_ε in $\mathscr{C}$ such that

$$\|(\lambda - L)h_\varepsilon - V\|_\pi \le \varepsilon, \qquad \|(\lambda - L^*)g_\varepsilon - V\|_\pi \le \varepsilon.$$

Let $u_\varepsilon = (\lambda - L)h_\varepsilon - V$, $w_\varepsilon = (\lambda - L^*)g_\varepsilon - V$ and keep in mind that $\|u_\varepsilon\|_\pi \le \varepsilon$, $\|w_\varepsilon\|_\pi \le \varepsilon$. Taking the scalar product with respect to h_ε on both sides of the identity $u_\varepsilon = (\lambda - L)h_\varepsilon - V$ and applying Schwarz inequality we obtain that $\|h_\varepsilon\|_\pi \le \lambda^{-1}(\|V\|_\pi + \varepsilon)$. A similar inequality holds for g_ε so that

$$\|h_\varepsilon\|_\pi \le \lambda^{-1}\big(\|V\|_\pi + \varepsilon\big), \qquad \|g_\varepsilon\|_\pi \le \lambda^{-1}\big(\|V\|_\pi + \varepsilon\big). \tag{4.7}$$

In the same way,

$$\|(\lambda - L)^{-1}W\|_\pi \le \lambda^{-1}\|W\|_\pi, \qquad \|(\lambda - S)^{-1}W\|_\pi \le \lambda^{-1}\|W\|_\pi \tag{4.8}$$

for any function W in $L^2(\pi)$.

Let $G_\varepsilon = (1/2)\{h_\varepsilon + g_\varepsilon\}$, $H_\varepsilon = (1/2)\{h_\varepsilon - g_\varepsilon\}$ and remark that the bounds (4.7) extend to G_ε, H_ε. From the equations for h_ε, g_ε, we obtain that

$$\begin{cases} (\lambda - S)G_\varepsilon - AH_\varepsilon = V + U_\varepsilon, \\ (\lambda - S)H_\varepsilon - AG_\varepsilon = W_\varepsilon, \end{cases}$$

where $U_\varepsilon = (1/2)\{u_\varepsilon + w_\varepsilon\}$, $W_\varepsilon = (1/2)\{u_\varepsilon - w_\varepsilon\}$. Of course, $\|U_\varepsilon\|_\pi \le \varepsilon$, $\|W_\varepsilon\|_\pi \le \varepsilon$.

Since $V = (\lambda - L)h_\varepsilon - u_\varepsilon$ and $\langle h_\varepsilon, (\lambda - L)h_\varepsilon\rangle_\pi = \langle h_\varepsilon, (\lambda - S)h_\varepsilon\rangle_\pi$, we have that

$$\langle V, (\lambda - L)^{-1}V\rangle_\pi = \langle h_\varepsilon, (\lambda - S)h_\varepsilon\rangle_\pi - \langle h_\varepsilon, u_\varepsilon\rangle_\pi - \langle(\lambda - L)^{-1}u_\varepsilon, V\rangle_\pi.$$

By (4.7) and since $\|u_\varepsilon\|_\pi \le \varepsilon$, the second term on the right-hand side is absolutely bounded by $\lambda^{-1}\varepsilon\{\|V\|_\pi + \varepsilon\}$. On the other hand, by (4.8), the third term is less than or equal to $\lambda^{-1}\varepsilon\|V\|_\pi$. Therefore,

$$\langle V, (\lambda - L)^{-1}V\rangle_\pi \ge \langle h_\varepsilon, (\lambda - S)h_\varepsilon\rangle_\pi - 2\lambda^{-1}\varepsilon\big\{\|V\|_\pi + \varepsilon\big\}.$$

Recall that $h_\varepsilon = G_\varepsilon + H_\varepsilon$ and that $(\lambda - S)H_\varepsilon = W_\varepsilon + AG_\varepsilon$. In particular, by the anti-symmetry of A,

$$\langle h_\varepsilon, (\lambda - S)h_\varepsilon\rangle_\pi = \langle H_\varepsilon, (\lambda - S)H_\varepsilon\rangle_\pi + 2\langle G_\varepsilon, W_\varepsilon\rangle_\pi + \langle G_\varepsilon, (\lambda - S)G_\varepsilon\rangle_\pi.$$

In view of (4.7), the second term on the right-hand side is absolutely bounded by $2\|G_\varepsilon\|_\pi \|W_\varepsilon\|_\pi \le 2\varepsilon\lambda^{-1}\{\|V\|_\pi + \varepsilon\}$. The third one can be rewritten as

$$\langle G_\varepsilon, V + AH_\varepsilon\rangle_\pi + \langle G_\varepsilon, U_\varepsilon\rangle_\pi$$

because $(\lambda - S)G_\varepsilon = V + AH_\varepsilon + U_\varepsilon$. By (4.7), the second term is less than or equal to $\varepsilon\lambda^{-1}\{\|V\|_\pi + \varepsilon\}$, while the first one is equal to

$$\big\langle(\lambda - S)^{-1}(V + AH_\varepsilon), V + AH_\varepsilon\big\rangle_\pi + \big\langle(\lambda - S)^{-1}U_\varepsilon, V + AH_\varepsilon\big\rangle_\pi.$$

By Schwarz inequality and by (4.8), the second term is absolutely bounded by $\lambda^{-1/2}\varepsilon\|V + AH_\varepsilon\|_{-1,\lambda} \le \lambda^{-1}\varepsilon + \varepsilon\|V + AH_\varepsilon\|^2_{-1,\lambda}$. Hence, the previous displayed formula is bounded below by

$$(1 - \varepsilon)\|V + AH_\varepsilon\|^2_{-1,\lambda} - \varepsilon\lambda^{-1}.$$

Putting all previous estimates together, we obtain that

$$\begin{aligned}\big\langle V, (\lambda - L)^{-1}V\big\rangle_\pi \ge (1 - \varepsilon)\big\{\|H_\varepsilon\|^2_{1,\lambda} + \|V + AH_\varepsilon\|^2_{-1,\lambda}\big\}\\ - 6\lambda^{-1}\varepsilon\{1 + \|V\|_\pi + \varepsilon\}.\end{aligned}$$

This concludes the proof of a slightly weaker version of (4.5), but strong enough for our needs. □

4.2 Bounds and Variational Formulas for the Variance

In this section, we present upper and lower bounds for the variance $\sigma^2(V)$ and we derive in Theorem 4.4 two variational formulas for $\sigma^2(V)$ under the assumptions (4.9), (4.10). Recall that f_λ stands for the solution of the resolvent equation (2.13).

Theorem 4.2 *Assume that V belongs to $L^2(\pi) \cap \mathscr{H}_{-1}$. Then,*

$$\sup_{g\in\mathscr{C}}\big\{2\langle V, g\rangle_\pi - \|g\|_1^2 - \|Ag\|^2_{-1}\big\} \le (1/2)\sigma^2(V) \le \inf_{h\in\mathscr{C}}\big\{\|h\|_1^2 + \|V + Ah\|^2_{-1}\big\}.$$

This result is a straightforward consequence of (4.1), Theorem 4.1 and the next lemma.

Lemma 4.3 *For any g in $\mathscr{C}$ and h in $L^2(\pi)$,*

$$\lim_{\lambda\to 0}\|g\|_{1,\lambda} = \|g\|_1, \qquad \lim_{\lambda\to 0}\|h\|_{-1,\lambda} = \|h\|_{-1}.$$

Proof The proof of the first identity is obvious and follows directly from the definition of the norm $\|\cdot\|_{1,\lambda}$. To show the second one, fix h in $L^2(\pi)$. For each f in $\mathscr{C}$,

$$\liminf_{\lambda\to 0}\|h\|^2_{-1,\lambda} \geq \liminf_{\lambda\to 0}\bigl\{2\langle h, f\rangle_\pi - \|f\|^2_{1,\lambda}\bigr\} = 2\langle h, f\rangle_\pi - \|f\|^2_1.$$

Taking the supremum over all f in $\mathscr{C}$, we conclude that $\liminf_{\lambda\to 0}\|h\|_{-1,\lambda} \geq \|h\|_{-1}$.

On the other hand, we claim that

$$\|h\|_{-1,\lambda} \leq \|h\|_{-1}$$

for all h in $L^2(\pi)$ and all $\lambda > 0$. Indeed, since $\|h\|^2_{-1,\lambda} \leq \lambda^{-1}\|h\|^2_\pi < \infty$, for any $\varepsilon > 0$ there exists f_ε in $\mathscr{C}$ such that

$$\begin{aligned}\|h\|^2_{-1,\lambda} &\leq 2\langle h, f_\varepsilon\rangle_\pi - \bigl\langle(\lambda - S)f_\varepsilon, f_\varepsilon\bigr\rangle_\pi + \varepsilon \\ &\leq 2\langle h, f_\varepsilon\rangle_\pi - \bigl\langle(-S)f_\varepsilon, f_\varepsilon\bigr\rangle_\pi + \varepsilon \leq \|h\|^2_{-1} + \varepsilon.\end{aligned}$$

Letting $\varepsilon \downarrow 0$, we conclude the proof of the claim and that of the lemma. □

To strengthen the conclusion of Theorem 4.2, we need further assumptions on V. Denote by f^*_λ the solution of the resolvent equation (2.13) with L^* in place of L. Assume that as $\lambda \downarrow 0$ the sequences

$$f_\lambda,\ f^*_\lambda \text{ converge in } \mathscr{H}_1 \text{ to limits denoted by } f_0,\ f^*_0, \text{ respectively.} \tag{4.9}$$

Theorem 4.4 *Assume that V belongs to $L^2(\pi)\cap\mathscr{H}_{-1}$, that* (4.9) *holds and that for any $\varepsilon > 0$ there exist h_ε, h^*_ε in $\mathscr{C}$ such that*

$$\begin{aligned}&\|Lh_\varepsilon + V\|_{-1} < \varepsilon \quad \textit{and} \quad \|h_\varepsilon - f_0\|_1 < \varepsilon, \\ &\bigl\|L^*h^*_\varepsilon + V\bigr\|_{-1} < \varepsilon \quad \textit{and} \quad \bigl\|h^*_\varepsilon - f^*_0\bigr\|_1 < \varepsilon.\end{aligned} \tag{4.10}$$

Then,

$$\begin{aligned}\sigma^2(V) &= 2\inf_{h\in\mathscr{C}}\bigl\{\|h\|^2_1 + \|V + Ah\|^2_{-1}\bigr\} \\ &= 2\sup_{g\in\mathscr{C}}\bigl\{2\langle V, g\rangle_\pi - \|g\|^2_1 - \|Ag\|^2_{-1}\bigr\}.\end{aligned}$$

Proof We start with the first identity, whose proof is similar to the one of Claim G in Theorem 4.1. In view of Theorem 4.2, we just need to prove the lower bound for $\sigma^2(V)$.

Fix $\varepsilon > 0$. According to assumption (4.10), there exists h_ε, g_ε in $\mathscr{C}$ such that $\|Lh_\varepsilon + V\|_{-1} \leq \varepsilon$, $\|L^*g_\varepsilon + V\|_{-1} \leq \varepsilon$, $\|h_\varepsilon - f_0\|_1 \leq \varepsilon$, $\|g_\varepsilon - f^*_0\|_1 \leq \varepsilon$. Let $u_\varepsilon = -Lh_\varepsilon - V$, $w_\varepsilon = -L^*g_\varepsilon - V$ and keep in mind that

$$\|u_\varepsilon\|_{-1} \leq \varepsilon, \qquad \|w_\varepsilon\|_{-1} \leq \varepsilon, \qquad \|h_\varepsilon\|_1 \leq \|f_0\|_1 + \varepsilon, \qquad \|g_\varepsilon\|_1 \leq \bigl\|f^*_0\bigr\|_1 + \varepsilon.$$

Let $G_\varepsilon = (1/2)\{h_\varepsilon + g_\varepsilon\}$, $H_\varepsilon = (1/2)\{h_\varepsilon - g_\varepsilon\}$. From the equations for h_ε, g_ε, we obtain that

$$\begin{cases} -SG_\varepsilon - AH_\varepsilon = V + U_\varepsilon, \\ -SH_\varepsilon - AG_\varepsilon = W_\varepsilon, \end{cases}$$

where $U_\varepsilon = (1/2)\{u_\varepsilon + w_\varepsilon\}$, $W_\varepsilon = (1/2)\{u_\varepsilon - w_\varepsilon\}$. Of course, $\|U_\varepsilon\|_{-1}$, $\|W_\varepsilon\|_{-1}$ are bounded by ε, and $\|G_\varepsilon\|_1$, $\|H_\varepsilon\|_1$ are less than or equal to $(1/2)\{\|f_0\|_1 + \|f_0^*\|_1 + 2\varepsilon\}$.

Recall from Claim C in Sect. 2.2 that the scalar product $\langle\cdot,\cdot\rangle_\pi$ has been extended to $\mathcal{H}_1 \times \mathcal{H}_{-1}$. Since f_λ converges in $\mathcal{H}_1$ to f_0, by the estimates on $\|u_\varepsilon\|_{-1}$, $\|h_\varepsilon\|_1$ and since $-Lh_\varepsilon = V + u_\varepsilon$,

$$\begin{aligned}(1/2)\sigma^2(V) &= \lim_{\lambda\to 0}\langle V, f_\lambda\rangle_\pi \ge \langle V, h_\varepsilon\rangle_\pi - \varepsilon\|V\|_{-1} \\ &\ge \langle V + u_\varepsilon, h_\varepsilon\rangle_\pi - \varepsilon\big\{\|V\|_{-1} + \|f_0\|_1 + \varepsilon\big\} \\ &= \|h_\varepsilon\|_1^2 - \varepsilon\big\{\|V\|_{-1} + \|f_0\|_1 + \varepsilon\big\}.\end{aligned}$$

Recall that $h_\varepsilon = G_\varepsilon + H_\varepsilon$ and that $-SH_\varepsilon = W_\varepsilon + AG_\varepsilon$. In particular, by the anti-symmetry of A,

$$\|h_\varepsilon\|_1^2 = \langle h_\varepsilon, -Sh_\varepsilon\rangle_\pi = \langle H_\varepsilon, -SH_\varepsilon\rangle_\pi + 2\langle G_\varepsilon, W_\varepsilon\rangle_\pi + \langle G_\varepsilon, -SG_\varepsilon\rangle_\pi. \quad (4.11)$$

The second term on the right-hand side of this identity is absolutely bounded by $2\|G_\varepsilon\|_1\|W_\varepsilon\|_{-1} \le \varepsilon\{\|f_0\|_1 + \|f_0^*\|_1 + 2\varepsilon\}$. The third one can be rewritten as

$$\|G_\varepsilon\|_1^2 = \|SG_\varepsilon\|_{-1}^2 = \|V + AH_\varepsilon + U_\varepsilon\|_{-1}^2.$$

Since $\|V + AH_\varepsilon\|_{-1}$ is equal to

$$\|U_\varepsilon + SG_\varepsilon\|_{-1} \le \varepsilon + \|G_\varepsilon\|_1 \le \varepsilon + (1/2)\big\{\|f_0\|_1 + \big\|f_0^*\big\|_1 + 2\varepsilon\big\},$$

and since $\|u + v\|_{-1}^2 \ge \|u\|_{-1}^2 - 2\|u\|_{-1}\|v\|_{-1}$,

$$\|V + AH_\varepsilon + U_\varepsilon\|_{-1}^2 \ge \|V + AH_\varepsilon\|_{-1}^2 - 2\varepsilon\big\{\|f_0\|_1 + \big\|f_0^*\big\|_1 + 2\varepsilon\big\}.$$

Assembling all previous estimates, we obtain that for every ε, there exists a function H_ε in $\mathcal{C}$ such that

$$(1/2)\sigma^2(V) \ge \|H_\varepsilon\|_1^2 + \|V + AH_\varepsilon\|_{-1}^2 - \varepsilon C_0,$$

where $C_0 = 4\{\|V\|_{-1} + \|f_0\|_1 + \|f_0^*\|_1 + 2\varepsilon\}$. This concludes the proof of the first statement of the theorem.

We now turn to the second identity. In view of Theorem 4.2, we just need to prove an upper bound for $\sigma^2(V)$. By definition of h_ε,

$$(1/2)\sigma^2(V) = \lim_{\lambda\to 0}\|f_\lambda\|_1^2 \le \|h_\varepsilon\|_1^2 + \varepsilon\big\{3\varepsilon + 2\|f_0\|_1\big\}.$$

By (4.11),

$$\|h_\varepsilon\|_1^2 = \|H_\varepsilon\|_1^2 + \|G_\varepsilon\|_1^2 + 2\langle G_\varepsilon, W_\varepsilon\rangle_\pi .$$

The last term is absolutely bounded by $\varepsilon\{\|f_0\|_1 + \|f_0^*\|_1 + 2\varepsilon\}$. On the one hand, in view of the equation relating H_ε and G_ε to V,

$$\begin{aligned}\|G_\varepsilon\|_1^2 + \|H_\varepsilon\|_1^2 &= \langle AH_\varepsilon + V + U_\varepsilon, G_\varepsilon\rangle_\pi + \langle AG_\varepsilon + W_\varepsilon, H_\varepsilon\rangle_\pi \\ &= \langle V, G_\varepsilon\rangle_\pi + \langle U_\varepsilon, G_\varepsilon\rangle_\pi + \langle W_\varepsilon, H_\varepsilon\rangle_\pi\end{aligned}$$

because A is anti-symmetric. By Schwarz inequality, the last two terms are absolutely bounded by $\varepsilon\{\|f_0\|_1 + \|f_0^*\|_1 + 2\varepsilon\}$. On the other hand, by the same relations,

$$\|G_\varepsilon\|_1^2 + \|H_\varepsilon\|_1^2 = \|G_\varepsilon\|_1^2 + \|SH_\varepsilon\|_{-1}^2 = \|G_\varepsilon\|_1^2 + \|AG_\varepsilon + W_\varepsilon\|_{-1}^2 .$$

Since $AG_\varepsilon = -SH_\varepsilon - W_\varepsilon$, $\|AG_\varepsilon\|_{-1} \le \|H_\varepsilon\|_1 + \|W_\varepsilon\|_{-1} \le \{\|f_0\|_1 + \|f_0^*\|_1 + 2\varepsilon\}$. In particular, the last term of the previous identity is bounded below by $\|AG_\varepsilon\|_{-1}^2 - 2\varepsilon\{\|f_0\|_1 + \|f_0^*\|_1 + 2\varepsilon\}$.

Therefore, writing $A = \|H_\varepsilon\|_1^2 + \|G_\varepsilon\|_1^2$ as $2A - A$, we obtain that $\|H_\varepsilon\|_1^2 + \|G_\varepsilon\|_1^2$ is bounded above by

$$2\langle V, G_\varepsilon\rangle_\pi - \|G_\varepsilon\|_1^2 - \|AG_\varepsilon\|_{-1}^2 + 4\varepsilon\big\{\|f_0\|_1 + \big\|f_0^*\big\|_1 + 2\varepsilon\big\}.$$

We have thus proved that for any $\varepsilon > 0$ there exists a function G_ε in the core $\mathscr{C}$ such that

$$(1/2)\sigma^2(V) \le 2\langle V, G_\varepsilon\rangle_\pi - \|G_\varepsilon\|_1^2 - \|AG_\varepsilon\|_{-1}^2 + C_1\varepsilon,$$

where $C_1 = 7\varepsilon\{\|f_0\|_1 + \|f_0^*\|_1 + 2\varepsilon\}$. This concludes the proof of the second identity of the theorem. □

The assumptions of Theorem 4.4 hold if the generator satisfies a sector condition.

Theorem 4.5 *Assume that the generators L, L^* satisfy a sector condition and that V belongs to $L^2(\pi) \cap \mathscr{H}_{-1}$. Then,* (4.9) *and* (4.10) *are in force.*

Proof Fix a function V in $L^2(\pi) \cap \mathscr{H}_{-1}$. We only prove the claim for L. Recall that we denote by f_λ the solution of the resolvent equation. By the sector condition, $\sup_{0<\lambda\le 1} \|Lf_\lambda\|_{-1} < \infty$. Hence, by Lemma 2.16 the sequence f_λ converges to some f_0 in $\mathscr{H}_1$. This proves (4.9). Moreover, by Claim 2 in Lemma 2.16, there exists a sequence $\{h_k : k \ge 1\}$, $h_k \in \mathscr{D}(L)$, such that $-Lh_k$, h_k converges strongly to V, f_0 in $\mathscr{H}_{-1}$, $\mathscr{H}_1$, respectively. Hence, for every $\varepsilon > 0$, there exists h_ε in $\mathscr{D}(L)$ such that

$$\|Lh_\varepsilon + V\|_{-1} < \varepsilon, \qquad \|h_\varepsilon - f_0\|_1 < \varepsilon.$$

Since $\mathscr{C}$ is a core for the generator L, there exists $g_\varepsilon \in \mathscr{C}$ such that $\|g_\varepsilon - h_\varepsilon\|_1 < \varepsilon$. On the other hand, by the sector condition, $\|Lg_\varepsilon - Lh_\varepsilon\|_{-1} < C_0\|g_\varepsilon - h_\varepsilon\|_1$ for some finite constant C_0. This proves the theorem. □

4.3 Variational Principles in the Graded Sector Context

Recall the formalism and the notation introduced in Sect. 2.7.4 on the graded sector condition. Assume that hypotheses (2.39), (2.40), (2.45), (2.50) are in force and that $\mathscr{A}_0$ is the subspace of $L^2(\pi)$ of constant functions.

Fix a function V of finite degree in $\mathscr{H}_{-1}$. We claim that V has finite triple norm:

$$|||V|||_{k,-1} < \infty \tag{4.12}$$

for all $k \geq 1$. Indeed, fix $k \geq 1$, assume that V belongs to $\mathscr{G}_n$, $n \geq 0$, recall the definition of the triple norm given just after (2.47) and note that

$$|||V|||_{k,-1} \leq C_0 \|V\|_{0,-1}$$

for some finite constant C_0 which depends only on n and k. Since V has finite degree, in the variational formula (2.43) which defines the $\|\cdot\|_{0,-1}$ norm of V, we may restrict the supremum to functions g in $\mathscr{C} \cap \mathscr{G}_n$. For such functions, by the estimates (2.45), (2.50), $\|g\|_1 \leq C_0 \|g\|_{0,1}$ for some finite constant C_0 which depends on n, β and γ. After replacing $\|g\|_{0,1}$ by $C_0^{-1}\|g\|_1$ in the variational formula, we obtain that $\|V\|_{0,-1} \leq C_0 \|V\|_{-1}$ which is finite because we assumed V to belong to $\mathscr{H}_{-1}$. This proves claim (4.12).

It follows from (4.12), Theorem 2.23, estimate (2.44) and Lemma 2.16 that conditions (4.9) are fulfilled. In this section, we prove that condition (4.10) also holds. Therefore, by Theorem 4.4 we have a variational formula for the variance $\sigma^2(V)$ in the graded sector context.

Recall from Corollary 2.24 that Lf belongs to $\mathscr{H}_{-1}$ for every function f in the domain $\mathscr{D}(L)$ such that $\sum_{n\geq 1} n^{2(\beta\vee\gamma)} \|\Pi_n f\|_{0,1}^2 < \infty$. The sum starts from $n=1$ because we assumed $\mathscr{A}_0$ to be the space of constant functions so that $\|\Pi_0 f\|_{0,1}^2 \leq C_0 \|\Pi_0 f\|_1^2 = 0$.

Lemma 4.6 *For every function f in the domain $\mathscr{D}(L)$ such that*

$$\sum_{n\geq 1} n^{2\alpha} \|\Pi_n f\|_{0,1}^2 < \infty,$$

where $\alpha = \max\{\beta, \gamma\}$, and for every $\varepsilon > 0$, there exists a function g_ε with finite degree in the core $\mathscr{C}$ such that

$$\|Lf - Lg_\varepsilon\|_{-1} < \varepsilon, \qquad \|f - g_\varepsilon\|_1 < \varepsilon.$$

A similar statement holds if we replace L by L^* everywhere.

Proof By the proof of Lemma 2.22, there exists a sequence $\{g_k : k \geq 1\}$ of functions of finite degree in the core $\mathscr{C}$ which converges to f in $L^2(\pi)$ and in $\mathscr{H}_1$, and such that

$$\lim_{k\to\infty} \sum_{n\geq 1} n^{2\alpha} \|\Pi_n\{g_k - f\}\|_{0,1}^2 = 0.$$

By (2.43),

$$\|Lf - Lg_k\|_{0,-1}^2 = \sum_{n\ge 1} \|\Pi_n\{Lf - Lg_k\}\|_{0,-1}^2.$$

By the proof of Theorem 2.23 this expression is less than or equal to

$$C_0 \sum_{n\ge 1} n^{2\alpha} \|\Pi_n\{f - g_k\}\|_{0,1}^2$$

for some finite constant C_0. This bound, together with (2.44) and the estimate of the previous paragraph shows that $\|Lf - Lg_k\|_{-1}$ vanishes as $k \uparrow \infty$.

Hence, we have obtained a sequence $\{g_k : k \ge 1\}$ of functions of finite degree in the core $\mathscr{C}$ which converges to f in $L^2(\pi)$ and in $\mathscr{H}_1$, and such that Lg_k converges to Lf in $\mathscr{H}_{-1}$. The same arguments apply to L^*. This proves the lemma. □

Theorem 4.7 *Suppose that V is a function of finite degree which belongs to $\mathscr{H}_{-1}$. Then,* (4.10) *holds.*

Proof Fix a function V in $\mathscr{H}_{-1}$ with finite degree. We only prove the claims regarding L. By Claim 2 in Lemma 2.16, there exists a sequence $\{h_k : k \ge 1\}$, $h_k \in \mathscr{D}(L)$, obtained as convex combinations of the solutions of the resolvent equation, such that $-Lh_k$, h_k converges strongly to V, f_0 in $\mathscr{H}_{-1}$, $\mathscr{H}_1$, respectively. Hence, for every $\varepsilon > 0$, there exists h_ε in $\mathscr{D}(L)$ such that

$$\|Lh_\varepsilon + V\|_{-1} < \varepsilon, \qquad \|h_\varepsilon - f_0\|_1 < \varepsilon.$$

Since V has finite degree, if f_λ, $0 < \lambda \le 1$, stands for the solution of the resolvent equation (2.13), by Lemma 2.21,

$$\sup_{0<\lambda\le 1} \sum_{n\ge 1} n^{2\alpha} \|\Pi_n f_\lambda\|_{0,1}^2 \le C_1$$

for some finite constant C_1 which depends on V and α. Since the sequence $\{h_k : k \ge 1\}$ is obtained as a convex combination of the solutions of the resolvent equation, the same estimate holds for $\{h_\varepsilon : 0 < \varepsilon \le 1\}$:

$$\sup_{0<\varepsilon\le 1} \sum_{n\ge 1} n^{2\alpha} \|\Pi_n h_\varepsilon\|_{0,1}^2 \le C_1.$$

To conclude the proof of the theorem it remains to apply Lemma 4.6. □

4.4 Estimates of the Variance

In this section we give some estimates of the variance of an additive functional of the Markov process $\{X_t : t \ge 0\}$. These bounds shall be useful in proving the superdiffusive behavior of a tracer particle in turbulent diffusion models of Sect. 12.6.

The first result is similar to Lemma 2.4.

Lemma 4.8 *Fix a function V in $L^2(\pi)$ and let f_λ be the solution of the resolvent equation* (2.13). *For every $t > 0$,*

$$\mathbb{E}_\pi\Big[\Big(\int_0^t V(X_s)\,ds\Big)^2\Big] \le 12t\langle f_{1/t}, V\rangle_\pi .$$

Proof Fix $\lambda > 0$ and recall the definition of the Dynkin martingale M_t^λ introduced just before (2.25). By this latter formula,

$$\int_0^t V(X_s)\,ds = M_t^\lambda + f_\lambda(X_0) - f_\lambda(X_t) + \lambda\int_0^t f_\lambda(X_s)\,ds.$$

By (2.16), the expectation of the quadratic variation of the martingale M_t^λ is equal to $2t\|f_\lambda\|_1^2$. Therefore,

$$\mathbb{E}_\pi\Big[\Big(\int_0^t V(X_s)\,ds\Big)^2\Big] \le 4\big\{2t\|f_\lambda\|_1^2 + 2\|f_\lambda\|_\pi^2 + (\lambda t)^2\|f_\lambda\|_\pi^2\big\}.$$

By (2.14), $\lambda\|f_\lambda\|_\pi^2 + \|f_\lambda\|_1^2$ is equal to $\langle f_\lambda, V\rangle_\pi$. Choosing $\lambda = 1/t$ we can therefore estimate the right-hand side of the previous equation by $12t\langle f_\lambda, V\rangle_\pi$ and the lemma follows. □

Proposition 4.9 *Fix a function V in $L^2(\pi)$ and denote by f_λ the solution of the resolvent equation* (2.13). *Suppose that there exist finite constants $\kappa < 1$, $C_0 > 0$ such that $\langle f_\lambda, V\rangle_\pi \ge C_0\lambda^{-\kappa}$ for all λ in $(0, 1]$. Then, for any $\kappa' \in (0, \kappa)$ there exists $C_1 > 0$ such that*

$$\int_0^t \mathbb{E}_\pi\Big[\Big(\int_0^s V(X_r)\,dr\Big)^2\Big]\,ds \ge C_1 t^{2+\kappa'}$$

for all t sufficiently large.

Proof Let

$$\varphi(t) = \int_0^t \mathbb{E}_\pi\Big[\Big(\int_0^s V(X_r)\,dr\Big)^2\Big]\,ds.$$

Of course, $\varphi(t)$ is a non-decreasing positive function. Moreover, by Schwarz inequality, $\varphi(t) \le (1/3)t^3\|V\|_\pi^2$. On the other hand, a long but straightforward computation gives that

$$\int_0^\infty e^{-\lambda t}\varphi(t)\,dt = \frac{2}{\lambda^3}\int_0^\infty e^{-\lambda t}\langle P_t V, V\rangle_\pi\,dt = \frac{2}{\lambda^3}\langle f_\lambda, V\rangle_\pi ,$$

where f_λ is the solution of the resolvent equation (2.13). Hence, by assumption,

$$\int_0^\infty e^{-\lambda t}\varphi(t)\,dt \ge C_0\lambda^{-3-\kappa}$$

for all $0<\lambda\le 1$ and some finite constant C_0.

The function $\varphi(t)$ satisfies therefore all assumptions of Lemma 4.10 below with $\rho=1$. In particular, for all $\kappa'<\kappa$, there exists a constant C_1 such that $\varphi(t)\ge C_1 t^{2+\kappa'}$ for all t sufficiently large. This concludes the proof of the proposition. □

We conclude this section with a Tauberian type estimate.

Lemma 4.10 *Suppose that $\varphi:[0,\infty)\to\mathbb{R}$ is a non-decreasing, positive function, for which there exist $\rho,\overline{C},\underline{c}>0$ and $\kappa\in(0,\rho)$ such that for all $t>0$, $\varphi(t)\le\overline{C}t^{2+\rho}$; and for all $\lambda\in(0,1)$,*

$$\int_0^{+\infty} e^{-\lambda t}\varphi(t)\,dt \ge \underline{c}\lambda^{-3-\kappa}.$$

Then, for any $\kappa'\in(0,\kappa)$ there exists $\tilde{c}>0$ such that $\varphi(t)\ge\tilde{c}t^{2+\kappa'}$ for all t sufficiently large.

Proof Fix $\kappa'<\kappa$ and choose $\gamma>0$ such that $(\kappa-2\gamma)/(1+\gamma)>\kappa'$. By assumption,

$$\underline{c}\lambda^{-3-\kappa}\le\int_0^\infty e^{-\lambda t}\varphi(t)\,dt=\frac{1}{\lambda}\int_0^\infty e^{-t}\varphi(t/\lambda)\,dt.$$

Last expression can be rewritten as

$$\frac{1}{\lambda}\int_0^{\lambda^{-\gamma}} e^{-t}\varphi(t/\lambda)\,dt+\frac{1}{\lambda}\int_{\lambda^{-\gamma}}^\infty e^{-t}\varphi(t/\lambda)\,dt.$$

Since φ is monotone and since $\varphi(t)\le\overline{C}t^{2+\rho}$ the previous sum can be estimated by

$$\frac{1}{\lambda}\varphi\big(\lambda^{-1-\gamma}\big)+\frac{\overline{C}}{\lambda^{3+\rho}}\int_{\lambda^{-\gamma}}^\infty e^{-t}t^{2+\rho}\,dt\le\frac{1}{\lambda}\varphi\big(\lambda^{-1-\gamma}\big)+\frac{c_1}{\lambda^{3+\rho}}e^{-(1/2)\lambda^{-\gamma}}$$

for some constant $c_1>0$ because the integral $\int_{[0,\infty)}\exp\{-t/2\}t^{2+\rho}\,dt$ is finite. Up to this point, we have proved that

$$\frac{\underline{c}}{\lambda^{3+\kappa}}\le\frac{1}{\lambda}\varphi\big(\lambda^{-1-\gamma}\big)+\frac{c_1}{\lambda^{3+\rho}}e^{-(1/2)\lambda^{-\gamma}}.$$

Since

$$\frac{c_1}{\lambda^{3+\rho}}e^{-(1/2)\lambda^{-\gamma}}\le\frac{\underline{c}}{2\lambda^{3+\kappa}}$$

for sufficiently small $\lambda > 0$, we obtain that $\varphi(\lambda^{-1-\gamma}) \geq C_0\lambda^{-2-\kappa}$ for some finite constant C_0 and all λ sufficiently small. Replacing λ by $t^{-1/(1+\gamma)}$, we deduce that $\varphi(t) \geq C_0 t^{2+\kappa_0}$ for all t sufficiently large, where $\kappa_0 = (\kappa - 2\gamma)/(1+\gamma) > \kappa'$. This concludes the proof of the lemma. □

4.5 Comments and References

A discussion on the references is presented at the end of Chap. 10.

References

Kesavan S (1989) Topics in functional analysis and applications. Wiley, New York

Part II
Simple Exclusion Processes

Chapter 5
The Simple Exclusion Process

Among the simplest and most widely studied interacting particle systems is the simple exclusion process. It represents the evolution of random walks on the lattice $\mathbb{Z}^d$ with a hard-core interaction which prevents more than one particle per site. To describe the dynamics, fix a *finite range* probability measure $p(\cdot)$ on $\mathbb{Z}^d$ *which does not charge the origin* and distribute particles on the lattice in such a way that each site is occupied by at most one particle. Particles evolve on $\mathbb{Z}^d$ as continuous-time random walks with translation-invariant transition probability $p(x, y) = p(y - x)$. Each time a particle tries to jump over a site already occupied, the jump is suppressed to respect the exclusion rule.

This informal description corresponds to a Feller process defined as follows. Consider the space $\mathbb{X} = \{0, 1\}^{\mathbb{Z}^d}$ endowed with the product topology which turns $\mathbb{X}$ into a metrizable, compact space. Denote the states of $\mathbb{X}$ by the Greek letters η, ξ, so that $\eta(x)$ is equal to 1 (resp. 0) if the site $x \in \mathbb{Z}^d$ is occupied (resp. vacant) for the configuration η. Let $C(\mathbb{X})$ be the collection of continuous functions on $\mathbb{X}$, regarded as a Banach space with norm

$$\|f\| = \sup_{\eta \in \mathbb{X}} \big| f(\eta) \big|,$$

and let $\mathscr{C}$ be the subset of cylinder functions, i.e., the continuous functions which depend on the configuration η only through a finite number of variables $\eta(x)$.

Let L be the operator defined on $\mathscr{C}$ by

$$(Lf)(\eta) = \sum_{x,z \in \mathbb{Z}^d} \eta(x)\big[1 - \eta(x+z)\big]p(z)\big[f\big(\sigma^{x,x+z}\eta\big) - f(\eta)\big], \tag{5.1}$$

where $\sigma^{x,y}\eta$ is the configuration obtained from η by interchanging the occupation variables $\eta(x)$, $\eta(y)$:

$$\big(\sigma^{x,y}\eta\big)(z) = \begin{cases} \eta(z) & \text{if } z \neq x, y, \\ \eta(y) & \text{if } z = x, \\ \eta(x) & \text{if } z = y. \end{cases} \tag{5.2}$$

T. Komorowski et al., *Fluctuations in Markov Processes*,
Grundlehren der mathematischen Wissenschaften 345,
DOI 10.1007/978-3-642-29880-6_5,

It is not difficult to check that the operator L defined on $\mathscr{C}$ is a Markov pre-generator, (Liggett, 1985, Definition I.2.1). By Liggett (1985, Theorem I.3.9) the closure of L, still denoted by L, is a Markov generator and $\mathscr{C}$ is a core for L. Denote by $\{S(t) : t \geq 0\}$ the strictly Markovian, Feller semigroup on $C(\mathbb{X})$ associated to the generator L through the Hille–Yosida theorem.

Let $D([0,\infty),\mathbb{X})$ be the space of r.c.l.l. trajectories $\eta : [0,\infty) \to \mathbb{X}$ and let $\{\Pi_t : t \geq 0\}$ be the canonical projections defined by $\Pi_t(\eta) = \eta_t$. We shall represent by $\mathscr{F}^o$ the smallest σ-algebra on $D([0,\infty),\mathbb{X})$ which turns the projections Π_s, $s \geq 0$, measurable; and by $\mathscr{F}^o_t$ the smallest σ-algebra relative to which all the mappings Π_s, $0 \leq s \leq t$, are measurable.

Denote by $\{\mathbb{P}_\eta : \eta \in \mathbb{X}\}$ the normal Markov process associated to the semigroup $\{S(t) : t \geq 0\}$. Recall from Chap. 2 that $\{\mathbb{P}_\eta : \eta \in \mathbb{X}\}$ is a family of probability measures defined on $(D([0,\infty),\mathbb{X}), \mathscr{F}^o)$ characterized by the properties

(1) $\mathbb{P}_\eta[\eta_0 = \eta] = 1$ for all $\eta \in \mathbb{X}$;
(2) For every $A \in \mathscr{F}^o$, the mapping $\eta \to \mathbb{P}_\eta[A]$ is measurable;
(3) For all $\eta \in \mathbb{X}$, $f \in C(\mathbb{X})$, $s, t \geq 0$,

$$\mathbb{E}_\eta\big[f(\eta_{t+s})|\mathscr{F}^o_s\big] = \big(S(t)f\big)(\eta_s), \quad \mathbb{P}_\eta \text{ a.s.},$$

where $\mathbb{E}_\eta$ represents the expectation with respect to $\mathbb{P}_\eta$.

Fix a cylinder function V. The purpose of this chapter is to obtain sufficient conditions which guarantee a central limit theorem for

$$\frac{1}{\sqrt{t}} \int_0^t V(\eta_s)\, ds. \tag{5.3}$$

The chapter is organized as follows. In Sect. 5.1 we present the main properties of exclusion processes and show that all assumptions formulated at the beginning of Chap. 2 are fulfilled for the exclusion process. In Sect. 5.2 we prove a central limit theorem for additive functionals of the exclusion process in the case where the probability measure p is symmetric. In Sect. 5.3, we show that the generator of the exclusion process associated to a probability measure p which has mean zero satisfies a sector condition. We deduce from this result a central limit theorem for (5.3) following the approach presented in Sect. 2.7.3. In Sects. 5.4, 5.5 and 5.6, we examine the case where the probability measure p has a non-vanishing mean. We show that in this case the L^2 spaces associated to the stationary states of the exclusion process can be decomposed as a direct sum of orthogonal spaces and that the generator of the process satisfies in dimension $d \geq 3$ the graded sector conditions stated in Sect. 2.7.4. This estimate provides a central limit theorem for (5.3) for local functions V in $\mathscr{H}_{-1}$. In the last section we state some results on transient Markov processes needed in this and in the following chapters.

5.1 Exclusion Processes

In this section, we show that all assumptions formulated at the beginning of Chap. 2 are fulfilled for the exclusion process. Throughout the second part of this book, we assume that the probability measure $p(\cdot)$ satisfies the following assumptions.

(1) To avoid degeneracies, we assume that the probability measure $p(\cdot)$ is irreducible in the sense that the set $\{x : p(x) + p(-x) > 0\}$ generates $\mathbb{Z}^d$: for any pair of sites x, y in $\mathbb{Z}^d$, there exists $M \geq 1$ and a sequence $x = x_0, \dots, x_M = y$ such that $p(x_i, x_{i+1}) + p(x_{i+1}, x_i) > 0$ for $0 \leq i \leq M - 1$, where $p(x, y) = p(y - x)$.
(2) We also suppose that the probability measure p has finite range: there exists R in $\mathbb{N}$ such that $p(z) = 0$ for all sites z such that $|z| \geq R$.
(3) We assume that p does not charge the origin: $p(0) = 0$.

Sometimes, we renormalize the transition probability in a slightly different way and we assume $\sum_x p(x) = d$ instead of 1.

We start this section by describing the stationary ergodic states of the simple exclusion process. The conservation of the total number of particles by the dynamics is reflected in the existence of a one-parameter family of invariant measures. For $0 \leq \alpha \leq 1$, denote by ν_α the Bernoulli product measure of parameter α. This means that under ν_α the variables $\{\eta(x), x \in \mathbb{Z}^d\}$ are independent with marginals given by

$$\nu_\alpha\bigl\{\eta(x) = 1\bigr\} = \alpha = 1 - \nu_\alpha\bigl\{\eta(x) = 0\bigr\}.$$

Proposition 5.1 *The Bernoulli measures $\{\nu_\alpha, 0 \leq \alpha \leq 1\}$ are invariant for simple exclusion processes.*

Proof By a simple change of variables, for any cylinder functions f, g and any bond $\{x, y\}$,

$$\begin{aligned} &\int f\bigl(\sigma^{x,y}\eta\bigr)g(\eta)\eta(x)\bigl[1 - \eta(y)\bigr]\nu_\alpha(d\eta) \\ &\quad = \int f(\eta)g\bigl(\sigma^{x,y}\eta\bigr)\eta(y)\bigl[1 - \eta(x)\bigr]\nu_\alpha(d\eta). \end{aligned} \tag{5.4}$$

This identity, the fact that $1 = \sum_{z\in\mathbb{Z}^d} p(z) = \sum_{z\in\mathbb{Z}^d} p(-z)$ and a change in the order of summation prove that

$$\int Lf \, d\nu_\alpha = 0$$

for all cylinder functions f. This proves the proposition in view of Liggett (1985, Proposition I.2.13). □

Notice that the family of invariant measures ν_α is parametrized by the density, for

$$E_{\nu_\alpha}\big[\eta(0)\big] = \nu_\alpha\big\{\eta(0) = 1\big\} = \alpha,$$

where E_{ν_α} stands for the expectation with respect to ν_α.

For a probability measure μ in $\mathbb{X}$, denote by $\mathbb{P}_\mu$ the measure on $(D([0,\infty),\mathbb{X}),\mathscr{F}^o)$ given by $\int \mathbb{P}_\eta\,\mu(d\eta)$. Expectation with respect to $\mathbb{P}_\mu$ is denoted by $\mathbb{E}_\mu$. For $0 \le \alpha \le 1$, denote by $(D([0,\infty),\mathbb{X}),\mathscr{F},\mathbb{P}_{\nu_\alpha},\{\mathscr{F}_t : t \ge 0\})$ the usual augmentation of the filtered space $(D([0,\infty),\mathbb{X}),\mathscr{F}^o,\mathbb{P}_{\nu_\alpha},\{\mathscr{F}_t^o : t \ge 0\})$. By Theorem 8.11 and Proposition 8.12 of Chap. 1 of Blumenthal and Getoor (1968), $\{\eta_t : t \ge 0\}$ is a strong Markov process with respect to the augmented filtration.

By Schwarz inequality, for any cylinder function f,

$$\big[S(t)f(\eta)\big]^2 \le S(t)f^2(\eta).$$

In particular, since ν_α is a stationary measure for any $\alpha \in [0,1]$,

$$\int \big[S(t)f(\eta)\big]^2 \nu_\alpha(d\eta) \le \int f(\eta)^2 \nu_\alpha(d\eta).$$

Therefore, the semigroup $S(t)$ extends to a Markov semigroup on $L^2(\nu_\alpha)$. Approximating a function in $L^2(\nu_\alpha)$ by cylinder functions, one can show that the semigroup $\{S(t) : t \ge 0\}$ is strongly continuous in $L^2(\nu_\alpha)$. Its generator L_{ν_α} is the closure of L in $L^2(\nu_\alpha)$. Since the density α remains fixed, to keep notation simple, we denote L_{ν_α} by L and by $\mathscr{D}(L)$ the domain of L in $L^2(\nu_\alpha)$.

Denote by L_{p^*} the infinitesimal generator in $L^2(\nu_\alpha)$ whose action on cylinder functions is defined by (5.1) with the probability measure $p^*(x) = p(-x)$ in place of p. Denote by L^* the adjoint of L in $L^2(\nu_\alpha)$. The first argument of the next result is taken from Aubin (1979).

Lemma 5.2 *The adjoint operator L^* is a generator and $L^* = L_{p^*}$. In particular, $\mathscr{C}$ is a common core for L and L^*.*

Proof We first show that L^* is closed. We claim that the space $\mathfrak{A} = \{(f, L^*f) : f \in \mathscr{D}(L^*)\}$ is the orthogonal of the space $\mathfrak{B} = \{(Lg, -g) : g \in \mathscr{D}(L)\}$: $\mathfrak{A} = \mathfrak{B}^\perp$. A simple computation shows that (f, L^*f), $f \in \mathscr{D}(L^*)$, is orthogonal to $\mathfrak{B}$. On the other hand, if a pair $(f,h) \in L^2(\nu_\alpha) \times L^2(\nu_\alpha)$ is orthogonal to $\mathfrak{B}$, $\langle f, Lg\rangle_{\nu_\alpha} = \langle h, g\rangle_{\nu_\alpha}$ for all $g \in \mathscr{D}(L)$, where $\langle\cdot,\cdot\rangle_{\nu_\alpha}$ stands for the inner product in $L^2(\nu_\alpha)$. This identity asserts that f belongs to $\mathscr{D}(L^*)$ and that $h = L^*f$, which proves the claim.

It is now easy to deduce that L^* is closed. Consider a sequence $\{f_n : n \ge 1\}$ of functions in $\mathscr{D}(L^*)$ such that f_n, L^*f_n converge in $L^2(\nu_\alpha)$ to f, h, respectively. By the claim, (f_n, L^*f_n) belongs to $\mathfrak{A} = \mathfrak{B}^\perp$. Since $\mathfrak{B}^\perp$ is closed, $(f,h) \in \mathfrak{B}^\perp = \mathfrak{A}$, so that $f \in \mathscr{D}(L^*)$ and $L^*f = h$.

By (5.4), for every cylinder function f, g,

$$\langle Lf, g\rangle_{\nu_\alpha} = \langle f, L_{p^*}g\rangle_{\nu_\alpha}. \tag{5.5}$$

Since $\mathscr{C}$ is a core for L, this identity can be extended to $f \in \mathscr{D}(L)$. Hence, $\mathscr{C} \subset \mathscr{D}(L^*)$ and $L^* g = L_{p^*} g$ for $g \in \mathscr{C}$. This proves that the domain $\mathscr{D}(L^*)$ is dense in $L^2(\nu_\alpha)$ and that 1 belongs to $\mathscr{D}(L^*)$ and $L^* 1 = L_{p^*} 1 = 0$.

Denote by $\{G_\lambda : \lambda > 0\}$, the resolvent associated to the generator L: $G_\lambda = (\lambda - L)^{-1}$ and keep in mind that G_λ is a bounded operator, $\|\lambda G_\lambda\| \le 1$. Let G_λ^*, $\lambda > 0$, be the adjoint of G_λ. It is easy to show that $\mathscr{D}(G_\lambda^*) = L^2(\nu_\alpha)$, $\|G_\lambda^*\| = \|G_\lambda\|$, $G_\lambda^*(L^2(\nu_\alpha)) \subset \mathscr{D}(L^*)$ and $G_\lambda^* = (\lambda - L^*)^{-1}$. In particular, the range of $\lambda - L^*$ is $L^2(\nu_\alpha)$ and L^* is dissipative, (Yosida, 1995), as for every $f \in \mathscr{D}(L^*)$,

$$\begin{aligned} \langle f, (-L^*) f \rangle_{\nu_\alpha} &= \lim_{\lambda \to 0} \langle f, (\lambda - L^*) f \rangle_{\nu_\alpha} = \lim_{\lambda \to 0} \langle G_\lambda^* (\lambda - L^*) f, (\lambda - L^*) f \rangle_{\nu_\alpha} \\ &= \lim_{\lambda \to 0} \langle (\lambda - L^*) f, G_\lambda (\lambda - L^*) f \rangle_{\nu_\alpha} \\ &= \lim_{\lambda \to 0} \langle (\lambda - L) G_\lambda (\lambda - L^*) f, G_\lambda (\lambda - L^*) f \rangle_{\nu_\alpha} \ge 0 \end{aligned}$$

because L is dissipative. This proves that L^* is a generator.

It remains to show that $L^* = L_{p^*}$. Denote by $\{G_{p^*,\lambda} : \lambda > 0\}$ the resolvent associated to the generator L_{p^*}. It is enough to show that $G_{p^*,\lambda} = G_\lambda^*$ since the latter operator is the resolvent of the generator L^*.

We claim that for every $f \in \mathscr{D}(L_{p^*})$, $G_\lambda^*(\lambda - L_{p^*}) f = f$. Indeed, assume first that f belongs to $\mathscr{C}$. In this case, by definition of G_λ^*, for every function $g \in L^2(\nu_\alpha)$,

$$\langle G_\lambda^*(\lambda - L_{p^*}) f, g \rangle_{\nu_\alpha} = \langle (\lambda - L_{p^*}) f, G_\lambda g \rangle_{\nu_\alpha}.$$

Since $G_\lambda g$ belongs to $\mathscr{D}(L)$ and since $\mathscr{C}$ is a core for L, approximating $G_\lambda g$ by a sequence $\{h_n : n \ge 1\}$ in $\mathscr{C}$ such that $h_n \to G_\lambda g$, $L h_n \to L G_\lambda g$, by (5.5) we obtain that the previous expression is equal to

$$\langle f, (\lambda - L) G_\lambda g \rangle_{\nu_\alpha} = \langle f, g \rangle_{\nu_\alpha},$$

which proves the claim for functions f in the core $\mathscr{C}$. Since $\mathscr{C}$ is also a core for the generator L_{p^*}, and since G_λ^* is a bounded operator, an approximation argument permits to extend the identity $G_\lambda^*(\lambda - L_{p^*}) f = f$ to functions f in the domain $\mathscr{D}(L_{p^*})$. This proves the claim.

Fix a function $h \in L^2(\nu_\alpha)$. Let $h_\lambda \in \mathscr{D}(L_{p^*})$ be the solution of $(\lambda - L_{p^*}) h_\lambda = h$, which exists because L_{p^*} is a generator. By the previous identity and by definition of $G_{p^*,\lambda}$,

$$G_\lambda^* h = G_\lambda^*(\lambda - L_{p^*}) h_\lambda = h_\lambda = G_{p^*,\lambda}(\lambda - L_{p^*}) h_\lambda = G_{p^*,\lambda} h,$$

which proves that $G_{p^*,\lambda} = G_\lambda^*$. □

The simple exclusion process is said to be symmetric if the probability measure p is symmetric, $p(z) = p(-z)$, mean zero if p has zero average, $\sum_z z p(z) = 0$, and asymmetric otherwise. In the symmetric case where $p(-x) = p(x)$, it follows from (5.5) that L is a symmetric operator in $L^2(\nu_\alpha)$, or equivalently, that ν_α is a reversible measure for the symmetric simple exclusion. The previous result shows that in fact L is self-adjoint.

Denote by $s(\cdot)$ (resp. $a(\cdot)$) the symmetric (resp. asymmetric) part of the probability measure p:

$$s(x) = (1/2)\{p(x) + p(-x)\}, \qquad a(x) = (1/2)\{p(x) - p(-x)\}. \tag{5.6}$$

We can decompose the operator L into a symmetric and anti-symmetric part as $L = S + A$ where, for a cylinder function f,

$$(Sf)(\eta) = \sum_{x,z\in\mathbb{Z}^d} \eta(x)\big(1-\eta(x+z)\big)s(z)\big[f\big(\sigma^{x,x+z}\eta\big) - f(\eta)\big],$$

$$(Af)(\eta) = \sum_{x,z\in\mathbb{Z}^d} \eta(x)\big(1-\eta(x+z)\big)a(z)\big[f\big(\sigma^{x,x+z}\eta\big) - f(\eta)\big].$$

By (5.4) and an elementary computation, for every cylinder function f, g,

$$\langle Sf, g\rangle_{\nu_\alpha} = \langle f, Sg\rangle_{\nu_\alpha}, \qquad \langle Af, g\rangle_{\nu_\alpha} = -\langle f, Ag\rangle_{\nu_\alpha},$$

which explains the terminology. Moreover, by the explicit expression for the adjoint L^* obtained in Lemma 5.2, $S = (1/2)(L + L^*)$ and $A = (1/2)(L - L^*)$.

Performing a change of variables $x' = x + z$, $z' = -z$, we obtain

$$(Sf)(\eta) = \sum_{x,z\in\mathbb{Z}^d} \eta(x+z)\big(1-\eta(x)\big)s(z)\big[f\big(\sigma^{x,x+z}\eta\big) - f(\eta)\big]$$

because $s(\cdot)$ is symmetric. Adding the two previous formulas for Sf and since $\sigma^{x,x+z}\eta = \eta$ unless $\eta(x)(1-\eta(x+z)) + \eta(x+z)(1-\eta(x)) = 1$, we deduce the simpler form

$$(Sf)(\eta) = (1/2) \sum_{x,z\in\mathbb{Z}^d} s(z)\big[f\big(\sigma^{x,x+z}\eta\big) - f(\eta)\big]$$

for the operator S. In fact, S is the generator of the simple exclusion process with probability measure $s(\cdot)$. Since $s(\cdot)$ is symmetric, by Lemma 5.2, S is a self-adjoint operator on $L^2(\nu_\alpha)$.

For a cylinder function f, denote by $D(f)$ the Dirichlet form of f:

$$\begin{aligned} D(f) &= \big\langle f, (-Lf)\big\rangle_{\nu_\alpha} = \big\langle f, (-Sf)\big\rangle_{\nu_\alpha} \\ &= (1/4) \sum_{x,z\in\mathbb{Z}^d} s(z) \int \big[f\big(\sigma^{x,x+z}\eta\big) - f(\eta)\big]^2 \nu_\alpha(d\eta). \end{aligned} \tag{5.7}$$

This formula holds also for functions f in the domain $\mathscr{D}(L)$ of the generator, and the series defined on the right-hand side converge absolutely. This can be proved by an approximation argument which can be found in Liggett (1985, Lemma IV.4.3).

Theorem 5.3 *For any $\alpha \in [0,1]$, ν_α is ergodic for L.*

Proof Let $f \in L^2(\nu_\alpha)$ such that $S(t)f = f$ for any $t \ge 0$. Then $f \in \mathscr{D}(L)$ and $Lf = 0$. Multiplying this last equation by f and integrating, by (5.7) for functions in the domain $\mathscr{D}(L)$ we obtain

$$\sum_{x,z} s(z) \int \big[f\big(\sigma^{x,x+z}\eta\big) - f(\eta)\big]^2 \nu_\alpha(d\eta) = 0.$$

By assumption the support of $s(\cdot)$ generates $\mathbb{Z}^d$. We deduce that for any $x, y \in \mathbb{Z}^d$

$$f\big(\sigma^{x,y}\eta\big) = f(\eta) \qquad \nu_\alpha\text{-a.e.}$$

By De Finetti's theorem we conclude that f is constant ν_α-a.e. □

5.2 Central Limit Theorems for Additive Functionals

Fix $0 < \alpha < 1$ and a mean zero cylinder function V. In this section, we prove a central limit theorem for the additive functional $t^{-1/2} \int_0^t V(\eta_s)\, ds$ in the context of mean zero asymmetric simple exclusion process.

Denote by $\mathscr{H}_1$, $\mathscr{H}_{-1}$ the Hilbert spaces introduced in Sect. 2.2 with the measure π replaced by ν_α. Note that the dependence of $\mathscr{H}_1$, $\mathscr{H}_{-1}$ on the parameter α is omitted, as α is kept fixed.

Theorem 5.4 *Assume that the jump rate p has mean zero: $\sum_{x\in\mathbb{Z}^d} x p(x) = 0$. Fix $0 < \alpha < 1$ and consider a mean zero cylinder function V in $\mathscr{H}_{-1}$. Then, under $\mathbb{P}_{\nu_\alpha}$,*

$$\frac{1}{\sqrt{t}} \int_0^t V(\eta_s)\, ds$$

converges in distribution, as $t \uparrow \infty$, to a mean zero Gaussian variable with variance $\sigma^2(V) = 2\lim_{\lambda\to 0} \|f_\lambda\|_1^2$, where f_λ is the solution of the resolvent equation (5.8). *In the symmetric case $\sigma^2(V) = 2\|V\|_{-1}^2$.*

Actually, the result is stronger. It follows from Theorem 2.7 that the law of $t^{-1/2} \int_0^t V(\eta_s)\, ds$ conditioned on $\mathscr{F}_0$ converges in $L^1(\mathbb{P}_{\nu_\alpha})$ to a Gaussian variable. This remark holds for all central limit theorems proved in this second part of the book and will not be stated again.

Consider the resolvent equation

$$\lambda f_\lambda - L f_\lambda = V, \quad \lambda > 0. \tag{5.8}$$

We have shown in Theorem 2.14 that a central limit theorem follows from a uniform bound on the $\mathscr{H}_{-1}$ norm of the solution of the resolvent equation:

$$\sup_{0<\lambda<1} \|L f_\lambda\|_{-1} < \infty. \tag{5.9}$$

Assume that the jump rate is symmetric: $p(-x) = p(x)$. In this case the generator L is self-adjoint in $L^2(\nu_\alpha)$. In particular, by Sect. 2.7.1, (5.9) and Theorem 5.4 are in force.

5.3 The Mean Zero Asymmetric Case

Assume now that the jump rate has mean zero: $\sum_{x\in\mathbb{Z}^d} xp(x) = 0$. In this case the generator L satisfies a sector condition:

Proposition 5.5 *Assume that the probability measure p has mean zero: $\sum_{x\in\mathbb{Z}^d} xp(x) = 0$. There exists a finite constant B, depending only on $p(\cdot)$, such that*

$$\left|\langle -Lf, g\rangle_{\nu_\alpha}\right| \le B\langle -Lf, f\rangle_{\nu_\alpha}^{1/2}\langle -Lg, g\rangle_{\nu_\alpha}^{1/2}$$

for all cylinder functions f, g.

Theorem 5.4 follows from this proposition, Theorem 2.14 and Sect. 2.7.3. Moreover, by Proposition 2.7, $\sigma^2(V) = 2\lim_{\lambda\to 0}\|f_\lambda\|_1^2$.

We now turn to the proof of Proposition 5.5, which relies on a decomposition of the generator in cycle generators, similar to the one presented in Sect. 3.3 for bistochastic centered random walks. The proof is divided into three steps. We first show in Lemma 5.6 that all mean zero finite-range probability measures on $\mathbb{Z}^d$ which do not charge the origin are convex combinations of cyclic probability measures. Then, in Lemma 5.7, we prove that a sector condition holds for a finite convex combination of generators if it holds individually. Finally, in Lemma 5.9 we conclude the proof of Proposition 5.5 showing that a sector condition holds for an exclusion process associated to a cyclic probability measure.

Recall from Sect. 3.3 the definition of a cycle C, of the mean zero probability measure p_C on $\mathbb{Z}^d$ associated to a cycle C and of the probability measure $p_{\mathbf{C},\mathbf{w}}(\cdot)$ associated to a finite collection of cycles $\mathbf{C} = \{C_1, \dots, C_m\}$ and to a probability measure $\mathbf{w} = \{w_1, \dots, w_m\}$ on $\{1, \dots, m\}$, $m \ge 1$.

Lemma 5.6 *Fix a finite-range, mean zero probability measure p on $\mathbb{Z}^d$ which does not charge the origin. There exists $m \ge 1$, finite cycles $\mathbf{C} = \{C_1, \dots, C_m\}$ and a probability measure $\mathbf{w} = \{w_1, \dots, w_m\}$ such that $p = p_{\mathbf{C},\mathbf{w}}(\cdot)$.*

Proof The proof is divided into three steps.

Claim 1 Fix a finite-range mean zero probability measure $p(\cdot)$ on $\mathbb{Z}^d$ which does not charge the origin. Then, there exists a mean zero probability $q(\cdot)$ taking only rational values and with the same support as $p(\cdot)$.

To prove this claim, denote by $S = \{x_1, \dots, x_n\}$ the support of $p(\cdot)$ and by $v_1, \dots, v_d$ the n-dimensional vectors $v_j = (\langle x_1, e_j\rangle, \dots, \langle x_n, e_j\rangle)$. Here $\{e_1, \dots, e_d\}$

stands for the canonical basis of $\mathbb{R}^d$ and $\langle \cdot, \cdot \rangle$ for the usual inner product in $\mathbb{R}^d$. Since $p(\cdot)$ has mean zero, $p = (p(x_1), \dots, p(x_n))$ is orthogonal to $v_1, \dots, v_d$.

Denote by $\mathbb{V}$ the vector space of the real numbers over the rationals. The real numbers $p(x_1), \dots, p(x_n)$ can be decomposed in terms of $\mathbb{Q}$-linearly independent reals: There exists $1 \le \ell \le n$, $\mathbb{Q}$-linearly independent real numbers $r_1, \dots, r_\ell$ and rational numbers $\{q_k(x_1), \dots, q_k(x_n)\}$, $1 \le k \le \ell$, such that

$$p(x) = \sum_{k=1}^{\ell} r_k q_k(x)$$

for each x in the support of $p(\cdot)$. Since $p(\cdot)$ has mean zero, by definition of $q_k(x)$,

$$0 = \sum_{x \in S} x p(x) = \sum_{x \in S} x \sum_{k=1}^{\ell} r_k q_k(x) = \sum_{k=1}^{\ell} r_k \sum_{x \in S} x q_k(x).$$

Since r_k are $\mathbb{Q}$-linearly independent and since x are integer-valued vectors, for $1 \le k \le \ell$,

$$\sum_{x \in S} x q_k(x) = 0.$$

One might think that the claim is proved because the coefficients $q_k(\cdot)$, $1 \le k \le \ell$, assume only rational values. However, nothing stops $q_k(\cdot)$ being negative, preventing us turning into a probability measure. We just know from the previous identity that the vectors $q_k = (q_k(x_1), \dots, q_k(x_n))$, $1 \le k \le \ell$, are orthogonal to $v_1, \dots, v_d$, as well as any linear combination of them. Since $p(\cdot)$ is a probability measure and since $p = \sum_{1 \le k \le \ell} r_k q_k$, there is one linear combination of these vectors which is strictly positive in the sense that all its coordinates are strictly positive. Approximating $r_1, \dots, r_\ell$ by rationals $m_1, \dots, m_\ell$ we obtain a vector $\tilde{q} = \sum_{1 \le k \le \ell} m_k q_k$ with rational, strictly positive entries and orthogonal to $v_1, \dots, v_d$ because each vector q_k is orthogonal. Thus a multiple q' of $\tilde{q}$ is a probability measure taking only rational values and such that

$$\sum_{x \in S} x q'(x) = 0$$

because q' is orthogonal to $v_1, \dots, v_d$. This proves the claim.

Claim 2 Every finite-range mean zero probability measure $p(\cdot)$ which does not charge the origin can be decomposed as $(1 - w)\tilde{p}(\cdot) + w q(\cdot)$ for some $0 < w \le 1$, where $\tilde{p}(\cdot)$ is a mean zero probability measure whose support is *strictly* contained in the support of $p(\cdot)$ and $q(\cdot)$ is a mean zero probability measure taking only rational values and whose support is equal to the support of $p(\cdot)$.

Fix such a probability measure $p(\cdot)$ and recall that we denote its support by S. By Claim 1, there exists a mean zero probability measure $q(\cdot)$ taking only rational values and with the same support as $p(\cdot)$. Let

$$w = \min_{x \in S} \frac{p(x)}{q(x)}.$$

Since both p and q are probability measures, $0 < w \le 1$. If $w = 1$, $p = q$ and the claim is proved. If $0 < w < 1$, let $\tilde{p}$ be defined by

$$\tilde{p}(x) = \frac{1}{1-w}\{p(x) - wq(x)\}.$$

By definition of w, $\tilde{p}$ is a probability measure whose support is strictly contained in the support of $p(\cdot)$. It has mean zero because both p and q have. This proves the claim.

Claim 3 Every finite-range mean zero probability measure $p(\cdot)$ which does not charge the origin and which assumes only rational values is a cyclic measure. This has been shown right after the definition of cyclic probability measures in Sect. 3.3.

We may now conclude the proof of the lemma. By Claims 2 and 3, every finite-range mean zero probability measure $p(\cdot)$ which does not charge the origin can be written as $(1-w)\tilde{p}(\cdot) + wq(\cdot)$, where $q(\cdot)$ is a cyclic probability measure and $\tilde{p}$ has a support strictly contained in the support of $p(\cdot)$. Repeating the same argument for $\tilde{p}$, after a finite number of steps, we decompose $p(\cdot)$ as a convex combinations of cyclic probability measures and a probability $p'(\cdot)$ concentrated on at most one site. Since any mean zero probability concentrated on one site is the Dirac measure on the origin and since we are assuming that $p(\cdot)$ puts no weight on the origin, $p'(\cdot)$ vanishes, which concludes the proof the lemma. □

For a cycle C, denote by L_C the cycle generator associated to the probability measure p_C:

$$(L_C f)(\eta) = \sum_{x,z \in \mathbb{Z}^d} \eta(x)\big[1 - \eta(x+z)\big] p_C(z) \big[f\big(\sigma^{x,x+z}\eta\big) - f(\eta)\big]. \tag{5.10}$$

Lemma 5.6 states that the generator L of an exclusion process associated to a mean zero probability measure p can be written as $L = \sum_{1 \le j \le m} w_j L_{C_j}$ for a positive integer m, a probability $\{w_1, \dots, w_m\}$ and cycles $\{C_1, \dots, C_m\}$.

Let L be a generator which can be decomposed as a convex combination of two generators, $L = \theta L_1 + (1-\theta) L_2$ for some $0 < \theta < 1$. Denote by $\langle \cdot, \cdot \rangle$ the scalar product in $L^2(\nu)$, where ν is a stationary state. The next result states that L satisfies a sector condition if both L_1, L_2 do.

Lemma 5.7 *Assume that there exist finite constants B_1, B_2 such that*

$$\big|\langle -L_j f, g\rangle\big| \le B_j \langle -L_j f, f\rangle^{1/2} \langle -L_j g, g\rangle^{1/2}$$

for $j = 1, 2$ *and all functions* f, g *in a common core* $\mathscr{C}$ *for* L, L_1, L_2. *Then,*

$$\big|\langle -Lf, g\rangle\big| \le B\langle -Lf, f\rangle^{1/2}\langle -Lg, g\rangle^{1/2}$$

for $B = B_1 + B_2$ *and all functions* f, g *in* $\mathscr{C}$.

Proof Fix two functions f, g in $\mathscr{C}$. By the triangle inequality and by assumption,

$$\begin{aligned}\big|\langle -Lf, g\rangle\big| &\le \theta\big|\langle -L_1 f, g\rangle\big| + (1-\theta)\big|\langle -L_2 f, g\rangle\big| \\ &\le \theta B_1\langle -L_1 f, f\rangle^{1/2}\langle -L_1 g, g\rangle^{1/2} \\ &\quad + (1-\theta) B_2\langle -L_2 f, f\rangle^{1/2}\langle -L_2 g, g\rangle^{1/2}.\end{aligned}$$

Since L_1, L_2 are generators, $\langle -L_j f, f\rangle$ are positive so that $\langle -L_1 f, f\rangle$, $\langle -L_2 f, f\rangle$ are bounded above by $\theta^{-1}\langle -Lf, f\rangle$, $(1-\theta)^{-1}\langle -Lf, f\rangle$, respectively. The previous expression is thus less than or equal to

$$\{B_1 + B_2\}\langle -Lf, f\rangle^{1/2}\langle -Lg, g\rangle^{1/2},$$

which concludes the proof of the lemma. □

In view of the previous two lemmas, Proposition 5.5 follows from a sector condition for generators associated to cycles. This estimate is based on the following general statement.

Lemma 5.8 *Fix a probability space* $(\Omega, \mathscr{F}, \mu)$ *and consider a family* $T_1, \dots, T_n$ *of measure-preserving bijections. Assume that* $T_1 \circ \cdots \circ T_n = I$, *where* I *is the identity and consider the generator* L *defined on* $L^2(\mu)$ *by*

$$(Lf)(\eta) = \frac{1}{n}\sum_{j=1}^{n}\big[f(T_j\eta) - f(\eta)\big].$$

For all functions f, g *in* $L^2(\mu)$,

$$\big|\langle Lf, g\rangle_\mu\big| \le 2n\langle -Lf, f\rangle_\mu^{1/2}\langle -Lg, g\rangle_\mu^{1/2}.$$

Proof We first compute the Dirichlet form $D(f) = \langle -Lf, f\rangle_\mu$ associated to the generator L. Since the operators $T_1, \dots, T_n$ are measure-preserving, an elementary computation shows that L^*, the adjoint of L with respect to μ, is given by

$$\big(L^* f\big)(\eta) = \frac{1}{n}\sum_{j=1}^{n}\big[f\big(T_j^{-1}\eta\big) - f(\eta)\big].$$

In particular, an elementary computation shows that

$$\langle -Lf, f\rangle_\mu = \frac{1}{2n}\sum_{j=1}^{n}\int\big\{f(T_j\eta) - f(\eta)\big\}^2\mu(d\eta). \tag{5.11}$$

We may turn now to the proof. Fix two functions f, g in $L^2(\mu)$. Since the operators are measure-preserving, after a change of variables $\eta = T_{j+1} \circ \cdots \circ T_n \xi$, we obtain that

$$\begin{aligned}
&\langle Lf, g\rangle_\mu \\
&\quad = \frac{1}{n}\sum_{j=1}^{n} \int \{f(T_j \circ \cdots \circ T_n \eta) - f(T_{j+1} \circ \cdots \circ T_n \eta)\} g(T_{j+1} \circ \cdots \circ T_n \eta)\, \mu(d\eta).
\end{aligned}$$

Since $T_1 \circ \cdots \circ T_n = I$, $\sum_{j=1}^{n} f(T_j \circ \cdots \circ T_n \eta) - f(T_{j+1} \circ \cdots \circ T_n \eta) = f(T_1 \circ \cdots \circ T_n \eta) - f(\eta) = 0$. The previous expression is therefore equal to

$$\begin{aligned}
&\frac{1}{n}\sum_{j=1}^{n} \int \{f(T_j \circ \cdots \circ T_n \eta) - f(T_{j+1} \circ \cdots \circ T_n \eta)\} \\
&\qquad \times \{g(T_{j+1} \circ \cdots \circ T_n \eta) - g(\eta)\}\, \mu(d\eta).
\end{aligned}$$

Applying Schwarz inequality, we obtain that this expression is less than or equal to

$$\begin{aligned}
&\frac{1}{2An}\sum_{j=1}^{n} \int \{f(T_j \circ \cdots \circ T_n \eta) - f(T_{j+1} \circ \cdots \circ T_n \eta)\}^2 \mu(d\eta) \\
&\quad + \frac{A}{2}\sum_{j=1}^{n}\sum_{k=j+1}^{n} \int \{g(T_k \circ \cdots \circ T_n \eta) - g(T_{k+1} \circ \cdots \circ T_n \eta)\}^2 \mu(d\eta)
\end{aligned}$$

for every $A > 0$. In the last step we replaced the difference $g(T_{j+1} \circ \cdots \circ T_n \eta) - g(\eta)$ by $\sum_{j+1 \le k \le n} g(T_k \circ \cdots \circ T_n \eta) - g(T_{k+1} \circ \cdots \circ T_n \eta)$ and applied Schwarz inequality. By a change of variables, we obtain that the previous sum is equal to

$$\frac{1}{2An}\sum_{j=1}^{n} \int \{f(T_j \eta) - f(\eta)\}^2 \mu(d\eta) + \frac{An}{2}\sum_{k=1}^{n} \int \{g(T_k \eta) - g(\eta)\}^2 \mu(d\eta).$$

In view of (5.11), this expression is bounded above by

$$\frac{1}{A}\langle -Lf, f\rangle_\mu + An^2\langle -Lg, g\rangle_\mu.$$

To conclude the proof of the lemma, it remains to minimize over A. □

Fix a cycle $C = \{0, x_1, \ldots, x_n = 0\}$ of length n and recall from (5.10) the expression of the generator L_C associated to the cycle C. It can be written as

$$L_C = \sum_{z \in \mathbb{Z}^d} L^0_{C+z},$$

where L^0_{C+z} is the generator of an exclusion process in which all particles not in the set $C+z$ are frozen while the ones in $C+z$ may jump from $z+x_j$ to $z+x_{j+1}$. The formal generator of this process is given by

$$\big(L^0_{C+z} f\big)(\eta) = \frac{1}{n}\sum_{j=0}^{n-1} \eta(z+x_j)\big[1-\eta(z+x_{j+1})\big]\big[f\big(\sigma^{z+x_j,z+x_{j+1}}\eta\big) - f(\eta)\big].$$

An elementary computation shows that the Dirichlet form associated to the generator L^0_{C+z} is given by

$$\big\langle -L^0_{C+z} f, f\big\rangle_{\nu_\alpha} = \frac{1}{4n}\sum_{j=0}^{n-1} E_{\nu_\alpha}\big[\big\{f\big(\sigma^{z+x_j,z+x_{j+1}}\eta\big) - f(\eta)\big\}^2\big].$$

Notice that the indicator function $\eta(z+x_j)[1-\eta(z+x_{j+1})]$ gave place to a factor $1/4$. Summing over z we obtain the full Dirichlet form:

$$\langle -L_C f, f\rangle_{\nu_\alpha} = \sum_{z\in\mathbb{Z}^d}\big\langle -L^0_{C+z} f, f\big\rangle_{\nu_\alpha} = \frac{1}{4n}\sum_{z\in\mathbb{Z}^d}\sum_{j=0}^{n-1} D_{z+x_j,z+x_{j+1}}(f), \tag{5.12}$$

where $D_{x,y}(f)$ stands for the piece of the Dirichlet form associated to jumps over the bond $\{x,y\}$:

$$D_{x,y}(f) = E_{\nu_\alpha}\big[\big\{f\big(\sigma^{x,y}\eta\big) - f(\eta)\big\}^2\big]. \tag{5.13}$$

Lemma 5.9 *For every cylinder function f, g,*

$$\big|\langle L_C f, g\rangle_{\nu_\alpha}\big| \le 4n^2 \langle -L_C f, f\rangle^{1/2}_{\nu_\alpha} \langle -L_C g, g\rangle^{1/2}_{\nu_\alpha}.$$

Proof Fix two cylinder functions f, g. We have just seen that $\langle L_C f, g\rangle_{\nu_\alpha}$ can be written as $\sum_{z\in\mathbb{Z}^d}\langle L^0_{C+z} f, g\rangle_{\nu_\alpha}$. By Lemma 5.10 below, this expression is less than or equal to

$$4n^2 \sum_{z\in\mathbb{Z}^d}\big\langle -L^0_{C+z} f, f\big\rangle^{1/2}_{\nu_\alpha}\big\langle -L^0_{C+z} g, g\big\rangle^{1/2}_{\nu_\alpha}.$$

By Schwarz inequality, this expression is bounded above by

$$\begin{aligned} &4n^2\Big\{\sum_{z\in\mathbb{Z}^d}\big\langle -L^0_{C+z} f, f\big\rangle_{\nu_\alpha}\sum_{z\in\mathbb{Z}^d}\big\langle -L^0_{C+z} g, g\big\rangle_{\nu_\alpha}\Big\}^{1/2} \\ &\quad = 4n^2\langle -L_C f, f\rangle^{1/2}_{\nu_\alpha}\langle -L_C g, g\rangle^{1/2}_{\nu_\alpha}. \end{aligned}$$

This concludes the proof of the lemma. □

Lemma 5.10 *For every cylinder function f, g,*

$$\left|\langle L_C^0 f, g\rangle_{\nu_\alpha}\right| \le 4n^2 \langle -L_C^0 f, f\rangle_{\nu_\alpha}^{1/2} \langle -L_C^0 g, g\rangle_{\nu_\alpha}^{1/2}.$$

Proof Recall that $C = \{0, x_1, \dots, x_n = 0\}$. For $0 \le j \le n-1$, denote by T_j the operator $\sigma^{x_j, x_{j+1}}$. Fix a configuration ξ in $\{0,1\}^C$ with a particle at the origin and observe that $T_{n-1} \circ \cdots \circ T_0 \xi$ is a configuration with a particle at the origin and where each other particle moved one step backward (two steps if the particle is originally at x_1). More precisely, if $\{0, x_{j_1}, \dots, x_{j_m}\}$, $j_1 < \cdots < j_m$, stands for the occupied sites of the configuration ξ, the occupied sites of $T_{n-1} \circ \cdots \circ T_0 \xi$ are

$$\begin{cases} \{0, x_{j_1 - 1}, \dots, x_{j_m - 1}\} & \text{if } x_{j_1} \neq x_1, \\ \{0, x_{j_2 - 1}, \dots, x_{j_m - 1}, x_{n-1}\} & \text{if } x_{j_1} = x_1. \end{cases}$$

In particular, for configurations ξ with a particle at the origin, $T^{n-1}\xi = \xi$ if $T = T_{n-1} \circ \cdots \circ T_0$.

By definition of the operators T_j, we may write the scalar product $\langle L_C^0 f, g\rangle_{\nu_\alpha}$ as

$$\frac{1}{n} \sum_{j=0}^{n-1} \int \eta(x_j)\big[1 - \eta(x_{j+1})\big]\big\{f(T_j \eta) - f(\eta)\big\} g(\eta) \nu_\alpha(d\eta).$$

The presence of the indicators $\eta(x_j)[1 - \eta(x_{j+1})]$ in the previous formula prevents us applying Lemma 5.8 directly. The proof, however, goes through.

Notice that we may take out the indicator $[1 - \eta(x_{j+1})]$ in the previous formula since the difference $f(T_j\eta) - f(\eta)$ vanishes if $\eta(x_{j+1}) = 1$. For $0 \le k \le n-2$, perform the change of variable $\eta = T_{j-1} \circ \cdots \circ T_0 \circ T^k \xi$ to rewrite the previous sum as

$$\begin{aligned} &\frac{1}{n(n-1)} \sum_{k=0}^{n-2} \sum_{j=0}^{n-1} \int \xi(0)\big\{f\big(T_j \circ \cdots \circ T_0 \circ T^k \xi\big) - f\big(T_{j-1} \circ \cdots \circ T_0 \circ T^k \xi\big)\big\} \\ &\quad \times g\big(T_{j-1} \circ \cdots \circ T_0 \circ T^k \xi\big) \nu_\alpha(d\xi). \end{aligned}$$

Observe that the indicator $\eta(x_j)$ is transformed in $\xi(0)$ by the change of variables. Since the configuration ξ has a particle at the origin, $\sum_{k=0}^{n-2} \sum_{j=0}^{n-1} \{f(T_j \circ \cdots \circ T_0 \circ T^k \xi) - f(T_{j-1} \circ \cdots \circ T_0 \circ T^k \xi)\} = f(T^{n-1}\xi) - f(\xi)$ vanishes. In particular, we may add $g(\xi)$ in the previous formula and rewrite it as

$$\begin{aligned} &\frac{1}{n(n-1)} \sum_{k=0}^{n-2} \sum_{j=0}^{n-1} \int \xi(0)\big\{f\big(T_j \circ \cdots \circ T_0 \circ T^k \xi\big) - f\big(T_{j-1} \circ \cdots \circ T_0 \circ T^k \xi\big)\big\} \\ &\quad \times \big\{g\big(T_{j-1} \circ \cdots \circ T_0 \circ T^k \xi\big) - g(\xi)\big\} \nu_\alpha(d\xi). \end{aligned}$$

It remains to repeat the arguments presented at the end of the proof of Lemma 5.8 to estimate the previous expression by

$$\frac{1}{2An}\sum_{j=0}^{n-1}\int\{f(T_j\eta)-f(\eta)\}^2\nu_\alpha(d\eta)$$
$$+\frac{An(n-1)^2}{2}\sum_{j=0}^{n-1}\int\{g(T_j\eta)-g(\eta)\}^2\nu_\alpha(d\eta)$$

for every $A>0$. To conclude the proof of the lemma it remains to optimize over A and to recall the expression (5.12) of the Dirichlet form $\langle -L_C^0 f, f\rangle_{\nu_\alpha}$. □

We conclude this section with an alternative proof of the sector condition for generators associated to cyclic measures. It gives a factor $3/2$ instead of 2.

Lemma 5.11 *Fix a cycle C of length n. There exists a finite universal constant C_0 such that*

$$\big|\langle -L_C f, g\rangle_{\nu_\alpha}\big| \le C_0 n^{3/2}\langle -L_C f, f\rangle_{\nu_\alpha}^{1/2}\langle -L_C g, g\rangle_{\nu_\alpha}^{1/2}$$

for every cylinder function f, g.

Proof Denote the cycle C by $(y_0,\dots,y_{n-1},y_n=y_0)$. Fix two cylinder functions f, g and recall the definition of the generators L_{C+z}^0 as well as the computations presented just before the statement of the lemma. By definition of L_{C+z}^0,

$$\langle L_C f, g\rangle_{\nu_\alpha} = \sum_{z\in\mathbb{Z}^d}\big\langle L_{C+z}^0 f, g\big\rangle_{\nu_\alpha}.$$

Fix a site z. Denote by $\mathscr{F}_z$ the σ-algebra generated by all occupation variables outside $C+z$ and the total number of particles in $C+z$:

$$\mathscr{F}_z=\sigma\bigg(\sum_{y\in C+z}\eta(y),\big\{\eta(y),\,y\notin C+z\big\}\bigg).$$

Denote by g_z the conditional expectation of g given $\mathscr{F}_z$:

$$g_z=E_{\nu_\alpha}[g|\mathscr{F}_z].$$

Since g_z depends on the variables $\{\eta(y_j),1\le j\le n\}$ only through their sum, $(L_{C+z}^0)^* g_z=0$ and

$$\big\langle L_{C+z}^0 f, g\big\rangle_{\nu_\alpha}=\big\langle L_{C+z}^0 f, g-g_z\big\rangle_{\nu_\alpha}=E_{\nu_\alpha}\big[E_{\nu_\alpha}\big[\big(L_{C+z}^0 f\big)\{g-g_z\}|\mathscr{F}_z\big]\big].$$

Observe that the expectation $E_{\nu_\alpha}[\cdot|\mathscr{F}_z]$ corresponds to an expectation with respect to the canonical measure on $C+z$ since all particles outside $C+z$ are frozen and since $\sum_{y\in C+z}\eta(y)$ is fixed. Hence, g_z is the expectation of g with respect to the canonical measure on $C+z$. It follows from the explicit formula for the generator L^0_{C+z} that the previous conditional expectation is equal to

$$\frac{1}{n}\sum_{j=0}^{n-1} E_{\nu_\alpha}\big[\eta(z+y_j)\big[1-\eta(z+y_{j+1})\big]\big(T^{z+y_j,z+y_{j+1}}f\big)(\eta)\big[g(\eta)-g_z(\eta)\big]|\mathscr{F}_z\big].$$

By Schwarz inequality, this expression is bounded above by

$$\frac{A}{2n}\sum_{j=0}^{n-1} E_{\nu_\alpha}\big[\big(T^{z+y_j,z+y_{j+1}}f\big)(\eta)^2|\mathscr{F}_z\big]+\frac{1}{2A}E_{\nu_\alpha}\big[\big[g(\eta)-g_z(\eta)\big]^2|\mathscr{F}_z\big]$$

for every $A>0$. Taking expectation with respect to ν_α, in view of (5.13), the first term becomes

$$\frac{A}{2n}\sum_{j=0}^{n-1} D_{z+y_j,z+y_{j+1}}(f).$$

On the other hand, the second term is the variance of g with respect to the canonical measure on $C+z$. Thus, by the spectral gap for the symmetric simple exclusion process on a finite torus, there exists a finite and universal constant B_0 which makes the second term in the penultimate expression bounded by

$$\frac{B_0 n^2}{2A}\sum_{j=0}^{n-1} E_{\nu_\alpha}\big[\big[g\big(\sigma^{z+y_j,z+y_{j+1}}\eta\big)-g(\eta)\big]^2|\mathscr{F}_z\big].$$

Taking expectation with respect to ν_α and recollecting all previous estimates, we obtain that

$$\big|\langle L^0_{C+z}f,g\rangle_{\nu_\alpha}\big|\le\frac{A}{2n}\sum_{j=0}^{n-1} D_{z+y_j,z+y_{j+1}}(f)+\frac{B_0 n^2}{2A}\sum_{j=0}^{n-1} D_{z+y_j,z+y_{j+1}}(g).$$

Summing over z and minimizing over A we conclude that

$$\big|\langle L_C f,g\rangle_{\nu_\alpha}\big|^2\le B_1 n\bigg\{\sum_{z\in\mathbb{Z}^d}\sum_{j=0}^{n-1} D_{z+y_j,z+y_{j+1}}(f)\bigg\}\bigg\{\sum_{z\in\mathbb{Z}^d}\sum_{j=0}^{n-1} D_{z+y_j,z+y_{j+1}}(g)\bigg\}$$

for some finite universal constant B_1. This concludes the proof of the lemma in view of the formula (5.12) for the Dirichlet form $\langle -L_C f,f\rangle_{\nu_\alpha}$. □

5.4 Duality

The proof of the central limit theorem for additive functionals of asymmetric exclusion processes relies on the decomposition of the space $L^2(\nu_\alpha)$ as a direct sum of orthogonal spaces, along the lines introduced in Sect. 2.7.4.

For each $n \geq 0$, denote by $\mathscr{E}_n$ the subsets of $\mathbb{Z}^d$ with n points and let $\mathscr{E} = \bigcup_{n\geq 0} \mathscr{E}_n$ be the class of finite subsets of $\mathbb{Z}^d$. For each A in $\mathscr{E}$, let Ψ_A be the cylinder function

$$\Psi_A(\eta) = \prod_{x\in A} \frac{\eta(x) - \alpha}{\sqrt{\chi(\alpha)}},$$

where $\chi(\alpha) = \alpha(1-\alpha)$. By convention, $\Psi_\varnothing = 1$. It is easy to check that $\{\Psi_A, A \in \mathscr{E}\}$ is an orthonormal basis of $L^2(\nu_\alpha)$. For each $n \geq 1$, denote by $\mathscr{A}_n$ the subspace of $L^2(\nu_\alpha)$ generated by $\{\Psi_A, A \in \mathscr{E}_n\}$, so that $L^2(\nu_\alpha) = \bigoplus_{n\geq 0} \mathscr{A}_n$.

Let $\mathscr{G}_n = \bigoplus_{0\leq k\leq n} \mathscr{A}_k$. Functions in $\mathscr{G}_n \setminus \mathscr{G}_{n-1}$, $\mathscr{G} = \bigcup_{n\geq 0} \mathscr{G}_n$ are said to have degree n, finite degree, respectively, while functions in $\mathscr{A}_n$ are said to be monomials of degree n. For $n \geq 0$, denote by Π_n the orthogonal projection on $\mathscr{A}_n$ so that

$$f = \sum_{n\geq 0} \Pi_n f \quad \text{and} \quad \Pi_n f \in \mathscr{A}_n, \quad n \geq 0, \ f \in L^2(\nu_\alpha).$$

Clearly, every cylinder function has finite degree and the set $\mathscr{C}$ of cylinder functions is closed under the projections $\{\Pi_n : n \geq 0\}$. In particular, condition (2.39) is fulfilled.

Consider a cylinder function f. Since $\{\Psi_A : A \in \mathscr{E}\}$ is a basis of $L^2(\nu_\alpha)$, there exists a finitely supported function $\mathfrak{f} : \mathscr{E} \to \mathbb{R}$ such that

$$f = \sum_{A\in\mathscr{E}} \mathfrak{f}(A)\Psi_A = \sum_{n\geq 0} \sum_{A\in\mathscr{E}_n} \mathfrak{f}(A)\Psi_A.$$

The function $\mathfrak{f}$ represents the Fourier coefficients of the cylinder function f and is denoted by $\mathfrak{F}f$ when we want to stress its dependence on f. $\mathfrak{f} : \mathscr{E} \to \mathbb{R}$ is a function of finite support because f is a cylinder function. Moreover, the Fourier coefficients $\mathfrak{f}(A)$ depend not only on f but also on the density α: $\mathfrak{f}(A) = \mathfrak{f}(\alpha, A)$.

Denote by $\mathfrak{C}$ the space of finitely supported functions $\mathfrak{f} : \mathscr{E} \to \mathbb{R}$, by μ the counting measure on $\mathscr{E}$ and by $\langle \cdot, \cdot \rangle_\mu$ the inner product in $L^2(\mu)$. For any two cylinder functions $f = \sum_{A\in\mathscr{E}} \mathfrak{f}(A)\Psi_A$, $g = \sum_{A\in\mathscr{E}} \mathfrak{g}(A)\Psi_A$,

$$\langle f, g\rangle_{\nu_\alpha} = \sum_{A,B\in\mathscr{E}} \mathfrak{f}(A)\mathfrak{g}(B)\langle \Psi_A, \Psi_B\rangle_{\nu_\alpha} = \sum_{A\in\mathscr{E}} \mathfrak{f}(A)\mathfrak{g}(A) = \langle \mathfrak{f}, \mathfrak{g}\rangle_\mu. \tag{5.14}$$

In particular, the map $\mathfrak{F} : L^2(\nu_\alpha) \to L^2(\mu)$ is an isomorphism.

To examine how the generator acts on the Fourier coefficients we need to introduce some notation. For a finite subset A of $\mathbb{Z}^d$ and x, y in $\mathbb{Z}^d$, denote by $A_{x,y}$, the set defined by

$$A_{x,y} = \begin{cases} (A\backslash\{x\}) \cup \{y\} & \text{if } x \in A,\, y \notin A, \\ (A\backslash\{y\}) \cup \{x\} & \text{if } x \notin A,\, y \in A, \\ A & \text{otherwise.} \end{cases} \tag{5.15}$$

Fix a cylinder function $f = \sum_{A\in\mathscr{E}} (\mathfrak{F}f)(A)\Psi_A$. We claim that

$$Lf = \sum_{A\in\mathscr{E}} (\mathfrak{L}_\alpha \mathfrak{F} f)(A)\Psi_A,$$

where

$$\mathfrak{L}_\alpha = \mathfrak{S} + (1-2\alpha)\mathfrak{N} + \sqrt{\chi(\alpha)}\mathfrak{J}_+ + \sqrt{\chi(\alpha)}\mathfrak{J}_- \tag{5.16}$$

and $\mathfrak{S}$, $\mathfrak{N}$, $\mathfrak{J}_+$, $\mathfrak{J}_-$ are the operators defined by

$$\begin{aligned} (\mathfrak{S}\mathfrak{f})(A) &= (1/2) \sum_{x,y\in\mathbb{Z}^d} s(y-x)\big[\mathfrak{f}(A_{x,y}) - \mathfrak{f}(A)\big], \\ (\mathfrak{N}\mathfrak{f})(A) &= \sum_{\substack{x\in A \\ y\notin A}} a(y-x)\big[\mathfrak{f}(A_{x,y}) - \mathfrak{f}(A)\big], \\ (\mathfrak{J}_+\mathfrak{f})(A) &= 2 \sum_{x,y\in A} a(y-x)\mathfrak{f}\big(A\setminus\{y\}\big), \\ (\mathfrak{J}_-\mathfrak{f})(A) &= 2 \sum_{x,y\notin A} a(y-x)\mathfrak{f}\big(A\cup\{y\}\big). \end{aligned} \tag{5.17}$$

In other words,

$$\mathfrak{F}L = \mathfrak{L}_\alpha\mathfrak{F}.$$

In general $\mathfrak{L}_\alpha$ is not the generator of a Markov process. However, in the particular case where the probability measure $p(\cdot)$ is symmetric, the operators $\mathfrak{N}$, $\mathfrak{J}_+$, $\mathfrak{J}_-$ vanish and $\mathfrak{L}_\alpha$ becomes $\mathfrak{S}$, which restricted to $\mathscr{E}_n$ corresponds to the generator of the Markov process in which n particles evolve on $\mathbb{Z}^d$ according to symmetric exclusion random walks. In fact, $\mathfrak{S}$ is the operator on the Fourier coefficients associated to the symmetric part of the generator since, by (5.17),

$$Sf = \sum_{A\in\mathscr{E}} (\mathfrak{S}\mathfrak{f})(A)\Psi_A, \quad \text{that is} \quad \mathfrak{F}S = \mathfrak{S}\mathfrak{F}. \tag{5.18}$$

To prove (5.16), we compute $L\Psi_A$ for finite subsets A of $\mathbb{Z}^d$. By definition of the generator L,

$$
\begin{aligned}
(L\Psi_A)(\eta) &= \sum_{x,y\in\mathbb{Z}^d} \eta(x)\big[1-\eta(y)\big]p(y-x)\big[\Psi_A\big(\sigma^{x,y}\eta\big)-\Psi_A(\eta)\big] \\
&= \sum_{x,y\in\mathbb{Z}^d} \eta(x)p(y-x)\big[\Psi_A\big(\sigma^{x,y}\eta\big)-\Psi_A(\eta)\big]
\end{aligned}
$$

because on the set $\eta(x)\eta(y)=1$, $\sigma^{x,y}\eta=\eta$. Moreover, $\Psi_A(\sigma^{x,y}\eta)-\Psi_A(\eta)$ vanishes unless $x\in A$, $y\notin A$ or $x\notin A$, $y\in A$. In these two cases, $\Psi_A(\sigma^{x,y}\eta)=\Psi_{A_{x,y}}(\eta)$. These remarks together with the elementary identity

$$
\eta(z)\Psi_B = \begin{cases} (1-\alpha)\Psi_B+\sqrt{\chi(\alpha)}\Psi_{B\setminus\{z\}} & \text{if } z\in B, \\ \alpha\Psi_B+\sqrt{\chi(\alpha)}\Psi_{B\cup\{z\}} & \text{if } z\notin B \end{cases}
$$

give that

$$
\begin{aligned}
L\Psi_A = {} & \sum_{\substack{x\in A\\ y\notin A}} s(y-x)[\Psi_{A_{x,y}}-\Psi_A] + (2\alpha-1)\sum_{\substack{x\in A\\ y\notin A}} a(y-x)[\Psi_{A_{x,y}}+\Psi_A] \\
& + 2\sqrt{\chi(\alpha)}\sum_{\substack{x\in A\\ y\notin A}} a(y-x)\Psi_{A\cup\{y\}} - 2\sqrt{\chi(\alpha)}\sum_{\substack{x\in A\\ y\notin A}} a(y-x)\Psi_{A\setminus\{x\}},
\end{aligned}
$$

from which (5.16) follows after a change of variables.

We claim that the operators $\mathfrak{N}$ and $\mathfrak{J}_+ + \mathfrak{J}_-$ are anti-symmetric with respect to the counting measure. More precisely, for any finitely supported functions $\mathfrak{f}$, $\mathfrak{g}:\mathscr{E}\to\mathbb{R}$,

$$
\langle \mathfrak{N}\mathfrak{f},\mathfrak{g}\rangle_\mu = -\langle \mathfrak{f},\mathfrak{N}\mathfrak{g}\rangle_\mu, \qquad \langle \mathfrak{J}_+\mathfrak{f},\mathfrak{g}\rangle_\mu = -\langle \mathfrak{f},\mathfrak{J}_-\mathfrak{g}\rangle_\mu. \tag{5.19}
$$

Moreover, for any finitely supported functions $\mathfrak{f}:\mathscr{E}_1\to\mathbb{R}$, $\mathfrak{g}:\mathscr{E}_0\to\mathbb{R}$,

$$
(\mathfrak{S}\mathfrak{g})(\varnothing)=0, \quad (\mathfrak{N}\mathfrak{g})(\varnothing)=0, \qquad (\mathfrak{J}_-\mathfrak{f})(\varnothing)=0, \qquad (\mathfrak{J}_+\mathfrak{g})\big(\{z\}\big)=0 \tag{5.20}
$$

for all z in $\mathbb{Z}$.

The proof of identities (5.19) is based on the following observation. Since a is anti-symmetric, $\sum_{z\in\mathbb{Z}^d} a(z)=0$. In particular, for any finite set A and any site y, $\sum_{x\in A} a(y-x) = -\sum_{x\notin A} a(y-x)$.

Fix two functions $\mathfrak{f}$, $\mathfrak{g}$ in $\mathfrak{C}$. By the previous observation and since $a(\cdot)$ is anti-symmetric, $\sum_{x\in A, y\notin A} a(y-x) = -\sum_{x\in A, y\in A} a(y-x) = 0$, so that $\sum_{A\in\mathscr{E}}\sum_{x\in A, y\notin A} a(y-x)\mathfrak{f}(A)\mathfrak{g}(A)=0$. Therefore, by definition of $\mathfrak{N}$,

$$
\langle \mathfrak{N}\mathfrak{f},\mathfrak{g}\rangle_\mu = \sum_{A\in\mathscr{E}}\sum_{\substack{x\in A\\ y\notin A}} a(y-x)\mathfrak{f}(A_{x,y})\mathfrak{g}(A).
$$

Changing variables $B=A_{x,y}$, $x'=y$, $y'=x$, since $a(\cdot)$ is anti-symmetric, we obtain that the previous sum is equal to

$$
-\sum_{A\in\mathscr{E}}\sum_{\substack{x\in A\\ y\notin A}} a(y-x)\mathfrak{f}(A)\mathfrak{g}(A_{x,y}) = -\langle \mathfrak{f},\mathfrak{N}\mathfrak{g}\rangle_\mu.
$$

This proves the first identity in (5.19). The second one is analogous. On the other hand, (5.20) follows immediately from the definition of the operators $\mathfrak{J}_-$, $\mathfrak{J}_+$ and the fact that $\sum_{y\in\mathbb{Z}^d} a(y-x) = 0$ for all x.

The basis $\{\Psi_A, A \in \mathscr{E}\}$ permitted to decompose the generator L of the simple exclusion process into four pieces: $\mathfrak{S}$, $\mathfrak{N}$, $\mathfrak{J}_+$ and $\mathfrak{J}_-$. We may now define the operators $\mathscr{N}$, $\mathscr{J}_+$, $\mathscr{J}_-$ on $L^2(\nu_\alpha)$ corresponding to $\mathfrak{N}$, $\mathfrak{J}_+$ and $\mathfrak{J}_-$: For a cylinder function f, let

$$(\mathscr{N} f)(\eta) = \sum_{A\in\mathscr{E}} (\mathfrak{N}\mathfrak{F} f)(A)\Psi_A(\eta), \qquad (\mathscr{J}_\pm f)(\eta) = \sum_{A\in\mathscr{E}} (\mathfrak{J}_\pm \mathfrak{F} f)(A)\Psi_A(\eta).$$

Of course, with these definitions, $\mathfrak{F}\mathscr{N} = \mathfrak{N}\mathfrak{F}$, $\mathfrak{F}\mathscr{J}_\pm = \mathfrak{J}_\pm\mathfrak{F}$.

The descriptions of a cylinder function and of the generator in terms of their Fourier coefficients have several applications. One of the first is the possibility to obtain simple criteria to describe cylinder functions with finite $\mathscr{H}_{-1}$ norm.

The Spaces $\mathscr{H}_1$ and $\mathscr{H}_1(\mathfrak{S})$ Denote by $\|\cdot\|_1 = \|\cdot\|_{\nu_\alpha,1}$ the norm of the Hilbert space $\mathscr{H}_1$ and by $\langle\cdot,\cdot\rangle_{\nu_\alpha,1}$ its scalar product:

$$\langle f, g\rangle_{\nu_\alpha,1} = \big\langle f, (-S)g\big\rangle_{\nu_\alpha}$$

for cylinder functions f, g. Since S preserves the degrees of monomials, the previous expression is equal to $\sum_{n\ge 0}\langle \Pi_n f, (-S)\Pi_n g\rangle_{\nu_\alpha}$ so that

$$\|f\|_1^2 = \sum_{n\ge 0} \|\Pi_n f\|_1^2 = \sum_{n\ge 1} \|\Pi_n f\|_1^2.$$

The last identity follows from the fact that $\mathscr{A}_0$ is the space of constants and from the identity $S1 = 0$.

An elementary computation, based on the change of variables $\xi = \sigma^{x,y}\eta$, shows that

$$\big\langle f, (-S)g\big\rangle_{\nu_\alpha} = (1/4) \sum_{x,y\in\mathbb{Z}^d} s(y-x) \int \big(T^{x,y} f\big)\big(T^{x,y} g\big)\, d\nu_\alpha,$$

where, for a cylinder function f,

$$\big(T^{x,y} f\big)(\eta) = f\big(\sigma^{x,y}\eta\big) - f(\eta). \tag{5.21}$$

The previous remarks can be formulated in terms of the Fourier coefficients. Let $\mathscr{H}_1(\mathfrak{S})$ be the Hilbert space induced by the set $\mathfrak{C}$ of finitely supported functions endowed with the scalar product

$$\langle \mathfrak{f}, \mathfrak{g}\rangle_{\mu,1} := \big\langle \mathfrak{f}, (-\mathfrak{L}_\alpha)\mathfrak{g}\big\rangle_\mu, \quad \mathfrak{f}, \mathfrak{g} \in \mathfrak{C}.$$

Since $\mathfrak{N}$ and $\mathfrak{J}_+ + \mathfrak{J}_-$ are anti-symmetric, the symmetric part of the operator $\mathfrak{L}_\alpha$ with respect to the counting measure μ is $\mathfrak{S}$. In particular,

$$\langle \mathfrak{f}, \mathfrak{g}\rangle_{\mu,1} = \big\langle \mathfrak{f}, (-\mathfrak{S})\mathfrak{g}\big\rangle_\mu.$$

An elementary computation shows that for any finitely supported functions $\mathfrak{f}$, $\mathfrak{g} : \mathscr{E} \to \mathbb{R}$,

$$\langle \mathfrak{f}, \mathfrak{g} \rangle_{\mu,1} = \langle \mathfrak{f}, (-\mathfrak{S})\mathfrak{g} \rangle_{\mu} = -(1/2) \sum_{x,y \in \mathbb{Z}^d} s(y-x) \sum_{A \in \mathscr{E}} \mathfrak{f}(A)\{\mathfrak{g}(A_{x,y}) - \mathfrak{g}(A)\}$$

$$= (1/4) \sum_{x,y \in \mathbb{Z}^d} s(y-x) \sum_{A \in \mathscr{E}} \{\mathfrak{f}(A_{x,y}) - \mathfrak{f}(A)\}\{\mathfrak{g}(A_{x,y}) - \mathfrak{g}(A)\}. \quad (5.22)$$

The last identity is obtained through a change of variables $B = A_{x,y}$.

Of course the Hilbert spaces $\mathscr{H}_1$, $\mathscr{H}_1(\mathfrak{S})$ are isomorphic since for any cylinder functions f, g, by (5.14) and (5.18),

$$\langle f, g \rangle_{\nu_\alpha,1} = \langle f, (-S)g \rangle_{\nu_\alpha} = \langle \mathfrak{F} f, -\mathfrak{F}(Sg) \rangle_{\mu} = \langle \mathfrak{F} f, (-\mathfrak{S})\mathfrak{F} g \rangle_{\mu} = \langle \mathfrak{F} f, \mathfrak{F} g \rangle_{\mu,1}.$$

Denote also by Π_n the restriction to $\mathscr{E}_n$ of a finitely supported function $\mathfrak{f} : \mathscr{E} \to \mathbb{R}$: $(\Pi_n \mathfrak{f})(A) = \mathfrak{f}(A)\mathbf{1}\{A \in \mathscr{E}_n\}$. With this definition, $\mathfrak{F}(\Pi_n f) = \Pi_n \mathfrak{F} f$ and

$$\|\mathfrak{f}\|_1^2 = \sum_{n \geq 0} \|\Pi_n \mathfrak{f}\|_1^2 = \sum_{n \geq 1} \|\Pi_n \mathfrak{f}\|_1^2 \quad (5.23)$$

because $\mathscr{E}_0 = \{\varnothing\}$.

The following alternative formula for the $\mathscr{H}_1$ norm of a function of degree n is sometimes useful. Fix a function $\mathfrak{f} : \mathscr{E}_n \to \mathbb{R}$. By definition of $A_{x,y}$, in (5.22) we may restrict the sum to sets A which either contain x and do not contain y or the opposite. In particular, by symmetry,

$$\|\mathfrak{f}\|_1^2 = (1/2) \sum_{x,y \in \mathbb{Z}^d} s(y-x) \sum_{\substack{A \in \mathscr{E}_n \\ A \ni x, A \not\ni y}} \{\mathfrak{f}(A_{x,y}) - \mathfrak{f}(A)\}^2.$$

Performing the change of variables $A' = A \cup \{y\}$, we get that

$$\|\mathfrak{f}\|_1^2 = (1/2) \sum_{x,y \in \mathbb{Z}^d} s(y-x) \sum_{\substack{A \in \mathscr{E}_{n+1} \\ A \ni x, A \ni y}} \{\mathfrak{f}(A \setminus \{x\}) - \mathfrak{f}(A \setminus \{y\})\}^2. \quad (5.24)$$

The Spaces $\mathscr{H}_{-1}$ and $\mathscr{H}_{-1}(\mathfrak{S})$ Recall from Sect. 2.2 that $\mathscr{H}_{-1}$ is the Hilbert space induced by cylinder functions endowed with the norm $\|\cdot\|_{-1}^2$ defined by

$$\|f\|_{-1}^2 = \sup_{g \in \mathscr{C}} \{2\langle f, g \rangle_{\nu_\alpha} - \langle g, g \rangle_{\nu_\alpha,1}\},$$

while $\mathscr{H}_{-1}(\mathfrak{S})$ is the Hilbert space induced by finitely supported functions $\mathfrak{f} : \mathscr{E} \to \mathbb{R}$ endowed with the norm $\|\cdot\|_{-1}^2$ defined by

$$\|\mathfrak{f}\|_{-1}^2 = \sup_{\mathfrak{g} \in \mathfrak{C}} \{2\langle \mathfrak{f}, \mathfrak{g} \rangle_{\mu} - \langle \mathfrak{g}, \mathfrak{g} \rangle_{\mu,1}\}.$$

Since for cylinder functions f, g, $\langle f, g\rangle_{\nu_\alpha} = \langle \mathfrak{F} f, \mathfrak{F} g\rangle_\mu$ and $\langle g, g\rangle_{\nu_\alpha,1} = \langle \mathfrak{F} g, \mathfrak{F} g\rangle_{\mu,1}$, we have that $\|f\|_{-1}^2 = \|\mathfrak{F} f\|_{-1}^2$. The spaces $\mathscr{H}_{-1}$ and $\mathscr{H}_{-1}(\mathfrak{S})$ are therefore isomorphic. Moreover, by (2.43),

$$\|\mathfrak{f}\|_{-1}^2 = \sum_{n\geq 0} \|\Pi_n \mathfrak{f}\|_{-1}^2. \tag{5.25}$$

Note that $\|\Pi_0 \mathfrak{f}\|_{-1} = \infty$ if $\Pi_0 \mathfrak{f} \neq 0$.

Fix a finitely supported function $\mathfrak{f} : \mathscr{E}_n \to \mathbb{R}$, $n \geq 1$. Since $\mathfrak{S}$ restricted to $\mathscr{E}_n$ corresponds to the generator of n symmetric random walks evolving on $\mathbb{Z}^d$ with an exclusion rule,

$$\|\mathfrak{f}\|_{-1}^2 = \big\langle \mathfrak{f}, (-\mathfrak{S})^{-1}\mathfrak{f}\big\rangle_\mu = \sum_{A,B\in\mathscr{E}_n} \mathfrak{f}(A)\mathfrak{G}_n(A,B)\mathfrak{f}(B),$$

where $\mathfrak{G}_n(A,B)$ is the Green function associated to $\mathfrak{S}$ restricted to $\mathscr{E}_n$.

By (5.45), n symmetric random walks evolving on $\mathbb{Z}^d$ with an exclusion rule form a transient Markov process if $nd \geq 3$. In particular, the Green function $\mathfrak{G}_n(A,B)$ is finite and we have the following result.

Lemma 5.12 *Fix $0 < \alpha < 1$. A mean zero cylinder function f belongs to $\mathscr{H}_{-1}$ if*

$$\sum_{A\in\mathscr{E}_n} (\mathfrak{F} f)(\alpha, A) = 0$$

for all n such that $nd \leq 2$.

Proof Fix $0 < \alpha < 1$ and a mean zero cylinder function f. We have just seen that

$$\|f\|_{-1}^2 = \|\mathfrak{F} f\|_{-1}^2 = \sum_{n\geq 1} \|\Pi_n \mathfrak{F} f\|_{-1}^2.$$

The sum starts from $n = 1$ because by assumption $(\Pi_0 \mathfrak{F} f)(\varnothing)$, which is the expectation of f with respect to ν_α, vanishes.

Recall that $\mathfrak{G}_n$ is the Green function associated to the Markov generator $\mathfrak{S}$ restricted to $\mathscr{E}_n$. By Landim (1998), $\mathfrak{G}_n(A,B)$ is uniformly bounded for $nd \geq 3$. In particular, in this range and since $\mathfrak{F} f$ is a finitely supported function,

$$\|\Pi_n \mathfrak{F} f\|_{-1}^2 = \sum_{A,B\in\mathscr{E}_n} \mathfrak{f}(A)\mathfrak{G}_n(A,B)\mathfrak{f}(B) < \infty,$$

where $\mathfrak{f} = \mathfrak{F} f$.

Fix now n such that $nd \leq 2$, a finitely supported function $\mathfrak{h} : \mathscr{E} \to \mathbb{R}$ and a set A_0 in $\mathscr{E}_n$. By assumption,

$$\sum_{A\in\mathscr{E}_n} \mathfrak{f}(A)\mathfrak{h}(A) = \sum_{A\in\mathscr{E}_n} \mathfrak{f}(A)\big\{\mathfrak{h}(A) - \mathfrak{h}(A_0)\big\}.$$

Since $\mathfrak{f}$ has finite support, by Schwarz inequality, the previous expression is bounded above by $C(\mathfrak{f}, n)\|\mathfrak{h}\|_1$ for some finite constant $C(\mathfrak{f}, n)$ depending only on f and n. This shows that $\|\Pi_n f\|_{-1}$ is finite for $nd \le 2$ and concludes the proof of the lemma. □

Fix a cylinder function $f = \sum_{A\in\mathscr{E}} \mathfrak{f}(A)\Psi_A$ and two densities α, β. Here the Fourier decomposition is performed with respect to the measure ν_α so that $\Psi_A(\eta) = \Psi_A(\alpha, \eta)$, $\mathfrak{f}(A) = \mathfrak{f}(\alpha, A)$. Denote by $|A|$ the total number of sites of a finite subset A of $\mathbb{Z}^d$. Since ν_β is a product measure, $E_{\nu_\beta}[\Psi_A(\alpha)] = \chi(\alpha)^{-|A|/2}(\beta-\alpha)^{|A|}$,

$$E_{\nu_\beta}[f] = \sum_{n\ge 0} \frac{(\beta-\alpha)^n}{\chi(\alpha)^{n/2}} \sum_{A\in\mathscr{E}_n} \mathfrak{f}(\alpha, A).$$

In particular, $E_{\nu_\alpha}[f] = \mathfrak{f}(\alpha, \varnothing)$ and more generally,

$$\frac{d^n}{d\beta^n} E_{\nu_\beta}[f]\Big|_{\beta=\alpha} = \frac{n!}{\chi(\alpha)^{n/2}} \sum_{A\in\mathscr{E}_n} \mathfrak{f}(\alpha, A), \quad n \ge 1.$$

The assumptions of the previous lemma can thus be restated as $E_{\nu_\alpha}[f] = 0$ in dimension $d \ge 3$ and

$$\frac{d^n}{d\beta^n} E_{\nu_\beta}[f]\Big|_{\beta=\alpha} = 0 \quad \text{for } n = 0,\ 1,\ 2 \text{ in dimension } 1,$$

$$\frac{d^n}{d\beta^n} E_{\nu_\beta}[f]\Big|_{\beta=\alpha} = 0 \quad \text{for } n = 0,\ 1 \text{ in dimension } 2.$$

We conclude this section with an example to illustrate that there might be a central limit theorem for an additive functional of a Markov process without having a solution of the Poisson equation in L^2.

Example 5.13 Assume that the probability measure $p(\cdot)$ is symmetric, fix a density $0 < \alpha < 1$ and consider the mean zero cylinder function $V(\eta) = \eta(0) - \alpha$ which can be written as $\sqrt{\chi(\alpha)}\Psi_{\{0\}}$. Taking Fourier coefficients, the Poisson equation $-Lf = V$ may be transformed into the elliptic equation $-\mathfrak{S}\mathfrak{f} = \mathfrak{V}$, where $\mathfrak{V}(A) = \sqrt{\chi(\alpha)}\mathbf{1}\{A = \{0\}\}$. This last transformation illustrates the utility of the duality introduced in this section. While the equation $-Lf = V$ is an equation on an infinite-dimensional space, $-\mathfrak{S}\mathfrak{f} = \mathfrak{V}$ is an equation on the finite-dimensional space $\mathbb{Z}^d$ because V has degree one and $\mathfrak{S}$ maps $\mathscr{E}_1$ into $\mathscr{E}_1$. Recall that $\mathfrak{S}$ restricted to $\mathscr{E}_1$ corresponds to the generator of a symmetric random walk on $\mathbb{Z}^d$ and that we denoted by $\mathfrak{G}_1(\cdot,\cdot)$ the Green kernel associated to $\mathfrak{S}$ restricted to $\mathscr{E}_1$. Since the symmetric random walk is transient in dimension $d \ge 3$, $-\mathfrak{S}\mathfrak{f} = \mathfrak{V}$ has a solution in $d \ge 3$ given by $\mathfrak{f}(x) = \sqrt{\chi(\alpha)}\mathfrak{G}_1(0, x)$. Here, to keep notation simple, we denoted the set $\{x\}$ simply by x so that $\mathfrak{G}_1(x, y)$ and $\mathfrak{f}(x)$ stand for $\mathfrak{G}_1(\{x\}, \{y\})$, $\mathfrak{f}(\{x\})$, respectively. It is well known that $\mathfrak{G}_1(0, x)$ decays as $|x|^{2-d}$ so that $\mathfrak{f}$ belongs to $L^2(\mu)$

only in dimension $d \geq 5$. Thus, the Poisson equation has a solution in $L^2(\nu_\alpha)$ only in dimension $d \geq 5$. Nevertheless, a central limit theorem holds for the time integral of V whenever V belongs to $\mathcal{H}_{-1}$. Since $\|V\|_{-1}^2 = \|\mathfrak{F}V\|_{-1}^2$ and in dimension $d \geq 3$ the latter expression is equal to

$$\sum_{x\in\mathbb{Z}^d} (\mathfrak{F}V)(x)\mathfrak{G}_1(x,y)(\mathfrak{F}V)(y) = \chi(\alpha)\mathfrak{G}_1(0,0),$$

a central limit theorem holds in dimension $d \geq 3$.

5.5 The Asymmetric Case, $\alpha = 1/2$

In this section and the next we extend the central limit theorem to asymmetric exclusion processes in dimension $d \geq 3$.

Theorem 5.14 *Assume that $d \geq 3$ and consider a mean zero cylinder function V in $\mathcal{H}_{-1}$. Then, under $\mathbb{P}_{\nu_\alpha}$,*

$$\frac{1}{\sqrt{t}}\int_0^t V(\eta_s)\,ds$$

converges in distribution, as $t \uparrow \infty$, to a mean zero Gaussian distribution with variance $\sigma^2(V) = 2\lim_{\lambda\to 0}\|f_\lambda\|_1^2$, where f_λ is the solution of the resolvent equation (5.8).

Recall from Sect. 2.7.4 the general approach to prove central limit theorems for additive functionals of Markov processes under a graded sector condition and the statement of Theorem 2.23. Keep also in mind the notation introduced in the previous section.

The operators S_0, B_0, L_+ and L_- defined in Sect. 2.7.4 correspond, respectively, to the operators S, $(1-2\alpha)\mathcal{N}$, $\sqrt{\chi(\alpha)}\mathcal{J}_+$ and $\sqrt{\chi(\alpha)}\mathcal{J}_-$ introduced in the previous section. We also already observed that condition (2.39) is in force.

Since $\mathfrak{N}$ and $\mathfrak{J}_- + \mathfrak{J}_+$ are anti-symmetric operators,

$$\langle \mathfrak{f}, (-\mathfrak{S})\mathfrak{f}\rangle_\mu = \langle \mathfrak{f}, (-\mathfrak{L}_\alpha)\mathfrak{f}\rangle_\mu \geq 0$$

for all finitely supported functions $\mathfrak{f} : \mathcal{E} \to \mathbb{R}$ and condition (2.40) is trivially satisfied with $C_0 = 1$. Moreover, the norms $\|\cdot\|_{0,1}$, $\|\cdot\|_{0,-1}$ introduced in Sect. 2.7.4 coincide with the norms $\|\cdot\|_1$, $\|\cdot\|_{-1}$ of the Hilbert spaces $\mathcal{H}_1$, $\mathcal{H}_{-1}$ defined in Sect. 5.2.

By Theorem 2.14 and Theorem 2.23, Theorem 5.14 holds if the generator satisfies the following two assumptions.

There exists $\beta < 1$ and a finite constant C_0 such that for all $n \geq 1$,

$$\langle \mathfrak{f}, (-\mathfrak{J}_+)\mathfrak{g}\rangle_\mu^2 \leq C_0 n^{2\beta}\langle \mathfrak{f}, (-\mathfrak{S})\mathfrak{f}\rangle_\mu \langle \mathfrak{g}, (-\mathfrak{S})\mathfrak{g}\rangle_\mu \tag{5.26}$$

for all finitely supported function $\mathfrak{f} : \mathscr{E}_{n+1} \to \mathbb{R}$, $\mathfrak{g} : \mathscr{E}_n \to \mathbb{R}$, and a similar inequality with $\mathfrak{J}_-$ in place of $\mathfrak{J}_+$.

There exist finite constants γ and C_0 such that for all $n \geq 1$,

$$\langle \mathfrak{f}, (-\mathfrak{N})\mathfrak{g} \rangle_\mu^2 \leq C_0 n^{2\gamma} \langle \mathfrak{f}, (-\mathfrak{S})\mathfrak{f} \rangle_\mu \langle \mathfrak{g}, (-\mathfrak{S})\mathfrak{g} \rangle_\mu \tag{5.27}$$

for all finitely supported functions $\mathfrak{f}, \mathfrak{g} : \mathscr{E}_n \to \mathbb{R}$.

In view of (5.20), the left-hand sides of (5.26) and (5.27) vanish for $n = 0$. This explains the ranges of n in both conditions.

Condition (5.26) is proved in Lemma 5.15 below in dimensions $d \geq 3$ with $\beta = 1/2$ and constitutes the main result of this section. To avoid proving the second estimate, we assume in this section that the density $\alpha = 1/2$, in which case the asymmetric part $(1 - 2\alpha)\mathfrak{N}$ vanishes.

Lemma 5.15 *Assume that $d \geq 3$. There exists a finite constant C_0, depending only on the probability $p(\cdot)$, such that for every $n \geq 1$,*

$$\langle \mathfrak{J}_+ \mathfrak{f}, \mathfrak{g} \rangle_\mu^2 \leq C_0 n \langle -\mathfrak{S}\mathfrak{f}, \mathfrak{f} \rangle_\mu \langle -\mathfrak{S}\mathfrak{g}, \mathfrak{g} \rangle_\mu$$

for any finitely supported functions $\mathfrak{f} : \mathscr{E}_n \to \mathbb{R}$, $\mathfrak{g} : \mathscr{E}_{n+1} \to \mathbb{R}$.

Since $-\mathfrak{J}_-$ is the adjoint of $\mathfrak{J}_+$, a similar bound holds for $\mathfrak{J}_-$.

Proof Fix $n \geq 1$ and two finitely supported functions $\mathfrak{f} : \mathscr{E}_n \to \mathbb{R}$, $\mathfrak{g} : \mathscr{E}_{n+1} \to \mathbb{R}$. By the explicit formula for $\mathfrak{J}_+$,

$$\langle \mathfrak{J}_+ \mathfrak{f}, \mathfrak{g} \rangle_\mu = 2 \sum_{A \in \mathscr{E}} \sum_{x,y \in A} a(y - x) \mathfrak{f}\big(A \setminus \{y\}\big) \mathfrak{g}(A).$$

Notice that the first sum is in fact carried over sets A in $\mathscr{E}_{n+1}$. Since $a(\cdot)$ is anti-symmetric, a change of variables $x' = y$, $y' = x$ permits to rewrite the previous scalar product as

$$\sum_{A \in \mathscr{E}_{n+1}} \sum_{x,y \in A} a(y - x) \big\{ \mathfrak{f}\big(A \setminus \{y\}\big) - \mathfrak{f}\big(A \setminus \{x\}\big) \big\} \mathfrak{g}(A).$$

Since $a(\cdot)$ is absolutely bounded by $s(\cdot)$, the elementary inequality $2ab \leq \gamma a^2 + \gamma^{-1} b^2$ gives that the previous expression is less than or equal to

$$\frac{\gamma}{2} \sum_{A \in \mathscr{E}_{n+1}} \sum_{x,y \in A} s(y - x) \big\{ \mathfrak{f}\big(A \setminus \{y\}\big) - \mathfrak{f}\big(A \setminus \{x\}\big) \big\}^2 + \frac{1}{2\gamma} \sum_{A \in \mathscr{E}_{n+1}} \mathfrak{g}(A)^2 W(A) \tag{5.28}$$

for all $\gamma > 0$, where $W : \mathscr{E} \to \mathbb{R}$ is the function defined by

$$W(A) = \sum_{x,y \in A} s(y - x). \tag{5.29}$$

By (5.24), the first expression in (5.28) is equal to $\gamma\langle -\mathfrak{S}\mathfrak{f},\mathfrak{f}\rangle_\mu$. By Lemma 5.18, the second term is bounded by $C_1\gamma^{-1}n\langle -\mathfrak{S}\mathfrak{g},\mathfrak{g}\rangle_\mu$ for some finite constant C_1 depending only on p. Therefore, (5.28) is less than or equal to

$$\gamma\langle -\mathfrak{S}\mathfrak{f},\mathfrak{f}\rangle_\mu + C_1 n\gamma^{-1}\langle -\mathfrak{S}\mathfrak{g},\mathfrak{g}\rangle_\mu$$

for every $\gamma > 0$. It remains to minimize over γ to conclude the proof of the lemma. □

The next result is a straightforward consequence of the lemma. Recall from (2.48) the definition of the triple norms. In terms of Fourier coefficients, these norms can be written as

$$|||\mathfrak{f}|||_k^2 = \sum_{n\ge 0}(n+1)^{2k}\|\Pi_n\mathfrak{f}\|^2, \qquad |||\mathfrak{f}|||_{k,1}^2 = \sum_{n\ge 1}(n+1)^{2k}\|\Pi_n\mathfrak{f}\|_1^2,$$
$$|||\mathfrak{f}|||_{k,-1}^2 = \sum_{n\ge 0}(n+1)^{2k}\|\Pi_n\mathfrak{f}\|_{-1}^2.$$

We explained in (5.23) why the second sum starts from $n = 1$.

Corollary 5.16 *There exists a finite constant C_0, depending only on the probability $p(\cdot)$, such that for all $n \ge 1$*

$$\|\mathfrak{J}_+\mathfrak{f}\|_{-1}^2 \le C_0 n\|\mathfrak{f}\|_1^2, \qquad \|\mathfrak{J}_-\mathfrak{g}\|_{-1}^2 \le C_0 n\|\mathfrak{g}\|_1^2$$

for all finitely supported functions $\mathfrak{f} : \mathscr{E}_n \to \mathbb{R}$, $\mathfrak{g} : \mathscr{E}_{n+1} \to \mathbb{R}$. In particular,

$$|||\mathfrak{J}_\pm\mathfrak{f}|||_{k,-1} \le C_0|||\mathfrak{f}|||_{k+1,1}$$

for any finitely supported function $\mathfrak{f} : \mathscr{E} \to \mathbb{R}$ and every $k \ge 0$.

Remark 5.17 Lemma 5.15 asserts that in dimension $d \ge 3$, there exists a finite constant C_0, depending only on the probability $p(\cdot)$, such that for every $n \ge 1$,

$$\langle \mathscr{J}_+ f, g\rangle_{\nu_\alpha}^2 \le C_0 n\langle -Lf, f\rangle_{\nu_\alpha}\langle -Lg, g\rangle_{\nu_\alpha}$$

for all cylinder functions $f \in \mathscr{C} \cap \mathscr{A}_n$, $g \in \mathscr{C} \cap \mathscr{A}_{n+1}$. This estimate, as well as the one stated in Lemma 5.18 below, can be extended by the usual approximation arguments to a larger class of functions. We leave the details to the reader.

We conclude this section with an estimate on two point correlation functions.

Lemma 5.18 *Let $W : \mathscr{E} \to \mathbb{R}$ be defined by $W(A) = \sum_{x,y\in A} s(y-x)$. There exists a finite constant C_0, depending only on $p(\cdot)$ such that*

$$\sum_{A\in\mathscr{E}_n} \mathfrak{g}(A)^2 W(A) \le C_0 n \langle -\mathfrak{S}\mathfrak{g}, \mathfrak{g}\rangle_\mu$$

for all $n \ge 1$ and all finitely supported functions $\mathfrak{g} : \mathscr{E}_n \to \mathbb{R}$.

Proof We need some sort of spectral gap since we have to estimate the square of $\mathfrak{g}$ multiplied by the function W by the $L^2(\mu)$ norm of the gradient of $\mathfrak{g}$. The idea is to reduce the problem to an estimate on symmetric random walks on $\mathbb{Z}^d$.

Inverting the order of summation, we may write the expression we want to estimate by

$$\sum_{x,y\in\mathbb{Z}^d} s(y-x)\rho\big(\{x,y\}\big)^2,$$

where, for each set $\{x,y\}$ in $\mathscr{E}_2$,

$$\rho\big(\{x,y\}\big)^2 = \sum_{\substack{A\in\mathscr{E}_n\\ A\ni y,x}} \mathfrak{g}(A)^2.$$

Note that ρ inherits a finite support from $\mathfrak{g}$.

Consider the irreducible, continuous-time random walk on $\mathbb{Z}^d$ with transition probability $p(x,y) = s(y-x)$. Since the process is transient in $d \ge 3$ and since $s(\cdot)$ has finite support, by Proposition 5.23, there exists a finite constant C_1, depending only on $s(\cdot)$, such that

$$\sum_{x\in\mathbb{Z}^d} s(x)F(x)^2 \le C_1 \sum_{x,y\in\mathbb{Z}^d} s(y-x)\big[F(y)-F(x)\big]^2$$

for all finitely supported functions $F : \mathbb{Z}^d \to \mathbb{R}$.

Since ρ has a finite support, replacing the origin by a site x_0 and taking $F(x) = \rho(\{x_0,x\})$, we obtain that

$$\sum_{x\in\mathbb{Z}^d} s(x-x_0)\rho\big(\{x_0,x\}\big)^2 \le C_1 \sum_{x,y\in\mathbb{Z}^d} s(y-x)\big\{\rho\big(\{x_0,y\}\big)-\rho\big(\{x_0,x\}\big)\big\}^2.$$

Summing over x_0, the left-hand side of the previous inequality becomes the expression we want to estimate. On the other hand, by the definition of ρ and by Schwarz inequality, we have that

$$\big\{\rho\big(\{x_0,y\}\big)-\rho\big(\{x_0,x\}\big)\big\}^2 \le \sum_{\substack{A\in\mathscr{E}_{n+1}\\ A\ni x_0,x,y}} \big\{\mathfrak{g}\big(A\setminus\{y\}\big)-\mathfrak{g}\big(A\setminus\{x\}\big)\big\}^2.$$

Summing over x_0 in the penultimate formula, we conclude that there exists a finite constant C_1, depending only on $s(\cdot)$, for which

$$\sum_{A\in\mathscr{E}_n} \mathfrak{g}(A)^2 \sum_{x,x_0\in A} s(x-x_0) \le C_1 \sum_{x,x_0,y\in\mathbb{Z}^d} s(y-x) \sum_{\substack{A\in\mathscr{E}_{n+1}\\ A\ni x_0,x,y}} \big\{\mathfrak{g}\big(A\setminus\{y\}\big)-\mathfrak{g}\big(A\setminus\{x\}\big)\big\}^2.$$

Since A has $n+1$ elements, summing over x_0, the right-hand side becomes

$$C_1 n \sum_{x,y\in\mathbb{Z}^d} s(y-x) \sum_{\substack{A\in\mathscr{E}_{n+1}\\ A\ni x,y}} \big\{\mathfrak{g}\big(A\setminus\{y\}\big)-\mathfrak{g}\big(A\setminus\{x\}\big)\big\}^2.$$

It follows from (5.24) that this expression is equal to $2C_1 n\langle -\mathfrak{S}\mathfrak{g},\mathfrak{g}\rangle_\mu$. □

5.6 The Asymmetric Case, $\alpha \neq 1/2$

In this section, we prove Theorem 5.14 in the case $\alpha\neq 1/2$. It remains to check the validity of (5.27). This assumption is not expected to hold because in the context of asymmetric simple exclusion processes, the operator $\mathfrak{N}$ is essentially a gradient, while $\mathfrak{S}$ is a Laplacian. This condition requires therefore that

$$\langle \mathfrak{f},\nabla\mathfrak{g}\rangle^2 \le C_0\langle\nabla\mathfrak{f},\nabla\mathfrak{f}\rangle\langle\nabla\mathfrak{g},\nabla\mathfrak{g}\rangle$$

which cannot hold because it would imply a Poincaré inequality in infinite volume, that is, a bound on the L^2 norm of a mean zero function in terms of the L^2 norm of its gradient: $\|\mathfrak{f}\|\le C_0'\|\nabla\mathfrak{f}\|$.

Nevertheless, if we return to the proof of Theorem 2.23 and recall that f_λ stands for the solution of the resolvent equation, we see that the bound

$$\|\Pi_n\mathfrak{N}\mathfrak{f}_\lambda\|_{-1}^2 \le C_0 n\|\Pi_n\mathfrak{F}V\|_{-1}^2, \tag{5.30}$$

or the one presented in Lemma 5.20 below, is sufficient to conclude the proof. This is the content of Theorem 5.19 below. Note that $\mathfrak{N}\mathfrak{f}_\lambda$ may not be defined. However, since the set of cylinder functions is a core for the generator, we may proceed as in (2.37) and approximate $\mathfrak{f}_\lambda$ by finitely supported functions.

Theorem 5.19 *Recall the notation introduced in Sect.* 2.7.4. *Suppose that the generator L satisfies hypotheses* (2.39), (2.40) *and* (2.45). *Fix a function V such that*

$$|||V|||_{k,-1}<\infty$$

for some $k \geq \beta$. *Let* f_λ *be the solution of the resolvent equation* (2.13), *and assume that there exist* $C_0 < \infty$, $\ell \geq 1$, $a, b > 0$, $\max\{a,b\} \leq k$, *such that*

$$\|B_0 \Pi_n f_\lambda\|_{0,-1}^2 \leq C_0 n^{2a} \|\Pi_n V\|_{0,-1}^2 + C_0 n^{2b} \sum_{j=n-\ell}^{n+\ell} \|\Pi_j f_\lambda\|_{0,1}^2$$

for all $n \geq 1$. *Then, for all* $j \leq k - m$, *where* $m = \max\{a, b, \beta\}$, *there exists a finite constant* C_j *such that*

$$\sup_{0<\lambda\leq 1} |||Lf_\lambda|||_{j,-1} < C_j |||V|||_{j+m,-1} < \infty.$$

In particular, $\sup_{0<\lambda\leq 1} \|Lf_\lambda\|_{0,-1} < \infty$, *and, by* (2.44), $\sup_{0<\lambda\leq 1} \|Lf_\lambda\|_{-1} < \infty$.

In the present context, since f_λ is essentially $(\nabla - \Delta)^{-1} V$, Δ being the Laplacian, inequality (5.30) states that

$$\langle \nabla(\nabla - \Delta)^{-1} V, (-\Delta)^{-1} \nabla(\nabla - \Delta)^{-1} V\rangle \leq C_0 \langle V, (-\Delta)^{-1} V\rangle.$$

Rewriting this estimate in Fourier variables, it is not difficult to show that such a bound should hold.

Lemma 5.20 *Let* f_λ *be the solution of the resolvent equation* (5.8). *There exists a finite constant* C_0, *depending only on* $p(\cdot)$, *such that*

$$\|\Pi_n \mathfrak{N} \mathfrak{f}_\lambda\|_{-1}^2 \leq C_0 n \|\Pi_n \mathfrak{F} V\|_{-1}^2 + C_0 n^3 \sum_{j=n-1}^{n+1} \|\Pi_j \mathfrak{f}_\lambda\|_1^2$$

for all $n \geq 1$. *In particular, for all* $k \geq 0$,

$$|||\mathfrak{N} \mathfrak{f}_\lambda|||_{k,-1}^2 \leq C_0 |||\mathfrak{F} V|||_{k+1,-1}^2 + C_0 |||\mathfrak{f}_\lambda|||_{k+2,1}^2.$$

Proof All constants appearing in this proof depend only on the probability measure p. Let $\mathfrak{w} = \mathfrak{F} V$, $\mathfrak{w}_1 = \mathfrak{w} + \sqrt{\chi(\alpha)} \{\mathfrak{J}_+ + \mathfrak{J}_-\} \mathfrak{f}_\lambda$ so that

$$\lambda \mathfrak{f}_\lambda - \{\mathfrak{S} + (1 - 2\alpha)\mathfrak{N}\} \mathfrak{f}_\lambda = \mathfrak{w}_1. \tag{5.31}$$

By Lemma 5.15 and Remark 5.17, there exists a finite constant C_0 such that

$$\|\Pi_n \mathfrak{w}_1\|_{-1}^2 \leq 2 \|\Pi_n \mathfrak{w}\|_{-1}^2 + C_0 n \sum_{j=n-1}^{n+1} \|\Pi_j \mathfrak{f}_\lambda\|_1^2 \tag{5.32}$$

for all $n \geq 1$.

The operator $\mathfrak{S} + (1 - 2\alpha)\mathfrak{N}$ does not change the degree of a function. We may therefore examine Eq. (5.31) on each set $\mathscr{E}_n$:

$$\lambda \Pi_n \mathfrak{f}_\lambda - \{\mathfrak{S} + (1 - 2\alpha)\mathfrak{N}\} \Pi_n \mathfrak{f}_\lambda = \Pi_n \mathfrak{w}_1.$$

Since n is fixed until estimate (5.39), we omit the operator Π_n in the following formulas.

The main idea of this proof is to approximate the operator $\mathfrak{S} + (1 - 2\alpha)\mathfrak{N}$ by a convolution operator that can be analyzed through Fourier transforms. Fix $n \geq 1$ and let $\mathscr{X}_n = (\mathbb{Z}^d)^n$. We consider a set A in $\mathscr{E}_n$ as an equivalent class of $n!$ sets of distinct points of $\mathbb{Z}^d$. A function $\mathfrak{f} : \mathscr{E}_n \to \mathbb{R}$ can be lifted into a symmetric function $\Xi\mathfrak{f}$ on $\mathscr{X}_n$, which vanishes on $\mathscr{X}_n \setminus \mathscr{E}_n$:

$$(\Xi\mathfrak{f})(x_1, \dots, x_n) = \begin{cases} \mathfrak{f}(\{x_1, \dots, x_n\}) & \text{if } x_i \neq x_j \text{ for } i \neq j, \\ 0 & \text{otherwise.} \end{cases} \tag{5.33}$$

Note that we use the same notation, $\mathscr{E}_n$, for two different sets, the finite subsets of $\mathbb{Z}^d$ with n points and the set of all n-tuples $(x_1, \dots, x_n) \in (\mathbb{Z}^d)^n$ such that $x_i \neq x_j$ for $i \neq j$. Clearly, for all $\mathfrak{f}$, $\mathfrak{g} \in L^2(\mu)$,

$$\sum_{A \in \mathscr{E}_n} \mathfrak{f}(A)\mathfrak{g}(A) = \frac{1}{n!} \sum_{\mathbf{x} \in \mathscr{X}_n} \Xi\mathfrak{f}(\mathbf{x}) \Xi\mathfrak{g}(\mathbf{x}).$$

The operators $\mathfrak{S}, \mathfrak{N}$ can also be extended in a natural way to $\mathscr{X}_n$. Denote by $\{\mathbf{e}_j, 1 \leq j \leq n\}$ the canonical basis of $\mathbb{R}^n$ and consider on $\mathscr{X}_n$ the operators $\mathfrak{S}^o$, $\mathfrak{N}^o$ defined by

$$\begin{aligned} \big(\mathfrak{S}^o\mathfrak{f}\big)(\mathbf{x}) &= \sum_{\substack{1 \leq j \leq n \\ z \in \mathbb{Z}^d}} s(z)\big\{\mathfrak{f}(\mathbf{x} + z\mathbf{e}_j) - \mathfrak{f}(\mathbf{x})\big\}, \\ \big(\mathfrak{N}^o\mathfrak{f}\big)(\mathbf{x}) &= \sum_{\substack{1 \leq j \leq n \\ z \in \mathbb{Z}^d}} a(z)\big\{\mathfrak{f}(\mathbf{x} + z\mathbf{e}_j) - \mathfrak{f}(\mathbf{x})\big\}. \end{aligned} \tag{5.34}$$

In this formula and below, $\mathbf{x} = (x_1, \dots, x_n)$ is an element of $\mathscr{X}_n$, so that each x_j belongs to $\mathbb{Z}^d$ and $\mathbf{x} + z\mathbf{e}_j = (x_1, \dots, x_{j-1}, x_j + z, x_{j+1}, \dots, x_n)$.

Denote by $\|\cdot\|_{\mathscr{X}_n,1}$ the $\mathscr{H}_1$ norm associated to the generator $\mathfrak{S}^o$: for each function $\mathfrak{f} : \mathscr{X}_n \to \mathbb{R}$,

$$\|\mathfrak{f}\|^2_{\mathscr{X}_n,1} = \frac{1}{n!} \sum_{\mathbf{x} \in \mathscr{X}_n} \mathfrak{f}(\mathbf{x})\big(-\mathfrak{S}^o\mathfrak{f}\big)(\mathbf{x}) \tag{5.35}$$

and denote by $\|\cdot\|_{\mathscr{X}_n,-1}$ its dual norm defined by

$$\|\mathfrak{f}\|^2_{\mathscr{X}_n,-1} = \frac{1}{n!} \sum_{\mathbf{x} \in \mathscr{X}_n} \mathfrak{f}(\mathbf{x})\big\{\big(-\mathfrak{S}^o\big)^{-1}\mathfrak{f}\big\}(\mathbf{x}). \tag{5.36}$$

Lifting the resolvent equation (5.31) to $\mathscr{X}_n$ and adding and subtracting $\mathfrak{S}^o\Xi\mathfrak{f}_\lambda + (1 - 2\alpha)\mathfrak{N}^o\Xi\mathfrak{f}_\lambda$, we obtain that

$$\lambda\Xi\mathfrak{f}_\lambda - \big\{\mathfrak{S}^o + (1 - 2\alpha)\mathfrak{N}^o\big\}\Xi\mathfrak{f}_\lambda = \mathfrak{w}_2, \tag{5.37}$$

where

$$\mathfrak{w}_2 = \Xi\mathfrak{w}_1 + \{\Xi\mathfrak{S} - \mathfrak{S}^o\Xi\}\mathfrak{f}_\lambda + (1-2\alpha)\{\Xi\mathfrak{N} - \mathfrak{N}^o\Xi\}\mathfrak{f}_\lambda.$$

We claim that $\mathfrak{w}_2$ has finite $\mathcal{H}_{-1}(\mathfrak{S}^o)$ norm. Indeed, for each $n \geq 1$, by (5.41) and Lemma 5.22 below, there exists a finite constant C_0 such that

$$\|\Xi\mathfrak{w}_1\|^2_{\mathcal{X}_n,-1} \leq \|\mathfrak{w}_1\|^2_{-1}, \qquad \|\Xi\mathfrak{S}\mathfrak{f}_\lambda - \mathfrak{S}^o\Xi\mathfrak{f}_\lambda\|^2_{\mathcal{X}_n,-1} \leq C_0 n^2 \|\mathfrak{f}_\lambda\|^2_1,$$
$$\|\Xi\mathfrak{N}\mathfrak{f}_\lambda - \mathfrak{N}^o\Xi\mathfrak{f}_\lambda\|^2_{\mathcal{X}_n,-1} \leq C_0 n^2 \|\mathfrak{f}_\lambda\|^2_1$$

so that

$$\|\mathfrak{w}_2\|^2_{\mathcal{X}_n,-1} \leq 2\|\mathfrak{w}_1\|^2_{-1} + C_0 n^2 \|\mathfrak{f}_\lambda\|^2_1. \tag{5.38}$$

It remains to examine the resolvent equation (5.37) through Fourier analysis. Let $\mathbb{T}_{n,d} = [-\pi,\pi]^{nd}$ and denote by $\widehat{\mathfrak{f}_\lambda} : (\mathbb{T}^d)^n \to \mathbb{C}$ the Fourier transform of $\Xi\mathfrak{f}_\lambda$:

$$\widehat{\mathfrak{f}_\lambda}(\mathbf{k}) = \sum_{\mathbf{x}\in\mathcal{X}_n} e^{i\mathbf{x}\cdot\mathbf{k}} (\Xi\mathfrak{f}_\lambda)(\mathbf{x}).$$

In this formula, $\mathbf{x}\cdot\mathbf{k} = \sum_{1\leq j\leq n} x_j \cdot k_j$. It follows from the resolvent equation (5.37) that $\widehat{\mathfrak{f}_\lambda}$ is the solution of

$$\lambda\widehat{\mathfrak{f}_\lambda}(\mathbf{k}) - \{\widehat{\mathfrak{S}^o}(\mathbf{k}) + (1-2\alpha)\widehat{\mathfrak{N}^o}(\mathbf{k})\}\widehat{\mathfrak{f}_\lambda}(\mathbf{k}) = \widehat{\mathfrak{w}_2}(\mathbf{k}),$$

where $\widehat{\mathfrak{S}^o}$, $\widehat{\mathfrak{N}^o}$ are the functions associated to the operators $\mathfrak{S}^o$, $\mathfrak{N}^o$:

$$-\widehat{\mathfrak{S}^o}(\mathbf{k}) = \sum_{\substack{1\leq j\leq n\\ z\in\mathbb{Z}^d}} s(z)\{1-\cos(k_j\cdot z)\}, \qquad -\widehat{\mathfrak{N}^o}(\mathbf{k}) = i\sum_{\substack{1\leq j\leq n\\ z\in\mathbb{Z}^d}} a(z)\sin(k_j\cdot z).$$

The $\mathcal{H}_{-1}(\mathfrak{S}^o)$ norm of a function $\mathfrak{v} : \mathcal{X}_n \to \mathbb{R}$ has a simple and explicit expression in terms of the Fourier transform:

$$\|\mathfrak{v}\|^2_{\mathcal{X}_n,-1} = \frac{-1}{n!(2\pi)^{nd}} \int_{\mathbb{T}_{n,d}} |\widehat{\mathfrak{v}}(\mathbf{k})|^2 \frac{1}{\widehat{\mathfrak{S}^o}(\mathbf{k})} d\mathbf{k}.$$

Since $\Xi\mathfrak{f}_\lambda$ is the solution of the resolvent equation (5.37), for every $\lambda > 0$,

$$\|\mathfrak{N}^o\Xi\mathfrak{f}_\lambda\|^2_{\mathcal{X}_n,-1} = \frac{-1}{n!(2\pi)^{nd}} \int_{\mathbb{T}_{n,d}} \left|\frac{\widehat{\mathfrak{N}^o}(\mathbf{k})}{\lambda - \widehat{\mathfrak{S}^o}(\mathbf{k}) - (1-2\alpha)\widehat{\mathfrak{N}^o}(\mathbf{k})}\right|^2 \frac{|\widehat{\mathfrak{w}_2}(\mathbf{k})|^2}{\widehat{\mathfrak{S}^o}(\mathbf{k})} d\mathbf{k}.$$

It follows from the explicit formulas for the functions $\widehat{\mathfrak{S}^o}$, $\widehat{\mathfrak{N}^o}$ and a Taylor expansion for $|\mathbf{k}|$ small that the previous expression is bounded by

$$\frac{-C_0}{n!(2\pi)^{nd}} \int_{\mathbb{T}_{n,d}} \frac{|\widehat{\mathfrak{w}_2}(\mathbf{k})|^2}{\widehat{\mathfrak{S}^o}(\mathbf{k})} d\mathbf{k} = C_0 \|\mathfrak{w}_2\|^2_{\mathcal{X}_n,-1}$$

for some finite constant C_0. We have thus proved that

$$\left\|\mathfrak{N}^o \Xi \mathfrak{f}_\lambda\right\|^2_{\mathscr{X}_n,-1} \le C_0 \|\mathfrak{w}_2\|^2_{\mathscr{X}_n,-1}. \tag{5.39}$$

We may now conclude the proof of Lemma 5.20. Fix $n \ge 1$. By (5.41), by the estimate presented just before (5.38) and by the inequality just derived, there exists a finite constant C_0, which may change from line to line, such that

$$\begin{aligned}
\|\Pi_n \mathfrak{N} \mathfrak{f}_\lambda\|^2_{-1} &= \|\mathfrak{N}\Pi_n \mathfrak{f}_\lambda\|^2_{-1} \le C_0 n \|\Xi \mathfrak{N}\Pi_n \mathfrak{f}_\lambda\|^2_{\mathscr{X}_n,-1} \\
&\le C_0 n\big\{n^2\|\Pi_n \mathfrak{f}_\lambda\|_1^2 + \left\|\mathfrak{N}^o \Xi \Pi_n \mathfrak{f}_\lambda\right\|^2_{\mathscr{X}_n,-1}\big\} \\
&\le C_0\big\{n^3\|\Pi_n \mathfrak{f}_\lambda\|_1^2 + n\|\Pi_n \mathfrak{w}_2\|^2_{\mathscr{X}_n,-1}\big\}.
\end{aligned}$$

In particular, by (5.38) and (5.32),

$$\begin{aligned}
\|\Pi_n \mathfrak{N} \mathfrak{f}_\lambda\|^2_{-1} &\le C_0\big\{n^3\|\Pi_n \mathfrak{f}_\lambda\|_1^2 + n\|\Pi_n \mathfrak{w}_1\|^2_{-1}\big\} \\
&\le C_0\left\{n^3 \sum_{j=n-1}^{n+1} \|\Pi_j \mathfrak{f}_\lambda\|_1^2 + n\|\Pi_n \mathfrak{w}\|^2_{-1}\right\},
\end{aligned}$$

which concludes the proof of the lemma. □

Lemma 5.21 *There exists a finite constant C_0, which depends only on p, such that for every $n \ge 1$ and every function $\mathfrak{f}: \mathscr{E}_n \to \mathbb{R}$ in $L^2(\mu) \cap \mathscr{H}_1(\mathfrak{S})$,*

$$\|\mathfrak{f}\|_1^2 \le \|\Xi \mathfrak{f}\|^2_{\mathscr{X}_n,1} \le C_0 n \|\mathfrak{f}\|_1^2.$$

Proof All constants in this lemma depend only on the probability measure p. The first inequality is elementary and follows from the explicit formulas for the respective $\mathscr{H}_1$ norms. To prove the second inequality, fix $n \ge 1$ and recall from (5.29) the definition of the function W. We also denote by W the lifted function ΞW. A simple computation shows that $\Xi \mathfrak{S}\mathfrak{f}(\mathbf{x}) - \mathfrak{S}^o \Xi \mathfrak{f}(\mathbf{x}) = W(\mathbf{x}) \Xi \mathfrak{f}(\mathbf{x})$. In particular,

$$\left|\Xi \mathfrak{S}\mathfrak{f}(\mathbf{x}) - \mathfrak{S}^o \Xi \mathfrak{f}(\mathbf{x})\right| \le W(\mathbf{x})\left|\Xi \mathfrak{f}(\mathbf{x})\right| \tag{5.40}$$

for every $\mathbf{x}$ in $\mathscr{E}_n$ and $\mathfrak{f}: \mathscr{E}_n \to \mathbb{R}$.

We are now in a position to prove the second bound. By definition,

$$\|\Xi \mathfrak{f}\|^2_{\mathscr{X}_n,1} = -\frac{1}{n!} \sum_{\mathbf{x} \in \mathscr{X}_n} (\Xi \mathfrak{f})(\mathbf{x}) \big(\mathfrak{S}^o \Xi \mathfrak{f}\big)(\mathbf{x}).$$

Since $\Xi \mathfrak{f}$ vanishes outside $\mathscr{E}_n$, we may restrict the sum to $\mathscr{E}_n$. Now, adding and subtracting $(\Xi \mathfrak{S}\mathfrak{f})(\mathbf{x})$ in this expression and recalling (5.40), we obtain that

$$\|\Xi \mathfrak{f}\|^2_{\mathscr{X}_n,1} \le \|\mathfrak{f}\|_1^2 + \frac{1}{n!} \sum_{\mathbf{x} \in \mathscr{E}_n} W(\mathbf{x}) \big\{\Xi \mathfrak{f}(\mathbf{x})\big\}^2 = \|\mathfrak{f}\|_1^2 + \sum_{A \in \mathscr{E}_n} W(A) \mathfrak{f}(A)^2.$$

By Lemma 5.18 and Remark 5.17, the second term of the previous formula is bounded by $C_0 n \|\mathfrak{f}\|_1^2$, which concludes the proof of the lemma. □

It follows from this result that there exists a finite constant C_0, which depends only on p, such that for every $n \geq 1$ and every function $\mathfrak{f} : \mathscr{E}_n \to \mathbb{R}$ in $L^2(\mu) \cap \mathscr{H}_{-1}(\mathfrak{S})$,

$$\frac{1}{C_0 n} \|\mathfrak{f}\|_{-1}^2 \leq \|\Xi \mathfrak{f}\|_{\mathscr{X}_n,-1}^2 \leq \|\mathfrak{f}\|_{-1}^2. \tag{5.41}$$

Lemma 5.22 *There exists a finite constant C_0, which depends only on p, such that for every $n \geq 1$ and every function $\mathfrak{f} : \mathscr{E}_n \to \mathbb{R}$ in $L^2(\mu) \cap \mathscr{H}_1(\mathfrak{S})$,*

$$\big\| \Xi \mathfrak{S} \mathfrak{f} - \mathfrak{S}^o \Xi \mathfrak{f} \big\|_{\mathscr{X}_n,-1}^2 \leq C_0 n^2 \|\mathfrak{f}\|_1^2, \qquad \big\| \Xi \mathfrak{N} \mathfrak{f} - \mathfrak{N}^o \Xi \mathfrak{f} \big\|_{\mathscr{X}_n,-1}^2 \leq C_0 n^2 \|\mathfrak{f}\|_1^2.$$

Proof We prove the first estimate and leave to the reader the details of the second. Fix $n \geq 1$ and a finitely supported function $\mathfrak{h} : \mathscr{X}_n \to \mathbb{R}$. We need to estimate the scalar product

$$\frac{1}{n!} \sum_{\mathbf{x} \in \mathscr{X}_n} \mathfrak{h}(\mathbf{x}) \big\{ \Xi \mathfrak{S} \mathfrak{f}(\mathbf{x}) - \mathfrak{S}^o \Xi \mathfrak{f}(\mathbf{x}) \big\} \tag{5.42}$$

in terms of the $\mathscr{H}_1(\mathfrak{S}^o)$ norm of $\mathfrak{h}$ and the $\mathscr{H}_1(\mathfrak{S})$ norm of $\mathfrak{f}$. There are two possible cases. Either $\mathbf{x}$ belongs to $\mathscr{E}_n$ or $\mathbf{x}$ does not belong to $\mathscr{E}_n$.

In the first case, by (5.40), the expression inside braces in the previous formula is absolutely bounded by $W(\mathbf{x}) |\Xi \mathfrak{f}(\mathbf{x})|$. Therefore, the corresponding piece in the previous formula is bounded above by

$$\frac{1}{n!} \sum_{\mathbf{x} \in \mathscr{E}_n} W(\mathbf{x}) \big|\mathfrak{h}(\mathbf{x})\big| \big|\Xi \mathfrak{f}(\mathbf{x})\big| \leq \frac{1}{2\ell} \frac{1}{n!} \sum_{\mathbf{x} \in \mathscr{E}_n} W(\mathbf{x}) \mathfrak{h}(\mathbf{x})^2 + \frac{\ell}{2} \sum_{A \in \mathscr{E}_n} W(A) \mathfrak{f}(A)^2$$

for every $\ell > 0$.

If $\mathbf{x}$ does not belong to $\mathscr{E}_n$, the corresponding piece of the scalar product can be written as

$$-\frac{1}{n!} \sum_{\substack{\mathbf{x} \in \mathscr{X}_n \setminus \mathscr{E}_n \\ z \in \mathbb{Z}^d, 1 \leq j \leq n}} s(z) \mathfrak{h}(\mathbf{x}) \Xi \mathfrak{f}(\mathbf{x} + z \mathbf{e}_j)$$

because in this case $\Xi \mathfrak{S} \mathfrak{f}(\mathbf{x}) = \Xi \mathfrak{f}(\mathbf{x}) = 0$. Since $\Xi \mathfrak{f}$ vanishes outside $\mathscr{E}_n$, it is implicit in the previous formula that the sum is restricted to all $\mathbf{x}$ such that $\mathbf{x} + z\mathbf{e}_j$ belongs to $\mathscr{E}_n$. Since $\mathbf{x} + z\mathbf{e}_j \in \mathscr{E}_n$ and $\mathbf{x} \notin \mathscr{E}_n$, $x_j = x_k$ for some k. In particular, since $2ab \leq \ell a^2 + \ell^{-1} b^2$ for every $\ell > 0$, a change of variables gives that the previous expression is bounded above by

$$\frac{1}{2n!\ell} \sum_{\mathbf{x} \in \mathscr{X}_n} \mathfrak{h}(\mathbf{x})^2 \widetilde{W}(\mathbf{x}) + \frac{\ell}{2n!} \sum_{\mathbf{x} \in \mathscr{X}_n} \Xi \mathfrak{f}(\mathbf{x})^2 W(\mathbf{x}),$$

where $\widetilde{W}(\mathbf{x}) = \sum_{j\neq k} \mathbf{1}\{x_j = x_k\}$. We may of course replace the sum over $\mathscr{X}_n$ by a sum over $\mathscr{E}_n$ in the second term, losing the factor $n!$.

Adding the two previous estimates, we obtain that the scalar product (5.42) is bounded above by

$$\frac{1}{2n!\ell} \sum_{\mathbf{x}\in\mathscr{X}_n} \mathfrak{h}(\mathbf{x})^2\big\{\widetilde{W}(\mathbf{x}) + W(\mathbf{x})\big\} + \ell \sum_{A\in\mathscr{E}_n} \mathfrak{f}(A)^2 W(A).$$

By Lemma 5.18, the second term is less than or equal to $C_0 n\ell\|\mathfrak{f}\|_1^2$ for some finite constant C_0. On the other hand, a simple adaptation of the proof of the same lemma gives that the first term is bounded by $C_0 n\ell^{-1}\|\mathfrak{h}\|^2_{\mathscr{X}_n,1}$. To conclude the proof, it remains to minimize over ℓ and to recall the variational formula for the $\mathscr{H}_{-1}$ norm of a function. □

5.7 Transient Markov Processes

In this section, we prove some estimates involving the Dirichlet form and the Green function of transient Markov processes which have their own interest and which were used in this chapter.

Consider a transient, irreducible Markov process $\{X_t, t\geq 0\}$ on a countable state space $\mathscr{E}$, reversible with respect to the counting measure. Denote by $r: \mathscr{E}\times\mathscr{E}\to\mathbb{R}_+$ the rate at which the process jumps from x to y and assume that $\sum_{y\in\mathscr{E}} r(x,y) < \infty$ for all x in $\mathscr{E}$. Since the process is reversible with respect to the counting measure, $r(x,y) = r(y,x)$ for all x, y in $\mathscr{E}$.

The generator of the Markov process reads

$$(Lf)(x) = \sum_{y\in\mathscr{E}} r(x,y)\big[f(y) - f(x)\big]$$

for every bounded function $f: \mathscr{E}\to\mathbb{R}$. Denote by $G(x,y)$ the Green function:

$$G(x,y) = \int_0^\infty p_t(x,y)\,dt,$$

which is finite because we assumed the process to be transient. Here $p_t(x,y)$ stands for the transition probability function of the Markov chain. For x in $\mathscr{E}$, denote by $\mathbb{P}_x$ the probability measure on the path space $D(\mathbb{R}_+,\mathscr{E})$ induced by the Markov chain X_t starting from x.

Proposition 5.23 *Let $V: \mathscr{E}\to\mathbb{R}_+$ be a positive function with compact support. Then, for all finitely supported $u: \mathscr{E}\to\mathbb{R}$,*

$$\sum_{y\in\mathscr{E}} u(y)^2 V(y) \leq (C_0/2)\sum_{x,y} r(x,y)\big\{u(y) - u(x)\big\}^2,$$

where

$$C_0 = C_0(V) = \sup_{x\in\mathscr{E}} \sum_{y\in\mathscr{E}} G(x,y)V(y).$$

Proof Notice that $C_0(V)$ is finite because V has compact support, the process is transient and $G(x,y) \le G(y,y)$ for all x, y in $\mathcal{E}$. Let $W(x) = \sum_{y\in\mathcal{E}} G(x,y)V(y)$ so that $LW = -V$ and

$$\sum_{y\in\mathcal{E}} u(y)^2 V(y) \le C_0 \sum_{y\in\mathcal{E}} \frac{u(y)^2}{W(y)} V(y) = -C_0 \sum_{y\in\mathcal{E}} \frac{u(y)^2}{W(y)} (LW)(y).$$

After a change of variables, since the process is reversible with respect to the counting measure, the previous expression becomes

$$\begin{aligned} &(C_0/2) \sum_{x,y\in\mathcal{E}} r(x,y) \left\{ \frac{u(y)^2}{W(y)} - \frac{u(x)^2}{W(x)} \right\} \{W(y) - W(x)\} \\ &= (C_0/2) \sum_{x,y\in\mathcal{E}} r(x,y) \left\{ u(y)^2 + u(x)^2 - \frac{u(y)^2}{W(y)} W(x) - \frac{u(x)^2}{W(x)} W(y) \right\}. \end{aligned}$$

Since $2ab \le a^2 + b^2$, the last expression is bounded above by

$$(C_0/2) \sum_{x,y\in\mathcal{E}} r(x,y) \{u(y) - u(x)\}^2,$$

which concludes the proof of the proposition. □

Consider a subset $\mathcal{E}_0$ of $\mathcal{E}$ and exclude all jumps to or from $\mathcal{R} = \mathcal{E} \setminus \mathcal{E}_0$. Consider on $\mathcal{E}_0$ the induced Markov chain $\{X_t^0 : t \ge 0\}$ with generator given by

$$(L_0 f)(x) = \sum_{y\in\mathcal{E}_0} r(x,y) \big[f(y) - f(x) \big]$$

for every bounded function $f : \mathcal{E}_0 \to \mathbb{R}$.

The first result states that the induced Markov chain is transient if the probability of hitting the set $\mathcal{R}$ starting from any site in $\mathcal{E}_0$ is less than one. More precisely, for each subset A of $\mathcal{E}$, denote by H_A the hitting time of A:

$$H_A = \inf\{t \ge 0 : X_t \in A\}.$$

Let

$$\theta(x) = \mathbb{P}_x[H_{\mathcal{R}} < \infty].$$

Lemma 5.24 *Assume that $\theta(x) < 1$ for all x in $\mathcal{E}_0$. Then the process is transient.*

Proof To prove that the process is transient, we will show that the function θ restricted to $\mathcal{E}_0$ is a bounded, non-constant, superharmonic function for $\{X_t^0 : t \ge 0\}$.

Fix x in $\mathscr{E}_0$. Conditioning on the first jump of the process, we obtain that θ solves the equation

$$\begin{cases} (L\theta)(x) = 0 & \text{for } x \text{ in } \mathscr{E}_0, \\ \theta(x) = 1 & \text{for } x \text{ in } \mathscr{R}. \end{cases}$$

Therefore, for x in $\mathscr{E}_0$,

$$\begin{aligned}(L_0\theta)(x) &= \sum_{y\in\mathscr{E}_0} r(x,y)\big\{\theta(y)-\theta(x)\big\} = -\sum_{y\in\mathscr{R}} r(x,y)\big\{\theta(y)-\theta(x)\big\} \\ &= -\big\{1-\theta(x)\big\}R(x),\end{aligned}$$

where $R(x) = \sum_{y\in\mathscr{R}} r(x,y)$. This identity shows that θ is a bounded superharmonic function for the Markov chain $\{X_t^0 : t \ge 0\}$. By irreducibility, R cannot vanish uniformly on $\mathscr{E}_0$. Since we assumed that $\theta(x) < 1$, $x \in \mathscr{E}_0$, $(L_0\theta)(x_0) < 0$ for some $x_0 \in \mathscr{E}_0$. Hence, θ is not constant, which proves the lemma. □

Denote by G_0 the Green function of the induced Markov chain. We now obtain estimates on G_0 assuming that

$$b = \sup_{x\in\mathscr{E}_0} \theta(x) < 1.$$

We proved in Lemma 5.24 that $-(L_0\theta)(x) = (1-\theta(x))R(x)$. By definition of b, this latter quantity is bounded below by $(1-b)R(x)$ so that $-(L_0\theta)(x) \ge (1-b)R(x)$. Applying Green's function on both sides of this inequality, we obtain that

$$\sum_{y\in\mathscr{E}_0} G_0(x,y)R(y) \le \frac{\theta(x)}{1-b} \tag{5.43}$$

for all x in $\mathscr{E}_0$.

Proposition 5.25 *Fix a positive function $U : \mathscr{E} \to \mathbb{R}_+$ supported on $\mathscr{E}_0$ (U vanishes on $\mathscr{R}$). Then,*

$$\begin{aligned}\sup_{x\in\mathscr{E}_0}\sum_{y\in\mathscr{E}_0} G_0(x,y)U(y) &\le \frac{1}{1-b}\sup_{x\in\mathscr{E}}\sum_{y\in\mathscr{E}} G(x,y)U(y) \\ &= \frac{1}{1-b}\sup_{x\in\mathscr{E}_0}\sum_{y\in\mathscr{E}_0} G(x,y)U(y).\end{aligned}$$

Moreover, for all x, y in $\mathscr{E}_0$,

$$G_0(x,y) \le G(x,y) + \frac{\theta(x)}{1-b}G(y,y), \qquad G_0(x,x) \le \frac{1}{1-b}G(x,x).$$

Proof Fix a positive function $U : \mathscr{E} \to \mathbb{R}_+$ supported on $\mathscr{E}_0$. Let

$$W(x) = \sum_{y \in \mathscr{E}} G(x, y)U(y) = \sum_{y \in \mathscr{E}_0} G(x, y)U(y).$$

The function W is non-negative and solves $LW = -U$. A computation shows that for x in $\mathscr{E}_0$,

$$(L_0 W)(x) = (LW)(x) - \sum_{y \in \mathscr{R}} r(x, y)\{W(y) - W(x)\} \le -U(x) + R(x)W(x).$$

The latter term is bounded above by $-U(x) + C_1 R(x)$, where $C_1 = \sup_{x \in \mathscr{E}_0} W(x)$. Thus, $U(x) \le -(L_0 W)(x) + C_1 R(x)$. Applying G_0 on both sides, in view of (5.43) we get that

$$(G_0 U)(x) \le W(x) + C_1 \sum_{y \in \mathscr{E}_0} G_0(x, y)R(y) \le W(x) + C_1 \frac{\theta(x)}{1-b}. \tag{5.44}$$

Taking the supremum over x and recalling the definition of C_1 we obtain that

$$\sup_{x \in \mathscr{E}_0} \sum_{y \in \mathscr{E}_0} G_0(x, y)U(y) \le \frac{1}{1-b} \sup_{x \in \mathscr{E}} \sum_{y \in \mathscr{E}} G(x, y)U(y).$$

Since U vanishes on the set $\mathscr{R}$, we may replace on the right-hand side $\mathscr{E}$ by $\mathscr{E}_0$ in both occurrences. This proves the first assertion of the proposition.

Fix z in $\mathscr{E}_0$ and take $U(y) = \mathbf{1}\{y = z\}$ in (5.44). Since $C_1 = \sup_{x \in \mathscr{E}_0} W(x) = \sup_{x \in \mathscr{E}_0} G(x, z) = G(z, z)$, we obtain that

$$G_0(x, z) = (G_0 U)(x) \le G(x, z) + \frac{\theta(x)}{1-b} G(z, z),$$

which is the second statement of the lemma. Taking $z = x$ in this inequality and recalling that $\theta(x) \le b$, we prove the second statement of the lemma. □

We now apply the previous results to the case of symmetric random walks on $\mathbb{Z}^d$, $d \ge 3$, for later applications in Chaps. 6 and 7. Fix a symmetric finite range probability p on $\mathbb{Z}^d$ and denote by $\{X_t : t \ge 0\}$, $\{X_t^j : t \ge 0\}$, $j = 1, 2$, independent copies of a random walk on $\mathbb{Z}^d$ which jumps from x to $x + y$ at rate $p(y)$. Let Y_t (resp. (Y_t^1, Y_t^2)) be the Markov chain on $\mathbb{Z}_*^d = \mathbb{Z}^d \setminus \{0\}$ (resp. $\{(x, y) \in \mathbb{Z}^d \times \mathbb{Z}^d : x \ne 0, y \ne 0, x \ne y\}$) induced by the process X_t (resp. (X_t^1, X_t^2)).

Lemma 5.26 *The Markov chains Y_t and (Y_t^1, Y_t^2) are transient.*

Proof It is well known that the random walk is transient in dimension $d \ge 3$. Exclude all jumps from and to the origin and consider the induced random walk. In this case,

$$\theta(x) = \mathbb{P}_x[H_{\{0\}} < \infty] < 1$$

for all $x \neq 0$. Moreover, since the probability has finite range,

$$b = \sup_{x \neq 0} \theta(x) = \max_{\substack{|x| \leq A \\ x \neq 0}} \theta(x) < 1.$$

Hence, the induced process is transient and the estimates derived in Proposition 5.25 are in force.

Let $\mathscr{R} = \{(x, y) \in \mathbb{Z}^d \times \mathbb{Z}^d : x \neq 0, y \neq 0, x \neq y\}$. Recall the definition of θ given at the beginning of the lemma. Then

$$\theta(x_1, x_2) = \mathbb{P}_{(x_1, x_2)}[H_{\mathscr{R}} < \infty] \leq \theta(x_1) + \theta(x_2) + \theta(x_1 - x_2).$$

Since $\sup_x \theta(x) < \infty$ and since $\theta(x)$ vanishes as $x \to \infty$, it follows from the previous bound that there exists $A > 0$ such that $\sup\{\theta(x_1, x_2) : |x_1| + |x_2| > A\} < 1$. To conclude the proof it remains to observe that $\theta(x_1, x_2) < 1$ for all (x_1, x_2) and that the set $\{(x_1, x_2) : |x_1| + |x_2| \leq A\}$ is finite. □

We conclude this section by showing the transience of n symmetric random walks evolving on $\mathbb{Z}^d$ with an exclusion rule whenever $nd \geq 3$.

Fix a subset A of $\mathbb{Z}^d$ with n points. Let $F : (\mathbb{Z}^d)^n \to \mathbb{R}$ be the bounded symmetric function given by $F(x_1, \dots, x_n) = \mathbf{1}\{\{x_1, \dots, x_n\} \subset A\}$. It is easy so check that F is positive definite in the sense of Liggett (1985, Section VIII.1). By Proposition VIII.1.7 of that book, for any configuration ξ with n particles,

$$\mathbb{P}_\xi[\eta_t \subset A] \leq \mathbb{P}_\xi[X_t \subset A], \tag{5.45}$$

where X_t represents n independent symmetric random walks. By the transience of independent random walks, the right-hand side is integrable, and so is the left-hand side, proving the transience of the symmetric random walks evolving with an exclusion rule.

5.8 Comments and References

The main reference for exclusion processes is Liggett (1985). We refer to Karr (1983); von Weizsäcker and Winkler (1979/1980, 1980); Winkler (1988) for generalizations of Lemma 5.6, which asserts that all finite range mean zero probability measures on $\mathbb{Z}^d$ which do not charge the origin are convex combinations of probability measures associated to cycles. A proof of this result appears in Xu (1993). The one presented here is a modification of Sued (2003). The sector condition for mean zero asymmetric exclusion processes stated in Proposition 5.5 and the general result presented in Lemma 5.8 are due to Varadhan (1995). The alternative proof presented in Lemma 5.11 seems to be new. Komoriya (1998) used the sector condition to prove the hydrodynamic behavior of asymmetric mean zero exclusion processes.

The duality method explained in Sect. 5.4 and the graded sector bound stated in Lemma 5.15, whose proof is based on the estimate of the two point correlation functions stated in Lemma 5.18, appeared in Landim and Yau (1997) and Sethuraman et al. (2000). The proof of Lemma 5.20, based on the removal of the hard core interaction, is due to Sethuraman et al. (2000). It has been used in Landim et al. (2004b) and Yau (2004) to prove the superdiffusivity of the second class particle in asymmetric exclusion processes in dimensions 1 and 2. Lemma 5.12 is due to Sethuraman and Xu (1996). The converse holds also: If a cylinder function belongs to $\mathcal{H}_{-1}$, then $\sum_{A\in\mathcal{E}_n}(\mathfrak{F}f)(\alpha, A)=0$ for all n such that $nd\le 2$. In this case an invariance principle is in force. We presented an alternative proof based on estimates of the Green function of n symmetric random walks evolving with exclusion obtained in Landim (1998). Sect. 5.7 is taken from Sethuraman et al. (2000).

Extensions We assumed throughout this chapter that the cylinder function V belongs to $\mathcal{H}_{-1}$, whose norm depends on the symmetric part of the generator. Lemma 5.12 presents, as we have seen, necessary and sufficient conditions for a cylinder function to belong to $\mathcal{H}_{-1}$. Sethuraman (2000) proved that in the mean zero case the variance $\sigma^2(V)$ is finite if and only if the conditions of Lemma 5.12 are fulfilled, and that in the asymmetric case in dimension $d\ge 3$ the variance is finite if and only if the cylinder function has mean zero. In all cases, when the variance is finite an invariance principle holds.

For a cylinder function V, let $\hat V:[0,1]\to\mathbb{R}$ be the smooth function defined by $\hat V(\beta)=E_{\nu_\beta}[V(\eta)]$. It follows from the previous paragraph that in the symmetric case and in the mean zero case the variance of a cylinder function V is finite if and only if V belongs to $\mathcal{H}_{-1}$, $\sigma^2(V)<\infty \Leftrightarrow \|V\|_{-1}<\infty$, which happens if and only if $\hat V^{(n)}(\alpha)=0$ for $nd\le 2$, where $\hat V^{(n)}$ represents the n-th derivative of $\hat V$. In the asymmetric case, in dimension $d\ge 3$, the variance of a cylinder function V is finite if and only if V belongs to $\mathcal{H}_{-1}$, which occurs if and only if V has mean zero, $\hat V(\alpha)=0$. In all these cases the invariance principle holds.

To complete the picture, it remains to understand what happens in the mean zero case for a cylinder function V such that $\hat V^{(n)}(\alpha)\neq 0$ for some n such that $nd\le 2$, and in the asymmetric case in dimensions 1 and 2.

The Symmetric Case The integral $\int_0^t \eta_s(0)ds$ represents the amount of time the origin stayed occupied in the interval $[0,t]$. For this reason, this additive functional is called the occupation time. Kipnis (1987) proved that in the symmetric case, under $\mathbb{P}_{\nu_\alpha}$, $\beta(d,t)^{-1}\int_0^t\{\eta_s(0)-\alpha\}ds$ converges to a mean zero Gaussian law with variance $\sigma^2(d,\alpha)$, where d represents the dimension, $\beta(1,t)=t^{3/4}$, $\beta(2,t)=\sqrt{t\log t}$, $\beta(d,t)=\sqrt{t}$ for $d\ge 3$, and $\sigma^2(d,\alpha)$ is known explicitly. The proof is carried out by decomposing the additive functional as the sum of a martingale and a negligible term.

Sethuraman (2000) extended this result to cylinder functions. Fix a mean zero function V such that $\hat V^{(1)}(\alpha)\neq 0$. In dimension 1, $N^{-3/4}\int_0^{Nt}V(\eta_s)ds$ converges to $B_{3/4}(Ct)$, for some constant $0<C<\infty$, where $B_{3/4}$ is the fractional Brownian motion with Hurst parameter 3/4. In dimension 2, $[N \log N]^{-1/2} \int_0^{Nt} V(\eta_s)ds$ converges to $B(Ct)$, for some constant $0 < C < \infty$, where B is the Brownian motion. Finally, in dimension 1, for mean zero cylinder functions such that $\hat{V}^{(1)}(\alpha) = 0$, $\hat{V}^{(2)}(\alpha) \neq 0$, Quastel et al. (2002) proved that $[N \log N]^{-1/2} \int_0^{Nt} V(\eta_s)ds$ converges to $B(Ct)$, where B is the Brownian motion and $C = [\chi(\alpha)^2/8\pi]\hat{V}^{(2)}(\alpha)^2$.

The Mean Zero Case In dimensions 1 and 2, Sethuraman (2000, 2006a) proved an invariance principle for coordinatewise increasing local functions provided the variance is finite. The proof relies on Newman and Wright's central limit theorem for stationary L^2 associated random variables, Newman (1983, 1984); Newman and Wright (1982). The general case under the assumption of the finiteness of the variances is open.

The Asymmetric Case As we shall see, the density α plays an important role in this context. We start from an identity derived in Sect. 2.5:

$$\mathbb{E}_{\nu_\alpha}\left[\left(\int_0^t \{\eta_s(0) - \alpha\}\, ds\right)^2\right] = 2\int_0^t (t-s)\big\{\mathbb{E}_{\nu_\alpha}\big[\eta_s(0)\eta_0(0)\big] - \alpha^2\big\}\, ds.$$

By Lemma 7.8, the expression inside braces on the right-hand side can be written in terms of the position of a second class particle:

$$\mathbb{E}_{\nu_\alpha}\big[\eta_s(0)\eta_0(0)\big] - \alpha^2 = \chi(\alpha)\mathbb{P}_{\nu_\alpha^1}[X_t = x],$$

where ν_α^1 is the product measure ν_α conditioned to have a particle at the origin, and $\chi(\alpha) = \alpha(1-\alpha)$ is the static compressibility of the simple exclusion process. Therefore,

$$\mathbb{E}_{\nu_\alpha}\left[\left(\int_0^t \{\eta_s(0) - \alpha\}\, ds\right)^2\right] = 2\chi(\alpha)\int_0^\infty (t-s)^+ \mathbb{P}_{\nu_\alpha^1}[X_s = 0]\, ds.$$

A simple computation shows that the drift of the second class particle under the equilibrium measure ν_α is $(1-2\alpha)\sum_x xp(x)$. In particular, if $\alpha \neq 1/2$, the second class particle has a drift and one expects the variance of the occupation time to be of order t, while for $\alpha = 1/2$ the variance should increase faster than linearly.

Assuming $\alpha \neq 1/2$, Seppäläinen and Sethuraman (2003) proved that in dimension 1 the time spent by the second class particle at the origin has finite expectation. They deduced from this result that the variance of any mean zero local function is finite and an invariance principle for the associated additive functional. The finiteness of the variance and the invariance principle were extended to dimension 2 by Bernardin (2004).

Central Limit Theorems for Functions Which Do Not Belong to $\mathscr{H}_{-1}$ We mentioned in Remark 2.5 the possibility to prove central limit theorems for functions which do not belong to $\mathscr{H}_{-1}$, but whose variance is finite.

The occupation time for equilibrium asymmetric exclusion processes in dimensions 1 and 2 with density $\alpha \neq 1/2$ provides such an example. On the one hand, as we have just seen, the variance is finite and the invariance principle holds. On the other hand, $\eta(0) - \alpha$ does not belong to $\mathscr{H}_{-1}$: Rewriting a local function f in the orthonormal basis Ψ, $f = \sum \mathfrak{f}(A)\Psi_A$, we get that

$$\|\eta(0) - \alpha\|_{-1}^2 = \sup_{f \in \mathscr{C}} \{2\langle f, \eta(0) - \alpha\rangle_{\nu_\alpha} - \|f\|_1^2\}$$

$$= \sup_{\mathfrak{f}} \left\{ 2\sqrt{\alpha(1-\alpha)}\mathfrak{f}(0) - (1/2)\sum_{j=1}^{d} \sum_{x \in \mathbb{Z}^d} [\mathfrak{f}(x+e_j) - \mathfrak{f}(x)]^2 \right\};$$

where the supremum is carried over all finitely supported functions $\mathfrak{f} : \mathbb{Z}^d \to \mathbb{R}$. We assumed the process to be nearest-neighbor: $p(x) = \sum_{1 \le j \le d} \{p_j \delta_{x,e_j} + (1 - p_j)\delta_{x,-e_j}\}$ and used the orthogonality properties of the spaces $\mathscr{A}_n$ to reduce the supremum to finitely supported functions $\mathfrak{f} : \mathscr{E}_1 \to \mathbb{R}$. In dimensions 1 and 2, it is easy to show, with a judicious choice of the test function $\mathfrak{f}$, that the supremum is equal to ∞.

The Asymmetric Case with $\alpha = 1/2$ In contrast, if the density $\alpha = 1/2$, the variance of a mean zero local function diverges: Bernardin (2004) proved that the occupation-time variance at the origin up to time t in dimension $d = 1$ in equilibrium with density $\alpha = 1/2$ is in a certain sense at least $t^{5/4}$. Sethuraman (2006b) extended this result to dimension 2, obtaining a lower bound of order $t \log\log t$.

Li and Mao (2008) proved that in dimension 1 the variance of the occupation time for the asymmetric exclusion process with density $\alpha = 1/2$ is bounded above by $O(t^{3/2})$.

$\mathscr{H}_{-1}$ Estimates We have seen the importance of $\mathscr{H}_{-1}$ estimates in the proof of the invariance principles. Sethuraman (2003) showed that the resolvent $\mathscr{H}_{-1}$ norm of a finite range exclusion process is equivalent to the resolvent $\mathscr{H}_{-1}$ norm of the nearest neighbor exclusion process with the same drift. A similar result holds for the environment process as seen from the tagged particle in dimension $d \geq 2$.

Zero-Range Processes Central limit theorems of additive functionals of symmetric zero-range processes are investigated in Sethuraman and Xu (1996) and Quastel et al. (2002).

References

Aubin JP (1979) Applied functional analysis: with exercises by Bernard Cornet and Jean-Michel Lasry. Wiley, New York/Chichester/Brisbane. Translated from the French by Carole Labrousse

Bernardin C (2004) Fluctuations in the occupation time of a site in the asymmetric simple exclusion process. Ann Probab 32(1B):855–879

Blumenthal RM, Getoor RK (1968) Markov processes and potential theory. Pure and applied mathematics, vol 29. Academic Press, New York

Karr AF (1983) Extreme points of certain sets of probability measures, with applications. Math Oper Res 8(1):74–85

Kipnis C (1987) Fluctuations des temps d'occupation d'un site dans l'exclusion simple symétrique. Ann Inst Henri Poincaré Probab Stat 23(1):21–35

Komoriya K (1998) Hydrodynamic limit for asymmetric mean zero exclusion processes with speed change. Ann Inst Henri Poincaré Probab Stat 34(6):767–797

Landim C (1998) Decay to equilibrium in L^∞ of finite interacting particle systems in infinite volume. Markov Process Relat Fields 4(4):517–534. I Brazilian School in Probability, Rio de Janeiro, 1997

Landim C, Yau HT (1997) Fluctuation-dissipation equation of asymmetric simple exclusion processes. Probab Theory Relat Fields 108(3):321–356

Landim C, Quastel J, Salmhofer M, Yau HT (2004b) Superdiffusivity of asymmetric exclusion process in dimensions one and two. Commun Math Phys 244(3):455–481

Li CM, Mao MC (2008) An upper bound on the occupation time for a simple exclusion process. Teor Mat Fiz 156(1):147–158

Liggett TM (1985) Interacting particle systems. Grundlehren der Mathematischen Wissenschaften [Fundamental principles of mathematical sciences], vol 276. Springer, New York

Newman CM (1983) A general central limit theorem for FKG systems. Commun Math Phys 91(1):75–80

Newman CM (1984) Asymptotic independence and limit theorems for positively and negatively dependent random variables. In: Inequalities in statistics and probability, Lincoln, NE, 1982. IMS lect notes monogr ser, vol 5. Inst Math Stat, Hayward, pp 127–140

Newman CM, Wright AL (1982) Associated random variables and martingale inequalities. Z Wahrscheinlichkeitstheor Verw Geb 59(3):361–371

Quastel J, Jankowski H, Sheriff J (2002) Central limit theorem for zero-range processes. Methods Appl Anal 9(3):393–406. Special issue dedicated to Daniel W Stroock and Srinivasa SR Varadhan on the occasion of their 60th birthday

Seppäläinen T, Sethuraman S (2003) Transience of second-class particles and diffusive bounds for additive functionals in one-dimensional asymmetric exclusion processes. Ann Probab 31(1):148–169

Sethuraman S (2000) Central limit theorems for additive functionals of the simple exclusion process. Ann Probab 28(1):277–302

Sethuraman S (2003) An equivalence of H_{-1} norms for the simple exclusion process. Ann Probab 31(1):35–62

Sethuraman S (2006a) Correction: "Central limit theorems for additive functionals of the simple exclusion process" [Ann Probab **28**(1):277–302 (2000); mr1756006]. Ann Probab 34(1):427–428

Sethuraman S (2006b) Superdiffusivity of occupation-time variance in 2-dimensional asymmetric exclusion processes with density $\rho = 1/2$. J Stat Phys 123(4):787–802

Sethuraman S, Xu L (1996) A central limit theorem for reversible exclusion and zero-range particle systems. Ann Probab 24(4):1842–1870

Sethuraman S, Varadhan SRS, Yau HT (2000) Diffusive limit of a tagged particle in asymmetric simple exclusion processes. Commun Pure Appl Math 53(8):972–1006

Sued M (2003) PhD dissertation, IMPA. http://www.preprint.impa.br/Shadows/SERIE_C/2003/22.html

Varadhan SRS (1995) Self-diffusion of a tagged particle in equilibrium for asymmetric mean zero random walk with simple exclusion. Ann Inst Henri Poincaré Probab Stat 31(1):273–285

von Weizsäcker H, Winkler G (1979/1980) Integral representation in the set of solutions of a generalized moment problem. Math Ann 246(1):23–32

von Weizsäcker H, Winkler G (1980) Noncompact extremal integral representations: some probabilistic aspects. In: Functional analysis: surveys and recent results, II. Proc second conf functional anal, Univ. Paderborn, Paderborn, 1979. Notas mat, vol 68. North-Holland, Amsterdam, pp 115–148

Winkler G (1988) Extreme points of moment sets. Math Oper Res 13(4):581–587

Xu L (1993) PhD dissertation, Courant Institute, New York University

Yau HT (2004) $(\log t)^{2/3}$ law of the two dimensional asymmetric simple exclusion process. Ann Math (2) 159(1):377–405

Yosida K (1995) Functional analysis. Classics in mathematics. Springer, Berlin. Reprint of the sixth 1980 edition

Chapter 6
Self-diffusion

In the previous chapter we examined the central limit theorem of additive functionals of the simple exclusion process. We investigate in this one the asymptotic behavior of a distinguished particle.

Recall the set-up introduced in Chap. 5. Fix a finite range probability measure p on $\mathbb{Z}^d$ such that $p(0) = 0$ and denote by $\{\eta_t : t \ge 0\}$ the exclusion process on $\mathbb{X} = \{0, 1\}^{\mathbb{Z}^d}$ associated to p. Consider a configuration η with a particle at the origin. Tag this particle and denote by Z_t, $t \ge 0$, its position at time t. Due to the presence of the other particles $\{Z_t : t \ge 0\}$ by itself is not a Markov process but the pair $\{(Z_t, \eta_t) : t \ge 0\}$ is a Markov process.

We are interested in this chapter in the asymptotic behavior of Z_t. If there were no other particles, Z_t would evolve as a continuous time random walk for which the law of large numbers, the central limit theorem and the corresponding invariance principle are well known. The interaction of the tagged particle with the other particles, called from now on the environment, clearly alters the dynamics of the tagged particle, and the study of its evolution requires new tools.

The Markov process (Z_t, η_t) has a very special feature which enables the analysis of the asymptotic behavior of the tagged particle. Consider the evolution of the environment as seen from the position of the tagged particle. This means that we set the origin to be at the position of the tagged particle, and observe from there the evolution of the remaining particles. To respect this rule, each time the tagged particle jumps by x, we translate the entire configuration by $-x$ to keep the tagged particle at the origin. A fundamental and quite particular property of this dynamics is that it possesses a very simple family of stationary states, the Bernoulli product measures with density $0 \le \alpha \le 1$ on $\mathbb{Z}^d_* = \mathbb{Z}^d \setminus \{0\}$, which we denote by ν^*_α.

A law of large numbers for the tagged particle starting from ν^*_α is easy to infer. The mean displacement of the tagged particle is $\mathfrak{m} = \sum_{x \in \mathbb{Z}^d} x p(x)$. Since ν^*_α is stationary, the site to which the tagged particle chooses to jump is empty with probability $1 - \alpha$. We thus expect a mean displacement in the stationary state to be equal to $(1 - \alpha)\mathfrak{m}$. This is in fact the content of the first main theorem in this chapter which states that, in the stationary state with density α, Z_t/t converges a.s. to $(1 - \alpha)\mathfrak{m}$ as $t \to \infty$.

T. Komorowski et al., *Fluctuations in Markov Processes*, Grundlehren der mathematischen Wissenschaften 345, DOI 10.1007/978-3-642-29880-6_6, © Springer-Verlag Berlin Heidelberg 2012

The second main result establishes a central limit theorem for the position of the tagged particle: $(Z_t - (1-\alpha)\mathfrak{m})/\sqrt{t}$ converges in distribution to a mean zero Gaussian random vector with a covariance matrix $D(\alpha)$ which depends on the surrounding density of particles. So far, this result has been proved in the mean zero case, and in the asymmetric case in dimension $d \ge 3$. It is conjectured to hold also in dimensions 1 and 2 in the asymmetric case, but there even the finiteness of the asymptotic variance has not yet been established.

6.1 The Exclusion Process as Seen from a Tagged Particle

Consider an exclusion process $\{\eta_t : t \ge 0\}$ on $\mathbb{X}$ satisfying the assumptions of Sect. 5.1. Denote by $\{\tau_x : x \in \mathbb{Z}^d\}$ the group of translations on $\mathbb{X}$ so that

$$(\tau_x \eta)(y) = \eta(x+y)$$

for x, y in $\mathbb{Z}^d$ and a configuration η in $\mathbb{X}$. The action of the translation group is naturally extended to functions and measures.

Consider a configuration η with a particle at the origin. Tag this particle and denote by Z_t, $t \ge 0$, its position at time t. Let $\{\xi_t : t \ge 0\}$ be the state of the process as seen from the position of this tagged particle, $\xi_t = \tau_{Z_t}\eta_t$, and notice that the origin stays always occupied, $\xi_t(0) = (\tau_{Z_t}\eta_t)(0) = \eta_t(Z_t) = 1$. In particular, we may consider ξ_t either as a configuration of $\mathbb{X}$ with a particle at the origin or as a configuration of $\mathbb{X}^* = \{0,1\}^{\mathbb{Z}^d_*}$, where $\mathbb{Z}^d_* = \mathbb{Z}^d \setminus \{0\}$. We adopt here the latter convention and denote by the Greek letter ξ the configurations of $\mathbb{X}^*$.

The evolution of ξ_t corresponds to a Feller process which we now describe. Assume that $\mathbb{X}^*$ is endowed with the product topology which turns $\mathbb{X}^*$ into a metrizable, compact space. Denote by $C(\mathbb{X}^*)$ the space of continuous functions on $\mathbb{X}^*$, regarded as a Banach space with norm

$$\|f\| = \sup_{\xi \in \mathbb{X}^*} \left| f(\xi) \right|,$$

and by $\mathscr{C} \subset C(\mathbb{X}^*)$ the space of functions which depend on the configuration η through a finite number of coordinates, called the cylinder functions.

Let $\{\theta_x : x \in \mathbb{Z}^d_*\}$ be the shift operators on $\mathbb{X}^*$ defined as follows:

$$(\theta_x \xi)(y) = \begin{cases} \xi(x) & \text{if } y = -x, \\ \xi(y+x) & \text{otherwise.} \end{cases}$$

This means that $\theta_x \xi$ stands for the configuration where the tagged particle, sitting at the origin, is first transferred to site x and then all the configuration is translated by $-x$.

Let $\mathscr{L}$ be the operator defined on $\mathscr{C}$ as the sum $\mathscr{L} = \mathscr{L}_0 + \mathscr{L}_\theta$, where $\mathscr{L}_0$, $\mathscr{L}_\theta$ are given by

$$\begin{aligned}(\mathscr{L}_0 f)(\xi) &= \sum_{x,y\in\mathbb{Z}^d_*} p(y-x)\xi(x)\big[1-\xi(y)\big]\big[f\big(\sigma^{x,y}\xi\big)-f(\xi)\big],\\ (\mathscr{L}_\theta f)(\xi) &= \sum_{z\in\mathbb{Z}^d_*} p(z)\big[1-\xi(z)\big]\big[f(\theta_z\xi)-f(\xi)\big].\end{aligned} \tag{6.1}$$

The first operator takes into account the jumps of the environment, while the second one corresponds to the jumps of the tagged particle.

It is not difficult to check that the operator $\mathscr{L}$ defined on $\mathscr{C}$ is a Markov pregenerator (Liggett, 1985, Definition I.2.1). By the proof of Liggett (1985, Theorem I.3.9) the closure of $\mathscr{L}$, still denoted by $\mathscr{L}$, is a Markov generator and $\mathscr{C}$ is a core for $\mathscr{L}$. Denote by $\{S(t) : t \ge 0\}$ the strictly Markovian, Feller semigroup on $C(\mathbb{X}^*)$ associated to the generator $\mathscr{L}$ through the Hille–Yosida theorem.

Let $D([0,\infty),\mathbb{X}^*)$ be the space of r.c.l.l. trajectories $\xi : [0,\infty) \to \mathbb{X}^*$ and let $\{\Pi_t : t \ge 0\}$ be the canonical projections defined by $\Pi_t(\xi) = \xi_t$. We shall represent by $\mathscr{F}^o$ the smallest σ-algebra on $D([0,\infty),\mathbb{X}^*)$ which turns the projections Π_s, $s \ge 0$, measurable; and by $\mathscr{F}^o_t$ the smallest σ-algebra relative to which all the mappings Π_s, $0 \le s \le t$, are measurable. Denote by $\{\mathbb{P}_\xi : \xi \in \mathbb{X}^*\}$ the normal Markov process associated to the semigroup $\{S(t) : t \ge 0\}$. Expectation with respect to $\mathbb{P}_\xi$ is represented by $\mathbb{E}_\xi$.

For $0 \le \alpha \le 1$, let ν^*_α be the Bernoulli product measure on $\mathbb{X}^*$ with marginals given by

$$\nu^*_\alpha\big\{\xi : \xi(x) = 1\big\} = \alpha$$

for x in $\mathbb{Z}^d_*$.

Proposition 6.1 *The Bernoulli measures* $\{\nu^*_\alpha, 0 \le \alpha \le 1\}$ *are invariant for the Markov process* $\{\xi_t : t \ge 0\}$.

Proof By a simple change of variables, for any cylinder functions f, g and any $z \in \mathbb{Z}^d_*$,

$$\int f(\theta_z\xi)g(\xi)\big[1-\xi(z)\big]\nu^*_\alpha(d\xi) = \int f(\xi)g(\theta_{-z}\xi)\big[1-\xi(-z)\big]\nu^*_\alpha(d\xi).$$

It follows from this identity and from (5.4), that

$$\langle g, \mathscr{L} f\rangle_{\nu^*_\alpha} = \langle \mathscr{L}_{p^*} g, f\rangle_{\nu^*_\alpha} \tag{6.2}$$

for all cylinder functions f, g. In this formula, $\langle\cdot,\cdot\rangle_{\nu^*_\alpha}$ represents the scalar product in $L^2(\nu^*_\alpha)$ and $\mathscr{L}_{p^*}$ the generator $\mathscr{L}$ defined above (6.1) with p^* in place of p, where $p^*(z) = p(-z)$. In particular, $\int \mathscr{L} f\, d\nu^*_\alpha = 0$ for any cylinder function f, which proves the proposition in view of Liggett (1985, Proposition I.2.13). □

Note that the probability measures ν_α^* are invariant for the operators $\mathscr{L}_0$, $\mathscr{L}_\theta$ taken individually if and only if the probability measure p is symmetric.

For a probability measure μ in $\mathbb{X}^*$, denote by $\mathbb{P}_\mu$ the measure on $(D([0,\infty),\mathbb{X}^*),\mathscr{F}^o)$ given by $\int \mathbb{P}_\xi\, \mu(d\xi)$. Expectation with respect to $\mathbb{P}_\mu$ is denoted by $\mathbb{E}_\mu$. For $0\le\alpha\le 1$, denote by $(D([0,\infty),\mathbb{X}^*),\mathscr{F},\mathbb{P}_{\nu_\alpha},\{\mathscr{F}_t : t\ge 0\})$ the usual augmentation of the filtered space $(D([0,\infty),\mathbb{X}^*),\mathscr{F}^o,\mathbb{P}_{\nu_\alpha},\{\mathscr{F}_t^o : t\ge 0\})$. By Theorem 8.11 and Proposition 8.12 of Chap. 1 of Blumenthal and Getoor (1968), $\{\xi_t : t\ge 0\}$ is a strong Markov process with respect to the augmented filtration.

As in the previous chapter, the semigroup $\{S(t) : t\ge 0\}$ extends to a Markov semigroup on $L^2(\nu_\alpha^*)$ whose generator $\mathscr{L}_{\nu_\alpha^*}$ is the closure of $\mathscr{L}$ in $L^2(\nu_\alpha^*)$. Since the density α remains fixed, to keep notation simple we denote $\mathscr{L}_{\nu_\alpha^*}$ by $\mathscr{L}$ and by $\mathscr{D}(\mathscr{L})$ the domain of $\mathscr{L}$ in $L^2(\nu_\alpha^*)$. We adopt the same convention with the operator $\mathscr{L}_{p^*}$, introduced above, and use the same notation $\mathscr{L}_{p^*}$ to represent the generator acting on $C(\mathbb{X}^*)$ and on $L^2(\nu_\alpha^*)$.

Denote by $\mathscr{L}^*$ the adjoint of $\mathscr{L}$ in $L^2(\nu_\alpha)$. The next result follows from the proof of Lemma 5.2 and from (6.2).

Lemma 6.2 *The adjoint operator $\mathscr{L}^*$ is a generator and $\mathscr{L}^*=\mathscr{L}_{p^*}$. In particular, $\mathscr{C}$ is a common core for $\mathscr{L}$ and $\mathscr{L}^*$, and $\mathscr{L}$ is self-adjoint with respect to each ν_α^* when p is symmetric.*

Recall that $s(\cdot)$ (resp. $a(\cdot)$) stands for the symmetric (resp. anti-symmetric) part of the probability p. For any cylinder function f

$$\langle f,(-\mathscr{L})f\rangle_{\nu_\alpha^*} =: \mathscr{D}(f)=\mathscr{D}_0(f)+\mathscr{D}_\theta(f), \tag{6.3}$$

where

$$\mathscr{D}_0(f)=(1/2)\sum_{x,y\in\mathbb{Z}_d^*} s(y-x)\int \xi(x)\big[1-\xi(y)\big]\big(T^{x,y}f\big)(\xi)^2\nu_\alpha^*(d\xi),$$

$$\mathscr{D}_\theta(f)=(1/2)\sum_{z\in\mathbb{Z}_d^*} s(z)\int \big[1-\xi(z)\big]\big(T^{z}f\big)(\xi)^2\nu_\alpha^*(d\xi).$$

In this formula and below, for a function g in $L^2(\nu_\alpha^*)$,

$$\big(T^{x,y}g\big)(\xi)=g\big(\sigma^{x,y}\xi\big)-g(\xi),\quad \big(T^{z}g\big)(\xi)=g(\theta_z\xi)-g(\xi).$$

Since $(T^{x,y}f)(\xi)$ vanishes unless $\xi(x)\ne\xi(y)$ and since $T^{x,y}f=T^{y,x}f$, we may rewrite $\mathscr{D}_0(f)$ as

$$\mathscr{D}_0(f)=(1/4)\sum_{x,y\in\mathbb{Z}_d^*} s(y-x)\int \big(T^{x,y}f\big)(\xi)^2\nu_\alpha^*(d\xi).$$

The explicit formulas for the Dirichlet forms written above hold also for functions f in the domain $\mathscr{D}(\mathscr{L})$ of the generator, and the series defined on the right-hand side converge absolutely. This can be proved by an approximation argument (cf. Liggett, 1985, Lemma IV.4.3).

Observe that, in contrast with what the notation suggests, in the case where the probability measure $p(\cdot)$ is not symmetric, $\mathscr{D}_0(f)$ is *not* equal to $\langle(-\mathscr{L}_0)f, f\rangle_{\nu_\alpha^*}$. Denote by $\mathscr{S}_0$ the generator $\mathscr{L}_0$ introduced in (6.1) with the probability measure $p(\cdot)$ replaced by its symmetric part $s(\cdot)$. A straightforward computation shows that

$$\mathscr{D}_0(f) = \big\langle(-\mathscr{S}_0)f, f\big\rangle_{\nu_\alpha^*} \neq \big\langle(-\mathscr{L}_0)f, f\big\rangle_{\nu_\alpha^*}, \quad f \in \mathscr{C}. \tag{6.4}$$

In particular, if p is not symmetric, $\mathscr{S}_0$ is *not* the symmetric part of the operator $\mathscr{L}_0$, even though the symmetric part of $\mathscr{L} = \mathscr{L}_0 + \mathscr{L}_\theta$ is $\mathscr{S}_0 + \mathscr{S}_\theta$, where $\mathscr{S}_\theta$ is the generator $\mathscr{L}_\theta$ defined in (6.1) with the probability measure p replaced by its symmetric part s.

We claim that the measures $\{\nu_\alpha^* : 0 \le \alpha \le 1\}$ are ergodic in all but one degenerate case.

Theorem 6.3 *Assume that $d \ge 2$ or assume that $d = 1$ and that $p(\cdot)$ is not nearest neighbor*: $\sum_{x \neq \pm 1} p(x) > 0$. *Then, for any $0 \le \alpha \le 1$, ν_α^* is ergodic for $\mathscr{L}$.*

Proof The proof follows closely the one of Theorem 5.3. Fix a function $f \in L^2(\nu_\alpha^*)$ invariant for the (L^2 extension) semigroup generated by $\mathscr{L}$. Then, f is in the domain of $\mathscr{L}$ and $\mathscr{L} f = 0$. By multiplying by f both sides of this equation and integrating, we obtain that

$$\begin{aligned}
&(1/4) \sum_{x,y \in \mathbb{Z}_d^*} s(y - x) \int \big[f\big(\sigma^{x,y}\xi\big) - f(\xi)\big]^2 \nu_\alpha^*(d\xi) \\
&\quad + (1/2) \sum_{x \in \mathbb{Z}_d^*} s(x) \int \big[1 - \xi(x)\big]\big[f(\theta_x \xi) - f(\xi)\big]^2 \nu_\alpha^*(d\xi) = 0.
\end{aligned}$$

Under our assumptions ($d \ge 2$ or $d = 1$ and p is not nearest neighbor), the support of $s(\cdot)$ generates $\mathbb{Z}_*^d$. Hence, for any $x, y \in \mathbb{Z}_*^d$

$$f\big(\sigma^{x,y}\xi\big) = f(\xi) \quad \nu_\alpha^*\text{-a.e.}$$

By De Finetti's theorem we conclude that f is constant ν_α^*-a.e. □

Remark 6.4 Without any further mention, we exclude from now on the degenerate one-dimensional case with only nearest neighbor jumps.

The main results of this chapter state a law of large numbers and a central limit theorem for the position of the tagged particle:

Theorem 6.5 *Fix $0 \le \alpha \le 1$. Then, as $t \uparrow \infty$, Z_t/t converges $\mathbb{P}_{\nu^*_\alpha}$-almost surely to $[1-\alpha]\mathfrak{m}$, where $\mathfrak{m} = \sum_{x\in\mathbb{Z}^d} xp(x)$.*

Theorem 6.6 *Assume that $\mathfrak{m} = 0$ or that $d \ge 3$. Then, under $\mathbb{P}_{\nu^*_\alpha}$,*

$$\frac{Z_t - (1-\alpha)\mathfrak{m}t}{\sqrt{t}}$$

converges in distribution, as $t \uparrow \infty$, to a mean zero Gaussian random vector with covariance matrix denoted by $D(\alpha)$.

As a quadratic form $D(\alpha)$ is strictly positive and finite: There exists a strictly positive and finite constant C_0, depending only on p, such that

$$C_0^{-1}\alpha(1-\alpha)|\mathfrak{a}|^2 \le \mathfrak{a}\cdot D(\alpha)\mathfrak{a} \le C_0(1-\alpha)|\mathfrak{a}|^2$$

for all $\mathfrak{a}$ in $\mathbb{R}^d$.

The matrix $D(\alpha)$ is called the self-diffusion matrix of the exclusion process. In this chapter, we adopt the convention that C_0 stands for a constant which depends only on the transition probability p and which may change from line to line.

Remark 6.7

(i) The result holds in the stronger form introduced in Definition 2.6. Conditioned on $\mathscr{F}_0$, the random vector $[Z_t - (1-\alpha)\mathfrak{m}t]/\sqrt{t}$ converges in $L^1(\mathbb{P}_{\nu^*_\alpha})$ to a mean zero Gaussian random vector with covariance matrix $D(\alpha)$.

(ii) Denote by Z_t^N, $N \ge 1$, $t \ge 0$, the rescaled position of the tagged particle speeded-up by N:

$$Z_t^N = \frac{Z_{tN} - tN(1-\alpha)\mathfrak{m}}{\sqrt{N}}.$$

An invariance principle holds: Under $\mathbb{P}_{\nu^*_\alpha}$, the sequence of processes $\{Z_t^N : t \ge 0\}$ converges in $D([0,\infty),\mathbb{R}^d)$ to a d-dimensional Brownian motion with diffusion matrix $D(\alpha)$.

(iii) It is conjectured that this theorem holds also in dimensions 1 and 2 in the case $\mathfrak{m} \ne 0$, but this is still an open problem. Actually, in these cases we are not even able to prove that the asymptotic variance is finite.

The proof of Theorem 6.6 follows the ideas discussed in Chap. 5 for additive functionals of simple exclusion processes. Some complications arise, however, due to the presence of the very *non-local* operators associated to the translations.

6.2 Elementary Martingales

It is useful to represent the position of the tagged particle in terms of elementary orthogonal martingales associated to the jumps of the process.

For z such that $p(z) > 0$ and for $0 \le s < t$, denote by $N^z_{[s,t]}$ the total number of jumps of the tagged particle from the origin to z in the time interval $[s,t]$. In the same way, for x, y in $\mathbb{Z}^d_*$ such that $p(y-x) > 0$, denote by $N^{x,y}_{[s,t]}$ the total number of jumps of a particle from x to y in the time interval $[s,t]$. Let $N^z_t = N^z_{[0,t]}$, $N^{x,y}_t = N^{x,y}_{[0,t]}$.

Lemma 6.8 *For x, y, z in $\mathbb{Z}^d_*$ such that $p(z) > 0$, $p(y-x) > 0$, let*

$$M^z_t = N^z_t - \int_0^t p(z)\big[1-\xi_s(z)\big]\,ds$$

$$M^{x,y}_t = N^{x,y}_t - \int_0^t p(y-x)\xi_s(x)\big[1-\xi_s(y)\big]\,ds.$$

$\{M^z_t : p(z) > 0\}$, $\{M^{x,y}_t : x, y \in \mathbb{Z}^d_*, p(y-x) > 0\}$ *are orthogonal martingales with quadratic variation* $\langle M^z\rangle_t$, $\langle M^{x,y}\rangle_t$ *given by*

$$\langle M^z\rangle_t = \int_0^t p(z)\big[1-\xi_s(z)\big]\,ds, \qquad \langle M^{x,y}\rangle_t = \int_0^t p(y-x)\xi_s(x)\big[1-\xi_s(y)\big]\,ds.$$

Proof Fix x, y, z in $\mathbb{Z}^d_*$ such that $p(z) > 0$, $p(y-x) > 0$. It is easy to check that $(\xi_t, N^z_t, N^{x,y}_t)$ is a Markov process on $\mathbb{X}^* \times \mathbb{Z} \times \mathbb{Z}$ with generator $\mathscr{L}_{x,y,z}$ given by

$$\begin{aligned}
&(\mathscr{L}_{x,y,z} f)(\xi,k,j) \\
&\quad = p(y-x)\xi(x)\big[1-\xi(y)\big]\big\{f\big(\sigma^{x,y}\xi,k,j+1\big) - f(\xi,k,j)\big\} \\
&\qquad + p(z)\big[1-\xi(z)\big]\big\{f(\theta_z\xi,k+1,j) - f(\xi,k,j)\big\} \\
&\qquad + \sum_{x',y' \in \mathbb{Z}^d_*} p(y'-x')\xi(x')\big[1-\xi(y')\big]\big\{f\big(\sigma^{x',y'}\xi,k,j\big) - f(\xi,k,j)\big\} \\
&\qquad + \sum_{z' \neq z} p(z')\big[1-\xi(z')\big]\big\{f(\theta_{z'}\xi,k,j) - f(\xi,k,j)\big\}.
\end{aligned}$$

In this formula the first sum is carried over all pairs (x', y') in $\mathbb{Z}^d_* \times \mathbb{Z}^d_*$ different from (x,y). Dynkin's formula applied to the functions $F_1(\xi,k,j) = k$, $F_2(\xi,k,j) = j$ shows that M^z_t, $M^{x,y}_t$ are martingales with quadratic variation as stated in the lemma.

To show that these martingales are orthogonal, we need to prove that $M^z_t M^{x,y}_t$ is a martingale. Let $W_z(\xi) = p(z)[1-\xi(z)]$, $W_{x,y}(\xi) = p(y-x)\xi(x)[1-\xi(y)]$. Dynkin's formula applied to $F(\xi,k,j) = kj$ gives that

$$N^z_t N^{x,y}_t - \int_0^t \big\{N^z_s W_{x,y}(\xi_s) + N^{x,y}_s W_z(\xi_s)\big\}\,ds$$

is a martingale. Rewriting the processes N_t^z, $N_t^{x,y}$ as $M_t^z + \int_0^t W_z(\xi_s)ds$, $M_t^{x,y} + \int_0^t W_{x,y}(\xi_s)ds$ and integrating by parts we obtain that

$$M_t^z M_t^{x,y} + \int_0^t \big\{M_s^z W_{x,y}(\xi_s) + M_s^{x,y} W_z(\xi_s)\big\}\, ds$$
$$+ \int_0^t W_{x,y}(\xi_s)\, ds \int_0^t W_z(\xi_s)\, ds - \int_0^t \big\{N_s^z W_{x,y}(\xi_s) + N_s^{x,y} W_z(\xi_s)\big\}\, ds$$

is a martingale. Expressing the martingales M^z, $M^{x,y}$ inside the first time integral in terms of the jump processes N^z, $N^{x,y}$ we obtain that

$$M_t^z M_t^{x,y} + \int_0^t \int_0^s W_z(\xi_r) dr\, W_{x,y}(\xi_s)\, ds$$
$$+ \int_0^t \int_0^s W_{x,y}(\xi_r) dr\, W_z(\xi_s)\, ds + \int_0^t W_{x,y}(\xi_s)\, ds \int_0^t W_z(\xi_s)\, ds$$

is a martingale. An integration by parts shows that the integrals cancel so that $M_t^z M_t^{x,y}$ is a martingale as claimed.

The same argument applies to any pair of distinct martingales in the set $\{M_t^z : p(z) > 0\} \cup \{M_t^{x,y} : x, y \in \mathbb{Z}_*^d, p(y-x) > 0\}$. This concludes the proof of the lemma. □

These martingales associated to the jumps of the Markov process are called the elementary martingales because any martingale given by Dynkin's formula can be written in terms of them: Fix a cylinder function $f : \mathbb{X}^* \to \mathbb{R}$. By Dynkin's formula,

$$M_t^f = f(\xi_t) - f(\xi_0) - \int_0^t (\mathscr{L} f)(\xi_s)\, ds \tag{6.5}$$

is a martingale. Since f is a cylinder function, rewriting $f(\xi_t) - f(\xi_0)$ as the sum of all differences arising from a jump to or from a site contained in the support of f in the time interval $[0, t]$, we obtain that

$$f(\xi_t) - f(\xi_0) = \sum_{x,y \in \mathbb{Z}_*^d} \int_0^t (T^{x,y} f)(\xi_{s-})\, dN_s^{x,y} + \sum_{z \in \mathbb{Z}_*^d} \int_0^t (T^z f)(\xi_{s-})\, dN_s^z$$

so that

$$M_t^f = \sum_{x,y \in \mathbb{Z}_*^d} \int_0^t (T^{x,y} f)(\xi_{s-})\, dM_s^{x,y} + \sum_{z \in \mathbb{Z}_*^d} \int_0^t (T^z f)(\xi_{s-})\, dM_s^z. \tag{6.6}$$

In (6.6) and below we set $M_t^z = 0$ and $M_t^{x,y} = 0$ if $p(z) = 0$, $p(y-x) = 0$.

By (2.16), we know that

$$\mathbb{E}_{\nu_\alpha^*}\big[(M_t^f)^2\big] = 2t\mathscr{D}(f). \tag{6.7}$$

We derive below this identity taking advantage of the representation of M_t^f in terms of the elementary martingales. As we shall see at the end of this section, this computation allows to describe the limits of $L^2(\mathbb{P}_{\nu_\alpha^*})$-Cauchy sequences of martingales M_t^f.

Since the elementary martingales are orthogonal, the quadratic variation $\langle M^f\rangle_t$ of the martingale M^f defined by (6.6) is given by

$$\langle M^f\rangle_t = \sum_{x,y\in\mathbb{Z}_*^d} p(y-x)\int_0^t \xi_s(x)\big[1-\xi_s(y)\big]\big(T^{x,y}f\big)(\xi_s)^2\,ds + \sum_{z\in\mathbb{Z}_*^d} p(z)\int_0^t \big[1-\xi_s(z)\big]\big(T^z f\big)(\xi_s)^2\,ds$$

so that

$$\frac{1}{t}\mathbb{E}_{\nu_\alpha^*}\big[\big(M_t^f\big)^2\big] = \sum_{x,y\in\mathbb{Z}_*^d} p(y-x)\int \xi(x)\big[1-\xi(y)\big]\big(T^{x,y}f\big)(\xi)^2\nu_\alpha^*(d\xi) + \sum_{z\in\mathbb{Z}_*^d} p(z)\int \big[1-\xi(z)\big]\big(T^z f\big)(\xi)^2\nu_\alpha^*(d\xi).$$

To obtain the expression of the Dirichlet form $\mathscr{D}(f)$ given in (6.3), we need to replace p by s and to remove the indicator $\xi(x)[1-\xi(y)]$ in the first sum. Since

$$\begin{aligned}&\big(T^{x,y}f\big)(\xi) = \big(T^{y,x}f\big)(\xi), \qquad \big(T^{x,y}f\big)\big(\sigma^{x,y}\xi\big) = -\big(T^{x,y}f\big)(\xi),\\ &\big(T^z f\big)(\theta_{-z}\xi) = -\big(T^{-z}f\big)(\xi),\end{aligned} \tag{6.8}$$

the change of variables $\xi' = \theta_z\xi$, $\xi'' = \sigma^{x,y}\xi$ permit to replace p by s. Observing now that

$$\big(T^{x,y}f\big)(\xi)^2\big\{\xi(x)\big[1-\xi(y)\big] + \xi(y)\big[1-\xi(x)\big]\big\} = \big(T^{x,y}f\big)(\xi)^2, \tag{6.9}$$

we obtain (6.7).

We claim that the representation (6.6) extends to functions in the domain $\mathscr{D}(\mathscr{L})$.

Lemma 6.9 *Fix a function u in the domain $\mathscr{D}(\mathscr{L})$. Then, the martingale M_t^u, defined by* (6.5) *with u in place of f, can be represented as in* (6.6):

$$M_t^u = \sum_{x,y\in\mathbb{Z}_*^d}\int_0^t \big(T^{x,y}u\big)(\xi_{s-})\,dM_s^{x,y} + \sum_{z\in\mathbb{Z}_*^d}\int_0^t \big(T^z u\big)(\xi_{s-})\,dM_s^z.$$

Proof Fix such a function u. Since u belongs to the domain of the generator and since the space of cylinder functions forms a core for the generator in $L^2(\nu_\alpha^*)$, there exists a sequence of cylinder functions $\{f_n : n\ge 1\}$ such that f_n, $\mathscr{L}f_n$ converges in

$L^2(\nu_\alpha^*)$ to u, $\mathscr{L}u$, respectively. In particular, the martingale $M_t^{f_n}$, defined by (6.5) with f_n in place of f, converges in $L^2(\nu_\alpha^*)$, as $n\uparrow\infty$, to the martingale M_t^u, defined by (6.5) with u in place of f. This holds of course for every $t>0$.

In view of (6.3), (6.6) with u in place of f defines a martingale in $L^2(\nu_\alpha^*)$ because the partial sums form a Cauchy sequence in $L^2(\nu_\alpha^*)$. Denote this martingale by m_t^u.

To show that $M_t^u = m_t^u$, it is enough to show that $M_t^{f_n}$ converges in $L^2(\nu_\alpha^*)$ to m_t^u. Since f_n is a cylinder function, the martingale $M_t^{f_n}$ can be represented through the elementary martingales M^z, $M^{x,y}$ by (6.6). Since the martingales M^z, $M^{x,y}$ are orthogonal, by the computations performed right after (6.7),

$$\frac{1}{t}\mathbb{E}_{\nu_\alpha^*}\big[\big(M_t^{f_n}-m_t^u\big)^2\big]=2\mathscr{D}(f_n-u).$$

This expression vanishes as $n\uparrow\infty$ by the choice of the sequence $\{f_n : n\ge 1\}$. This concludes the proof of the lemma. □

6.3 The Spaces $\mathscr{H}_1$ and $\mathscr{H}_{-1}$

In this section, we examine the $L^2(\mathbb{P}_{\nu_\alpha^*})$ limits of the martingales M_t^f, where f is a cylinder function. As in Sect. 2.2, denote by $\mathscr{H}_1$ the Hilbert space generated by the space $\mathscr{C}$ of cylinder functions endowed with the scalar product $\langle f,(-\mathscr{L})g\rangle_{\nu_\alpha^*}$. Let $\|\cdot\|_1$ be the norm in $\mathscr{H}_1$. We have seen in (6.3) that

$$\|f\|_1^2=\mathscr{D}_0(f)+\mathscr{D}_\theta(f)$$

for functions f in $\mathscr{C}$. This identity extends to the domain $\mathscr{D}(\mathscr{L})$ because $\mathscr{C}$ forms a core for $\mathscr{L}$.

Denote by $\mathscr{H}_{-1}$ the dual space of $\mathscr{H}_1$ defined as in (2.6) with ν_α^* in place of π. Recall that the $\mathscr{H}_{-1}$ norm is defined by the variational formula

$$\|f\|_{-1}^2=\sup_{g\in\mathscr{C}}\big\{2\langle f,g\rangle_{\nu_\alpha^*}-\|g\|_1^2\big\},\quad f\in L^2\big(\nu_\alpha^*\big).$$

As in (2.8), (2.9), it is easy to check from this variational formula that for every function f in $\mathscr{H}_1$ and every function g in $L^2(\nu_\alpha^*)\cap\mathscr{H}_{-1}$

$$\big|\langle f,g\rangle_{\nu_\alpha^*}\big|\le\|f\|_1\|g\|_{-1}.$$

The same variational formula permits to show that a function in $L^2(\nu_\alpha^*)$ belongs to $\mathscr{H}_{-1}$ if and only if there exists a finite constant C_1 such that

$$\langle f,g\rangle_{\nu_\alpha^*}\le C_1\|g\|_1 \tag{6.10}$$

for every g in $\mathscr{C}$. In this case, $\|f\|_{-1}\le C_1$.

Denote by $\mathbb{L}^2(\nu_\alpha^*)$ the space of sequences $\Psi = \{\Psi_z : \mathbb{X}^* \to \mathbb{R};\, p(z) > 0\} \times \{\Psi_{x,y} : \mathbb{X}^* \to \mathbb{R};\, x, y \in \mathbb{Z}_*^d,\, p(y-x) > 0\}$ of $L^2(\nu_\alpha^*)$ functions such that

$$\sum_{z\in\mathbb{Z}_*^d} p(z) E_{\nu_\alpha^*}\big[[1-\xi(z)]\Psi_z(\xi)^2\big] + \sum_{x,y\in\mathbb{Z}_*^d} p(y-x) E_{\nu_\alpha^*}\big[\xi(x)[1-\xi(y)]\Psi_{x,y}(\xi)^2\big]$$

is finite. $\mathbb{L}^2(\nu_\alpha^*)$ is endowed with the scalar product $\langle\cdot,\cdot\rangle$ defined by

$$\begin{aligned}\langle \Psi, \Phi\rangle = &\sum_{z\in\mathbb{Z}_*^d} p(z) E_{\nu_\alpha^*}\big[[1-\xi(z)]\Psi_z(\xi)\Phi_z(\xi)\big] \\ &+ \sum_{x,y\in\mathbb{Z}_*^d} p(y-x) E_{\nu_\alpha^*}\big[\xi(x)[1-\xi(y)]\Psi_{x,y}(\xi)\Phi_{x,y}(\xi)\big].\end{aligned}$$

A function u in the domain $\mathscr{D}(\mathscr{L})$ induces a sequence in $\mathbb{L}^2(\nu_\alpha^*)$, denoted by Ψ^u and given by $\Psi_z^u = T^z u$, $\Psi_{x,y}^u = T^{x,y}u$. Moreover, for Ψ in $\mathbb{L}^2(\nu_\alpha^*)$, M^Ψ defined by (6.6), with Ψ_z, $\Psi_{x,y}$ in place of $T^z f$, $T^{x,y} f$, defines a square integrable martingale such that $\mathbb{E}_{\nu_\alpha^*}[(M_t^\Psi)^2] = t\langle\Psi,\Psi\rangle$.

In view of (6.8), (6.9), denote by $\mathbb{L}_0^2(\nu_\alpha^*)$ the closed subspace of $\mathbb{L}^2(\nu_\alpha^*)$ composed by all sequences Ψ such that ν_α^*-a.s.

$$\begin{aligned}&\Psi_{x,y}(\xi) = \Psi_{y,x}(\xi), \qquad \Psi_{x,y}\big(\sigma^{x,y}\xi\big) = -\Psi_{x,y}(\xi),\\ &\Psi_{x,y}(\xi)^2\big\{\xi(x)[1-\xi(y)] + \xi(y)[1-\xi(x)]\big\} = \Psi_{x,y}(\xi)^2, \\ &\Psi_z(\theta_{-z}\xi) = -\Psi_{-z}(\xi).\end{aligned} \tag{6.11}$$

Repeating the computation performed right after (6.7) and taking advantage of the relations (6.11), we obtain that

$$\begin{aligned} t^{-1}\mathbb{E}_{\nu_\alpha^*}\big[\big(M_t^\Psi\big)^2\big] &= \langle\Psi,\Psi\rangle \\ &= (1/2)\sum_{x,y\in\mathbb{Z}_*^d} s(y-x)\int \Psi_{x,y}(\xi)^2\nu_\alpha^*(d\xi) \\ &\quad + \sum_{z\in\mathbb{Z}_*^d} s(z)\int \big[1-\xi(z)\big]\Psi_z(\xi)^2\nu_\alpha^*(d\xi).\end{aligned} \tag{6.12}$$

Lemma 6.10 *Consider a sequence of cylinder functions $\{f_n : n \ge 1\}$ which forms a Cauchy sequence in $\mathscr{H}_1$. Then, there exists Ψ in $\mathbb{L}_0^2(\nu_\alpha^*)$ such that $M_t^{f_n}$ converges to M_t^Ψ in $L^2(\mathbb{P}_{\nu_\alpha^*})$ for all $t \ge 0$.*

Proof A sequence $\{f_n : n \ge 1\}$ of cylinder functions forms a Cauchy sequence in $\mathscr{H}_1$ if and only if $\{\Psi^{f_n} : n \ge 1\}$ forms a Cauchy sequence in $\mathbb{L}^2(\nu_\alpha^*)$. In particular, Ψ^{f_n} converges in $\mathbb{L}^2(\nu_\alpha^*)$ to some Ψ which belongs to $\mathbb{L}_0^2(\nu_\alpha^*)$ because this space is closed. By (6.12), $M_t^{f_n}$ converges to M_t^Ψ in $\mathbb{L}^2(\nu_\alpha^*)$. □

6.4 Law of Large Numbers

In this section, we prove Theorem 6.5. Recall the definition of the jump processes $\{N_t^z : t \geq 0\}$ defined in Sect. 6.2. The position at time t of the tagged particle is obtained by summing over the number of jumps multiplied by their size:

$$Z_t = \sum_{z \in \mathbb{Z}_*^d} z N_t^z = \sum_{z \in \mathbb{Z}_*^d} z M_t^z + \int_0^t \tilde{V}(\xi_s)\, ds, \tag{6.13}$$

where $\tilde{V}$ is the cylinder function given by

$$\tilde{V}(\xi) = \sum_{x \in \mathbb{Z}_*^d} x p(x)\big[1 - \xi(x)\big].$$

Under the stationary measure $\mathbb{P}_{\nu_\alpha^*}$, the quadratic variation of the vector-valued martingale $M_t = \sum_{x \in \mathbb{Z}_*^d} x M_t^x$ is bounded by $C_0 t$. In particular, as $t \uparrow \infty$, M_t/t converges to 0-a.s., (cf. Feller, 1971, Theorem VII.9.3). On the other hand, since $\mathbb{P}_{\nu_\alpha^*}$ is ergodic, as $t \uparrow \infty$, $t^{-1} \int_0^t \tilde{V}(\xi_s)\, ds$ converges a.s. to $E_{\nu_\alpha^*}[\tilde{V}] = \mathfrak{m}(1-\alpha)$. This proves the theorem.

6.5 Central Limit Theorem

In this section, we prove Theorem 6.6 following the strategy presented in Chap. 2. Recall the definition of Z_t, the position of the tagged particle. By (6.13),

$$Z_t - (1-\alpha)t\mathfrak{m} = \sum_{z \in \mathbb{Z}_*^d} z M_t^z + \int_0^t V(\xi_s)\, ds,$$

where V is the mean zero cylinder function

$$V(\xi) = \tilde{V}(\xi) - (1-\alpha)\mathfrak{m} = \sum_{x \in \mathbb{Z}^d} x p(x)\big\{\alpha - \xi(x)\big\}. \tag{6.14}$$

We have seen in Chap. 2 that the proof of a central limit theorem for additive functionals of Markov processes relies on bounds on $\mathcal{H}_{-1}$. Fix a vector $\mathfrak{a} \in \mathbb{R}^d$ and let $V_\mathfrak{a} = \mathfrak{a} \cdot V$. Denote by u_λ the solution of the resolvent equation

$$\lambda u_\lambda - \mathscr{L} u_\lambda = V_\mathfrak{a}. \tag{6.15}$$

In this section, we prove Theorem 6.6 assuming that for every vector $\mathfrak{a} \in \mathbb{R}^d$,

$$V_\mathfrak{a} \in \mathcal{H}_{-1} \quad \text{and} \quad \sup_{0<\lambda\leq 1} \|\mathscr{L} u_\lambda\|_{-1} < \infty. \tag{6.16}$$

By Lemma 2.16, it follows from these conditions that

$$\lim_{\lambda\to 0} \lambda\langle u_\lambda, u_\lambda\rangle_{\nu_\alpha^*} = 0 \quad \text{and} \quad u_\lambda \text{ converges in } \mathscr{H}_1 \text{ as } \lambda \downarrow 0. \tag{6.17}$$

The strategy of the proof of Theorem 6.6 relies on the ideas presented in Chap. 2. The goal is to represent the additive functional $\int_0^t V_\mathfrak{a}(\xi_s)ds$ as the sum of a martingale m_t and a negligible term and then to use the central limit theorem for the martingale $M_t + m_t$, where $M_t = \sum_z z\cdot \mathfrak{a}M_t^z$.

For $\lambda > 0$, let m_t^λ be the martingale

$$m_t^\lambda = u_\lambda(\xi_t) - u_\lambda(\xi_0) - \int_0^t (\mathscr{L}u_\lambda)(\xi_s)\,ds.$$

By Lemma 6.9, the martingale m_t^λ can be represented as

$$\sum_{x,y\in\mathbb{Z}_*^d} \int_0^t \Psi_{x,y}^\lambda(\xi_{s-})\,dM_s^{x,y} + \sum_{z\in\mathbb{Z}_*^d} \int_0^t \Psi_z^\lambda(\xi_{s-})\,dM_s^z,$$

where $\Psi_{x,y}^\lambda = T^{x,y}u_\lambda$, $\Psi_z^\lambda = T^z u_\lambda$. The resolvent equation permits to write the position of the tagged particle Z_t as

$$Z_t\cdot\mathfrak{a} - (1-\alpha)t(\mathfrak{m}\cdot\mathfrak{a}) = M_t + m_t^\lambda + R_t^\lambda, \tag{6.18}$$

where M_t is the martingale $\sum_z (z\cdot\mathfrak{a})M_t^z$ and where the remainder R_t^λ is given by

$$R_t^\lambda = u_\lambda(\xi_0) - u_\lambda(\xi_t) + \lambda\int_0^t u_\lambda(\xi_s)\,ds.$$

We first claim that

Lemma 6.11 *For every $t > 0$, the martingale m_t^λ converges in $L^2(\mathbb{P}_{\nu_\alpha^*})$ to some martingale m_t as $\lambda \downarrow 0$.*

Proof Since the sequence u_λ converges in $\mathscr{H}_1$ as $\lambda\downarrow 0$ and since $\mathscr{C}$ forms a core for the generator $\mathscr{L}$, by Lemma 6.10 the martingale m_t^λ converges in $L^2(\mathbb{P}_{\nu_\alpha^*})$ to a martingale $m_t = M_t^\Psi$ associated to a sequence Ψ in $\mathbb{L}_0^2(\nu_\alpha^*)$. □

It follows from this lemma that the remainder R_t^λ appearing in (6.18) converges in $L^2(\mathbb{P}_{\nu_\alpha^*})$ as $\lambda\downarrow 0$ so that

$$Z_t\cdot\mathfrak{a} - (1-\alpha)t(\mathfrak{m}\cdot\mathfrak{a}) = M_t + m_t + R_t \tag{6.19}$$

where m_t is a martingale in $L^2(\mathbb{P}_{\nu_\alpha^*})$.

Lemma 6.12 *$t^{-1/2}R_t$ vanishes in $L^2(\mathbb{P}_{\nu_\alpha^*})$ as $t\uparrow\infty$.*

Proof The proof of this lemma is similar to the one of Lemma 2.10 and relies on (6.17). □

Since both martingales M_t and m_t are written in terms of the elementary martingales, the quadratic variation of the sum is easy to compute and equals

$$\begin{aligned}\langle M+m\rangle_t = &\sum_{x,y\in\mathbb{Z}^d_*} p(y-x)\int_0^t \xi_s(x)\big[1-\xi_s(y)\big]\Psi_{x,y}(\xi_s)^2\,ds\\ &+\sum_{z\in\mathbb{Z}^d_*} p(z)\int_0^t \big[1-\xi_s(z)\big]\big\{\mathfrak{a}\cdot z+\Psi_z(\xi_s)\big\}^2\,ds.\end{aligned}$$

By the ergodic theorem under $\mathbb{P}_{\nu^*_\alpha}$, $t^{-1}\langle M+m\rangle_t$ converges a.s. and in $L^1(\mathbb{P}_{\nu^*_\alpha})$. Therefore, by Theorem 2.1, $t^{-1/2}\{M_t+m_t\}$, and therefore $[Z_t\cdot\mathfrak{a}-(1-\alpha)t(\mathfrak{m}\cdot\mathfrak{a})]/\sqrt{t}$, converges in distribution to a mean zero Gaussian variable with variance $D(\alpha)$ satisfying

$$\begin{aligned}\mathfrak{a}\cdot D(\alpha)\mathfrak{a} = &\sum_{x,y\in\mathbb{Z}^d_*} p(y-x)\int \xi(x)\big[1-\xi(y)\big]\Psi_{x,y}(\xi)^2\nu^*_\alpha(d\xi)\\ &+\sum_{z\in\mathbb{Z}^d_*} p(z)\int \big[1-\xi(z)\big]\big\{\mathfrak{a}\cdot z+\Psi_z(\xi)\big\}^2\nu^*_\alpha(d\xi).\end{aligned}$$

Since Ψ belongs to $\mathbb{L}^2_0(\nu^*_\alpha)$, an analogous computation to the one presented just after (6.7) shows that

$$\begin{aligned}\mathfrak{a}\cdot D(\alpha)\mathfrak{a} = (1/2)&\sum_{x,y\in\mathbb{Z}^d_*} s(y-x)\int \Psi_{x,y}(\xi)^2\nu^*_\alpha(d\xi)\\ &+\sum_{z\in\mathbb{Z}^d_*} s(z)\int \big[1-\xi(z)\big]\big\{\mathfrak{a}\cdot z+\Psi_z(\xi)\big\}^2\nu^*_\alpha(d\xi).\end{aligned}\tag{6.20}$$

Nothing in principle prevents the asymptotic variance $D(\alpha)$ from vanishing or being $+\infty$ and the proof of the central limit theorem would be incomplete without a strictly positive lower bound and a finite upper bound. Such bounds are derived in Sect. 6.8 below. This concludes the proof of Theorem 6.6 under the assumptions (6.16) on the solution of the resolvent equation. The purpose of the following sections is to show that (6.16) holds if $\mathfrak{m}=0$ or if $d\geq 3$.

We close this section by showing that the function $V_\mathfrak{a}$ belongs to $\mathcal{H}_{-1}$ if $\mathfrak{m}=0$ and that the solution of the resolvent equation satisfies the assumptions (6.16) provided $p(-x)=p(x)$, proving the central limit theorem for the tagged particle in the symmetric case.

Lemma 6.13 *Assume that* $\mathfrak{m}=0$. *Then, the cylinder function* $V_\mathfrak{a}$ *introduced in* (6.14) *belongs to* $\mathcal{H}_{-1}$ *and* $\|V_\mathfrak{a}\|^2_{-1}\leq C_0\chi(\alpha)|\mathfrak{a}|^2$.

Proof In view of (6.10) and the remarks thereafter, to show that $V_{\mathfrak{a}}$ belongs to $\mathscr{H}_{-1}$, we need to prove that there exists a finite constant C_0 such that

$$\langle f, V_{\mathfrak{a}} \rangle_{\nu_\alpha^*} \leq C_0 \sqrt{\chi(\alpha)} \| f \|_1 |\mathfrak{a}|$$

for every cylinder function f.

Fix a cylinder function f. In the mean zero case $\mathfrak{m} = 0$, $V_{\mathfrak{a}}$ can be rewritten as

$$\sum_{x \in \mathbb{Z}^d} (\mathfrak{a} \cdot x) p(x) \big\{ \xi(e) - \xi(x) \big\},$$

where e is any fixed site of $\mathbb{Z}_*^d$. Since $s(\cdot)$ generates $\mathbb{Z}_*^d$, for each x such that $p(x) > 0$, there exists a path $x = y_0, \dots, y_n = e$ going from x to e avoiding the origin and such that $s(y_{i+1} - y_i) > 0$. Since

$$\big\langle \xi(e) - \xi(x), f \big\rangle_{\nu_\alpha^*} = \sum_{i=0}^{n-1} \big\langle \xi(y_{i+1}) - \xi(y_i), f \big\rangle_{\nu_\alpha^*},$$

performing the change of variables $\xi' = \sigma^{y_i, y_{i+1}} \xi$, we may rewrite this sum as

$$-(1/2) \sum_{i=0}^{n-1} \big\langle \xi(y_{i+1}) - \xi(y_i), T^{y_i, y_{i+1}} f \big\rangle_{\nu_\alpha^*}.$$

Clearly, this expression is less than or equal to

$$\frac{1}{2} n A \alpha(1-\alpha) + \frac{1}{4A} \sum_{i=0}^{n-1} E_{\nu_\alpha^*} \big[\big(T^{y_i, y_{i+1}} f \big)^2 \big]$$

for every $A > 0$. Summing over all x such that $p(x) > 0$, we get that

$$\langle V_{\mathfrak{a}}, f \rangle_{\nu_\alpha^*} \leq C_0 A \alpha(1-\alpha) |\mathfrak{a}|^2 + \frac{C_0}{A} \mathscr{D}_0(f)$$

for all $A > 0$ and some finite constant C_0. To conclude, it remains to minimize the previous expression with respect to A. □

The previous lemma shows that the first condition in (6.16) holds in the mean zero case. On the other hand, since the process is reversible in the symmetric case, the second assumption in (6.16) follows from Sect. 2.7.1.

The Spaces $\mathscr{H}_{0,1}$ and $\mathscr{H}_{0,-1}$ Note that in the previous lemma only the piece of the Dirichlet form $\mathscr{D}_0$ associated to jumps of the environment were used. This part plays an important role in the sequel and deserves a special notation. Recall that we denote by $\mathscr{S}_0$ the generator $\mathscr{L}_0$ introduced in (6.1) with the probability measure p replaced by its symmetric part s, and that $\mathscr{D}_0(f) = \langle (-\mathscr{S}_0) f, f \rangle_{\nu_\alpha^*}$ for all cylinder functions f.

Let $\mathscr{H}_{0,1}$ be the Hilbert space generated by the cylinder functions $\mathscr{C}$ endowed with the scalar product $\langle f, (-\mathscr{S}_0)g\rangle_{\nu_\alpha^*}$. Denote also by $\mathscr{H}_{0,-1}$ the dual space of $\mathscr{H}_{0,1}$, given by (2.6) with ν_α^* in place of π. Clearly,

$$\|f\|_{0,1} \le \|f\|_1 \quad \text{so that} \quad \|f\|_{-1} \le \|f\|_{0,-1} \tag{6.21}$$

for all cylinder functions f.

The proof of Lemma 6.13 shows in fact that $V_\mathfrak{a}$ belongs to $\mathscr{H}_{0,-1}$ and that

$$\|V_\mathfrak{a}\|_{0,-1}^2 \le C_0 \chi(\alpha)|\mathfrak{a}|^2 \tag{6.22}$$

for all $\mathfrak{a}$ in $\mathbb{R}^d$.

6.6 The Mean Zero Asymmetric Case

In this section, we examine the mean zero case $\sum_x p(x)x = 0$. In the previous section, we have reduced the proof of the central limit theorem to the inspection of the conditions stated in (6.16). The first assumption was proved in Lemma 6.13. In view of Sect. 2.7.3, to derive the second assumption in (6.16), it is enough to prove that the generator $\mathscr{L}$ satisfies a sector condition. This is the content of the main result of this section, whose proof relies on the decomposition of a mean zero probability p in cycle probability measures, as seen in Sect. 5.3. The argument is similar to the one of Proposition 5.5 but the presence of the highly non-local shift operator $\mathscr{L}_\theta$ introduces further complications.

Proposition 6.14 *There exists a finite constant C_0 depending only on the probability measure p such that*

$$\langle f, (-\mathscr{L})g\rangle_{\nu_\alpha^*}^2 \le C_0 \langle f, (-\mathscr{L})f\rangle_{\nu_\alpha^*} \langle g, (-\mathscr{L})g\rangle_{\nu_\alpha^*}$$

for all cylinder functions f, g.

Proof By Lemma 5.6 and Lemma 5.7, we may assume that p is a cycle probability measure:

$$p(x) = \frac{1}{n} \sum_{j=0}^{n-1} \mathbf{1}\{x = y_j - y_{j-1}\},$$

for some $n \ge 1$, the length of the cycle, and some set $C = \{y_0, \dots, y_{n-1}, y_n = y_0\}$. We may of course assume that the cycle is irreducible in the sense that $y_i \ne y_j$ if $i \ne j$, $0 \le i, j \le n-1$.

Let $C + x$ be the cycle $\{y_0 + x, \dots, y_{n-1} + x, y_n + x\}$. Since the cycle C is irreducible, there are exactly n cycles of the form $C + x$ which intersect the origin: $C - y_0, \dots, C - y_{n-1}$.

For a cycle $C+x$ which does not intersect the origin, let $\mathscr{L}^0_{C+x}$ be the generator defined by

$$\big(\mathscr{L}^0_{C+x} f\big)(\xi) = \frac{1}{n}\sum_{k=0}^{n-1} \xi(y_k + x)\big[1 - \xi(y_{k+1} + x)\big]\big\{f\big(\sigma^{y_k+x, y_{k+1}+x}\xi\big) - f(\xi)\big\}.$$

In contrast, for the cycle $C - y_j$, $0 \le j \le n-1$, let $\mathscr{L}^0_{C-y_j}$ be the generator defined by

$$\begin{aligned}&\big(\mathscr{L}^0_{C-y_j} f\big)(\xi)\\ &\quad = \frac{1}{n}\sum_{\substack{0\le k\le n-1\\ k\ne j-1, j}} \xi(y_k - y_j)\big[1 - \xi(y_{k+1} - y_j)\big]\big\{f\big(\sigma^{y_k-y_j, y_{k+1}-y_j}\xi\big) - f(\xi)\big\}.\end{aligned}$$

Notice that we suppressed the jump from $y_{j-1} - y_j$ to the origin, because the origin is always occupied, and from the origin to $y_{j+1} - y_j$ because this jump does not appear in the generator associated to the environment. Note also that in the formulas of the generators $\mathscr{L}^0_{C+x}$, $\mathscr{L}^0_{C-y_j}$, we may remove the factors $\xi(y_k + x)$, $\xi(y_k - y_j)$ without modifying the generators.

For the probability measure p obtained from the cycle C, the generator associated to the jumps of the tagged particle is written as

$$(\mathscr{L}_\theta f)(\xi) = \frac{1}{n}\sum_{k=0}^{n-1}\big[1 - \xi(y_{k+1} - y_k)\big]\big\{f(\theta_{y_{k+1}-y_k}\xi) - f(\xi)\big\}.$$

Let $\mathscr{L}_1 = \sum_{x\notin C}\mathscr{L}^0_{C-x}$ be the piece of the generator $\mathscr{L}$ whose cycles do not intersect the origin and let $\mathscr{L}_2 = \mathscr{L}_\theta + \sum_{0\le j\le n-1}\mathscr{L}^0_{C-y_j}$ be the remaining part of the generator. By Lemma 5.9,

$$\big\langle f, (-\mathscr{L}_1)g\big\rangle^2_{\nu^*_\alpha} \le 16 n^4 \big\langle f, (-\mathscr{L}_1)f\big\rangle_{\nu^*_\alpha}\big\langle g, (-\mathscr{L}_1)g\big\rangle_{\nu^*_\alpha}. \tag{6.23}$$

We may of course replace $\mathscr{L}_1$ by $\mathscr{L}$ on the right-hand side since we will be adding only non-negative terms.

It remains to prove a similar bound for the generator $\mathscr{L}_2$. The generator $\mathscr{L}_2$ involves $n(n-2)$ measure-preserving transformations coming from $\sum_{0\le j\le n-1}\mathscr{L}^0_{C-y_j}$ and n additional measure-preserving transformations coming from the piece $\mathscr{L}_\theta$. Denote by $T_1, \dots, T_{n(n-1)}$ these transformations. We claim that there is a permutation s of $\{1, \dots, n(n-1)\}$ such that $T_{s(n(n-1))} \circ \dots \circ T_{s(1)}\xi = \xi$ provided ξ has a hole in a specific site.

A rigorous proof of this property is lengthy and requires too much notation. We prefer to convince the reader with an example. Consider the cycle $\{0, -e_2, e_1, 0\}$ in $\mathbb{Z}^2$. In this case, the generators $\mathscr{L}_3 = \sum_{0\le j\le n-1}\mathscr{L}^0_{C-y_j}$, $\mathscr{L}_\theta$ can be written as

$$3(\mathscr{L}_3 f)(\xi) = \big[1 - \xi(e_1)\big]\big\{f\big(\sigma^{-e_2, e_1}\xi\big) - f(\xi)\big\}$$

$$
\begin{aligned}
&+\big[1-\xi(-e_1-e_2)\big]\big\{f\big(\sigma^{-e_1,-e_1-e_2}\xi\big)-f(\xi)\big\}\\
&+\big[1-\xi(e_2)\big]\big\{f\big(\sigma^{e_1+e_2,e_2}\xi\big)-f(\xi)\big\},\\
3(\mathscr{L}_\theta f)(\xi)=&\big[1-\xi(-e_2)\big]\big\{f(\theta_{-e_2}\xi)-f(\xi)\big\}\\
&+\big[1-\xi(-e_1)\big]\big\{f(\theta_{-e_1}\xi)-f(\xi)\big\}\\
&+\big[1-\xi(e_1+e_2)\big]\big\{f(\theta_{e_1+e_2}\xi)-f(\xi)\big\}.
\end{aligned}
$$

The six measure-preserving transformations are σ^{-e_2,e_1}, $\sigma^{-e_1,-e_1-e_2}$, $\sigma^{e_1+e_2,e_2}$, $\theta_{-e_2}, \theta_{-e_1}, \theta_{e_1+e_2}$. We have to estimate $\langle f, (-\mathscr{L}_3-\mathscr{L}_\theta)g\rangle_{\nu_\alpha^*}$. This expression consists of six terms. To fix ideas, consider the first one which is multiplied by $1-\xi(e_1)$. A simple computation shows that

$$\theta_{-e_1}\circ\sigma^{-e_1,-e_1-e_2}\circ\theta_{e_1+e_2}\circ\sigma^{e_1+e_2,e_2}\circ\theta_{-e_2}\circ\sigma^{-e_2,e_1}\xi=\xi,$$

provided $\xi(e_1)=0$. Similar identities hold for the other five cases. We may therefore write the generator $\mathscr{L}_\theta+\mathscr{L}_3$ as the generator L in the statement of Lemma 5.8 and deduce that

$$\big\langle f,(-\mathscr{L}_3-\mathscr{L}_\theta)g\big\rangle_{\nu_\alpha^*}^2\le 16n^2\big\langle f,(-\mathscr{L}_3-\mathscr{L}_\theta)f\big\rangle_{\nu_\alpha^*}\big\langle g,(-\mathscr{L}_3-\mathscr{L}_\theta)g\big\rangle_{\nu_\alpha^*},$$

where $n=6$. Here again we may replace the generator $\mathscr{L}_3+\mathscr{L}_\theta$ by $\mathscr{L}$ on the right-hand side. This bound together with (6.23) proves the sector condition for the mean zero exclusion process as seen from the tagged particle. □

6.7 Duality

If $\mathfrak{m}\neq 0$, the proof of the central limit theorem for the position of the tagged particle relies on the dual spaces and operators introduced in Sect. 5.4. The presence of the tagged particle and the corresponding shifts operators create new difficulties.

For $n\ge 0$, denote by $\mathscr{E}_n^*$ the subsets of $\mathbb{Z}_*^d$ with n points and let $\mathscr{E}^*=\bigcup_{n\ge 0}\mathscr{E}_n^*$. For each A in $\mathscr{E}^*$, let Ψ_A be the cylinder function

$$\Psi_A(\xi)=\prod_{x\in A}\frac{\xi(x)-\alpha}{\sqrt{\chi(\alpha)}},$$

where $\chi(\alpha)=\alpha(1-\alpha)$. By convention, $\Psi_\varnothing=1$. The class $\{\Psi_A, A\in\mathscr{E}^*\}$ forms an orthonormal basis of $L^2(\nu_\alpha^*)$. For each $n\ge 1$, denote by $\mathscr{A}_n$ the subspace of $L^2(\nu_\alpha^*)$ generated by $\{\Psi_A, A\in\mathscr{E}_n^*\}$, so that $L^2(\nu_\alpha^*)=\bigoplus_{n\ge 0}\mathscr{A}_n$. Let $\mathscr{G}_n=\bigoplus_{0\le k\le n}\mathscr{A}_k$. Functions in $\mathscr{G}_n\setminus\mathscr{G}_{n-1}$, $\mathscr{G}=\bigcup_{n\ge 0}\mathscr{G}_n$ are said to have degree n, finite degree, respectively, while functions in $\mathscr{A}_n$ are said to be monomials of degree n.

Consider a cylinder function f. Since $\{\Psi_A : A\in\mathscr{E}^*\}$ is a basis of $L^2(\nu_\alpha^*)$, there exists a finitely supported function $\mathfrak{f}:\mathscr{E}^*\to\mathbb{R}$ such that

$$f=\sum_{A\in\mathscr{E}^*}\mathfrak{f}(A)\Psi_A=\sum_{n\ge 0}\sum_{A\in\mathscr{E}_n^*}\mathfrak{f}(A)\Psi_A.$$

The function $\mathfrak{f}$ represents the Fourier coefficients of the cylinder function f and is denoted by $\mathfrak{F}f$ when we want to stress its dependence on f. $\mathfrak{f}$: $\mathscr{E}^* \to \mathbb{R}$ is a function of finite support because f is a cylinder function. Moreover, the Fourier coefficients $\mathfrak{f}(A)$ depend not only on f but also on the density α: $\mathfrak{f}(A) = \mathfrak{f}(\alpha, A)$. Note finally that $\mathfrak{f}(\varnothing) = E_{\nu_\alpha^*}[f]$.

Denote by Π_n the orthogonal projection on $\mathscr{A}_n$, or the restriction to $\mathscr{E}_n^*$ of a finitely supported function $\mathfrak{f} : \mathscr{E}^* \to \mathbb{R}$: $(\Pi_n \mathfrak{f})(A) = \mathfrak{f}(A)\mathbf{1}\{A \in \mathscr{E}_n^*\}$. With this definition, $\mathfrak{F}(\Pi_n f) = \Pi_n \mathfrak{F} f$.

Denote by $\mathfrak{C}$ the space of finitely supported functions $\mathfrak{f} : \mathscr{E}^* \to \mathbb{R}$, by $\mu_\star$ the counting measure on $\mathscr{E}^*$ and by $\langle \cdot, \cdot \rangle_{\mu_\star}$ the scalar product in $L^2(\mu_\star)$. For any two cylinder functions $f = \sum_{A\in\mathscr{E}^*} \mathfrak{f}(A)\Psi_A$, $g = \sum_{A\in\mathscr{E}^*} \mathfrak{g}(A)\Psi_A$,

$$\langle f, g\rangle_{\nu_\alpha^*} = \sum_{A,B\in\mathscr{E}^*} \mathfrak{f}(A)\mathfrak{g}(B)\langle \Psi_A, \Psi_B\rangle_{\nu_\alpha^*} = \sum_{A\in\mathscr{E}^*} \mathfrak{f}(A)\mathfrak{g}(A) = \langle \mathfrak{f}, \mathfrak{g}\rangle_{\mu_\star}.$$

In particular, the map $\mathfrak{F} : L^2(\nu_\alpha^*) \to L^2(\mu_\star)$ is an isomorphism.

To examine how the generator acts on the Fourier coefficients, recall the definition of the set $A_{x,y}$ introduced in (5.15). To represent the shift operator, for A in $\mathscr{E}^*$, denote by $\theta_y A$ the set defined by

$$\theta_y A = \begin{cases} A + y & \text{if } -y \notin A, \\ (A + y)_{0,y} & \text{if } -y \in A, \end{cases} \tag{6.24}$$

where $B+z$ is the set $\{x+z; x \in B\}$. Therefore, if $-y$ belongs to A, we first translate A by y (obtaining a new set which contains the origin) and then we remove the origin and we add the site y. Of course, $\theta_y : \mathscr{E}_n^* \to \mathscr{E}_n^*$ is a one-to-one function and a straightforward computation shows that $\Psi_A(\theta_y\xi) = \Psi_{\theta_y A}(\xi)$ for every configuration $\xi \in \mathbb{X}^*$.

Fix a cylinder function $f = \sum_{A\in\mathscr{E}^*} (\mathfrak{F}f)(A)\Psi_A$. A straightforward computation shows that

$$\mathscr{L} f = \sum_{A\in\mathscr{E}^*} (\mathfrak{L}_{*,\alpha}\mathfrak{F}f)(A)\Psi_A,$$

where

$$\mathfrak{L}_{*,\alpha} = \mathfrak{L}_{0,\alpha} + \mathfrak{L}_{\theta,\alpha}$$

and

$$\begin{aligned}\mathfrak{L}_{0,\alpha} &= \mathfrak{S}_\star + (1-2\alpha)\mathfrak{N}_\star + \sqrt{\chi(\alpha)}\mathfrak{J}_{0,+} + \sqrt{\chi(\alpha)}\mathfrak{J}_{0,-}, \\ \mathfrak{L}_{\theta,\alpha} &= \alpha\mathfrak{L}_{\theta,p,1} + (1-\alpha)\mathfrak{L}_{\theta,p,2} + \sqrt{\chi(\alpha)}\mathfrak{J}_{\theta,p,+} + \sqrt{\chi(\alpha)}\mathfrak{J}_{\theta,p,-}.\end{aligned}$$

The operators $\mathfrak{S}_\star$, $\mathfrak{N}_\star$, $\mathfrak{J}_{0,+}$, $\mathfrak{J}_{0,-}$ are given by

$$(\mathfrak{S}_\star\mathfrak{f})(A) = (1/2) \sum_{x,y\in\mathbb{Z}_*^d} s(y-x)\big[\mathfrak{f}(A_{x,y}) - \mathfrak{f}(A)\big],$$

$$
\begin{aligned}
(\mathfrak{N}_\star\mathfrak{f})(A) &= \sum_{\substack{x\in A, y\notin A\\ y\neq 0}} a(y-x)\big[\mathfrak{f}(A_{x,y}) - \mathfrak{f}(A)\big],\\
(\mathfrak{J}_{0,+}\mathfrak{f})(A) &= 2\sum_{x,y\in A} a(y-x)\mathfrak{f}\big(A\setminus\{y\}\big),\\
(\mathfrak{J}_{0,-}\mathfrak{f})(A) &= 2\sum_{\substack{x,y\notin A\\ x,y\neq 0}} a(y-x)\mathfrak{f}\big(A\cup\{y\}\big)
\end{aligned}
\tag{6.25}
$$

for any function $\mathfrak{f}:\mathscr{E}^*\to\mathbb{R}$ of finite support. Note that the generator $\mathfrak{S}_\star$ defined above is different from the generator $\mathfrak{S}_\star$ of Sect. 5.4 because the summation is carried over sites in $\mathbb{Z}^d_*$. The operators $\mathfrak{L}_{\theta,p,1}$, $\mathfrak{L}_{\theta,p,2}$, $\mathfrak{J}_{\theta,p,+}$, $\mathfrak{J}_{\theta,p,-}$ are defined as follows:

$$
\begin{aligned}
(\mathfrak{L}_{\theta,p,1}\mathfrak{f})(A) &= \sum_{x\in A} p(x)\big\{\mathfrak{f}(\theta_{-x}A) - \mathfrak{f}(A)\big\},\\
(\mathfrak{L}_{\theta,p,2}\mathfrak{f})(A) &= \sum_{x\notin A} p(x)\big\{\mathfrak{f}(\theta_{-x}A) - \mathfrak{f}(A)\big\},\\
(\mathfrak{J}_{\theta,p,+}\mathfrak{f})(A) &= \sum_{x\in A} p(x)\big\{\mathfrak{f}\big(A\backslash\{x\}\big) - \mathfrak{f}\big([\theta_{-x}A]\backslash\{-x\}\big)\big\},\\
(\mathfrak{J}_{\theta,p,-}\mathfrak{f})(A) &= \sum_{x\notin A} p(x)\big\{\mathfrak{f}\big(A\cup\{x\}\big) - \mathfrak{f}\big([\theta_{-x}A]\cup\{-x\}\big)\big\}
\end{aligned}
\tag{6.26}
$$

for any function $\mathfrak{f}:\mathscr{E}^*\to\mathbb{R}$ of finite support. In these equations the origin never appears in the summation because $p(0)=0$. We denote by $\mathfrak{L}_{\theta,q,1}$, $\mathfrak{L}_{\theta,q,2}$, $\mathfrak{J}_{\theta,q,+}$, $\mathfrak{J}_{\theta,q,-}$ the operators $\mathfrak{L}_{\theta,p,1}$, $\mathfrak{L}_{\theta,p,2}$, $\mathfrak{J}_{\theta,p,+}$, $\mathfrak{J}_{\theta,p,-}$ defined above with p replaced by q, where $q=p^*$, s or a.

Up to this point we have proved that

$$\mathfrak{F}\mathscr{L} = \mathfrak{L}_{*,\alpha}\mathfrak{F}.$$

To proceed, we rewrite the operator $\mathfrak{L}_{*,\alpha}$ as a sum of operators which are either symmetric or anti-symmetric. Notice first that $\mathfrak{S}_\star$ is a symmetric operator in $L^2(\mu_\star)$ which preserves the degree of a function. In contrast with the previous chapter, we shall see that it does not carry all the symmetric part of the operator $\mathfrak{L}_{*,\alpha}$.

The operators $\mathfrak{N}_\star$, $\mathfrak{L}_{\theta,p,1}$ and $\mathfrak{L}_{\theta,p,2}$ also preserves the degree of functions. The special role played by the origin introduces some complications: $\mathfrak{N}_\star$ is not anti-symmetric as its notation suggests, but it carries a piece which is compensated by $\mathfrak{L}_{\theta,p,1}$ and $\mathfrak{L}_{\theta,p,2}$. More precisely, let $U:\mathscr{E}^*\to\mathbb{R}$ be given by

$$U(A) = \sum_{x\in A} a(x).$$

Note that U vanishes on sets A which do not contain points close to the origin. Denote by $\mathfrak{N}^*_\star$, $\mathfrak{L}^*_{\theta,p,1}$ and $\mathfrak{L}^*_{\theta,p,2}$ the adjoints of $\mathfrak{N}_\star$, $\mathfrak{L}_{\theta,p,1}$ and $\mathfrak{L}_{\theta,p,2}$, respectively, in

$L^2(\mu_\star)$. An elementary computation shows that for every finitely supported function $\mathfrak{f}: \mathscr{E}^* \to \mathbb{R}$

$$\begin{aligned}
(\mathfrak{N}_\star^* \mathfrak{f})(A) &= -(\mathfrak{N}_\star \mathfrak{f})(A) - 2U(A)\mathfrak{f}(A),\\
(\mathfrak{L}_{\theta,p,1}^* \mathfrak{f})(A) &= (\mathfrak{L}_{\theta,p^*,1} \mathfrak{f})(A) - 2U(A)\mathfrak{f}(A),\\
(\mathfrak{L}_{\theta,p,2}^* \mathfrak{f})(A) &= (\mathfrak{L}_{\theta,p^*,2} \mathfrak{f})(A) + 2U(A)\mathfrak{f}(A).
\end{aligned}$$

The derivations of these identities rely on the facts that $\sum_{x\in A} a(x) = -\sum_{x\notin A} a(x)$, $\sum_{x,y\in A} a(y-x) = 0$ because a is anti-symmetric.

It follows from the previous computation that the symmetric and anti-symmetric parts of the operator $\mathfrak{L}_{\theta,p,1} - \mathfrak{N}_\star$ are $\mathfrak{L}_{\theta,s,1}$ and $\mathfrak{L}_{\theta,a,1} - \mathfrak{N}_\star$, respectively, while the symmetric and anti-symmetric parts of the operator $\mathfrak{L}_{\theta,p,2} + \mathfrak{N}_\star$ are $\mathfrak{L}_{\theta,s,2}$ and $\mathfrak{L}_{\theta,a,2} + \mathfrak{N}_\star$, respectively. It is therefore natural to rewrite the operator $(1-2\alpha)\mathfrak{N}_\star + \alpha\mathfrak{L}_{\theta,p,1}(1-\alpha)\mathfrak{L}_{\theta,p,2}$ as

$$\alpha(\mathfrak{L}_{\theta,p,1} - \mathfrak{N}_\star) + (1-\alpha)(\mathfrak{L}_{\theta,p,2} + \mathfrak{N}_\star).$$

Furthermore, $\mathfrak{L}_{\theta,s,1}$ and $\mathfrak{L}_{\theta,s,2}$ are symmetric operators in $L^2(\mu_\star)$ which preserve the degrees of functions, while the operators $\mathfrak{L}_{\theta,a,1} - \mathfrak{N}_\star$, $\mathfrak{L}_{\theta,a,2} + \mathfrak{N}_\star$ are anti-symmetric operators in $L^2(\mu_\star)$ which preserve the degrees of functions.

We turn now to the operators which do not preserve the degree of a function: $\mathfrak{J}_{0,+}$, $\mathfrak{J}_{0,-}$, $\mathfrak{J}_{\theta,p,+}$, $\mathfrak{J}_{\theta,p,-}$. The adjoint of these operators can be written as

$$\begin{aligned}
(\mathfrak{J}_{0,+}^* \mathfrak{f})(A) &= -(\mathfrak{J}_{0,-} \mathfrak{f})(A) - 2\sum_{y\notin A} a(y)\mathfrak{f}(A\cup\{y\}),\\
(\mathfrak{J}_{0,-}^* \mathfrak{f})(A) &= -(\mathfrak{J}_{0,+} \mathfrak{f})(A) - 2\sum_{y\in A} a(y)\mathfrak{f}(A\setminus\{y\}),\\
(\mathfrak{J}_{\theta,p,+}^* \mathfrak{f})(A) &= (\mathfrak{J}_{\theta,p^*,-} \mathfrak{f})(A) + 2\sum_{y\notin A} a(y)\mathfrak{f}(A\cup\{y\}),\\
(\mathfrak{J}_{\theta,p,-}^* \mathfrak{f})(A) &= (\mathfrak{J}_{\theta,p^*,+} \mathfrak{f})(A) + 2\sum_{y\in A} a(y)\mathfrak{f}(A\setminus\{y\}).
\end{aligned}$$

Note that the indices $\pm$ on the left-hand side are changed to $\mp$ on the right-hand side. It follows from these identities that

$$\mathfrak{J}_{\theta,s,+}^* = \mathfrak{J}_{\theta,s,-}, \quad \{\mathfrak{J}_{0,+} + \mathfrak{J}_{\theta,a,+}\}^* = -\{\mathfrak{J}_{0,-} + \mathfrak{J}_{\theta,a,-}\}.$$

Therefore, $\mathfrak{J}_{\theta,s,+} + \mathfrak{J}_{\theta,s,-}$ is symmetric in $L^2(\mu_\star)$, while $\mathfrak{J}_{0,+} + \mathfrak{J}_{0,-} + \mathfrak{J}_{\theta,a,+} + \mathfrak{J}_{\theta,a,-}$ is anti-symmetric.

Let $\mathfrak{J}_+ = \mathfrak{J}_{0,+} + \mathfrak{J}_{\theta,p,+}$, $\mathfrak{J}_- = \mathfrak{J}_{0,-} + \mathfrak{J}_{\theta,p,-}$. With this notation we have that

$$\mathfrak{L}_{*,\alpha} = \mathfrak{S}_\star + \alpha\{\mathfrak{L}_{\theta,p,1} - \mathfrak{N}_\star\} + (1-\alpha)\{\mathfrak{L}_{\theta,p,2} + \mathfrak{N}_\star\} + \sqrt{\chi(\alpha)}\{\mathfrak{J}_+ + \mathfrak{J}_-\}.$$

The Spaces $\mathfrak{H}_{0,1}$ and $\mathfrak{H}_{0,-1}$ Denote by $\mathfrak{H}_{0,1}$ the Hilbert space generated by the finitely supported functions endowed with the scalar product $\langle \mathfrak{f}, (-\mathfrak{S}_\star)\mathfrak{g}\rangle_{\mu_\star}$. We use the same notation $\|\cdot\|_{0,1}$ to denote the norm of the Hilbert space $\mathfrak{H}_{0,1}$ and the norm of the Hilbert space $\mathscr{H}_{0,1}$ introduced at the end of Sect. 6.5. An elementary computation shows that

$$\|\mathfrak{f}\|_{0,1}^2 = \langle \mathfrak{f}, (-\mathfrak{S}_\star)\mathfrak{f}\rangle_{\mu_\star} = \frac{1}{4} \sum_{x,y\in\mathbb{Z}_*^d} s(y-x) \sum_{A\in\mathscr{E}^*} \{\mathfrak{f}(A_{x,y}) - \mathfrak{f}(A)\}^2 \tag{6.27}$$

for all finitely supported functions $\mathfrak{f}: \mathscr{E}^* \to \mathbb{R}$.

Let $\mathfrak{H}_{0,-1}$ be the dual space of $\mathfrak{H}_{0,1}$, given by (2.6) with $\mu_\star$, $\mathfrak{C}$ in place of π, $\mathscr{C}$. Here also $\|\cdot\|_{0,-1}$ stands for the norm of the $\mathfrak{H}_{0,-1}$ and the norm of the $\mathscr{H}_{0,-1}$.

Fix a cylinder function f in $\mathscr{C}$ and recall that we denote by $\mathscr{S}_0$ the generator $\mathscr{L}_0$ with the probability measure p replaced by its symmetric part s. An elementary computation shows that $\mathfrak{F}\mathscr{S}_0 f = \mathfrak{S}_\star\mathfrak{F} f$ so that

$$\mathfrak{F}\mathscr{S}_0 = \mathfrak{S}_\star\mathfrak{F}.$$

Hence, since $\mathfrak{F}$ is an isomorphism from $L^2(\nu_\alpha^*)$ to $L^2(\mu_\star)$, for any cylinder function f,

$$\|f\|_{0,1}^2 = \langle f, (-\mathscr{S}_0)f\rangle_{\nu_\alpha^*} = \langle \mathfrak{f}, (-\mathfrak{S}_\star)\mathfrak{f}\rangle_{\mu_\star} = \|\mathfrak{f}\|_{0,1}^2, \tag{6.28}$$

where $\mathfrak{f} = \mathfrak{F} f$. Therefore $\mathfrak{F}$ is also an isomorphism from $\mathscr{H}_{0,1}$ to $\mathfrak{H}_{0,1}$. By duality these identities extend to $\mathscr{H}_{0,-1}$, $\mathfrak{H}_{0,-1}$. For every function $f \in L^2(\nu_\alpha^*)$,

$$\|f\|_{0,-1}^2 = \|\mathfrak{F} f\|_{0,-1}^2. \tag{6.29}$$

The isomorphism from $L^2(\nu_\alpha^*)$ to $L^2(\mu_\star)$ and the relation $\mathfrak{F}\mathscr{L} = \mathfrak{L}_{*,\alpha}\mathfrak{F}$ give also that

$$\langle f, (-\mathscr{L})f\rangle_{\nu_\alpha^*} = \langle \mathfrak{F} f, (-\mathfrak{L}_{*,\alpha})\mathfrak{F} f\rangle_{\mu_\star} \tag{6.30}$$

for all cylinder functions f, an identity needed later.

6.8 The Asymmetric Case in Dimension $d \geq 3$

In this section, we prove Theorem 6.6 for asymmetric exclusion processes in dimension $d \geq 3$. We have seen in Sect. 6.5 that the proof is reduced to the inspection of conditions (6.16).

This section is divided in two parts. We first prove that the cylinder function V_α belongs to $\mathscr{H}_{0,-1}$ in dimensions $d \geq 3$. This part is based on the estimates of the Green function of transient Markov processes presented in Sect. 5.7. In the second part of the section we prove that the second condition in (6.16) holds following the strategy presented in Sect. 2.7.4.

Lemma 6.15 *Assume that* $d \geq 3$. *Then, the cylinder function* $V_{\mathfrak{a}}$ *introduced in* (6.14) *belongs to* $\mathscr{H}_{0,-1}$ *and* $\| V_{\mathfrak{a}} \|_{0,-1}^2 \leq C_0 \alpha(1-\alpha)|\mathfrak{a}|^2$.

Proof The cylinder function $V_{\mathfrak{a}}$ can be written in the orthonormal basis $\{\Psi_A : A \in \mathscr{E}^*\}$ as $-\sqrt{\chi(\alpha)} \sum_{x \in \mathbb{Z}^d} x \cdot \mathfrak{a} p(x) \Psi_{\{x\}}$ and therefore belongs to the space $\mathscr{A}_1$ of functions of degree 1. Since functions of different degrees are orthogonal both in $L^2(\nu_\alpha^*)$ and in $\mathscr{H}_{0,1}$, in the variational formula which defines the $\| \cdot \|_{0,-1}$ norm we may restrict the supremum to cylinder functions of degree 1:

$$\| V_{\mathfrak{a}} \|_{0,-1}^2 = \sup_{f \in \mathscr{C}} \bigl\{ 2\langle V_{\mathfrak{a}}, f \rangle_{\nu_\alpha^*} - \| f \|_{0,1}^2 \bigr\} = \sup_{f \in \mathscr{C} \cap \mathscr{A}_1} \bigl\{ 2\langle V_{\mathfrak{a}}, f \rangle_{\nu_\alpha^*} - \| f \|_{0,1}^2 \bigr\}.$$

Denote the Fourier coefficients of the cylinder function f by $\mathfrak{f} : \mathscr{E}_1^* \to \mathbb{R}$ so that $f = \sum_x \mathfrak{f}(\{x\}) \Psi_{\{x\}}$. With this notation the previous variational formula can be written as

$$\sup_{\mathfrak{f}} \biggl\{ -2\sqrt{\chi(\alpha)} \sum_x x \cdot \mathfrak{a} p(x) \mathfrak{f}\bigl(\{x\}\bigr) - D_0(\mathfrak{f}) \biggr\}.$$

In this formula the supremum is carried over all finitely supported functions $\mathfrak{f} : \mathscr{E}_1^* \to \mathbb{R}$ and $D_0(\mathfrak{f})$ corresponds to the Dirichlet form of a symmetric random walk on $\mathbb{Z}_*^d$ in which a jump from x to y occurs at rate $(1/2)s(y-x)$:

$$D_0(\mathfrak{f}) = \frac{1}{4} \sum_{x,y \in \mathbb{Z}_*^d} s(y-x) \bigl\{ \mathfrak{f}\bigl(\{y\}\bigr) - \mathfrak{f}\bigl(\{x\}\bigr) \bigr\}^2.$$

By Schwarz inequality, the linear term in the variational formula is less than or equal to

$$A\chi(\alpha) \sum_x (x \cdot \mathfrak{a})^2 p(x) + \frac{1}{A} \sum_x p(x) \mathfrak{f}\bigl(\{x\}\bigr)^2$$

for all $A > 0$. By Proposition 5.23, since the transition probability p has finite range, the second term is less than or equal to

$$\frac{1}{A} \sup_{y \in \mathbb{Z}_*^d} G_0(y,y) D_0(\mathfrak{f}),$$

where $G_0(x,y)$ stands for the Green function of the random walk with Dirichlet form D_0 defined above. By Proposition 5.25, the Green function G_0 can be estimated by the Green function of the random walk on $\mathbb{Z}^d$ which jumps from x to y at rate $(1/2)s(y-x)$. The previous expression is thus less than or equal to $C_0 A^{-1} D_0(\mathfrak{f})$, for some finite constant C_0 depending only on p. Choosing $A = C_0$ we conclude the proof of the lemma. □

We now turn to the proof of the second condition in (6.16). Let $\mathfrak{u}_\lambda = \mathfrak{F} u_\lambda$, $\lambda > 0$. Since $\mathfrak{F}\mathscr{L} = \mathfrak{L}_{*,\alpha}\mathfrak{F}$, applying the operator $\mathfrak{F}$ on both sides of the resolvent equation (6.15), we obtain that

$$\lambda \mathfrak{u}_\lambda - \mathfrak{L}_{*,\alpha} \mathfrak{u}_\lambda = \mathfrak{V}_{\mathfrak{a}}, \tag{6.31}$$

where $\mathfrak{V}_\mathfrak{a} = \mathfrak{F} V_\mathfrak{a}$. Since $\mathfrak{L}_{*,\alpha}\mathfrak{u}_\lambda = \mathfrak{F}\mathscr{L} u_\lambda$, by (6.29) $\|\mathscr{L} u_\lambda\|_{0,-1} = \|\mathfrak{L}_{*,\alpha}\mathfrak{u}_\lambda\|_{0,-1}$. Thus, in view of (6.21), to prove the second condition in (6.16) it is enough to show that

$$\sup_{0<\lambda\le 1} \|\mathfrak{L}_{*,\alpha}\mathfrak{u}_\lambda\|_{0,-1} < \infty. \tag{6.32}$$

We place ourselves in the set-up of Sect. 2.7.4, where $\mu_\star$ stands for π and we have the following correspondence between the operators:

$$B_0 \to \alpha\{\mathfrak{L}_{\theta,p,1} - \mathfrak{N}_\star\} + (1-\alpha)\{\mathfrak{L}_{\theta,p,2} + \mathfrak{N}_\star\},$$

$$S_0 \to \mathfrak{S}_\star, \quad L_+ \to \sqrt{\chi(\alpha)}\mathfrak{J}_+, \quad L_- \to \sqrt{\chi(\alpha)}\mathfrak{J}_-.$$

By Lemma 2.23, (6.32) holds if we show that conditions (2.39), (2.40), (2.45) and (2.50) are in force. The fist condition is trivially satisfied. In the present context the other three conditions read as follows.

There exists a finite constant C_0 such that

$$0 \le \langle \mathfrak{f}, -\mathfrak{S}_\star \mathfrak{f}\rangle_{\mu_\star} \le C_0 \big\langle \mathfrak{f}, (-\mathfrak{L}_{*,\alpha})\mathfrak{f}\big\rangle_{\mu_\star} \tag{6.33}$$

for all functions $\mathfrak{f}: \mathscr{E}^* \to \mathbb{R}$ of finite support.

There exists $\beta < 1$ and a finite constant C_0 such that for all $n \ge 0$,

$$\begin{aligned} \langle \mathfrak{J}_+ \mathfrak{f}, \mathfrak{g}\rangle_{\mu_\star}^2 &\le C_0 (n+1)^{2\beta} \langle -\mathfrak{S}_\star \mathfrak{f}, \mathfrak{f}\rangle_{\mu_\star} \langle -\mathfrak{S}_\star \mathfrak{g}, \mathfrak{g}\rangle_{\mu_\star} \\ \langle \mathfrak{f}, \mathfrak{J}_- \mathfrak{g}\rangle_{\mu_\star}^2 &\le C_0 (n+1)^{2\beta} \langle -\mathfrak{S}_\star \mathfrak{f}, \mathfrak{f}\rangle_{\mu_\star} \langle -\mathfrak{S}_\star \mathfrak{g}, \mathfrak{g}\rangle_{\mu_\star} \end{aligned} \tag{6.34}$$

for any finitely supported functions $\mathfrak{f}: \mathscr{E}_n^* \to \mathbb{R}$, $\mathfrak{g}: \mathscr{E}_{n+1}^* \to \mathbb{R}$.

As we have seen in Sect. 5.6, condition (2.50) is not expected to hold, but can be replaced by the following one. There exists a finite constant C_0 such that

$$\|\Pi_n \mathfrak{N}_\star \mathfrak{u}_\lambda\|_{0,-1}^2 \le C_0 n \|\mathfrak{V}_\mathfrak{a}\|_{0,-1}^2 + C_0 n^3 \sum_{j=n-1}^{n+1} \|\Pi_j \mathfrak{u}_\lambda\|_{0,1}^2 \tag{6.35}$$

for all $n \ge 1$; and identical bounds with $\mathfrak{L}_{\theta,p,j}$, $j = 1, 2$, in place of $\mathfrak{N}_\star$, because in our context the asymmetric part of the generator which keeps the degree is the operator $\alpha\{\mathfrak{L}_{\theta,p,1} - \mathfrak{N}_\star\} + (1-\alpha)\{\mathfrak{L}_{\theta,p,2} + \mathfrak{N}_\star\}$.

In the remaining part of this section we prove assertions (6.33), (6.34), (6.35).

Condition (6.33) This condition is straightforward since the operators $\mathfrak{S}_\star$, $\mathfrak{L}_{*,\alpha}$ are generators and therefore non-positive: For every finitely supported function $\mathfrak{f}: \mathscr{E}^* \to \mathbb{R}$, by (6.28), (6.30),

$$0 \le \langle -\mathscr{S}_0 f, f\rangle_{\nu_\alpha^*} = \langle -\mathfrak{S}_\star \mathfrak{f}, \mathfrak{f}\rangle_{\mu_\star},$$

$$\langle -\mathfrak{S}_\star \mathfrak{f}, \mathfrak{f}\rangle_{\mu_\star} = \langle -\mathscr{S}_0 f, f\rangle_{\nu_\alpha^*} \le \langle -\mathscr{L} f, f\rangle_{\nu_\alpha^*} = \big\langle \mathfrak{f}, (-\mathfrak{L}_{*,\alpha})\mathfrak{f}\big\rangle_{\mu_\star},$$

where f is the cylinder function $f(\xi) = \sum_{A\in\mathscr{E}^*} \mathfrak{f}(A)\Psi_A(\xi)$.

Condition (6.34) We start with a graded sector condition on the symmetric diagonal operators.

Lemma 6.16 *Fix $j = 1, 2$. There exists a constant C_0, depending only on $p(\cdot)$, such that for all $n \geq 1$*

$$\langle -\mathfrak{L}_{\theta,s,j}\mathfrak{f}, \mathfrak{f}\rangle_{\mu_\star} \leq C_0 n \langle -\mathfrak{S}_\star \mathfrak{f}, \mathfrak{f}\rangle_{\mu_\star}$$

for all finitely supported functions $\mathfrak{f} : \mathcal{E}_n^* \to \mathbb{R}$.

Proof Set $j = 1$ and fix $n \geq 1$ and a finitely supported function $\mathfrak{f} : \mathcal{E}_n^* \to \mathbb{R}$. An elementary computation shows that

$$\langle -\mathfrak{L}_{\theta,s,1}\mathfrak{f}, \mathfrak{f}\rangle_{\mu_\star} = (1/2) \sum_{z \in \mathbb{Z}_*^d} s(z) \sum_{A \ni z} \big\{\mathfrak{f}(\theta_{-z}A) - \mathfrak{f}(A)\big\}^2.$$

Since the sum over z is finite, we need to estimate for each z

$$\sum_{A \in \mathcal{E}_n^*} \big\{\mathfrak{f}(\theta_{-z}A) - \mathfrak{f}(A)\big\}^2$$

in terms of the Dirichlet form $\langle -\mathfrak{S}_\star \mathfrak{f}, \mathfrak{f}\rangle_{\mu_\star}$ in which only exchanges of sites are allowed.

We can assume without loss of generality that $z = (1, 0, \dots, 0)$, the unit vector in the direction of the first coordinate axis. The problem is now reduced to the following: we are given a function $\mathfrak{f} : \mathcal{E}_n^* \to \mathbb{R}$. We think of $\mathcal{E}_n^*$ as a graph with edges $\mathbb{E}_n^*$, defined as

$$\mathbb{E}_n^* = \big\{(A, A_{x,y}) : s(y - x) > 0\big\}.$$

The Dirichlet form can be written as

$$\langle -\mathfrak{S}_\star \mathfrak{f}, \mathfrak{f}\rangle_{\mu_\star} = \frac{1}{2} \sum_{e \in \mathbb{E}^*{}_n} s(e) \big|(\delta \mathfrak{f})(e)\big|^2$$

where $(\delta \mathfrak{f})(e) = f(A_{x,y}) - f(A)$, if $e = (A, A_{x,y})$ and $s(e) = s(y - x)$.

We are claiming an estimate of the form

$$\sum_{A \in \mathcal{E}_n^*} \big|\mathfrak{f}(\theta_z A) - \mathfrak{f}(A)\big|^2 \leq C_0 n \langle -\mathfrak{S}_\star \mathfrak{f}, \mathfrak{f}\rangle_{\mu_\star}$$

with a constant C_0 independent of n.

It is clear that for any set A one can move from A to $\theta_{-z}A$ along edges of the graph $\mathcal{E}_n^*$, using only the edges in $\mathbb{E}_n^*$. We will verify that for every $A \in \mathcal{E}_n^*$ we can assign a set of edges $\mathbb{E}^*{}_A \subset \mathbb{E}_n^*$ such that, (i) for every A one can use the edges of $\mathbb{E}^*{}_A$ to go from A to $\theta_{-z}A$, (ii) for any A there are at most n edges in $\mathbb{E}^*{}_A$ and (iii) the subsets $\{\mathbb{E}^*{}_A\}$ are mutually disjoint as A varies over $\mathscr{E}_n^*$. Then it is easy to see that

$$|\mathfrak{f}(\theta_{-z}A) - \mathfrak{f}(A)|^2 = \Big| \sum_{e \in \mathbb{E}^*{}_A} (\delta \mathfrak{f})(e) \Big|^2 \leq n \sum_{e \in \mathbb{E}^*{}_A} |(\delta \mathfrak{f})(e)|^2$$

and summing over $A \in \mathscr{E}_n^*$, because $\mathbb{E}^*{}_A$ are disjoint, we can establish the lemma.

To construct the paths from A to $\theta_{-z}A$ we totally order the points of $\mathbb{Z}_*^d$ by lexicographic ordering. We say that $z = (z_1, \dots, z_d) \in \mathbb{Z}_*^d$ is *positive* if, either $z_1 > 0$; or $z_1 = 0, z_2 = 0, \dots, z_{j-1} = 0$ and $z_j > 0$ for some $2 \leq j \leq d$. The total ordering declares $y > x$ if $y - x$ is *positive*. Let the set $A \in \mathscr{E}_n^*$ consist of the n points $(x_1, x_2, \dots, x_n)$ of $\mathbb{Z}_*^d$. We can assume that they are ordered so that $x_1 > x_2 > \cdots > x_n$. Then $\theta_{-z}A = (x_1^*, \dots, x_n^*)$ where $x_j^* = x_j - z$ unless $x_j = z$ in which case $x_j^* = x_j - 2z = -z$. Assume without loss of generality that $z < 0$. We use the edges in $\mathbb{E}_n^*$ to shift successively each x_i to x_i^* starting from x_1 and proceeding in order and ending with shifting x_n. Any edge that is used goes from some A_1 to $A_2 = \sigma^{x_i, x_i^*} A_1$. Since the shifts were made in lexicographic order we can determine without ambiguity which points of A_1 have already been shifted and which have not been. In other words the paths from any two different A to the corresponding $\theta_{-z}A$ do not share a common edge. It is also clear that exactly n edges are used. Any edge corresponds to $x_j^* - x_j = -z$ or $-2z$.

We have thus proved that

$$\langle -\mathfrak{L}_{\theta,s,j}\mathfrak{f}, \mathfrak{f}\rangle_{\mu_\star} \leq n \langle -\hat{\mathfrak{S}}_\star \mathfrak{f}, \mathfrak{f}\rangle_{\mu_\star}$$

where $\hat{\mathfrak{S}}_\star$ allows jumps of size $-2z$. It is however an easy matter to show that

$$\langle -\hat{\mathfrak{S}}_\star \mathfrak{f}, \mathfrak{f}\rangle_{\mu_\star} \leq C_0 \langle -\mathfrak{S}_\star \mathfrak{f}, \mathfrak{f}\rangle_{\mu_\star}$$

for some finite constant C_0 depending only on the probability $p(\cdot)$. This concludes the proof of the lemma for $j = 1$. The proof for $j = 2$ is identical. □

It follows from the variational formula for the $\mathfrak{H}_{0,-1}$ norm, from the symmetry of the operators $\mathfrak{L}_{\theta,s,j}$ and from the previous lemma that there exists a finite constant C_0 such that for all $n \geq 1$,

$$\|\mathfrak{L}_{\theta,s,j}\mathfrak{f}\|_{0,-1}^2 \leq C_0 n^2 \|\mathfrak{f}\|_{0,1}^2 \tag{6.36}$$

for $j = 1, 2$ and all finitely supported functions $\mathfrak{f} : \mathscr{E}_n^* \to \mathbb{R}$.

We are now in a position to show that the off-diagonal operators $\mathfrak{J}_+$, $\mathfrak{J}_-$ satisfy the graded sector condition (6.34) with $\beta = 1/2$.

Lemma 6.17 *Assume that $d \geq 3$. There exists a finite constant C_0, depending only on the probability $p(\cdot)$, such that for every $n \geq 0$,*

$$\langle \mathfrak{J}_+ \mathfrak{f}, \mathfrak{g}\rangle_{\mu_\star}^2 \leq C_0 n \langle -\mathfrak{S}_\star \mathfrak{f}, \mathfrak{f}\rangle_{\mu_\star} \langle -\mathfrak{S}_\star \mathfrak{g}, \mathfrak{g}\rangle_{\mu_\star}$$

for any finitely supported functions $\mathfrak{f} : \mathcal{E}_n^* \to \mathbb{R}$, $\mathfrak{g} : \mathcal{E}_{n+1}^* \to \mathbb{R}$. *A similar bound holds for* $\mathfrak{J}_-$.

Proof To fix ideas we prove the estimate for the operator $\mathfrak{J}_{\theta,s,+}$. The proof for $\mathfrak{J}_{\theta,a,+}$ is identical and the one for $\mathfrak{J}_{0,+}$ is similar to the proof of Lemma 5.15.

Fix $n \geq 0$ and consider two finitely supported functions $\mathfrak{f} : \mathcal{E}_n^* \to \mathbb{R}$, $\mathfrak{g} : \mathcal{E}_{n+1}^* \to \mathbb{R}$. From the explicit form of the operator $\mathfrak{J}_{\theta,s,+}$, we obtain that $\mathfrak{J}_{\theta,s,+}\mathfrak{f}(\{x\}) = 0$ for all $x \in \mathbb{Z}_*^d$. We may therefore assume that $n \geq 1$. In this case,

$$\big|\langle \mathfrak{J}_{\theta,s,+}\mathfrak{f}, \mathfrak{g}\rangle_{\mu_\star}\big| \leq \sum_{A \in \mathcal{E}_{n+1}^*} \sum_{y \in A} s(y) \big|\mathfrak{f}\big(A \setminus \{y\}\big) - f\big([\theta_{-y} A] \setminus \{-y\}\big)\big| \big|\mathfrak{g}(A)\big|.$$

The elementary inequality $2ab \leq \lambda a^2 + \lambda^{-1} b^2$ gives that the previous expression is less than or equal to

$$(\lambda/2) \sum_{A \in \mathcal{E}_{n+1}^*} \sum_{y \in A} s(y) \big\{\mathfrak{f}\big(A \setminus \{y\}\big) - f\big([\theta_{-y} A] \setminus \{-y\}\big)\big\}^2 + (1/2\lambda) \sum_{y \in \mathbb{Z}_*^d} s(y) \mathfrak{g}_1(y)^2 \tag{6.37}$$

for every $\lambda > 0$, where $\mathfrak{g}_1(y)^2 = \sum_{A \ni y} \mathfrak{g}(A)^2$. We estimate separately these two expressions.

The change of variables $B = A \setminus \{y\}$ permits to rewrite the first term in (6.37) as $\lambda \langle (-\mathfrak{L}_{\theta,s,2})\mathfrak{f}, \mathfrak{f}\rangle_{\mu_\star}$. By Lemma 6.16, this term is bounded by $C_0 \lambda n \langle (-\mathfrak{S}_\star)\mathfrak{f}, \mathfrak{f}\rangle_{\mu_\star}$.

On the other hand, since s generates an irreducible transient Markov process on $\mathbb{Z}_*^d$ and since $s(\cdot)$ has finite range, by Propositions 5.23 and 5.25,

$$\sum_{y \in \mathbb{Z}_*^d} s(y) \mathfrak{g}_1(y)^2 \leq C_0 \sum_{x, y \in \mathbb{Z}_*^d} s(y - x) \big\{\mathfrak{g}_1(y) - \mathfrak{g}_1(x)\big\}^2.$$

A change of variables $B = A_{x,y}$ gives that

$$\mathfrak{g}_1(y) = \Big\{ \sum_{A \ni x, y} \mathfrak{g}(A)^2 + \sum_{A \ni x, A \not\ni y} \mathfrak{g}(A_{x,y})^2 \Big\}^{1/2}.$$

Since

$$\mathfrak{g}_1(x) = \Big\{ \sum_{A \ni x, y} \mathfrak{g}(A)^2 + \sum_{A \ni x, A \not\ni y} \mathfrak{g}(A)^2 \Big\}^{1/2},$$

by Schwarz inequality,

$$\big\{\mathfrak{g}_1(y) - \mathfrak{g}_1(x)\big\}^2 \leq \sum_{A \ni x, A \not\ni y} \big\{\mathfrak{g}(A_{x,y}) - \mathfrak{g}(A)\big\}^2.$$

Therefore, in view of (6.27),

$$\sum_{y \in \mathbb{Z}_*^d} s(y) \mathfrak{g}_1(y)^2 \leq C_0 \big\langle (-\mathfrak{S}_\star)\mathfrak{g}, \mathfrak{g}\big\rangle$$

for some finite constant C_0 depending only on $p(\cdot)$.

In view of (6.37), to conclude the proof of the lemma, it remains to optimize on $\lambda > 0$. □

It follows from the variational formula for the $\mathfrak{H}_{0,-1}$ norm and from this lemma that there exists a finite constant C_0 such that for every $n \ge 1$,

$$\|\mathfrak{J}_\pm \mathfrak{f}\|_{0,-1}^2 \le C_0 n \|\mathfrak{f}\|_{0,1}^2 \tag{6.38}$$

for all finitely supported functions $\mathfrak{f} : \mathscr{E}_n^* \to \mathbb{R}$.

Condition (6.35) There are three diagonal operators: $\mathfrak{L}_{\theta,p,1}$, $\mathfrak{L}_{\theta,p,2}$ and $\mathfrak{N}_\star$. It follows from (6.36) that $\mathfrak{L}_{\theta,s,j}$, $j = 1, 2$, satisfies trivially Condition (6.35). The next lemma shows that $\mathfrak{L}_{\theta,a,1}$ also satisfies a graded sector condition.

Lemma 6.18 *Assume that $d \ge 3$. There exists a finite constant C_0, depending only on the probability $p(\cdot)$, such that for every $n \ge 0$,*

$$\langle \mathfrak{L}_{\theta,a,1}\mathfrak{f}, \mathfrak{g}\rangle_{\mu_\star}^2 \le C_0 n \langle -\mathfrak{S}_\star \mathfrak{f}, \mathfrak{f}\rangle_{\mu_\star} \langle -\mathfrak{S}_\star \mathfrak{g}, \mathfrak{g}\rangle_{\mu_\star}$$

for any finitely supported functions $\mathfrak{f}$, $\mathfrak{g} : \mathscr{E}_n^* \to \mathbb{R}$.

Proof The proof is similar and slightly simpler than the one of Lemma 6.17. Details are left to the reader which should remember that $|a(y)| \le s(y)$. □

It follows from the variational formula for the $\mathfrak{H}_{0,-1}$ norm and from this lemma that there exists a finite constant C_0 such that for all $n \ge 0$,

$$\|\mathfrak{L}_{\theta,a,1}\mathfrak{f}\|_{0,-1}^2 \le C_0 n \|\mathfrak{f}\|_{0,1}^2 \tag{6.39}$$

for all finitely supported functions $\mathfrak{f} : \mathscr{E}_n^* \to \mathbb{R}$. In particular, Condition (6.35) holds also for $\mathfrak{L}_{\theta,a,1}$ and $\mathfrak{L}_{\theta,p,1}$.

The proof of the previous lemma does not apply to $\mathfrak{L}_{\theta,a,2}$ because in the definition of this operator the sum is carried over sites not in A. Actually, such an estimate is not expected to hold for $\mathfrak{L}_{\theta,a,2}$ and for $\mathfrak{N}_\star$, as explained in the beginning of Sect. 5.6.

The proof of Condition (6.35) for the operators $\mathfrak{N}_\star$, $\mathfrak{L}_{\theta,a,2}$ is similar to the one of Lemma 5.20 and relies on the comparison of the operators $\mathfrak{S}_\star$, $\mathfrak{N}_\star$, $\mathfrak{L}_{\theta,a,2}$ with generators associated to the evolution of independent randoms walks which can be examined through Fourier analysis.

Lemma 6.19 *There exists a finite constant C_0 depending only on $p(\cdot)$ such that for all $n \ge 1$,*

$$\|\Pi_n \mathfrak{N}_\star \mathfrak{u}_\lambda\|_{0,-1}^2 \leq C_0 n \|\mathfrak{V}_a\|_{0,-1}^2 + C_0 n^3 \sum_{j=n-1}^{n+1} \|\Pi_j \mathfrak{u}_\lambda\|_{0,1}^2.$$

An identical bound holds if we replace $\mathfrak{N}_\star$ by $\mathfrak{L}_{\theta,a,2}$.

Proof The proof of this lemma is similar to the one of Lemma 5.20 and divided into several steps. Let $\mathfrak{w}_1 = \mathfrak{V}_a + \sqrt{\chi(\alpha)}\{\mathfrak{J}_+ + \mathfrak{J}_-\}\mathfrak{u}_\lambda + \alpha \mathfrak{L}_{\theta,p,1}\mathfrak{u}_\lambda + (1-\alpha)\mathfrak{L}_{\theta,s,2}\mathfrak{u}_\lambda$ so that

$$\lambda \mathfrak{u}_\lambda - \big\{\mathfrak{S}_\star + (1-2\alpha)\mathfrak{N}_\star + (1-\alpha)\mathfrak{L}_{\theta,a,2}\big\}\mathfrak{u}_\lambda = \mathfrak{w}_1. \tag{6.40}$$

By (6.36), (6.38), (6.39), there exists a finite constant C_0 such that

$$\|\Pi_n \mathfrak{w}_1\|_{0,-1}^2 \leq 2\|\Pi_n \mathfrak{V}_a\|_{0,-1}^2 + C_0 n^2 \sum_{j=n-1}^{n+1} \|\Pi_j \mathfrak{u}_\lambda\|_{0,1}^2 \tag{6.41}$$

for all $n \geq 1$.

The operator $\mathfrak{S}_\star + (1-2\alpha)\mathfrak{N}_\star + (1-\alpha)\mathfrak{L}_{\theta,a,2}$ does not change the degree of a function. We may therefore examine equation (6.40) on each set $\mathcal{E}_n^{o*}$:

$$\lambda \mathfrak{u}_{\lambda,n} - \big\{\mathfrak{S}_\star + (1-2\alpha)\mathfrak{N}_\star + (1-\alpha)\mathfrak{L}_{\theta,a,2}\big\}\mathfrak{u}_{\lambda,n} = \Pi_n \mathfrak{w}_1,$$

where $\mathfrak{u}_{\lambda,n} = \Pi_n \mathfrak{u}_\lambda$. Since n is fixed until estimate (6.44), we omit the operator Π_n in the following formulas.

The main idea of this proof is to approximate the operator $\mathfrak{S}_\star + (1-2\alpha)\mathfrak{N}_\star + (1-\alpha)\mathfrak{L}_{\theta,a,2}$ by a convolution operator which can be analyzed through Fourier transforms. Fix $n \geq 1$ and let $\mathcal{X}_n = (\mathbb{Z}^d)^n$. Note that we include the origin of $\mathbb{Z}^d$ in $\mathcal{X}_n$. We consider a set A in $\mathcal{E}_n^*$ as an equivalent class of $n!$ sets of distinct points of $\mathbb{Z}^d$. A function $\mathfrak{f} : \mathcal{E}_n^* \to \mathbb{R}$ can be lifted into a symmetric function $\Xi^\star \mathfrak{f}$ on $\mathcal{X}_n$ which vanishes on $\mathcal{X}_n \setminus \mathcal{E}_n^*$:

$$\big(\Xi^\star \mathfrak{f}\big)(x_1, \dots, x_n) = \begin{cases} \mathfrak{f}(\{x_1, \dots, x_n\}) & \text{if } x_i \neq x_j \text{ for } i \neq j \text{ and } x_i \neq 0, \\ 0 & \text{otherwise.} \end{cases}$$

The operators $\mathfrak{S}_\star$, $\mathfrak{N}_\star$, $\mathfrak{L}_{\theta,a,2}$ can also be extended in a natural way to $\mathcal{X}_n$. Consider on $\mathcal{X}_n$ the operators $\mathfrak{S}^o$, $\mathfrak{N}^o$ defined in the previous chapter by (5.34) and $\mathfrak{L}_\theta^o$ defined by

$$\big(\mathfrak{L}_\theta^o \mathfrak{f}\big)(\mathbf{x}) = \sum_{z \in \mathbb{Z}^d} a(z)\big\{\mathfrak{f}(x_1 - z, \dots, x_n - z) - \mathfrak{f}(\mathbf{x})\big\}.$$

In this formula and below, $\mathbf{x} = (x_1, \dots, x_n)$ is an element of $\mathcal{X}_n$, so that each x_j belongs to $\mathbb{Z}^d$. Recall the definition of the norms $\|\cdot\|_{\mathcal{X}_n,1}$ and $\|\cdot\|_{\mathcal{X}_n,-1}$ defined by (5.35) and (5.36).

Lifting the resolvent equation (6.40) to $\mathcal{X}_n$ and adding and subtracting $\mathfrak{S}^o \Xi^\star \mathfrak{u}_\lambda + (1-2\alpha)\mathfrak{N}^o \Xi^\star \mathfrak{u}_\lambda + (1-\alpha)\mathfrak{L}_\theta^o \Xi^\star \mathfrak{u}_\lambda$, we obtain that

$$\lambda \Xi^\star \mathfrak{u}_\lambda - \big\{\mathfrak{S}^o + (1-2\alpha)\mathfrak{N}^o + (1-\alpha)\mathfrak{L}_\theta^o\big\}\Xi^\star \mathfrak{u}_\lambda = \mathfrak{w}_2, \tag{6.42}$$

where $\mathfrak{w}_2$ is equal to $\Xi^\star \mathfrak{w}_1$ plus the remainder

$$\begin{aligned}&\big\{\Xi^\star \mathfrak{S}_\star - \mathfrak{S}^o \Xi^\star\big\}\mathfrak{u}_\lambda + (1-2\alpha)\big\{\Xi^\star \mathfrak{N}_\star - \mathfrak{N}^o \Xi^\star\big\}\mathfrak{u}_\lambda\\ &\quad + (1-\alpha)\big\{\Xi^\star \mathfrak{L}_{\theta,a,2} - \mathfrak{L}_\theta^o \Xi^\star\big\}\mathfrak{u}_\lambda .\end{aligned}$$

We claim that $\mathfrak{w}_2$ has finite $\mathscr{H}_{-1}(\mathscr{X}_n)$ norm. Indeed, for each $n \geq 1$, by (6.46) and Lemma 6.21 below, there exists a finite constant C_0 such that

$$\begin{aligned}&\big\| \Xi^\star \Pi_n \mathfrak{w}_1 \big\|^2_{\mathscr{X}_n,-1} \leq \|\Pi_n \mathfrak{w}_1\|^2_{0,-1},\\ &\big\| \Xi^\star \mathfrak{S}_\star \Pi_n \mathfrak{u}_\lambda - \mathfrak{S}^o \Xi^\star \Pi_n \mathfrak{u}_\lambda \big\|^2_{\mathscr{X}_n,-1} \leq C_0 n^2 \|\Pi_n \mathfrak{u}_\lambda\|^2_{0,1},\\ &\big\| \Xi^\star \mathfrak{N}_\star \Pi_n \mathfrak{u}_\lambda - \mathfrak{N}^o \Xi^\star \Pi_n \mathfrak{u}_\lambda \big\|^2_{\mathscr{X}_n,-1} \leq C_0 n^2 \|\Pi_n \mathfrak{u}_\lambda\|^2_{0,1},\\ &\big\| \Xi^\star \mathfrak{L}_{\theta,a,2} \Pi_n \mathfrak{u}_\lambda - \mathfrak{L}^o_\theta \Xi^\star \Pi_n \mathfrak{u}_\lambda \big\|^2_{\mathscr{X}_n,-1} \leq C_0 n^2 \|\Pi_n \mathfrak{u}_\lambda\|^2_{0,1},\end{aligned}$$

so that

$$\|\Pi_n \mathfrak{w}_2\|^2_{\mathscr{X}_n,-1} \leq 2\|\Pi_n \mathfrak{w}_1\|^2_{0,-1} + C_0 n^2 \|\Pi_n \mathfrak{u}_\lambda\|^2_{0,1} \tag{6.43}$$

for some finite constant C_0.

It remains to examine the resolvent equation (6.42) through Fourier analysis. Let $\mathbb{T}_{n,d} = [-\pi,\pi]^{nd}$ and denote by $\widehat{\mathfrak{u}}_\lambda : \mathbb{T}_{n,d} \to \mathbb{C}$ the Fourier transform of $\Xi^\star \mathfrak{u}_\lambda$:

$$\widehat{\mathfrak{u}}_\lambda(\mathbf{k}) = \sum_{\mathbf{x}\in\mathscr{X}_n} e^{i\mathbf{x}\cdot\mathbf{k}} \big(\Xi^\star \mathfrak{u}_\lambda\big)(\mathbf{x}).$$

In this formula, $\mathbf{x}\cdot\mathbf{k} = \sum_{1\leq j\leq n} x_j \cdot k_j$. It follows from the resolvent equation (6.42) that $\widehat{\mathfrak{u}}_\lambda$ is the solution of

$$\lambda \widehat{\mathfrak{u}}_\lambda(\mathbf{k}) - \big\{\widehat{\mathfrak{S}^o}(\mathbf{k}) + (1-2\alpha)\widehat{\mathfrak{N}^o}(\mathbf{k}) + (1-\alpha)\widehat{\mathfrak{L}^o_\theta}(\mathbf{k})\big\}\widehat{\mathfrak{u}}_\lambda(\mathbf{k}) = \widehat{\mathfrak{w}}_2(\mathbf{k}),$$

where $\widehat{\mathfrak{S}^o}$, $\widehat{\mathfrak{N}^o}$, $\widehat{\mathfrak{L}^o_\theta}$ are the functions associated to the operators $\mathfrak{S}^o$, $\mathfrak{N}^o$, $\mathfrak{L}^o_\theta$:

$$\begin{aligned}-\widehat{\mathfrak{S}^o}(\mathbf{k}) &= \sum_{\substack{1\leq j\leq n\\ z\in\mathbb{Z}^d}} s(z)\big\{1-\cos(k_j\cdot z)\big\},\\ -\widehat{\mathfrak{N}^o}(\mathbf{k}) &= i \sum_{\substack{1\leq j\leq n\\ z\in\mathbb{Z}^d}} a(z)\sin(k_j\cdot z),\\ \widehat{\mathfrak{L}^o_\theta}(\mathbf{k}) &= i \sum_{z\in\mathbb{Z}^d} a(z)\sin\bigg(\sum_{j=1}^n k_j\cdot z\bigg).\end{aligned}$$

The $\mathscr{H}_{-1}(\mathscr{X}_n)$ norm of a function $\mathfrak{v} : \mathscr{X}_n \to \mathbb{R}$ has a simple and explicit expression in terms of the Fourier transform:

$$\|\mathfrak{v}\|^2_{\mathscr{X}_n,-1} = \frac{-1}{n!(2\pi)^{nd}} \int_{\mathbb{T}_{n,d}} d\mathbf{k} \big|\widehat{\mathfrak{v}}(\mathbf{k})\big|^2 \frac{1}{\widehat{\mathfrak{S}^o}(\mathbf{k})}.$$

Since $\Xi^\star \mathfrak{u}_\lambda$ is the solution of the resolvent equation (6.42), for every $\lambda > 0$,

$$\begin{aligned}
&\big\|\mathfrak{N}^o \Xi^\star \mathfrak{u}_\lambda\big\|^2_{\mathscr{X}_n,-1} \\
&= \frac{-1}{n!(2\pi)^{nd}} \int_{\mathbb{T}_{n,d}} \left| \frac{\widehat{\mathfrak{N}^o}(\mathbf{k})}{\lambda - \widehat{\mathfrak{S}^o}(\mathbf{k}) - (1-2\alpha)\widehat{\mathfrak{N}^o}(\mathbf{k}) - (1-\alpha)\widehat{\mathfrak{L}^o_\theta}(\mathbf{k})} \right|^2 \frac{|\widehat{\mathfrak{w}}_2(\mathbf{k})|^2}{\widehat{\mathfrak{S}^o}(\mathbf{k})} d\mathbf{k}.
\end{aligned}$$

It follows from the explicit formulae for the functions $\widehat{\mathfrak{S}^o}$, $\widehat{\mathfrak{N}^o}$, $\widehat{\mathfrak{L}^o_\theta}$ and a Taylor expansion for $|\mathbf{k}|$ small that the previous expression is bounded by

$$\frac{-C_0}{n!(2\pi)^{nd}} \int_{\mathbb{T}_{n,d}} \frac{|\widehat{\mathfrak{w}}_2(\mathbf{k})|^2}{\widehat{\mathfrak{S}^o}(\mathbf{k})} d\mathbf{k} = C_0 \|\mathfrak{w}_2\|^2_{\mathscr{X}_n,-1}$$

for some finite constant C_0. We have thus proved that

$$\big\|\mathfrak{N}^o \Xi^\star \mathfrak{u}_\lambda\big\|^2_{\mathscr{X}_n,-1} \leq C_0 \|\mathfrak{w}_2\|^2_{\mathscr{X}_n,-1}. \tag{6.44}$$

In the same way, for every $\lambda > 0$, $\|\mathfrak{L}^o_\theta \Xi^\star \mathfrak{u}_\lambda\|^2_{\mathscr{X}_n,-1}$ is equal to

$$\begin{aligned}
&\frac{-1}{n!(2\pi)^{nd}} \int_{\mathbb{T}_{n,d}} \left| \frac{\widehat{\mathfrak{L}^o_\theta}(\mathbf{k})}{\lambda - \widehat{\mathfrak{S}^o}(\mathbf{k}) - (1-2\alpha)\widehat{\mathfrak{N}^o}(\mathbf{k}) - (1-\alpha)\widehat{\mathfrak{L}^o_\theta}(\mathbf{k})} \right|^2 \frac{|\widehat{\mathfrak{w}}_2(\mathbf{k})|^2}{\widehat{\mathfrak{S}^o}(\mathbf{k})} d\mathbf{k} \\
&\quad \leq \frac{-C_0}{n!(2\pi)^{nd}} \int_{\mathbb{T}_{n,d}} \frac{|\widehat{\mathfrak{w}}_2(\mathbf{k})|^2}{\widehat{\mathfrak{S}^o}(\mathbf{k})} d\mathbf{k} = C_0 \|\mathfrak{w}_2\|^2_{\mathscr{X}_n,-1}.
\end{aligned}$$

for some finite constant C_0 so that

$$\big\|\mathfrak{L}^o_\theta \Xi^\star \mathfrak{u}_\lambda\big\|^2_{\mathscr{X}_n,-1} \leq C_0 \|\mathfrak{w}_2\|^2_{\mathscr{X}_n,-1}.$$

We may now conclude the proof of Lemma 6.19. Fix $n \geq 1$. By (6.46), Lemma 6.21 and (6.44), there exists a finite constant C_0, which may change from line to line, such that

$$\begin{aligned}
\|\Pi_n \mathfrak{N}_\star \mathfrak{u}_\lambda\|^2_{0,-1} = \|\mathfrak{N}_\star \Pi_n \mathfrak{u}_\lambda\|^2_{0,-1} &\leq C_0 n \big\|\Xi^\star \mathfrak{N}^o \Pi_n \mathfrak{u}_\lambda\big\|^2_{\mathscr{X}_n,-1} \\
&\leq C_0 n \big\{n^2 \|\Pi_n \mathfrak{u}_\lambda\|^2_{0,1} + \big\|\mathfrak{N}^o \Xi^\star \Pi_n \mathfrak{u}_\lambda\big\|^2_{\mathscr{X}_n,-1}\big\} \\
&\leq C_0 \big\{n^3 \|\Pi_n \mathfrak{u}_\lambda\|^2_{0,1} + n \|\Pi_n \mathfrak{w}_2\|^2_{\mathscr{X}_n,-1}\big\}.
\end{aligned}$$

In particular, by (6.43) and (6.41),

$$\|\Pi_n \mathfrak{N}_\star \mathfrak{u}_\lambda\|^2_{0,-1} \leq C_0 \big\{n^3 \|\Pi_n \mathfrak{u}_\lambda\|^2_{0,1} + n \|\Pi_n \mathfrak{w}_1\|^2_{0,-1}\big\}$$

$$\leq C_0 \left\{ n^3 \sum_{j=n-1}^{n+1} \|\Pi_j \mathfrak{u}_\lambda\|_{0,1}^2 + n \|\Pi_n \mathfrak{V}_\mathfrak{a}\|_{0,-1}^2 \right\}.$$

A similar set of inequalities holds for $\mathfrak{L}_{\theta,a,2}$ in place of $\mathfrak{N}_\star$. This concludes the proof of the lemma. □

Estimates on Liftings We close this section with some results used in the previous proof.

Lemma 6.20 *There exists a finite constant C_0 such that for any $n \geq 1$ and any finitely supported function $\mathfrak{f} : \mathscr{E}_n^* \to \mathbb{R}$ in $\mathfrak{H}_{0,1}$,*

$$\|\mathfrak{f}\|_{0,1}^2 \leq \left\| \Xi^\star \mathfrak{f} \right\|_{\mathscr{X}_n,1}^2 \leq C_0 n \|\mathfrak{f}\|_{0,1}^2.$$

Proof The first inequality is elementary and follows from the explicit formulae for the respective H_1 norms. The only difference between the two expressions is that some gradients which are present in the $\mathscr{H}_1(\mathscr{X}_n)$ norm do not appear in $\mathfrak{H}_{0,1}$ norm.

To prove the second inequality, let

$$W(A) = \sum_{x,y \in A} s(y-x) + \sum_{x \in A} s(x).$$

We also denote by W the lifted function $\Xi^\star W$. A simple computation shows that there exists a finite constant C_0 such that

$$\left| \Xi^\star \mathfrak{S}_\star \mathfrak{f}(\mathbf{x}) - \mathfrak{S}^o \Xi^\star \mathfrak{f}(\mathbf{x}) \right| \leq C_0 W(\mathbf{x}) \left| \Xi^\star \mathfrak{f}(\mathbf{x}) \right| \tag{6.45}$$

for every $\mathbf{x}$ in $\mathscr{E}_n^*$ and $\mathfrak{f} : \mathscr{E}_n^* \to \mathbb{R}$.

We are now in a position to prove the second bound. By definition,

$$\left\| \Xi^\star \mathfrak{f} \right\|_{\mathscr{X}_n,1}^2 = -\frac{1}{n!} \sum_{\mathbf{x} \in \mathscr{X}_n} (\Xi^\star \mathfrak{f})(\mathbf{x}) (\mathfrak{S}^o \Xi^\star \mathfrak{f})(\mathbf{x}).$$

Since $\Xi^\star \mathfrak{f}$ vanishes outside $\mathscr{E}_n^*$, we may restrict the sum to $\mathscr{E}_n^*$. Now, adding and subtracting $(\Xi^\star \mathfrak{S}_\star \mathfrak{f})(\mathbf{x})$ in this expression and recalling (6.45), we obtain that

$$\begin{aligned} \left\| \Xi^\star \mathfrak{f} \right\|_{\mathscr{X}_n,1}^2 &\leq \|\mathfrak{f}\|_{0,1}^2 + \frac{C_0}{n!} \sum_{\mathbf{x} \in \mathscr{E}_n^*} W(\mathbf{x}) \left\{ \Xi^\star \mathfrak{f}(\mathbf{x}) \right\}^2 \\ &= \|\mathfrak{f}\|_{0,1}^2 + C_0 \sum_{A \in \mathscr{E}_n^*} W(A) \mathfrak{f}(A)^2. \end{aligned}$$

By Lemma 6.22, the second term of the previous formula is bounded by $C_0 n \|\mathfrak{f}\|_{0,1}^2$, which concludes the proof of the lemma. □

It follows from this result and from the variational formulas for the H_{-1} norms that there exists a finite constant C_0 such for any $n \geq 1$ and any finitely supported function $\mathfrak{f}: \mathscr{E}_n^* \to \mathbb{R}$

$$\frac{1}{C_0 n} \|\mathfrak{f}\|_{0,-1}^2 \leq \left\| \Xi^\star \mathfrak{f} \right\|_{\mathscr{X}_n,-1}^2 \leq \|\mathfrak{f}\|_{0,-1}^2. \tag{6.46}$$

Lemma 6.21 *There exists a finite constant C_0 depending only on $p(\cdot)$ such that*

$$\left\| \Xi^\star \mathfrak{S}_\star \mathfrak{f} - \mathfrak{S}^o \Xi^\star \mathfrak{f} \right\|_{\mathscr{X}_n,-1}^2 \leq C_0 n^2 \|\mathfrak{f}\|_{0,1}^2,$$

$$\left\| \Xi^\star \mathfrak{N}_\star \mathfrak{f} - \mathfrak{N}^o \Xi^\star \mathfrak{f} \right\|_{\mathscr{X}_n,-1}^2 \leq C_0 n^2 \|\mathfrak{f}\|_{0,1}^2,$$

$$\left\| \Xi^\star \mathfrak{L}_{\theta,a,2} \mathfrak{f} - \mathfrak{L}_\theta^o \Xi^\star \mathfrak{f} \right\|_{\mathscr{X}_n,-1}^2 \leq C_0 n^2 \|\mathfrak{f}\|_{0,1}^2,$$

for all $n \geq 1$ and all functions $\mathfrak{f}: \mathscr{E}_n^* \to \mathbb{R}$.

Proof We prove the first and the third estimates and leave to the reader the details of the second. Fix $n \geq 1$ and a finitely supported function $\mathfrak{h}: \mathscr{X}_n \to \mathbb{R}$. We need to estimate the scalar product

$$\frac{1}{n!} \sum_{\mathbf{x} \in \mathscr{X}_n} \mathfrak{h}(\mathbf{x}) \left\{ \Xi^\star \mathfrak{S}_\star \mathfrak{f}(\mathbf{x}) - \mathfrak{S}^o \Xi^\star \mathfrak{f}(\mathbf{x}) \right\} \tag{6.47}$$

in terms of the $\mathscr{H}_1(\mathscr{X}_n)$ norm of $\mathfrak{h}$ and the $\mathfrak{H}_{0,1}$ norm of $\mathfrak{f}$. There are two possible cases. Either $\mathbf{x}$ belongs to $\mathscr{E}_n^*$ or $\mathbf{x}$ does not belong to $\mathscr{E}_n^*$.

In the first case, by (6.45), the expression inside braces in the previous formula is absolutely bounded by $C_0 W(\mathbf{x}) |\Xi^\star \mathfrak{f}(\mathbf{x})|$ for some finite constant C_0. Therefore, the corresponding piece in the previous formula is bounded above by

$$\frac{C_0}{n!} \sum_{\mathbf{x} \in \mathscr{E}_n^*} W(\mathbf{x}) \left| \mathfrak{h}(\mathbf{x}) \right| \left| \Xi^\star \mathfrak{f}(\mathbf{x}) \right| \leq \frac{C_0}{\ell n!} \sum_{\mathbf{x} \in \mathscr{E}_n^*} W(\mathbf{x}) \mathfrak{h}(\mathbf{x})^2 + C_0 \ell \sum_{A \in \mathscr{E}_n^*} W(A) \mathfrak{f}(A)^2$$

for every $\ell > 0$.

If $\mathbf{x}$ does not belong to $\mathscr{E}_n^*$, the corresponding piece of the scalar product writes

$$-\frac{1}{n!} \sum_{\substack{\mathbf{x} \in \mathscr{X}_n \setminus \mathscr{E}_n^* \\ z \in \mathbb{Z}^d, 1 \leq j \leq n}} s(z) \mathfrak{h}(\mathbf{x}) \Xi^\star \mathfrak{f}(\mathbf{x} + z\mathbf{e}_j)$$

because in this case $\Xi^\star \mathfrak{S}_\star \mathfrak{f}(\mathbf{x}) = \Xi^\star \mathfrak{f}(\mathbf{x}) = 0$. In this formula, $\{\mathbf{e}_j, 1 \leq j \leq n\}$ stands for the canonical basis of $\mathbb{R}^n$ so that $\mathbf{x} + z\mathbf{e}_j = (x_1, \ldots, x_{j-1}, x_j + z, x_{j+1}, \ldots, x_n)$. Since $\Xi^\star \mathfrak{f}$ vanishes outside $\mathscr{E}_n^*$, it is implicit in the previous formula that the sum is restricted to all $\mathbf{x}$ such that $\mathbf{x} + z\mathbf{e}_j$ belongs to $\mathscr{E}_n^*$. Since $\mathbf{x} + z\mathbf{e}_j \in \mathscr{E}_n^*$ and $\mathbf{x} \notin \mathscr{E}_n^*$, either $x_j = x_k$ for some k or $x_j = 0$. In particular, since $2ab \le \ell a^2 + \ell^{-1} b^2$ for every $\ell > 0$, a change of variables shows that the previous sum is bounded above by

$$\frac{1}{n!\ell} \sum_{\mathbf{x} \in \mathscr{X}_n} \mathfrak{h}(\mathbf{x})^2 \widetilde{W}(\mathbf{x}) + \frac{\ell}{n!} \sum_{\mathbf{x} \in \mathscr{X}_n} \Xi^\star \mathfrak{f}(\mathbf{x})^2 W(\mathbf{x}), \tag{6.48}$$

where $\widetilde{W}(\mathbf{x}) = \sum_{j \neq k} \mathbf{1}\{x_j = x_k\} + \sum_j \mathbf{1}\{x_j = 0\}$. We may of course replace the sum over $\mathscr{X}_n$ by a sum over $\mathscr{E}_n^*$ in the second term, losing the factor $n!$.

Adding together all previous estimates, we obtain that the scalar product (6.47) is bounded above by

$$\frac{C_0}{n!\ell} \sum_{\mathbf{x} \in \mathscr{X}_n} \mathfrak{h}(\mathbf{x})^2 \big\{ \widetilde{W}(\mathbf{x}) + W(\mathbf{x}) \big\} + C_0 \ell \sum_{A \in \mathscr{E}_n^*} \mathfrak{f}(A)^2 W(A).$$

By Lemma 6.22, the second term is less than or equal to $C_0 n \ell \|\mathfrak{f}\|_{0,1}^2$ for some finite constant C_0. On the other hand, a simple adaptation of the proof of the same lemma gives that the first term is bounded by $C_1 n \ell^{-1} \|\mathfrak{h}\|_{\mathscr{X}_n,1}^2$. To conclude the proof, it remains to minimize over ℓ and to recall the variational formula for the H_{-1} norm of a function.

We turn now to the third estimate. As above, fix a finitely supported function $\mathfrak{h} : \mathscr{X}_n \to \mathbb{R}$. Observe that $\Xi^\star \mathfrak{L}_{\theta,a,2} \mathfrak{f}(\mathbf{x}) = \mathfrak{L}_\theta^o \Xi^\star \mathfrak{f}(\mathbf{x})$ if $\mathbf{x}$ belongs to $\mathscr{E}_n^*$. On the other hand, if $\mathbf{x} \notin \mathscr{E}_n^*$, this difference is equal to $-\sum_z a(z) \mathfrak{f}(x_1 - z, \dots, x_n - z)$. Therefore

$$\begin{aligned} \frac{1}{n!} \sum_{\mathbf{x} \in \mathscr{X}_n} & \mathfrak{h}(\mathbf{x}) \big\{ \Xi^\star \mathfrak{L}_{\theta,a,2} \mathfrak{f}(\mathbf{x}) - \mathfrak{L}_\theta^o \Xi^\star \mathfrak{f}(\mathbf{x}) \big\} \\ &= \frac{-1}{n!} \sum_{\substack{\mathbf{x} \notin \mathscr{E}_n^* \\ z \in \mathbb{Z}^d}} \mathfrak{h}(\mathbf{x}) a(z) \Xi^\star \mathfrak{f}(x_1 - z, \dots, x_n - z). \end{aligned}$$

Since $\Xi^\star \mathfrak{f}$ vanishes in $(\mathscr{E}_n^*)^c$ and since $\mathbf{x}$ does not belong to $\mathscr{E}_n^*$, we must have $x_i = 0$ for some i for the previous term to be different from 0. Introducing the indicator function $\sum_{1 \le i \le n} \mathbf{1}\{x_i = 0\}$ and proceeding as in the first part of the proof, we estimate the absolute value of the previous sum by (6.48) This concludes the proof of the lemma. □

Lemma 6.22 *Let $W : \mathscr{E}^* \to \mathbb{R}$ be defined by $W(A) = \sum_{x,y \in A} s(y - x) + \sum_{x \in A} s(x)$. There exists a finite constant C_0, depending only on $p(\cdot)$ such that*

$$\sum_{A \in \mathscr{E}_n^*} \mathfrak{g}(A)^2 W(A) \le C_0 n \langle -\mathfrak{S}_\star \mathfrak{g}, \mathfrak{g} \rangle_{\mu_\star}$$

for all $n \ge 1$ and all finitely supported function $\mathfrak{g} : \mathscr{E}_n^* \to \mathbb{R}$.

The proof of this lemma is similar to the one of Lemma 5.18 and left to the reader. Instead of employing the transience of the symmetric random walk on $\mathbb{Z}^d$, $d \geq 3$, one uses the transience of the symmetric walk which avoids the origin stated in Lemma 5.26 and the estimates of the Green function presented in Proposition 5.25.

6.9 The Self-diffusion Matrix

In this section, we obtain a finite upper bound and a strictly positive lower bound for the diffusion matrix $D(\alpha)$ seen as a quadratic form.

Proposition 6.23 *There exists a strictly positive and finite constant C_0, depending only on p, such that*

$$C_0^{-1}\alpha(1-\alpha)|\mathfrak{a}|^2 \leq \mathfrak{a} \cdot D(\alpha)\mathfrak{a} \leq C_0(1-\alpha)|\mathfrak{a}|^2$$

for all $\mathfrak{a}$ in $\mathbb{R}^d$.

The proof of this proposition is divided into several steps. We first derive a variational formula for the self-diffusion matrix in the symmetric case, denoted by $D_s(\alpha)$.

Lemma 6.24 *Assume that $p(x) = p(-x)$. For all $\mathfrak{a}$ in $\mathbb{R}^d$,*

$$\mathfrak{a} \cdot D_s(\alpha)\mathfrak{a} = \inf_u \Big\{ \sum_{z \in \mathbb{Z}_*^d} s(z) \int [1-\xi(z)]\big\{z \cdot \mathfrak{a} + \big(T^z u\big)(\xi)\big\}^2 \nu_\alpha^*(d\xi)$$
$$+ (1/2) \sum_{x,y \in \mathbb{Z}_*^d} s(y-x) \int \big(T^{x,y}u\big)(\xi)^2 \nu_\alpha^*(d\xi)\Big\},$$

where the infimum is taken over all cylinder functions u.

Proof Fix $\mathfrak{a}$ in $\mathbb{R}^d$. By (6.19), Lemma 6.11 and Lemma 6.12, the asymptotic variance of $t^{-1/2}Z_t \cdot \mathfrak{a}$ is equal to $\lim_{t\to\infty}\lim_{\lambda\to 0} t^{-1}\mathbb{E}_{\nu_\alpha^*}[(M_t + m_t^\lambda)^2]$. By the representation of the martingales M_t, m_t^λ in terms of the elementary martingales, this expectation is equal to

$$\sum_{z \in \mathbb{Z}_*^d} s(z) \int [1-\xi(z)]\big\{z \cdot \mathfrak{a} + \big(T^z u_\lambda\big)(\xi)\big\}^2 \nu_\alpha^*(d\xi)$$
$$+ (1/2) \sum_{x,y \in \mathbb{Z}_*^d} s(y-x) \int \big(T^{x,y}u_\lambda\big)(\xi)^2 \nu_\alpha^*(d\xi),$$

where u_λ is the solution of the resolvent equation (6.15). Expand the square $\{z \cdot \mathfrak{a} - (T^z u_\lambda)(\xi)\}^2$. The contribution of the term which does not depend on u_λ is equal to

$(1-\alpha)\mathfrak{a}\cdot\sigma^2\mathfrak{a}$, where σ^2 is the symmetric matrix with entries $\sigma^2_{i,j}=\sum_x x_i x_j p(x)$, since $s=p$. A change of variables and the symmetry of $s(\cdot)$ show that the cross term is equal to

$$\begin{aligned}2\sum_{z\in\mathbb{Z}^d_*} s(z)z\cdot\mathfrak{a}\int\big[1-\xi(z)\big]\big\{u_\lambda(\theta_z\xi)-u_\lambda(\xi)\big\}\nu^*_\alpha(d\xi)\\ =-4\int V_\mathfrak{a}(\xi)u_\lambda(\xi)\nu^*_\alpha(d\xi).\end{aligned}\tag{6.49}$$

Therefore, by the explicit formula for the Dirichlet forms obtained right after (6.3), the asymptotic variance of $t^{-1/2}Z_t\cdot\mathfrak{a}$ is equal to

$$(1-\alpha)\mathfrak{a}\cdot\sigma^2\mathfrak{a}-2\lim_{\lambda\to 0}\big\{2\langle V_\mathfrak{a},u_\lambda\rangle_{\nu^*_\alpha}-\|u_\lambda\|_1^2\big\}.$$

By (2.24) and Sect. 2.7.1, both $\langle V_\mathfrak{a},u_\lambda\rangle_{\nu^*_\alpha}$ and $\|u_\lambda\|_1^2$ converge to $\|V_\mathfrak{a}\|_{-1}^2$ as $\lambda\downarrow 0$. Therefore,

$$\mathfrak{a}\cdot D_s(\alpha)\mathfrak{a}=(1-\alpha)\mathfrak{a}\cdot\sigma^2\mathfrak{a}-2\|V_\mathfrak{a}\|_{-1}^2.\tag{6.50}$$

It remains to recall the variational formula for the $\mathcal{H}_{-1}$ norm of a cylinder function and to repeat the computations presented at the beginning of the proof in the opposite order to conclude the lemma. □

The second step in the proof of Proposition 6.23 consists in obtaining a lower bound for the self-diffusion coefficient in the symmetric case. For sake of completeness, we present also an upper bound.

Lemma 6.25 *Assume that $p(x)=p(-x)$. There exists a strictly positive constant C_0, depending only on p, such that*

$$C_0\alpha(1-\alpha)\mathfrak{a}\cdot\sigma^2\mathfrak{a}\le\mathfrak{a}\cdot D_s(\alpha)\mathfrak{a}\le(1-\alpha)\mathfrak{a}\cdot\sigma^2\mathfrak{a}$$

for all $\mathfrak{a}$ in $\mathbb{R}^d$.

Proof The upper bound follows from (6.50). In view of this identity, to prove the lower bound it is enough to show that

$$2\|V_\mathfrak{a}\|_{-1}^2\le C_1(1-\alpha)\mathfrak{a}\cdot\sigma^2\mathfrak{a}\tag{6.51}$$

for some constant $C_1<1$.

We claim that there exists a finite constant C_0 such that

$$\begin{aligned}\sup_f\big\{2\langle f,V_\mathfrak{a}\rangle_{\nu^*_\alpha}-\mathscr{D}_\theta(f)\big\}\le(1/2)(1-\alpha)\mathfrak{a}\cdot\sigma^2\mathfrak{a},\\ \sup_f\big\{2\langle f,V_\mathfrak{a}\rangle_{\nu^*_\alpha}-\mathscr{D}_0(f)\big\}\le C_0\alpha(1-\alpha)\mathfrak{a}\cdot\sigma^2\mathfrak{a},\end{aligned}\tag{6.52}$$

where the supremum is carried over all cylinder functions f.

To prove the first inequality, recall (6.49) and apply Schwarz inequality to obtain that $2\langle f, V_{\mathfrak{a}}\rangle_{\nu_\alpha^*}$ is bounded above by

$$(1/2)(1-\alpha)\sum_{x\in\mathbb{Z}_*^d} s(x)(x\cdot\mathfrak{a})^2 + (1/2)\sum_{x\in\mathbb{Z}_*^d} s(x)\int\big(T^x f\big)(\xi)^2\nu_\alpha^*(d\xi).$$

The first term is equal to $(1/2)(1-\alpha)\mathfrak{a}\cdot\sigma^2\mathfrak{a}$, while the second term is equal to $\mathscr{D}_\theta(f)$. This proves the first inequality in (6.52). The second inequality follows from (6.22) and the strict ellipticity of the matrix σ^2.

We are now in a position to prove (6.51). Since

$$\|V_{\mathfrak{a}}\|_{-1}^2 = \sup_f\big\{2\langle f, V_{\mathfrak{a}}\rangle_{\nu_\alpha^*} - \mathscr{D}_\theta(f) - \mathscr{D}_0(f)\big\},$$

we may split $2\langle f, V_{\mathfrak{a}}\rangle_{\nu_\alpha^*}$ optimally in two pieces and recall (6.52) to obtain (6.51) with $C_1 = 2C_0\alpha(1+2C_0\alpha)^{-1}$. Some elementary algebra permits to conclude. □

Proof of Proposition 6.23 We first derive the upper bound which is easier. In view of (6.19) and Lemma 6.12, to obtain an upper bound on the variance of $t^{-1/2}Z_t\cdot\mathfrak{a}$, we just need to estimate the variances of the martingales M_t and m_t. On the one hand, by definition (6.18) of the martingale M_t and by the explicit formulas for the quadratic variations of the elementary martingales M_t^z,

$$\mathbb{E}_{\nu_\alpha^*}\big[M_t^2\big] \le C_0 t(1-\alpha)|\mathfrak{a}|^2.$$

On the other hand, by Lemma 6.11,

$$\mathbb{E}_{\nu_\alpha^*}\big[m_t^2\big] = \lim_{\lambda\to 0}\mathbb{E}_{\nu_\alpha^*}\big[\big(m_t^\lambda\big)^2\big] = \lim_{\lambda\to 0} 2t\|u_\lambda\|_1^2,$$

where u_λ is the solution of the resolvent equation (6.15). By (2.15) and by Lemma 6.13, this last expression is less than or equal to $2t\|V_{\mathfrak{a}}\|_{-1}^2 \le C_0 t\alpha(1-\alpha)|\mathfrak{a}|^2$. This concludes the proof of the upper bound.

We now turn to the lower bound. In view of the previous lemma, it is enough to show that the self-diffusion coefficient of the asymmetric exclusion process is bounded below by the self-diffusion coefficient of the symmetric process: $D(\alpha) \ge D_s(\alpha)$.

Recall formula (6.20) of the diffusion coefficient $D(\alpha)$. Since the sequence Ψ is the limit in $\mathbb{L}^2(\nu_\alpha^*)$ of Ψ^λ,

$$\begin{aligned}\mathfrak{a}\cdot D(\alpha)\mathfrak{a} \ge \inf_u\bigg\{&\sum_{z\in\mathbb{Z}_*^d} s(z)\int\big[1-\xi(z)\big]\big\{z\cdot\mathfrak{a} + \big(T^z u\big)(\xi)\big\}^2\nu_\alpha^*(d\xi)\\ &+ (1/2)\sum_{x,y\in\mathbb{Z}_*^d} s(y-x)\int\big(T^{x,y}u\big)(\xi)^2\nu_\alpha^*(d\xi)\bigg\},\end{aligned}$$

where the infimum is taken over all cylinder functions u. By Lemma 6.24, the right-hand side is just $\mathfrak{a}\cdot D_s(\alpha)\mathfrak{a}$. This concludes the proof of the proposition. □

6.10 Comments and References

The invariance of the Bernoulli measures ν_α^* for the process seen from the tagged particle was proved by Spitzer (1970). The ergodicity of the measures ν_α^* and the law of large numbers stated in Theorem 6.5 are due to Saada (1987). Kipnis and Varadhan (1986) proved the invariance principle for a tagged particle in symmetric exclusion processes and showed that the diffusion matrix is non-degenerate. This results has been extended to the mean zero case by Varadhan (1995), and to the asymmetric case in dimension $d \geq 3$ by Sethuraman et al. (2000). We refer to Sect. 5.8 for comments and references on the duality method, on the estimates of the asymmetric diagonal and off-diagonal terms of the generator, and on the removal of the hard core interaction used in the proof of Lemma 6.19. We followed here (Landim et al., 2001), where the reader can find a variational formula for the diffusion coefficient in the asymmetric case.

A central limit theorem for a tagged particle has been proved along the lines described in this chapter in several different contexts.

Nearest Neighbor, Asymmetric Case in Dimension 1 In the totally asymmetric case, $p(1) = 1$, Kesten, cited in Spitzer (1970), observed that under the stationary state $\mathbb{P}_{\nu_\alpha^*}$ the position of the tagged particle, Z_t, is a Markov process with mean one exponential jump rates. It follows from this remark that Z_t/t converges a.s. to $1 - \alpha$ and that $[Z_t - (1-\alpha)t]/\sqrt{t}$ converges in distribution to a mean zero Gaussian random variable with variance $(1-\alpha)^2$. Kipnis (1986) extended this result to the asymmetric, nearest neighbor case, $p(-1) + p(1) = 1$, $p(1) \neq 1/2$, proving that Z_t/t converges a.s. to $(1-\alpha)\mathfrak{m}$ and that $[Z_t - (1-\alpha)\mathfrak{m}t]/\sqrt{t}$ converges in distribution to a mean zero Gaussian random variable with a strictly positive variance.

Hard Spheres Tanemura (1989) considers a reversible system of infinitely many hard balls with the same diameter moving on $\mathbb{R}^d$ by random jumps under the hard core condition. The jump rates depend on the configuration. The author proves a central limit theorem for a tagged particle provided the density of particles is sufficiently small. This latter condition is imposed to ensure the ergodicity of the process. Osada (1998a,b) proves an invariance principle for a tagged particle among infinitely many hard core Brownian balls and shows that the self-diffusion matrix is strictly positive.

Exclusion Processes and Related Models Carlson et al. (1993a,b) proves an invariance principle for a tagged particle in a one-dimensional exclusion process where a particle jumps from site x to site $x + y$ at rate $c(|y|)$ if all sites between x and $x + y$ are occupied, where c is a non-increasing function. When c decays slowly they show that the variance of the Brownian motion diverges as the density increases to one.

An invariance principle for a tagged particle in exclusion processes with degenerate rates is proved in Bertini and Toninelli (2004) and Toninelli and Biroli (2004). The self-diffusion matrix is shown to be always positive for cooperative models.

Sethuraman (2007) examines the central limit theorem for a tagged particle in asymmetric zero-range processes.

Mechanical Systems A classical problem in mathematical physics consists in deriving Brownian motion from Hamiltonian principles. There is a huge bibliography on the subject. The martingale approximation presented in this chapter has been used by Szász and Tóth (1987a,b) to study the following problem. They consider a one-dimensional system of a Brownian particle of fixed mass interacting via elastic collisions with an infinite ideal gas of bath particles of mass 1. Letting the mass of the Brownian particle increase at an appropriate rate as the scaling parameter changes, the authors proved convergence to Brownian motion.

We prove in this book convergence to Brownian motion. Certain models, such as the tagged particle in the one-dimensional nearest neighbor symmetric exclusion process or the tagged particle in an exclusion process with long range jumps, exhibit different asymptotic behavior.

Convergence to Fractional Brownian Motion Consider the tagged particle in the one-dimensional nearest neighbor symmetric exclusion process. This is the case not covered by Theorem 6.3 and not examined in this chapter. Arratia (1983) proved that $X_t/t^{1/4}$ converges in distribution to a normal random variable with mean 0 and variance $(1-\alpha)/\alpha\sqrt{2/\pi}$. Based on the approach developed by Brox and Rost (1984) to prove the equilibrium fluctuations for the density field, Rost and Vares (1985) showed that the finite dimensional distributions of the tagged particle converge to the finite dimensional distributions of a fractional Brownian motion of Hurst parameter $1/4$. Peligrad and Sethuraman (2008) completed the proof of the functional central limit theorem by deriving the tightness of the process. Jara and Landim (2006, 2008) proved the convergence to a Gaussian process of the finite dimensional distributions of the position of the tagged particle starting from a local equilibrium state. In the second article the authors considered an exclusion process with bond disorder. This is one of the few examples where a non-equilibrium fluctuation has been proven. With similar methods Gonçalves and Jara (2008) examined the tagged particle in the exclusion process with variable diffusion coefficient.

Subdiffusive Behavior Shiga (1988) examined the evolution of a tagged particle in exclusion models on $\mathbb{Z}$ in which more than one particle jumps simultaneously. He proved that the position of the tagged particle is asymptotically Gaussian with variance equal to $Ct^{1/a}$, where $a \in (1,2]$ depends on the rates at which particles jump.

Convergence to Lévy Processes Jara (2009) proved that a tagged particle in an equilibrium exclusion process with long range jumps converges to a symmetric, α-stable Lévy process. When the initial state is a local equilibrium, the tagged particle converges to the solution of a martingale problem involving the solution of the hydrodynamic equation.

Rod Dynamics In the exclusion model, a "rod" is a large particle which occupies several lattice sites. In the symmetric case Alexander and Lebowitz (1994) prove that the rescaled position of the rod converges to a Brownian motion and obtain by computer simulations how the variance depends on the length of the rod. In the asymmetric case Alexander and Lebowitz (1990, 1994) show that the velocity of the rod increases with its length, which is counter-intuitive since we would expect the size to restrain the displacement.

Einstein Relation Consider particles evolving in time and observe the displacement of one them, called the tagged particle. Denote by $X(t)$ its position at time t and assume that this displacement process converges weakly to a centered d-dimensional Brownian motion with covariance matrix D, when space and time are appropriately scaled. Perturb the process by adding a small external force of intensity f felt only by the tagged particle. Under reasonable conditions, it is expected that the tagged particle converges in the same time scale to a Brownian motion with the same covariance matrix, but with a drift of the form Mf, where M is called the mobility. The Einstein relation predicts that the diffusivity and the mobility are related by the equation $M = (1/2kT)D$, where T is the temperature and k the Boltzmann constant.

Lebowitz and Rost (1994) proved the validity of Einstein relation for a tagged particle in a system of interacting diffusions, for a random walk with random conductances, and for a particle in a random potential whose velocity is governed by a Langevin equation.

Loulakis (2002, 2005) computed the mobility of the tagged particle in the symmetric and in the mean zero asymmetric simple exclusion process in dimension $d \geq 3$. While the Einstein relation holds for the symmetric simple exclusion process, it fails to hold in the mean zero asymmetric case.

Landim et al. (1998a) considered on $\mathbb{Z}$ a tagged particle jumping with rate $p > 1/2$ to the right and rate $1 - p$ to the left, evolving in a sea of particles moving according to the nearest neighbor symmetric simple exclusion process. Denote by X_t the position of the tracer particle at time t. Assuming that the tracer particle starts at the origin, while the remaining particles start from a product measure with density α, the authors prove that $X_t/\sqrt{t}$ converges in probability to $v_\alpha(p)$, and that

$$\lim_{p \downarrow 1/2} \frac{v_\alpha(p)}{p - (1-p)} = \frac{1-\alpha}{\alpha}\sqrt{\frac{2}{\pi}}.$$

When $p = 1/2$, Arratia (1983) proved that $X_t/t^{1/4}$ converges in distribution to a normal random variable with mean 0 and variance $(1-\alpha)/\alpha\sqrt{2/\pi}$. This establishes the Einstein relation for this process. In contrast with the previous models, the drift is not rescaled in this last example.

Stability of the Effective Diffusion Consider a tagged particle in the exclusion process on a d-dimensional torus of length N with K particles. Denote by $\sigma^2_{N,K}$ the effective diffusion of the limiting Brownian motion. Landim et al. (2002) proved that in the symmetric case, $\sigma^2_{N,K}$ converges, as $N \uparrow \infty$ and $K/N^d \to \alpha$, to the effective diffusion of the Brownian motion obtained as limit of a tagged particle in the exclusion process on $\mathbb{Z}^d$ under the stationary measure ν_α. The argument reduces essentially to the proof of the Liouville D-property, discussed in Chap. 1, for symmetric exclusion processes. Jara (2006) extends the result to mean zero exclusion processes in every dimension and to asymmetric exclusion processes in dimension $d \geq 3$. Caputo and Ioffe (2003) examined this problem for random walks with random conductances. In the same spirit, Owhadi (2003) investigated the approximation of the effective conductivity of bounded elliptic random operators by periodization.

References

Alexander FJ, Lebowitz JL (1990) Driven diffusive systems with a moving obstacle: a variation on the Brazil nuts problem. J Phys A, Math Gen 23:L375–L381

Alexander FJ, Lebowitz JL (1994) On the drift and diffusion of a rod in a lattice fluid. J Phys A, Math Gen 27:683–696

Arratia R (1983) The motion of a tagged particle in the simple symmetric exclusion system on **Z**. Ann Probab 11(2):362–373

Bertini L, Toninelli C (2004) Exclusion processes with degenerate rates: convergence to equilibrium and tagged particle. J Stat Phys 117(3–4):549–580

Blumenthal RM, Getoor RK (1968) Markov processes and potential theory. Pure and applied mathematics, vol 29. Academic Press, New York

Brox T, Rost H (1984) Equilibrium fluctuations of stochastic particle systems: the role of conserved quantities. Ann Probab 12(3):742–759

Caputo P, Ioffe D (2003) Finite volume approximation of the effective diffusion matrix: the case of independent bond disorder. Ann Inst Henri Poincaré Probab Stat 39(3):505–525

Carlson JM, Grannan ER, Swindle GH (1993a) A limit theorem for tagged particles in a class of self-organizing particle systems. Stoch Process Appl 47(1):1–16

Carlson JM, Grannan ER, Swindle GH, Tour J (1993b) Singular diffusion limits of a class of reversible self-organizing particle systems. Ann Probab 21(3):1372–1393

Feller W (1971) An introduction to probability theory and its applications, vol II, 2nd edn. Wiley, New York

Gonçalves P, Jara M (2008) Scaling limits of a tagged particle in the exclusion process with variable diffusion coefficient. J Stat Phys 132(6):1135–1143

Jara M (2006) Finite-dimensional approximation for the diffusion coefficient in the simple exclusion process. Ann Probab 34(6):2365–2381

Jara M (2009) Nonequilibrium scaling limit for a tagged particle in the simple exclusion process with long jumps. Commun Pure Appl Math 62(2):198–214

Jara MD, Landim C (2006) Nonequilibrium central limit theorem for a tagged particle in symmetric simple exclusion. Ann Inst Henri Poincaré Probab Stat 42(5):567–577

Jara MD, Landim C (2008) Quenched non-equilibrium central limit theorem for a tagged particle in the exclusion process with bond disorder. Ann Inst Henri Poincaré Probab Stat 44(2):341–361

Kipnis C (1986) Central limit theorems for infinite series of queues and applications to simple exclusion. Ann Probab 14(2):397–408

Kipnis C, Varadhan SRS (1986) Central limit theorem for additive functionals of reversible Markov processes and applications to simple exclusions. Commun Math Phys 104(1):1–19

Landim C, Olla S, Volchan SB (1998a) Driven tracer particle in one-dimensional symmetric simple exclusion. Commun Math Phys 192(2):287–307

Landim C, Olla S, Varadhan SRS (2001) Symmetric simple exclusion process: regularity of the self-diffusion coefficient. Commun Math Phys 224(1):307–321. Dedicated to Joel L Lebowitz

Landim C, Olla S, Varadhan SRS (2002) Finite-dimensional approximation of the self-diffusion coefficient for the exclusion process. Ann Probab 30(2):483–508

Lebowitz JL, Rost H (1994) The Einstein relation for the displacement of a test particle in a random environment. Stoch Process Appl 54(2):183–196

Liggett TM (1985) Interacting particle systems. Grundlehren der Mathematischen Wissenschaften [Fundamental principles of mathematical sciences], vol 276. Springer, New York

Loulakis M (2002) Einstein relation for a tagged particle in simple exclusion processes. Commun Math Phys 229(2):347–367

Loulakis M (2005) Mobility and Einstein relation for a tagged particle in asymmetric mean zero random walk with simple exclusion. Ann Inst Henri Poincaré Probab Stat 41(2):237–254

Osada H (1998a) An invariance principle for Markov processes and Brownian particles with singular interaction. Ann Inst Henri Poincaré Probab Stat 34(2):217–248

Osada H (1998b) Positivity of the self-diffusion matrix of interacting Brownian particles with hard core. Probab Theory Relat Fields 112(1):53–90

Owhadi H (2003) Approximation of the effective conductivity of ergodic media by periodization. Probab Theory Relat Fields 125(2):225–258

Peligrad M, Sethuraman S (2008) On fractional Brownian motion limits in one dimensional nearest-neighbor symmetric simple exclusion. ALEA Lat Am J Probab Math Stat 4:245–255

Rost H, Vares ME (1985) Hydrodynamics of a one-dimensional nearest neighbor model. In: Particle systems, random media and large deviations, Brunswick, Maine, 1984. Contemp math, vol 41. Am Math Soc, Providence, pp 329–342

Saada E (1987) A limit theorem for the position of a tagged particle in a simple exclusion process. Ann Probab 15(1):375–381

Sethuraman S (2007) On diffusivity of a tagged particle in asymmetric zero-range dynamics. Ann Inst Henri Poincaré Probab Stat 43(2):215–232

Sethuraman S, Varadhan SRS, Yau HT (2000) Diffusive limit of a tagged particle in asymmetric simple exclusion processes. Commun Pure Appl Math 53(8):972–1006

Shiga T (1988) Tagged particle motion in a clustered random walk system. Stoch Process Appl 30(2):225–252

Spitzer F (1970) Interaction of Markov processes. Adv Math 5:246–290

Szász D, Tóth B (1987a) A dynamical theory of Brownian motion for the Rayleigh gas. In: Proceedings of the symposium on statistical mechanics of phase transitions—mathematical and physical aspects, Trebon, 1986, vol 47, pp 681–693

Szász D, Tóth B (1987b) Towards a unified dynamical theory of the Brownian particle in an ideal gas. Commun Math Phys 111(1):41–62

Tanemura H (1989) Ergodicity for an infinite particle system in $\mathbf{R}^d$ of jump type with hard core interaction. J Math Soc Jpn 41(4):681–697

Toninelli C, Biroli G (2004) Dynamical arrest, tracer diffusion and kinetically constrained lattice gases. J Stat Phys 117(1–2):27–54

Varadhan SRS (1995) Self-diffusion of a tagged particle in equilibrium for asymmetric mean zero random walk with simple exclusion. Ann Inst Henri Poincaré Probab Stat 31(1):273–285

Chapter 7
Equilibrium Fluctuations of the Density Field

The techniques presented in the first part of the book have a wide range of applications. They have been used, for instance, to prove the hydrodynamic limit of non-gradient interacting particle systems. To illustrate this fact, we depart in this chapter from the main stream of the book and consider the fluctuations of a scalar random field instead of the fluctuations of a particle or an additive functional. More precisely, we examine the equilibrium space-time fluctuations of the density field of simple exclusion processes. As the dynamics conserve the total number of particles, the fluctuations of the density field evolve in a longer time scale than the other fluctuation fields, yielding to an autonomous equation in a proper scaling limit.

The general framework of the fluctuations of scalar random fields with a conserved quantity is the following. Let Y be a stationary random scalar field on $\mathbb{Z}^d$ with a distribution ν that is invariant under translations by $x \in \mathbb{Z}^d$. Under some reasonable conditions the distribution μ_ε of the rescaled field

$$Y^\varepsilon(\varepsilon x) = \varepsilon^{-d/2}\{Y(x) - \alpha\}, \quad x \in \mathbb{Z}^d,$$

where α is the mean $E_\nu[Y(x)]$, will converge, as $\varepsilon \to 0$, to white noise μ with variance

$$E_\mu\big[\xi(x)\xi(y)\big] = \chi\delta(y-x)$$

for some positive constant χ.

If there is a Markovian evolution of the field $Y(\cdot)$ with ν as invariant measure then we have a space time process $Y(x,t)$ with distribution $\mathbb{P}_\nu$, with marginals ν for any fixed time. If we now do a diffusive space-time rescaling

$$Y^\varepsilon(\varepsilon x, t) = \varepsilon^{-d/2}\{Y(x, \varepsilon^{-2}t) - \alpha\},$$

the scaled process $Y^\varepsilon(\varepsilon x, t)$ with distribution $\mathbb{P}_\nu$ is expected to converge to an Ornstein–Uhlenbeck type fluctuation process $\xi(x,t)$ with distribution $\mathbb{P}$, satisfying

$$d\xi(x,t) = (\mathscr{A}\xi)(x,t)\,dt + d\beta(x,t)$$

T. Komorowski et al., *Fluctuations in Markov Processes*,
Grundlehren der mathematischen Wissenschaften 345,
DOI 10.1007/978-3-642-29880-6_7,

expressed in the weak form as a linear stochastic differential equation

$$d\xi_t(G) = \xi_t(\mathscr{A}G)\,dt + d\beta_G(t).$$

Usually $\mathscr{A}$ takes the form of an elliptic second order differential operator with constant coefficients

$$\mathscr{A}G = \sum_{i,j=1}^{d} D_{i,j}\partial_{x_i}\partial_{x_j}G$$

where $D_{i,j}$ is a symmetric positive definite matrix. The fluctuation–dissipation relation asserts that the family of Brownian motions $\beta_G(\cdot)$, that depend linearly on the test function G, satisfy

$$\mathbb{E}\big[\beta_G(t)^2\big] = \chi t \int_{\mathbb{R}^d} \nabla G \cdot D\nabla G dx,$$

where $\mathbb{E}$ stands for the expectation with respect to $\mathbb{P}$.

The proof usually follows the martingale methods. We can write a martingale decomposition (Itô's formula) of the form

$$dY_t^\varepsilon(G) = \Psi\big(Y_t^\varepsilon, G\big)\,dt + dM_{\varepsilon,G}(t).$$

We only have to check that

$$\lim_{\varepsilon\to 0} \Psi\big(Y_t^\varepsilon, G\big) = Y_t(\mathscr{A}G) \quad \text{and} \quad \lim_{\varepsilon\to 0} \mathbb{E}_\nu\big[M_{\varepsilon,G}(t)^2\big] = \chi t \int_{\mathbb{R}^d} \nabla G \cdot D\nabla G\,dx.$$

A central limit theorem for martingales ensures that any limit of $M_{\varepsilon,G}(t)$ is a Brownian motion. Except for some technical issues this is the basic proof.

However there are models where something quite different happens. The term $\Psi(Y_t^\varepsilon, G)$ becomes big but its integral

$$\int_0^t \Psi\big(Y_s^\varepsilon, G\big)\,ds$$

stays finite. $\Psi(Y_s^\varepsilon, G)$ is represented as a sum $\phi_\varepsilon^{(1)} + \phi_\varepsilon^{(2)}$. The big piece $\phi_\varepsilon^{(1)}$ is transformed, by the martingale method presented in the previous chapters, into Brownian noise and combines with the noise $M_{\varepsilon,G}(t)$ to provide the new Brownian motion $\tilde{\beta}_G(t)$. The remaining part $\phi_\varepsilon^{(2)}$ stays finite and becomes

$$\int_0^t Y_t(\tilde{\mathscr{A}}G)\,ds$$

where $\tilde{\mathscr{A}}$ again is of the same form, i.e., a second order elliptic operator. This splitting has to be done by a carefully constructed decomposition and a new formula provided for the coefficients $\tilde{D}_{i,j}$ that defines $\tilde{\mathscr{A}}$. The simple exclusion models studied in the previous chapters provide examples of both situations. The symmetric case is an easy textbook case, while the asymmetric versions exhibit the more complex behavior alluded to earlier.

The Density Field Let us consider the simple exclusion process on $\mathbb{Z}^d$ as defined in Chap. 5. We consider this dynamics in a stationary state with density α. For any $\varepsilon > 0$, let us define the density field Y^ε

$$Y^\varepsilon = \varepsilon^{d/2} \sum_{x \in \mathbb{Z}^d} \big[\eta(x) - \alpha\big]\delta_{\varepsilon x},$$

where δ_u is the Dirac measure concentrated at u. The density field Y^ε is to be thought of as density fluctuation over a spatial scale of size ε^{-1} and can be expressed in the weak form as

$$Y^\varepsilon(G) = \varepsilon^{d/2} \sum_{x \in \mathbb{Z}^d} G(\varepsilon x)\big[\eta(x) - \alpha\big],$$

for any continuous test function $G : \mathbb{R}^d \to \mathbb{R}$ which decays sufficiently fast at infinity. A computation reveals

$$\Psi^\varepsilon(G) = LY^\varepsilon(G) = \varepsilon^{d/2} \sum_{x,y \in \mathbb{Z}^d} p(y-x)\eta(x)\big[1 - \eta(y)\big]\big[G(\varepsilon y) - G(\varepsilon x)\big]. \quad (7.1)$$

If we assume that $p(\cdot)$ is symmetric,

$$\begin{aligned} LY^\varepsilon(G) &= \varepsilon^{d/2} \sum_{x,y \in \mathbb{Z}^d} p(y-x)\eta(x)\big[G(\varepsilon y) - G(\varepsilon x)\big] \\ &\simeq \varepsilon^{2+(d/2)} \sum_{x \in \mathbb{Z}^d} (\mathscr{A}G)(\varepsilon x)\big[\eta(x) - \alpha\big], \end{aligned}$$

where $\mathscr{A}$ is the elliptic operator

$$\mathscr{A} = (1/2) \sum_{i,j=1}^{d} \sigma_{i,j} \partial_{x_i} \partial_{x_j} \quad \text{with } \sigma_{i,j} = \sum_{x \in \mathbb{Z}^d} p(x) x_i x_j \quad (7.2)$$

the covariance matrix of $p(\cdot)$. Therefore the density fluctuation Y^ε has an approximate dissipation that equals $\varepsilon^2 \mathscr{A} Y^\varepsilon$.

Hence, if we define the martingale $M_t^\varepsilon(G)$ by

$$Y_t^\varepsilon(G) = Y_0^\varepsilon(G) + \int_0^t LY_s^\varepsilon(G)\, ds + M_t^\varepsilon(G),$$

where Y_t^ε is the density field at time t, namely

$$Y_t^\varepsilon = \varepsilon^{d/2} \sum_{x \in \mathbb{Z}^d} \big[\eta_t(x) - \alpha\big]\delta_{\varepsilon x},$$

the previous computation shows that in the symmetric case

$$Y_t^\varepsilon(G) \simeq Y_0^\varepsilon(G) + \varepsilon^2 \int_0^t Y_s^\varepsilon(\mathscr{A} G)\,ds + M_t^\varepsilon(G).$$

Speeding up time by the factor ε^{-2} takes us to diffusive scaling, replacing L by $\varepsilon^{-2}L$, and now the dissipation will be approximately $\mathscr{A} Y^\varepsilon$. There is noise that is built in, in the Poisson process generating the jumps, which scales in the diffusive scale to a Brownian noise, leading to an Ornstein–Uhlenbeck process for the density fluctuations in the diffusive scale. The equilibrium fluctuations are that of white noise

$$\mathbb{E}\big[Y_t(G)^2\big] = \mathbb{E}\big[Y_0(G)^2\big] = \chi(\alpha) \int_{\mathbb{R}^d} G(x)^2\,dx,$$

and the evolution of the fluctuations in the diffusive scale takes the form

$$dY_t(G) = Y_t(\mathscr{A} G)\,dt + d\beta_G(t)$$

with

$$\mathbb{E}\big[\beta_G(t)^2\big] = \chi(\alpha) \int_0^t ds \int_{\mathbb{R}^d} \sum_{i,j=1}^d \sigma_{i,j}(\partial_{x_i} G)(\partial_{x_j} G)\,dx.$$

If we drop the assumption of symmetry on $p(\cdot)$, the situation is much more complex. Let us still suppose that $p(\cdot)$ has mean zero:

$$\sum_{z\in\mathbb{Z}^d} zp(z) = 0.$$

We cannot proceed beyond (7.1). The rescaling, which should still be by a factor of ε^{-2}, leads us to a term of magnitude ε^{-1}:

$$\begin{aligned}\Psi^\varepsilon(G) &= \varepsilon^{(d/2)-2} \sum_{x,y\in\mathbb{Z}^d} p(y-x)\eta(x)\big[1-\eta(y)\big]\big[G(\varepsilon y) - G(\varepsilon x)\big] \\ &\simeq \varepsilon^{(d/2)-1} \sum_{j=1}^d \sum_{x\in\mathbb{Z}^d} (\partial_{x_j} G)(\varepsilon x)\tau_x w_j,\end{aligned} \tag{7.3}$$

where w_j is the current in the j-th direction given by

$$w_j(\eta) = \frac{1}{2} \sum_{z\in\mathbb{Z}^d} z_j \big\{p(z)\eta(0)\big[1-\eta(z)\big] - p(-z)\eta(z)\big[1-\eta(0)\big]\big\}. \tag{7.4}$$

In the reversible case the current $w_j(\eta)$ is of a special form

$$w_j(\eta) = \frac{1}{2} \sum_{z\in\mathbb{Z}^d} z_j p(z)\big[\eta(0) - \eta(z)\big]$$

allowing for another summation by parts, getting rid of the factor ε^{-1} and showing that $\Psi^\varepsilon(G) \simeq Y_t^\varepsilon(\mathscr{A}G)$ for $\mathscr{A}$ given by (7.2). More generally if we have a *gradient system*, i.e., if the current takes the form

$$w_j(\eta) = \sum_{z\in\mathbb{Z}^d} r_j(z)\big[h_j(\eta) - \tau_z h_j(\eta)\big]$$

for some finitely supported function r_j and some mean zero cylinder function h_j, then

$$\Psi^\varepsilon(G) \simeq \varepsilon^{d/2} \sum_{x\in\mathbb{Z}^d} (\mathscr{A}G)(\varepsilon x)\big[\eta(x) - \alpha\big],$$

where $\mathscr{A}G = \sum_{i,j} c_i(\alpha) r_i(z) z_j \partial_{x_i}\partial_{x_j} G$ and the constant $c_j(\alpha)$ is calculated as

$$c_j(\alpha) = \frac{d}{d\alpha} E_{\nu_\alpha}\big[h_j(\eta)\big].$$

In general if $w_j(\eta)$ is not of gradient type then we show that $w_j(\eta)$ can be replaced by $\sum_z c(w_j, z)[\eta(z) - \eta(0)]$ for a suitable choice of $c(w_j, z)$. In all these cases, a replacement permits to get rid of the troublesome factor ε^{-1} and yields that $\Psi^\varepsilon(G) \simeq Y_t^\varepsilon(\mathscr{A}G)$ for some second order elliptic differential operator $\mathscr{A}$ of the form

$$\mathscr{A} = \frac{1}{2}\sum_{i,j=1}^{d} D_{i,j}(\alpha)\partial_{x_i}\partial_{x_j}. \tag{7.5}$$

Finally when $\mathfrak{m} = \sum_z z p(z) \neq 0$, it turns out that if we speed up time by ε^{-1}, then there is a limit for the fluctuations and it is

$$dY_t + (1 - 2\alpha)\sum_{j=1}^{d} \mathfrak{m}_j \partial_{x_j} Y_t = 0.$$

In that scaling the Poisson noise becomes ineffective and disappears leaving just a translation. If we denote $v = (1 - 2\alpha)\mathfrak{m}$, then we can center the translations and consider the centered fluctuation field

$$Z_t^\varepsilon = \varepsilon^{d/2} \sum_{x\in\mathbb{Z}^d} \big[\eta_t(x) - \alpha\big]\delta_{\varepsilon[x - vt]}. \tag{7.6}$$

The previous computation shows that on the time scale ε^{-1} the centered fluctuation field Z^ε does not evolve, while on the time scale ε^{-2}, Z^ε will have a scaling limit whose analysis is very similar to the one for the mean zero asymmetric case. In view of (7.3), in order to carry out all the steps we need to understand the behavior of

$$\int_0^t \sum_{x\in V} \tau_x f(\eta_s)\, ds$$

over large times, where V is the box $\{x \in \mathbb{Z}^d : |x_i| \le \varepsilon^{-1}, 1 \le i \le d\}$. In the time scale ε^{-1}, we can show that

$$\varepsilon^{d/2} \int_0^t \sum_{x \in V} \tau_x \big\{ f(\eta_{s\varepsilon^{-1}}) - c(f,\alpha)\eta_{s\varepsilon^{-1}}(0) \big\}\, ds$$

is negligible if we choose $c(f,\alpha)$ to be

$$c(f,\alpha) = \frac{d}{d\alpha} E_{\nu_\alpha}\big[f(\eta)\big].$$

In the time scale which matters, ε^{-2}, the main result is Theorem 7.1 below, known as the fluctuation–dissipation theorem. There are two distinct types of fluctuations one can handle. If g is a cylinder function, then fluctuations of the form

$$\varepsilon^{(d/2)-1} \int_0^t \sum_{x \in \mathbb{Z}^d} G(\varepsilon x)(\tau_x Lg)(\eta_{s\varepsilon^{-2}})\, ds$$

satisfy a central limit theorem and converge to a Brownian motion, since this time integral can be written as

$$M^\varepsilon(t) + \varepsilon^{(d/2)+1} \sum_{x \in \mathbb{Z}^d} G(\varepsilon x)(\tau_x g)(\eta_{t\varepsilon^{-2}}) - \varepsilon^{(d/2)+1} \sum_{x \in \mathbb{Z}^d} G(\varepsilon x)(\tau_x g)(\eta_0),$$

where $M^\varepsilon(t)$ is a martingale with finite quadratic variation.

On the other hand, if $h = \sum_z b(z)[\eta(z) - \eta(0)]$, i.e., if h is of the *gradient* form for some b supported on a finite subset of $\mathbb{Z}^d_*$, a summation by parts is possible and we reduce the fluctuations of

$$\begin{aligned}
&\varepsilon^{(d/2)-1} \int_0^t \sum_{x \in \mathbb{Z}^d} G(\varepsilon x)(\tau_x h)(\eta_{s\varepsilon^{-2}})\, ds \\
&\quad \simeq -\varepsilon^{d/2} \int_0^t \sum_{x,z \in \mathbb{Z}^d} b(z)(z \cdot \nabla G)(\varepsilon x)\big[\eta_{s\varepsilon^{-2}}(x) - \alpha\big]\, ds
\end{aligned}$$

to the fluctuations of the density field.

The analysis of non-gradient systems depends therefore on our ability to write any w that satisfies some conditions as $w = Lg + h$, with $h = \sum_z b(z)[\eta(z) - \eta(0)]$. While g may not be a cylinder function, what we do show is that it is possible to approximate w well enough by $Lg + h$ with some local g. In fact, the fluctuation–dissipation theorem asserts that if w is a cylinder function which is orthogonal to all functions of degree one, then there are coefficients $b(z, w)$ such that the fluctuation of $w - \sum_z b(z,w)[\eta(z) - \eta(0)]$ is well approximated by fluctuations of the first type.

Theorem 7.1 *Assume that $d \geq 3$. Fix $T > 0$, a cylinder function w in $\mathscr{G}_2^\infty = \bigoplus_{n\geq 2} \mathscr{A}_n$ and a smooth function $G : \mathbb{R}^d \to \mathbb{R}$ with compact support. Denote by $\mathscr{S}$ the support of the probability measure $s(\cdot)$: $\mathscr{S} = \{x \in \mathbb{Z}^d : s(x) > 0\}$. There exist coefficients $\{b(z,\alpha) : z \in \mathscr{S}\}$ and a sequence of cylinder functions u_m such that*

$$\limsup_{m\to\infty} \limsup_{\varepsilon\to 0} \mathbb{E}_{\nu_\alpha}\Big[\sup_{0\leq t\leq T} \Big(\varepsilon^{(d/2)-1} \int_0^t \sum_{x\in\mathbb{Z}^d} G(\varepsilon x)\tau_x W_m(\eta_{s\varepsilon^{-2}})\, ds \Big)^2 \Big] = 0,$$

where

$$W_m(\eta) = W_{\alpha,m}(\eta) = w - Lu_m + \sum_{z\in\mathbb{Z}^d} b(z,\alpha)\big\{\eta(z) - \eta(0)\big\}.$$

The coefficients $b(z,\alpha)$ can be expressed in terms of the Fourier coefficients of the solution of the resolvent equation $\lambda u_\lambda - Lu_\lambda = w$. We prove in Chap. 8 that they depend smoothly on the density α. Moreover, in the non-gradient mean zero case it is not difficult to obtain from this result and from formula (7.3) an expression for the diffusion matrix $D_{i,j}(\alpha)$ appearing in (7.5) in terms of the coefficients $b_j(z,\alpha)$ associated to the currents w_j, introduced in (7.4), by the previous theorem. A similar result holds in the asymmetric case, one just needs to compute Ψ^ε in this case.

The study of fluctuations therefore depends on controlling space-time variances of the form

$$\mathbb{E}_{\nu_\alpha}\Big[\Big(\int_0^t \sum_{x\in V} \tau_x w(\eta_s)\, ds \Big)^2\Big]$$

for large times t and large volumes V. The natural way to control such objects is to invert the generator of the process L, i.e., to solve the equation

$$Lu = \sum_{z\in V} \tau_z w \tag{7.7}$$

in the Hilbert space $\mathscr{H}_{-1}$ induced by the asymptotic space-time variances, perhaps with $u = \sum_{z\in V} \tau_z u_0$ for some cylinder function u_0. Our goal is to show that we can approximate, in a proper weak sense, the solution of Eq. (7.7) by such functions in a diffusive space-time scaling limit.

In contrast with the previous chapters, where we solved the equation $Lu = w$ in $\mathscr{H}_{-1}$ for cylinder functions u, w, there is now a spatial sum. This ingredient modifies the scalar product between cylinder functions and consequently the dual relations and the associated $\mathscr{H}_1$ and $\mathscr{H}_{-1}$ spaces. We introduce in Sect. 7.1 the new framework. In Sect. 7.2 we characterize the functions w such that the corresponding asymptotic space-time variance is finite and then show that any such function can be approximated in $\mathscr{H}_{-1}$ by a function in the domain of the generator. This result leads to the proof of the fluctuation–dissipation theorem in Sect. 7.3. In Sect. 7.4 we show that the fluctuations of the density field can be understood as part of the linear response theory and are related to the fluctuations of a special particle, called the second class particle. We conclude this chapter by presenting some properties of the diverse parts of the generator in Sect. 7.5.

We hope that we have convinced the reader that the fluctuation–dissipation theorem is the main step in the proof of the equilibrium fluctuations of the density field. A complete proof of this result for gradient models can be found in Chap. 11 of Kipnis and Landim (1999) and in Chang et al. (2001) for asymmetric simple exclusion processes in dimension $d \geq 3$.

7.1 Duality

Consider the simple exclusion process defined in Chap. 5. For mean zero cylinder functions f, g, define the scalar product $\langle\langle \cdot, \cdot \rangle\rangle$ by

$$\langle\langle f, g \rangle\rangle := \sum_{x \in \mathbb{Z}^d} \langle \tau_x f, g \rangle_{\nu_\alpha},$$

where $\langle \cdot \rangle_{\nu_\alpha}$ stands for the expectation with respect to the measure ν_α. That $\langle\langle f, g \rangle\rangle$ is in fact an inner product can be seen by the relation

$$\langle\langle f, g \rangle\rangle = \lim_V \frac{1}{|V|} E_{\nu_\alpha} \left[\sum_{x \in V} \tau_x f \sum_{y \in V} \tau_y g \right],$$

where the limit is carried over an increasing family of finite sets V which eventually contains all sites of $\mathbb{Z}^d$. Since $\langle\langle f - \tau_x f, g \rangle\rangle = 0$ for all x in $\mathbb{Z}^d$ and mean zero cylinder functions f, g, this scalar product is only positive semi-definite. We will show later that these *gradients* are the only elements of the kernel of $\langle\langle \cdot, \cdot \rangle\rangle$. Denote by $\mathscr{L}^2 = \mathscr{L}^2(\nu_\alpha)$ the Hilbert space generated by the mean zero cylinder functions and the inner product $\langle\langle \cdot, \cdot \rangle\rangle$.

In order to prove the fluctuation–dissipation theorem we need to analyze the resolvent equation $\lambda f_\lambda - L f_\lambda = W$. As in the previous chapters, duality will be an important tool to reduce the problem to a random walk estimate, similar to the ones encountered before.

Recall from Sect. 5.4 the definition of the orthonormal basis $\{\Psi_A : A \in \mathscr{E}\}$ of $L^2(\nu_\alpha)$. Fix two mean zero cylinder functions f, g and write them in the basis $\{\Psi_A, A \in \mathscr{E}\}$:

$$f = \sum_{A \in \mathscr{E}} \mathfrak{f}(A) \Psi_A = \sum_{n \geq 1} \sum_{A \in \mathscr{E}_n} \mathfrak{f}(A) \Psi_A, \qquad g = \sum_{A \in \mathscr{E}} \mathfrak{g}(A) \Psi_A = \sum_{n \geq 1} \sum_{A \in \mathscr{E}_n} \mathfrak{g}(A) \Psi_A.$$

The sum starts from $n = 1$ because f and g have mean zero so that $\mathfrak{f}(\varnothing) = \mathfrak{g}(\varnothing) = 0$. An elementary computation shows that

$$\langle\langle f, g \rangle\rangle = \sum_{x \in \mathbb{Z}^d} \sum_{n \geq 1} \sum_{A \in \mathscr{E}_n} \mathfrak{f}(A) \mathfrak{g}(A + x),$$

where $B+z$ is the set $\{x+z : x \in B\}$.

We say that two finite subsets A, B of $\mathbb{Z}^d$ are equivalent if one is the translation of the other. This equivalence relation is denoted by $\sim$ so that $A \sim B$ if $A = B + x$ for some x in $\mathbb{Z}^d$. Let $\tilde{\mathscr{E}}_n$ be the quotient of $\mathscr{E}_n$ with respect to this equivalence relation: $\tilde{\mathscr{E}}_n = \mathscr{E}_n/\sim$, $\tilde{\mathscr{E}} = \mathscr{E}/\sim$. If $\mathfrak{f}\colon \mathscr{E}_n \to \mathbb{R}$ is a summable function,

$$\sum_{A\in\mathscr{E}_n} \mathfrak{f}(A) = \sum_{\tilde{A}\in\tilde{\mathscr{E}}_n} \tilde{\mathfrak{f}}(\tilde{A})$$

where, for any equivalence class $\tilde{A}$ and summable function $\mathfrak{f}\colon \mathscr{E} \to \mathbb{R}$,

$$\tilde{\mathfrak{f}}(\tilde{A}) = \sum_{z\in\mathbb{Z}^d} \mathfrak{f}(A+z),$$

A being any representative from $\tilde{A}$. In particular, for two mean zero cylinder functions f, g,

$$\langle\langle f, g\rangle\rangle = \sum_{x,z\in\mathbb{Z}^d} \sum_{n\geq 1} \sum_{\tilde{A}\in\tilde{\mathscr{E}}_n} \mathfrak{f}(A+z)\mathfrak{g}(A+x+z) = \sum_{n\geq 1} \sum_{\tilde{A}\in\tilde{\mathscr{E}}_n} \tilde{\mathfrak{f}}(\tilde{A})\tilde{\mathfrak{g}}(\tilde{A}).$$

We say that a function $\mathfrak{f}\colon \mathscr{E} \to \mathbb{R}$ is translation invariant if $\mathfrak{f}(A+x) = \mathfrak{f}(A)$ for all sets A in $\mathscr{E}$ and all sites x of $\mathbb{Z}^d$. Of course, functions $\tilde{\mathfrak{f}}$ on $\tilde{\mathscr{E}}$ are the same as translation invariant functions on $\mathscr{E}$. Fix a subset A of $\mathbb{Z}^d$ with n points. There are n sets in the equivalence class of A which contain the origin. Therefore, summing a translation invariant function $\mathfrak{f}$ over all equivalence classes $\tilde{A}$ in $\tilde{\mathscr{E}}_n$ is the same as summing $\mathfrak{f}$ over all sets B in $\mathscr{E}_n$ which contain the origin and dividing by n:

$$\sum_{\tilde{A}\in\tilde{\mathscr{E}}_n} \mathfrak{f}(\tilde{A}) = \frac{1}{n} \sum_{\substack{A\in\mathscr{E}_n \\ A\ni 0}} \mathfrak{f}(A)$$

provided that $\mathfrak{f}(A) = \mathfrak{f}(A+x)$ for all A, x. Let $\mathscr{E}^*$ be the class of all finite subsets of $\mathbb{Z}^d_* = \mathbb{Z}^d\setminus\{0\}$ and let $\mathscr{E}^*_n$ be the class of all subsets of $\mathbb{Z}^d_*$ with n points. Then, we may write

$$\langle\langle f, g\rangle\rangle = \sum_{n\geq 1} \frac{1}{n} \sum_{\substack{A\in\mathscr{E}_n \\ A\ni 0}} \tilde{\mathfrak{f}}(A)\tilde{\mathfrak{g}}(A) = \sum_{n\geq 0} \frac{1}{n+1} \sum_{A\in\mathscr{E}^*_n} \tilde{\mathfrak{f}}\big(A\cup\{0\}\big)\tilde{\mathfrak{g}}\big(A\cup\{0\}\big).$$

In conclusion, if for a finitely supported function $\mathfrak{f} : \mathscr{E} \to \mathbb{R}$, we define $\mathbb{M}\mathfrak{f} : \mathscr{E}^* \to \mathbb{R}$ by

$$(\mathbb{M}\mathfrak{f})(A) = \tilde{\mathfrak{f}}\big(A\cup\{0\}\big) = \sum_{z\in\mathbb{Z}^d} \mathfrak{f}\big(\big[A\cup\{0\}\big]+z\big), \tag{7.8}$$

for every mean zero cylinder functions f, g, we have that

$$\langle\langle f, g\rangle\rangle = \sum_{n\geq 0}\frac{1}{n+1}\sum_{A\in\mathscr{E}_n^*}\mathbb{M}\mathfrak{f}(A)\mathbb{M}\mathfrak{g}(A) = \sum_{n\geq 0}\frac{1}{n+1}\langle \Pi_n\mathbb{M}\mathfrak{f}, \Pi_n\mathbb{M}\mathfrak{g}\rangle_{\mu_\star}, \quad (7.9)$$

where $\mu_\star$ is the counting measure on $\mathscr{E}^*$, $\langle\cdot,\cdot\rangle_{\mu_\star}$ the scalar product on $L^2(\mu_\star)$, and $\Pi_n\mathfrak{h}$ the restriction of a function $\mathfrak{h}:\mathscr{E}^*\to\mathbb{R}$ to $\mathscr{E}_n^*$: $(\Pi_n\mathfrak{h})(A) = \mathfrak{h}(A)\mathbf{1}\{A\in\mathscr{E}_n^*\}$.

We call $(\mathbb{M}\mathfrak{F}f)(A)$, $A\in\mathscr{E}^*$, the $\mathbb{M}$-coefficients of a mean zero cylinder function f and $\mathfrak{f}(A) = (\mathfrak{F}f)(A)$, $A\in\mathscr{E}$, the Fourier coefficients. The $\mathbb{M}$-coefficients of a mean zero cylinder function f, g are denoted by the symbols $\mathbf{f}$, $\mathbf{g}$, while the Fourier coefficients are denoted by the symbols $\mathfrak{f}$, $\mathfrak{g}$. Observe that $\mathbb{M}$ transforms a function on $\mathscr{E}_n$ into a function on $\mathscr{E}_{n-1}^*$.

Remark 7.2 Recall the definition of $\theta_x A$, $A\in\mathscr{E}^*$, $x\in\mathbb{Z}_*^d$, given in (6.24). Not every function $\mathbf{f}:\mathscr{E}^*\to\mathbb{R}$ is the image by $\mathbb{M}$ of some function $\mathfrak{f}:\mathscr{E}\to\mathbb{R}$ since $(\mathbb{M}\mathfrak{f})(A) = (\mathbb{M}\mathfrak{f})(\theta_{-z}A)$ for all z in A. In contrast, if $\mathbf{f}:\mathscr{E}^*\to\mathbb{R}$ is a finitely supported function satisfying

$$\mathbf{f}(A) = \mathbf{f}(\theta_{-z}A) \quad \text{for all } z\in A, \quad (7.10)$$

define $\mathfrak{f}:\mathscr{E}\to\mathbb{R}$ by

$$\mathfrak{f}(B) = \begin{cases} |B|^{-1}\mathbf{f}(B\setminus\{0\}) & \text{if } B\ni 0,\\ 0 & \text{otherwise.}\end{cases} \quad (7.11)$$

An elementary computation shows that $\mathbb{M}\mathfrak{f} = \mathbf{f}$. With this choice, which is natural but not unique, $\mathfrak{f}(0) = \mathbf{f}(\varnothing)$.

Remark 7.3 The transformation $\mathbb{M}$ maps $\mathscr{E}_n$ into $\mathscr{E}_{n-1}^*$ lowering the degree of a function by one. Thus, the translations in the inner product $\langle\langle\cdot,\cdot\rangle\rangle$ effectively reduce the degree by one while replacing the space $\mathbb{Z}^d$ by $\mathbb{Z}_*^d$.

Formula (7.9) shows also that a mean zero cylinder function f is in the kernel of the inner product $\langle\langle\cdot,\cdot\rangle\rangle$ if and only if $\mathbb{M}\mathfrak{f}$ vanishes, i.e., if and only if

$$\sum_{x\in\mathbb{Z}^d}\mathfrak{f}(A+x) = 0$$

for all finite subsets A such that $|A|\geq 1$. Examples of such functions are the translations of a cylinder function: $f - \tau_x f$.

To examine how the generator L acts on the $\mathbb{M}$-coefficients $\mathbb{M}\mathfrak{f}$ of a cylinder function f, recall the definition of the operator $\mathfrak{L}_\alpha$, introduced in (5.16), which permitted to commute the operator $\mathfrak{F}$ with the generator L, $\mathfrak{F}L = \mathfrak{L}_\alpha\mathfrak{F}$. In the present context, where the scalar product of two cylinder functions f, g is expressed in terms of their $\mathbb{M}$-coefficients $\mathbb{M}\mathfrak{F}f$, $\mathbb{M}\mathfrak{F}g$, as shown in (7.9), we need to find an operator $\mathfrak{L}_{\star,\alpha}$ for which $\mathbb{M}\mathfrak{F}L = \mathfrak{L}_{\star,\alpha}\mathbb{M}\mathfrak{F}$. We first consider the symmetric part of $\mathfrak{L}_\alpha$.

Fix a function f of degree $n \geq 1$ and denote by $\mathfrak{f}$ its Fourier coefficients. Recall the definition of the operator $\mathfrak{S}$ introduced in (5.17). An elementary computation, based on the fact that

$$\sum_{z \in \mathbb{Z}^d} \mathfrak{f}\big([B \cup \{y\}] + z\big) = (\mathbb{M}\mathfrak{f})(\theta_{-y} B)$$

for all subsets B of $\mathbb{Z}^d_*$, sites $y \notin B$ and finitely supported functions $\mathfrak{f} : \mathscr{E} \to \mathbb{R}$, shows that

$$\mathbb{M}\mathfrak{S}\mathfrak{f} = \mathfrak{L}_{\star,s}\mathbb{M}\mathfrak{f}, \tag{7.12}$$

where $\mathfrak{L}_{\star,s} = \mathfrak{S}_\star + \mathfrak{L}_{\theta,s,2}$, and $\mathfrak{S}_\star$, $\mathfrak{L}_{\theta,s,2}$ are the operators defined by

$$(\mathfrak{S}_\star \mathbf{f})(A) = (1/2) \sum_{x,y \in \mathbb{Z}^d_*} s(y-x)\big[\mathbf{f}(A_{x,y}) - \mathbf{f}(A)\big],$$

$$(\mathfrak{L}_{\theta,s,2}\mathbf{f})(A) = \sum_{x \notin A} s(x)\big[\mathbf{f}(\theta_{-x} A) - \mathbf{f}(A)\big].$$

The operators $\mathfrak{S}_\star$ and $\mathfrak{L}_{\theta,s,2}$ appeared in (6.25) and (6.26) in the previous chapter when we defined the operator $\mathfrak{L}_{*,\alpha}$.

This computation should be understood as follows. We introduced an equivalence relation in $\mathscr{E}$ when we decided not to distinguish between a set and its translations. This is the same as assuming that all sets contain the origin. If n particles evolve as exclusion random walks on $\mathbb{Z}^d$, one of them fixed to be at the origin, two things may happen. Either one of the particles which is not at the origin jumps or the particle we assumed to be at the origin jumps. In the first case, this is just a jump on $\mathbb{Z}^d_*$ and is taken care of by the generator $\mathfrak{S}_\star$. In the second case, however, since we are imposing the origin to be always occupied, we need to translate back the configuration to the origin. This part corresponds to the operator $\mathfrak{L}_{\theta,s,2}$.

We now turn to the asymmetric part of the operator $\mathfrak{L}_\alpha$. Let $\mathfrak{L}_{\star,\alpha}$ be given by

$$\mathfrak{L}_{\star,\alpha} = \mathfrak{L}_{\star,s} + (1-2\alpha)\mathfrak{L}_{\star,a} + \sqrt{\chi(\alpha)}\{\mathfrak{J}_{\star,+} + \mathfrak{J}_{\star,-}\},$$

where $\mathfrak{L}_{\star,a} = \mathfrak{N}_\star + \mathfrak{L}_{\theta,a,2}$, and, for a finitely supported function $\mathbf{f} : \mathscr{E}^* \to \mathbb{R}$ and $A \in \mathscr{E}^*$,

$$(\mathfrak{N}_\star \mathbf{f})(A) = \sum_{\substack{x \in A, y \notin A \\ y \neq 0}} a(y-x)\big\{\mathbf{f}(A_{x,y}) - \mathbf{f}(A)\big\}$$

$$(\mathfrak{L}_{\theta,a,2}\mathbf{f})(A) = \sum_{\substack{y \notin A \\ y \neq 0}} a(y)\big\{\mathbf{f}(\theta_{-y} A) - \mathbf{f}(A)\big\},$$

$$(\mathfrak{J}_{\star,+}\mathbf{f})(A) = 2 \sum_{x \in A, y \in A} a(y-x)\mathbf{f}\big(A \setminus \{y\}\big)$$

$$+2\sum_{x\in A}a(x)\big\{\mathbf{f}\big(A\backslash\{x\}\big)-\mathbf{f}\big(\theta_{-x}\big[A\backslash\{x\}\big]\big)\big\},$$

$$(\mathfrak{J}_{\star,-}\mathbf{f})(A)=2\sum_{\substack{x\notin A,y\notin A\\x,y\neq 0}}a(y-x)\mathbf{f}\big(A\cup\{y\}\big).$$

A long but simple computation shows that

$$\mathbb{M}\mathfrak{L}_\alpha\mathfrak{f}=\mathfrak{L}_{\star,\alpha}\mathbb{M}\mathfrak{f}$$

for all finitely supported functions $\mathfrak{f}:\mathscr{E}\to\mathbb{R}$. All operators in the decomposition of $\mathfrak{L}_{\star,\alpha}$ also appear in the definition of the operator $\mathfrak{L}_{*,\alpha}$ introduced just before (6.25): $\mathfrak{J}_{\star,+}=\mathfrak{J}_{0,+}+2\mathfrak{J}_{\theta,a,+}$ and $\mathfrak{J}_{\star,-}=\mathfrak{J}_{0,-}$.

Some Hilbert Spaces In view of (7.9), it is natural to introduce the Hilbert space $L^2_\star(\mu_\star)$ generated by the finitely supported functions $\mathbf{f}:\mathscr{E}^*\to\mathbb{R}$ endowed with the scalar product $\langle\cdot,\cdot\rangle_{\star,\mu_\star}$ defined by

$$\langle\mathbf{f},\mathbf{g}\rangle_{\star,\mu_\star}=\sum_{n\ge 0}\frac{1}{n+1}\sum_{A\in\mathscr{E}_n^*}\mathbf{f}(A)\mathbf{g}(A)=\sum_{n\ge 0}\frac{1}{n+1}\langle\Pi_n\mathbf{f},\Pi_n\mathbf{g}\rangle_{\mu_\star}.$$

The norm of $L^2_\star(\mu_\star)$ is denoted by $\|\cdot\|_{\star,\mu_\star}$. By (7.9), $\mathbb{M}\mathfrak{F}$ is an isomorphism from $\mathscr{L}^2$ to $L^2_\star(\mu_\star)$.

Denote by $\mathscr{I}_n$, $n\ge 0$, the closed subspace of $L^2_\star(\mu_\star)$ of all functions $\mathbf{f}:\mathscr{E}_n^*\to\mathbb{R}$ for which (7.10) holds and let

$$\mathscr{I}=\bigoplus_{n\ge 0}\mathscr{I}_n. \tag{7.13}$$

More generally, let $L^2_k(\mu_\star)$, $k\ge 0$, be the Hilbert space generated by the finitely supported functions $\mathbf{f}:\mathscr{E}^*\to\mathbb{R}$ endowed with the scalar product $\langle\cdot,\cdot\rangle_{k,\mu_\star}$ defined by

$$\langle\mathbf{f},\mathbf{g}\rangle_{k,\mu_\star}=\sum_{n\ge 0}(n+1)^{2k}\sum_{A\in\mathscr{E}_n^*}\mathbf{f}(A)\mathbf{g}(A)=\sum_{n\ge 0}(n+1)^{2k}\langle\Pi_n\mathbf{f},\Pi_n\mathbf{g}\rangle_{\mu_\star}.$$

The norm of $L^2_k(\mu_\star)$ is denoted by $|||\cdot|||_{k,\mu_\star}$.

Recall from the end of Sect. 6.7 the definition of the Hilbert spaces $\mathfrak{H}_{0,1}$, $\mathfrak{H}_{0,-1}$ associated to the symmetric operator $\mathfrak{S}_\star$. The norms of $\mathfrak{H}_{0,1}$, $\mathfrak{H}_{0,-1}$ are represented by $\|\cdot\|_{0,1}$ and $\|\cdot\|_{0,-1}$, respectively. Denote by $\mathfrak{H}_{k,1}$, $\mathfrak{H}_{k,-1}$, $k\ge 0$, the graded Hilbert spaces generated by the finitely supported functions $\mathbf{f}:\mathscr{E}^*\to\mathbb{R}$ and the norms

$$\begin{aligned}|||\mathbf{f}|||^2_{k,1}&=\sum_{n\ge 0}(n+1)^{2k}\|\Pi_n\mathbf{f}\|^2_{0,1},\\|||\mathbf{f}|||^2_{k,-1}&=\sum_{n\ge 0}(n+1)^{2k}\|\Pi_n\mathbf{f}\|^2_{0,-1}.\end{aligned} \tag{7.14}$$

7.2 Approximations in $\mathscr{H}_{0,-1}$

The main goal of this section is to show that finitely supported functions $\mathbf{f}: \mathscr{E}^* \to \mathbb{R}$ can be approximated in $\mathfrak{H}_{k,-1}$ by finitely supported functions in the image of the operator $\mathfrak{L}_{\star,\alpha}$. We start by proving that all finitely supported functions $\mathbf{f}: \mathscr{E}^* \to \mathbb{R}$ such that $\mathbf{f}(\varnothing) = 0$ are in $\mathfrak{H}_{k,-1}$.

Lemma 7.4 *A finitely supported function* $\mathbf{f}: \mathscr{E}^* \to \mathbb{R}$ *belongs to* $\mathfrak{H}_{k,-1}$, $k \geq 1$, *provided* $\mathbf{f}(\varnothing) = 0$.

The previous result states that $\mathbb{M}\mathfrak{F}f$ belongs to $\mathfrak{H}_{k,-1}$ for any cylinder function f orthogonal to the functions of degree one. The degree has to be greater than or equal to 2 because by (6.27) and (7.8) $\mathbb{M}\mathfrak{F}f$ vanishes in $\mathfrak{H}_{0,1}$ for a cylinder function f in $\mathscr{A}_1$. Hence, $\mathbb{M}\mathfrak{F}f$ belongs to $\mathfrak{H}_{0,-1}$, $f \in \mathscr{A}_1$, only if $\mathbb{M}\mathfrak{F}f(\varnothing) = 0$, i.e., only if f vanishes in $\mathscr{L}^2$.

Proof The proof of this lemma is similar to the one of Lemma 6.15. Fix a finitely supported function $\mathbf{f}: \mathscr{E}^* \to \mathbb{R}$ such that $\mathbf{f}(\varnothing) = 0$. In view of the explicit formula (7.14) for the $\mathfrak{H}_{k,-1}$ norm, it is enough to show that for every $n \geq 1$

$$\|\Pi_n \mathbf{f}\|_{0,-1}^2 = \sup_{\mathbf{g}} \big\{ 2\langle \Pi_n \mathbf{f}, \mathbf{g}\rangle_{\mu_\star} - \langle \mathbf{g}, (-\mathfrak{S}_\star)\mathbf{g}\rangle_{\mu_\star} \big\} < \infty,$$

where the supremum is carried over all finitely supported functions $\mathbf{g}: \mathscr{E}_n^* \to \mathbb{R}$. The linear term is bounded by

$$\frac{1}{\gamma} \sum_{A \in \mathscr{E}_n^*} \mathbf{f}(A)^2 + \gamma \sum_{A \in \mathscr{E}_n^*} \mathbf{1}\{\mathbf{f}(A) \neq 0\} \mathbf{g}(A)^2$$

for every $\gamma > 0$. By Lemma 5.26, n particles evolving on $\mathbb{Z}_*^d$, $d \geq 3$, as symmetric random walks with exclusion is a transient process. Hence, by Proposition 5.23, the second term is bounded above by $C_0 \langle \mathbf{g}, (-\mathfrak{S}_\star)\mathbf{g}\rangle_{\mu_\star}$ for some finite constant C_0 depending only on the size of the support of $\Pi_n \mathbf{f}$. To conclude the proof of the lemma it remains to set $\gamma = C_0^{-1}$. □

Recall the definition of the set $\mathscr{I} \subset L^2_\star(\mu_\star)$ introduced in (7.13). Recall also the set-up of Sect. 2.7.4 with the measure π and the sets $\mathscr{C}$, $L^2(\pi)$ replaced by the measure $\mu_\star$ and the sets $\mathfrak{C}$, $\mathscr{I}$; and with the following correspondence between operators:

$$S_0 \to \mathfrak{S}_\star, \quad B_0 \to (1 - 2\alpha)\mathfrak{L}_{\star,a} + \mathfrak{L}_{\theta,s,2},$$
$$L_+ \to \sqrt{\chi(\alpha)}\mathfrak{J}_{\star,+}, \quad L_- \to \sqrt{\chi(\alpha)}\mathfrak{J}_{\star,-}.$$

The hypotheses of Sect. 2.7.4 hold. Assumption (2.39) is clearly satisfied. By Lemma 7.11, the operator $\mathfrak{L}_{\star,a}$ is anti-symmetric in $L^2_\star(\mu_\star)$. By Corollary 7.12,

$\mathfrak{J}_{\star,-} + \mathfrak{J}_{\star,+}$ is anti-symmetric in $\mathscr{I}$. Hence, since by Lemma 7.10 $\mathfrak{L}_{\theta,s,2}$ is a non-negative, symmetric operator in $L^2_\star(\mu_\star)$, for every finitely supported function $\mathbf{f}$ in $\mathscr{I}$,

$$0 \le \langle \mathbf{f}, (-\mathfrak{S}_\star)\mathbf{f}\rangle_{\star,\mu_\star} \le \langle \mathbf{f}, [-(\mathfrak{S}_\star + \mathfrak{L}_{\theta,s,2})]\mathbf{f}\rangle_{\star,\mu_\star} = \langle \mathbf{f}, (-\mathfrak{L}_{\star,\alpha})\mathbf{f}\rangle_{\star,\mu_\star}.$$

Assumption (2.40) is therefore in force. Finally, by Lemma 7.14, condition (2.50) with $\beta = 1/2$ is fulfilled.

Lemma 7.5 *Fix a finitely supported function* $\mathbf{w} : \mathscr{E}^* \to \mathbb{R}$ *in* $\mathscr{I}$ *and such that* $\mathbf{w}(\varnothing) = 0$. *For each* $\lambda > 0$, *there exists a function* $\mathbf{u}_\lambda$ *in* $\mathscr{I}$ *which solves the resolvent equation*

$$\lambda \mathbf{u}_\lambda - \mathfrak{L}_{\star,\alpha}\mathbf{u}_\lambda = \mathbf{w}. \tag{7.15}$$

Proof This assertion requires a proof because $\mathfrak{L}_{\star,\alpha}$ is not the generator of a Markov process. The obvious approach also does not work due to the presence of the translations in the scalar product. One would proceed as follows. Since $\mathbf{w}$ belongs to $\mathscr{I}$, $\mathfrak{w}$ given by (7.11) with $\mathbf{w}$ replacing $\mathbf{f}$ is such that $\mathbb{M}\mathfrak{w} = \mathbf{w}$. Let w be the cylinder function $w = \sum_A \mathfrak{w}(A)\Psi_A$, consider the resolvent equation

$$\lambda u_\lambda - L u_\lambda = w,$$

and set $\mathbf{u}_\lambda = \mathbb{M}\mathfrak{F}u_\lambda$. It is not clear, however, that u_λ belongs to $\mathscr{L}^2$ and that we may define $\mathbb{M}\mathfrak{F}u_\lambda$. In the proof below we examine (7.15) directly without relying on the lifted resolvent equation.

Since $\mathbf{w}$ is a finitely supported function, there exists $n_0 \ge 1$ such that $\mathbf{w}(A) = 0$ for all $A \notin \bigcup_{j=0}^{n_0} \mathscr{E}_j^*$. Let Π_n^+, $n \ge 1$, be the projection on $\bigcup_{j=0}^{n} \mathscr{E}_j^*$: $\Pi_n^+ = \sum_{0\le j\le n} \Pi_j$ and let $\mathfrak{M}_n = \Pi_n^+(\mathfrak{J}_{\star,+} + \mathfrak{J}_{\star,-})\Pi_n^+$. We first prove the existence of a solution in $L^2_\star(\mu_\star)$ of the truncated resolvent equation

$$\lambda \mathbf{u}_{\lambda,n} - \{\mathfrak{L}_{\star,s} + (1-2\alpha)\mathfrak{L}_{\star,a} + \sqrt{\chi(\alpha)}\mathfrak{M}_n\}\mathbf{u}_{\lambda,n} = \mathbf{w} \tag{7.16}$$

for each $n \ge n_0$.

Fix $n \ge n_0$. By Lemma 7.9, the operators $\mathfrak{L}_{\star,s}$, $\mathfrak{L}_{\star,a}$ and $\mathfrak{M}_n$ are bounded in $\{\Pi_n^+\mathbf{f} : \mathbf{f} \in L^2_\star(\mu_\star)\}$. There exists, in particular, a solution for λ large enough. Since $\mathbf{w}$ belongs to $\mathscr{I}$, by (7.30), $\mathbf{u}_{\lambda,n}$ also belongs to $\mathscr{I}$. On the other hand, $\mathfrak{L}_{\star,s}$ is a symmetric, negative, semi-definite operator while, by Lemma 7.11 and Corollary 7.12, $\mathfrak{L}_{\star,a}$ and $\mathfrak{J}_{\star,+} + \mathfrak{J}_{\star,-}$ are anti-symmetric in $\mathscr{I}$. Since Π_n^+ is a projection, for any $\mathbf{f}$ in $L^2_\star(\mu_\star)$, $\langle \mathbf{f}, \mathfrak{M}_n\mathbf{f}\rangle_{\star,\mu_\star} = \langle \Pi_n^+\mathbf{f}, (\mathfrak{J}_{\star,+} + \mathfrak{J}_{\star,-})\Pi_n^+\mathbf{f}\rangle_{\star,\mu_\star} = 0$, so that $\mathfrak{M}_n$ is also anti-symmetric in $\mathscr{I}$. Therefore, taking the scalar product on both sides of the previous identity with respect to $\mathbf{u}_{\lambda,n}$ we obtain by Schwarz inequality that

$$\lambda \|\mathbf{u}_{\lambda,n}\|_{\star,\mu_\star} \le \|\mathbf{w}\|_{\star,\mu_\star}.$$

By the proof of Proposition I.2.8(b) in Liggett (1985), there exists a solution to (7.16) for every $\lambda > 0$.

Up to this point we have proved the existence of a solution of Eq. (7.16) which belongs to $\mathscr{I}$. The previous estimate shows that the sequence $\{\mathbf{u}_{\lambda,n}, n \geq 1\}$ is uniformly bounded in $L^2_\star(\mu_\star)$ for each $\lambda > 0$. Moreover, in view of the discussion presented before the statement of this lemma and by Lemma 2.21, since $\mathbf{w}$ is finitely supported, for every $k \geq 1$, there exists a finite constant C_k, depending only on $\mathbf{w}$ and p such that

$$\lambda \sum_{j\geq 0} (j+1)^{2k} \|\Pi_j \mathbf{u}_{\lambda,n}\|^2_{\mu_\star} \leq C_k$$

uniformly over n. Let $\mathbf{u}_\lambda$ be a limit point of the sequence $\{\mathbf{u}_{\lambda,n}, n \geq 1\}$. $\mathbf{u}_\lambda$ inherits the previous bound and belongs therefore to the domain of the operators $\mathfrak{L}_{\star,s}$, $\mathfrak{L}_{\star,a}$, $\mathfrak{J}_{\star,\pm}$. Furthermore, taking scalar products with finitely supported functions, it is easy to show that any limit point of this sequence is a solution of Eq. (7.15). Finally, $\mathbf{u}_\lambda$ belongs to $\mathscr{I}$ because each function $\mathbf{u}_{\lambda,n}$ belongs and $\mathscr{I}$ is closed. This proves the lemma. □

Let $\mathbf{u}_\lambda$ be the solution of the resolvent equation (7.15). Since the assumptions of Lemma 2.21 are in force, for each $k \geq 0$, there exists a finite constant C_k such that

$$\sup_{0<\lambda\leq 1} \left\{\lambda |||\mathfrak{u}_\lambda|||_{k,\mu_\star} + |||\mathfrak{u}_\lambda|||_{k,1}\right\} \leq C_k |||\mathfrak{w}|||_{k,-1}. \tag{7.17}$$

Moreover, by Lemma 7.15 and Theorem 5.19, for all $k \geq 1$, there exists a finite constant C_k such that

$$\sup_{0<\lambda\leq 1} |||\mathfrak{L}_{\star,\alpha}\mathfrak{u}_\lambda|||_{k,-1} \leq C_k |||\mathfrak{w}|||_{k+2,-1}. \tag{7.18}$$

We are now in a position to prove the main result of this section.

Theorem 7.6 *Fix a finitely supported function* $\mathbf{w} : \mathscr{E}^* \to \mathbb{R}$ *in* $\mathscr{I}$ *and such that* $\mathbf{w}(\varnothing) = 0$. *For each* $\varepsilon > 0$ *and* $k \geq 0$, *there exists a finitely supported function* $\mathbf{h} : \mathscr{E}_* \to \mathbb{R}$ *such that*

$$|||\mathfrak{L}_{\star,\alpha}\mathbf{h} + \mathbf{w}|||_{k,-1} \leq \varepsilon.$$

We may take $\mathbf{h}$ *in* $\mathscr{I}$ *and* $\mathbf{h}(\varnothing) = 0$.

Proof Fix a finitely supported function $\mathbf{w} : \mathscr{E}^* \to \mathbb{R}$ in $\mathscr{I}$ and such that $\mathbf{w}(\varnothing) = 0$. By the proof of Lemma 2.16, there exists a sequence $\{\mathbf{v}_j : j \geq 1\}$, obtained as a convex combination of $\mathbf{u}_\lambda$, for which $\mathfrak{L}_{\star,\alpha}\mathbf{v}_j$ converges strongly in $\mathfrak{H}_{k,-1}$ to $\mathbf{w}$:

$$\lim_{j\to 0} |||\mathfrak{L}_{\star,\alpha}\mathbf{v}_j + \mathbf{w}|||_{k,-1} = 0.$$

It remains to show that for each fixed j and $\varepsilon > 0$, there exists a finitely supported function $\mathbf{v}$ on $\mathscr{E}^*$ such that

$$|||\mathfrak{L}_{\star,\alpha}\mathbf{v}_j - \mathfrak{L}_{\star,\alpha}\mathbf{v}|||_{k,-1} \leq \varepsilon.$$

To prove the existence of such a function $\mathbf{v}$, assume that $\mathbf{v}$ vanishes on $\bigcup_{j\ge n}\mathscr{E}_j^*$ and recall the decomposition of the operator $\mathfrak{L}_{\star,\alpha}$ to write that the left-hand side of the previous inequality is bounded above by

$$\begin{aligned}&\big|\!\big|\!\big|\mathfrak{J}_{\star,+}\big(\mathbf{v}-\Pi_n^+\mathbf{v}_j\big)\big|\!\big|\!\big|_{k,-1}+\big|\!\big|\!\big|\mathfrak{J}_{\star,-}\big(\mathbf{v}-\Pi_n^+\mathbf{v}_j\big)\big|\!\big|\!\big|_{k,-1}+\big|\!\big|\!\big|\mathfrak{L}_{\star,a}\big(\mathbf{v}-\Pi_n^+\mathbf{v}_j\big)\big|\!\big|\!\big|_{k,-1}\\&\quad+\big|\!\big|\!\big|\mathfrak{L}_{\star,s}\big(\mathbf{v}-\Pi_n^+\mathbf{v}_j\big)\big|\!\big|\!\big|_{k,-1}+\big|\!\big|\!\big|\mathfrak{L}_{\star,\alpha}\big(I-\Pi_n^+\big)\mathbf{v}_j\big|\!\big|\!\big|_{k,-1}.\end{aligned}\tag{7.19}$$

We estimate each term on the right-hand side separately. By (7.31), Lemma 7.14 and Lemma 7.13, there exists a finite constant C_0 depending only on the probability measure p such that

$$\begin{aligned}\big|\!\big|\!\big|\mathfrak{J}_{\star,+}\big(\mathbf{v}-\Pi_n^+\mathbf{v}_j\big)\big|\!\big|\!\big|_{k,-1}^2&=\sum_{\ell=1}^{n}(\ell+1)^{2k}\big\|\mathfrak{J}_{\star,+}\Pi_\ell\{\mathbf{v}-\mathbf{v}_j\}\big\|_{0,-1}^2\\&\le C_0\sum_{\ell=1}^{n+1}(\ell+1)^{2k+1}\big\|\Pi_\ell\{\mathbf{v}-\mathbf{v}_j\}\big\|_{0,1}^2\\&\le C_0\sum_{\ell=1}^{n+1}(\ell+1)^{2k+2}\big\|\Pi_\ell\{\mathbf{v}-\mathbf{v}_j\}\big\|_{\mu_\star}^2.\end{aligned}$$

The second term of (7.19) is estimated in the same way. By Lemma 7.13,

$$\begin{aligned}\big|\!\big|\!\big|\mathfrak{L}_{\star,a}\big(\mathbf{v}-\Pi_n^+\mathbf{v}_j\big)\big|\!\big|\!\big|_{k,-1}^2&=\sum_{\ell=1}^{n}(\ell+1)^{2k}\big\|\mathfrak{L}_{\star,a}\Pi_\ell(\mathbf{v}-\mathbf{v}_j)\big\|_{0,-1}^2\\&\le C_0\sum_{\ell=1}^{n}(\ell+1)^{2k+1}\big\|\Pi_\ell(\mathbf{v}-\mathbf{v}_j)\big\|_{\mu_\star}^2\end{aligned}$$

for some finite constant C_0. The fourth term is estimated by exactly the same arguments. Finally, since $\mathbf{v}_j$ is a convex combination of the solutions of the resolvent equation (7.15), by Lemmas 7.10, 7.14 and 7.15,

$$\begin{aligned}\big|\!\big|\!\big|\mathfrak{L}_{\star,\alpha}\big(I-\Pi_n^+\big)\mathbf{v}_j\big|\!\big|\!\big|_{k,-1}^2&\le C_0\sum_{\ell>n}(\ell+1)^{2k}\|\mathfrak{L}_{\star,\alpha}\Pi_\ell\mathbf{v}_j\|_{0,-1}^2\\&\le C_0\sum_{\ell\ge n}(\ell+1)^{2k+3}\|\Pi_\ell\mathbf{v}_j\|_{0,1}^2\\&\quad+C_0\sum_{\ell\ge n}(\ell+1)^{2k+1}\|\Pi_\ell\mathbf{w}\|_{0,-1}^2\end{aligned}$$

for some finite constant C_0 which changes at each line.

For $\varepsilon>0$ fixed, since $\mathbf{w}$ is finitely supported, by (7.17) there exists $n_0>0$ large enough for the last quantity to be bounded by ε. For this fixed n_0, find a finitely supported function $\mathbf{v} : \bigcup_{n \le n_0} \mathscr{E}_n^* \to \mathbb{R}$ for which all previous expressions are bounded by ε, which is possible because $\mathbf{v}_j$ belongs to $L^2_k(\mu_\star)$.

It remains to check that we may take $\mathbf{v}$ in $\mathscr{I}$ with $\mathbf{v}(\varnothing) = 0$. The first property follows from (7.30) which asserts that the operators $\mathfrak{L}_{\star,s}$, $\mathfrak{L}_{\star,a}$, $\mathfrak{J}_{\star,+}$, $\mathfrak{J}_{\star,-}$ map the closed subspace $\mathscr{I}$ of $L^2_\star(\mu_\star)$ into $\mathscr{I}$. In particular, the solutions of the resolvent equations, as well as their convex combinations, belong to $\mathscr{I}$ so that $\mathbf{v}$ can be taken in $\mathscr{I}$.

The second requirement follows from the fact that $(\mathfrak{L}_\star \mathbf{g})(\varnothing) = 0$ for any finitely supported function $\mathbf{g}$ in $\mathscr{I}$, where $\mathfrak{L}_\star$ stands for any of the four operators $\mathfrak{L}_{\star,s}$, $\mathfrak{L}_{\star,a}$, $\mathfrak{J}_{\star,+}$, $\mathfrak{J}_{\star,-}$ and from the fact that $(\mathfrak{J}_{\star,+}\mathbf{g})(\{x\}) = 0$ for all x in $\mathbb{Z}^d_*$. These two properties show that we may set the value of $\mathbf{v}$ at $\varnothing$ to be 0 without changing $\mathfrak{L}_{\star,\alpha}\mathbf{v}$. □

7.3 The Fluctuation–Dissipation Theorem

The next theorem asserts that a cylinder function f in $\mathscr{G}_2^\infty$ has a finite space-time variance in the diffusive scaling. Denote by $\|G\|_{L^2(\mathbb{R}^d)}$ the L^2 norm of a function $G : \mathbb{R}^d \to \mathbb{R}$:

$$\|G\|^2_{L^2(\mathbb{R}^d)} = \int_{\mathbb{R}^d} G(x)^2 \, dx.$$

Theorem 7.7 *Assume that $d \ge 3$. Fix $T > 0$, a smooth function $G : [0,T] \times \mathbb{R}^d \to \mathbb{R}$ with compact support and a cylinder function f in $\mathscr{G}_2^\infty$. There exists a finite constant C_0, depending only on p, such that*

$$\begin{aligned} &\limsup_{\varepsilon \to 0} \mathbb{E}_{\nu_\alpha}\Bigg[\sup_{0 \le t \le T}\bigg(\varepsilon^{(d/2)-1}\int_0^t \sum_{x \in \mathbb{Z}^d} G(s, \varepsilon x)\tau_x f(\eta_{s\varepsilon^{-2}})\, ds\bigg)^2\Bigg] \\ &\quad \le C_0 \int_0^T \|G(s,\cdot)\|^2_{L^2(\mathbb{R}^d)}\, ds\, \|\mathbb{M}\mathfrak{f}\|^2_{0,-1}. \end{aligned} \tag{7.20}$$

Proof Fix a smooth function $G : [0,T] \times \mathbb{R}^d \to \mathbb{R}$ with compact support and a cylinder function f in $\mathscr{G}_2^\infty$. By Lemma 2.4,

$$\begin{aligned} &\mathbb{E}_{\nu_\alpha}\Bigg[\sup_{0 \le t \le T}\bigg(\varepsilon^{(d/2)-1}\int_0^t \sum_{x \in \mathbb{Z}^d} G(s, \varepsilon x) f(\tau_x \eta_{s\varepsilon^{-2}})\, ds\bigg)^2\Bigg] \\ &\quad \le 24\varepsilon^d \int_0^T \|g(s,\cdot)\|^2_{-1}\, ds, \end{aligned}$$

where $g(s,\eta)$ is the time-dependent cylinder function given by $g(s,\eta) = \sum_x G(s,\varepsilon x)\, \tau_x f$. The factor ε^{-1} multiplying the sum over x has been canceled with the factor ε^{-2} speeding up the process.

Since the spaces $\mathscr{H}_{-1}$ and $\mathscr{H}_{-1}(\mathfrak{S})$, introduced in Sect. 5.4, are isomorphic, by (5.25) the previous expression is equal to

$$24\varepsilon^d \int_0^T \|\mathfrak{g}(s,\cdot)\|_{-1}^2 \, ds = 24\varepsilon^d \int_0^T \sum_{n\geq 2} \|\Pi_n \mathfrak{g}(s,\cdot)\|_{-1}^2 \, ds,$$

where $\mathfrak{g}(s,\cdot) = \mathfrak{F}g(s,\cdot)$. The sum starts at $n=2$ because we assumed f to be orthogonal to $\mathscr{A}_1$. Recall from Sect. 5.6 the definitions of the space $\mathscr{X}_n = (\mathbb{Z}^d)^n$, and of the operators Ξ, $\mathfrak{S}^o$, presented in (5.33), (5.34), respectively. By (5.41), the previous expression is less than or equal to

$$C_0 \varepsilon^d \int_0^T \sum_{n\geq 2} n \|\Xi \Pi_n \mathfrak{g}(s,\cdot)\|_{\mathscr{X}_n,-1}^2 \, ds$$

for some finite constant C_0. Denote by $\mathfrak{G}_n^o(\cdot,\cdot)$ the Green function associated to the generator $\mathfrak{S}^o$ on $\mathscr{X}_n$. $\mathfrak{G}_n^o$ corresponds to the Green operator of n independent random walks with jump rates $s(\cdot)$, defined in (5.6), evolving on $\mathbb{Z}^d$. By (5.36), the previous expression can be written as

$$C_0 \varepsilon^d \int_0^T \sum_{n\geq 2} \frac{1}{(n-1)!} \sum_{\mathbf{x},\mathbf{y}\in\mathscr{X}_n} \mathfrak{g}_n^o(s,\mathbf{x}) \mathfrak{G}_n^o(\mathbf{x},\mathbf{y}) \mathfrak{g}_n^o(s,\mathbf{y}) \, ds,$$

where $\mathfrak{g}_n^o(s,\cdot) = \Xi \Pi_n \mathfrak{g}(s,\cdot)$.

Replacing $\mathfrak{g}(s,\cdot)$ by its value, we obtain that the expectation appearing in (7.20) is bounded above by

$$C_0 \varepsilon^d \int_0^T \sum_{z,w\in\mathbb{Z}^d} G(s,\varepsilon z) G(s,\varepsilon w) A(z,w) \, ds, \tag{7.21}$$

where

$$A(w,z) = \sum_{n\geq 2} \frac{1}{(n-1)!} \sum_{\mathbf{x},\mathbf{y}\in\mathscr{X}_n} (\Xi \mathfrak{f}_n)(\mathbf{x} - z\mathbf{1}) \mathfrak{G}_n^o(\mathbf{x},\mathbf{y}) (\Xi \mathfrak{f}_n)(\mathbf{y} - w\mathbf{1}),$$

$\mathbf{1}$ represents the n-dimensional vector with all coordinates equal to 1 and $\mathfrak{f}_n = \Pi_n \mathfrak{F} f$. A change of variables shows that $A(w,z) = A(0, z-w)$. We may rewrite (7.21) as

$$\begin{aligned} &C_0 \varepsilon^d \int_0^T \sum_{z,w\in\mathbb{Z}^d} \big\{G(s,\varepsilon[z+w]) - G(s,\varepsilon w)\big\} G(s,\varepsilon w) A(0,z) \, ds \\ &+ C_0 \varepsilon^d \int_0^T \sum_{w\in\mathbb{Z}^d} G(s,\varepsilon w)^2 \, ds \sum_{z\in\mathbb{Z}^d} A(0,z). \end{aligned} \tag{7.22}$$

We claim that the first term is bounded above by $C(f,G)\varepsilon^{1/2}$ for some finite constant $C(f,G)$ depending only on f and G. Indeed, since G has a bounded derivative, the difference $G(s,\varepsilon[z+w]) - G(s,\varepsilon w)$ is absolutely bounded by $C_0\varepsilon^{1/2}|z|^{1/2}$ for some finite constant C_0, whose value may change from line to line. For $\varepsilon|z|\le 1$, this estimate follows from the boundedness of the derivative of G, while for $\varepsilon|z|>1$, it follows from the boundedness of G. Hence, the first line of the previous expression is less than or equal to

$$C_0 T\varepsilon^{1/2}\sum_{z\in\mathbb{Z}^d}|z|^{1/2}\big|A(0,z)\big|.$$

After some change of variables, this sum becomes

$$C_0 T\varepsilon^{1/2}\sum_{n\ge 2}\frac{1}{(n-1)!}\sum_{z\in\mathbb{Z}^d}|z|^{1/2}\sum_{\mathbf{x},\mathbf{y}\in\mathscr{X}_n}(\Xi\mathfrak{f}_n)(\mathbf{x})\mathfrak{G}^o_n(\mathbf{0},\mathbf{y}-\mathbf{x}-z\mathbf{1})(\Xi\mathfrak{f}_n)(\mathbf{y}).$$

Since f is a cylinder function, $\mathfrak{f}_n$ vanishes for sufficiently large n and $\Xi\mathfrak{f}_n$ is a finitely supported function for all n. The previous expression is thus bounded by

$$C_0 T\varepsilon^{1/2}\sup_{\mathbf{x}\in\Gamma(f)}\sum_{z\in\mathbb{Z}^d}|z|^{1/2}\mathfrak{G}^o_n(\mathbf{0},\mathbf{x}-z\mathbf{1}),$$

where the supremum is carried over the set $\Gamma(f)$ of all sites which can be written as the difference of two points in the support of $\Xi\mathfrak{f}$: $\Gamma(f)=\{\mathbf{y}-\mathbf{x},\,\Xi\mathfrak{f}(\mathbf{x})\Xi\mathfrak{f}(\mathbf{y})\ne 0\}$. Since $\mathfrak{G}^o_n(0,\mathbf{z})$ is the Green function of n independent random walks on $\mathbb{Z}^d$, it decays as $|\mathbf{z}|^{2-nd}$. The sum over z is thus finite, uniformly over $\mathbf{x}$, as soon as $(n-1)d>5/2$, an inequality which is fulfilled because we are assuming $d\ge 3$ and $n\ge 2$.

We now turn to the second term of (7.22). After a change of variables, the sum $\sum_z A(0,z)$ becomes

$$\sum_{n\ge 2}\frac{1}{(n-1)!}\sum_{z\in\mathbb{Z}^d}\sum_{\mathbf{x},\mathbf{y}\in\mathscr{X}_n}(\Xi\mathfrak{f}_n)(\mathbf{x})\mathfrak{G}^o_n(\mathbf{x},\mathbf{y}-z\mathbf{1})(\Xi\mathfrak{f}_n)(\mathbf{y}).$$

Fix $n\ge 2$. For a finitely supported function $\mathfrak{f}:\mathscr{X}_n\to\mathbb{R}$, define $\mathbb{M}^o\mathfrak{f}:\mathscr{X}_{n-1}\to\mathbb{R}$ by

$$\big(\mathbb{M}^o\mathfrak{f}\big)(x_1,\dots,x_{n-1})=\sum_{z\in\mathbb{Z}^d}\mathfrak{f}(z,x_1+z,\dots,x_{n-1}+z).$$

Clearly, for every finitely supported function $\mathfrak{f}:\mathscr{E}_n\to\mathbb{R}$,

$$\mathbb{M}^o\Xi\mathfrak{f}=\Xi^\star\mathbb{M}\mathfrak{f}.\tag{7.23}$$

A computation, similar to the one which yielded (7.9) and (7.12), shows that for each $n \geq 2$, and finitely supported function $\mathfrak{f}, \mathfrak{g} : \mathscr{X}_n \to \mathbb{R}$,

$$\sum_{z \in \mathbb{Z}^d} \sum_{\mathbf{x} \in \mathscr{X}_n} \mathfrak{f}(\mathbf{x} - z\mathbf{1})(\mathfrak{S}^o \mathfrak{g})(\mathbf{x}) = \sum_{\mathbf{x} \in \mathscr{X}_{n-1}} (\mathbb{M}^o \mathfrak{f})(\mathbf{x})(\mathfrak{L}^o_{\star,s} \mathbb{M}^o \mathfrak{g})(\mathbf{x}), \tag{7.24}$$

where $\mathfrak{L}^o_{\star,s}$ is the operator defined on the finitely supported functions $\mathfrak{f} : \mathscr{X}_{n-1} \to \mathbb{R}$ by

$$(\mathfrak{L}^o_{\star,s} \mathfrak{f})(\mathbf{x}) = \sum_{\substack{1 \leq j \leq n-1 \\ z \in \mathbb{Z}^d}} s(z)\{\mathfrak{f}(\mathbf{x} + z\mathbf{e}_j) - \mathfrak{f}(\mathbf{x})\} + \sum_{z \in \mathbb{Z}^d} s(z)\{\mathfrak{f}(\mathbf{x} - z\mathbf{1}) - \mathfrak{f}(\mathbf{x})\}.$$

In this formula, $\{\mathbf{e}_j : 1 \leq j \leq n\}$ stands for the canonical basis of $\mathbb{R}^n$, and $\mathbf{x} + z\mathbf{e}_j$ for $(x_1, \dots, x_{j-1}, x_j + z, x_{j+1}, \dots, x_n)$ if $\mathbf{x} = (x_1, \dots, x_n)$, $x_i \in \mathbb{R}^d$. In contrast with (7.9), there is no factor $1/n$ on the right-hand side of (7.24) because we are setting here the first coordinate to be the origin.

For $m \geq 1$, denote by $\| \cdot \|_{\mathscr{X}_m,\theta,1}$ the $\mathscr{H}_1$ norm associated to the generator $\mathfrak{L}^o_{\star,s}$: for each finitely supported function $\mathfrak{f} : \mathscr{X}_m \to \mathbb{R}$,

$$\|\mathfrak{f}\|^2_{\mathscr{X}_m,\theta,1} = \frac{1}{m!} \sum_{\mathbf{x} \in \mathscr{X}_m} \mathfrak{f}(\mathbf{x})(-\mathfrak{L}^o_{\star,s} \mathfrak{f})(\mathbf{x}).$$

A computation gives that

$$\|\mathfrak{f}\|^2_{\mathscr{X}_m,\theta,1} = \frac{1}{2} \sum_{\substack{1 \leq j \leq m \\ z \in \mathbb{Z}^d}} s(z)[\mathfrak{f}(\mathbf{x} + z\mathbf{e}_j) - \mathfrak{f}(\mathbf{x})]^2 + \frac{1}{2} \sum_{z \in \mathbb{Z}^d} s(z)\{\mathfrak{f}(\mathbf{x} - z\mathbf{1}) - \mathfrak{f}(\mathbf{x})\}^2.$$

Denote by $\| \cdot \|^2_{\mathscr{X}_m,\theta,-1}$ the dual norm defined by

$$\|\mathfrak{f}\|^2_{\mathscr{X}_m,\theta,-1} = \frac{1}{m!} \sum_{\mathbf{x} \in \mathscr{X}_m} \mathfrak{f}(\mathbf{x})(\mathfrak{G}^{\star,o}_m \mathfrak{f})(\mathbf{x}),$$

where $\mathfrak{G}^{\star,o}_m$ is the Green function associated to the generator $\mathfrak{L}^o_{\star,s}$ restricted to $\mathscr{X}_m$. It follows from (7.24) that for each pair of finitely supported functions $\mathfrak{f}, \mathfrak{g} : \mathscr{X}_m \to \mathbb{R}$,

$$\sum_{z \in \mathbb{Z}^d} \sum_{\mathbf{x},\mathbf{y} \in \mathscr{X}_m} \mathfrak{f}(\mathbf{x}) \mathfrak{G}^o_m(\mathbf{x}, \mathbf{y} - z\mathbf{1}) \mathfrak{g}(\mathbf{y}) = \sum_{\mathbf{x},\mathbf{y} \in \mathscr{X}_{m-1}} (\mathbb{M}^o \mathfrak{f})(\mathbf{x}) \mathfrak{G}^{\star,o}_{m-1}(\mathbf{x}, \mathbf{y})(\mathbb{M}^o \mathfrak{g})(\mathbf{y}).$$

Hence, by (7.23),

$$\begin{aligned}
\sum_z A(0, z) &= \sum_{n \geq 2} \frac{1}{(n-1)!} \sum_{\mathbf{x},\mathbf{y} \in \mathscr{X}_{n-1}} (\mathbb{M}^o \Xi \mathfrak{f}_n)(\mathbf{x}) \mathfrak{G}^{\star,o}_{n-1}(\mathbf{x}, \mathbf{y})(\mathbb{M}^o \Xi \mathfrak{f}_n)(\mathbf{y}) \\
&= \sum_{n \geq 2} \|\mathbb{M}^o \Xi \mathfrak{f}_n\|^2_{\mathscr{X}_{n-1},\theta,-1} = \sum_{n \geq 2} \|\Xi^\star \mathbb{M} \mathfrak{f}_n\|^2_{\mathscr{X}_{n-1},\theta,-1}.
\end{aligned}$$

By (7.33) and since $\mathbb{M}\Pi_n \mathfrak{f} = \Pi_{n-1}\mathbb{M}\mathfrak{f}$, this sum is less than or equal to

$$\sum_{n\geq 1} \|\Pi_n \mathbb{M}\mathfrak{f}\|_{0,-1}^2 = \|\mathbb{M}\mathfrak{f}\|_{0,-1}^2 .$$

Therefore, the second line in (7.22) is bounded above by

$$C_0 \|\mathbb{M}\mathfrak{f}\|_{0,-1}^2 \varepsilon^d \int_0^T \sum_{w\in\mathbb{Z}^d} G(s,\varepsilon w)^2 \, ds,$$

which concludes the proof of the lemma. □

Fix a cylinder function w in $\mathscr{G}_2^\infty$ and let $\mathbf{w} = \mathbb{M}\mathfrak{F}w$ be its $\mathbb{M}$-coefficients. For each $\lambda > 0$, let $\mathbf{u}_\lambda(\alpha,\cdot)\colon \mathscr{E}^* \to \mathbb{R}$ be the solution of the resolvent equation (7.15). In Sect. 8.5, we prove the existence of a subsequence λ_k for which the sequence $\mathbf{u}_{\lambda_k}(\alpha,\{z\})$ converges, as $k \uparrow \infty$, uniformly in α in $[0,1]$, to some limit, denoted by $D_z(\alpha)$:

$$D_z(\alpha) = \lim_{k\to\infty} \mathbf{u}_{\lambda_k}\big(\alpha,\{z\}\big)$$

for each z in $\mathbb{Z}_*^d$. We are now in a position to prove the fluctuation–dissipation theorem.

Proof of Theorem 7.1 Fix $T > 0$, a continuous function $G : \mathbb{R}^d \to \mathbb{R}$ with compact support, and a mean zero cylinder function w orthogonal to $\mathscr{A}_1$. Let $b(z,\alpha) = a(z)D_z(\alpha)$ so that

$$W_m = w - Lu_m + \sqrt{\chi(\alpha)} \sum_{z\in\mathbb{Z}^d} a(z) D_z(\alpha)\{\Psi_z - \Psi_0\}.$$

Since w belongs to $\mathscr{G}_2^\infty$, by Theorem 7.6 with $k = 0$, there exists a sequence of finitely supported functions $\mathbf{v}_m : \mathscr{E}^* \to \mathbb{R}$ such that $\mathbf{v}_m(\varnothing) = 0$, $\mathbf{v}_m$ belongs to $\mathscr{I}$ and

$$\lim_{m\to\infty} \|\mathbf{w} - \mathfrak{L}_{\star,\alpha}\mathbf{v}_m\|_{0,-1} = 0.$$

Since $\mathbf{v}_m$ satisfies (7.10), in view of (7.11), there exists a finitely supported function $\mathfrak{v}_m : \mathscr{E} \to \mathbb{R}$ such that $\mathbb{M}\mathfrak{v}_m = \mathbf{v}_m$. Moreover, $\mathfrak{v}_m(A) \neq 0$ only if A contains the origin and $\mathfrak{v}_m(\{0\}) = 0$ because $\mathbf{v}_m(\varnothing) = 0$. Let v_m be the cylinder function defined by $v_m = \sum_{A\in\mathscr{E}} \mathfrak{v}_m(A)\Psi_A$ and let $\hat{W}_m$ be the cylinder function in $\mathscr{G}_2^\infty$ defined by

$$\hat{W}_m = w - \{Lv_m - \Pi_1 Lv_m\}.$$

By Theorem 7.7,

$$\begin{aligned}
&\limsup_{\varepsilon\to 0} \mathbb{E}_{\nu_\alpha}\Big[\sup_{0\le t\le T}\Big(\varepsilon^{(d/2)-1}\int_0^t \sum_{x\in\mathbb{Z}^d} G(\varepsilon x)\tau_x \hat{W}_m(\eta_{s\varepsilon^{-2}})\,ds\Big)^2\Big] \\
&\quad\le C_0 T\,\|G\|^2_{L^2(\mathbb{R}^d)}\|\mathbb{M}\mathfrak{F}\hat{W}_m\|^2_{0,-1}
\end{aligned}\tag{7.25}$$

for some finite constant C_0. An elementary computation based on (5.16) gives that for every cylinder function f,

$$\begin{aligned}
\Pi_1(Lf) = &-\frac{1}{2}\sum_{x,y\in\mathbb{Z}^d} s(y-x)\big\{\mathfrak{f}(y)-\mathfrak{f}(x)\big\}\{\Psi_y-\Psi_x\} \\
&-\frac{1-2\alpha}{2}\sum_{x,y\in\mathbb{Z}^d} a(y-x)\big\{\mathfrak{f}(y)+\mathfrak{f}(x)\big\}\{\Psi_y-\Psi_x\} \\
&+\sqrt{\chi(\alpha)}\sum_{x,y\in\mathbb{Z}^d} a(y-x)\mathfrak{f}(x,y)\{\Psi_y-\Psi_x\}
\end{aligned}$$

if $\mathfrak{f}(\cdot)$ represents the Fourier coefficients of f. To obtain the difference $\Psi_y-\Psi_x$ in the second line we used the fact that $\sum_x a(x)=0$. Since $\mathfrak{v}_m(\{x\})=0$ for every x in $\mathbb{Z}^d$, $\mathfrak{v}_m(A)\neq 0$ only if A contains the origin, and $\mathfrak{v}_m(A)=|A|^{-1}\mathbf{v}_m(A\setminus\{0\})$, we have that

$$\Pi_1(Lv_m)=\sqrt{\chi(\alpha)}\sum_{x\in\mathbb{Z}^d} a(x)\mathbf{v}_m(x)\{\Psi_x-\Psi_0\},$$

which vanishes in $\mathscr{L}^2$ so that $\mathbb{M}\mathfrak{F}\Pi_1(Lv_m)=0$. In particular, $\mathbb{M}\mathfrak{F}\hat{W}_m=\mathbf{w}-\mathfrak{L}_{\star,\alpha}\mathbf{v}_m$ and the right-hand side of (7.25) is bounded above by

$$C_0 T\,\|G\|^2_{L^2(\mathbb{R}^d)}\|\mathbf{w}-\mathfrak{L}_{\star,\alpha}\mathbf{v}_m\|^2_{0,-1}.$$

Since this expression vanishes as $m\uparrow\infty$, all we need to prove is that

$$\limsup_{m\to\infty}\limsup_{\varepsilon\to 0} \mathbb{E}_{\nu_\alpha}\Big[\sup_{0\le t\le T}\Big(\varepsilon^{(d/2)-1}\int_0^t \sum_{x\in\mathbb{Z}^d} G(\varepsilon x)\tau_x \tilde{W}_m(\eta_{s\varepsilon^{-2}})\,ds\Big)^2\Big]=0,$$

where

$$\tilde{W}_m=\Pi_1 Lv_m-\sqrt{\chi(\alpha)}\sum_{z\in\mathbb{Z}^d} a(z)D_z(\alpha)\{\Psi_z-\Psi_0\}.$$

By the explicit formula for $\Pi_1 Lv_m$ obtained above,

$$\tilde{W}_m=\sqrt{\chi(\alpha)}\sum_{x\in\mathbb{Z}^d} a(x)\big\{\mathbf{v}_m(x)-D_x(\alpha)\big\}\{\Psi_x-\Psi_0\}.$$

Since ν_α is a stationary state, by Schwarz inequality, the previous expectation is bounded above by

$$\varepsilon^{d-2}T^2 E_{\nu_\alpha}\Big[\Big(\sum_{x\in\mathbb{Z}^d} G(\varepsilon x)\tau_x \tilde{W}_m(\eta)\Big)^2\Big].$$

A change of variables shows that this expression is equal to

$$\varepsilon^{d-2}\chi(\alpha)T^2\sum_{x\in\mathbb{Z}^d}\Big(\sum_{y\in\mathbb{Z}^d} a(y)\big\{G\big(\varepsilon[x-y]\big)-G(\varepsilon x)\big\}\big\{\mathbf{v}_m(y)-D_y(\alpha)\big\}\Big)^2.$$

A Taylor expansion shows that this expression converges, as $\varepsilon\downarrow 0$, to

$$\chi(\alpha)T^2\int_{\mathbb{R}^d}\big\{(\nabla G)(x)\cdot R_m\big\}^2\,dx,$$

where $R_m=(R_m^1,\dots,R_m^d)$, $R_m^j=\sum_{y\in\mathbb{Z}^d}a(y)y_j\{\mathbf{v}_m(y)-D_y(\alpha)\}$. By the definition of $D_y(\alpha)$, $R_m^j\to 0$ as $m\uparrow\infty$ because $\mathbf{v}_m$ is constructed as convex combinations of $\mathfrak{u}_{\lambda_k}$, the solution of the resolvent equation (7.15). The integral therefore tends to 0. □

7.4 The Second Class Particle

The macroscopic evolution of the density field can be reinterpreted as a central limit theorem for a special type of random walk, called the second class particle. This is also generally known as linear response.

Given two continuous functions $F,G:\mathbb{R}^d\to\mathbb{R}$ with compact support, we can compute the correlation

$$\lim_{\varepsilon\to 0}\mathbb{E}_{\nu_\alpha}\big[Y_t^\varepsilon(G)Y_t^\varepsilon(F)\big]=\chi(\alpha)\int_{\mathbb{R}^d}F(q)G(q)\,dq,$$

where $\chi(\alpha)=\alpha(1-\alpha)$ is the static compressibility of the exclusion process. Hence, for any $t\geq 0$, the sequence of random fields $Y_t^\varepsilon(\cdot)$ converges in law to the centered Gaussian random field with covariance $\chi(\alpha)\delta(q-q')$, which is the standard white noise on $\mathbb{R}^d$ multiplied by $\sqrt{\chi(\alpha)}$.

More generally, as we have seen in the introduction to this chapter, in the mean zero case, $\mathfrak{m}=\sum_x xp(x)=0$, the rescaled density field Y_t^ε converges to the Ornstein–Uhlenbeck process Y_t satisfying

$$dY(x,t)=(\mathscr{A}Y)(x,t)\,dt+d\beta(x,t) \tag{7.26}$$

where $\mathscr{A}$ is the differential operator $\mathscr{A} = (1/2)\nabla \cdot D(\alpha)\nabla$ and β is the white noise, known as the Gaussian free field, with space-time covariances given by

$$\mathbb{E}\big[\beta_G(t)\beta_H(s)\big] = \chi(\alpha)\min\{s,t\}\int_{\mathbb{R}^d} \nabla G \cdot D(\alpha)\nabla H\,dx, \tag{7.27}$$

for smooth functions G, H with compact support.

Similarly, in the asymmetric case, $\mathfrak{m} = \sum_x xp(x) \neq 0$, in dimension $d \geq 3$, the centered rescaled density field Z_t^ε defined in (7.6) converges in the diffusive scale to the Ornstein–Uhlenbeck process Y_t satisfying (7.26) for a white noise β with covariances given by (7.27). The diffusion matrix $D(\alpha)$ in the asymmetric case is however different from the one obtained in the mean zero case.

The centered density field Z_t^ε coincides with the density field Y_t^ε if the mean displacement $\mathfrak{m}$ vanishes. We may therefore examine simultaneously the mean zero case and the asymmetric case in dimension $d \geq 3$ by considering the centered density field Z_t^ε.

It follows from the previous results that the correlations of the density fields Z_t^ε converge:

$$\lim_{\varepsilon\to 0}\mathbb{E}_{\nu_\alpha}\big[Z_{t+s}^\varepsilon(G)Z_s^\varepsilon(F)\big] = \chi(\alpha)\int_{\mathbb{R}^d}\int_{\mathbb{R}^d} F(y)P(x-y,t)G(x)\,dy\,dx, \tag{7.28}$$

where

$$P(x,t) = \frac{1}{(2\pi t|D(\alpha)|)^{d/2}}\exp\Big\{-\frac{1}{2t}x\cdot D(\alpha)^{-1}x\Big\},$$

and $|D(\alpha)|$ is the determinant of $D(\alpha)$.

On the other hand, computing the time correlations and keeping in mind that the process is translation invariant, we obtain

$$\begin{aligned}
&\mathbb{E}_{\nu_\alpha}\big[Z_{t+s}^\varepsilon(G)Z_s^\varepsilon(F)\big]\\
&\quad = \varepsilon^d\sum_{x,y\in\mathbb{Z}^d} G(\varepsilon x)F(\varepsilon y)\mathbb{E}_{\nu_\alpha}\big[\big\{\eta_{\varepsilon^{-2}t}\big(x+\varepsilon^{-2}vt\big)-\alpha\big\}\big\{\eta_0(y)-\alpha\big\}\big]\\
&\quad = \varepsilon^d\sum_{x,y\in\mathbb{Z}^d} G(\varepsilon x)F(\varepsilon y)R\big(\varepsilon^{-2}vt+x-y,\varepsilon^{-2}t\big),
\end{aligned}$$

where $R(x,t) = \mathbb{E}_{\nu_\alpha}[\eta_t(x)\eta_0(0)] - \alpha^2$ and v is the velocity $v = (1-2\alpha)\mathfrak{m}$. Therefore, in view of (7.28), in a weak sense,

$$\lim_{\varepsilon\to 0}\varepsilon^{-d}R\big(\varepsilon^{-1}z+vt\varepsilon^{-2},\varepsilon^{-2}t\big) = \chi(\alpha)P(z,t).$$

In particular,

$$\frac{1}{2}D_{i,j}(\alpha) = \lim_{t\to\infty}\frac{1}{2\chi t}\sum_{x\in\mathbb{Z}^d} x_ix_jR(x+vt,t). \tag{7.29}$$

The correlation function $R(x,t)$ may be expressed in terms of the probability transition of a *second class particle*, an extra particle added to the exclusion process which has the same jump rates of the other particles. Its dynamics however differs from the others. When an original (first class) particle attempts to jump to the site occupied by the second class particle, the two particles exchange site. In contrast, when the second class particle attempts to jump to a site occupied by a first class particle, the jump is suppressed. The evolution of the first class particles is therefore unaffected by the presence of the second class particle. In other words, the first class particles evolve as the original exclusion process, as well as the superposition of the first class particles and the second class particle.

In the symmetric case, the dynamics of the exclusion process can be seen as an exchange dynamics: the occupation variables $\eta(x)$ and $\eta(y)$ exchange at rate $p(x-y)=p(y-x)$. It is clear that in this case the second class particle moves like a symmetric random walk with rate p, unaffected by the presence of the other particles.

If p is not symmetric, the motion of the second class particle is altered by the presence of the other particles. That the second class particle has a drift equal to $v=(1-2\alpha)\mathfrak{m}$ is easy to understand. The second class particle moves for two reasons: when it tries to jump and when a first class particle attempts to jump to the site occupied by the second class particle. In the first case, the jump is performed if the site chosen by the second class particle is empty, which happens with probability $1-\alpha$, since the density of first class particles is α. These jumps entail a drift equal to $(1-\alpha)\mathfrak{m}$. In the second case, the second class particle jumps to the site occupied by the first class particle. Since the density of first class particles is α, this second type of jump causes a drift equal to $-\alpha\mathfrak{m}$, showing that the total drift of the second class particle is $(1-2\alpha)\mathfrak{m}$.

The next result shows that the asymptotic evolution of the second class particle is closely related to the equilibrium fluctuations of the density. Let ν_α^1 be the Bernoulli measure ν_α conditioned to have a particle at the origin. Tag the particle at the origin and let it evolve as a second class particle. Let X_t, $t\ge 0$, be the position at time t of the second class particle. There is an exact relation between the transition probability of the second class particle and the correlation function $R(x,t)$.

Lemma 7.8 *For all $t\ge 0$, $x\in\mathbb{Z}^d$,*

$$\mathbb{P}_{\nu_\alpha^1}[X_t=x]=\chi(\alpha)^{-1}R(x,t).$$

Proof Fix $t\ge 0$ and $x\in\mathbb{Z}^d$. By definition,

$$\mathbb{E}_{\nu_\alpha}\big[\eta_t(x)\eta_0(0)\big]=\int\eta(0)\mathbb{E}_\eta\big[\eta_t(x)\big]\nu_\alpha(d\eta).$$

Denote by $\sigma_0\eta$ the configuration which coincides with η outside the origin and which has no particle at the origin: $(\sigma_0\eta)(x)=\eta(x)$, $x\neq 0$, and $(\sigma_0\eta)(0)=0$. The previous integral can be rewritten as

$$\int \eta(0)\big\{\mathbb{E}_\eta\big[\eta_t(x)\big] - \mathbb{E}_{\sigma_0\eta}\big[\eta_t(x)\big]\big\}\nu_\alpha(d\eta) + \int \eta(0)\mathbb{E}_{\sigma_0\eta}\big[\eta_t(x)\big]\nu_\alpha(d\eta).$$

Replace the measure ν_α by the measure ν_α^1 in the first integral and perform the change of variables $\xi = \eta - \mathfrak{d}_0$, where $\mathfrak{d}_x$ is the configuration with only one particle at x, to rewrite the previous expression as

$$\alpha \int \big\{\mathbb{E}_\eta\big[\eta_t(x)\big] - \mathbb{E}_{\sigma_0\eta}\big[\eta_t(x)\big]\big\}\nu_\alpha^1(d\eta) + \frac{\alpha}{1-\alpha}\int \big[1-\xi(0)\big]\mathbb{E}_\xi\big[\eta_t(x)\big]\nu_\alpha(d\xi).$$

The expression inside braces in the first integral is equal to $\mathbb{P}_\eta[X_t = x]$, while the second integral is equal to $\alpha(1-\alpha)^{-1}\{\alpha - \mathbb{E}_{\nu_\alpha}[\eta_t(x)\eta_0(0)]\}$.

Therefore, up to this point we have showed that

$$\mathbb{E}_{\nu_\alpha}\big[\eta_t(x)\eta_0(0)\big] = \alpha\mathbb{P}_{\nu_\alpha^1}[X_t = x] + \frac{\alpha}{1-\alpha}\big\{\alpha - \mathbb{E}_{\nu_\alpha}\big[\eta_t(x)\eta_0(0)\big]\big\}.$$

It remains to reorder the terms to conclude the proof of the lemma. □

It follows from this result and from (7.29) that the diffusion matrix $D(\alpha)$ is the asymptotic covariance matrix of the rescaled process $\varepsilon[X_{\varepsilon^{-2}t} - vt\varepsilon^{-2}]$.

7.5 Estimates on the Operators $\mathfrak{L}_{\theta,s,2}$, $\mathfrak{L}_{\star,a}$ and $\mathfrak{J}_{\star,\pm}$

In this section, we prove some estimates involving the operators $\mathfrak{L}_{\star,s}$, $\mathfrak{L}_{\star,a}$ and $\mathfrak{J}_{\star,\pm}$. To avoid long sentences when a property holds for some of them, we represent these operators by the symbol $\mathfrak{L}_\star$. Recall that all functions $\mathbf{f} : \mathscr{E}^* \to \mathbb{R}$ which come from a cylinder function f through the transformation $\mathbb{M}$ of the Fourier coefficients $\mathfrak{F}f$ of f are such that $\mathbf{f}(\theta_{-z}A) = \mathbf{f}(A)$ for all z in A. Recall also the definition of the spaces $\mathscr{I}_n$ given in (7.13).

A simple computation shows that the space $\mathscr{I}$ is left invariant by the operators $\mathfrak{L}_\star$: For every $n \geq 1$ and every finitely supported function $\mathbf{f}$ in $\mathscr{I}_n$,

$$\mathfrak{L}_{\star,s}\mathbf{f} \in \mathscr{I}_n, \quad \mathfrak{L}_{\star,a}\mathbf{f} \in \mathscr{I}_n, \quad \mathfrak{J}_{\star,-}\mathbf{f} \in \mathscr{I}_{n-1}, \quad \mathfrak{J}_{\star,+}\mathbf{f} \in \mathscr{I}_{n+1}. \tag{7.30}$$

This claim can be proved in two different ways: either by a direct computation or by reconstructing a cylinder function f from $\mathbf{f}$. More precisely, to prove that $\mathfrak{L}_{\star,s}\mathbf{f}$ belongs to $\mathscr{I}_n$ if $\mathbf{f}$ belongs to $\mathscr{I}_n$, let $\mathfrak{f}$ be given by (7.11) so that $\mathbb{M}\mathfrak{f} = \mathbf{f}$. By (7.12), $\mathbb{M}\mathfrak{S}\mathfrak{f} = \mathfrak{L}_{\star,s}\mathbf{f}$, which proves that $\mathfrak{L}_{\star,s}\mathbf{f}$ belongs to $\mathscr{I}_n$.

We turn now to an elementary identity to illustrate the fact that the space $\mathscr{I}$ carries some special properties. For every $\mathbf{f} : \mathscr{E}_1^* \to \mathbb{R}$,

$$(\mathfrak{J}_{\star,-}\mathbf{f})(\varnothing) = -2\sum_{x\neq 0} a(x)\mathbf{f}\big(\{x\}\big).$$

In particular, $(\mathfrak{J}_{\star,-}\mathbf{f})(\varnothing) = 0$ for all $\mathbf{f}$ in $\mathscr{I}_1$ because in this space $\mathbf{f}(\{x\}) = \mathbf{f}(\{-x\})$ and $a(\cdot)$ is anti-symmetric. In contrast, $(\mathfrak{J}_{\star,+}\mathbf{g})(\{x\}) = 0$ for all functions $\mathbf{g} : \mathscr{E}_0^* \to \mathbb{R}$ so that, for all $\mathbf{f}$ in $\mathscr{I}_1$ and all $\mathbf{g} : \mathscr{E}_0^* \to \mathbb{R}$,

$$\mathfrak{J}_{\star,-}\mathbf{f} = 0, \qquad \mathfrak{J}_{\star,+}\mathbf{g} = 0. \tag{7.31}$$

We turn now to the proof of some estimates involving the operators $\mathfrak{L}_\star$. The first lemma states that for each $n \ge 1$ the operators $\mathfrak{L}_\star$ are bounded in $L^2(\mathscr{E}_n^*) = \{\mathbf{f} : \mathscr{E}_n^* \to bbR\} \cap L^2_\star(\mu_\star)$.

Lemma 7.9 *There exists a finite constant C_0 such that*

$$\|\mathfrak{L}_\star \mathbf{f}\|^2_{\mu_\star} \le C_0 n^2 \|\mathbf{f}\|^2_{\mu_\star}$$

for each $\mathbf{f}$ *in* $L^2(\mathscr{E}_n^*)$, *where the operator* $\mathfrak{L}_\star$ *stands for* $\mathfrak{L}_{\star,s}$, $\mathfrak{L}_{\star,a}$, $\mathfrak{J}_{\star,+}$ *or* $\mathfrak{J}_{\star,-}$.

Proof We prove the estimate for $\mathfrak{J}_{\star,-}$, the other ones being elementary. Fix a function $\mathbf{f} : \mathscr{E}_n^* \to \mathbb{R}$ in $L^2_\star(\mu_\star)$ and keep in mind that $\mathfrak{J}_{\star,-}\mathbf{f}$ maps $\mathscr{E}_{n-1}^*$ in $\mathbb{R}$. By the explicit expression for $\mathfrak{J}_{\star,-}$,

$$\|\mathfrak{J}_{\star,-}\mathbf{f}\|^2_{\mu_\star} = 4 \sum_{A \in \mathscr{E}_{n-1}^*} \Big\{ \sum_{\substack{x,y \notin A \\ x,y \neq 0}} a(y-x)\mathbf{f}\big(A \cup \{y\}\big) \Big\}^2 .$$

Since $\sum_{x \in \mathbb{Z}^d} a(x) = 0$, the previous expression is less than or equal to

$$8 \sum_{A \in \mathscr{E}_{n-1}^*} \Big\{ \sum_{\substack{y \notin A \\ y \neq 0}} a(y)\mathbf{f}\big(A \cup \{y\}\big) \Big\}^2 + 8 \sum_{A \in \mathscr{E}_{n-1}^*} \Big\{ \sum_{\substack{y \notin A, y \neq 0 \\ x \in A}} a(y-x)\mathbf{f}\big(A \cup \{y\}\big) \Big\}^2 .$$

By Schwarz inequality, since $a(\cdot)$ is absolutely bounded and $\sum_{x \in \mathbb{Z}^d} |a(x)|$ is finite, this expression is less than or equal to

$$C_0 n \sum_{A \in \mathscr{E}_{n-1}^*} \sum_{\substack{y \notin A \\ y \neq 0}} \mathbf{f}\big(A \cup \{y\}\big)^2$$

for some finite constant C_0. To sum over A in $\mathscr{E}_{n-1}^*$ and over $y \neq 0$, $y \notin A$ is the same as to sum over B in $\mathscr{E}_n^*$ with a multiplicity factor n because all sets are counted n times. The previous expression is thus equal to

$$C_0 n^2 \sum_{A \in \mathscr{E}_n^*} \mathbf{f}(A)^2,$$

which concludes the proof of the lemma. □

The next result asserts that the operator $\mathfrak{L}_{\theta,s,2}$ is symmetric and non-negative in $L^2_\star(\mu_\star)$. The proof is elementary and left to the reader. The last assertion is Lemma 6.16.

Lemma 7.10 *For every finitely supported function* $\mathbf{f}, \mathbf{g} : \mathscr{E}_n^* \to \mathbb{R}$,

$$\langle \mathbf{f}, \mathfrak{L}_{\theta,s,2}\mathbf{g}\rangle_{\mu_\star} = \langle \mathfrak{L}_{\theta,s,2}\mathbf{f}, \mathbf{g}\rangle_{\mu_\star}.$$

Moreover,

$$\big\langle \mathbf{f}, (-\mathfrak{L}_{\theta,s,2})\mathbf{f}\big\rangle_{\mu_\star} = \frac{1}{2}\sum_{A\in\mathscr{E}_n^*}\sum_{x\notin A} s(x)\big[\mathbf{f}(\theta_{-x}A) - \mathbf{f}(A)\big]^2,$$

$$\text{and}\quad \big\langle \mathbf{f}, (-\mathfrak{L}_{\theta,s,2})\mathbf{f}\big\rangle_{\mu_\star} \le C_0 n \|\mathbf{f}\|_{0,1}^2$$

for some finite constant C_0 *which depends only on the probability* $p(\cdot)$.

It follows from the next statement that the operators $\mathfrak{L}_{\star,a}$ and $\mathfrak{J}_{\star,+} + \mathfrak{J}_{\star,-}$ are anti-symmetric on $\mathscr{I}$.

Lemma 7.11 *For every* $n \ge 1$ *and every finitely supported function* $\mathbf{u}, \mathbf{v} : \mathscr{E}_n^* \to \mathbb{R}$

$$\langle \mathfrak{L}_{\star,a}\mathbf{u}, \mathbf{v}\rangle_{\mu_\star} = -\langle \mathbf{u}, \mathfrak{L}_{\star,a}\mathbf{v}\rangle_{\mu_\star}.$$

For every finitely supported function $\mathbf{f}, \mathbf{g}$ *in* $\mathscr{I}_{n-1}$, $\mathscr{I}_n$ *respectively,*

$$\frac{1}{n+1}\langle \mathfrak{J}_{\star,+}\mathbf{f}, \mathbf{g}\rangle_{\mu_\star} = -\frac{1}{n}\langle \mathbf{f}, \mathfrak{J}_{\star,-}\mathbf{g}\rangle_{\mu_\star}.$$

Proof The first identity is elementary and relies on the fact that $\sum_{x,y\in A} a(y-x) = 0$. Note, however, that both pieces of the operator $\mathfrak{L}_{\star,a}$ are needed.

The proof of the second statement is more demanding. Fix finitely supported functions $\mathbf{f}, \mathbf{g}$ in $\mathscr{I}_{n-1}$, $\mathscr{I}_n$, respectively. By the explicit form of $\mathfrak{J}_{\star,+}$,

$$\begin{aligned}\langle \mathbf{g}, \mathfrak{J}_{\star,+}\mathbf{f}\rangle_{\mu_\star} &= 2\sum_{A\in\mathscr{E}_n^*}\sum_{x,y\in A} a(y-x)\mathbf{g}(A)\mathbf{f}\big(A\setminus\{y\}\big) \\ &\quad + 2\sum_{A\in\mathscr{E}_n^*}\sum_{x\in A} a(x)\mathbf{g}(A)\big\{\mathbf{f}\big(A\setminus\{x\}\big) - \mathbf{f}\big(\theta_{-x}\big[A\setminus\{x\}\big]\big)\big\}.\end{aligned}$$

Since $\theta_{-x}[A\setminus\{x\}] = \theta_{-x}A\setminus\{-x\}$ and since $\mathbf{g}(\theta_{-x}A) = \mathbf{g}(A)$ for x in A because $\mathbf{g}$ belongs to $\mathscr{I}_n$, a change of variables $B = \theta_{-x}A$, $x' = -x$ in the second part of the second term permits to rewrite the second term on the right-hand side as

$$4\sum_{x\neq 0} a(x)\sum_{\substack{A\in\mathscr{E}_n^*\\ A\ni x}} \mathbf{g}(A)\mathbf{f}\big(A\setminus\{x\}\big)$$

because $a(-x) = -a(x)$. We claim that

$$\sum_{x\neq 0} a(x) \sum_{\substack{A\in\mathscr{E}_n^*\\ A\ni x}} \mathbf{g}(A)\mathbf{f}\big(A\setminus\{x\}\big)$$

$$= \frac{1}{n-1} \sum_{x,y\neq 0} a(y-x) \sum_{\substack{A\in\mathscr{E}_n^*\\ A\ni x,y}} \mathbf{g}(A)\mathbf{f}\big(A\setminus\{y\}\big). \tag{7.32}$$

We conclude the proof of the lemma assuming (7.32), whose proof is presented at the end. This identity, which states that $A = (n-1)^{-1}B$ can be written as $A = n^{-1}A + n^{-1}B$. It follows from identity (7.32) written in the latter way and the previous expression for $\langle \mathbf{g}, \mathfrak{J}_{\star,+}\mathbf{f}\rangle_{\mu_\star}$ that

$$\langle \mathbf{g}, \mathfrak{J}_{\star,+}\mathbf{f}\rangle_{\mu_\star} = 2\Big(1+\frac{1}{n}\Big) \sum_{A\in\mathscr{E}_n^*} \sum_{x,y\in A} a(y-x)\mathbf{g}(A)\mathbf{f}\big(A\setminus\{y\}\big)$$

$$+ 2\Big(1+\frac{1}{n}\Big) \sum_{A\in\mathscr{E}_n^*} \sum_{x\in A} a(x)\mathbf{g}(A)\mathbf{f}\big(A\setminus\{x\}\big).$$

The first term of the right-hand side, which can be written as

$$2\Big(1+\frac{1}{n}\Big) \sum_{y\neq 0} \sum_{\substack{A\in\mathscr{E}_n^*\\ A\ni y}} \mathbf{g}(A)\mathbf{f}\big(A\setminus\{y\}\big) \sum_{x\in A} a(y-x),$$

is equal to

$$-2\Big(1+\frac{1}{n}\Big) \sum_{y\neq 0} a(y) \sum_{\substack{A\in\mathscr{E}_n^*\\ A\ni y}} \mathbf{g}(A)\mathbf{f}\big(A\setminus\{y\}\big)$$

$$-2\Big(1+\frac{1}{n}\Big) \sum_{x,y\neq 0} a(y-x) \sum_{\substack{A\in\mathscr{E}_n^*\\ A\ni y,\, A\not\ni x}} \mathbf{g}(A)\mathbf{f}\big(A\setminus\{y\}\big)$$

because $\sum_{x\in A} a(y-x) = -a(y) - \sum_{x\neq 0, x\notin A} a(y-x)$. The first term of this formula cancels with the second one in the last expression for $\langle \mathbf{g}, \mathfrak{J}_{\star,+}\mathbf{f}\rangle_{\mu_\star}$. Therefore,

$$\langle \mathbf{g}, \mathfrak{J}_{\star,+}\mathbf{f}\rangle_{\mu_\star} = -2\Big(1+\frac{1}{n}\Big) \sum_{x,y\neq 0} a(y-x) \sum_{\substack{A\in\mathscr{E}_n^*\\ A\ni y,\, A\not\ni x}} \mathbf{g}(A)\mathbf{f}\big(A\setminus\{y\}\big).$$

To conclude the proof of the lemma, it remains to change variables $B = A\setminus\{y\}$ and to recall the definition of the operator $\mathfrak{J}_{\star,-}$.

We turn now to the proof of (7.32). Since for y in A, $\mathbf{g}(A)=\mathbf{g}(\theta_{-y}A)$ and since $|A|=n$, the left-hand side of (7.32) is equal to

$$\frac{1}{n}\sum_{x\neq 0} a(x) \sum_{\substack{A\in\mathscr{E}_n^*\\ A\ni x}} \sum_{\substack{y\in A\cup\{0\}\\ y\neq x}} \mathbf{g}(\theta_{-y}A)\mathbf{f}\big(A\setminus\{x\}\big)$$

$$=\frac{1}{n}\sum_{\substack{x,y\neq 0\\ y\neq x}} a(x) \sum_{\substack{A\in\mathscr{E}_n^*\\ A\ni x,y}} \mathbf{g}(\theta_{-y}A)\mathbf{f}\big(A\setminus\{x\}\big)+\frac{1}{n}\sum_{x\neq 0} a(x) \sum_{\substack{A\in\mathscr{E}_n^*\\ A\ni x}} \mathbf{g}(A)\mathbf{f}\big(A\setminus\{x\}\big).$$

Notice that the second term on the right-hand side is precisely the original one. Consider the first term. Perform a change of variables $B=\theta_{-y}A$, rewrite $(\theta_y B)\setminus\{x\}$ as $\theta_y(B\setminus\{x-y\})$ and recall that $\mathbf{f}(\theta_y(B\setminus\{x-y\}))=\mathbf{f}(B\setminus\{x-y\})$ if $-y$ belongs to B because $\mathbf{f}$ is in $\mathscr{I}_{n-1}$, to rewrite this expression as

$$\frac{1}{n}\sum_{\substack{x,y\neq 0\\ y\neq x}} a(x) \sum_{\substack{A\in\mathscr{E}_n^*\\ A\ni x-y,-y}} \mathbf{g}(A)\mathbf{f}\big(A\setminus\{x-y\}\big).$$

A change of variables $x'=x-y$, $y'=-y$, shows that this expression is equal to

$$\frac{1}{n}\sum_{x,y\neq 0} a(x-y) \sum_{\substack{A\in\mathscr{E}_n^*\\ A\ni x,y}} \mathbf{g}(A)\mathbf{f}\big(A\setminus\{x\}\big).$$

To prove (7.32), it remains to recollect all previous identities. □

Corollary 7.12 *The operator $\mathfrak{J}_{\star,+}+\mathfrak{J}_{\star,-}$ is anti-symmetric in $\mathscr{I}$:*

$$\big\langle\mathbf{f},(\mathfrak{J}_{\star,+}+\mathfrak{J}_{\star,-})\mathbf{g}\big\rangle_{\star,\mu_\star}=-\big\langle(\mathfrak{J}_{\star,+}+\mathfrak{J}_{\star,-})\mathbf{f},\mathbf{g}\big\rangle_{\star,\mu_\star}$$

for all finitely supported functions $\mathbf{f}$, $\mathbf{g}$ in $\mathscr{I}$. The same statement remains in force if $\mathfrak{J}_{\star,+}+\mathfrak{J}_{\star,-}$ is replaced by $\Pi_n^+(\mathfrak{J}_{\star,+}+\mathfrak{J}_{\star,-})\Pi_n^+$ for every $n\geq 1$, where $\Pi_n^+=\sum_{0\leq j\leq n}\Pi_j$.

The proof of Corollary 7.12 is elementary and left to the reader, one needs only to recall identities (7.31). The next result states that $\mathfrak{L}_{\star,a}$ and $\mathfrak{L}_{\star,s}$ are bounded operators from $L^2(\mathscr{E}_n^*)$ to $\mathfrak{H}_{0,-1}(\mathscr{E}_n^*)$.

Lemma 7.13 *There exists a finite constant C_0, depending only on the probability measure $p(\cdot)$, such that for each $n\geq 1$ and for any finitely supported functions $\mathbf{f}$, $\mathbf{g}:\mathscr{E}_n^*\to\mathbb{R}$,*

$$\langle\mathfrak{L}_\star\mathbf{f},\mathbf{g}\rangle_{\mu_\star}\leq C_0\sqrt{n}\|\mathbf{g}\|_{0,1}\|\mathbf{f}\|_{\mu_\star},$$

where $\mathfrak{L}_\star$ represents the operators $\mathfrak{N}_\star$, $\mathfrak{L}_{\theta,a,2}$, $\mathfrak{L}_{\theta,s,2}$ and $\mathfrak{S}_\star$. In particular,

$$\|\mathfrak{L}_\star\mathbf{f}\|_{0,-1}\leq C_0\sqrt{n}\|\mathbf{f}\|_{\mu_\star}\quad\textit{and}\quad\|\mathbf{f}\|_{0,1}\leq C_0\sqrt{n}\|\mathbf{f}\|_{\mu_\star}.$$

Proof We prove the lemma for $\mathfrak{L}_\star = \mathfrak{L}_{\theta,s,2}$. Fix $n \ge 1$ and two finitely supported functions $\mathbf{f}, \mathbf{g} : \mathscr{E}_n^* \to \mathbb{R}$. Since $\mathfrak{L}_{\theta,s,2}$ is a symmetric operator, $\langle \mathfrak{L}_{\theta,s,2}\mathbf{f}, \mathbf{g}\rangle_{\mu_\star} = \langle \mathbf{f}, \mathfrak{L}_{\theta,s,2}\mathbf{g}\rangle_{\mu_\star}$. By Schwarz inequality and since $\sum_{\notin A} s(x) \le 1$, this scalar product is bounded above by

$$\gamma \sum_{A\in\mathscr{E}_n^*} \mathbf{f}(A)^2 + \frac{1}{\gamma} \sum_{A\in\mathscr{E}_n^*} \sum_{x\notin A} s(x)\big[\mathbf{g}(\theta_{-x}A) - \mathbf{g}(A)\big]^2$$

for every $\gamma > 0$. The second term is a multiple of $\langle \mathbf{g}, \mathfrak{L}_{\theta,s,2}\mathbf{g}\rangle_{\mu_\star}$, which, by Lemma 6.16, is bounded by $C_0 n \|\mathbf{g}\|_{0,1}^2$. To conclude the proof of the lemma we minimize over γ. □

We conclude this section with the estimates needed in order to prove that the solution $\mathbf{u}_\lambda$ of the resolvent equation (7.15) satisfies the bound

$$\sup_{0<\lambda\le 1} \|\mathbf{u}_\lambda\|_{0,-1}^2 < \infty.$$

The first result follows from Lemma 6.17 since $\mathfrak{J}_{\star,+} = \mathfrak{J}_{0,+} + 2\mathfrak{J}_{\theta,a,+}$ and $\mathfrak{J}_{\star,-} = \mathfrak{J}_{0,-}$.

Lemma 7.14 *There exists a finite constant C_0 depending only on the transition probability p such that*

$$\langle \mathfrak{J}_{\star,+}\mathbf{h}, \mathbf{g}\rangle_{\mu_\star}^2 \le C_0 n \|\mathbf{h}\|_{0,1}^2 \|\mathbf{g}\|_{0,1}^2,$$
$$\langle \mathfrak{h}, \mathfrak{J}_{\star,-}\mathbf{g}\rangle_{\mu_\star}^2 \le C_0 n \|\mathbf{h}\|_{0,1}^2 \|\mathbf{g}\|_{0,1}^2$$

for all $n \ge 1$ and all finite supported functions $\mathbf{h}$: $\mathscr{E}_n^* \to \mathbb{R}$, $\mathbf{g}$: $\mathscr{E}_{n+1}^* \to \mathbb{R}$. In particular,

$$\|\mathfrak{J}_{\star,+}\mathbf{h}\|_{0,-1}^2 \le C_0 n \|\mathbf{h}\|_{0,1}^2, \qquad \|\mathfrak{J}_{\star,-}\mathbf{g}\|_{0,-1}^2 \le C_0 n \|\mathbf{g}\|_{0,1}^2.$$

The second result follows from Lemma 6.19 since $\mathfrak{L}_{\star,a} = \mathfrak{N}_\star + \mathfrak{L}_{\theta,a,2}$.

Lemma 7.15 *Fix a finitely supported function $\mathbf{w} : \mathscr{E}^* \to \mathbb{R}$ in $\mathscr{I}$ and such that $\mathbf{w}(\varnothing) = 0$. Let $\mathbf{u}_\lambda$ be the solution of the resolvent equation* (7.15). *There exists a finite constant C_0 depending only on $p(\cdot)$ such that for all $n \ge 1$,*

$$\|\Pi_n \mathfrak{L}_{\star,a}\mathbf{u}_\lambda\|_{0,-1}^2 \le C_0 n \|\mathbf{w}\|_{0,-1}^2 + C_0 n^3 \sum_{j=n-1}^{n+1} \|\Pi_j \mathbf{u}_\lambda\|_{0,1}^2.$$

The proof of the last result of this section is similar to the one of Lemma 6.20.

Lemma 7.16 *There exists a finite constant C_0, depending only on p, such that for every $n \ge 1$ and every function $\mathfrak{f} : \mathscr{E}_n^* \to \mathbb{R}$ in $\mathfrak{H}_{0,1}$,*

$$\|\mathfrak{f}\|_{0,1}^2 \le \big\|\Xi^\star \mathfrak{f}\big\|_{\mathscr{X}_n,\theta,1}^2 \le C_0 n \|\mathfrak{f}\|_{0,1}^2.$$

It follows from this result that

$$\frac{1}{C_0 n} \|\mathfrak{f}\|_{0,-1}^2 \le \left\| \Xi^\star \mathfrak{f} \right\|_{\mathscr{X}_n,\theta,-1}^2 \le \|\mathfrak{f}\|_{0,-1}^2 . \tag{7.33}$$

7.6 Comments and References

The *bulk diffusion* of the density in systems with conservation laws is a classic problem in the study of hydrodynamic limits. We refer to Kipnis and Landim (1999) for complete references on the problem in equilibrium. We look here only at small perturbations (fluctuations) from equilibrium in the exclusion processes. As we have seen, by the attractive properties of the exclusion processes, these fluctuation can be *traced* by the movement of a *second class particle*. In the symmetric case the problem is trivial, as the second class particle moves exactly like a symmetric simple random walk. In the asymmetric case, the problem falls in the class of *non-gradient systems* (cf. Kipnis and Landim, 1999, Chap. 7). In this chapter, we followed (Landim et al., 2004a).

The Fluctuation–Dissipation Theorem The main step in the derivation of equilibrium fluctuations is the so-called fluctuation–dissipation theorem, also referred to as the Boltzmann–Gibbs principle in the gradient case, which allows the replacement of local fields by the density field in the fluctuations regime. Brox and Rost (1984) stated the Boltzmann–Gibbs principle, presented a proof in the case of one-dimensional zero range processes, and deduced the equilibrium fluctuations of this model. Spohn (1986) derived the equilibrium fluctuations of interacting Brownian motions. De Masi et al. (1986) proved the Boltzmann–Gibbs principle for speed change exclusion processes and deduced the equilibrium fluctuations.

Chang (1994) proposed a general method to prove the Boltzmann–Gibbs principle in equilibrium. He deduced from this result the equilibrium fluctuations of gradient interacting particle systems. The method was extended by Lu (1994) to non-gradient models. The fluctuation–dissipation decomposition of the density current for asymmetric exclusion processes in dimension $d \ge 3$ was first proved in Landim and Yau (1997). Sellami (1999) examined the equilibrium fluctuations of generalized exclusion processes and Chang et al. (2001) proved the equilibrium fluctuations for asymmetric exclusion processes in dimension $d \ge 3$. Olla and Tremoulet (2003) considered the equilibrium fluctuations for interacting Ornstein–Uhlenbeck particles and Benois et al. (2003) proved the equilibrium fluctuations for lattices gases.

Hydrodynamic Limit of Non-gradient Models The replacement of the current by the sum of a gradient of type $\eta(0) - \eta(z)$ and a cylinder function in the range of the generator is one of the main steps in the proof of the hydrodynamic limit of a large class of interacting particles systems. It has its origins in the articles (Quastel, 1992; Varadhan, 1994a). Kipnis and Landim (1999) presents the method and discusses the literature up to 1999. Recent developments concern the hydrodynamic behavior of systems in random environment, Faggionato and Martinelli (2003); Quastel (2006), and dynamics with more than one conserved quantity, Sasada (2010).

Diffusive Behavior of the Asymmetric Exclusion Process Similar arguments to the ones presented in this chapter permitted to derive the so-called Navier–Stokes equations for stochastic lattice gases. We refer to Esposito et al. (1994, 1996); Landim et al. (1996, 1997).

Second Class Particle in the Asymmetric Exclusion Process in Dimensions 1 and 2 van Beijeren et al. (1985) conjectured that the diffusion coefficient of the second class particle in the asymmetric exclusion process diverges as $t^{1/3}$ in dimension 1 and $(\log t)^{2/3}$ in dimension 2. Landim et al. (2004b) proved that the diffusion coefficient diverges, in the resolvent sense, at least as fast as $t^{1/4}$ in dimension 1 and as $\sqrt{\log t}$ in dimension 2. The method relies on two ingredients: estimates of the resolvent associated to the generator restricted to low degree functions in terms of the resolvent associated to the full generator, and a comparison between the resolvent associated to the generator of the asymmetric exclusion process restricted to functions of a fixed degree with the resolvent associated to the generator of free particles. Yau (2004) obtained a lower bound of the correct order in dimension 2 for nearest neighbor asymmetric exclusion processes totally asymmetric in the first direction and symmetric in the second one. In dimension 1, a lower bound of the correct order is a simple consequence of the scaling limit for the two-point function established in Ferrari and Spohn (2006) for nearest neighbor asymmetric exclusion processes. Using the resolvent method of Landim et al. (2004b) and the estimates of the $\mathscr{H}_{-1}$ norms obtained in Sethuraman (2003), Quastel and Valko (2007) extend the lower bound of Ferrari and Spohn (2006) to general finite-range asymmetric simple exclusion processes. The convergence of the second class particle properly rescaled is an open problem.

Non-equilibrium Fluctuations Ravishankar (1992) prove the non-equilibrium fluctuations for symmetric simple exclusion processes, a model in which the Boltzmann–Gibbs principle is not needed. Chang and Yau (1992) extended the method introduced in Chang (1994) to prove the Boltzmann–Gibbs principle to the non-equilibrium context in dimension 1. Landim et al. (2008) proved the stationary non-equilibrium fluctuations for boundary driven exclusion processes. Non-equilibrium fluctuations in higher dimension is still an open problem.

References

van Beijeren H, Kutner R, Spohn H (1985) Excess noise for driven diffusive systems. Phys Rev Lett 54:2026–2029

Benois O, Esposito R, Marra R (2003) Equilibrium fluctuations for lattice gases. Ann Inst Henri Poincaré Probab Stat 39(5):743–777

Brox T, Rost H (1984) Equilibrium fluctuations of stochastic particle systems: the role of conserved quantities. Ann Probab 12(3):742–759

Chang CC (1994) Equilibrium fluctuations of gradient reversible particle systems. Probab Theory Relat Fields 100(3):269–283

Chang CC, Yau HT (1992) Fluctuations of one-dimensional Ginzburg-Landau models in nonequilibrium. Commun Math Phys 145(2):209–234

Chang CC, Landim C, Olla S (2001) Equilibrium fluctuations of asymmetric simple exclusion processes in dimension $d \geq 3$. Probab Theory Relat Fields 119(3):381–409

De Masi A, Presutti E, Spohn H, Wick WD (1986) Asymptotic equivalence of fluctuation fields for reversible exclusion processes with speed change. Ann Probab 14(2):409–423

Esposito R, Marra R, Yau HT (1994) Diffusive limit of asymmetric simple exclusion. Rev Math Phys 6(5A):1233–1267. Special issue dedicated to Elliott H Lieb

Esposito R, Marra R, Yau HT (1996) Navier-Stokes equations for stochastic particle systems on the lattice. Commun Math Phys 182(2):395–456

Faggionato A, Martinelli F (2003) Hydrodynamic limit of a disordered lattice gas. Probab Theory Relat Fields 127(4):535–608

Ferrari PL, Spohn H (2006) Scaling limit for the space-time covariance of the stationary totally asymmetric simple exclusion process. Commun Math Phys 265(1):1–44

Kipnis C, Landim C (1999) Scaling limits of interacting particle systems. Grundlehren der Mathematischen Wissenschaften [Fundamental principles of mathematical sciences], vol 320. Springer, Berlin

Landim C, Yau HT (1997) Fluctuation-dissipation equation of asymmetric simple exclusion processes. Probab Theory Relat Fields 108(3):321–356

Landim C, Olla S, Yau HT (1996) Some properties of the diffusion coefficient for asymmetric simple exclusion processes. Ann Probab 24(4):1779–1808

Landim C, Olla S, Yau HT (1997) First-order correction for the hydrodynamic limit of asymmetric simple exclusion processes in dimension $d \geq 3$. Commun Pure Appl Math 50(2):149–203

Landim C, Olla S, Varadhan SRS (2004a) On viscosity and fluctuation-dissipation in exclusion processes. J Stat Phys 115(1–2):323–363

Landim C, Quastel J, Salmhofer M, Yau HT (2004b) Superdiffusivity of asymmetric exclusion process in dimensions one and two. Commun Math Phys 244(3):455–481

Landim C, Milanés A, Olla S (2008) Stationary and nonequilibrium fluctuations in boundary driven exclusion processes. Markov Process Relat Fields 14(2):165–184

Liggett TM (1985) Interacting particle systems. Grundlehren der Mathematischen Wissenschaften [Fundamental principles of mathematical sciences], vol 276. Springer, New York

Lu SL (1994) Equilibrium fluctuations of a one-dimensional nongradient Ginzburg-Landau model. Ann Probab 22(3):1252–1272

Olla S, Tremoulet C (2003) Equilibrium fluctuations for interacting Ornstein-Uhlenbeck particles. Commun Math Phys 233(3):463–491

Quastel J (1992) Diffusion of color in the simple exclusion process. Commun Pure Appl Math 45(6):623–679

Quastel J (2006) Bulk diffusion in a system with site disorder. Ann Probab 34(5):1990–2036

Quastel J, Valko B (2007) $t^{1/3}$ superdiffusivity of finite-range asymmetric exclusion processes on $\mathbb{Z}$. Commun Math Phys 273(2):379–394

Ravishankar K (1992) Fluctuations from the hydrodynamical limit for the symmetric simple exclusion in $\mathbf{Z}^d$. Stoch Process Appl 42(1):31–37

Sasada M (2010) Hydrodynamic limit for two-species exclusion processes. Stoch Process Appl 120(4):494–521

Sellami S (1999) Equilibrium density fluctuations of a one-dimensional non-gradient reversible model: the generalized exclusion process. Markov Process Relat Fields 5(1):21–51

Sethuraman S (2003) An equivalence of H_{-1} norms for the simple exclusion process. Ann Probab 31(1):35–62

Spohn H (1986) Equilibrium fluctuations for interacting Brownian particles. Commun Math Phys 103(1):1–33

Varadhan SRS (1994a) Nonlinear diffusion limit for a system with nearest neighbor interactions II. In: Asymptotic problems in probability theory, stochastic models and diffusions on fractals. Pitman res notes math ser, vol 283. Wiley, New York, pp 75–128

Yau HT (2004) $(\log t)^{2/3}$ law of the two dimensional asymmetric simple exclusion process. Ann Math (2) 159(1):377–405

Chapter 8
Regularity of the Asymptotic Variance

Consider the exclusion process $\{\eta_s : s \geq 0\}$ associated to a finite range probability measure $p(\cdot)$ on $\mathbb{Z}^d$. Fix $0 < \alpha < 1$ and a mean zero cylinder function V in the Hilbert space $\mathscr{H}_{-1}$ introduced in Sect. 5.2. By Theorems 5.4 and 5.14, if the probability measure p has mean zero or if the dimension d is greater than 2, the additive functional $t^{-1/2}\int_0^t V(\eta_s)ds$ converges to a mean zero Gaussian variable with variance $\sigma^2(V) = 2\lim_{\lambda\to 0}\|f_\lambda\|_1^2$, where f_λ is the solution of the resolvent equation (5.8). In this chapter, we examine the dependence of the variance $\sigma^2(V)$ on the density α.

We assumed in Theorems 5.4 and 5.14 that the cylinder function V belongs to the Hilbert space $\mathscr{H}_{-1}$ which clearly depends on the parameter α. To stress this dependence, in this chapter we sometimes denote $\mathscr{H}_{-1}$ by $\mathscr{H}_{-1}(\alpha)$. We also required the cylinder function V to have mean zero with respect to the product measure ν_α. Therefore, to investigate the problem stated in the previous paragraph we have to consider cylinder functions V which have mean zero with respect to all measures ν_α and which belong to all spaces $\mathscr{H}_{-1}(\alpha)$, $0 < \alpha < 1$. According to Lemma 5.12, the function $V(\eta) = \eta(1) - \eta(0)$ is a cylinder function which satisfies these assumptions.

Since we are interested in the dependence of the variance $\sigma^2(V)$ on the density α, it is natural to weaken the previous assumptions to include in our considerations cylinder functions which also depend on the parameter α, as $\eta(0) - \alpha$ which has mean zero with respect to all measures ν_α and, by Lemma 5.12, belongs to $\mathscr{H}_{-1}(\alpha)$ in dimension $d \geq 3$.

Consider, therefore, a family $V(\alpha, \cdot)$, $0 \leq \alpha \leq 1$, of cylinder functions. Assume that, for each $0 \leq \alpha \leq 1$, $V(\alpha, \cdot)$ has mean zero with respect to ν_α and belongs to the Hilbert space $\mathscr{H}_{-1}(\alpha)$. Denote by $\sigma^2(\alpha, V)$ the variance of the central limit theorem for $t^{-1/2}\int_0^t V(\alpha, \eta_s)ds$. The main result of this chapter states that $\sigma^2(\alpha, V)$ is a smooth function of α under further assumptions on $V(\alpha, \eta)$.

T. Komorowski et al., *Fluctuations in Markov Processes*,
Grundlehren der mathematischen Wissenschaften 345,
DOI 10.1007/978-3-642-29880-6_8,

Recall from Sect. 5.4 the definition of the orthonormal basis $\{\Psi_A : A \in \mathcal{E}\}$ of $L^2(\nu_\alpha)$. Let $\mathfrak{V}(\alpha, A)$, $A \in \mathcal{E}$, be the Fourier coefficients of $V(\alpha, \cdot)$ so that

$$V(\alpha, \eta) = \sum_{A \in \mathcal{E}} \mathfrak{V}(\alpha, A)\Psi_A.$$

For each $0 \le \alpha \le 1$, $\mathfrak{V}(\alpha, \cdot) : \mathcal{E} \to \mathbb{R}$ is a finitely supported function. We shall assume that

$$\text{the functions } \mathfrak{V}(\alpha, \cdot),\ 0 \le \alpha \le 1,\ \text{have a common finite support.} \tag{H1}$$

To prove smoothness in α of the asymptotic variance $\sigma^2(\alpha, V)$, we clearly need to assume that

$$\text{for each } A \text{ in } \mathcal{E},\ \mathfrak{V}(\cdot, A) \text{ is an infinitely differentiable function.} \tag{H2}$$

Denote by $\mathfrak{V}^{(j)}(\alpha, \cdot)$ the j-th derivative of $\mathfrak{V}(\alpha, \cdot)$ in α:

$$\mathfrak{V}^{(j)}(\alpha, A) = \frac{d^j}{d\alpha^j}\mathfrak{V}(\alpha, A)\cdot$$

The last assumption requires $\mathfrak{V}(\alpha, \cdot)$, as well as its derivatives, to belong to the Hilbert space $\mathcal{H}_{-1}(\mathfrak{S})$ introduced in Sect. 5.4: For each $j \ge 0, 0 \le \alpha \le 1, \mathfrak{V}^{(j)}(\alpha, \cdot)$ belongs to $\mathcal{H}_{-1}(\mathfrak{S})$ and

$$\sup_{0 \le \alpha \le 1} \left\| \mathfrak{V}^{(j)}(\alpha, \cdot) \right\|_{-1} < \infty. \tag{H3}$$

Theorem 8.1 *Fix a function $V(\alpha, \eta)$ satisfying hypotheses* (H1)–(H3). *If the probability measure p is symmetric, the limiting variance $\sigma^2(\alpha, V)$, as function of α, is of class C^∞ in the interval* $[0, 1]$. *If the probability measure has mean zero or if $d \ge 3$, the limiting variance is of class C^∞ in the open interval* $(0, 1)$.

8.1 The Resolvent Equation

Consider the resolvent equation associated to V: for $\lambda > 0$, denote by f_λ the solution of the resolvent equation:

$$\lambda f_\lambda(\alpha, \cdot) - L f_\lambda(\alpha, \cdot) = V(\alpha, \cdot).$$

We use the dual representation to carry out the estimates. Recall the notation introduced in Sect. 5.4 and let $\mathfrak{f}_\lambda = \mathfrak{F} f_\lambda \in L^2(\mu)$ which clearly depends on α. Applying $\mathfrak{F}$ on both sides of the resolvent equation, we obtain

$$\lambda \mathfrak{f}_\lambda(\alpha, \cdot) - \mathfrak{L}_\alpha \mathfrak{f}_\lambda(\alpha, \cdot) = \mathfrak{V}(\alpha, \cdot). \tag{8.1}$$

Theorems 5.4 and 5.14 give an explicit formula for the variance in terms of the solution of the resolvent equation:

$$(1/2)\sigma^2(V) = \lim_{\lambda\to 0} \|f_\lambda\|_1^2 = \lim_{\lambda\to 0} \langle V, f_\lambda\rangle_{\nu_\alpha} = \lim_{\lambda\to 0} \langle \mathfrak{V}, \mathfrak{f}_\lambda\rangle_\mu$$
$$= \lim_{\lambda\to 0} \sum_{n\geq 1} \sum_{A\in\mathscr{E}_n} \mathfrak{V}(\alpha, A)\mathfrak{f}_\lambda(\alpha, A).$$

The sum starts from $n = 1$ because $\mathfrak{V}(\alpha, \phi) = 0$. Moreover, all but a finite number of terms vanish because the cylinder functions $V(\alpha, \cdot)$ have a finite common support. In view of this identity, to prove Theorem 8.1 in the asymmetric case we just need to show that for every $\varepsilon > 0$ there exists a subsequence $\lambda_k \downarrow 0$ for which the sequence of functions $\sum_{A\in\mathscr{E}_n} \mathfrak{V}(\cdot, A)\mathfrak{f}_{\lambda_k}(\cdot, A)$ converges uniformly in the interval $[\varepsilon, 1-\varepsilon]$ to a smooth function for each $n \geq 1$.

To prove the existence of such a subsequence, it is enough to show that

$$\text{the functions } \mathfrak{f}_\lambda(\cdot, A) \text{ are smooth for each } \lambda > 0,\ A \in \mathscr{E} \tag{8.2}$$

and that

$$\sup_{0<\lambda\leq 1} \sup_{\varepsilon\leq\alpha\leq 1-\varepsilon} \left| \frac{d^j}{d\alpha^j} \sum_{A\in\mathscr{E}_n} \mathfrak{V}(\alpha, A)\mathfrak{f}_\lambda(\alpha, A)\right| < \infty$$

for each $\varepsilon > 0$, $j \geq 0$, $n \geq 1$. To prove this bound, it is enough to show that

$$\sup_{0<\lambda\leq 1} \sup_{\varepsilon\leq\alpha\leq 1-\varepsilon} \left| \sum_{A\in\mathscr{E}_n} \mathfrak{V}^{(j)}(\alpha, A)\mathfrak{f}_\lambda^{(k)}(\alpha, A)\right| < \infty$$

for each $\varepsilon > 0$, $j, k \geq 0$, $n \geq 1$. Here, $\mathfrak{g}^{(j)}$ stands for the j-th derivative of a function $\mathfrak{g} : [0, 1] \times \mathscr{E} \to \mathbb{R}$ with respect to the first coordinate. To prove Theorem 8.1 in the symmetric case we need to extend this estimate up to the boundary of the interval $[0, 1]$.

In the following sections we prove that $\mathfrak{f}_\lambda^{(k)}(\alpha, \cdot)$ belongs to $L^2(\mu)$ for each $\lambda > 0$, $k \geq 1$, $0 \leq \alpha \leq 1$. We assume this property for the moment and we claim that

$$\left| \sum_{A\in\mathscr{E}_n} \mathfrak{V}^{(j)}(\alpha, A)\mathfrak{f}_\lambda^{(k)}(\alpha, A)\right| \leq C_0 \left\| \Pi_n \mathfrak{f}_\lambda^{(k)}(\alpha, \cdot)\right\|_1 \tag{8.3}$$

for some finite constant C_0 depending only on j, the probability measure p and V.

Fix $j, k \geq 0$ and assume first that $nd \leq 2$. Since $V(\alpha, \cdot)$ belongs to $\mathscr{H}_{-1}(\alpha)$, by Lemma 5.12, $\sum_{A\in\mathscr{E}_n} \mathfrak{V}(\alpha, A) = 0$. In particular,

$$\sum_{A\in\mathscr{E}_n} \mathfrak{V}^{(j)}(\alpha, A) = 0.$$

Denote by $S(\mathfrak{V})$ the common support of the cylinder functions $V(\alpha,\cdot)$: $S(\mathfrak{V}) = \{A \in \mathscr{E} : \mathfrak{V}(\alpha, A) \neq 0 \text{ for some } \alpha\}$. By the previous identity,

$$\begin{aligned}\sum_{A\in\mathscr{E}_n} & \mathfrak{V}^{(j)}(\alpha, A)\mathfrak{f}_\lambda^{(k)}(\alpha, A) \\ &= \frac{1}{M}\sum_{A,B\in\mathscr{E}_n\cap S(\mathfrak{V})} \mathfrak{f}_\lambda^{(k)}(\alpha, A)\big\{\mathfrak{V}^{(j)}(\alpha, A) - \mathfrak{V}^{(j)}(\alpha, B)\big\} \\ &= \frac{1}{M}\sum_{A,B\in\mathscr{E}_n\cap S(\mathfrak{V})} \mathfrak{V}^{(j)}(\alpha, A)\big\{\mathfrak{f}_\lambda^{(k)}(\alpha, A) - \mathfrak{f}_\lambda^{(k)}(\alpha, B)\big\},\end{aligned}$$

where M stands for the cardinality of $\mathscr{E}_n \cap S(\mathfrak{V})$. By assumptions (H1), (H2) and by Schwarz inequality, the square of this expression is bounded by

$$C_0 \sum_{A,B\in\mathscr{E}_n\cap S(\mathfrak{V})} \big\{\mathfrak{f}_\lambda^{(k)}(\alpha, A) - \mathfrak{f}_\lambda^{(k)}(\alpha, B)\big\}^2.$$

In this formula and below, C_0 represents a finite constant which depends only on j, V and p, and which may change from line to line. Since $S(\mathfrak{V})$ is a finite set, constructing paths from A to B and applying Schwarz inequality, we obtain that the previous expression is less than or equal to

$$C_0 \sum_{A\in\mathscr{E}_n}\sum_{B\sim A} \big\{\mathfrak{f}_\lambda^{(k)}(\alpha, B) - \mathfrak{f}_\lambda^{(k)}(\alpha, A)\big\}^2.$$

The second sum is carried over all sets B which may be obtained from A by moving a single particle: $A \sim B$ if $B = A_{x,y}$ for some pair $\{x, y\}$ such that $x \in A$, $y \notin A$, $s(y-x) > 0$. It is easy to see from the explicit formula (5.24) for the $\mathscr{H}_1$ norm, that the previous expression is bounded by $C_0\|\varPi_n \mathfrak{f}_\lambda^{(k)}(\alpha,\cdot)\|_1^2$. This proves Claim (8.3) in the case $nd \le 2$.

Assume now that $nd > 2$. In this case, by (5.45), the evolution of n symmetric exclusion particles in $\mathbb{Z}^d$ is transient. In particular, by Proposition 5.23 there exists a finite constant $C(p)$, depending only on the probability measure $p(\cdot)$, such that for any finitely supported function $\mathfrak{g} : \mathscr{E}_n \to \mathbb{R}$ and any set A in $\mathscr{E}_n$,

$$\mathfrak{g}(A)^2 \le C(p) \sum_{x,y\in\mathbb{Z}^d}\sum_{A\in\mathscr{E}_n} s(y-x)\big\{\mathfrak{g}(A_{x,y}) - \mathfrak{g}(A)\big\}^2 = 4C(p)\big\|\varPi_n \mathfrak{g}(\alpha,\cdot)\big\|_1^2.$$

This estimate can be extended to any function $\mathfrak{g}$ in $L^2(\mu)$. Therefore, since $\mathfrak{V}(\cdot,\cdot)$ is a finitely supported function, smooth in the first coordinate, and since $\mathfrak{f}_\lambda^{(k)}(\alpha,\cdot)$ belongs to $L^2(\mu)$ for each $k \ge 0$, $\lambda > 0$, $0 \le \alpha \le 1$,

$$\Big|\sum_{A\in\mathscr{E}_n} \mathfrak{V}^{(j)}(\alpha, A)\mathfrak{f}_\lambda^{(k)}(\alpha, A)\Big| \le C(j, V) \sum_{A\in\mathscr{E}_n\cap S(\mathfrak{V})} \big|\mathfrak{f}_\lambda^{(k)}(\alpha, A)\big|$$

$$\leq C(j, V, p)\big\|\Pi_n \mathfrak{f}_\lambda^{(k)}(\alpha, \cdot)\big\|_1$$

because $S(\mathfrak{V})$ is a finite set. This proves Claim (8.3) in the case $nd > 2$.

In view of (8.3), to prove Theorem 8.1 in the asymmetric case we need to show that $\{\mathfrak{f}_\lambda(\cdot, A) : \lambda > 0, A \in \mathscr{E}\}$ are smooth functions and that

$$\sup_{\varepsilon \leq \alpha \leq 1-\varepsilon} \sup_{0<\lambda\leq 1} \big\|\mathfrak{f}_\lambda^{(j)}(\alpha, \cdot)\big\|_1 < \infty \tag{8.4}$$

for all $\varepsilon > 0$, $j \geq 0$. In Sect. 8.2, we prove this bound with $\varepsilon = 0$ for symmetric simple exclusion processes. In Sect. 8.3 we consider the mean zero case and in Sect. 8.4 asymmetric processes in dimension $d \geq 3$.

8.2 The Symmetric Case

Recall the definition of the operator $\mathfrak{L}_\alpha$ introduced in (5.16). In the symmetric case, the operator $\mathfrak{L}_\alpha = \mathfrak{S}$ does not depend on the parameter α and the proof of (8.2), (8.4) with $\varepsilon = 0$ is elementary.

We say that a function $\mathfrak{h} : [0, 1] \to L^2(\mu)$ is differentiable at α if $\gamma^{-1}[\mathfrak{h}(\alpha+\gamma) - \mathfrak{h}(\alpha)]$ converges, as $\gamma \to 0$, strongly in $L^2(\mu)$ to some function denoted by $\mathfrak{h}^{(1)}(\alpha)$.

By assumption, for each $A \in \mathscr{E}$, $\mathfrak{V}(\cdot, A)$ is infinitely differentiable, and the cylinder functions $\mathfrak{V}(\alpha, \cdot)$, $0 \leq \alpha \leq 1$, have a common finite support. In particular, $\alpha \mapsto \mathfrak{V}(\alpha, \cdot)$ is infinitely differentiable, $\mathfrak{V}^{(j)}(\alpha, \cdot)$ belongs to $L^2(\mu)$ for each $j \geq 0$ and we may examine the resolvent equation for $\mathfrak{V}^{(j)}(\alpha, \cdot)$. For $\lambda > 0$, $j \geq 1$, $0 \leq \alpha \leq 1$, denote by $\mathfrak{g}_{\lambda,j}(\alpha)$ the solution of the resolvent equation

$$\lambda \mathfrak{g}_{\lambda,j}(\alpha) - \mathfrak{S}\mathfrak{g}_{\lambda,j}(\alpha) = \mathfrak{V}^{(j)}(\alpha). \tag{8.5}$$

Recall that $\mathfrak{f}_\lambda(\alpha)$ stands for the solution of (8.1). We claim that $\alpha \mapsto \mathfrak{f}_\lambda(\alpha, \cdot)$ is infinitely differentiable and that $\mathfrak{f}_\lambda^{(j)}(\alpha) = \mathfrak{g}_{\lambda,j}(\alpha)$ for all $j \geq 1$. In particular, $\mathfrak{f}_\lambda^{(j)}(\alpha)$ belongs to $L^2(\mu)$ for $j \geq 1$, $\lambda > 0$, $0 \leq \alpha \leq 1$, as claimed in the previous section.

Indeed, let us first prove the claim for $j = 1$. For $h \neq 0$, let $\mathfrak{v}_h = h^{-1}\{\mathfrak{f}_\lambda(\alpha+h) - \mathfrak{f}_\lambda(\alpha)\}$, $\mathfrak{V}_h = h^{-1}\{\mathfrak{V}(\alpha+h) - \mathfrak{V}(\alpha)\}$. By assumption, $\mathfrak{V}_h$ converges to $\mathfrak{V}^{(1)}(\alpha)$ in $L^2(\mu)$ as $h \to 0$. Since

$$\lambda\big\{\mathfrak{v}_h - \mathfrak{g}_{\lambda,1}(\alpha)\big\} - \mathfrak{S}\big\{\mathfrak{v}_h - \mathfrak{g}_{\lambda,1}(\alpha)\big\} = \big\{\mathfrak{V}_h - \mathfrak{V}^{(1)}(\alpha)\big\},$$

taking the scalar product on both sides with respect to $\mathfrak{v}_h - \mathfrak{g}_{\lambda,1}(\alpha)$, we obtain that $\mathfrak{v}_h$ converges to $\mathfrak{g}_{\lambda,1}(\alpha)$ in $L^2(\mu)$ as $h \to 0$. Therefore, $\mathfrak{f}_\lambda(\alpha)$ is differentiable and its derivative is equal to $\mathfrak{g}_{\lambda,1}(\alpha)$. An induction argument permits to extend the statement to $j \geq 2$.

We have just shown that $\mathfrak{f}_\lambda(\alpha)$ is a smooth function of α. Since $\mathfrak{f}_\lambda^{(j)}(\alpha)$ is the solution of the resolvent equation (8.5), taking the scalar product on both sides with respect to $\mathfrak{f}_\lambda^{(j)}$ and applying Schwarz inequality, we obtain that

$$\big\|\mathfrak{f}_\lambda^{(j)}(\alpha)\big\|_1 \leq \big\|\mathfrak{V}^{(j)}(\alpha)\big\|_{-1}$$

for all $j \geq 1$, $\lambda > 0$, $0 \leq \alpha \leq 1$. Hence (8.4) with $\varepsilon = 0$ follows from assumption (H3) and the proof of Theorem 8.1 is complete in the symmetric case.

8.3 The Mean Zero Case

In this section, we assume that the probability p has mean zero: $\sum_x xp(x) = 0$ and that it does not charge the origin. We have seen in Sect. 5.3 that such probabilities can be written as convex combinations of cyclic probability measures and that the generators of the simple exclusion process associated to such probability measures satisfy a sector condition.

Recall the definition of the operator $\mathfrak{L}_\alpha$ introduced in (5.16). Since the coefficients of $\mathfrak{L}_\alpha$ are not smooth at the boundary of $[0, 1]$, we reparametrize the family of equations by $\alpha = \sin^2 t$, $t \in [0, \pi/2]$, to get

$$\mathfrak{L}(t) = \mathfrak{S} + \cos(2t)\mathfrak{N} + \sin t \cos t \mathfrak{J},$$

where $\mathfrak{J} = \mathfrak{J}_+ + \mathfrak{J}_-$. Consider the resolvent equation

$$\lambda \mathfrak{g}_\lambda(t) - \mathfrak{L}(t)\mathfrak{g}_\lambda(t) = \mathfrak{W}(t), \tag{8.6}$$

where $\mathfrak{W}(t) = \mathfrak{V}(\alpha(t))$. Of course, since $\mathfrak{L}(t) = \mathfrak{L}_{\alpha(t)}$, $\mathfrak{g}_\lambda(t) = \mathfrak{f}_\lambda(\alpha(t))$.

To prove that the functions $\{\mathfrak{f}_\lambda(\cdot, A) : \lambda > 0, A \in \mathscr{E}\}$ are smooth in any compact interval of $(0, 1)$, it is enough to show a similar statement for the functions $\{\mathfrak{g}_\lambda(\cdot, A) : \lambda > 0, A \in \mathscr{E}\}$. On the other hand, to prove (8.4) for some $\varepsilon > 0$, it suffices to show that

$$\sup_{0 \leq t \leq \pi/2} \sup_{0 < \lambda \leq 1} \left\| \mathfrak{g}_\lambda^{(j)}(t, \cdot) \right\|_1 < \infty \tag{8.7}$$

for all $j \geq 0$.

We now begin the proof that $\mathfrak{g}_\lambda$ is a sequence of smooth functions with derivatives in $\mathscr{H}_1$. Fix $\mathfrak{f}$ in $L^2(\mu)$. Denote by $\mathfrak{h}_\lambda(t)$, $0 \leq t \leq \pi/2$, the solution of the resolvent equation

$$\lambda \mathfrak{h}_\lambda(t) - \mathfrak{L}(t)\mathfrak{h}_\lambda(t) = \mathfrak{f}.$$

It is by now well known that

$$\sup_{0 < \lambda \leq 1} \left\{ \lambda \left\| \mathfrak{h}_\lambda(t) \right\|_\mu^2 + \left\| \mathfrak{h}_\lambda(t) \right\|_1^2 \right\} \leq C_0 \|\mathfrak{f}\|_{-1}^2 \tag{8.8}$$

for a finite constant C_0 independent of t.

Lemma 8.2 *Suppose that $\mathfrak{U} : [0, 1] \to L^2(\mu)$ is a differentiable function such that*

$$\sup_{0 \leq t \leq \pi/2} \left\| \mathfrak{U}(t) \right\|_{-1} < \infty, \qquad \sup_{0 \leq t \leq \pi/2} \left\| \mathfrak{U}'(t) \right\|_{-1} < \infty.$$

Let $\mathfrak{h}_\lambda(t)$ be the solution of the resolvent equation

$$\lambda \mathfrak{h}_\lambda(t) - \mathfrak{L}(t)\mathfrak{h}_\lambda(t) = \mathfrak{U}(t). \tag{8.9}$$

Then, for each $0 < \lambda \leq 1$, $0 \leq t \leq \pi/2$, $\mathfrak{h}_\lambda(t)$ is differentiable and its derivative $\mathfrak{h}'_\lambda(t)$ is the solution of

$$\lambda \mathfrak{h}'_\lambda(t) - \mathfrak{L}(t)\mathfrak{h}'_\lambda(t) = \mathfrak{U}'(t) + \mathfrak{L}'(t)\mathfrak{h}_\lambda(t), \tag{8.10}$$

where

$$\mathfrak{L}'(t) = -2\sin(2t)\mathfrak{N} + \cos(2t)\mathfrak{J}.$$

Proof Fix $0 < \lambda \leq 1$, $0 \leq t \leq \pi/2$ and denote by $\mathfrak{v}_\lambda(t)$ the solution of equation (8.10). We have that

$$\lambda \big\| \mathfrak{v}_\lambda(t) \big\|_\mu^2 + \big\| \mathfrak{v}_\lambda(t) \big\|_1^2 \leq \big\| \mathfrak{U}'(t) + \mathfrak{L}'(t)\mathfrak{h}_\lambda(t) \big\|_{-1}^2.$$

By Lemma 8.3 and (8.8), the right-hand side is bounded above by $C_0\{\|\mathfrak{U}'(t)\|_{-1}^2 + \|\mathfrak{U}(t)\|_{-1}^2\}$ for some finite constant C_0 independent of t and λ. Therefore,

$$\lambda \big\| \mathfrak{v}_\lambda(t) \big\|_\mu^2 + \big\| \mathfrak{v}_\lambda(t) \big\|_1^2 \leq C_0\big\{ \big\| \mathfrak{U}'(t) \big\|_{-1}^2 + \big\| \mathfrak{U}(t) \big\|_{-1}^2 \big\}. \tag{8.11}$$

For $h \neq 0$, let $a_h = h^{-1}\{\cos 2(t+h) - \cos 2t\}$, $b_h = h^{-1}\{\sin(t+h)\cos(t+h) - \sin t \cos t\}$, $\mathfrak{v}_h = h^{-1}\{\mathfrak{h}_\lambda(t+h) - \mathfrak{h}_\lambda(t)\}$, $\mathfrak{U}_h = h^{-1}\{\mathfrak{U}(t+h) - \mathfrak{U}(t)\}$ so that

$$\begin{aligned} &\lambda\big\{\mathfrak{v}_h - \mathfrak{v}_\lambda(t)\big\} - \mathfrak{L}(t)\big\{\mathfrak{v}_h - \mathfrak{v}_\lambda(t)\big\} \\ &\quad = \big\{\mathfrak{U}_h - \mathfrak{U}'(t)\big\} + h a_h \mathfrak{N}\big\{\mathfrak{v}_h - \mathfrak{v}_\lambda(t)\big\} \\ &\qquad + h b_h \mathfrak{J}\big\{\mathfrak{v}_h - \mathfrak{v}_\lambda(t)\big\} + \varepsilon_1(h)\mathfrak{N}\mathfrak{v}_\lambda(t) + \varepsilon_2(h)\mathfrak{J}\mathfrak{v}_\lambda(t), \end{aligned}$$

where $\varepsilon_1(h) = a_h - a_0(t) + h a_h$, $\varepsilon_2(h) = b_h - b_0(t) + h b_h$, $a_0(t) = -2\sin(2t)$, $b_0(t) = \cos(2t)$. Take the scalar product on both sides of this equation with respect to $\mathfrak{v}_h - \mathfrak{v}_\lambda(t)$ and recall that $\mathfrak{N}$ and $\mathfrak{J}$ are asymmetric operators to get that

$$\begin{aligned} &\lambda \big\| \mathfrak{v}_h - \mathfrak{v}_\lambda(t) \big\|_\mu^2 + \big\| \mathfrak{v}_h - \mathfrak{v}_\lambda(t) \big\|_1^2 \\ &\quad = \big\langle \mathfrak{U}_h - \mathfrak{U}'(t), \mathfrak{v}_h - \mathfrak{v}_\lambda(t) \big\rangle_\mu \\ &\qquad + \varepsilon_1(h)\big\langle \mathfrak{N}\mathfrak{v}_\lambda(t), \mathfrak{v}_h - \mathfrak{v}_\lambda(t) \big\rangle_\mu + \varepsilon_2(h)\big\langle \mathfrak{J}\mathfrak{v}_\lambda(t), \mathfrak{v}_h - \mathfrak{v}_\lambda(t) \big\rangle_\mu. \end{aligned}$$

By Schwarz inequality and Lemma 8.3,

$$\lambda \big\| \mathfrak{v}_h - \mathfrak{v}_\lambda(t) \big\|_\mu^2 + \big\| \mathfrak{v}_h - \mathfrak{v}_\lambda(t) \big\|_1^2 \leq \frac{1}{\lambda} \big\| \mathfrak{U}_h - \mathfrak{U}'(t) \big\|_\mu^2 + \varepsilon(h) \big\| \mathfrak{v}_\lambda(t) \big\|_1^2,$$

where $\varepsilon(h)$ vanishes as $h \to 0$. It remains to recall the a-priori estimate (8.11) to conclude the proof of the lemma. □

Proof of Theorem 8.1 in the mean zero case By assumption (H3), for every $j \ge 0$,

$$\sup_{0\le t\le \pi/2} \left\| \mathfrak{W}^{(j)}(t) \right\|_{-1} < \infty. \tag{8.12}$$

Hence, by (8.8),

$$\sup_{0\le t\le \pi/2} \sup_{0<\lambda\le 1} \left\| \mathfrak{g}_\lambda(t) \right\|_1 < \infty. \tag{8.13}$$

By Lemma 8.2, $\mathfrak{g}_\lambda(t)$ is differentiable and its derivative $\mathfrak{g}_\lambda^{(1)}(t)$ satisfies the resolvent equation

$$\lambda \mathfrak{g}_\lambda^{(1)}(t) - \mathfrak{L}(t)\mathfrak{g}_\lambda^{(1)}(t) = \mathfrak{W}^{(1)}(t) + \mathfrak{L}^{(1)}(t)\mathfrak{g}_\lambda(t),$$

where $\mathfrak{L}^{(j)}(t)$ stands for the j-th derivative of $\mathfrak{L}(t)$. Let $\mathfrak{W}_1(t,\lambda) = \mathfrak{W}^{(1)}(t) + \mathfrak{L}^{(1)}(t)\mathfrak{g}_\lambda(t)$. By (8.12), (8.13) and Lemma 8.3,

$$\sup_{0\le t\le \pi/2} \sup_{0<\lambda\le 1} \left\| \mathfrak{W}_1(t,\lambda) \right\|_{-1} < \infty.$$

To iterate the argument, we just need to prove by induction the existence of constants $\{a_{n,i}, n \ge 1, 0 \le i < n\}$ such that

$$\lambda \mathfrak{g}_\lambda^{(j)}(t) - \mathfrak{L}(t)\mathfrak{g}_\lambda^{(j)}(t) = \mathfrak{W}^{(j)}(t) + \sum_{i=0}^{j-1} a_{j,i}\mathfrak{L}^{(j-i)}(t)\mathfrak{g}_\lambda^{(i)}(t).$$

This is elementary and left to the reader. This procedure yields (8.7), which concludes the proof of the theorem. □

We conclude this section with the proof of a sector condition for the asymmetric operators $\mathfrak{N}$ and $\mathfrak{J}$.

Lemma 8.3 *There exists a finite constant C_0 depending only on the probability measure p such that for all finitely supported functions $\mathfrak{f}, \mathfrak{g} : \mathcal{E} \to \mathbb{R}$,*

$$\langle \mathfrak{N}\mathfrak{f}, \mathfrak{g}\rangle_\mu \le C_0 \|\mathfrak{f}\|_1 \|\mathfrak{g}\|_1, \qquad \langle \mathfrak{J}\mathfrak{f}, \mathfrak{g}\rangle_\mu \le C_0 \|\mathfrak{f}\|_1 \|\mathfrak{g}\|_1.$$

In particular, there exists a finite constant C_0 such that $\|\mathfrak{N}\mathfrak{f}\|_{-1} \le C_0\|\mathfrak{f}\|_1$ and $\|\mathfrak{J}\mathfrak{f}\|_{-1} \le C_0\|\mathfrak{f}\|_1$ for all finitely supported functions $\mathfrak{f} : \mathcal{E} \to \mathbb{R}$.

Proof In view of Lemmas 5.6 and 5.7, it is enough to prove the lemma for operators associated to cyclic probability measures. Fix, therefore, an irreducible cycle $C = (y_0, y_1, \dots, y_{\ell-1}, y_\ell = y_0)$ and assume that the probability measure p appearing in the definition (5.1) of the generator of the exclusion process is the cyclic probability measure p_C introduced in (3.18).

Recall the explicit formula of the operator $\mathfrak{N}$ given in (5.17). Since $2a(x) = p_C(x) - p_C(-x)$ and since the probability measure p_C^* defined by $p_C^*(x) = p_C(-x)$

is also a cyclic probability measure, it is enough to prove the sector condition for the operator $\mathfrak{N}$ with a replaced by p_C.

The proof is similar to the one of Lemma 5.8. Since $A_{x,y} = A$ if $x, y \in A$, in the definition of $\mathfrak{N}$, we may replace the sum carried over $y \notin A$ by a sum carried over $y \in \mathbb{Z}^d$. In this case, in view of (3.18) and after the replacement of a by p_C, the operator $\mathfrak{N}$ becomes

$$(\mathfrak{N}\mathfrak{f})(A) = \frac{1}{\ell} \sum_{i=0}^{\ell-1} \sum_{x \in A} \big[\mathfrak{f}(A_{x,x+y_{i+1}-y_i}) - \mathfrak{f}(A)\big].$$

Let $T_{x,i} : \mathcal{E} \to \mathcal{E}$, $0 \le i \le \ell - 1$, $x \in \mathbb{Z}^d$, be the operator $T_{x,i} A = A_{x+y_i, x+y_{i+1}}$ and let $T_x = T_{x,\ell-1} \circ \cdots \circ T_{x,0}$. Note that $T_x^{\ell-1} A = A$ if T_x^0 is the identity and $T_x^{j+1} = T_x \circ T_x^j$, $j \ge 0$. With this notation, the Dirichlet form can be written as

$$\big\langle(-\mathfrak{S}\mathfrak{f}), \mathfrak{f}\big\rangle_\mu = \frac{1}{8} \sum_{i=0}^{\ell-1} \sum_{x \in \mathbb{Z}^d} \sum_{A \in \mathcal{E}} \big[\mathfrak{f}(T_{x,i} A) - \mathfrak{f}(A)\big]^2$$

plus a similar term in which the cycle $(y_0, y_1, \dots, y_{n-1}, y_n = y_0)$ is replaced by its adjoint $(y_0, y_{n-1}, \dots, y_1, y_0)$.

Fix finitely supported functions $\mathfrak{f}, \mathfrak{g} : \mathcal{E} \to \mathbb{R}$. In view of the explicit formula for $\mathfrak{N}$ presented above and a change of variables,

$$\langle \mathfrak{N}\mathfrak{f}, \mathfrak{g}\rangle_\mu = \frac{1}{\ell} \sum_{i=0}^{\ell-1} \sum_{x \in \mathbb{Z}^d} \sum_{A \in \mathcal{E}} \mathbf{1}\{x + y_i \in A\}\big[\mathfrak{f}(T_{x,i} A) - \mathfrak{f}(A)\big]\mathfrak{g}(A).$$

Perform a change of variables to rewrite the previous sum as

$$\frac{1}{\ell(\ell-1)} \sum_{j=0}^{\ell-2} \sum_{i=0}^{\ell-1} \sum_{x \in \mathbb{Z}^d} \sum_{A \in \mathcal{E}} \mathbf{1}\{x + y_0 \in A\} \mathfrak{g}\big(T_{x,i-1} \circ \cdots \circ T_{x,0} \circ T_x^j A\big)$$
$$\times \big[\mathfrak{f}\big(T_{x,i} \circ \cdots \circ T_{x,0} \circ T_x^j A\big) - \mathfrak{f}\big(T_{x,i-1} \circ \cdots \circ T_{x,0} \circ T_x^j A\big)\big].$$

Since $T_x^{\ell-1}$ is the identity, if we had $\mathfrak{g}(A)$ instead of $\mathfrak{g}(T_{x,i-1} \circ \cdots \circ T_{x,0} \circ T_x^j A)$ the expression would vanish due to the presence of the telescopic sum in the variables i, j, in which the first and the last term coincide. Therefore, we may rewrite the previous sum as

$$\frac{1}{\ell(\ell-1)} \sum_{j=0}^{\ell-2} \sum_{i=0}^{\ell-1} \sum_{x \in \mathbb{Z}^d} \sum_{A \in \mathcal{E}} \mathbf{1}\{x + y_0 \in A\} \big[\mathfrak{g}\big(T_{x,i-1} \circ \cdots \circ T_{x,0} \circ T_x^j A\big) - \mathfrak{g}(A)\big]$$
$$\times \big[\mathfrak{f}\big(T_{x,i} \circ \cdots \circ T_{x,0} \circ T_x^j A\big) - \mathfrak{f}\big(T_{x,i-1} \circ \cdots \circ T_{x,0} \circ T_x^j A\big)\big].$$

To conclude the proof it remains to proceed as in the end of the proof of Lemma 5.8. This argument shows that the operator $\mathfrak{N}$ satisfies the sector condition claimed in the statement of the lemma. Since the symmetric part $\mathfrak{S}$ of the generator $\mathfrak{L}(t)$ clearly satisfies a similar sector condition, the sector condition for the operator $\mathfrak{J}$ follows from the one of $\mathfrak{N}$, the one of $\mathfrak{L}(t)$, proved in Lemma 5.9, and the isomorphism $\mathfrak{F}$ introduced in the beginning of Sect. 5.4. □

8.4 The Asymmetric Case in $d \geq 3$

In this section, we examine the asymmetric case in dimension $d \geq 3$. The proof relies on the graded sector estimates obtained in Lemmas 5.15 and 5.20 and on Lemma 2.21 which provides bounds for the triple norms of the solution of a resolvent equation.

Recall the definition of the triple norms $|||\cdot|||_{k,a}$, $k \geq 0$, $a = -1, 1, \mu$, introduced in (2.48) and denote by $\mathscr{H}_{k,a}$ the Hilbert space generated by finitely supported functions $\mathfrak{f} : \mathscr{E} \to \mathbb{R}$ endowed with the norm $|||\cdot|||_{k,a}$. The next result follows from Lemma 5.15 and Lemma 2.21.

Lemma 8.4 *Fix $k \geq 0$ and $\mathfrak{f}$ in $L^2(\mu) \cap \mathscr{H}_{k,-1}$. Denote by $\mathfrak{h}_\lambda(t)$, $0 \leq t \leq \pi/2$, $\lambda > 0$, the solution of the resolvent equation*

$$\lambda \mathfrak{h}_\lambda(t) - \mathfrak{L}(t)\mathfrak{h}_\lambda(t) = \mathfrak{f}. \tag{8.14}$$

Then,

$$\sup_{0<\lambda\leq 1} \left\{\lambda \left|\left|\left| \mathfrak{h}_\lambda(t) \right|\right|\right|_{k,\mu}^2 + \left|\left|\left| \mathfrak{h}_\lambda(t) \right|\right|\right|_{k,1}^2 \right\} \leq C(k) |||\mathfrak{f}|||_{k,-1}^2$$

for a finite constant $C(k)$ independent of t.

The proof of the next result is similar to the one of Lemma 8.2 but involves the triple norms. We give the full proof instead of referring to one of that lemma.

Lemma 8.5 *Suppose that $\mathfrak{U} : [0, 1] \to L^2(\mu)$ is a differentiable function such that*

$$\sup_{0\leq t\leq \pi/2} \left|\left|\left| \mathfrak{U}(t) \right|\right|\right|_{3,-1} < \infty, \qquad \sup_{0\leq t\leq \pi/2} \left|\left|\left| \mathfrak{U}'(t) \right|\right|\right|_{1,-1} < \infty.$$

Let $\mathfrak{h}_\lambda(t)$ be the solution of the resolvent equation

$$\lambda \mathfrak{h}_\lambda(t) - \mathfrak{L}(t)\mathfrak{h}_\lambda(t) = \mathfrak{U}(t). \tag{8.15}$$

Then, for each $0 < \lambda \leq 1$, $0 \leq t \leq \pi/2$, $\mathfrak{h}_\lambda(t)$ is differentiable and its derivative $\mathfrak{h}'_\lambda(t)$ is the solution of

$$\lambda \mathfrak{h}'_\lambda(t) - \mathfrak{L}(t)\mathfrak{h}'_\lambda(t) = \mathfrak{U}'(t) + \mathfrak{L}'(t)\mathfrak{h}_\lambda(t). \tag{8.16}$$

Proof Fix $0 < \lambda \leq 1$, $0 \leq t \leq \pi/2$. Denote by $\mathfrak{v}_\lambda(t)$ the solution of (8.16). By Lemma 2.21,

$$\lambda \left\| \mathfrak{v}_\lambda(t) \right\|_{1,\mu}^2 + \left\| \mathfrak{v}_\lambda(t) \right\|_{1,1}^2 \leq C_0 \left\| \mathfrak{U}'(t) + \mathfrak{L}'(t)\mathfrak{h}_\lambda(t) \right\|_{1,-1}^2$$

for some finite constant C_0, which may change from line to line. By Lemma 8.6 below, the right-hand side is bounded above by

$$C_0 \left\{ \left\| \mathfrak{U}'(t) \right\|_{1,-1}^2 + \left\| \mathfrak{U}(t) \right\|_{2,-1}^2 + \left\| h_\lambda(t) \right\|_{3,1}^2 \right\}.$$

By Lemma 8.4, $\|h_\lambda(t)\|_{3,1}$ is bounded above by $\|\mathfrak{U}(t)\|_{3,-1}$. Hence,

$$\lambda \left\| \mathfrak{v}_\lambda(t) \right\|_{1,\mu}^2 + \left\| \mathfrak{v}_\lambda(t) \right\|_{1,1}^2 \leq C_0 \left\{ \left\| \mathfrak{U}'(t) \right\|_{1,-1}^2 + \left\| \mathfrak{U}(t) \right\|_{3,-1}^2 \right\}. \tag{8.17}$$

For $h \neq 0$, let $a_h = h^{-1}\{\cos 2(t+h) - \cos 2t\}$, $b_h = h^{-1}\{\sin(t+h)\cos(t+h) - \sin t \cos t\}$, $\mathfrak{v}_h = h^{-1}\{\mathfrak{h}_\lambda(t+h) - \mathfrak{h}_\lambda(t)\}$, $\mathfrak{U}_h = h^{-1}\{\mathfrak{U}(t+h) - \mathfrak{U}(t)\}$ so that

$$\begin{aligned}
&\lambda\{\mathfrak{v}_h - \mathfrak{v}_\lambda(t)\} - \mathfrak{L}(t)\{\mathfrak{v}_h - \mathfrak{v}_\lambda(t)\} \\
&\quad = \{\mathfrak{U}_h - \mathfrak{U}'(t)\} + h a_h \mathfrak{N}\{\mathfrak{v}_h - \mathfrak{v}_\lambda(t)\} \\
&\qquad + h b_h \mathfrak{J}\{\mathfrak{v}_h - \mathfrak{v}_\lambda(t)\} + \varepsilon_1(h)\mathfrak{N}\mathfrak{v}_\lambda(t) + \varepsilon_2(h)\mathfrak{J}\mathfrak{v}_\lambda(t),
\end{aligned}$$

where $\varepsilon_1(h) = a_h - a_0(t) + h a_h$, $\varepsilon_2(h) = b_h - b_0(t) + h b_h$, $a_0(t) = -2\sin(2t)$, $b_0(t) = \cos(2t)$. Take the scalar product on both sides of this equation with respect to $\mathfrak{v}_h - \mathfrak{v}_\lambda(t)$ and recall that $\mathfrak{N}$ and $\mathfrak{J}$ are asymmetric operators to get that

$$\begin{aligned}
&\lambda \left\| \mathfrak{v}_h - \mathfrak{v}_\lambda(t) \right\|_\mu^2 + \left\| \mathfrak{v}_h - \mathfrak{v}_\lambda(t) \right\|_1^2 \\
&\quad = \left\langle \mathfrak{U}_h - \mathfrak{U}'(t), \mathfrak{v}_h - \mathfrak{v}_\lambda(t) \right\rangle_\mu \\
&\qquad + \varepsilon_1(h) \left\langle \mathfrak{N}\mathfrak{v}_\lambda(t), \mathfrak{v}_h - \mathfrak{v}_\lambda(t) \right\rangle_\mu + \varepsilon_2(h) \left\langle \mathfrak{J}\mathfrak{v}_\lambda(t), \mathfrak{v}_h - \mathfrak{v}_\lambda(t) \right\rangle_\mu.
\end{aligned}$$

By Schwarz inequality and Lemma 8.7, we obtain that

$$\begin{aligned}
&(\lambda/4) \left\| \mathfrak{v}_h - \mathfrak{v}_\lambda(t) \right\|_\mu^2 + \left\| \mathfrak{v}_h - \mathfrak{v}_\lambda(t) \right\|_1^2 \\
&\quad \leq \frac{1}{\lambda} \left\{ \left\| \mathfrak{U}_h - \mathfrak{U}'(t) \right\|_\mu^2 + 4\varepsilon_1(h)^2 \left\| \mathfrak{v}_\lambda(t) \right\|_{1,\mu}^2 + 4\varepsilon_2(h)^2 \left\| \mathfrak{v}_\lambda(t) \right\|_{1,\mu}^2 \right\}.
\end{aligned}$$

By (8.17), the right-hand side is less than or equal to

$$\frac{1}{\lambda} \left\| \mathfrak{U}_h - \mathfrak{U}'(t) \right\|_\mu^2 + \frac{C_0 \varepsilon(h)}{\lambda^2} \left\{ \left\| \mathfrak{U}'(t) \right\|_{1,-1}^2 + \left\| \mathfrak{U}(t) \right\|_{3,-1}^2 \right\},$$

where $\varepsilon(h)$ vanishes as $h \to 0$. This proves the lemma. □

Proof of Theorem 8.1 By assumptions (H1)–(H3), for every k, $j \geq 0$,

$$\sup_{0 \leq t \leq \pi/2} \left\| \mathfrak{W}^{(j)}(t) \right\|_{k,-1} < \infty. \tag{8.18}$$

Hence, by Lemma 8.4, for all $k \geq 0$,

$$\sup_{0 \leq t \leq \pi/2} \sup_{0 < \lambda \leq 1} \left|\left|\left| \mathfrak{g}_\lambda(t) \right|\right|\right|_{k,1} < \infty. \tag{8.19}$$

By Lemma 8.5, $\mathfrak{g}_\lambda(t)$ is differentiable and its derivative $\mathfrak{g}_\lambda^{(1)}(t)$ satisfies the resolvent equation

$$\lambda \mathfrak{g}_\lambda^{(1)}(t) - \mathfrak{L}(t)\mathfrak{g}_\lambda^{(1)}(t) = \mathfrak{W}^{(1)}(t) + \mathfrak{L}^{(1)}(t)\mathfrak{g}_\lambda(t),$$

where $\mathfrak{L}^{(j)}(t)$ stands for the j-th derivative of $\mathfrak{L}(t)$. Let $\mathfrak{W}_1(t,\lambda) = \mathfrak{W}^{(1)}(t) + \mathfrak{L}^{(1)}(t)\mathfrak{g}_\lambda(t)$. By (8.18), (8.19) and Lemma 8.6,

$$\sup_{0 \leq t \leq \pi/2} \sup_{0 < \lambda \leq 1} \left|\left|\left| \mathfrak{W}_1(t,\lambda) \right|\right|\right|_{k,-1} < \infty$$

for all $k \geq 0$. The rest of the proof is exactly as in the mean zero case. □

We conclude this section with two estimates on the operators $\mathfrak{N}$, $\mathfrak{J}$.

Lemma 8.6 *Fix $\mathfrak{f}$ in $L^2(\mu)$ such that $|||\mathfrak{f}|||_{k,-1} < \infty$ for all $k \geq 0$. Let $\mathfrak{h}_\lambda$ be the solution of the resolvent equation* (8.14). *For each $k \geq 0$, there exists a finite constant C_k, depending only on k and on the probability measure p, such that*

$$|||\mathfrak{J}\mathfrak{h}_\lambda|||_{k,-1} \leq C_k |||\mathfrak{h}_\lambda|||_{k+1,1}, \qquad |||\mathfrak{N}\mathfrak{h}_\lambda|||_{k,-1} \leq C_k\big\{|||\mathfrak{f}|||_{k+1,-1} + |||\mathfrak{h}_\lambda|||_{k+2,1}\big\}.$$

Proof The first assertion is a restatement of Corollary 5.16, while the second one is the content of Lemma 5.20. □

Lemma 8.7 *There exists a finite constant C_0 depending only on the probability measure p such that*

$$\|\mathfrak{N}\mathfrak{f}\|_\mu \leq 2n\|\mathfrak{f}\|_\mu \quad \text{and} \quad \|\mathfrak{J}_\pm \mathfrak{f}\|_\mu \leq 2(n+1)\|\mathfrak{f}\|_\mu$$

for all $\mathfrak{f}: \mathcal{E}_n \to \mathbb{R}$ in $L^2(\mu)$.

Proof Fix a function in $\mathfrak{f}: \mathcal{E}_n \to \mathbb{R}$ in $L^2(\mu)$. By the explicit expression of the operator $\mathfrak{N}$ given in (5.17) and by Schwarz inequality,

$$\|\mathfrak{N}\mathfrak{f}\|_\mu^2 \leq n \sum_{A \in \mathcal{E}_n} \sum_{\substack{x \in A \\ y \notin A}} s(y-x)\big\{f(A_{x,y}) - f(A)\big\}^2$$

because $|a(z)| \leq s(z)$ and $\sum_{x\in A, y\notin A} s(y-x) \leq |A| = n$. It remains to bound the square by $2\{f(A_{x,y})^2 + f(A)^2\}$ and to perform a change of variables $B = A_{x,y}$ to conclude the proof of the first claim.

The proof of the second assertion is similar. One just needs to rewrite the operator $\mathfrak{J}_-$ as $(\mathfrak{J}_-\mathfrak{f})(A) = -2\sum_{y\notin A, x\in A} a(y-x)\mathfrak{f}(A\cup\{y\})$, which is allowed because $\sum_z a(z) = 0$. The details are left to the reader. □

8.5 Regularity of the Diffusion Coefficients

The method presented in the previous sections permit also to show that the self-diffusion and the bulk diffusion coefficients, introduced in Chaps. 6 and 7, depend smoothly on the density. In this section, we examine the bulk diffusion whose regularity was used in the proof of the fluctuation–dissipation theorem.

Recall the set-up and the notation introduced in Chap. 7 and that we are considering exclusion processes in dimension $d \ge 3$. Fix a finitely supported function $\mathbf{w}\colon \mathscr{E}^* \to \mathbb{R}$ in $\mathscr{I}$ and such that $\mathbf{w}(\varnothing) = 0$. Since by Lemma 5.26 a random walk on $\mathbb{Z}_*^d$, $d \ge 3$, is transient, by Proposition 5.23, $\mathbf{w}$ belongs to $\mathfrak{H}_{0,-1}$ and therefore to $\mathfrak{H}_{k,-1}$:

$$|||\mathbf{w}|||_{k,-1} < \infty \quad \text{for all } k \ge 1. \tag{8.20}$$

For each $\lambda > 0$, let $\mathbf{u}_\lambda$ be the solution of the resolvent equation

$$\lambda \mathbf{u}_\lambda - \mathfrak{L}_{\star,\alpha} \mathbf{u}_\lambda = \mathbf{w},$$

whose existence is asserted by Lemma 7.5. In this section, we prove that for each $\delta > 0$, there exists a subsequence λ_k such that for each $z \in \mathbb{Z}_*^d$, $\mathbf{u}_{\lambda_k}(\cdot, z)$ converges uniformly $[\delta, 1-\delta]$, as well as all its derivatives, to a smooth function $D_z(\cdot)$.

To prove the existence of such subsequence it is enough to show that $\mathbf{u}_\lambda(\alpha, \{z\})$ are smooth functions of α for each $\lambda > 0$ and each z, and that

$$\sup_{\lambda > 0} \sup_{\delta \le \alpha \le 1-\delta} \left| \mathbf{u}_\lambda^{(j)}(\alpha, \{z\}) \right| < \infty,$$

where $\mathbf{u}_\lambda^{(j)}(\alpha, \{z\})$ stands for the j-th derivative of $\mathbf{u}_\lambda$.

By Proposition 5.23 again,

$$\left| \mathbf{u}_\lambda^{(j)}(\alpha, \{z\}) \right| \le C_0 \left\| \mathbf{u}_\lambda^{(j)}(\alpha, \cdot) \right\|_{0,1}$$

for some finite constant C_0 depending only on $p(\cdot)$. With the iterated use of (7.17), we will prove that, for any k and j,

$$\sup_{\lambda > 0} \sup_{\delta \le \alpha \le 1-\delta} \left|\left|\left| \mathbf{u}_\lambda^{(j)}(\alpha, \cdot) \right|\right|\right|_{k,1} < \infty.$$

Since the coefficients of $\mathfrak{L}_{\star,\alpha}$ are not smooth at the boundary of $[0,1]$, we reparametrize by $\alpha = \sin^2 t$, $t \in [0, 2\pi]$, as done in the previous sections. We obtain in this way the operator $\mathfrak{L}_\star(t)$ given by

$$\mathfrak{L}_\star(t) = \mathfrak{L}_{\star,s} + \left(\cos^2 t - \sin^2 t\right) \mathfrak{L}_{\star,a} + (\sin t \cos t)\{\mathfrak{J}_{\star,+} + \mathfrak{J}_{\star,-}\}.$$

Consider the equation

$$\lambda \mathbf{v}_\lambda(t) - \mathfrak{L}_\star(t) \mathbf{v}_\lambda(t) = \mathbf{w}. \tag{8.21}$$

Since $\mathbf{w}$ does not depend on α, we have $\mathbf{u}_\lambda(\alpha(t)) = \mathbf{v}_\lambda(t)$. So if we prove that $|||\mathbf{v}_\lambda^{(j)}(t)|||_{k,1}$ are uniformly bounded, we obtain the boundedness in the same norm for $\mathbf{u}_\lambda(\alpha)$ for α in the interior of $[0, 1]$.

Differentiating formally $\mathfrak{L}_\star(t)$ in t, we obtain

$$\mathfrak{L}'_\star(t) = -4(\sin t \cos t)\mathfrak{L}_{\star,a} + \left(\cos^2 t - \sin^2 t\right)\{\mathfrak{J}_{\star,+} + \mathfrak{J}_{\star,-}\}.$$

By (8.20) and Lemma 8.5, $\mathbf{v}_\lambda(t)$ is differentiable in t, and its derivative $\mathbf{v}'_\lambda(t)$ satisfies

$$\lambda \mathbf{v}'_\lambda(t) - \mathfrak{L}_\star(t)\mathbf{v}'_\lambda(t) = \mathfrak{L}'_\star(t)\mathbf{v}_\lambda(t). \tag{8.22}$$

Recall the estimate (7.18). By (7.17) and Lemma 7.14, this bound holds for the operators $\mathfrak{L}_{\star,s}$ and $\mathfrak{J}_{\star,\pm}$ in place of $\mathfrak{L}_{\star,\alpha}$. It also holds therefore for the operator $\mathfrak{L}_{\star,a}$. In particular, by the explicit form of $\mathfrak{L}'_\star(t)$ and since $\mathbf{v}_\lambda(t)$ solves the resolvent equation (8.21),

$$\left|\left|\left|\mathfrak{L}'_\star(t)\mathbf{v}_\lambda(t)\right|\right|\right|_{k,-1} \le C_k |||\mathbf{w}|||_{k+2,-1}.$$

If we now apply (7.17) to Eq. (8.22), we obtain a bound for $|||\mathbf{v}'_\lambda(t)|||_{k,1}$ uniform in t and λ. The argument can be iterated exactly as before, yielding similar bounds for all the derivatives $\mathbf{v}_\lambda^{(j)}(t)$.

One can extend the regularity up to the boundary and prove the existence of a subsequence λ_k for which $\mathbf{u}_{\lambda_k}(\cdot, z)$ converges uniformly in $[0, 1]$, as well as all its derivatives, to a smooth function $D_z(\cdot)$. We refer to Landim et al. (2004a).

8.6 Comments and References

Varadhan (1994b) proved that the self-diffusion coefficient is Lipschitz continuous in dimension $d \ge 3$. In this chapter, we followed (Landim et al., 2001), who proved smoothness of the self-diffusion coefficient for symmetric simple exclusion processes, and for asymmetric exclusion processes in dimension $d \ge 3$. The method was extended in several directions.

Diffusion Coefficient Landim et al. (2004a) proved the regularity of the diffusion coefficient of asymmetric exclusion processes in dimension $d \ge 3$. Bernardin (2002) proved the smoothness of the diffusion coefficient of a non-gradient exclusion process, reversible with respect to Bernoulli product measures. Sued (2005) extended the regularity of the diffusion coefficient to mean zero exclusion processes. Nagahata (2005, 2006, 2007) proved the regularity of the diffusion coefficient for several lattice gas models. In the last article the reference measure is not a product and substantial modifications of the original arguments are required. For example, the basis used is formed by the functions $\prod_{x\in A} \eta(x)$, for finite subsets A of $\mathbb{Z}^d$.

We mention also that Beltrán (2005) proved the smoothness, with respect to the drift, of the asymptotic variance in the central limit theorem for additive functionals of one-dimensional nearest neighbor reversible asymmetric exclusion processes.

Tracer Particle Carlson et al. (1993a) examined the self-diffusion coefficient of a one-dimensional simple exclusion process where a particle jumps from site x to site $x + y$ at rate $c(|y|)$ if all sites between x and $x + y$ are occupied. They proved that the self-diffusion diverges as the density increases to one if the non-increasing function $c(\cdot)$ has a sufficiently slow decay.

References

Beltrán J (2005) Regularity of diffusion coefficient for nearest neighbor asymmetric simple exclusion on $\mathbb{Z}$. Stoch Process Appl 115(9):1451–1474

Bernardin C (2002) Regularity of the diffusion coefficient for lattice gas reversible under Bernoulli measures. Stoch Process Appl 101(1):43–68

Carlson JM, Grannan ER, Swindle GH (1993a) A limit theorem for tagged particles in a class of self-organizing particle systems. Stoch Process Appl 47(1):1–16

Landim C, Olla S, Varadhan SRS (2001) Symmetric simple exclusion process: regularity of the self-diffusion coefficient. Commun Math Phys 224(1):307–321. Dedicated to Joel L Lebowitz

Landim C, Olla S, Varadhan SRS (2004a) On viscosity and fluctuation-dissipation in exclusion processes. J Stat Phys 115(1–2):323–363

Nagahata Y (2005) Regularity of the diffusion coefficient matrix for the lattice gas with energy. Ann Inst Henri Poincaré Probab Stat 41(1):45–67

Nagahata Y (2006) Regularity of the diffusion coefficient matrix for generalized exclusion process. Stoch Process Appl 116(6):957–982

Nagahata Y (2007) Regularity of the diffusion coefficient matrix for lattice gas reversible under Gibbs measures with mixing condition. Commun Math Phys 273(3):637–650

Sued M (2005) Regularity properties of the diffusion coefficient for a mean zero exclusion process. Ann Inst Henri Poincaré Probab Stat 41(1):1–33

Varadhan SRS (1994b) Regularity of self-diffusion coefficient. In: The Dynkin festschrift. Progr probab, vol 34. Birkhäuser Boston, Boston, pp 387–397

Part III
Diffusions in Random Environments

Chapter 9
Diffusions in Random Environments

9.1 Diffusions with Periodic Coefficients

We start with a simple example of a one-dimensional diffusion with periodic coefficients whose generator is in a divergence form. It is convenient to think about the state space of such a process as a one-dimensional torus. In this setting the normalized Lebesgue measure is invariant and the generator satisfies the spectral gap estimate. The central limit theorem is a consequence of Theorem 2.14.

More precisely, assume that $a, U : \mathbb{R} \to \mathbb{R}$ are two 1-periodic functions that are C^2 smooth and such that $a(x) \ge c_0$ for all $x \in \mathbb{R}$ and some positive constant c_0. Suppose that $\{X_t^x, t \ge 0\}$ is a diffusion that starts at x with the generator

$$\mathscr{L} f(x) := \frac{1}{2} e^{U(x)} \frac{d}{dx}\left(e^{-U(x)} a(x) \frac{df}{dx}\right), \quad f \in C_c^2(\mathbb{R}). \tag{9.1}$$

For $m \ge 1$ we use the notation $C^m(\mathbb{R})$, $C_c^m(\mathbb{R})$ (resp. $C_c^\infty(\mathbb{R})$) for the spaces of m times continuously differentiable functions and its subspace consisting of compactly supported functions (resp. the subspace of infinitely differentiable compactly supported functions). This notation will be extended in an obvious way to an arbitrary dimension d. Process $\{X_t^x, t \ge 0\}$ is the unique solution of the Itô stochastic differential equation

$$dX_t^x = V\big(X_t^x\big)\,dt + a^{1/2}\big(X_t^x\big)\,dw_t, \quad X_0^x = x, \tag{9.2}$$

where

$$V(x) := -(1/2)\big[U'(x)a(x) + a'(x)\big]$$

and $\{w_t, t \ge 0\}$ is a standard, one-dimensional Brownian motion given over a probability space $(\Sigma, \mathscr{A}, \mathbb{P})$. Under regularity assumptions made about the coefficients the solution to (9.2) is unique and has the strong Markov property, see e.g. Theorem 5.3.4, p. 112 of Friedman (1975). The corresponding probability transitions have densities with respect to the Lebesgue measure $p_t(x, y)$, which are strictly

T. Komorowski et al., *Fluctuations in Markov Processes*,
Grundlehren der mathematischen Wissenschaften 345,
DOI 10.1007/978-3-642-29880-6_9, © Springer-Verlag Berlin Heidelberg 2012

positive and C^1 smooth jointly in (t, x, y) variables and of C^2 class in both x and y. The transition probability semigroup for the diffusion equals

$$P_t f(x) = \int_{\mathbb{R}} p_t(x, y) f(y)\, dy, \quad f \in B_b(\mathbb{R}).$$

Here $B_b(\mathbb{R}^d)$ is the space of bounded and measurable Borel functions on $\mathbb{R}^d$.

It follows from (9.2) that the trajectory of the particle can be written as a sum of the additive functional

$$\int_0^t V\big(X_s^x\big) ds$$

and the martingale

$$\int_0^t a^{1/2}\big(X_s^x\big) dw_s.$$

Note also that

$$\int_{\mathbb{R}} \mathscr{L} f(x) e^{-U(x)} dx = 0, \quad \forall f \in C_c^2(\mathbb{R})$$

thus $e^{-U(x)}dx$ is an invariant measure of infinite mass. This prevents us from applying directly the results of Chap. 2. To deal with this issue we work with the diffusion induced by X_t^x on the one-dimensional torus $\mathbb{T} := \mathbb{R}/\sim$, where the equivalence relation $x \sim y$, for arbitrary $x, y \in \mathbb{R}$, means that x and y are equal mod 1, or equivalently $x - y \in \mathbb{Z}$. Denote by $\dot{x} \in \mathbb{T}$ the equivalence class corresponding to a given $x \in \mathbb{R}$.

The process $\{\dot{X}_t^{\dot{x}}, t \ge 0\}$ is Markovian on $\mathbb{T}$ and its transition probability densities, with respect to the normalized Lebesgue measure on the torus, and the generator are given by

$$p_t(\dot{x}, \dot{y}) = \sum_{y \in \dot{y}} p_t(x, y) > 0, \quad \dot{x}, \dot{y} \in \mathbb{T} \tag{9.3}$$

and

$$Lf(\dot{x}) = \frac{1}{2} e^{U(\dot{x})} D\big(e^{-U(\dot{x})} a(\dot{x}) Df(\dot{x})\big), \quad f \in C^2(\mathbb{T}).$$

We shall identify 1-periodic functions with the functions on the torus $\mathbb{T}$ and when there is no danger of confusion we denote them by the same symbols. By D^m, where for simplicity $D^1 = D$, we denote the m-th derivative of a given function on the torus. Let $\{\dot{P}_t, t \ge 0\}$ be the transition semigroup corresponding to $\{\dot{X}_t^{\dot{x}}, t \ge 0\}$.

Let π be the probability measure on $\mathbb{T}$ given by

$$\pi(d\dot{x}) := \frac{1}{Z} e^{-U(\dot{x})} d\dot{x},$$

with

$$Z := \int_{\mathbb{T}} e^{-U(\dot{x})} d\dot{x}.$$

Note that

$$\int_{\mathbb{T}} Lf d\pi = 0$$

for all f belonging to $C^2(\mathbb{T})$—the class of functions on the torus that are C^2 smooth. The above implies π is an invariant measure for $\{\dot{P}_t, t \geq 0\}$ and the semigroup can be extended to a strongly continuous semigroup of contractions on $L^2(\pi)$. Since $C^2(\mathbb{T})$ is dense in $L^2(\mathbb{T})$ and invariant under the semigroup, it is a core of the generator. Suppose that a Borel set $A \subset \mathbb{T}$ satisfies $1_A = \dot{P}_t 1_A$ for some $t > 0$. Then,

$$0 = 1_{A^c}(\dot{x}) \dot{P}_t 1_A(\dot{x}) = \int_A 1_{A^c}(\dot{x}) p_t(\dot{x}, \dot{y}) d\dot{y}, \quad \forall \dot{x} \in \mathbb{T}^d.$$

This, together with (9.3), implies that A is of trivial Lebesgue, thus also, π measure. Hence we conclude ergodicity of π under $\{\dot{P}_t, t \geq 0\}$. As an immediate consequence of the ergodic theorem and the law of large numbers for martingales, see e.g. Theorem VII.9.3 of Feller (1971), we obtain that for $t \to +\infty$

$$\frac{1}{t} X_t^x = \frac{1}{t}\left[x + \int_0^t V(\dot{X}_s^{\dot{x}}) ds + \int_0^t a^{1/2}(\dot{X}_s^{\dot{x}}) dw_s\right]$$

tends to

$$\int_{\mathbb{T}} V(\dot{x}) \pi(d\dot{x}) = \frac{-1}{2Z} \int_{\mathbb{T}} D\big(a(\dot{x}) e^{-U(\dot{x})}\big) d\dot{x} = 0$$

for a.e. x and $\mathbb{P}$ a.s. realization of $\{X_t^x, t \geq 0\}$. The last equality follows from integration by parts.

In order to prove the central limit theorem we look for the solution χ of *the cell problem*

$$-L\chi = V. \tag{9.4}$$

The above equation reads

$$-\frac{1}{2} e^{U(\dot{x})} D\big(e^{-U(\dot{x})} a(\dot{x}) D\chi(\dot{x})\big) = \frac{1}{2} e^{U(\dot{x})} D\big(e^{-U(\dot{x})} a(\dot{x})\big),$$

hence

$$D\chi(\dot{x}) = -1 + C_1 e^{U(\dot{x})} a^{-1}(\dot{x}) \tag{9.5}$$

for some constant C_1. It has to be chosen in such a way that

$$\int_{\mathbb{T}} D\chi(\dot{x}) d\dot{x} = 0,$$

which is consistent with the property of a gradient, due to the integration by parts formula. This, in turn yields

$$C_1 = \left(\int_{\mathbb{T}} e^{U(\dot{x})} a^{-1}(\dot{x}) d\dot{x} \right)^{-1}.$$

From (9.5) we deduce

$$\chi(\dot{x}) = -\dot{x} + C_1 \int_0^{\dot{x}} e^{U(y)} a^{-1}(y) dy + C,$$

where C is another constant, which we choose in such a way that $\int_{\mathbb{T}} \chi(\dot{x}) d\dot{x} = 0$. Note that

$$-\mathscr{L}\chi(x) = V(x). \tag{9.6}$$

Applying Itô's formula to $\chi(X_t^x)$ we conclude that

$$\chi(X_t^x) = \chi(x) + \int_0^t \mathscr{L}\chi(X_s^x) ds + \int_0^t a^{1/2}(X_s^x) D\chi(X_s^x) dw_s.$$

Combining the above with (9.6) and (9.2) we obtain

$$X_t^x = x + \chi(x) - \chi(X_t^x) + \int_0^t a^{1/2}(X_s^x)\left[D\chi(X_s^x) + 1\right] dw_s.$$

Denote by M_t the martingale term appearing on the right-hand side. Its quadratic variation satisfies

$$\frac{1}{t}\langle M \rangle_t = \frac{1}{t} \int_0^t a(\dot{X}_s^{\dot{x}})\left[D\chi(\dot{X}_s^{\dot{x}}) + 1\right]^2 ds.$$

By the ergodic theorem we obtain that the limit of the above expression, as $t \to +\infty$, equals

$$\bar{a} := \int_{\mathbb{T}} a(\dot{x})\left[C_1 e^{U(\dot{x})} a^{-1}(\dot{x})\right]^2 \pi(d\dot{x}) = \frac{1}{Z} C_1, \tag{9.7}$$

for $\mathbb{P}$ a.s. realization of $\{X_t^x, t \ge 0\}$ and a.e. $x \in \mathbb{R}$. The martingale central limit theorem (see Theorem 2.1) implies that

$$\lim_{t \to +\infty} \int_{\mathbb{T}} \left| E_{\mathbb{P}} f(X_t^{\dot{x}}/\sqrt{t}) - \int_{\mathbb{R}} f(y) \Phi_{\bar{a}}(y) dy \right| d\dot{x} = 0$$

for any $f \in C_b(\mathbb{R})$. Here $\Phi_{\bar{a}}$ is a density of a zero mean normal distribution with the variance given by (9.7). We conclude therefore that if $\{X_t, t \ge 0\}$ is a diffusion with the generator (9.1) and the initial data whose law is absolutely continuous with respect to the Lebesgue measure then

$$\lim_{t \to +\infty} E_{\mathbb{P}} \left| E_{\mathbb{P}}\left[f(X_t/\sqrt{t}) \mid \mathscr{F}_0\right] - \int_{\mathbb{R}} f(y) \Phi_{\bar{a}}(y) dy \right| = 0.$$

The multidimensional case can be handled quite similarly. The only significant difference comes from the fact that in a higher dimension we cannot in general write an explicit solution to the cell problem (9.4). To deal with this difficulty we use some basic estimates from the multidimensional calculus.

Assume that $a(\cdot)$ is a $d\times d$ matrix valued function defined on $\mathbb{R}^d$, whose symmetric part $a^s(x):=1/2[a(x)+a^t(x)]$ is uniformly positive definite, i.e. $a^s(x)\ge c_0 I_d$ for some $c_0>0$ and all $x\in\mathbb{R}^d$. Here I_d is the $d\times d$ identity matrix. Furthermore we suppose that $a(x), U(x)$ are C^2 smooth and 1-periodic in all variables.

Denote by $\{X_t^x, t\ge 0\}$ an $\mathbb{R}^d$-valued diffusion that starts at x with the generator

$$\mathscr{L}f(x)=\frac{1}{2}e^{U(x)}\sum_{i,j=1}^{d}\partial_i\bigl(e^{-U(x)}a_{i,j}(x)\partial_j f(x)\bigr),\quad f\in C_c^2(\mathbb{R}^d). \tag{9.8}$$

It can be viewed as the solution of the stochastic differential equation

$$dX_t^x=V(X_t^x)dt+c(X_t^x)dw_t,\quad X_0^x=x, \tag{9.9}$$

where $c(x)$ is the square root of $a^s(x)$ and the drift $V=(V_1,\dots,V_d)$ is given by

$$V_k(x):=\frac{1}{2}e^{U(x)}\sum_{j=1}^{d}\partial_j\bigl(e^{-U(x)}a_{j,k}(x)\bigr),\quad k=1,\dots,d. \tag{9.10}$$

The process $\{w_t, t\ge 0\}$ is a standard, d-dimensional Brownian motion over the probability space $(\Sigma,\mathscr{A},\mathbb{P})$. Analogously with the one-dimensional situation we identify a d-dimensional torus $\mathbb{T}^d$ with $\mathbb{R}^d/\sim$, where $x\sim y$ iff $x-y\in\mathbb{Z}^d$. We also maintain the conventions of denoting by $\dot{x}$ the equivalence class of x and identifying the notation for 1-periodic functions and their counterparts induced on $\mathbb{T}^d$. The partial derivative on the torus D_k can be introduced as the partial of the respective periodic function in the direction of the k-th vector of the canonical base.

Let $\pi(d\dot{x}):=Z^{-1}e^{-U(\dot{x})}d\dot{x}$ and let

$$L_0^2(\pi):=\bigl[f\in L^2(\pi):\langle f\rangle_\pi=0\bigr].$$

For each direction $\ell\in\mathbb{R}^d$ we formulate the corresponding cell problem

$$-L\chi_\ell(\dot{x})=V_\ell(\dot{x}), \tag{9.11}$$

where $V_\ell(\dot{x}):=V(\dot{x})\cdot\ell$. Here L is the generator of the transition semigroup $\{\dot{P}_t, t\ge 0\}$ corresponding to the diffusion $\{\dot{X}_t^{\dot{x}}, t\ge 0\}$ on $\mathbb{T}^d$. The operator is the closure in $L^2(\pi)$ of

$$Lf(\dot{x})=\frac{1}{2}e^{U(\dot{x})}\sum_{k,l=1}^{d}D_k\bigl(e^{-U(\dot{x})}a_{kl}(\dot{x})D_l f(\dot{x})\bigr),\quad f\in C^2(\mathbb{T}^d). \tag{9.12}$$

Here $C^m(\mathbb{T}^d)$ denotes the class of all m times, continuously differentiable functions on the torus. Using the classical Poincare inequality one can show, see the proof below, the following *spectral gap estimate*: there exists a constant $\lambda_0 > 0$ such that

$$\lambda_0\big(\|f\|_\pi^2 - \langle f, 1\rangle_\pi^2\big) \le \|\nabla f\|_\pi^2, \quad \forall f \in C^1\big(\mathbb{T}^d\big). \tag{9.13}$$

Analogously with the one-dimensional situation we can argue that

$$\dot{P}_t\big(C^2\big(\mathbb{T}^d\big)\big) \subset C^2\big(\mathbb{T}^d\big), \quad \forall t \ge 0$$

and, since π is invariant $\langle \dot{P}_t f, 1\rangle_\pi = 0$ for all $f \in L_0^2(\pi)$. Inequality (9.13) implies that for any $f \in C^2(\mathbb{T}^d) \cap L_0^2(\pi)$ we have

$$\begin{aligned}\frac{d}{dt}\|\dot{P}_t f\|_\pi^2 &= \langle L\dot{P}_t f, \dot{P}_t f\rangle_\pi = -\frac{1}{2Z}\sum_{k,l=1}^{d}\int_{\mathbb{T}^d} e^{-U} a_{k,l} D_k \dot{P}_t f D_l \dot{P}_t f d\dot{x} \\ &\le -c_0\|\nabla \dot{P}_t f\|_\pi^2 \le -c_0\lambda_0\|f\|_\pi^2.\end{aligned}$$

In consequence

$$\|\dot{P}_t f\|_\pi \le e^{-c_0\lambda_0 t/2}\|f\|_\pi, \quad \forall t \ge 0.$$

This estimate can be easily extended by density to the entire $L_0^2(\pi)$. Therefore

$$\chi_\ell = \int_0^{+\infty} \dot{P}_t V_\ell dt$$

is well defined and solves uniquely (9.11) in $L_0^2(\mathbb{T}^d)$. Standard regularity results from the theory of elliptic partial differential equations, see e.g. Chap. 8 of Gilbarg and Trudinger (1983), imply that $\chi_\ell \in C^2(\mathbb{T}^d)$. From this point on the proof of the central limit theorem for $\{X_t^x/\sqrt{t}, t \ge 0\}$ goes along the same lines as in the one-dimensional setting. We conclude therefore the following.

Theorem 9.1 *Suppose that $\{X_t, t \ge 0\}$ is a diffusion with the generator given by* (9.8) *such that the law of X_0 is absolutely continuous with respect to the Lebesgue measure. Then, for any $f \in C_b(\mathbb{R}^d)$ we have*

$$\lim_{t\to+\infty} E_{\mathbb{P}}\left| E_{\mathbb{P}}\big[f(X_t/\sqrt{t})\big|\mathscr{F}_0\big] - \int_{\mathbb{R}^d} f(y)\Phi_{\bar{a}}(y)dy\right| = 0,$$

where $\Phi_{\bar{a}}$ is the density of a d-dimensional, zero mean normal random vector with the covariance matrix $\bar{a}$ determined by the quadratic form

$$\bar{a}\ell\cdot\ell = \int_{\mathbb{T}^d} a(\ell + \nabla\chi_\ell)\cdot(\ell + \nabla\chi_\ell)d\pi, \quad \ell \in \mathbb{R}^d. \tag{9.14}$$

Proof of estimate (9.13) We prove it first in case $d = 1$. It is clear that

$$\langle f^2 \rangle_\pi = \frac{1}{2} \int_{\mathbb{T}} \int_{\mathbb{T}} \left[f(\dot{x}) - f(\dot{y}) \right]^2 e^{-U(\dot{x}) - U(\dot{y})} d\dot{x} d\dot{y}$$

$$= \int_{\mathbb{T}} \int_{\mathbb{T}} \left[\int_0^1 t^{-1/4} t^{1/4} (\dot{x} - \dot{y}) f'\big(t\dot{x} + (1-t)\dot{y}\big) dt \right]^2 e^{-U(\dot{x}) - U(\dot{y})} d\dot{x} d\dot{y}. \tag{9.15}$$

Using Cauchy–Schwartz inequality we can estimate the utmost right-hand side by

$$\int_0^1 \frac{dt}{\sqrt{t}} \int_0^1 \sqrt{t} dt \left\{ \int_{\mathbb{T}} \int_{\mathbb{T}} (\dot{x} - \dot{y})^2 \left[f'\big(t\dot{x} + (1-t)\dot{y}\big) \right]^2 e^{-U(\dot{x}) - U(\dot{y})} d\dot{x} d\dot{y} \right\}. \tag{9.16}$$

Then changing variables according to $\dot{z}' := t\dot{x} + (1-t)\dot{y}$, $\dot{y}' := \dot{y}$ and dropping primes in our notation we obtain, due to $|\dot{x} - \dot{y}| \le 1$, that (9.16) can be estimated by

$$2 \int_0^1 \frac{dt}{\sqrt{t}} \int_{\mathbb{T}} \int_{\mathbb{T}} \left[f'(\dot{z}) \right]^2 e^{-U(t^{-1}(\dot{z} - (1-t)\dot{y})) - U(\dot{y})} d\dot{z} d\dot{y}. \tag{9.17}$$

Since U is bounded, we can find C such that

$$U\big(t^{-1}\big(\dot{z} - (1-t)\dot{y}\big)\big) + U(\dot{y}) \ge U(\dot{z}) - C, \quad \forall \dot{y}, \dot{z} \in \mathbb{T}$$

and the inequality follows for $d = 1$. The generalization of the above argument to an arbitrary dimension d is straightforward. We replace $f(\dot{x}) - f(\dot{y})$ by

$$\int_0^1 \nabla f\big(t\dot{x} + (1-t)\dot{y}\big) \cdot (\dot{x} - \dot{y}) dt$$

and perform the calculation as before. □

9.2 Remark About the Quasi-periodic Case

Before dealing with diffusions with random coefficients we briefly discuss the quasi-periodic case. What distinguishes it from the previously considered periodic situation is the fact that the solution of the cell problem need not exist in the L^2 space corresponding to the invariant measure. A similar situation occurs for diffusions with random coefficients and as it turns out the quasi-periodic case can be treated within the same framework. We shall pursue this subject in the following sections. Here we limit the scope of our discussion only to the description of the set-up needed to deal with quasi-periodic coefficients. To simplify our presentation we assume also that $d = 1$.

Suppose that $N \ge 1$ is an integer, $a(\cdot)$, $U(\cdot)$ are two functions that belong to $C^2(\mathbb{T}^N)$ and $a(\omega) \ge c_0$ for all $\omega \in \mathbb{T}^N$ and some $c_0 > 0$. Assume also that the components of $\lambda = (\lambda_1, \dots, \lambda_N) \in \mathbb{T}^N$, are rationally independent, i.e. for any rationals

$r_1, \ldots, r_N$ such that $\sum_{j=1}^N r_j\lambda_j = 0$ we have $r_1 = \cdots = r_N = 0$. For any $x, l \in \mathbb{R}$ we let $x_l := lx$. Define *quasi-periodic* functions

$$\tilde{a}(x) := a(\dot{x}_{\lambda_1}, \ldots, \dot{x}_{\lambda_N}) \quad \text{and} \quad \tilde{U}(x) := U(\dot{x}_{\lambda_1}, \ldots, \dot{x}_{\lambda_N}), \qquad x \in \mathbb{R}.$$

The process $\{X_t^x, t \geq 0\}$ is the diffusion on $\mathbb{R}$ that starts at x with the generator

$$\mathscr{L} f(x) := \frac{1}{2} e^{\tilde{U}(x)} \frac{d}{dx}\left(e^{-\tilde{U}(x)} \tilde{a}(x) \frac{df(x)}{dx}\right), \quad f \in C_c^2(\mathbb{R}). \tag{9.18}$$

Define a diffusion on $\mathbb{T}^N$ by $\eta_t := (\dot{X}_t^{x,\lambda_1}, \ldots, \dot{X}_t^{x,\lambda_N})$, where $X_t^{x,\lambda} := \lambda X_t^x$. Here, as we recall, $\dot{x}$ denotes the mod 1 projection of x onto $\mathbb{T}$. The generator of $\{\eta_t, t \geq 0\}$ equals

$$Lf(\omega) = \frac{1}{2} e^{U(\omega)} D\big(e^{-U(\omega)} a(\omega) Df(\omega)\big), \quad f \in C^2(\mathbb{T}^N), \tag{9.19}$$

$\omega = (\omega_1, \ldots, \omega_N) \in \mathbb{T}^N$, where D is defined by the formula

$$Df(\omega) := \sum_{j=1}^N \lambda_j \partial_j f(\omega),$$

and ∂_j is the partial in the direction ω_j. The transition semigroup $\{\dot{P}_t, t \geq 0\}$ of the diffusion has Feller property, i.e. $\dot{P}_t(C(\mathbb{T}^N)) \subset C(\mathbb{T}^N)$ (see e.g. Proposition 8.2.4 in Ethier and Kurtz, 1986), and $C^2(\mathbb{T}^N)$ is a core of its generator. On $\mathbb{T}^N$ define a probability measure

$$\pi(d\omega) := \frac{1}{Z} e^{-U(\omega)} d\omega,$$

where Z is a normalizing factor and $d\omega$ is the normalized Lebesgue measure on the torus. Since

$$\int_{\mathbb{T}^N} Lf d\pi = 0, \quad \forall f \in C^2(\mathbb{T}^N)$$

the measure π is invariant and the semigroup can be extended to $L^2(\pi)$.

Suppose that $f \in C^1(\mathbb{T}^N)$. Then

$$Df(\omega) = i \sum_{k \in \mathbb{Z}^N} k \cdot \lambda \hat{f}(k) \exp\{2\pi i k \cdot \omega\}, \tag{9.20}$$

where

$$\hat{f}(k) := \int_{\mathbb{T}^N} e^{-2\pi i k \cdot \omega} f(\omega) d\omega, \quad k \in \mathbb{Z}^N$$

are the Fourier coefficients of f. Let H^1 be the completion of the space consisting of functions f for which $\hat{f}(0)=0$ in the norm

$$\|f\|_{H^1}^2 := \sum_{k\in\mathbb{Z}^N} (k\cdot\lambda)^2 |\hat{f}(k)|^2.$$

Operator D can be continuously extended to the entire H^1. Since the components of λ are rationally independent $Df=0$ implies that $\hat{f}(k)$ for all $k\in\mathbb{Z}^N\setminus\{0\}$, thus f is constant π a.s.

Integration by parts show that

$$\langle f,(-L)f\rangle_\pi = \frac{1}{2}Z\langle a(Df)^2\rangle_\pi, \quad f\in C^2(\mathbb{T}^N). \tag{9.21}$$

Recall that $\langle\cdot\rangle_\pi$ and $\langle\cdot,\cdot\rangle_\pi$ are the expectation with respect to π and the scalar product in $L^2(\pi)$, respectively. From (9.21) we deduce that $D(L)\subset H^1$. Furthermore, for any $f\in D(L)$ that satisfies $Lf=0$ we conclude, from the above formula, that $Df=0$. This in turn implies that f is constant π a.s. Hence, measure π has to be ergodic.

In analogy with the periodic case we formulate the cell problem (9.4) for the corrector, which leads to the equality

$$D\chi = F, \tag{9.22}$$

where

$$F = -1 + \langle e^U a^{-1}\rangle_\pi^{-1} e^U a^{-1}$$

(cf. (9.5)). From (9.22) we obtain that the Fourier coefficients of χ satisfy

$$\hat{\chi}(k) := \frac{\hat{F}(k)}{ik\cdot\lambda} \quad \text{for } k\neq 0.$$

We also let $\hat{\chi}(0)=0$. In contrast with the periodic case this time one cannot guarantee, without putting additional restrictions on a and U, that $\sum_k |\hat{\chi}(k)|^2 < +\infty$. The solution of the cell problem (9.4) need not exist in $L^2(\pi)$. As a result the argument made in the periodic setting cannot be directly applied. We explain how to deal with this problem in the following sections. At this point we only remark that it suffices to work with the corrector that is an object whose gradient $D\chi$ belongs to $L^2(\pi)$. This situation occurs also for diffusions with random and stationary coefficients that will be examined later on. In fact, as we have already mentioned at the beginning of this section, it turns out, see Remark 9.5 below, that the diffusions with quasi-periodic coefficients can be treated within this framework as well.

9.3 Diffusions with Stationary Coefficients

Suppose that $\{\tilde{a}(x;\omega), x\in\mathbb{R}^d\}$ and $\{\tilde{U}(x;\omega), x\in\mathbb{R}^d\}$ are $d\times d$ matrix and scalar valued random fields respectively, given over a probability space $(\Omega,\mathcal{F},\mathbb{Q})$. For a fixed $\omega \in \Omega$ denote by $\{X_t^{x,\omega}, t \geq 0\}$ a diffusion starting at x whose generator is given by

$$\mathscr{L}_\omega f(x) := \frac{1}{2} e^{\tilde{U}(x;\omega)} \sum_{k,l=1}^{d} \partial_{x_k} \big(e^{-\tilde{U}(x;\omega)} \tilde{a}_{kl}(x;\omega) \partial_{x_l} f(x)\big), \quad f \in C_c^2(\mathbb{R}^d), \quad (9.23)$$

where $\tilde{a}(x;\omega) = [\tilde{a}_{kl}(x;\omega)]$, $k, l = 1, \ldots, d$. In what follows we present a general framework that allows to prove the central limit theorem when the coefficient fields are stationary and ergodic. The main idea is to represent the trajectory of the diffusion as a sum of an additive functional of a Markov process, taking values in Ω—the space of possible "realizations" of the environment—and a martingale with stationary and ergodic increments. For a diffusion whose generator is in a divergence form we can identify an invariant and ergodic probability measure corresponding to such a process that is absolutely continuous with respect to $\mathbb{Q}$. This in turn allows us to apply the techniques developed in Chap. 2 very much in the same spirit as it has been done for random walks in random environment, see Chap. 3.

9.3.1 Preliminaries on Stationary Environments

We assume that (Ω, d) is a Polish metric space, i.e. the metric d is complete and separable. Denote by $\mathscr{F}$ its Borel σ-algebra and by $\mathscr{C}(\Omega)$, resp. $\mathscr{C}_b(\Omega)$, the space of all continuous, resp. bounded continuous functions on Ω.

Suppose that $\{\tau_x, x \in \mathbb{R}^d\}$ is a measurable group of transformations acting on Ω, the so-called *shifts* $\tau_x : \Omega \to \Omega$: i.e. $\tau_x \circ \tau_y = \tau_{x+y}$ for all $x, y \in \mathbb{R}^d$ and for each $A \in \mathscr{F}$ the mapping $(x, \omega) \mapsto 1_A(\tau_x \omega)$ is jointly measurable with respect to the σ-algebra $\mathscr{B}(\mathbb{R}^d) \otimes \mathscr{F}$. This in particular implies that $\tau_x(A) \in \mathscr{F}$ for each $A \in \mathscr{F}$ and $x \in \mathbb{R}^d$.

Let $\mathbb{Q}$ be a Borel probability measure on $(\Omega, \mathscr{F})$ that is *invariant* under the group, i.e. $\mathbb{Q} \circ \tau_x = \mathbb{Q}$ for all $x \in \mathbb{R}^d$. Furthermore, its action is *ergodic* and *stochastically continuous*: i.e.

(1) $\mathbb{Q}[A] = 0$, or 1 for any event A such that $\mathbb{Q}[A \Delta \tau_x(A)] = 0$ for all $x \in \mathbb{R}^d$ and
(2) for any $\delta > 0$ and $f \in B(\Omega)$ we have

$$\lim_{h \to 0} \mathbb{Q}\big(\omega : \big|f(\tau_h \omega) - f(\omega)\big| \geq \delta\big) = 0,$$

respectively. Here Δ denotes the symmetric difference of sets and $B(\Omega)$ is the space of all bounded, Borel, real valued functions on Ω.

For any $f \in L^2(\mathbb{Q})$ we let $T_x f := f \circ \tau_x$. The family $\{T_x, x \in \mathbb{R}^d\}$ forms a d-parameter group of unitary operators on $L^2(\mathbb{Q})$. Stochastic continuity implies that the group is strongly continuous. On the other hand from hypothesis (1) we conclude that the group of operators is ergodic in the following sense: any $f \in L^2(\mathbb{Q})$ that satisfies $T_x f = f$ for all $x \in \mathbb{R}^d$ has to be $\mathbb{Q}$-a.s. constant.

Remark 9.2 Suppose that $\{(\tilde{a}(x), \tilde{U}(x)), x \in \mathbb{R}^d\}$ is a random field defined over some probability space $(\tilde{\Omega}, \tilde{\mathscr{F}}, \tilde{\mathbb{Q}})$, where each $\tilde{a}(x)$ is a $d \times d$ random matrix and $\tilde{U}(x)$ is a scalar valued random variable. We assume furthermore that the realizations of the field are almost surely C^m-regular componentwise for some $m \geq 0$. When $m = 0$ the above simply means that the fields are continuous.

The field is called *stationary* if for any integer $k \geq 1$ and $x_1, \dots, x_k, h \in \mathbb{R}^d$ the law of

$$\big(\tilde{a}(x_1 + h), \tilde{U}(x_1 + h), \dots, \tilde{a}(x_k + h), \tilde{U}(x_k + h)\big)$$

and that of

$$\big(\tilde{a}(x_1), \tilde{U}(x_1), \dots, \tilde{a}(x_k), \tilde{U}(x_k)\big)$$

are identical.

Let Ω be the space of all C^m-regular mappings $\omega = (a, U) : \mathbb{R}^d \to \mathbb{R}^{d^2+1}$. It is a Polish metric space, when equipped with the standard Fréchet metric

$$\mathrm{d}(\omega_1, \omega_2) := \sum_{k=1}^{+\infty} \frac{1}{2^k} \cdot \frac{\|\omega_1 - \omega_2\|_{k,m}}{1 + \|\omega_1 - \omega_2\|_{k,m}}$$

for any $\omega_1, \omega_2 \in \Omega$ and

$$\|\omega\|_{k,m} := \sum_{l_1, l_2 = 1}^{d} \|a_{l_1, l_2}\|_{k,m} + \|U\|_{k,m}.$$

Here

$$\|f\|_{k,m} := \sup_{|x| \leq k} \Big[\big|f(x)\big| + \sum_{|i| = m} \big|\partial_i^m f(x)\big| \Big]$$

and the summation extends over all non-negative integer valued multi-indices $i = (i_1, \dots, i_d)$, with $|i| := i_1 + \cdots + i_d$ and $\partial_i^m := \partial_{x_1}^{i_1} \dots \partial_{x_d}^{i_d}$. The probability measure $\mathbb{Q}$ is defined as the law of the random field over Ω, i.e. it is given by

$$\mathbb{Q}[A] = \mathbb{P}\big[\big(\tilde{a}(\cdot), \tilde{U}(\cdot)\big) \in A\big]$$

for A belonging to $\mathscr{F}$—the σ-algebra of Borel subsets of Ω. The group of measure $\mathbb{Q}$ preserving shifts on Ω is given by $\tau_x \omega(\cdot) := \omega(\cdot + x)$ for any $x \in \mathbb{R}^d$. It is easy to see that the action of the group is jointly measurable and stochastically continuous with respect to $\mathbb{Q}$. We say that the random field is *ergodic* if the measure is ergodic under the group action.

9.3.2 Spaces of Smooth Functions

The generators of the group $\{T_x, x \in \mathbb{R}^d\}$ correspond to differentiations (in $L^2(\mathbb{Q})$) in the canonical directions e_k and are denoted by

$$D_k : \mathscr{D}(D_k) \to L^2(\mathbb{Q}), \quad k = 1, \dots, d.$$

Since T_x are unitary the generators are anti-self-adjoint and we have the following *integration by parts* formula

$$\langle D_k f, g \rangle_{\mathbb{Q}} = -\langle f, D_k g \rangle_{\mathbb{Q}}, \quad \forall f, g \in \mathscr{D}(D_k),\ k = 1, \dots, d. \tag{9.24}$$

For a fixed positive integer k denote by $H_0^k(\Omega)$ the subspace of $L^2(\mathbb{Q})$ that is an intersection of the domains of operators $D^m := D_1^{m_1} \dots D_d^{m_d}$ corresponding to all integer multi-indices $m = (m_1, \dots, m_d)$ with $|m| = k$. On this space we introduce a semi-norm given by

$$\|f\|_{H^k}^2 := \sum_{|m|=k} \|D^m f\|_{\mathbb{Q}}^2, \quad f \in H_0^k(\Omega) \tag{9.25}$$

and consider the equivalence relation: $f \sim g$ iff $f - g$ is constant. The quotient space, denoted by the same symbol, is pre-Hilbert. Let $H^k(\Omega)$ be its completion with respect to the norm defined in (9.25). We identify $H^0(\Omega)$ with $L^2(\mathbb{Q})$. Differentiation operators D^m, $|m| = k$ extend continuously to $H^k(\Omega)$.

By $L_d^2(\mathbb{Q})$ we denote the space of d-dimensional random vectors whose components belong to $L^2(\mathbb{Q})$. We define *the abstract gradient operator* $\nabla : H^1(\Omega) \to L_d^2(\mathbb{Q})$ by letting $\nabla := (D_1, \dots, D_d)$.

Suppose that f is a random element taking values in some measurable space. The corresponding stationary random field shall be denoted by

$$\tilde{f}(x; \omega) = f(\tau_x \omega), \quad x \in \mathbb{R}^d.$$

Let $C^k(\Omega)$ be the subspace of $H^k(\Omega)$ consisting of those random variables for which the corresponding random field possesses a C^k regular modification. By $C_b^k(\Omega)$ we denote its subspace consisting of those f for which

$$\|f\|_{k,\infty} := \sum_{|m| \le k} \|D^m f\|_\infty < +\infty.$$

Here $\|\cdot\|_\infty$ denotes the essential supremum norm on Ω. Let

$$C^\infty(\Omega) := \bigcap_{k \ge 1} C^k(\Omega) \quad \text{and} \quad C_b^\infty(\Omega) := \bigcap_{k \ge 1} C_b^k(\Omega).$$

Proposition 9.3 *The space $C_b^\infty(\Omega)$ is dense in $H^k(\Omega)$ for any $k \ge 0$.*

Proof Suppose that $g \in H^k(\Omega)$. Let $\phi \in C_c^\infty(\mathbb{R}^d)$ be such that $\int_{\mathbb{R}^d} \phi(x)dx = 1$. Let $\delta > 0$ and

$$g_\delta(\omega) := \delta^{-d} \int_{\mathbb{R}^d} \phi\left(\frac{x}{\delta}\right) \tilde{g}(x;\omega)dx.$$

Then,

$$\lim_{\delta \to 0+} \left\| D^m g_\delta - D^m g \right\|_{\mathbb{Q}} = \lim_{\delta \to 0+} \left\| \left(D^m g\right)_\delta - D^m g \right\|_{\mathbb{Q}} = 0$$

for any multi-index $|m| \le k$ and the conclusion of the proposition follows. □

Remark 9.4 Suppose that $\{(\tilde{a}(x), \tilde{U}(x)), x \in \mathbb{R}^d\}$ is a C^m-regular, stationary, random field as considered in Example 9.2. Let $(\Omega, \mathscr{F}, \mathbb{Q})$ and $\{\tau_x, x \in \mathbb{R}^d\}$ be the probability space and the group of transformations constructed in that example. Define $f(\omega) := \omega(0)$ for $\omega \in \Omega$. Let $a(\omega)$ and $U(\omega)$ be the $d \times d$ matrix and scalar components of $f(\omega)$, respectively. Both U and all the entries of $a(\omega)$ belong to $C^m(\Omega)$. The random field

$$\left\{\left(a(\tau_x\omega), U(\tau_x\omega)\right), x \in \mathbb{R}^d\right\}, \tag{9.26}$$

defined over $(\Omega, \mathscr{F}, \mathbb{Q})$, has an identical law with that of $\{(\tilde{a}(x), \tilde{U}(x)), x \in \mathbb{R}^d\}$. In fact, since any statement involving statistical properties of the field depends only on its law, with no loss of generality we may and will assume that the random fields of coefficients corresponding to a diffusion given by (9.23) is of the form (9.26).

9.3.3 Itô Equations with Stationary Coefficients

We suppose that a random variable U and matrix $a = [a_{k,l}]$, $k, l = 1, \dots, d$ satisfy the following hypotheses:

(R1) Both the random variable and entries of the matrix belong to $C_b^2(\Omega)$;
(R2) a is *uniformly positive definite*, i.e. there exists a (deterministic) positive constant $c_0 > 0$ such that

$$\sum_{kl=1}^{d} a_{k,l}(\omega)\xi_k\xi_l \ge c_0|\xi|^2, \quad \forall \xi \in \mathbb{R}^d, \mathbb{Q}\text{-a.s.}; \tag{9.27}$$

(R3) The probability space $(\Omega, \mathscr{F}, \mathbb{Q})$ and the group of shift transformations $\{\tau_x, x \in \mathbb{R}^d\}$ satisfy assumptions formulated in Sect. 9.3.1.

Let $\{(\tilde{a}(x), \tilde{U}(x)), x \in \mathbb{R}^d\}$ be the random field given by (9.26). We define a random vector $V = (V_1, \dots, V_d)$ by

$$V_k := \frac{1}{2}\sum_{l=1}^{d}(-a_{kl}D_lU + D_la_{kl}), \quad k = 1, \dots, d. \tag{9.28}$$

Let c be the square root of the symmetric part of a and

$$\tilde{V}(x;\omega) := V(\tau_x\omega), \quad \tilde{c}(x;\omega) := c(\tau_x\omega). \tag{9.29}$$

Suppose that $\{w_t = (w_t^1, \dots, w_t^d), t \ge 0\}$ is a d-dimensional, standard Brownian motion over a probability space $(\Sigma, \mathscr{A}, \mathbb{P})$. For a given $\omega \in \Omega$ we define a process $\{X_t^{x,\omega}, t \ge 0\}$ as a solution of the Itô stochastic differential equation

$$\begin{aligned} dX_t^{x,\omega} &= \tilde{V}\big(X_t^{x,\omega};\omega\big)dt + \tilde{c}\big(X_t^{x,\omega};\omega\big)dw_t, \\ X_0^{x,\omega} &= x. \end{aligned} \tag{9.30}$$

A simple calculation shows that the generator of the process is given by (9.23). By *a diffusion in a random environment* we understand the stochastic process $\{X_t^{x,\omega}, t \ge 0\}$ defined over the product probability space $(\Omega \times \Sigma, \mathscr{F} \otimes \mathscr{A}, \mathbb{Q} \otimes \mathbb{P})$. As in the case of random walks we omit writing the superscript corresponding to the starting point if $x = 0$. The superscript denoting the random argument ω will also be omitted when its value is obvious from the context.

Remark 9.5 Diffusions with periodic, as well as quasi-periodic coefficients, considered in Sects. 9.1 and 9.2 are included in the above framework. Assume that we are given C^2 smooth, 1-periodic, $d \times d$ matrix valued function $a(x)$, with the uniformly positive symmetric part, and a scalar function $U(x)$. Let $c(x)$ be the square root of the symmetric part of $a(x)$ and let $V_k(x)$ be given by (9.10). Define $\Omega := \mathbb{T}^d$ and set $\mathbb{Q}$ to be the normalized Lebesgue measure on Ω. The shift transformation $\tau_x : \Omega \to \Omega$ is given by

$$\tau_x(\omega) := \dot{x} \oplus \omega \quad \text{for } (x,\omega) \in \mathbb{R}^d \times \Omega.$$

Here $\oplus$ denotes the usual modulo summation on $\mathbb{T}^d$. Spaces $H^k(\Omega)$ coincide with the usual Sobolev spaces of periodic functions having square integrable weak derivatives up to order k. For a given x we denote by $[x]$ and (x) the vectors consisting of the integer parts of x and their remainders $(x) := x - [x]$ correspondingly. We also let

$$\tilde{V}(x;\omega) = V\big(\tau_x(\omega)\big), \qquad \tilde{c}(x;\omega) := c\big(\tau_x(\omega)\big). \tag{9.31}$$

Then, $\{X_t^x, t \ge 0\}$—the solution of (9.9)—satisfies

$$X_t^x = X_t^{(x)} + [x] \tag{9.32}$$

and

$$X_t^{x+\omega} = X_t^{x,\omega} + \omega, \quad \forall (x,\omega) \in \mathbb{R}^d \times \mathbb{T}^d, \tag{9.33}$$

where $X_t^{x,\omega}$ solves (9.30). With some abuse of notation, we identify $\omega \in \mathbb{T}^d$ with the corresponding vector in $[0,1)^d$, so $x + \omega$ is well defined in $\mathbb{R}^d$.

In the quasi-periodic setting, discussed in Sect. 9.2, we let $\Omega := \mathbb{T}^N$ with $\mathbb{Q}$ the respective normalized product Lebesgue measure. For $\omega = (\omega_1, \ldots, \omega_N)$ and $x \in \mathbb{R}$ we let

$$\tau_x \omega := (\dot{x}_{\lambda_1} \oplus \omega_1, \ldots, \dot{x}_{\lambda_N} \oplus \omega_N),$$

where $x_\lambda := \lambda x$ for a given $\lambda \in \mathbb{R}$ and $(\lambda_1, \ldots, \lambda_N) \in \mathbb{T}^N$ is a vector with rationally independent coordinates. The generator D of the corresponding unitary group $\{T_x, x \in \mathbb{R}\}$ is given by (9.20). We let $X_t^{x,\omega}$ be the solution of (9.30) where $\tilde{V}(x)$, $\tilde{c}(x)$ correspond to

$$V = 1/2(-aDU + DU) \quad \text{and} \quad c = a^{1/2}.$$

9.4 Environment Process and Its Properties

The *environment process* corresponding to diffusion $\{X_t^\omega, t \geq 0\}$ is an Ω-valued process over $(\Sigma, \mathscr{A}, \mathbb{P})$ defined by

$$\eta_t^\omega := \tau_{X_t^\omega} \omega, \quad t \geq 0. \tag{9.34}$$

Non-degeneracy and boundedness of the coefficients of the diffusion imply, see e.g. Theorem 6.4.6, p. 142 of Friedman (1975), the existence of continuous transition probability densities $p_t^\omega(x, y)$ with respect to the d-dimensional Lebesgue measure.

Lemma 9.6 *We have*

$$p_t^{\tau_z \omega}(x, y) = p_t^\omega(x + z, y + z), \quad \textit{for all } t > 0,\ x, y, z \in \mathbb{R}^d,\ \omega \in \Omega. \tag{9.35}$$

Proof Since $\tilde{V}(x; \tau_y \omega) = \tilde{V}(x + y; \omega)$ and $\tilde{c}(x; \tau_y \omega) = \tilde{c}(x + y; \omega)$, by the uniqueness of solutions to (9.30), we obtain that $X_{t,x}^{\tau_y \omega} + y = X_{t,x+y}^\omega$. The respective transition probability functions $P_t^\omega(x, \cdot)$ satisfy therefore

$$P_t^{\tau_y \omega}(x, A) = P_t^\omega(x + y, A + y), \quad \forall A \in \mathscr{B}(\mathbb{R}^d),\ \omega \in \Omega. \tag{9.36}$$

The equality for transition probability densities follows from their continuity. □

Proposition 9.7 *The process $\{\eta_t, t \geq 0\}$ is Markovian. Its transition probability semigroup is given by*

$$P_t f(\omega) = \int p_t^\omega(0, x) \tilde{f}(x; \omega) dx \quad \textit{for all } f \in B(\Omega),\ \omega \in \Omega. \tag{9.37}$$

Proof The proof is almost the same as for random walks, see Lemma 3.1. It uses Kolmogorov–Chapman equations for $p_t^{\omega}(x, y)$ and homogeneity of transition probability densities, see Lemma 9.6. □

In the next step we identify the invariant probability measure for the environment process and its generator. We can easily find an invariant (non-probabilistic) measure for the diffusion process corresponding to a fixed ω. Denote by $\{P_t^{\omega}, t \geq 0\}$ the respective transition probability semigroup. Suppose that $m_{\omega}(dx) := e^{-\tilde{U}(x;\omega)}dx$. A direct calculation, see (9.23), shows that

$$\frac{d}{dt}_{|t=0} \int P_t^{\omega} f(x) m_{\omega}(dx) = \int \mathscr{L}_{\omega} f(x) m_{\omega}(dx) = 0 \tag{9.38}$$

for all $f \in C_c^2(\mathbb{R}^d)$. Hence, m_{ω} is an invariant measure for the semigroup, non-probabilistic $\mathbb{Q}$ a.s., see Theorem 8.2.5, p. 373 of Ethier and Kurtz (1986).

Proposition 9.8 *Let $Z := \langle e^{-U} \rangle_{\mathbb{Q}}$. Then,*

$$\pi(d\omega) := \frac{1}{Z} e^{-U(\omega)} \mathbb{Q}(d\omega) \tag{9.39}$$

is an invariant and ergodic probability measure for the semigroup $\{P_t, t \geq 0\}$. Therefore, it extends to a strongly continuous semigroup of contractions on $L^2(\pi)$. The set $C_b^2(\Omega)$ is a core of its generator, given by the formula

$$Lf = \frac{1}{2} e^U \sum_{k,l=1}^{d} D_k \big(e^{-U} a_{k,l} D_l f\big), \quad \forall f \in C_b^2(\Omega). \tag{9.40}$$

Next, we describe the adjoint semigroup and show that $C_b^2(\Omega)$ is also a core of its generator. Suppose that $\{\hat{X}_t^{x,\omega}, t \geq 0\}$ is a diffusion that starts at $x \in \mathbb{R}^d$ and corresponds to the generator whose coefficient matrix is the transpose of $\tilde{a}(x)$, i.e.

$$\hat{\mathscr{L}}_{\omega} f(x) := \frac{1}{2} e^{\tilde{U}(x;\omega)} \sum_{k,l=1}^{d} \partial_{x_k} \big(e^{-\tilde{U}(x;\omega)} \tilde{a}_{l,k}(x;\omega) \partial_{x_l} f(x)\big), \quad f \in C_c^2(\mathbb{R}^d). \tag{9.41}$$

Let $\{\hat{\eta}_t, t \geq 0\}$ be the environment process that corresponds to this diffusion. Using Proposition 9.8 we conclude that π is also an invariant and ergodic measure for its transition semigroup $\{\hat{P}_t, t \geq 0\}$.

Proposition 9.9 *The operator $\hat{P}_t$ is the $L^2(\pi)$-adjoint to P_t. In consequence, the set $C_b^2(\Omega)$ is a core of the dual of the generator L^*, given by the formula*

$$L^* f = \frac{1}{2} e^U \sum_{k,l=1}^{d} D_k \big(e^{-U} a_{l,k} D_l f\big), \quad \forall f \in C_b^2(\Omega).$$

Due to the technical nature of this and the previous proposition we postpone their proofs until Sect. 9.7.

Let a' be the transpose of a matrix a and

$$a^s = 1/2(a + a'), \qquad a^\dagger := 1/2(a - a')$$

be its respective symmetric and anti-symmetric parts. An immediate consequence of the results formulated above is the following.

Corollary 9.10 *The set $C_b^2(\Omega)$ is a common core of L, L^* and the respective Dirichlet form. The operator L satisfies sector condition* (2.36) *and its symmetric and anti-symmetric parts are correspondingly equal to*

$$Sf = \frac{1}{2} e^U \sum_{k,l=1}^{d} D_k\big(e^{-U} a_{k,l}^s D_l f\big) \tag{9.42}$$

and

$$Af = \frac{1}{2} e^U \sum_{k,l=1}^{d} D_k\big(e^{-U} a_{k,l}^\dagger D_l f\big), \quad f \in C_b^2(\Omega). \tag{9.43}$$

Remark 9.11 (Isomorphism of $\mathcal{H}_1$ and $H^1(\Omega)$) Note that

$$\|f\|_1^2 = \frac{1}{2}\big\langle a^s \nabla f, \nabla f\big\rangle_\pi, \quad f \in C_b^2(\Omega), \tag{9.44}$$

where

$$\langle f, g\rangle_\pi := \sum_{i=1}^{d} \langle f_i, g_i\rangle_\pi$$

for any vectors $f = (f_1, \dots, f_d)$, $g = (g_1, \dots, g_d)$ with components from $L^2(\pi)$.

We introduce the spaces $\mathcal{H}_1$ and $\mathcal{H}_{-1}$, as in Sect. 2.2. It can be easily seen that $\mathcal{H}_1$ is isomorphic to $H^1(\Omega)$ defined in Sect. 9.3.2. We can define therefore the differentiation operators $D_k : \mathcal{H}_1 \to L^2(\pi)$, $k = 1, \dots, d$. Denote $\nabla := (D_1, \dots, D_d)$. Then,

$$\big\langle(-L)f, g\big\rangle_\pi = \frac{1}{2}\langle a\nabla f, \nabla g\rangle_\pi, \quad \forall g \in \mathcal{H}_1, f \in D(L), \tag{9.45}$$

and formula (9.44) extends to the entire $\mathcal{H}_1$.

9.5 Martingale Decomposition and Central Limit Theorem

Equation (9.30) yields the decomposition of the trajectory into a martingale and an additive functional of the environment process

$$X_t = \mathfrak{M}_t + \int_0^t V(\eta_s)\,ds, \tag{9.46}$$

where

$$\mathfrak{M}_t := \int_0^t c(\eta_s)\,dw_s$$

and $V = (V_1, \ldots, V_d)$ is given by (9.28).

Proposition 9.12 *$V_p \in L^2(\pi) \cap \mathscr{H}_{-1}$ for each $p = 1, \ldots, d$.*

Proof By (9.44) we obtain

$$\langle V_p, f\rangle_\pi = \frac{1}{2}\langle a\nabla f \cdot e_p\rangle_\pi \le C\|f\|_1, \quad \forall f \in C_b^2(\Omega) \tag{9.47}$$

for some constant $C > 0$. The conclusion follows from density of $C_b^2(\Omega)$ in $\mathscr{H}_1$. □

Let $\lambda > 0$ and $p = 1, \ldots, d$. The solution of the resolvent equation

$$(\lambda - L)\chi_\lambda^{(p)} = V_p \tag{9.48}$$

is called *the λ-corrector* in the direction e_p.

From Corollary 9.10 and the assumptions made about the coefficients we conclude that generator L satisfies the sector condition. We can apply therefore the results of Sect. 2.7.3 and conclude that $\chi_\lambda^{(p)} \in \mathscr{H}_1$,

$$\lim_{\lambda\to 0+} \lambda\|\chi_\lambda^{(p)}\|_\pi^2 = 0 \quad \text{and} \quad \lim_{\lambda\to 0+} \chi_\lambda^{(p)} = \chi^{(p)} \quad \text{in } \mathscr{H}_1. \tag{9.49}$$

The limit $\chi^{(p)}$ is called *the corrector* corresponding to e_p. It need not belong to $L^2(\pi)$. Since it is an element of $\mathscr{H}_1$, we have $D_q\chi^{(p)} \in L^2(\pi)$ for $q = 1, \ldots, d$.

Proposition 9.13 *The following decomposition holds*

$$\int_0^t V(\eta_s)\,ds = m_{t,\lambda} + R_{t,\lambda}, \tag{9.50}$$

where $\{m_{t,\lambda}, t \ge 0\}$ and $\{R_{t,\lambda}, t \ge 0\}$ are processes with components given by

$$m^p_{t,\lambda} := \int_0^t c(\eta_s)\nabla\chi^{(p)}_\lambda(\eta_s)\cdot dw_s,$$

$$R^p_{t,\lambda} := \chi^{(p)}_\lambda(\eta_0) - \chi^{(p)}_\lambda(\eta_t) + \lambda\int_0^t \chi^{(p)}_\lambda(\eta_s)\,ds, \quad p = 1,\dots,d.$$

Proof Proposition 9.8 implies that for any $\lambda > 0$ there exists a sequence $\{f_n, n \ge 1\}$ of elements of $C^2_b(\Omega)$ such that $Lf_n \to L\chi^{(p)}_\lambda$ and $f_n \to \chi^{(p)}_\lambda$ in $L^2(\pi)$, as $n \to +\infty$. Using Itô formula for $\tilde{f}_n(X_t) = f_n \circ \tau_{X_t}$ we obtain

$$\begin{aligned} f_n(\eta_t) - f_n(\eta_0) &= \tilde{f}_n\big(X^\omega_t\big) - \tilde{f}_n\big(X^\omega_0\big) \\ &= \int_0^t \mathscr{L}_\omega \tilde{f}_n\big(X^\omega_s\big)ds + \int_0^t \tilde{c}\big(X^\omega_s\big)\nabla_x \tilde{f}_n\big(X^\omega_s\big)\cdot dw_s. \end{aligned} \tag{9.51}$$

Note that $\mathscr{L}_\omega \tilde{f}_n(X^\omega_s) = Lf_n(\eta_s)$ and

$$\tilde{c}\big(X^\omega_s\big)\nabla_x f_n\big(X^\omega_s\big) = c(\eta_s)\nabla f_n(\eta_s).$$

Letting $n \to +\infty$ in (9.51) we obtain

$$\chi^{(p)}_\lambda(\eta_t) - \chi^{(p)}_\lambda(\eta_0) = \int_0^t L\chi^{(p)}_\lambda(\eta_s)ds + m^p_{t,\lambda}.$$

Formula (9.50) follows then from (9.48). □

We let $\lambda \to 0+$ in (9.50). Using the argument from the proofs of Lemmas 2.9 and 2.10 we obtain

$$\int_0^t V(\eta_s)\,ds = m_t + R_t, \tag{9.52}$$

where $m_t = (m^1_t, \dots, m^d_t)$, given by

$$m^p_t := \int_0^t c(\eta_s)\nabla\chi^{(p)}(\eta_s)\cdot dw_s, \quad p = 1,\dots,d,$$

is an $\mathbb{R}^d$-valued, square integrable martingale and $R_t = (R^1_t, \dots, R^d_t)$ satisfies

$$\lim_{t\to+\infty} \frac{\langle E_{\mathbb{P}}|R_t|^2\rangle_\pi}{t} = 0. \tag{9.53}$$

Formula (9.46) can be rewritten therefore in the form

$$X_t = M_t + R_t, \tag{9.54}$$

where $M_t = (M^1_t, \dots, M^d_t)$ is an $\mathbb{R}^d$-valued, square integrable martingale given by $M_t := \mathfrak{M}_t + m_t$. We have

$$M^p_t = \int_0^t c(\eta_s)\big(e_p + \nabla\chi^{(p)}(\eta_s)\big)\cdot dw_s, \quad p = 1,\dots,d.$$

The quadratic covariation of the martingale equals

$$\langle M^p, M^q\rangle_t = \int_0^t a^s(\eta_s)\big(e_p + \nabla\chi^{(p)}(\eta_s)\big)\cdot\big(e_q + \nabla\chi^{(q)}(\eta_s)\big)ds, \quad p,q = 1,\dots,d.$$

Let $\Phi_{\bar{a}}$ denote the density of a d-dimensional Gaussian random vector with zero mean and the covariance matrix $\bar{a} = [\bar{a}_{p,q}]$, where

$$\bar{a}_{p,q} := \big\langle a^s\big(e_p + \nabla\chi^{(p)}\big)\cdot\big(e_q + \nabla\chi^{(q)}\big)\big\rangle_\pi. \tag{9.55}$$

From Proposition 9.8 and the ergodic theorem we conclude that

$$\lim_{t\to+\infty}\left\langle\left|\frac{E_{\mathbb{P}}\langle M^p, M^q\rangle_t}{t} - \bar{a}_{p,q}\right|\right\rangle_\pi = 0. \tag{9.56}$$

By virtue of Theorem 2.26 the martingale $\{M_t, t\ge 0\}$ satisfies the central limit theorem. In fact taking into account its statement we can formulate the following definition.

Definition 9.14 We say that random variables $X_t/\sqrt{t}$ satisfy the central limit theorem in probability with respect to the environment if for any $f\in C_b(\mathbb{R}^d)$ we have

$$\lim_{t\to+\infty}\left\langle\left|E_{\mathbb{P}} f\left(\frac{X_t}{\sqrt{t}}\right) - \int_{\mathbb{R}^d} f(y)\Phi_{\bar{a}}(y)dy\right|\right\rangle_{\mathbb{Q}} = 0. \tag{9.57}$$

Summarizing our considerations in this section, we have proved the following.

Theorem 9.15 *Suppose that a random matrix a and variable U satisfy the assumptions* (R1)–(R3). *Then, $X_t/\sqrt{t}$ satisfy the central limit theorem in probability with respect to the environment.*

Remark 9.16 Using (9.28) and (9.48) we can write

$$\begin{aligned}0 &= \big\langle(\lambda - L)\chi_\lambda^{(p)} + V_p, \chi_\lambda^{(q)}\big\rangle_\pi\\ &= \lambda\big\langle\chi_\lambda^{(p)}, \chi_\lambda^{(q)}\big\rangle_\pi + \frac{1}{2}\big\langle a\big(e_p + \nabla\chi_\lambda^{(p)}\big), \nabla\chi_\lambda^{(q)}\big\rangle_\pi.\end{aligned}$$

Letting $\lambda\to 0+$ we obtain that

$$0 = \big\langle a\big(e_p + \nabla\chi^{(p)}\big), \nabla\chi^{(q)}\big\rangle_\pi.$$

Using (9.55), we conclude an alternative formula for the limiting covariance matrix

$$\bar{a}_{p,q} = \big\langle a\big(e_p + \nabla\chi^{(p)}\big)\cdot e_q\big\rangle_\pi, \quad p,q = 1,\dots,d. \tag{9.58}$$

Remark 9.17 From (9.53) and (9.56) it follows that the covariance matrix can be also characterized by the limit

$$\bar{a}_{p,q} = \lim_{t\to+\infty} \frac{\langle E_{\mathbb{P}}(X_t^p X_t^q)\rangle_\pi}{t}, \qquad p,q = 1,\dots,d.$$

9.6 Homogenization of Solutions of Parabolic Partial Differential Equations

The question of the central limit theorem for a diffusion in a random environment is closely related to the problem of convergence of solutions of partial differential equations with random, fast oscillating coefficients. More precisely, suppose that $\{X_t^{x,\omega}, t \ge 0\}$ is a diffusion given by (9.30). For a function $u_0 \in C_b(\mathbb{R}^d)$ consider a solution of the Cauchy problem

$$\begin{aligned} &\partial_t u^{(\varepsilon)}(t,x) = \mathscr{L}_\omega^{(\varepsilon)} u^{(\varepsilon)}(t,x), \qquad \text{for } (t,x) \in (0,+\infty)\times\mathbb{R}^d, \\ &u^{(\varepsilon)}(0,x) = u_0(x), \end{aligned} \tag{9.59}$$

where

$$\mathscr{L}_\omega^{(\varepsilon)} f(x) := \frac{1}{2} e^{\tilde{U}(x/\varepsilon;\omega)} \sum_{k,l=1}^d \partial_{x_k}\left[e^{-\tilde{U}(x/\varepsilon;\omega)} \tilde{a}_{k,l}\left(\frac{x}{\varepsilon};\omega\right)\partial_{x_l} f(x)\right] \tag{9.60}$$

for $f \in C_b^2(\mathbb{R}^d)$. It is well known, see e.g. Sect. 1.7 of Friedman (1964), that there exists a unique solution to (9.59) that has continuous and bounded derivatives: one in t and two in x, in $(0,+\infty)\times\mathbb{R}^d$ and is continuous in its closure. Let

$$X_{t,\varepsilon}^{x,\omega} := \varepsilon X_{t/\varepsilon^2}^{x/\varepsilon,\omega}. \tag{9.61}$$

It solves the Itô stochastic differential equation

$$\begin{aligned} dX_{t,\varepsilon}^{x,\omega} &= \frac{1}{\varepsilon}\tilde{V}\left(\frac{X_{t,\varepsilon}^{x,\omega}}{\varepsilon};\omega\right)dt + \tilde{c}\left(\frac{X_{t,\varepsilon}^{x,\omega}}{\varepsilon};\omega\right)dw_t, \\ X_{0,\varepsilon}^{x,\omega} &= x, \end{aligned} \tag{9.62}$$

with $\tilde{V}(x;\omega)$ and $\tilde{c}(x;\omega)$ given by (9.29). The generator of the process is given by (9.60). Using Itô formula we find that $u^{(\varepsilon)}(t,x)$ can be expressed with the help of the diffusion process as follows:

$$u^{(\varepsilon)}(t,x;\omega) = E_{\mathbb{P}}\big[u_0\big(X_{t,\varepsilon}^{x,\omega}\big)\big] \tag{9.63}$$

(see Theorem 5.3, p. 149 of Friedman 1975). Since $\tilde{V}(x+y;\omega) = \tilde{V}(x;\tau_y\omega)$ for all $x, y \in \mathbb{R}^d$ from uniqueness of solutions of (9.62) we conclude that for $\mathbb{P}$ a.s.

realization of diffusion and $\mathbb{Q}$ a.s. ω

$$X_{t,\varepsilon}^{x,\omega} = x + X_{t,\varepsilon}^{\tau_{x/\varepsilon}\omega}, \quad \forall (t,x) \in [0,+\infty) \times \mathbb{R}^d. \tag{9.64}$$

We maintain our convention of omitting the superscript if the diffusion starts at 0.

Define $\bar{u}(t,x)$ as the solution of the Cauchy problem for a diffusion with constant coefficients

$$\begin{aligned} \partial_t \bar{u}(t,x) &= \frac{1}{2} \sum_{p,q=1}^{d} \bar{a}_{p,q} \partial^2_{x_p,x_q} \bar{u}(t,x), \\ \bar{u}(0,x) &= u_0(x), \end{aligned} \tag{9.65}$$

with $\bar{a}_{p,q}$, given by (9.55). It is well known, see Sect. 1.2 of Friedman (1964), that the unique bounded solution of (9.65) is given by the formula

$$\bar{u}(t,x) = \int_{\mathbb{R}^d} u_0(x + \sqrt{t}y) \Phi_{\bar{a}}(y) dy, \tag{9.66}$$

where $\Phi_{\bar{a}}(y)$ is the density of a normal vector of zero mean and the covariance matrix $\bar{a} = [\bar{a}_{pq}]$

9.6.1 Random Coefficient Case

The following result is a direct consequence of Theorem 9.15.

Theorem 9.18 *Suppose that $u_0 \in C_b(\mathbb{R}^d)$ and assumptions* (R1)–(R3) *hold. Then,*

$$\lim_{\varepsilon \to 0+} \langle |u^{(\varepsilon)}(t,x) - \bar{u}(t,x)| \rangle_{\mathbb{Q}} = 0, \quad \forall (t,x) \in [0,+\infty) \times \mathbb{R}^d. \tag{9.67}$$

If $u_0 \in L^p(\mathbb{R}^d)$ for some $p \in [1,+\infty)$, then

$$\lim_{\varepsilon \to 0+} \langle \| u^{(\varepsilon)}(t) - \bar{u}(t) \|^p_{L^p(\mathbb{R}^d)} \rangle_{\mathbb{Q}} = 0, \quad \forall t \ge 0. \tag{9.68}$$

Proof Using (9.63) and (9.64) we can rewrite the expression under the limit in (9.67) in the form

$$\langle |u_0(x + X_{t,\varepsilon}^{\tau_{x/\varepsilon}\omega}) - \bar{u}(t,x)| \rangle_{\mathbb{Q}} = \langle |u_0(x + \varepsilon X^{\omega}_{t/\varepsilon^{-2}}) - \bar{u}(t,x)| \rangle_{\mathbb{Q}} \to 0,$$

as $\varepsilon \to 0+$, by virtue of Theorem 9.15 and formula (9.66).

Now we prove (9.68). Assume first that the support of $u_0(\cdot)$ is contained in a ball of radius $K > 0$, centered at 0. In light of the already proved formula (9.67) it suffices only to show that for any $\rho > 0$ there exists $R > 0$ such that

$$\limsup_{\varepsilon \to 0+} \left\langle \int_{[|x| \ge R]} |u^{(\varepsilon)}(t,x)|^p dx \right\rangle_{\mathbb{Q}} < \rho. \tag{9.69}$$

The formula follows then from (9.67) and a straightforward observation that

$$\lim_{R\to+\infty}\int_{[|x|\ge R]}\left|\bar{u}(t,x)\right|^p dx=0.$$

Using (9.63) for any $\rho>0$ and $R>2K$ the expression under the limit in (9.69) can be rewritten as:

$$\int_{[|x|\ge R]}\left\langle \left|E_{\mathbb{P}}\left[u_0\left(x+X_{t,\varepsilon}^{\tau_{x/\varepsilon}\omega}\right),\left|X_{t,\varepsilon}^{\tau_{x/\varepsilon}\omega}\right|\ge R-2K\right]\right|^p\right\rangle_{\mathbb{Q}} dx.$$

Applying Jensen inequality the above expression can be estimated by

$$\begin{aligned}&\int_{\mathbb{R}^d}\left\langle E_{\mathbb{P}}\left[\left|u_0\left(x+X_{t,\varepsilon}^{\tau_{x/\varepsilon}\omega}\right)\right|^p,\left|X_{t,\varepsilon}^{\tau_{x/\varepsilon}\omega}\right|\ge R-2K\right]\right\rangle_{\mathbb{Q}} dx\\&\quad=\left\langle E_{\mathbb{P}}\left[\int_{\mathbb{R}^d}\left|u_0\left(x+X_{t,\varepsilon}^{\omega}\right)\right|^p dx,\left|X_{t,\varepsilon}^{\omega}\right|\ge R-2K\right]\right\rangle_{\mathbb{Q}}\\&\quad=\|u_0\|_{L^p(\mathbb{R}^d)}^p\mathbb{Q}\otimes\mathbb{P}\left[\left|X_{t,\varepsilon}^{\omega}\right|\ge R-2K\right]<\rho,\end{aligned}$$

provided that R is sufficiently large. The last estimate is a consequence of tightness of the laws of $X_{t,\varepsilon}^{\omega}$, as $\varepsilon\to+\infty$, for a fixed $t>0$. Hence, (9.69) follows.

To extend (9.68) to an arbitrary $u_0\in L^p(\mathbb{R}^d)$ note that for any $\rho>0$ there exists $u_1\in C_c(\mathbb{R}^d)$ such that

$$e^{2\|U\|_\infty}\|u_0-u_1\|_{L^p(\mathbb{R}^d)}^p<\rho/3^p\quad\text{and}\quad\left\|\bar{u}(t)-\bar{u}_1(t)\right\|_{L^p(\mathbb{R}^d)}^p<\rho/3^p.$$

Here $\bar{u}_1(t,x)$ is given by (9.66), where u_0 has been replaced by u_1. We denote by $u_1^{(\varepsilon)}(t,x;\omega)$ the respective solution of (9.59). Then,

$$\left\langle\left\|u^{(\varepsilon)}(t)-u_1^{(\varepsilon)}(t)\right\|_{L^p(\mathbb{R}^d)}^p\right\rangle_{\mathbb{Q}}\le e^{2\|U\|_\infty}\left\langle\left\|u^{(\varepsilon)}(t)-u_1^{(\varepsilon)}(t)\right\|_{L^p(m_\omega^{(\varepsilon)})}^p\right\rangle_{\mathbb{Q}},$$

where $m_\omega^{(\varepsilon)}(dx):=e^{-\tilde{U}(x/\varepsilon;\omega)}dx$ is an invariant measure under $X_{t,\varepsilon}^{x,\omega}$, see (9.38). Using this fact and (9.63) the right-hand side can be further estimated by

$$\begin{aligned}&e^{2\|U\|_\infty}\left\langle\int_{\mathbb{R}^d}E_{\mathbb{P}}\left|u_0\left(X_{t,\varepsilon}^{x,\omega}\right)-u_1\left(X_{t,\varepsilon}^{x,\omega}\right)\right|^p m_\omega^{(\varepsilon)}(dx)\right\rangle_{\mathbb{Q}}\\&\quad=e^{2\|U\|_\infty}\|u_0-u_1\|_{L^p(m_\omega^{(\varepsilon)})}^p\le e^{4\|U\|_\infty}\|u_0-u_1\|_{L^p(\mathbb{R}^d)}^p<\rho/3^p.\end{aligned}$$

Hence, by the already proved formula (9.68) for solutions with the compactly supported initial data we have

$$\limsup_{\varepsilon\to 0+}\big\langle\|u^{(\varepsilon)}(t)-\bar{u}(t)\|^p_{L^p(\mathbb{R}^d)}\big\rangle_{\mathbb{Q}}$$
$$\le 3^p\limsup_{\varepsilon\to 0+}\big\{\big\langle\|u^{(\varepsilon)}(t)-u_1^{(\varepsilon)}(t)\|^p_{L^p(\mathbb{R}^d)}\big\rangle_{\mathbb{Q}}+\|\bar{u}_1(t)-\bar{u}(t)\|^p_{L^p(\mathbb{R}^d)}\big\}<\rho$$

for all $\rho>0$. Thus (9.68) follows. □

Since the limiting procedure described in the above theorem removes inhomogeneities appearing in the formulation of Eq. (9.59), leading to a constant coefficient (thus homogeneous) Eq. (9.65), it is sometimes called *homogenization.* The central limit theorem for diffusions with random coefficients allows us to conclude homogenization results for the solutions of the corresponding scaled Kolmogorov equations. We return to this issue, taking a more analytic approach, in Chap. 14.

9.6.2 Periodic Case

Assume that $u^{(\varepsilon)}(t,x)$ is the solution of

$$\partial_t u^{(\varepsilon)}(t,x) = \mathscr{L}^{(\varepsilon)}u^{(\varepsilon)}(t,x) \tag{9.70}$$

with

$$\mathscr{L}^{(\varepsilon)}f(x):=\frac{1}{2}e^{U(x/\varepsilon)}\sum_{k,l=1}^{d}\partial_{x_k}\left[e^{-U(x/\varepsilon)}a_{k,l}\left(\frac{x}{\varepsilon}\right)\partial_{x_l}f(x)\right],\quad f\in C^2(\mathbb{R}^d), \tag{9.71}$$

where coefficients $a(x)=[a_{k,l}(x)]$, $k,l=1,\dots,d$ and $U(x)$ are 1-periodic functions satisfying the same conditions as in Sect. 9.1. We suppose furthermore that the initial condition $u_0(\cdot)$ belongs to $L^p(\mathbb{R}^d)$ for some $p\in[1,+\infty)$.

The solution $u^{(\varepsilon)}(t,x)$ can be represented as

$$u^{(\varepsilon)}(t,x)=E_{\mathbb{P}}\big[u_0\big(\varepsilon X_{t/\varepsilon^2}^{x/\varepsilon}\big)\big]=E_{\mathbb{P}}\big[u_0\big(\varepsilon X_{t/\varepsilon^2}^{(x/\varepsilon)}+\varepsilon[x/\varepsilon]\big)\big], \tag{9.72}$$

where the last equality follows from (9.32).

On the other hand for $\omega\in\mathbb{T}^d$ consider the process $X_t^{x,\omega}$ that is the solution of the stochastic differential equation with "randomized" coefficients, i.e. it solves (9.30) with coefficients that are random fields obtained from periodic functions via (9.31). The solution of the corresponding parabolic equation (9.59) is given by

$$u^{(\varepsilon)}(t,x;\omega)=E_{\mathbb{P}}\big[u_0\big(\varepsilon X_{t/\varepsilon^2}^{x/\varepsilon,\omega}\big)\big]$$

and from (9.31) and the second equality in (9.72) we conclude that

$$u^{(\varepsilon)}(t,x;\omega)=E_{\mathbb{P}}\big[u_0\big(\varepsilon X_{t/\varepsilon^2}^{(x/\varepsilon+\omega)}+\varepsilon[x/\varepsilon+\omega]-\varepsilon\omega\big)\big]. \tag{9.73}$$

The results of Theorem 9.18, translated to the periodic case, read as follows

$$\lim_{\varepsilon\to 0+}\int_{\mathbb{T}^d}\left|u^{(\varepsilon)}(t,x;\omega)-\bar{u}(t,x)\right|d\omega=0,\quad \forall (t,x)\in[0,+\infty)\times\mathbb{R}^d \tag{9.74}$$

and

$$\lim_{\varepsilon\to 0+}\int_{\mathbb{R}^d}\int_{\mathbb{T}^d}\left|u^{(\varepsilon)}(t,x;\omega)-\bar{u}(t,x)\right|^p dx d\omega=0,\quad \forall t\ge 0 \tag{9.75}$$

for any $p\in[1,+\infty)$. In what follows we show that from this result the classical homogenization of parabolic equations with periodic coefficients can be concluded.

Theorem 9.19 *If $u_0\in L^p(\mathbb{R}^d)$ for some $p\in[1,+\infty)$, then*

$$\lim_{\varepsilon\to 0+}\left\|u^{(\varepsilon)}(t)-\bar{u}(t)\right\|_{L^p(\mathbb{R}^d)}=0,\quad \forall t\ge 0. \tag{9.76}$$

Here $\bar{u}(t)$ is given by (9.66).

Proof Using a density argument it suffices only to show (9.76) for $u_0\in C_c(\mathbb{R}^d)$. An elementary consideration invoking the L^p continuity of $\bar{u}(t,x)$ in the x variable and invariance of the Lebesgue measure under spatial translations reduces the proof to showing that

$$\lim_{\varepsilon\to 0+}\int_{\mathbb{R}^d}\int_{\mathbb{T}^d}\left|u^{(\varepsilon)}(t,x+\varepsilon\omega)-\bar{u}(t,x)\right|^p dx d\omega=0,\quad \forall t\ge 0. \tag{9.77}$$

From the uniform continuity of u_0, equalities (9.72) and (9.73), we conclude that

$$\begin{aligned}
&\lim_{\varepsilon\to 0+}\int_{\mathbb{R}^d}\int_{\mathbb{T}^d}\left|u^{(\varepsilon)}(t,x;\omega)-u^{(\varepsilon)}(t,x+\varepsilon\omega)\right|^p dx d\omega\\
&\quad\le \lim_{\varepsilon\to 0+}\int_{\mathbb{R}^d}\int_{\mathbb{T}^d}E_{\mathbb{P}}\Big|u_0\big(\varepsilon X_{t/\varepsilon^2}^{(x/\varepsilon+\omega)}+\varepsilon[x/\varepsilon+\omega]-\varepsilon\omega\big)\\
&\qquad-u_0\big(\varepsilon X_{t/\varepsilon^2}^{(x/\varepsilon+\omega)}+x\big)\Big|^p dx d\omega=0.
\end{aligned}$$

Combining the above with (9.75) we conclude (9.77). □

9.7 Proofs of Propositions 9.8 and 9.9

This section is devoted to the proofs of technical results characterizing the generator of the environment process and its adjoint announced in Sect. 9.4.

9.7.1 Proof of Proposition 9.8

Lemma 9.20 *We have $P_t(B(\Omega)) \subset C_b^2(\Omega)$ for any $t > 0$. In addition, for a multi-index $m = (m_1, \dots, m_d)$ with $|m| \le 2$*

$$D^m P_t f(\omega) = \int_{\mathbb{R}^d} \nabla^m p_t^{\omega}(0, y) f(\tau_y \omega) dy, \quad \forall f \in B(\Omega), \quad \mathbb{Q} \text{ a.s.,} \tag{9.78}$$

where the gradient operator acts in the x variable of $p_t^{\omega}(x, y)$.

Proof According to (9.37) we have

$$P_t f(\tau_x \omega) = \int_{\mathbb{R}^d} p_t^{\tau_x \omega}(0, y) f(\tau_{y+x} \omega) dy = \int_{\mathbb{R}^d} p_t^{\omega}(x, y) f(\tau_y \omega) dy = P_t^{\omega} \tilde{f}(x). \tag{9.79}$$

Recall that $\tilde{f}(x; \omega) = f(\tau_x \omega)$. The second equality follows from homogeneity of transition probability densities, see (9.35), and the substitution $y := y + x$. The regularity of $P_t f$ and formula (9.78) are easy consequences of differentiability of $p_t^{\omega}(x, y)$ in the x variable and Gaussian bounds of the respective derivatives: for any $T > 0$ we have

$$\left|\nabla_x^m p_t^{\omega}(x, y)\right| \le \frac{C}{t^{(d+|m|)/2}} \exp\left\{-\frac{|x - y|^2}{Ct}\right\}, \quad \forall t \in [0, T], x, y \in \mathbb{R}^d$$

for any multi-index m satisfying $|m| = 0, 1, 2$. The constant $C > 0$ does not depend on m, ω, but may depend on T, see e.g. Theorems 6.4.5 and 6.4.7 of Friedman (1975). □

For any $f \in C_b^2(\Omega)$ and $t \ge 0$ we can write

$$P_t^{\omega} \tilde{f}(x) = \tilde{f}(x) + \int_0^t \mathcal{L}_{\omega} P_s^{\omega} \tilde{f}(x) ds \tag{9.80}$$

and $P_t f = P_t^{\omega} \tilde{f}(0)$. Here $\mathcal{L}_{\omega}$ is given by (9.23). From Lemma 9.20 we deduce that $P_t f \in C_b^2(\Omega)$. In addition it follows from (9.79) that $P_t^{\omega} \tilde{f} \in C_b^2(\mathbb{R}^d)$ and

$$\left\| P_t^{\omega} \tilde{f} \right\|_{C_b^2(\mathbb{R}^d)} \le \| P_t f \|_{2,\infty}, \quad \mathbb{Q}\text{-a.s.}$$

Using (9.80) and the dominated convergence theorem, we conclude that the limit

$$L P_t f(\omega) := \lim_{h \to 0+} \frac{P_{h+t} f(\omega) - P_t f(\omega)}{h} = \mathcal{L}_{\omega} P_t^{\omega} \tilde{f}(0),$$

understood in $L^2(\mathbb{Q})$, exists for any $t \ge 0$. The above equality implies that

$$L P_t f(\omega) = \frac{1}{2} e^U \sum_{k,l=1}^{d} D_k\left(a_{k,l} e^{-U} D_l P_t f\right), \quad \forall f \in C_b^2(\Omega). \tag{9.81}$$

Using this formula we deduce that

$$\frac{d}{dt}\langle P_t f\rangle_\pi = \langle L P_t f\rangle_\pi = \frac{1}{2Z}\sum_{k,l=1}^{d}\big\langle D_k\big(a_{kl}e^{-U}D_l P_t f\big)\big\rangle_{\mathbb{Q}} = 0. \tag{9.82}$$

This implies the invariance of measure π and, as a consequence, allows to extend the semigroup to $L^2(\pi)$. To show its strong continuity it suffices only to prove that

$$\lim_{h\to 0+}\|P_h f - f\|_\pi = 0 \quad \text{for any } f \in C_b^\infty(\Omega). \tag{9.83}$$

From (9.80)

$$\big|P_h f(\omega) - f(\omega)\big| \le h\max\big\{\|c\|_\infty, \|b\|_\infty\big\}\|f\|_{C_b^2(\Omega)}, \quad \forall h > 0.$$

The limit in (9.83) follows from the dominated convergence theorem.

To show ergodicity suppose that $A \in \mathscr{F}$ satisfies $P_t 1_A = 1_A$ for some $t > 0$. From (9.37) we have

$$0 = \langle 1_{A^c}, P_t 1_A\rangle_{\mathbb{Q}} = \int_{\mathbb{R}^d}\big\langle 1_{A^c}(\omega)p_t^\omega(0,x)1_A(\tau_x\omega)\big\rangle_{\mathbb{Q}}dx.$$

Since $p_t^\omega(0,x) > 0$ for all $x \in \mathbb{R}^d$ and $\mathbb{Q}$-a.s. ω we conclude that $\langle 1_{A^c}1_A\circ\tau_x\rangle_{\mathbb{Q}} = 0$ for all $x \in \mathbb{R}^d$. This however means that A is invariant under the group of spatial shifts, thus it has to be $\mathbb{Q}$-trivial thanks to ergodicity of the group action. Since π is equivalent to $\mathbb{Q}$ this, in turn implies that A is π-trivial and ergodicity of the invariant measure follows.

The formula for Lf—the generator of $\{P_t, t \ge 0\}$—can be obtained from (9.81), substituting there $t = 0$. From Lemma 9.20 we conclude that $C_b^2(\Omega)$ is invariant under the semigroup. Therefore, it is a core of L, see Proposition 1.3.3 of Ethier and Kurtz (1986).

9.7.2 Proof of Proposition 9.9

The following result permits to identify a formula for the transition probability density of the diffusion that corresponds to the formal adjoint given by (9.41).

Lemma 9.21 *The transition probability density of $\{\hat{X}_t^{x,\omega}, t \ge 0\}$ equals*

$$\hat{p}_t^\omega(x,y) = \exp\big\{\tilde{U}(x) - \tilde{U}(y)\big\}p_t^\omega(y,x), \quad \forall x,y \in \mathbb{R}^d,\ t > 0. \tag{9.84}$$

Proof Suppose that f belongs to $C_b^2(\mathbb{R}^d)$—the space of functions twice continuously differentiable with bounded derivatives. Then

$$v_\omega(t,x) := \int_{\mathbb{R}^d} p_t^\omega(y,x)f(y)dy$$

is the unique solution in $C_b^{1,2}([0,+\infty)\times\mathbb{R}^d)$—the space of functions that are continuously differentiable, once in t, twice in x with bounded derivatives—of the Cauchy problem

$$\partial_t v_\omega(t,x)=\mathscr{L}_\omega^* v_\omega(t,x),$$
$$v_\omega(0,x)=f(x),$$

where

$$\mathscr{L}_\omega^* f(x):=\frac{1}{2}\sum_{k,l=1}^d \partial_{x_k}\big\{e^{-\tilde U(x;\omega)}\tilde a_{l,k}(x;\omega)\partial_{x_l}\big[e^{\tilde U(x;\omega)}f(x)\big]\big\}, \tag{9.85}$$

see e.g. p. 149 of Friedman (1975). Let

$$u_\omega(t,x):=e^{\tilde U(x;\omega)}v_\omega(t,x)=\int_{\mathbb{R}^d} e^{\tilde U(x;\omega)}p_t^\omega(y,x)f(y)dy. \tag{9.86}$$

It satisfies the Cauchy problem

$$\begin{aligned}\partial_t u_\omega(t,x)&=\hat{\mathscr{L}}_\omega u_\omega(t,x),\\ u_\omega(0,x)&=e^{\tilde U(x;\omega)}f(x),\end{aligned} \tag{9.87}$$

with $\hat{\mathscr{L}}_\omega$ given by formula (9.41). Using the fundamental solution of the above equation, see Theorem 6.4.6 of Friedman (1975), we can represent $u_\omega(t,x)$ as follows

$$u_\omega(t,x)=\int_{\mathbb{R}^d}\hat p_t^\omega(x,y)e^{\tilde U(y;\omega)}f(y)dy.$$

Comparing the above formula with (9.86) we obtain (9.84). □

For any $f,g\in B(\Omega)$ we have

$$\begin{aligned}\langle P_t f,g\rangle_\pi&=\frac{1}{Z}\left\langle\int_{\mathbb{R}^d}p_t^\omega(0,y)f(\tau_y\omega)g(\omega)e^{-U(\omega)}dy\right\rangle_{\mathbb{Q}}\\ &=\frac{1}{Z}\left\langle\int_{\mathbb{R}^d}p^{\tau_{-y}\omega}(t,0,y)f(\omega)g(\tau_{-y}\omega)e^{-U(\tau_{-y}\omega)}dy\right\rangle_{\mathbb{Q}}.\end{aligned} \tag{9.88}$$

Using homogeneity of transition probability densities and substituting $y:=-y$ we can write that the utmost right-hand side of (9.88) equals

$$\frac{1}{Z}\left\langle\int_{\mathbb{R}^d}e^{U(\omega)-U(\tau_y\omega)}p^\omega(t,y,0)f(\omega)g(\tau_y\omega)e^{-U(\omega)}dy\right\rangle_{\mathbb{Q}}$$
$$\stackrel{(9.84)}{=}\frac{1}{Z}\left\langle\int_{\mathbb{R}^d}\hat p^\omega(t,0,y)f(\omega)g(\tau_y\omega)e^{-U(\omega)}dy\right\rangle_{\mathbb{Q}}=\langle f,\hat P_t g\rangle_\pi.$$

□

9.8 One-Dimensional Case

In dimension $d = 1$ we have an explicit formula for the limiting variance. In fact the random matrix a reduces to a random variable such that

$$c_0^{-1} \geq a(\omega) \geq c_0$$

for some constant $c_0 > 0$ and $\mathbb{Q}$ a.s. ω. We shall also assume that $a, U \in C_b^2(\Omega)$. Since the gradient operator has only one component we denote it by D.

Theorem 9.22 *Under the above conditions we have*

$$D\chi = \frac{e^U a^{-1}}{\langle e^U a^{-1} \rangle_{\mathbb{Q}}} - 1. \tag{9.89}$$

The limiting variance appearing in Theorem 9.15 *equals*

$$\bar{a} = \left[\langle e^U a^{-1} \rangle_{\mathbb{Q}} \langle e^{-U} \rangle_{\mathbb{Q}} \right]^{-1}. \tag{9.90}$$

Proof Denote by ψ the right-hand side of (9.89). A simple calculation shows that

$$\frac{1}{2} e^U D\left(e^{-U} a\psi\right) = -V.$$

Hence,

$$(\lambda - L)\chi_\lambda = -\frac{1}{2} e^U D\left(e^{-U} a\psi\right).$$

Multiplying both sides of the above equality by an arbitrary $f \in L^2(\pi) \cap \mathscr{H}_1$ and integrating with respect to the measure π (see (9.39)) we obtain

$$\lambda \langle \chi_\lambda, f \rangle_\pi + \frac{1}{2} \langle aD\chi_\lambda, Df \rangle_\pi = \frac{1}{2} \langle a\psi, Df \rangle_\pi.$$

Since $D\chi_\lambda \to D\chi$ and $\lambda\chi_\lambda \to 0$, strongly in $L^2(\pi)$, as $\lambda \to 0+$, we obtain

$$\langle aD\chi, Df \rangle_\pi = \langle a\psi, Df \rangle_\pi, \quad \forall f \in L^2(\pi) \cap \mathscr{H}_1.$$

In consequence,

$$e^U a^{-1}(D\chi - \psi) \equiv \text{const}, \quad \mathbb{Q} \text{ a.s.}$$

Note however that $\langle D\chi - \psi \rangle_{\mathbb{Q}} = 0$ thus the constant on the right-hand side has to vanish and (9.89) follows. Formula (9.90) follows directly from (9.89) and (9.55). □

Remark 9.23 Note that from the Cauchy–Schwartz inequality we obtain

$$\langle a^{-1/2} \rangle_{\mathbb{Q}}^2 = \langle a^{-1/2} e^{U/2} e^{-U/2} \rangle_{\mathbb{Q}}^2 \leq \langle e^U a^{-1} \rangle_{\mathbb{Q}} \langle e^{-U} \rangle_{\mathbb{Q}}.$$

On the other hand, Jensen inequality gives

$$\langle a^{-1/2}\rangle_{\mathbb{Q}}^{-2} \le \langle a^{1/2}\rangle_{\mathbb{Q}}^{2} \le \langle a\rangle_{\mathbb{Q}}.$$

Summarizing, we have proved that

$$\bar{a} \le \langle a\rangle_{\mathbb{Q}}.$$

In fact equality can only hold when a is constant. The limiting variance is therefore always bounded from above by the average of the diffusivity coefficient.

Remark 9.24 Let us consider again the case of a diffusion with quasi-periodic coefficients discussed in Sect. 9.2. Suppose that X_t^{ω} is the diffusion that starts at 0 and corresponds to the generator (9.18). We can use Remark 9.5 together with Theorem 9.15 and conclude that for any density $u_*(\omega)$ with respect to the normalized Lebesgue measure $d\omega$ on $\mathbb{T}^N$ and $f \in C_b(\mathbb{R})$ we have

$$\lim_{t\to+\infty}\int_{\mathbb{T}^N}\left[E_{\mathbb{P}} f\left(X_t^{\omega}/\sqrt{t}\right) - \int_{\mathbb{R}^d} f(y)\Phi_{\bar{a}}(y)dy\right]^2 u_*(\omega)d\omega = 0.$$

Here

$$\bar{a} = \left[\int_{\mathbb{T}^N} e^{U(\omega)}a^{-1}(\omega)d\omega \int_{\mathbb{T}^N} e^{-U(\omega)}d\omega\right]^{-1}. \tag{9.91}$$

9.9 Diffusions with Time Dependent Coefficients

In this section we consider the situation when the diffusion coefficients depend on the temporal variable. Some complication comes from the fact that the generator of the respective environment process does not satisfy the sector condition. We can however consider it as a bounded perturbation of a normal operator, that corresponds to the generator of a suitable constant coefficient diffusion, and apply the results of Sect. 2.7.5. The arguments are for the most part analogous to those made in the time independent case. We discuss only those points where more significant modifications are required.

9.9.1 Space-Time Stationary Environments

Suppose that $\{\tau_{t,x}, (t,x) \in \mathbb{R}^{d+1}\}$ is a measurable $d+1$ parameter group of transformations of the space Ω preserving a Borel probability measure $\mathbb{Q}$. In the same way as in Sect. 9.3.1 we define a $d+1$ parameter strongly continuous group of unitary operators $T_{t,x}f = f\circ\tau_{t,x}$ on $L^2(\mathbb{Q})$. The generator corresponding to $e_0 = (1,\dots,0)$

shall be denoted by D_0, while the generators corresponding to the other directions e_i shall be denoted by D_i, $i=1,\dots,d$.

Suppose that k_1,k_2 are non-negative integers. Spaces $C^{k_1,k_2}(\Omega)$, resp. $C_b^{k_1,k_2}(\Omega)$, consist of random variables that are k_1 times continuously differentiable, resp. boundedly and continuously differentiable, in e_0 and k_2 times differentiable in the spatial directions. On this space we define, in the obvious way, the norm $\|\cdot\|_{k_1,k_2,\infty}$. Let

$$C^\infty(\Omega):=\bigcap_{k_1,k_2\ge 0} C^{k_1,k_2}(\Omega) \quad \text{and} \quad C_b^\infty(\Omega):=\bigcap_{k_1,k_2\ge 0} C_b^{k_1,k_2}(\Omega).$$

We can also introduce the spaces $H^{k_1,k_2}(\Omega)$ consisting of those elements that possess respectively k_1 and k_2 derivatives in e_0 and the spatial directions e_p, $p=1,\dots,d$. The norm on $H^{k_1,k_2}(\Omega)$ is given by

$$\|f\|^2_{H^{k_1,k_2}} := \sum_{|m|=k_2} \|D_0^{k_1} D^m f\|^2_{\mathbb{Q}}, \tag{9.92}$$

where $D^m = D_1^{m_1}\dots D_d^{m_d}$ for a multi-index $m=(m_1,\dots,m_d)$. Analogously as in Proposition 9.3 we deduce that $C_b^\infty(\Omega)$ is dense in $H^{k_1,k_2}(\Omega)$ for any $k_1,k_2\ge 0$.

We assume that the entries of the random matrix a and random variable U belong to $C_b^{1,2}(\Omega)$, cf. condition (R1) of Sect. 9.3.3 and that the matrix is uniformly positive definite in the sense of condition (R2). Condition (R3) is replaced by an analogous hypothesis concerning the group of time-space shifts. Furthermore we assume that:

(R4) U does not depend on shifts in the t direction, i.e. $U\circ\tau_{t,0}=U$ for all $t\in\mathbb{R}$.

Let $\tilde a(t,x;\omega)$ and $\tilde U(x;\omega)$ be given by $a(\tau_{t,x}\omega)$ and $U(\tau_{0,x}\omega)$ respectively. For any (s,x) we define $\{X_t^{s,x,\omega}, t\ge s\}$ as the solution to

$$dX_t^{s,x,\omega} = \tilde V\big(t,X_t^{s,x,\omega};\omega\big)dt + \tilde c\big(t,X_t^{s,x,\omega};\omega\big)dw_t^s, \qquad X_s^{s,x,\omega}=x. \tag{9.93}$$

Here $\{w_t^s, t\ge s\}$ is the Brownian motion given by

$$w_t^s := w_{t-s}, \quad t\ge s,$$

where $\{w_t, t\ge 0\}$ is a standard, d-dimensional Brownian motion. We shall omit, as usual, writing the subscripts corresponding to the initial data if $(s,x)=(0,0)$. From uniqueness of solutions to (9.93) and space-time stationarity of the coefficients it follows that

$$X_t^{s,x,\tau_{h,y}\omega} + y = X_{t+h}^{s+h,x+y,\omega} \quad \text{for all } t\ge s,\ x,y,h\in\mathbb{R}^d,\ \omega\in\Omega,\ \mathbb{P}\text{ a.s.}$$

Define the environment process over $(\Sigma,\mathscr{A},\mathbb{P})$ by

$$\eta_t^\omega := \tau_{t,X_t^\omega}\omega, \quad t\ge 0. \tag{9.94}$$

Most of the facts concerning this process follow from the respective properties of diffusions with time dependent coefficients in the same way as for their time independent counterparts, see Sect. 9.4. We shall list them below without proofs. The arguments presented in the time independent case require only minor adjustments and we leave them as an exercise.

Suppose that $s < t$. Let $p^{\omega}_{s,t}(x, y)$, $x, y \in \mathbb{R}^d$ be the transition probability density functions corresponding to the process given by (9.93). Since

$$p^{\tau_{h,z}\omega}_{s,t}(x, y) = p^{\omega}_{s+h,t+h}(x+z, y+z) \tag{9.95}$$

for all $(h, z) \in \mathbb{R}^{1+d}$ we conclude the following.

Proposition 9.25 *The process* $\{\eta_t, t \ge 0\}$ *is Markovian. Its transition probability semigroup is given by*

$$P_t f(\omega) = \int p^{\omega}_{0,t}(0, x) f(\tau_{t,x}\omega) dx \quad \textit{for all } f \in B(\Omega),\ \omega \in \Omega. \tag{9.96}$$

The probability measure

$$\pi(d\omega) := \frac{1}{Z} e^{-U(\omega)} \mathbb{Q}(d\omega),$$

with $Z := \langle e^{-U} \rangle_{\mathbb{Q}}$, *is invariant and ergodic for the semigroup.*

In analogy with the time independent situation the transition probability semigroup extends to a strongly continuous semigroup of contractions on $L^2(\pi)$. Using Itô's formula we conclude the formula for its generator. For $f \in C^{1,2}_b(\Omega)$ we have

$$\begin{aligned} Lf(\omega) &= (\partial_t + \mathscr{L}_{\omega}) \tilde{f}(0,0) \\ &= D_0 f(\omega) + \frac{1}{2} e^{U(\omega)} \sum_{k,l=1}^{d} D_k \big(e^{-U(\omega)} a_{k,l}(\omega) D_l f(\omega) \big). \end{aligned} \tag{9.97}$$

Here $a_{k,l}(\omega)$, $k, l = 1, \ldots, d$ are the entries of the matrix $a(\omega)$. Adapting the respective part of the proof of Proposition 9.8 we conclude that

$$P_t\big(B(\Omega)\big) \subset C^{1,2}_b(\Omega), \quad \forall t > 0.$$

Thus $C^{1,2}_b(\Omega)$ is a core of the generator.

To describe the adjoint semigroup $\{P^*_t, t \ge 0\}$ we introduce the transition probability densities $\hat{p}^{\omega}_{s,t}(y, x)$ that correspond to the time inhomogeneous diffusion with the generator $\partial_t + \hat{\mathscr{L}}_{\omega}$, where

$$\hat{\mathscr{L}}_{\omega} f(t, x) := \frac{1}{2} e^{\tilde{U}(x;\omega)} \sum_{k,l=1}^{d} \partial_{x_k} \big(e^{-\tilde{U}(x;\omega)} \tilde{a}_{l,k}(-t, x; \omega) \partial_{x_l} f(t, x) \big)$$

for $f \in C_b^{1,2}(\mathbb{R}^{1+d})$. Repeating the calculations done in the proof of Proposition 9.9 we obtain the following.

Proposition 9.26 *For any $t \ge 0$ we have*

$$P_t^* f(\omega) = \int \hat{p}_{0,t}^{\omega}(0, x) f(\tau_{-t,x}\omega) dx, \quad f \in L^2(\pi). \tag{9.98}$$

Also, a straightforward modification of the proof of Lemma 9.21 allows us to conclude that for any $s < t$, $x, y \in \mathbb{R}^d$ we have

$$\hat{p}_{s,t}^{\omega}(x, y) = \exp\{\tilde{U}(x) - \tilde{U}(y)\} p_{-t,-s}^{\omega}(y, x). \tag{9.99}$$

Therefore, we conclude the following.

Corollary 9.27 *The space $C_b^{1,2}(\Omega)$ is a core of L^* and for any $f \in C_b^{1,2}(\Omega)$*

$$L^* f = -D_0 f + \frac{1}{2} e^U \sum_{k,l=1}^{d} D_k\big(e^{-U} a_{l,k} D_l f\big). \tag{9.100}$$

The symmetric and anti-symmetric parts of the generator equal respectively

$$Sf = \frac{1}{2} e^U \sum_{k,l=1}^{d} D_k\big(e^{-U} a_{k,l}^s D_l f\big) \tag{9.101}$$

and

$$Af = D_0 f + \frac{1}{2} e^U \sum_{k,l=1}^{d} D_k\big(e^{-U} a_{k,l}^{\dagger} D_l f\big), \tag{9.102}$$

where a^s, $a^{\dagger}$ denote the symmetric and anti-symmetric parts of matrix a.

We can introduce the spaces $\mathscr{H}_1$, $\mathscr{H}_{-1}$ as in Sect. 2.2. Note however that, in contrast with the time independent situation, the generator L need not satisfy the sector condition. Indeed, any f that is invariant under the subgroup of spatial shifts $\{\tau_{0,x}, x \in \mathbb{R}^d\}$ but depending on temporal shifts $\{\tau_{t,0}, t \in \mathbb{R}\}$ satisfies $Sf = 0$ but $Af \neq 0$. This, of course, contradicts the sector condition.

Remark 9.28 Similarly to the time independent case, discussed in Remark 9.11, one can see that the space $\mathscr{H}_1$ is isomorphic with $H^{0,1}(\Omega)$. Using this isomorphism we can define D_k, $k = 1, \dots, d$ on $\mathscr{H}_1$.

9.9.2 Central Limit Theorem

From (9.93) we obtain decomposition (9.46). Using the argument from the proof of Proposition 9.12 we immediately conclude that $V_p \in \mathscr{H}_{-1}$ for each $p = 1, \dots, d$. For any $\lambda > 0$ we can define $\chi_\lambda^{(p)}$ belonging to $D(L)$—the λ-corrector in the direction e_p—as the unique solution of the resolvent equation (9.48). Since $C_b^{1,2}(\Omega)$ is a core of the generator L formula (9.50) holds. For a given constant $c_0 > 0$ we let

$$S_0 f = \frac{1}{2} c_0 e^U \sum_{k=1}^{d} D_k\left(e^{-U} D_k f\right)$$

and

$$Bf := \frac{1}{2} e^U \sum_{k,l=1}^{d} D_k\left[e^{-U}(a_{k,l} - c_0 \delta_{k,l}) D_l f\right]$$

for $f \in C_b^{1,2}(\Omega)$. Let also $r_t^\omega(x, y)$ be the transition probability density for the diffusion corresponding to the generator

$$\mathscr{S}_0^\omega f(x) = \frac{1}{2} c_0 e^{\tilde{U}(x;\omega)} \sum_{k=1}^{d} \partial_{x_k}\left(e^{-\tilde{U}(x;\omega)} \partial_{x_k} f(x)\right), \quad f \in C_b^2\left(\mathbb{R}^d\right).$$

On $C_b^{1,2}(\Omega)$ we can write that $L = L_0 + B$, where $L_0 := D_0 + S_0$. The symmetric and anti-symmetric parts of L_0 on $C_b^{1,2}(\Omega)$ are given by S_0 and D_0 correspondingly. The closures of L_0 and S_0 are the generators of the respective semigroups $\{R_t, t \geq 0\}$ and $\{Q_t, t \geq 0\}$ given by

$$R_t f(\omega) = \int_{\mathbb{R}^d} r_t^\omega(0, y) f(\tau_{t,y}\omega) dy$$

and

$$Q_t f(\omega) = \int_{\mathbb{R}^d} r_t^\omega(0, y) f(\tau_{0,y}\omega) dy, \quad f \in L^2(\pi), t \geq 0$$

while the closure of D_0 generates the unitary group $\{T_{t,0}, t \in \mathbb{R}\}$. One can easily verify the commutation relation

$$[Q_t, T_{s,0}] = 0, \quad \forall t \geq 0, s \in \mathbb{R}.$$

This in turn implies that the resolvent operators corresponding to the semigroups commute, from which we deduce the commutation of the spectral resolutions of their generators, see Sect. 2.7.5. Thus, L_0 is normal in the sense of the definition used in that section.

Since

$$\begin{aligned}|\langle f, Bg\rangle_\pi| &= \frac{1}{2}|\langle (a - c_0 I)\nabla f, \nabla g\rangle_\pi| \\ &\le (\|a\|_\infty / c_0)\langle f, (-L_0) f\rangle_\pi|^{1/2}\langle g, (-L_0) g\rangle_\pi^{1/2}, \quad \forall f, g \in C_b^{1,2}(\Omega)\end{aligned} \tag{9.103}$$

condition (2.56) holds with $\mathscr{C} = C_b^{1,2}(\Omega)$. Verification of the remaining hypotheses made in Sect. 2.7.5 is straightforward. As a consequence we obtain the $\mathscr{H}_1$ convergence of $\chi_\lambda^{(p)}$ to $\chi^{(p)}$, as $\lambda \to 0+$. Thus, we conclude the following result.

Theorem 9.29 *Under the assumptions made at the beginning of Sect.* 9.9.1, *random variables $X_t/\sqrt{t}$ satisfy the central limit theorem in probability with respect to the environment.*

9.10 Comments and References

The central limit theorem for diffusions with periodic, almost periodic, or random stationary coefficients is a well-established subject that can be approached from both the probabilistic and analytic points of view. In the latter case, the respective question concerns proving that the limit of the Green functions corresponding to the random operators with fast varying coefficients is the Green function of an operator with constant coefficients (the so-called G-limit). Since the limiting operator does not depend any longer on the inhomogeneities of the medium the procedure of taking this limit is referred to as a *homogenization*. There exists a vast literature concerning the subject. We refer the reader to monographs Bensoussan et al. (1978) and Zhikov et al. (1994) and references therein. More detailed discussion of analytic aspects of homogenization is presented in Chap. 14.

The first proofs of the central limit theorem for diffusions with stationary and ergodic coefficients are due to Kozlov, Papanicolaou-Varadhan and Osada. In Kozlov (1979) and Papanicolaou and Varadhan (1981) diffusions with generators in a divergence form have been considered. It has been shown there that the respective generators G-converge to the limiting constant coefficient "homogenized" operator. The results contained in this chapter are essentially covered by Osada (1983) although the argument presented here is quite different. The proof in Osada (1983) has more analytic flavor and relies on estimates of the Hölder modulus of continuity of the fundamental solutions corresponding to the parabolic differential operator $\partial_t - \mathscr{L}^\omega$, where $\mathscr{L}^\omega$ is given by (9.23).

The homogenization method can be applied to deal with other classes of random operators. It is possible to incorporate random elliptic operators, whose principal part is in a divergence form and contains a zero order potential term. In this case one can use Feynman–Kac formula to identify the limit. This question has been considered in Lejay (2001), see also Campillo et al. (2001); Iftimie et al. (2008);

Pardoux and Piatnitski (2006) for further results on the subject. The method can also be used to prove a non-central limit theorem for processes whose generators are given by non-local operators with random characteristics, e.g. the solutions of stochastic different equations driven by a Levy noise, whose jump measure is either periodic or random, see Franke (2007); Rhodes and Vargas (2009).

The proof of homogenization for diffusions with time dependent coefficients contained in Sect. 9.9 comes essentially from Landim et al. (1998b), see also Rhodes (2007, 2008).

A central limit theorem for diffusions in random environments, whose generators are not in a divergence form, is quite difficult to prove due to the fact that in general an invariant measure, absolutely continuous with respect to the probability measure corresponding to the "randomness" of the environment, is not known. In Papanicolaou and Varadhan (1982) the central limit theorem is proved for diffusions whose generator is symmetric and without a drift term. The corresponding diffusion is then a martingale and the question of showing the central limit theorem can be reduced to the problem of proving the law of large numbers for the quadratic covariation of those martingales. See also Lawler (1982) for the discrete case of a random walk in random environment. Somewhat special in this context is the case of spatial dimension $d = 1$. Then one can characterize the conditions for the existence of an invariant measure in terms of the properties of coefficients. The central limit theorem can be obtained using the method based on the environment as seen from the particle process, described in this chapter, see Schmitz (2009). In a higher dimension an invariance principle has been shown by a different approach in case of isotropic diffusions in a random environment that are small perturbations of Brownian motion, when the space dimension is three or more, see Sznitman and Zeitouni (2006). Results establishing the existence of an invariant measure and the central limit theorem can be obtained for non-divergence form diffusions using coupling techniques, provided that the coefficients are sufficiently strongly mixing and there exists a deterministic drift, sufficiently large so that the diffusion trajectory in the future explores the regions where it has not been in the past. Mixing of the environment assures then sufficient decorrelation properties of the path so the central limit theorem follows. This type of problem has been considered e.g. in Komorowski and Krupa (2004); Schmitz (2006); Shen (2002, 2003).

References

Bensoussan A, Lions JL, Papanicolaou G (1978) Asymptotic analysis for periodic structures. Studies in mathematics and its applications, vol 5. North-Holland, Amsterdam

Campillo F, Kleptsyna M, Piatnitski A (2001) Homogenization of random parabolic operator with large potential. Stoch Process Appl 93(1):57–85

Ethier SN, Kurtz TG (1986) Markov processes: characterization and convergence. Wiley series in probability and mathematical statistics: probability and mathematical statistics. Wiley, New York

Feller W (1971) An introduction to probability theory and its applications, vol II, 2nd edn. Wiley, New York

Franke B (2007) A functional non-central limit theorem for jump-diffusions with periodic coefficients driven by stable Lévy-noise. J Theor Probab 20(4):1087–1100

Friedman A (1964) Partial differential equations of parabolic type. Prentice-Hall, Englewood Cliffs

Friedman A (1975) Stochastic differential equations and applications, vol 1. Probability and mathematical statistics, vol 28. Academic Press [Harcourt Brace Jovanovich Publishers], New York

Gilbarg D, Trudinger NS (1983) Elliptic partial differential equations of second order. Grundlehren der Mathematischen Wissenschaften [Fundamental principles of mathematical sciences], vol 224, 2nd edn. Springer, Berlin

Iftimie B, Pardoux É, Piatnitski A (2008) Homogenization of a singular random one-dimensional PDE. Ann Inst Henri Poincaré Probab Stat 44(3):519–543

Komorowski T, Krupa G (2004) On stationarity of Lagrangian observations of passive tracer velocity in a compressible environment. Ann Appl Probab 14(4):1666–1697

Kozlov SM (1979) The averaging of random operators. Mat Sb (NS) 109 (151):188–202, 327

Landim C, Olla S, Yau HT (1998b) Convection-diffusion equation with space-time ergodic random flow. Probab Theory Relat Fields 112(2):203–220

Lawler GF (1982) Weak convergence of a random walk in a random environment. Commun Math Phys 87(1):81–87

Lejay A (2001) Homogenization of divergence-form operators with lower-order terms in random media. Probab Theory Relat Fields 120(2):255–276

Osada H (1983) Homogenization of diffusion processes with random stationary coefficients. In: Probability theory and mathematical statistics, Tbilisi, 1982. Lecture notes in math, vol 1021. Springer, Berlin, pp 507–517

Papanicolaou GC, Varadhan SRS (1981) Boundary value problems with rapidly oscillating random coefficients. In: Random fields, vols I, II, Esztergom, 1979. Colloq math soc János Bolyai, vol 27. North-Holland, Amsterdam, pp 835–873

Papanicolaou GC, Varadhan SRS (1982) Diffusions with random coefficients. In: Statistics and probability: essays in honor of CR Rao. North-Holland, Amsterdam, pp 547–552

Pardoux E, Piatnitski A (2006) Homogenization of a singular random one dimensional PDE. In: Multi scale problems and asymptotic analysis. GAKUTO internat ser math sci appl, vol 24. Gakkōtosho, Tokyo, pp 291–303

Rhodes R (2007) On homogenization of space-time dependent and degenerate random flows. Stoch Process Appl 117(10):1561–1585

Rhodes R (2008) On homogenization of space-time dependent and degenerate random flows. II. Ann Inst Henri Poincaré Probab Stat 44(4):673–692

Rhodes R, Vargas V (2009) Scaling limits for symmetric Itô-Lévy processes in random medium. Stoch Process Appl 119(12):4004–4033

Schmitz T (2006) Diffusions in random environment and ballistic behavior. Ann Inst Henri Poincaré Probab Stat 42(6):683–714

Schmitz T (2009) On the equivalence of the static and dynamic points of view for diffusions in a random environment. Stoch Process Appl 119(8):2501–2522

Shen L (2002) Asymptotic properties of certain anisotropic walks in random media. Ann Appl Probab 12(2):477–510

Shen L (2003) On ballistic diffusions in random environment. Ann Inst Henri Poincaré Probab Stat 39(5):839–876

Sznitman AS, Zeitouni O (2006) An invariance principle for isotropic diffusions in random environment. Invent Math 164(3):455–567

Zhikov VV, Kozlov SM, Oleĭnik OA (1994) Homogenization of differential operators and integral functionals. Springer, Berlin. Translated from the Russian by GA Yosifian [GA Iosif'yan]

Chapter 10
Variational Principles for the Limiting Variance

In Chap. 9 we have shown a central limit theorem for a class of diffusions with stationary and ergodic coefficients. Formula (9.55) expressing the limiting covariance matrix of the particle trajectory involves taking the limit of the resolvent (at the bottom of the spectrum of the generator), which makes it not very useful in actual computations. In this chapter we derive upper and lower bounds for the quadratic form corresponding to the limiting covariance matrix. They are stated in terms of variational principles formulated using the language of stationary random vector fields.

Somewhat similar bounds have already been obtained in Chap. 4 for the variance of an additive functional of a Markov process. Recall that functionals of this type are given by integrals over the path. They are invariant therefore under time reversal transformation $s \mapsto t - s$. On the other hand, the displacement of a particle considered in Chap. 9, defined as $X_t - X_0$, is anti-symmetric with respect to such a transformation.

10.1 Spaces of Vector Fields

We begin our considerations by introducing some spaces of random vectors that will later appear in the formulations of variational principles. Let $L^2_d(\mathbb{Q})$ be the space of all d-dimensional random vectors with square integrable components. For $\mathbf{f}, \mathbf{g} \in L^2_d(\mathbb{Q})$ we introduce the scalar product $\langle \mathbf{f}, \mathbf{g}\rangle_{\mathbb{Q}} := \sum_{k=1}^d \langle f_k, g_k\rangle_{\mathbb{Q}}$ and the corresponding norm $\|\mathbf{f}\|_{\mathbb{Q}} := \langle \mathbf{f}, \mathbf{f}\rangle_{\mathbb{Q}}^{1/2}$. Gradient operator $\nabla = (D_1, \dots, D_d)$ extends to an isometric embedding of $H^1(\Omega)$ into $L^2_d(\mathbb{Q})$. The range of this operator shall be denoted by H_g and the fields belonging to this space shall be referred to as *gradient fields*.

Denote by H_c *the space of constant fields* in $L^2_d(\mathbb{Q})$, i.e. $H_c := \text{span}\{e_1, \dots, e_d\}$. We introduce also the space of all *divergence free* (or *solenoidal*), zero mean fields H_{div} that is defined as the $L^2_d(\mathbb{Q})$-orthogonal complement of $H_c \oplus H_g$. The name of the space is justified by the following.

T. Komorowski et al., *Fluctuations in Markov Processes*,
Grundlehren der mathematischen Wissenschaften 345,
DOI 10.1007/978-3-642-29880-6_10, © Springer-Verlag Berlin Heidelberg 2012

Proposition 10.1 *Suppose that* $\mathbf{f} = (f_1, \dots, f_d)$ *belongs to* H_{div} *and its components are from* $H^1(\Omega)$. *Then,*

$$\nabla \cdot \mathbf{f} := \sum_{k=1}^{d} D_k f_k = 0. \tag{10.1}$$

Proof We conclude, using integration by parts formula (9.24), that

$$0 = \langle \mathbf{f}, \nabla g \rangle_{\mathbb{Q}} = -\langle \nabla \cdot \mathbf{f}, g \rangle_{\mathbb{Q}}, \quad \forall g \in H^1(\Omega)$$

which proves (10.1). □

The space H_{div} can also be described as the closure of fields that are given as a divergence of an anti-symmetric random matrix. More precisely, let $b = [b_{pq}]$ be a $d \times d$ random matrix whose components belong to $C_b^1(\Omega)$. We define its divergence $\nabla \cdot b$ as the vector field whose k-th component equals $\sum_{p=1}^{d} D_p b_{kp}$. Let $A(d)$ be the space of all $d \times d$ anti-symmetric random matrices with entries belonging to $C_b^1(\Omega)$, i.e. $b \in A(d)$, when $b_{pq} = -b_{qp}$ and $b_{pq} \in C_b^1(\Omega)$ for all $p, q = 1, \dots, d$.

Proposition 10.2 H_{div} *is the closure of* $H_{div}^{(0)} := [\nabla \cdot b : b \in A(d)]$ *in* $L_d^2(\mathbb{Q})$.

Proof Suppose that $\mathbf{f} \in H_{div}$ and $\hat{\mathbf{f}} = (\hat{f}_1, \dots, \hat{f}_d)$ is the spectral measure that is the Fourier transform of the homogeneous random field (see Sect. 10.4 below)

$$\mathbf{f}(\tau_x \omega) = \int_{\mathbb{R}^d} e^{i x \cdot \lambda} \hat{\mathbf{f}}(d\lambda; \omega).$$

Condition (10.1) is equivalent to (see Theorem 10.14 below)

$$\sum_{p=1}^{d} \lambda_p \hat{f}_p(d\lambda) = 0. \tag{10.2}$$

For any $\delta > 0$ we let

$$b_{pq}^{(\delta)} := i \int_{[|\lambda| \ge \delta]} |\lambda|^{-2} \big[\lambda_q \hat{f}_p(d\lambda) - \lambda_p \hat{f}_q(d\lambda)\big].$$

Note that $b_{pq}^{(\delta)} = -b_{qp}^{(\delta)}$ and $b_{pq}^{(\delta)} \in H^1(\Omega)$ for any $p, q = 1, \dots, d$. Using (10.2) we obtain

$$\begin{aligned}\sum_{p=1}^{d} D_p b_{pq}^{(\delta)} &= -\sum_{p=1}^{d} \int_{[|\lambda| \ge \delta]} \lambda_p |\lambda|^{-2} \big[\lambda_q \hat{f}_p(d\lambda) - \lambda_p \hat{f}_q(d\lambda)\big] \\ &= \int_{[|\lambda| \ge \delta]} \hat{f}_q(d\lambda)\end{aligned}$$

which tends to f_q in $L^2(\mathbb{Q})$, as $\delta \to 0+$. Using regularization we can in fact approximate b in $H^1(\Omega)$ by anti-symmetric matrices from $A(d)$, see Proposition 9.3, and the conclusion of the proposition follows. □

Assume that $a(\cdot)$ and $U(\cdot)$ are respectively matrix and scalar valued random fields satisfying the hypotheses (R1)–(R3) of Sect. 9.3.3. Denote by P_g the orthogonal projection onto H_g. For a given $\mathbf{f} \in L^2_d(\mathbb{Q})$ we let $\mathbf{g} := P_g\mathbf{f}$. Its spectral measure $\hat{\mathbf{g}} = (\hat{g}_1, \dots, \hat{g}_d)$ is given by

$$\hat{g}_j(A) = \sum_{p=1}^{d} \int_A \frac{\lambda_j \lambda_p}{|\lambda|^2} \hat{f}_p(d\lambda), \quad \forall A \in \mathscr{B}(\mathbb{R}^d),\ j = 1, \dots, d.$$

By convention we admit that the value of the integrand, appearing on the right-hand side, at 0 equals 0.

Let $\mathfrak{L} : L^2_d(\mathbb{Q}) \to L^2_d(\mathbb{Q})$ be a bounded linear operator given by

$$\mathfrak{L}\mathbf{f} := e^{-U} a\mathbf{f}, \quad \forall \mathbf{f} \in L^2_d(\mathbb{Q}). \tag{10.3}$$

The operator is coercive, i.e. there exists $c_* > 0$ such that

$$\langle \mathfrak{L}\mathbf{f}, \mathbf{f}\rangle_{\mathbb{Q}} \ge c_* \|\mathbf{f}\|^2_{\mathbb{Q}}, \quad \forall \mathbf{f} \in L^2_d(\mathbb{Q}). \tag{10.4}$$

Its $L^2_d(\mathbb{Q})$ adjoint is given by $\mathfrak{L}^*\mathbf{f} := e^{-U} a'\mathbf{f}$, where, as we recall, a' is the transpose of a. The symmetric and anti-symmetric parts are respectively given by $\mathfrak{L}^s\mathbf{f} = e^{-U} a^s\mathbf{f}$ and $\mathfrak{L}^a\mathbf{f} = e^{-U} a^\dagger\mathbf{f}$ for $\mathbf{f} \in L^2_d(\mathbb{Q})$. Here a^s and $a^\dagger$ are the corresponding symmetric and anti-symmetric parts of matrix a. Note that $\mathfrak{L}^s$ is a linear automorphism of $L^2_d(\mathbb{Q})$ that satisfies

$$\frac{1}{2Z}\langle \mathfrak{L}^s \nabla f, \nabla f\rangle_{\mathbb{Q}} = \|f\|_1^2, \quad \forall f \in \mathscr{H}_1, \tag{10.5}$$

where $Z = \langle e^{-U}\rangle_{\mathbb{Q}}$. Its inverse is given by $(\mathfrak{L}^s)^{-1}\mathbf{f} = e^U (a^s)^{-1}\mathbf{f}$. The corrector fields $\chi^{(1)}, \dots, \chi^{(d)}$ can be characterized as follows.

Proposition 10.3 *For each $p = 1, \dots, d$ the field $\mathbf{h}_p := \nabla\chi^{(p)}$ is the only solution in $L^2_d(\mathbb{Q})$ of the system of equations*

$$\begin{aligned} &P_g\mathfrak{L}(e_p + \mathbf{h}_p) = 0,\\ &(I - P_g)\mathbf{h}_p = 0. \end{aligned} \tag{10.6}$$

Proof We verify first that $\mathbf{h}_p$ satisfies (10.6). We recall (see (9.49)) that $\chi^{(p)}$ belongs to $\mathscr{H}_1$, so its gradient is well defined as an element of $L^2_d(\mathbb{Q})$ and of course $\mathbf{h}_p \in H_g$. The second equation of (10.6) thus follows. Let $\mathbf{h}_{p,\lambda} := \nabla\chi^{(p)}_\lambda$. The resolvent equation (9.48) written in the weak form reads

$$\lambda\langle\chi_\lambda^{(p)},\phi\rangle_\pi+\frac{1}{2Z}\langle\mathfrak{L}(e_p+\mathbf{h}_{p,\lambda}),\nabla\phi\rangle_{\mathbb{Q}}=0,\quad\forall\phi\in C_b^1(\Omega).\tag{10.7}$$

Thanks to (9.49) we conclude that $\mathbf{h}_{p,\lambda}\to\mathbf{h}_p$ in $L_d^2(\mathbb{Q})$ and $\lambda\chi_\lambda^{(p)}\to 0$ in $L^2(\mathbb{Q})$, as $\lambda\to 0+$. In consequence,

$$\langle\mathfrak{L}(e_p+\mathbf{h}_p),\mathbf{g}\rangle_{\mathbb{Q}}=0,\quad\forall\mathbf{g}\in H_g.\tag{10.8}$$

Hence, the first equation of (10.6) holds.

We prove the uniqueness part. For this purpose suppose that $\mathbf{h}_p^*\in H_g$ is another solution of the system (10.6). Then both $\mathbf{h}_p$ and $\mathbf{h}_p^*$ satisfy (10.8). Let $\delta\mathbf{h}_p:=\mathbf{h}_p^*-\mathbf{h}_p$. As a consequence of (10.8) and the coercivity of $\mathfrak{L}$ we obtain

$$0=\langle\mathfrak{L}\delta\mathbf{h}_p,\delta\mathbf{h}_p\rangle_{\mathbb{Q}}\overset{(10.4)}{\ge}c_*\|\delta\mathbf{h}_p\|^2_{L_d^2(\mathbb{Q})}$$

and $\delta\mathbf{h}_p=0$. □

For any vector $\ell=(l_1,\dots,l_d)\in\mathbb{R}^d$ we let $\chi_\ell:=\sum_{p=1}^d\chi^{(p)}l_p$. The following simple corollary of Proposition 10.3 holds.

Corollary 10.4 *For each $\ell\in\mathbb{R}^d$ the field $\mathbf{h}_\ell:=\nabla\chi_\ell$ is the unique solution of the system of equations*

$$\begin{aligned}&P_g\mathfrak{L}(\ell+\mathbf{h}_\ell)=0,\\&(I-P_g)\mathbf{h}_\ell=0.\end{aligned}\tag{10.9}$$

Define a bilinear form:

$$\bar{a}(\ell_1,\ell_2):=\frac{1}{Z}\langle\mathfrak{L}(\ell_1+\nabla\chi_{\ell_1}),\ell_2+\nabla\chi_{\ell_2}\rangle_{\mathbb{Q}},\quad\ell_1,\ell_2\in\mathbb{R}^d.\tag{10.10}$$

The limiting covariance matrix $\bar{a}=[\bar{a}_{p,q}]$, $p,q=1,\dots,d$ (see (9.55)) can be written as

$$\bar{a}_{p,q}=\frac{1}{2}\big[\bar{a}(e_p,e_q)+\bar{a}(e_q,e_p)\big].\tag{10.11}$$

10.2 Upper Bound

Suppose first that random matrix a is symmetric and coercive. Then $\mathfrak{L}$ given by (10.3) is symmetric and coercive as well. Introduce a scalar product

$$\ll\mathbf{f}_1,\mathbf{f}_2\gg:=\langle\mathfrak{L}\mathbf{f}_1,\mathbf{f}_2\rangle_{\mathbb{Q}},\quad\forall\mathbf{f}_1,\mathbf{f}_2\in L_d^2(\mathbb{Q}),$$

with the respective Hilbert space norm $|||\cdot|||$.

Theorem 10.5 *For any vector $\ell \in \mathbb{R}^d$ we have*

$$\bar{a}(\ell) := \bar{a}(\ell, \ell) = \frac{1}{Z} \min_{\mathbf{f} \in H_g + \ell} |||\mathbf{f}|||^2. \tag{10.12}$$

Proof It is clear that the minimum on the right-hand side of (10.12) is attained at $\mathbf{f} := \ell - \tilde{P}\ell$, where $\tilde{P}$ is the orthogonal projection onto H_g in the scalar product introduced above. Vector $\tilde{P}\ell$ satisfies (10.9), which is the Euler–Lagrange equation for the quadratic functional appearing on the utmost right-hand side of (10.12). Uniqueness of solutions of (10.9) implies that $\mathbf{h}_\ell = \tilde{P}\ell$ and the conclusion of the theorem follows from formula (10.10). In addition, we deduce that

$$\bar{a}(\ell) = \frac{1}{Z} |||\ell - \tilde{P}\ell|||^2. \tag{10.13}$$

□

In the non-symmetric case we need to first conduct a symmetrization procedure and then follow the argument made for symmetric matrices. For that purpose consider the adjoint Euler–Lagrange system

$$\begin{aligned} P_g \mathfrak{L}^* \left(\ell + \mathbf{h}_\ell^*\right) &= 0, \\ (I - P_g)\mathbf{h}_\ell^* &= 0. \end{aligned} \tag{10.14}$$

The existence and uniqueness of solutions can be justified in the same way as has been done for (10.6). We just replace the random matrix $a(\cdot)$ by its transpose $a'(\cdot)$ and repeat word by word the argument used there.

Lemma 10.6 *For any $\ell \in \mathbb{R}^d$ we have*

$$\bar{a}(\ell) = \frac{1}{Z} \langle \mathfrak{L}^*(\ell + \mathbf{f}), \ell + \mathbf{g} \rangle_{\mathbb{Q}}, \tag{10.15}$$

where $\mathbf{f}, \mathbf{g}$ can be equal to either $\mathbf{h}_\ell$, or $\mathbf{h}_\ell^*$.

Proof Since both $\mathbf{h}_\ell$ and $\mathbf{h}_\ell^*$ belong to H_g we have

$$P_g \mathbf{h}_\ell^* = \mathbf{h}_\ell^*, \qquad P_g \mathbf{h}_\ell = \mathbf{h}_\ell.$$

From (10.14) and the above equalities we further conclude that

$$\begin{aligned} \langle \mathfrak{L}^*(\ell + \mathbf{h}_\ell^*), \ell + \mathbf{h}_\ell^* \rangle_{\mathbb{Q}} &= \langle \mathfrak{L}^*(\ell + \mathbf{h}_\ell^*), \ell \rangle_{\mathbb{Q}} \\ &= \langle \mathfrak{L}^*(\ell + \mathbf{h}_\ell^*), \ell + \mathbf{h}_\ell \rangle_{\mathbb{Q}} \\ &= \langle \ell + \mathbf{h}_\ell^*, \mathfrak{L}(\ell + \mathbf{h}_\ell) \rangle_{\mathbb{Q}}. \end{aligned} \tag{10.16}$$

Repeating the argument, this time using (10.9) instead of (10.14), we obtain that the utmost right-hand side of (10.16) can be written as

$$\langle \ell, \mathfrak{L}(\ell + \mathbf{h}_\ell) \rangle_{\mathbb{Q}} = \langle \ell + \mathbf{h}_\ell, \mathfrak{L}(\ell + \mathbf{h}_\ell) \rangle_{\mathbb{Q}} = Z\bar{a}(\ell)$$

and the conclusion of the lemma follows. □

Let us define

$$\mathbf{g} := \mathfrak{L}(\ell + \mathbf{h}_\ell) \quad \text{and} \quad \mathbf{g}^* := \mathfrak{L}^*(\ell + \mathbf{h}_\ell^*) \tag{10.17}$$

and furthermore

$$\mathbf{g}^{\pm} := \frac{1}{2}(\mathbf{g} \pm \mathbf{g}^*), \qquad \mathbf{h}^{\pm} := \frac{1}{2}(\mathbf{h}_\ell \pm \mathbf{h}_\ell^*). \tag{10.18}$$

Using this notation and Lemma 10.6 we write

$$\begin{aligned} \bar{a}(\ell) &= \frac{1}{4}\left[\langle \mathbf{g} - \mathbf{g}^*, \mathbf{h}_\ell - \mathbf{h}_\ell^* \rangle_{\mathbb{Q}} + \langle \mathbf{g} + \mathbf{g}^*, \mathbf{h}_\ell + \mathbf{h}_\ell^* + 2\ell \rangle_{\mathbb{Q}}\right] \\ &= \langle \mathbf{g}^-, \mathbf{h}^- \rangle_{\mathbb{Q}} + \langle \mathbf{g}^+, \mathbf{h}^+ + \ell \rangle_{\mathbb{Q}}. \end{aligned} \tag{10.19}$$

From (10.17), (10.18) and the definitions of $\mathfrak{L}^s$ and $\mathfrak{L}^a$ we obtain the following system of equations

$$\begin{aligned} \mathbf{g}^+ &:= \mathfrak{L}^s(\mathbf{h}^+ + \ell) + \mathfrak{L}^a \mathbf{h}^-, \\ \mathbf{g}^- &:= \mathfrak{L}^s \mathbf{h}^- + \mathfrak{L}^a(\mathbf{h}^+ + \ell). \end{aligned}$$

Using a block matrix notation, where each entry corresponds to an action of an appropriate operator on a copy of $L_d^2(\mathbb{Q})$, we can express $\mathbf{b} := [\mathbf{g}^+, \mathbf{h}^-]$ by $\mathbf{a} := [\mathbf{h}^+ + \ell, \mathbf{g}^-]$ as follows

$$\mathbf{b}^T = \mathfrak{S}\mathbf{a}^T. \tag{10.20}$$

Column vectors $\mathbf{a}^T$ and $\mathbf{b}^T$ are the transposes of the row vectors $\mathbf{a}$ and $\mathbf{b}$, and $\mathfrak{S}$, mapping $L_{2d}^2(\mathbb{Q}) := L_d^2(\mathbb{Q}) \oplus L_d^2(\mathbb{Q})$ into itself, given by

$$\mathfrak{S} := \begin{pmatrix} \mathfrak{L}^s - \mathfrak{L}^a(\mathfrak{L}^s)^{-1}\mathfrak{L}^a & \mathfrak{L}^a(\mathfrak{L}^s)^{-1} \\ -(\mathfrak{L}^s)^{-1}\mathfrak{L}^a & (\mathfrak{L}^s)^{-1} \end{pmatrix} \tag{10.21}$$

is a bounded and symmetric operator. Note that $\mathfrak{S}$ is also coercive. Indeed, for $\mathbf{h} := [\mathbf{f}, \mathbf{g}] \in L_{2d}^2(\mathbb{Q})$ and any $\alpha \in (0, 1)$ we can write

$$\begin{aligned} \langle \mathfrak{S}\mathbf{h}^T, \mathbf{h} \rangle_{\mathbb{Q}} &= \langle \mathfrak{L}^s \mathbf{f}, \mathbf{f} \rangle_{\mathbb{Q}} \\ &\quad + (1 - \alpha^2)\langle (\mathfrak{L}^s)^{-1}\mathbf{g}, \mathbf{g} \rangle_{\mathbb{Q}} - (\alpha^{-2} - 1)\langle (\mathfrak{L}^s)^{-1}\mathfrak{L}^a \mathbf{f}, \mathfrak{L}^a \mathbf{f} \rangle_{\mathbb{Q}} \end{aligned}$$

$$+\left\langle (\mathfrak{L}^s)^{-1}\left(\alpha \mathbf{g} - \alpha^{-1}\mathfrak{L}^a \mathbf{f}\right), \left(\alpha \mathbf{g} - \alpha^{-1}\mathfrak{L}^a \mathbf{f}\right)\right\rangle_{\mathbb{Q}}. \tag{10.22}$$

Suppose that α is chosen sufficiently close to 1 so that

$$\left(\alpha^{-2} - 1\right)\left\langle (\mathfrak{L}^s)^{-1}\mathfrak{L}^a \mathbf{f}, \mathfrak{L}^a \mathbf{f}\right\rangle_{\mathbb{Q}} \le \frac{1}{2}\langle \mathfrak{L}^s \mathbf{f}, \mathbf{f}\rangle_{\mathbb{Q}}, \quad \forall \mathbf{f} \in L^2_d(\mathbb{Q}).$$

The right-hand side of (10.22) can then be estimated from below by

$$\frac{1}{2}\langle \mathfrak{L}^s \mathbf{f}, \mathbf{f}\rangle_{\mathbb{Q}} + \left(1-\alpha^2\right)\left\langle (\mathfrak{L}^s)^{-1}\mathbf{g}, \mathbf{g}\right\rangle_{\mathbb{Q}} \ge \beta \|\mathbf{h}\|^2_{\mathbb{Q}}$$

for some $\beta > 0$.

Equality (10.19) implies that

$$\bar{a}(\ell) = \frac{1}{Z}\langle \mathfrak{S}\mathbf{a}^T, \mathbf{a}\rangle_{\mathbb{Q}}.$$

Obviously, we have $(I - P_g)\mathbf{h}^+ = 0$. From definitions (10.17) and Eqs. (10.9) and (10.14) we get

$$P_g \mathbf{g}^- = \frac{1}{2}\left(P_g \mathbf{g} + P_g \mathbf{g}^*\right) = 0.$$

Similarly,

$$P_g \mathbf{g}^+ = 0 \quad \text{and} \quad (I - P_g)\mathbf{h}^- = 0. \tag{10.23}$$

Let

$$\mathfrak{P} := \begin{pmatrix} P_g & 0 \\ 0 & I - P_g \end{pmatrix}$$

be an orthogonal projection of $L^2_{2d}(\mathbb{Q})$ onto $\mathscr{K}$—the direct product of H_g and its orthogonal complement $H_g^{\perp}$. It is the closure of

$$\mathscr{K}_0 := \left[\mathbf{h} : \mathbf{h} = [\nabla\phi, \nabla \cdot b + u], \phi \in C^1_b(\Omega), b \in A(d), u \in \mathbb{R}^d\right].$$

Equation (10.23) can then be rewritten as $\mathfrak{P}\mathfrak{S}\mathbf{a}^T = 0$. From this equation and the symmetry of $\mathfrak{S}$ we conclude that the quadratic form $\mathbf{h} \mapsto \langle \mathfrak{S}\mathbf{h}^T, \mathbf{h}\rangle_{\mathbb{Q}}$ achieves at $\mathbf{a}$ its minimum, on the linear manifold $\ell_0 + \mathscr{K}$, where $\ell_0 := [\ell, 0] \in \mathbb{R}^{2d}$ (see e.g. Proposition 6 p. 70 of Zeidler 1995). Summarizing, we conclude the following.

Theorem 10.7 *For any $\ell \in \mathbb{R}^d$ let $\ell_0 := [\ell, 0] \in \mathbb{R}^{2d}$. Then,*

$$\begin{aligned} \bar{a}(\ell) &= \frac{1}{Z}\min\left[\langle \mathfrak{S}\mathbf{h}^T, \mathbf{h}\rangle_{\mathbb{Q}} : \mathbf{h} - \ell_0 \in \mathscr{K}\right] \\ &= \frac{1}{Z}\inf\left[\langle \mathfrak{S}(\mathbf{h} + \ell_0)^T, \mathbf{h}\rangle_{\mathbb{Q}} : \mathbf{h} \in \mathscr{K}_0\right]. \end{aligned} \tag{10.24}$$

The second equality in (10.24) follows directly from the definition of the space $\mathscr{K}_0$ and Proposition 10.2. We can further reformulate the above variational principle taking first the infimum over the divergence free fields. For that purpose we will need the following auxiliary result.

Lemma 10.8 *Suppose that the components of* $\mathbf{g}$ *belong to* $H^1(\Omega)$ *and*

$$f := 1/2e^U \nabla \cdot (\mathfrak{L}^s \mathbf{g}).$$

Then, f *belongs to* $\mathscr{H}_{-1}$ *and*

$$\|f\|_{-1}^2 = \frac{1}{2Z} \inf\big[\langle \mathfrak{L}^s \mathbf{h}, \mathbf{h}\rangle_{\mathbb{Q}} : \mathbf{h} \in \mathscr{C}\big], \tag{10.25}$$

where

$$\mathscr{C} = \big[\mathbf{h} : \mathbf{h} = \mathbf{g} - (\mathfrak{L}^s)^{-1}(\nabla \cdot b + u), b \in A(d), u \in \mathbb{R}^d\big].$$

Proof Let $\phi \in C_b^1(\Omega)$. A simple calculation shows that

$$2\langle f, \phi\rangle_\pi - \|\phi\|_1^2 = \frac{1}{2Z}\big[-2\langle \mathfrak{L}^s \mathbf{g}, \nabla\phi\rangle_{\mathbb{Q}} - \langle \mathfrak{L}^s \nabla\phi, \nabla\phi\rangle_{\mathbb{Q}}\big]. \tag{10.26}$$

On $L_d^2(\mathbb{Q})$ introduce a scalar product $\ll \mathbf{f}_1, \mathbf{f}_2 \gg := \langle \mathfrak{L}^s \mathbf{f}_1, \mathbf{f}_2\rangle_{\mathbb{Q}}$ with the respective norm denoted by $|||\cdot|||$. The right-hand side of (10.26) can then be written as

$$\frac{1}{2Z}\big[|||\mathbf{g}|||^2 - |||\mathbf{g} - \nabla\phi|||^2\big].$$

The maximum of this expression over $\phi \in C_b^1(\Omega)$ equals $(2Z)^{-1}|||\tilde{P}\mathbf{g}|||^2$, where $\tilde{P}$ is the orthogonal projection onto H_g, in $\ll \cdot,\cdot \gg$. Since $(\mathfrak{L}^s)^{-1}(H_{div} \oplus H_c)$ is the orthogonal complement of H_g under this scalar product we obtain

$$\|f\|_{-1}^2 = \frac{1}{2Z}|||\tilde{P}\mathbf{g}|||^2 = \frac{1}{2Z}\inf\big[|||\mathbf{g} - \mathbf{h}|||^2 : \mathbf{h} \in (\mathfrak{L}^s)^{-1}(H_{div} \oplus H_c)\big].$$

Equality (10.25) follows from the definition of the $|||\cdot|||$ norm and Proposition 10.2. □

Let

$$V_s := \frac{1}{2}e^U \nabla \cdot (e^{-U} a^s) \tag{10.27}$$

and

$$V_a := \frac{1}{2}e^U \nabla \cdot (e^{-U} a^\dagger). \tag{10.28}$$

Suppose that L is the generator of the environment process defined in (9.40) and S, A are its respective symmetric and anti-symmetric parts, see Corollary 9.10. As a consequence of the above lemma and Theorem 10.7 we conclude.

Theorem 10.9 *For any $\ell \in \mathbb{R}^d$ we have*

$$\bar{a}(\ell) = a(\ell) + 2 \inf_{\phi \in C_b^2(\Omega)} \left[\| V_a \cdot \ell + A\phi \|_{-1}^2 + \|\phi\|_1^2 - 2\langle V_s \cdot \ell, \phi \rangle_\pi \right]. \qquad (10.29)$$

Here $a(\ell) := \langle a\ell, \ell \rangle_\pi$.

Proof It suffices only to note that the expression appearing under infimum on the utmost right-hand side of (10.24) equals

$$\frac{1}{Z}\left[\langle \mathfrak{L}^s \mathbf{h}, \mathbf{h} \rangle_{\mathbb{Q}} + \langle \mathfrak{L}^s \mathbf{g}, \mathbf{g} \rangle_{\mathbb{Q}} \right], \qquad (10.30)$$

where $\mathbf{h} := \nabla\phi + \ell$ and

$$\mathbf{g} := (\mathfrak{L}^s)^{-1}\left[\mathfrak{L}^a(\nabla\phi + \ell) - \nabla \cdot b - u \right].$$

The formula (10.29) is obtained by first taking the infimum over $b \in A(d)$, $u \in \mathbb{R}^d$ and then by applying Lemma 10.8. □

10.3 Lower Bound

Similarly to what has been done in the previous section we start our consideration with an additional assumption that $a(\cdot)$ is symmetric.

Theorem 10.10 *Suppose that $a(\cdot)$ is symmetric and $\mathfrak{L}$ is given by* (10.3). *Then,*

$$\bar{a}(\ell) = \frac{1}{Z} \max\left\{ 2\langle \ell, \mathbf{f} \rangle_{\mathbb{Q}} - \langle \mathfrak{L}^{-1}\mathbf{f}, \mathbf{f} \rangle_{\mathbb{Q}} : \mathbf{f} \in H_c \oplus H_{div} \right\}, \quad \forall \ell \in \mathbb{R}^d. \qquad (10.31)$$

Proof Introduce a scalar product $\ll \cdot, \cdot \gg$ and the respective norm $|||\cdot|||$ as in the proof of Theorem 10.5. Substituting $\mathfrak{L}\mathbf{f}$ for $\mathbf{f}$ we deduce that the right-hand side of (10.31) equals

$$\frac{1}{Z} \max\left\{ 2 \ll \ell, \mathbf{f} \gg - |||\mathbf{f}|||^2 : \mathbf{f} \perp_{\mathfrak{L}} H_g \right\}. \qquad (10.32)$$

Here $\perp_{\mathfrak{L}}$ denotes orthogonality under scalar product $\ll \cdot, \cdot \gg$. The expression in (10.32) equals

$$\frac{1}{Z}\left\{ |||\ell|||^2 - \min\left[|||\ell - \mathbf{f}|||^2 : \mathbf{f} \perp_{\mathfrak{L}} H_g \right] \right\} = \frac{1}{Z} |||Q\ell|||^2,$$

where Q is the orthogonal projection, in the scalar product $\ll \cdot, \cdot \gg$, onto the orthogonal complement of H_g. The conclusion of the theorem follows from (10.13). □

When $a(\cdot)$ is non-symmetric we carry out a symmetrization procedure similar to the one described in the previous section. We can write the inverse of $\mathfrak{S}$, see (10.21), using the matrix notation

$$\mathfrak{S}^{-1} = \begin{pmatrix} (\mathfrak{L}^s)^{-1} & -(\mathfrak{L}^s)^{-1}\mathfrak{L}^a \\ \mathfrak{L}^a(\mathfrak{L}^s)^{-1} & \mathfrak{L}^s - \mathfrak{L}^a(\mathfrak{L}^s)^{-1}\mathfrak{L}^a \end{pmatrix}.$$

Let $\mathbf{h} := [\mathbf{f}, \mathbf{g}]$ and denote

$$\mathfrak{J}_\ell \mathbf{h} := 2\langle \ell, \mathbf{f}\rangle_{\mathbb{Q}} - \langle \mathfrak{S}^{-1}\mathbf{h}^T, \mathbf{h}\rangle_{\mathbb{Q}}.$$

Recall that $\mathscr{K}$ is the direct product in $L^2_{2d}(\mathbb{Q})$ of H_g and the range of $(I - P_g)$. Note that its orthogonal complement $\mathscr{K}^\perp$ in $L^2_{2d}(\mathbb{Q})$ is the direct product of the range of $(I - P_g)$ and H_g. Calculations identical with those done in the proof of Theorem 10.10 lead to the following.

Theorem 10.11 *For any vector $\ell \in \mathbb{R}^d$ we have*

$$\begin{aligned} \bar{a}(\ell) &= \frac{1}{Z}\max[\mathfrak{J}_\ell \mathbf{h} : \mathbf{h} \perp \mathscr{K}] \\ &= \frac{1}{Z}\sup\bigl[\mathfrak{J}_\ell \mathbf{h} : \mathbf{h} = [u + \nabla\cdot b, \nabla\phi], u \in \mathbb{R}^d, b \in A(d), \phi \in C^1_b(\Omega)\bigr]. \end{aligned} \tag{10.33}$$

We compute first the supremum over the divergence free fields and then obtain the following.

Theorem 10.12 *For any $\ell \in \mathbb{R}^d$ we have*

$$\bar{a}(\ell) = a(\ell) + 2\sup_{\phi\in C^2_b(\Omega)}\bigl[2\langle V_a\cdot\ell, \phi\rangle_\pi - \|\phi\|_1^2 - \|V_s\cdot\ell - A\phi\|_{-1}^2\bigr]. \tag{10.34}$$

Proof For $\mathbf{h} = [u + \nabla\cdot b, \nabla\phi]$ we obtain

$$\mathfrak{J}_\ell \mathbf{h} = \langle \mathfrak{L}^s\ell, \ell\rangle_{\mathbb{Q}} + 2\langle \mathfrak{L}^a\nabla\phi, \ell\rangle_{\mathbb{Q}} - \langle \mathfrak{L}^s\nabla\phi, \nabla\phi\rangle_{\mathbb{Q}} - \langle \mathfrak{L}^s\tilde{\phi}_\ell, \tilde{\phi}_\ell\rangle_{\mathbb{Q}}, \tag{10.35}$$

where

$$\tilde{\phi}_\ell := (\mathfrak{L}^s)^{-1}(\mathfrak{L}^s\ell + \mathfrak{L}^a\nabla\phi) + (\mathfrak{L}^s)^{-1}(u + \nabla\cdot b).$$

Taking supremum over $b \in A(d)$ and $u \in \mathbb{R}^d$ in the last term on the right-hand side of (10.35) we obtain, due to Lemma 10.8, that

$$\begin{aligned} \mathfrak{J}_\ell \mathbf{h} &= \langle \mathfrak{L}^s\ell, \ell\rangle_{\mathbb{Q}} + 2\langle \mathfrak{L}^a\nabla\phi, \ell\rangle_{\mathbb{Q}} - \langle \mathfrak{L}^s\nabla\phi, \nabla\phi\rangle_{\mathbb{Q}} - 2Z\|V_s\ell + A\phi\|_{-1}^2 \\ &= Z\langle a^s\ell, \ell\rangle_\pi - 4Z\langle \phi, V_a\cdot\ell\rangle_\pi - 2Z\|\phi\|_1^2 - 2Z\|V_s\ell + A\phi\|_{-1}^2 \end{aligned}$$

and the conclusion of the theorem follows. □

As an application of the above result we show that the limiting covariance matrix of a random diffusion obtained in Theorem 9.15 is strictly positive definite.

Theorem 10.13 *There exists a constant $C > 0$ such that*

$$\bar{a}(\ell) \geq C|\ell|^2, \quad \forall \ell \in \mathbb{R}^d. \tag{10.36}$$

Proof Substituting $\phi = 0$ into (10.34) we obtain

$$\bar{a}(\ell) \geq a(\ell) - 2\|V_s \cdot \ell\|_{-1}^2. \tag{10.37}$$

From the definitions of V_s and the $\|\cdot\|_{-1}$ norm (see (10.27) and (9.44)) we obtain that the right-hand side of (10.37) equals

$$\inf_{\phi \in C_b^1(\Omega)} \langle a(\ell - \nabla\phi), \ell - \nabla\phi \rangle_\pi .$$

Using uniform ellipticity and the supremum bound on U, see hypotheses (R1) and (R2) made in Sect. 9.3.3, we conclude that the above expression is estimated from below by

$$\frac{c_0}{Z} e^{-\|U\|_\infty} \inf_{\phi \in C_b^1(\Omega)} \|\ell - \nabla\phi\|_{\mathbb{Q}}^2 = \frac{c_0}{Z} e^{-\|U\|_\infty} |\ell|^2.$$

□

10.4 Spectral Representation of Homogeneous Fields

The results presented here come from Chaps. 2 and 3 of monograph (Yaglom, 1962). Suppose that $(\Omega, \mathcal{F}, \mathbb{Q})$ is a probability space and $\{\tau_x, x \in \mathbb{R}^d\}$ is a group of measure preserving transformations as in Sect. 9.3.1 and $\hat{R}(d\xi) = [\hat{R}_{pq}(d\xi)]$ is a non-negative, hermitian $d \times d$ matrix valued Borel measure. A (possibly) complex vector valued set function $Z(\cdot) = (Z_1(\cdot), \ldots, Z_d(\cdot))$ defined on $\mathcal{B}(\mathbb{R}^d)$ and taking values in $L_d^2(\mathbb{Q})$ is called a *stochastic spectral measure* if:

(i) $Z(\emptyset) = 0$,
(ii) $Z(A \cup B) = Z(A) + Z(B)$ for any $A, B \in \mathcal{B}(\mathbb{R}^d)$ such that $A \cap B = \emptyset$,
(iii) $\langle Z_p(A) Z_q^*(B) \rangle_{\mathbb{Q}} = \hat{R}_{pq}(A \cap B)$ for any $A, B \in \mathcal{B}(\mathbb{R}^d)$, $p, q = 1, \ldots, d$.

Measure $\hat{R}(\cdot)$ is called the *structure measure* of $Z(\cdot)$.

For any Borel measurable function $g : \mathbb{R}^d \to \mathbb{C}^d$ such that

$$\int_{\mathbb{R}^d} \hat{R}(d\xi) g(\xi) \cdot g(\xi) = \sum_{p,q=1}^d \int_{\mathbb{R}^d} \hat{R}_{pq}(d\xi) g_q(\xi) g_p^*(\xi) < +\infty$$

we can define an element of $L^2(\mathbb{Q})$:

$$\int_{\mathbb{R}^d} g(\xi) \cdot Z(d\xi) := \lim_{n \to +\infty} \frac{1}{n} \sum_{p=1}^d \sum_{m_p^{(1)}, m_p^{(2)} \in \mathbb{Z}^d} \left(m_p^{(1)} + i m_p^{(2)}\right) Z_p\left(A^n_{m_p^{(1)}, m_p^{(2)}}\right)$$

called *a stochastic integral* of g with respect to spectral measure $Z(\cdot)$. Here

$$A^n_{m_p^{(1)},m_p^{(2)}} := \left[\frac{m_p^{(1)}}{n} \le \operatorname{Re} g < \frac{m_p^{(1)}+1}{n}, \frac{m_p^{(2)}}{n} \le \operatorname{Im} g < \frac{m_p^{(2)}+1}{n}\right]$$

and the limit is understood in the mean square sense. In addition, the following isometry property holds

$$\left\langle \left| \int g(\xi) Z(d\xi) \right|^2 \right\rangle_{\mathbb{Q}} = \int_{\mathbb{R}^d} \hat{R}(d\xi) g(\xi) \cdot g(\xi).$$

Suppose that $\mathbf{f} = (f_1, \ldots, f_d) : \Omega \to \mathbb{C}^d$ is a complex valued, square integrable random vector such that $\langle f_p \rangle_{\mathbb{Q}} = 0$, $p = 1, \ldots, d$. Define $\tilde{f}(x;\omega); = \mathbf{f}(\tau_x \omega)$ and the $d \times d$ covariance matrix of the field $R(x) = [R_{pq}(x)]$, where $R_{pq}(x) = \langle \tilde{f}_p(x) \tilde{f}_q(0) \rangle_{\mathbb{Q}}$, $x \in \mathbb{R}^d$.

The following result is a simple multidimensional extension of results of Sect. 3.16 of Yaglom (1962) concerning the Fourier integral representation of a one-dimensional, scalar valued random field.

Theorem 10.14 *Suppose that $\tilde{f}(x;\omega); = \mathbf{f}(\tau_x\omega)$ is a stationary, square integrable, random vector field as discussed above. Then,*

(i) *there exists a unique, $d \times d$ non-negative, Hermitian matrix valued, Borel measure $\hat{R}(d\xi) = [\hat{R}_{pq}(d\xi)]$ such that*

$$R_{pq}(x) = \int_{\mathbb{R}^d} e^{ix\cdot\xi} \hat{R}_{pq}(d\xi), \quad p, q = 1, \ldots, d,$$

(ii) *there exists a stochastic spectral measure $\hat{\mathbf{f}}(\cdot) = (\hat{f}_1(\cdot), \ldots, \hat{f}_d(\cdot))$ such that*

$$\tilde{f}(x) = \int_{\mathbb{R}^d} e^{ix\cdot\xi} \hat{\mathbf{f}}(d\xi), \quad x \in \mathbb{R}^d.$$

Its structure measure equals $\hat{R}(\cdot) = [\hat{R}_{pq}(\cdot)]$,

(iii) *for any $p, q = 1, \ldots, d$ the component f_p of the field belongs to the domain of D_q if and only if*

$$\int_{\mathbb{R}^d} \xi_q^2 \hat{R}_{pp}(d\xi) < +\infty$$

and

$$D_q f_p \circ \tau_x = i \int_{\mathbb{R}^d} \xi_q e^{ix\cdot\xi} \hat{f}_p(d\xi), \quad \forall x \in \mathbb{R}^d.$$

Definition 10.15 The numerical measure $r(d\xi) := \operatorname{tr} \hat{R}(d\xi)$ is called the *energy spectrum* of the field $\tilde{f}(x)$.

10.5 Comments and References

The symmetrization procedure leading to the variational principles stated in Theorems 10.7 and 10.11 is explained in Cherkaev and Gibiansky (1994); Fannjiang and Papanicolaou (1994) see also Fannjiang and Papanicolaou (1996); Fannjiang (1998) and Avellaneda and Majda (1991). The principles formulated in Cherkaev and Gibiansky (1994) concern a general elliptic operator while in Fannjiang and Papanicolaou (1994) it is assumed that the symmetric part of the random matrix is of the form $a^s = c_0 I_d$ for some $c_0 > 0$. The formulations contained in Theorems 10.9 and 10.12 appear to be new, although the case when a^s is a positive multiplicity of the identity matrix can be found in Chap. 4 of Olla (1994b). A somewhat different upper bound for the limiting covariance corresponding to a diffusion whose generator is given by (9.23) is contained in Sect. A.8 of Rhodes (2006).

References

Avellaneda M, Majda AJ (1991) An integral representation and bounds on the effective diffusivity in passive advection by laminar and turbulent flows. Commun Math Phys 138(2):339–391

Cherkaev AV, Gibiansky LV (1994) Variational principles for complex conductivity, viscoelasticity, and similar problems in media with complex moduli. J Math Phys 35(1):127–145

Fannjiang A (1998) Boundary value problem with unbounded, fast oscillatory random flows. J Differ Equ 143(2):327–359

Fannjiang A, Papanicolaou G (1994) Convection enhanced diffusion for periodic flows. SIAM J Appl Math 54(2):333–408

Fannjiang A, Papanicolaou G (1996) Diffusion in turbulence. Probab Theory Relat Fields 105(3):279–334

Olla S (1994b) Lectures on homogenization of diffusion processes in random fields. Publications de l'école doctorale de l'école polytechnique

Rhodes R (2006) Homogénéisation en milieu aléatoire. These de l'universite de provence, ecole doctorale de mathématiques et informatique de Marseille

Yaglom AM (1962) An introduction to the theory of stationary random functions. Prentice-Hall, Englewood Cliffs. Revised English edition. Translated and edited by Richard A Silverman

Zeidler E (1995) Applied functional analysis: main principles and their applications. Applied mathematical sciences, vol 109. Springer, New York

Chapter 11
Diffusions with Divergence Free Drifts

11.1 Passive Tracer Model

A *passive tracer model* is one of the simplest mathematical models of particle transport in an incompressible turbulent flow. The trajectory of a particle is the solution of an Itô stochastic differential equation

$$\begin{aligned} dX_t^{x,\omega} &= \tilde{V}\big(X_t^{x,\omega};\omega\big)dt + dw_t, \\ X_0^{x,\omega} &= x, \end{aligned} \tag{11.1}$$

where $\tilde{V}:\mathbb{R}^d \times \Omega \to \mathbb{R}^d$ is a random vector field, defined over a probability space $(\Omega, \mathscr{F}, \mathbb{Q})$ satisfying the incompressibility condition

$$\sum_{p=1}^{d} \partial_{x_p} \tilde{V}_p(x;\omega) = 0,$$

and $\{w_t, t \ge 0\}$ is a standard d-dimensional Brownian motion given over another probability space $(\Sigma, \mathscr{A}, \mathbb{P})$. The Brownian motion and random velocity appearing above correspond to two physical processes that are responsible for the particle transport: the diffusion, due to the interaction of the tracer particle with the fluid molecules, and the convection by a turbulent flow. Process $\{X_t^{x,\omega}, t \ge 0\}$ is a random diffusion considered over the corresponding product probability space. In agreement with the notational conventions adopted in previous sections we shall omit writing superscript x when the particle starts at the origin.

11.2 Properties of the Flow and the Definition of the Stream Matrix

Field $\tilde{V}(x;\omega)$, appearing in (11.1), is assumed to be stationary and of the form

$$\tilde{V}(x;\omega) = V(\tau_x \omega). \tag{11.2}$$

T. Komorowski et al., *Fluctuations in Markov Processes*,
Grundlehren der mathematischen Wissenschaften 345,
DOI 10.1007/978-3-642-29880-6_11, © Springer-Verlag Berlin Heidelberg 2012

The group $\{\tau_x, x \in \mathbb{R}^d\}$ of measure $\mathbb{Q}$ preserving transformations fulfills the hypotheses made in Sect. 9.3.1. The components of the random vector $V = (V_1, \dots, V_d)$ belong to $L^2(\mathbb{Q})$ and the field is *incompressible* (or *divergence free*), i.e.

$$\nabla \cdot V = \sum_{p=1}^{d} D_p V_p \equiv 0.$$

The above equality holds in the weak sense, i.e.

$$\langle V, \nabla\phi\rangle_{\mathbb{Q}} = 0, \quad \forall\phi \in C^\infty(\Omega).$$

A *stream matrix* of the flow is a $d \times d$ matrix valued field $\tilde{b}(x;\omega) = [\tilde{b}_{lm}(x;\omega)]$, $l, m = 1, \dots, d$, with entries differentiable in the L^2 sense and such that

$$\tilde{V}_m(x;\omega) = \sum_{l=1}^{d} \partial_{x_l} \tilde{b}_{lm}(x;\omega), \tag{11.3}$$

$$\tilde{b}_{lm}(x;\omega) = -\tilde{b}_{ml}(x;\omega), \quad l, m = 1, \dots, d. \tag{11.4}$$

The following proposition (that is a special case of the classical Helmholtz theorem) states that a solenoidal random vector field is the divergence of an anti-symmetric matrix valued potential.

Proposition 11.1 *Suppose that $V = (V_1, \dots, V_d)$ is incompressible. Then, there exists an $H^1(\Omega)$ valued stream matrix $\{\tilde{b}(x), x \in \mathbb{R}^d\}$. In addition, we can choose it in such a way that its increments are stationary, i.e. tensor field $\{\partial_{x_k} \tilde{b}_{lm}(x), k, l, m = 1, \dots, d, x \in \mathbb{R}^d\}$ is stationary, with $\tilde{b}(0) = 0$. When the components of the vector field are locally Hölder $\mathbb{Q}$ a.s. then we can find a stream matrix whose entries are C^1 smooth with locally Hölder continuous partials, $\mathbb{Q}$ a.s.*

Proof For each x we can define a $d \times d$ matrix $\tilde{b}(x) = [\tilde{b}_{lm}(x)]$, $l, m = 1, \dots, d$, with entries in $H^1(\Omega)$, by the formula

$$\tilde{b}_{lm}(x;\omega) := i \int_{\mathbb{R}^d} \frac{e^{i\xi\cdot x} - 1}{|\xi|^2} \big[\xi_l \hat{V}_m(d\xi;\omega) - \xi_m \hat{V}_l(d\xi;\omega)\big]. \tag{11.5}$$

Here $\hat{V}(d\xi;\omega) = (\hat{V}_1(d\xi;\omega), \dots, \hat{V}_d(d\xi;\omega))$ is the spectral measure corresponding to $\tilde{V}(x;\omega)$, see Sect. 10.4. Note that $\partial_{x_k} \tilde{b}_{lm}(x) = c_{k,l,m}(\tau_x\omega)$, where

$$c_{k,l,m}(\omega) = \int_{\mathbb{R}^d} |\xi|^{-2} \xi_m \big[\xi_m \hat{V}_l(d\xi;\omega) - \xi_l \hat{V}_m(d\xi;\omega)\big].$$

Thus the field $\tilde{b}(x)$ has stationary increments. It can also be easily checked that

$$\sum_{m=1}^{d} \partial_{x_m} \tilde{b}_{lm}(x;\omega) := \sum_{m=1}^{d} \int_{\mathbb{R}^d} \frac{e^{i\xi\cdot x}}{|\xi|^2} \xi_m \big[\xi_m \hat{V}_l(d\xi;\omega) - \xi_l \hat{V}_m(d\xi;\omega)\big]. \tag{11.6}$$

Using the relation $\sum_{m=1}^{d} \xi_m \hat{V}_m(d\xi) = 0$ we conclude that the right-hand side of (11.6) equals

$$\int_{\mathbb{R}^d} e^{i\xi\cdot x} \hat{V}_l(d\xi;\omega) = \tilde{V}_l(x;\omega), \qquad \forall l = 1,\dots,d.$$

When the field is locally Hölderian we can find a suitably regular modification of the stream matrix. For that purpose note that $\tilde{b}_{lm}(x;\omega)$ is a solution, in the weak sense, of

$$\Delta_x \tilde{b}_{lm}(x;\omega) = \partial_{x_m} \tilde{V}_l(x;\omega) - \partial_{x_l} \tilde{V}_m(x;\omega), \quad \mathbb{Q} \text{ a.s.}$$

see Chap. 8 of Gilbarg and Trudinger (1983). Using regularity results from the theory of partial differential equations, see Theorems 8.17 and 8.33 of Gilbarg and Trudinger (1983), we can find $\tilde{b}_{lm}(x;\omega)$ that are continuously differentiable, with locally Hölder derivatives. This modification determines a random field $\{\tilde{b}(x), x \in \mathbb{R}^d\}$ whose entries are differentiable and satisfy (11.3) and (11.4). □

A matrix $b = [b_{lm}]$, $l, m = 1,\dots,d$ is called a *stationary stream matrix* of V if

(1) $b_{lm} = -b_{ml}$ for $l, m = 1,\dots,d$,
(2) its entries belong to $H^1(\Omega)$ and

$$V_l = \sum_{m=1}^{d} D_m b_{lm}, \quad l = 1,\dots,d. \tag{11.7}$$

The matrix is said to be C^k-*bounded* if its entries belong to $C_b^k(\Omega)$.

Given a random vector field V we can define a non-negative, Hermitian matrix valued, Borel measure $\hat{R}(d\xi) = [\hat{R}_{p,q}(d\xi)]$ by

$$\langle \hat{V}_p(d\xi)\hat{V}_q^*(d\xi')\rangle_{\mathbb{Q}} = \delta(\xi - \xi')\hat{R}_{p,q}(d\xi)d\xi', \quad p, q = 1,\dots,d. \tag{11.8}$$

The *energy spectrum* of the field is then defined as $r(d\xi) := \operatorname{tr} \hat{R}(d\xi)$. Since the field is square integrable it is a finite Borel measure. By the *Péclet number* of the stationary flow associated with V we understand

$$\mathrm{Pe} := \int_{\mathbb{R}^d} \frac{r(d\xi)}{|\xi|^2}. \tag{11.9}$$

Suppose that $\mathrm{Pe} < +\infty$. Then the entries of a stationary stream matrix can be defined by the formula

$$b_{lm} := i \int_{\mathbb{R}^d} \frac{1}{|\xi|^2} \big[\xi_l \hat{V}_m(d\xi) - \xi_m \hat{V}_l(d\xi)\big]. \tag{11.10}$$

Finiteness of the Péclet number ensures that the stochastic integrals appearing above converge in the L^2 sense. Indeed, calculations, similar to the ones done in the proof of Proposition 11.1, yield (11.7) and an estimate

$$\sum_{l,m=1}^{d} \|b_{lm}\|_{\mathbb{Q}}^2 \le 2\mathrm{Pe}. \tag{11.11}$$

11.3 Central Limit Theorem for a Diffusion with Bounded Stream Matrix

In the case when the field possesses a C^2-bounded, stationary stream matrix b the generator of the diffusion corresponding to a "frozen environment" ω can be rewritten in a divergence form

$$\mathscr{L}_\omega f(x) = \frac{1}{2}\sum_{k,l=1}^{d} \partial_{x_k}\big[\big(\delta_{k,l} + 2\tilde{b}_{k,l}(x;\omega)\big)\partial_{x_l} f(x)\big], \quad f \in C_b^2(\mathbb{R}^d).$$

The central limit theorem for $X_t/\sqrt{t}$, can be viewed as a special case of Theorem 9.15 corresponding to $U = 0$, $a = I + 2b$ and $\pi = \mathbb{Q}$. Space $C_b^2(\Omega)$ is a core of the respective generator L and

$$Lf = \frac{1}{2}\sum_{k,l=1}^{d} D_k\big[(\delta_{k,l} + 2b_{k,l})D_l f\big], \quad f \in C_b^2(\Omega). \tag{11.12}$$

According to the results of Sect. 9.5 the λ-correctors $\chi_\lambda^{(p)}$ converge to $\chi^{(p)}$ in $\mathscr{H}_1$, as $\lambda \to 0+$. Therefore, we have the following.

Theorem 11.2 *Suppose that the random flow is given by* (11.2) *and it possesses a C^2-bounded, stationary stream matrix. Then, $\{X_t/\sqrt{t}, t \ge 0\}$ satisfy the central limit theorem in probability with respect to the environment. The limiting covariance matrix $\bar{a} = [\bar{a}_{pq}]$, $p, q = 1, \ldots, d$, is given by*

$$\bar{a}_{pq} = \delta_{p,q} + \langle \nabla\chi^{(p)}, \nabla\chi^{(q)}\rangle_{\mathbb{Q}}. \tag{11.13}$$

In addition, this matrix coincides with the asymptotic covariance of trajectories, i.e.

$$\lim_{t\to+\infty} \frac{1}{t}\langle E_{\mathbb{P}}[X_t^p X_t^q]\rangle_{\mathbb{Q}} = \bar{a}_{p,q}. \tag{11.14}$$

The covariance matrix of the limiting normal distribution can be expressed using the variational principles (see Sects. 10.2 and 10.3). Combining the conclusions of Theorems 10.9 and 10.12 we state the following.

Theorem 11.3 *For an arbitrary* $\ell \in \mathbb{R}^d$ *we have* $\bar{a}(\ell) = |\ell|^2 + 2\bar{a}_e(\ell)$, *where*

$$\begin{aligned}\bar{a}_e(\ell) :=& \inf_{\phi \in C_b^1(\Omega)} \left[\left\| V \cdot (\ell + \nabla\phi) \right\|_{-1}^2 + \|\phi\|_1^2 \right] \\ =& \sup_{\phi \in C_b^1(\Omega)} \left[2\langle V \cdot \ell, \phi \rangle_{\mathbb{Q}} - \|\phi\|_1^2 - \|V \cdot \nabla\phi\|_{-1}^2 \right]. \end{aligned} \tag{11.15}$$

Proof The formulas claimed above follow directly from (10.29) and (10.31). One should use the fact that $V_a = V$, $V_s = 0$, $A\phi = V \cdot \nabla\phi$ and $a(\ell) = |\ell|^2$. □

11.4 Convection Enhanced Diffusions

For a given $\ell = (l_1, \ldots, l_d)$ we let $V_\ell := V \cdot \ell$. We define also the λ-corrector in the direction ℓ as

$$\chi_{\lambda,\ell} := \sum_{p=1}^{d} \chi_\lambda^{(p)} l_p$$

and analogously we introduce χ_ℓ. From the results contained in Sect. 9.5 we deduce that $\lim_{\lambda \to 0+} \chi_{\lambda,\ell} = \chi_\ell$, strongly in $\mathscr{H}_1$. Since

$$\lambda\chi_{\lambda,\ell} - L\chi_{\lambda,\ell} = V_\ell$$

we have $\chi_\ell \neq 0$, provided that $V_\ell \neq 0$. In addition

$$2\|\chi_\ell\|_1^2 = \|\nabla\chi_\ell\|_{\mathbb{Q}}^2. \tag{11.16}$$

Using (11.13) and (11.16), we conclude that for any $\ell \in \mathbb{R}^d$, for which $\|V_\ell\|_{\mathbb{Q}} \neq 0$

$$\bar{a}(\ell) = |\ell|^2 + 2\|\chi_\ell\|_1^2 > |\ell|^2.$$

Thus, the asymptotic variance is greater (in the sense of partial ordering of the respective symmetric matrices) than the variance of the underlying Brownian motion. This phenomenon is sometimes referred to as *convection enhanced diffusion.*

The situation discussed above should be contrasted with the case when the anti-symmetric part of the diffusion generator is not present. Suppose that a diffusion $\{X_t^x, t \geq 0\}$ is given by (9.30), where the anti-symmetric part of the respective random matrix $a(\omega)$ vanishes. The limiting variance is characterized by variational principle (10.34). In this case $V_s = (1/2)e^U \nabla \cdot (e^{-U} a)$ and $V_a = 0$. Hence

$$\bar{a}(\ell) = a(\ell) - 2\|V_s \cdot \ell\|_{-1}^2 < a(\ell) = \langle a\ell, \ell \rangle_\pi, \tag{11.17}$$

provided that $V_s \cdot \ell \neq 0$. In the perhaps simplest possible situation when $a = I_d$ the trajectory $\{X_t^x, t \geq 0\}$ satisfies

$$dX_t^x = -\nabla_x \tilde{U}\left(X_t^x\right)dt + dw_t, \quad X_0^x = x,$$

where $\tilde{U}(x;\omega) = U(\tau_x\omega)$, i.e. the drift is a gradient of a stationary potential. Define the derivative of the potential in the direction of ℓ as $D_\ell U := \ell \cdot \nabla U$. From (11.17) we get

$$\bar{a}(\ell) = |\ell|^2 - 2\|D_\ell U\|_{-1}^2 < |\ell|^2,$$

provided $\|D_\ell U\|_{-1} \neq 0$. We conclude therefore that a random drift that is the gradient of a stationary potential diminishes, rather than enhances, a diffusion.

11.5 Time Dependent Flows with Finite Péclet Number

The results of Sect. 11.3 can be generalized to diffusions with time dependent, divergence free, stationary drifts admitting unbounded realizations. A sufficient condition for the central limit theorem can be formulated in terms of the spectral measure of the drift. Suppose that the particle trajectory satisfies an Itô stochastic differential equation

$$\begin{aligned} dX_t^{s,x} &= \tilde{V}\big(t, X_t^{s,x};\omega\big)dt + dw_{t,s}, \\ X_s^{s,x} &= x, \end{aligned} \tag{11.18}$$

where

$$\tilde{V}(t,x;\omega) = V(\tau_{t,x}\omega) \tag{11.19}$$

and $V \in L_d^2(\mathbb{Q})$ is of zero mean. Here $\{\tau_{t,x}, (t,x) \in \mathbb{R}^{d+1}\}$ is an ergodic group of $\mathbb{Q}$ preserving transformations of Ω, as introduced in Sect. 9.9.1. For a given $s \in \mathbb{R}$ the Brownian motion $\{w_{t,s}, t \geq s\}$ is defined by $w_{t,s} := w_{t-s}$, where $\{w_t, t \geq 0\}$ is a d-dimensional, standard Brownian motion.

Suppose that the components of V belong to $H^{0,m}(\Omega)$ for some $m > d/2+1$. Furthermore we assume that the field $\tilde{V}(t,x;\omega)$ is continuous in both (t,x) variables and incompressible, i.e.

$$\sum_{i=1}^{d} \partial_{x_i} \tilde{V}_i(t,x;\omega) \equiv 0 \tag{11.20}$$

for $\mathbb{Q}$ a.s. ω. In Proposition 11.7 below we prove that under the above hypotheses, the solutions of (11.18) can be defined for all times for $\mathbb{Q}$ a.s. ω.

Suppose that $\hat{V}(d\xi)$ is the stochastic measure corresponding to the stationary field $\tilde{V}(0,x)$. Then $\hat{R}(d\xi) = [\hat{R}_{p,q}(d\xi)]$, given by (11.8), is a non-negative definite, Hermitian matrix valued, Borel measure that is the Fourier transform of the covariance matrix of the field, i.e.

$$R(0,x) = \int_{\mathbb{R}^d} e^{ix\cdot\xi}\,\hat{R}(d\xi),$$

where $R(t,x) = [R_{p,q}(t,x)]$, $p,q = 1,\dots,d$, and

$$R_{p,q}(t,x) := \langle \tilde{V}_p(t,x)\tilde{V}_q(0,0)\rangle_{\mathbb{Q}}, \quad \forall (t,x) \in \mathbb{R}^{1+d},$$

see Sect. 10.4. The Péclet number Pe of the flow, defined by (11.9), is assumed to be finite. Then, formula (11.10) defines a stationary stream matrix $b(\omega)$ such that

$$\tilde{V}(t,x;\omega) = \nabla \cdot b(\tau_{t,x}\omega). \tag{11.21}$$

Process $\eta_t := \tau_{t,X_t}\omega$ is Markovian with $\mathbb{Q}$ an invariant and ergodic measure, see Sect. 11.6.3 below. Denote by L its generator. For a given $p = 1,\dots,d$ and $\lambda > 0$ let $\chi_\lambda^{(p)}$ be the solution of the resolvent equation

$$(\lambda - L)\chi_\lambda^{(p)} = V_p. \tag{11.22}$$

The main result of this section is the following.

Theorem 11.4 *Suppose that the field $\tilde{V}(t,x)$ satisfies the assumptions made above. Then, $\chi_\lambda^{(p)}$ converge to $\chi^{(p)}$, weakly in $\mathcal{H}_1$, as $\lambda \to 0+$. In addition, the energy identity holds, i.e.*

$$\frac{1}{2}\|\chi^{(p)}\|_1^2 = \langle V_p, \chi^{(p)}\rangle_{\mathbb{Q}} \quad \textit{for each } p = 1,\dots,d. \tag{11.23}$$

As a direct consequence of the above result and Theorem 2.17 we conclude the following.

Corollary 11.5 *Under the hypotheses of the above theorem $\{X_t/\sqrt{t}, t \ge 0\}$ satisfies a central limit theorem, in probability with respect to the environment. The limiting covariance matrix is given by* (11.13). *In addition, formula* (11.14) *holds.*

11.6 Proof of Theorem 11.4

The main difficulty lies in the fact that, due to unboundedness of V, none of the sufficient conditions for the generator of the environment process, formulated in Sect. 2.7, holds. Using truncation of the drift we can verify however that any $\mathcal{H}_1$-weak limit of the solutions of the resolvent equation (11.22) can be used to define a corrector field with stationary increments, see (11.48), that solves, in a weak sense, the Kolmogorov equation corresponding to (11.18), see Proposition 11.17 below. In Sect. 11.6.6 we apply this fact to conclude the energy identity (given by formula (11.23)). Uniqueness of an $\mathcal{H}_1$-weak limit is a consequence of the fact that the energy identity holds for any convex combinations of possible weak limit points. Finally, to conclude the central limit theorem we use Theorem 2.17 together with Remark 2.13.

11.6.1 Notation

Here we introduce some of the notation frequently used throughout the proof of the theorem. Let B_R, $\bar{B}_R$ denote a ball of radius R centered at 0 and its closure. For $t<0$ we shall write $C_{t,R}$, $\bar{C}_{t,R}$ and $S_{t,R}$ to denote an open cylinder $(t,0)\times B_R$, its closure and lateral boundary $(t,0)\times\partial B_R$, respectively. In the special case when $R=+\infty$ we shall write C_t to denote the slab $(t,0)\times\mathbb{R}^d$ and $\mathbb{H}$ for the open half-space $[(t,x):t<0]$.

For $p\in[1,+\infty)$, a non-negative integer $k\ge 0$ and a region $D\subset\mathbb{R}^d$ we define the Sobolev space $W_p^k(D)$ consisting of functions having k generalized derivatives integrable with p-th power. It is the completion of $C^\infty(D)$ in the norm

$$\|f\|_{W_p^k(D)}:=\left\{\|f\|^p_{L^p(D)}+\sum_{|m|\le k}\|\partial^m f\|^p_{L^p(D)}\right\}^{1/p},\quad f\in C^\infty(D),\tag{11.24}$$

see Sect. 7.5 of Gilbarg and Trudinger (1983). According to Theorem 7.26 of Gilbarg and Trudinger (1983) if $k>d/p$, then for any $R>0$ we can embed $W_p^k(B_R)$ in $C_b^m(B_R)$, provided $m=0,\dots,[k-d/p]$.

In the time dependent case, for any of the regions E defined in the foregoing we denote by $C^{k,m}(E)$ the space of all functions having k derivatives in t and m derivatives in x equipped with the norm that is the sum of the suprema of the respective derivatives. We can also define the space of functions having k generalized derivatives in t and resp. m derivatives in x that are L^p integrable. It shall be denoted by $W_p^{k,m}(E)$ with the respective norm $\|\cdot\|_{W_p^{k,m}(E)}$. Let $W_{p,loc}^{k,m}(E)$ be the space consisting of those functions that belong to $W_p^{k,m}(E_1)$ for any relatively compact, cylindrical region E_1 such that $\bar{E}_1\subset E$.

11.6.2 Statements of Some Technical Results

In this subsection we state a few facts concerning the global existence and regularity of solutions of (11.18) as well as a possibility of approximation of divergence free flows by bounded flows. Due to the rather technical nature of such results we postpone the presentation of their proofs until Sect. 11.7.

The following result makes it possible to approximate a divergence free, finite Péclet number field by bounded, smooth, incompressible fields.

Proposition 11.6 *Let $V=(V_1,\dots,V_d)$ be a divergence free, random vector with a finite Péclet number, whose components belong to $H^{0,m}(\Omega)$ for some $m>d/2+1$. Then, there exists a family of anti-symmetric, random matrices $\{b^{(n)},n\ge 1\}$ whose entries belong to $C_b^\infty(\Omega)$ and such that $V^{(n)}=\nabla\cdot b^{(n)}$ satisfies:*

(i)

$$\lim_{n\to+\infty}\left\|V^{(n)}-V\right\|_{\mathbb{Q}}=0,$$

(ii) *for any* $R, T > 0$

$$\lim_{n\to+\infty}\int_0^T \|\tilde{V}^{(n)}(t,\cdot;\omega)-\tilde{V}(t,\cdot;\omega)\|^2_{C^1(B_R)}dt=0, \tag{11.25}$$

both $\mathbb{Q}$ *a.s. and in the* $L^1(\mathbb{Q})$ *sense.*

Our second result deals with the issue of global pathwise existence and uniqueness for the solutions of (11.18). The assumed regularity of the drift (its realizations are C^1 smooth in x by Sobolev embedding) guarantees local existence and uniqueness property of solutions via a general theory of stochastic differential equations, see e.g. Sect. IV.3 of Ikeda and Watanabe (1989). Since we have not placed any restriction on the growth of the coefficients, it is a priori possible that a solution may explode in finite time. As it turns out however this is not the case when the drift is incompressible.

To simplify our presentation assume that the starting point of the diffusion is at $(0,0)$ and we shall omit it in our notation. For $(\omega,\zeta)\in\Omega\times\Sigma$ let $\{X^{\omega}_{t,n}(\zeta), t\ge 0\}$ be the solution of (11.18) corresponding to the regularized drift $\tilde{V}^{(n)}(t,x;\omega)$. Let $\{X^{\omega}_t(\zeta), t\in[0,\mathfrak{e}^{\omega})\}$ be the solution corresponding to $\tilde{V}(t,x;\omega)$ that exists up to a possibly finite explosion time $\mathfrak{e}^{\omega}(\zeta)$.

Proposition 11.7 *We have* $\mathfrak{e}^{\omega}(\zeta)=\infty$, $\mathbb{Q}\otimes\mathbb{P}$ *a.s. in* (ω,ζ).

In addition, the trajectories corresponding to the regularized drifts approximate the solution of (11.18) as seen from the following.

Proposition 11.8 *For any* $T>0$ *we have*

$$\lim_{n\to+\infty}\sup_{t\in[0,T]}|X_{t,n}-X_t|=0, \quad \mathbb{Q}\otimes\mathbb{P}\text{-a.s.} \tag{11.26}$$

Pathwise uniqueness implies the Markov property of solutions of (11.18) for a fixed ω, see e.g. Sect. 7.1 of Øksendal (2003). Denote by $P^{\omega}_{s,t}(x,\cdot)$ the transition probabilities of $X^{s,x,\omega}_t$. Our final result of this section concerns the existence of transition probability densities and their positivity.

Proposition 11.9 *For each* $s<t$ *and* $x\in\mathbb{R}^d$ *transition probability* $P^{\omega}_{s,t}(x,\cdot)$ *is absolutely continuous with respect to the Lebesgue measure. Its density satisfies*

$$p^{\omega}_{s,t}(x,y)>0 \quad \text{a.e. in } y \tag{11.27}$$

for $\mathbb{Q}$ *a.s.* ω.

11.6.3 Properties of the Environment Process

Suppose that $\eta_t=\tau_{t,X_t}\omega$ is the environment process corresponding to the diffusion $\{X_t, t\ge 0\}$. For any f belonging to the space of bounded and measurable functions

$B_b(\Omega)$ and $t > 0$ we define

$$P_t f(\omega) := E_{\mathbb{P}} f(\eta_t) = \int_{\mathbb{R}^d} p^{\omega}_{0,t}(0, y) f(\tau_{t,y}\omega) dy. \tag{11.28}$$

From (11.18) it can be deduced that $\lim_{t\to 0+}\langle E_{\mathbb{P}}|X_t|^2\rangle_{\mathbb{Q}} = 0$. Hence

$$\lim_{t\to 0+} \|P_t f - f\|_{\mathbb{Q}} = 0 \quad \text{for any } f \in C_b(\Omega). \tag{11.29}$$

Suppose that $X_{t,n}$ is the diffusion that corresponds to $\tilde{V}^{(n)}(t,x)$ as in Proposition 11.6. Let $\eta_t^{(n)} = \tau_{t,X_{t,n}}\omega$ be the respective environment process and $\{P_t^{(n)}, t \ge 0\}$ its transition semigroup of contractions on $L^2(\mathbb{Q})$. From Proposition 11.8 we immediately conclude that

$$P_t f = \lim_{n\to+\infty} P_t^{(n)} f \quad \text{strongly in } L^2(\mathbb{Q}), \quad \forall f \in L^2(\mathbb{Q}). \tag{11.30}$$

In addition, for any such f we obtain that

$$\langle P_t f\rangle_{\mathbb{Q}} = \lim_{n\to+\infty} \langle P_t^{(n)} f\rangle_{\mathbb{Q}} = \langle f\rangle_{\mathbb{Q}},$$

which in turn implies that $\mathbb{Q}$ is invariant for $\{P_t, t \ge 0\}$ and the semigroup extends to a strongly continuous semigroup of contractions on $L^2(\mathbb{Q})$.

From (11.27) we can see that

$$P_t f > 0, \quad \mathbb{Q} \text{ a.s.}$$

for any $t > 0$ and $f \ge 0$ that is not identically equal to 0. Using this property one can easily argue ergodicity of $\mathbb{Q}$, exactly in the same way as it was done in Sect. 9.4. Summarizing the above considerations we can formulate the following.

Proposition 11.10 *The process $\{\eta_t, t \ge 0\}$ is Markovian with $\{P_t, t \ge 0\}$ its transition semigroup. In addition, measure $\mathbb{Q}$ is invariant and ergodic, and the semigroup extends to a strongly continuous semigroup of contractions on $L^2(\mathbb{Q})$. For each $t \ge 0$ we have $P_t = \lim_{n\to+\infty} P_t^{(n)}$ strongly in $L^2(\mathbb{Q})$.*

Denote by L, $L^{(n)}$ the generators corresponding to $\{P_t, t \ge 0\}$ and $\{P_t^{(n)}, t \ge 0\}$, respectively. For any $g \in L^2(\mathbb{Q})$ and $\lambda > 0$ denote by $G_\lambda^{(n)} g$ the solution of the resolvent equation $(\lambda - L^{(n)}) f = g$ given by

$$G_\lambda^{(n)} g = \int_0^{+\infty} e^{-\lambda t} P_t^{(n)} g dt. \tag{11.31}$$

Since $\{P_t^{(n)}, t \ge 0\}$ is also a semigroup of contractions on any $L^p(\mathbb{Q})$ space, for $p \in [1, +\infty]$, we obtain

$$\left\| G_\lambda^{(n)} g \right\|_{L^p(\mathbb{Q})} \le \frac{1}{\lambda} \|g\|_{L^p(\mathbb{Q})}, \quad \forall n \ge 1. \tag{11.32}$$

The above estimate holds also for G_λ—the resolvent corresponding to L.

Lemma 11.11 *Suppose that $g \in L^2(\mathbb{Q})$ and $\lambda > 0$. Then,*

$$\lim_{n\to+\infty} \left\| G_\lambda^{(n)} g - G_\lambda g \right\|_{\mathbb{Q}} = 0. \tag{11.33}$$

In addition, $G_\lambda^{(n)} g$ and $G_\lambda g$ belong to $H^{0,1}(\Omega)$, and

$$\lim_{n\to+\infty} \nabla G_\lambda^{(n)} g = \nabla G_\lambda g \tag{11.34}$$

weakly in $L_d^2(\mathbb{Q})$.

Proof Formula (11.33) follows directly from (11.31), the corresponding formula for $G_\lambda g$ and the strong convergence of the respective semigroups, see Proposition 11.16.

Since $G_\lambda^{(n)} g$ belongs to $D(L^{(n)})$ it is in $H^{0,1}(\Omega)$ (see Remark 9.28). Multiplying the resolvent equation by $G_\lambda^{(n)} g$ and integrating with respect to $\mathbb{Q}$ we obtain

$$\lambda \left\| G_\lambda^{(n)} g \right\|_{\mathbb{Q}}^2 + \frac{1}{2} \left\| \nabla G_\lambda^{(n)} g \right\|_{\mathbb{Q}}^2 = \left\langle g, G_\lambda^{(n)} g \right\rangle_{\mathbb{Q}}. \tag{11.35}$$

From this and (11.32) it follows that $\{\nabla G_\lambda^{(n)} g, n \geq 1\}$ is bounded in $L_d^2(\mathbb{Q})$. Weak convergence of the sequence and the fact that $G_\lambda g \in H^1(\Omega)$ follow from (11.33). □

11.6.4 Properties of the $\mathscr{H}_1$-Norm

One can easily verify that $C_b^{1,2}(\Omega) \subset D(L)$. From Itô's formula it follows that

$$Lf = D_0 f + (1/2)\Delta f + V \cdot \nabla f, \quad f \in C_b^{1,2}(\Omega).$$

From here we deduce that for f from $C_b^{1,2}(\Omega)$ we have

$$\|\nabla f\|_{\mathbb{Q}}^2 = 2\|f\|_1^2. \tag{11.36}$$

This formula can be extended to the entire $\mathscr{H}_1$ in case when $V \in L^\infty(\mathbb{Q})$.

Proposition 11.12 *Suppose that $\|V\|_\infty < +\infty$. Then, $C_b^{1,2}(\Omega)$ is a core of L and (11.36) holds on $\mathscr{H}_1$.*

Proof The set $C_b^{1,2}(\Omega)$ is clearly invariant under the semigroup

$$R_t f(\omega) := \int q_t(x) f(\tau_{t,x}\omega) dx, \tag{11.37}$$

where

$$q_t(x) = \frac{1}{(2\pi t)^{d/2}} \exp\left\{-\frac{|x|^2}{2t}\right\}, \quad t > 0. \tag{11.38}$$

According to Proposition 3.3 of Ethier and Kurtz (1986) it is a core of the generator L_0 of this semigroup. By an application of the Itô formula we obtain that

$$L_0 f = D_0 f + (1/2)\Delta f, \quad \forall f \in C_b^{1,2}(\Omega).$$

On that set we define the operator $Bf := V \cdot \nabla f$. Suppose that $\lambda > 0$. Note that

$$\langle(\lambda - B)f, f\rangle_{\mathbb{Q}} = \lambda \| f \|_{\mathbb{Q}}^2,$$

hence in particular we get $\|(\lambda - B)f\|_{\mathbb{Q}} \geq \lambda \| f \|_{\mathbb{Q}}$. Operator B is therefore dissipative in the sense of Ethier and Kurtz (1986), see the definition on p. 11.

Suppose that $\alpha < 1$ is arbitrary. Using an elementary inequality $\sqrt{ab} \leq (a+b)/2$, holding for all $a, b > 0$ we obtain

$$\begin{aligned} \|Bf\|_{\mathbb{Q}} &= \|V \cdot \nabla f\|_{\mathbb{Q}} \leq \sqrt{2}\|V\|_\infty \langle f, (-L_0)f\rangle_{\mathbb{Q}}^{1/2} \\ &\leq \alpha \|L_0 f\|_{\mathbb{Q}}/2 + \|f\|_{\mathbb{Q}} \|V\|_\infty^2/\alpha, \quad \forall f \in C_b^{1,2}(\Omega). \end{aligned}$$

Therefore B extends to the domain of L_0 and, by virtue of Theorem 7.1, p. 37 of Ethier and Kurtz (1986), the closure of $L_0 + B$, which we denote by the same symbol, is a generator of a contraction semigroup on $L^2(\mathbb{Q})$ with $C_b^{1,2}(\Omega)$ as its a core. This operator agrees with L on $C_b^{1,2}(\Omega)$. The latter, as a generator of a strongly continuous semigroup, being a closed operator has to be an extension of $L_0 + B$. In fact, since L is also dissipative (as a generator of a contraction semigroup), by virtue of Proposition 4.1, p. 21 of Ethier and Kurtz (1986), it coincides with $L_0 + B$. The conclusion of the proposition then follows. □

Remark 11.13 The above argument shows that if $\|V\|_\infty < +\infty$ then for any $k \geq 1$, $m \geq 2$ the set $C_b^{k,m}(\Omega)$ is a core of L.

When $V \in L_d^2(\mathbb{Q})$ one can establish an inequality between the $\mathcal{H}_1$-norm and the L^2 norm of a gradient.

Lemma 11.14 *Suppose that $V \in L_d^2(\mathbb{Q})$. Then $\mathcal{H}_1 \subset H^{0,1}(\Omega)$ and*

$$\|\nabla f\|_{\mathbb{Q}}^2 \leq 2\|f\|_1^2, \quad f \in \mathcal{H}_1. \tag{11.39}$$

Proof Suppose first that $f \in D(L)$ and $f = G_\lambda g$. Consider a regularizing sequence $\{V^{(n)}, n \geq 1\}$, as in Proposition 11.6. Letting $n \to +\infty$ we obtain from (11.35) that

$$\lambda\|G_\lambda g\|_{\mathbb{Q}}^2+\frac{1}{2}\|\nabla G_\lambda g\|_{\mathbb{Q}}^2\le\langle g,G_\lambda g\rangle_{\mathbb{Q}}=\lambda\|G_\lambda g\|_{\mathbb{Q}}^2+\|G_\lambda g\|_1^2. \tag{11.40}$$

Estimate (11.39) follows from (11.40) and the fact that $D(L)$ is dense in $\mathscr{H}_1$. □

Since the components of V satisfy (11.7) we deduce the following.

Corollary 11.15 *For each $l=1,\dots,d$ we have $V_l\in\mathscr{H}_{-1}$ and*

$$\big|\langle V_l,f\rangle_{\mathbb{Q}}\big|=\big|\langle b_l,\nabla f\rangle_{\mathbb{Q}}\big|\le\sqrt{2}\|b_l\|_{\mathbb{Q}}\|f\|_1,\quad\forall f\in\mathscr{H}_1. \tag{11.41}$$

11.6.5 Construction of the Corrector Field

In the next step we construct the corrector field $\{\chi^{(p)}(t,x),(t,x)\in\overline{\mathbb{H}}\}$ corresponding to a given spatial direction e_p, $p=1,\dots,d$. It is a random (non-stationary) field, over the probability space $(\Omega,\mathscr{F},\mathbb{Q})$, that satisfies the backward Kolmogorov equation

$$(\partial_t+\mathscr{L}_{t,\omega})\chi^{(p)}(t,x)=-\tilde{V}_p(t,x),$$

where $\mathscr{L}_{t,\omega}$ is the generator of the diffusion, given by (11.49). We have already mentioned that the field is not stationary but its absolute moment satisfies

$$\lim_{a\to+\infty}\frac{1}{a}\langle\big|\chi^{(p)}\big(ta^2,xa\big)\big|\rangle_{\mathbb{Q}}=0$$

in the weak sense in (t,x), see Proposition 11.17. Both of these properties will be crucial in the proof of the energy identity presented in Sect. 11.6.6.

The starting point for the construction of the corrector is an $\mathscr{H}_1$-weak limit point of $\chi_\lambda^{(p)}$, as $\lambda\to0+$. To simplify the notation we shall suppress superscript p from our subsequent notation. Using the resolvent equation (11.22) and estimate (11.41) we obtain

$$\lambda\|\chi_\lambda\|_{\mathbb{Q}}^2+\|\chi_\lambda\|_1^2\le\sqrt{2}\|b_p\|_{\mathbb{Q}}\|\chi_\lambda\|_1.$$

Hence, from (11.11) we get

$$\|\chi_\lambda\|_1\le2\sqrt{\mathrm{Pe}},\quad\forall\lambda>0. \tag{11.42}$$

Let χ be an $\mathscr{H}_1$-weak limit of $\{\chi_{\lambda_n},n\ge1\}$, where $\lambda_n\to0+$, as $n\to+\infty$. Let

$$\mathbf{f}=(f_1,\dots,f_d):=\nabla\chi\in L_d^2(\mathbb{Q})$$

and let $\hat{\mathbf{f}}(d\xi):=(\hat{f}_1(d\xi),\dots,\hat{f}_d(d\xi))$ be the spectral measure corresponding to the spectral resolution of $\mathbf{f}$ with respect to the subgroup $\{\tau_{0,x},x\in\mathbb{R}^d\}$. Note that $\hat{f}_q(\{0\})=0$ for any $q=1,\dots,d$. Indeed, $\hat{f}_q(\{0\})$ is the orthogonal projection of f_q

onto the subspace Π of $L^2(\mathbb{Q})$ consisting of the elements that are invariant under spatial shifts. Therefore for any $g \in \Pi$ we have

$$\langle D_q \chi_{\lambda_n}, g\rangle_{\mathbb{Q}} = 0, \quad \forall n, q.$$

Thus, letting $n \to +\infty$, we obtain that $\langle f_q, g\rangle_{\mathbb{Q}} = 0$, which in turn implies that 0 cannot be an atom of the spectral measure $\hat{f}_q$. As a result

$$\theta(x) := -i \sum_{q=1}^{d} \int_{\mathbb{R}^d} \frac{e^{ix\cdot\xi} - 1}{|\xi|^2} \xi_q \hat{f}_q(d\xi) \tag{11.43}$$

is well defined as function from $\mathbb{R}^d$ to $L^2(\mathbb{Q})$. Here we adopt the convention that the integrand equals 0 for $\xi = 0$. Note that from the definition of the field it follows that $\xi_k \hat{f}_l = \xi_l \hat{f}_k$. Using this and (11.43) we conclude that

$$\theta(\cdot;\omega) \in W^1_{2,loc}\big(\mathbb{R}^d\big) \quad \text{and} \quad \partial_{x_k}\theta(x;\omega) = \tilde{f}_k(0,x;\omega) \quad \text{for } \mathbb{Q} \text{ a.s. } \omega, \tag{11.44}$$

where $\tilde{f}_k(t,x;\omega) = f_k(\tau_{t,x}\omega)$.

Lemma 11.16 *We have*

$$\lim_{|x|\to+\infty} \frac{\langle\theta^2(x)\rangle_{\mathbb{Q}}}{|x|^2} = 0. \tag{11.45}$$

Proof Denote by $\mu(d\xi)$ the energy spectrum of $\nabla_x\theta(x) = \mathbf{f} \circ \tau_{0,x}$. As we have already observed it has no atom at 0. Therefore

$$\limsup_{|x|\to+\infty} \frac{\langle\theta^2(x)\rangle_{\mathbb{Q}}}{|x|^2} \le \limsup_{|x|\to+\infty} \int_{\mathbb{R}^d} \frac{|e^{ix\cdot\xi} - 1|^2}{|x|^2|\xi|^2} \mu(d\xi) = 0. \tag{11.46}$$

The last equality follows from the Lebesgue dominated convergence theorem. □

On $\overline{\mathbb{H}}$ we define a non-stationary *corrector field* by letting

$$\chi(0,x) := \theta(x) \tag{11.47}$$

and

$$\chi(t,x) := \int_{\mathbb{R}^d} q_{-t}(x-y)\theta(y)dy + \int_{C_t} q_{s-t}(x-y)\big[\tilde{V}(s,y)\cdot\tilde{\mathbf{f}}(s,y) + \tilde{V}_p(s,y)\big]dsdy \tag{11.48}$$

for $(t,x) \in \mathbb{H}$. Note that both integrals appearing in (11.48) converge, $\mathbb{Q}$ a.s.

As it turns out, see the proposition below, the corrector field is both a mild and a weak solution of the backward Kolmogorov equation corresponding to the diffusion given by (11.18). More precisely, let

$$\mathscr{L}_{t,\omega} f(t,x) := \frac{1}{2}\Delta_x f(t,x) + \tilde{V}(t,x;\omega)\cdot\nabla_x f(t,x) \tag{11.49}$$

for $f \in C^{1,2}(\mathbb{R}^{d+1})$. Suppose that $u_0(x;\omega)$ is a measurable random field and $\lambda \in \mathbb{R}$. We say that $u(t,x)$ is *a mild solution* of

$$\begin{aligned} &(\partial_t + \mathscr{L}_{t,\omega} - \lambda)u(t,x) = -\tilde{V}_p(t,x), \quad (t,x)\in\mathbb{H} \\ &u(0,x) = u_0(x), \end{aligned} \tag{11.50}$$

if

$$\begin{aligned} &\int_{\mathbb{R}^d} q_{-t}(x-y)\big|u_0(y)\big|dy + \int_{C_t} q_{s-t}(x-y)\big[\big|\tilde{V}_p(s,y)\big| + \big|\tilde{V}(s,y)\cdot\nabla_y u(s,y)\big| \\ &\quad + |\lambda|\big|u(s,y)\big|\big]dsdy < +\infty \end{aligned}$$

for all $(t,x)\in\mathbb{H}$ and

$$u(t,x) = \mathscr{G}_\lambda(u_0)(t,x), \quad \mathbb{Q} \text{ a.s.}, \tag{11.51}$$

where

$$\begin{aligned} \mathscr{G}_\lambda(u_0)(t,x) := &\int_{\mathbb{R}^d} q_{-t}(x-y)u_0(y)dy \\ &+ \int_{C_t} q_{s-t}(x-y)\big[\tilde{V}(s,y)\cdot\nabla_y u(s,y) \\ &+ \tilde{V}_p(s,y) - \lambda u(s,y)\big]dsdy. \end{aligned} \tag{11.52}$$

We say that $u(t,x)$ solves (11.50) in a weak sense if

$$\int_{C_{t,R}} (u^2 + |\nabla_x u|^2)dsdx < +\infty, \quad \forall t, R > 0 \tag{11.53}$$

and for any $v \in C_c^{1,1}(\mathbb{R}^{d+1})$

$$\begin{aligned} &\int_{\mathbb{R}^d} u_0(x)v(0,x)dx + \int_{\mathbb{H}}\left[u\partial_t v + \frac{1}{2}\nabla_x u\cdot\nabla_x v + (\tilde{V}\cdot\nabla_x u + \lambda u)v\right]dtdx \\ &\quad = \int_{\mathbb{H}} \tilde{V}_p v dtdx, \quad \mathbb{Q} \text{ a.s.} \end{aligned} \tag{11.54}$$

Proposition 11.17

(i) *For each* $\lambda > 0$ *the field* $\tilde{\chi}_\lambda(t,x) := \chi_\lambda \circ \tau_{t,x}$ *is both a mild and a weak solution of* (11.50), *with* $u_0(x)$ *given by* (11.47),

(ii) *for* $\lambda = 0$ *the same holds for* $\chi(t,x)$. *Moreover* $\chi \in W^{1,1}_{2,loc}(\mathbb{H})$ *and*

$$\nabla_x \chi(t,x) = \nabla \chi \circ \tau_{t,x}, \quad a.e.\ in\ \mathbb{H}, \tag{11.55}$$

$$\begin{aligned}\int_{\mathbb{H}} \partial_t \chi v dt dx = &- \int_{\mathbb{H}} \chi \partial_t v dt dx \\ &+ \int_{\mathbb{R}^d} \theta(x) v(0,x) dt dx, \quad \forall v \in C^{1,1}_c(\mathbb{R}^{d+1}), \quad \mathbb{Q}\ a.s.,\end{aligned} \tag{11.56}$$

(iii) *for any* $\phi \in C_c^\infty(\mathbb{H})$ *we have*

$$\lim_{a\to+\infty} \frac{1}{a} \int_{\mathbb{H}} \phi(t,x) \langle |\chi(a^2 t, ax)| \rangle_{\mathbb{Q}} dt dx = 0. \tag{11.57}$$

Proof (i) Suppose first that the components of V belong to $C^{1,1}_b(\Omega)$. Then, by Remark 11.13, one can find a sequence $\{f_n, n \ge 1\} \subset C^{1,2}_b(\Omega)$ such that

$$\lim_{n\to+\infty} \big(\|f_n - \chi_\lambda\|_{H^{0,1}} + \|g_n - V_p\|_{\mathbb{Q}} \big) = 0, \tag{11.58}$$

where $g_n := (\lambda - L) f_n$. Using Itô's formula we conclude that $\tilde{f}_n(t,x) := f_n \circ \tau_{t,x}$ satisfies

$$(\lambda - \partial_t - \mathscr{L}_\omega) \tilde{f}_n(t,x) = \tilde{g}_n(t,x) \tag{11.59}$$

in the classical sense. Since the coefficients of (11.59) are C^1 smooth and bounded and $\tilde{f}_n$ is also bounded, using the classical theory of partial differential equations, see e.g. Theorem 16, p. 29 of Friedman (1964), we infer that

$$\tilde{f}_n(t,x) = \mathscr{G}_\lambda\big(\tilde{f}_n(0,\cdot)\big)(t,x), \quad \forall (t,x) \in \mathbb{H}. \tag{11.60}$$

Letting $n \to +\infty$ and using (11.58) we deduce that for each (t,x) Eq. (11.51) holds $\mathbb{Q}$ a.s. Generalization to an unbounded V can be achieved by approximating the field by smooth, bounded vector fields as in Proposition 11.6. The same argument allows us to deduce that $\tilde{\chi}_\lambda(t,x)$ is a weak solution. Since such solutions, in case of parabolic equations with continuous coefficients whose principal part is in a divergence form, are locally Hölder continuous for a.s. ω (see Theorem 5.1.1, p. 476, of Ladyženskaja et al. 1968) we can choose a modification of the random field $\tilde{\chi}_\lambda(t,x)$ that is continuous in $\mathbb{H}$. In this way we can ensure that equality (11.51) holds on this set, $\mathbb{Q}$ a.s. We have proved therefore that $\tilde{\chi}_\lambda(t,x)$ is a mild solution of (11.50).

(ii) We prove (11.55) first. Differentiating both sides of (11.48) with respect to x_k we obtain $\partial_{x_k} \chi(t,x) = I(t,x) + II(t,x)$, where

$$
\begin{aligned}
I(t,x) &:= \int_{\mathbb{R}^d} \partial_{x_k} q_{-t}(x-y)\theta(y)dy,\\
II(t,x) &:= \int_{C_t} \partial_{x_k} q_{s-t}(x-y)\big[\tilde{V}(s,y)\cdot \tilde{\mathbf{f}}(s,y) + \tilde{V}_p(s,y)\big]dsdy.
\end{aligned}
\tag{11.61}
$$

Passage with differentiation under the integration can be substantiated using the Lebesgue dominated convergence theorem. In case of the first integral appearing in (11.48) it suffices to verify that for any $K>0$, $t<0$ we have

$$
\int_{\mathbb{R}^d} \sup_{|x|\le K} \big|\partial_{x_k} q_{-t}(x-y)\big|\langle|\theta(y)|\rangle_{\mathbb{Q}} dy \le C\int_{\mathbb{R}^d} \sup_{|x|\le K} \big|\partial_{x_k} q_{-t}(x-y)y\big|dy < +\infty,
$$

which holds thanks to (11.45). A similar argument can be made also for the second integral appearing there.

Integrating by parts the expression on the right-hand side of the formula for $I(t,x)$ (note that the boundary term vanishes due to (11.46)) we obtain

$$
\partial_{x_k}\chi(t,x) = \int_{\mathbb{R}^d} q_{-t}(x-y)\tilde{f}_k(y)dy + II(t,x).
$$

Since $\tilde{\chi}_\lambda(t,x)$ is a mild solution of (11.50), differentiating both sides of (11.51) with respect to x_k yields

$$
\begin{aligned}
D_k\chi_{\lambda_n}\circ\tau_{t,x}(\omega) &= \int_{\mathbb{R}^d} q_{-t}(x-y)D_k\chi_{\lambda_n}\circ\tau_{0,y}(\omega)dy\\
&\quad + \int_{C_t} \partial_{x_k} q_{s-t}(x-y)\big[\tilde{V}(s,y)\cdot\nabla\chi_{\lambda_n}\circ\tau_{s,y}(\omega)\\
&\quad + \tilde{V}_p(s,y) - \lambda_n\tilde{\chi}_{\lambda_n}(s,y;\omega)\big]dsdy.
\end{aligned}
\tag{11.62}
$$

Here, the sequence $\{\lambda_n, n\ge 1\}$ is selected as in the definition of χ. Since $\nabla\chi_{\lambda_n}$ and $\lambda_n\chi_{\lambda_n}$ converge to $\mathbf{f}$, weakly in $L^2_d(\mathbb{Q})$, and 0, strongly in $L^2(\mathbb{Q})$, respectively we can take weak limits in $L^1(\mathbb{Q})$ on both sides of (11.62) and deduce (11.55).

The fact that $\chi(t,x)$ is a mild solution follows from the definition of the field, cf. (11.48) and formula (11.55). To show that it is also a weak solution note that, as an obvious consequence of (11.44) and (11.55), it satisfies (11.53), $\mathbb{Q}$ a.s. Let us consider $u_\rho(t,x) := \chi * \phi_\rho(t,x)$, where $\phi_\rho(t,x) := \rho^{-d-2}\phi(t\rho^{-2}, x\rho^{-1})$, $\rho>0$ and ϕ is a C^∞ probability density function, supported in $(-1,1)\times B_1$. To make sense of the convolution we suppose that $\chi(t,x)\equiv 0$ for $t>0$. Observe that $u_\rho(t,x)$ converges in $L^2_{loc}(\mathbb{H})$, as $\rho\to 0+$, together with $\nabla_x u_\rho$, to χ and $\nabla_x\chi$, respectively. On the other hand, $u_\rho(t,x)$ is a classical solution of the Cauchy problem

$$
\begin{aligned}
&\partial_t u_\rho(t,x) + \frac{1}{2}\Delta_x u_\rho(t,x) = f_\rho(t,x), \quad (t,x)\in\mathbb{H},\\
&u_\rho(0,x) = \chi * \phi_\rho(0,x),
\end{aligned}
\tag{11.63}
$$

where $f_\rho := -(\tilde{V} \cdot \nabla_x \tilde{\chi} + \tilde{V}_p) * \phi_\rho$. Letting $\rho \to 0+$ and applying (11.55) we conclude that χ is a weak solution of (11.50) with $\lambda = 0$.

Using a priori bounds for solutions of a heat equation, see Theorem 8, p. 459 of Il'in et al. (2002), we deduce that for any $T, R > 0$ there exists $C > 0$, independent of ω and $\rho > 0$, such that

$$\|u_\rho\|^2_{W_2^{1,2}(C_{T/2,R/2})} \le C\big[\|f_\rho\|^2_{L^2(C_{T,R})} + \|u_\rho\|^2_{L^2(C_{T,R})} + \big\|u_\rho(0,\cdot)\big\|^2_{W_2^1(B_R)}\big]. \quad (11.64)$$

Letting $\rho \to 0+$ we obtain that $\chi \in W^{1,1}_{2,loc}(\mathbb{H})$. Formula (11.56) can be concluded from the integration by parts formula for $u_\rho(t,x)$, after taking the limit as $\rho \to 0+$.

(iii) Using (11.48) we obtain that the expression under the limit in (11.57) can be estimated by $I_a + II_a$, where

$$I_a := \frac{1}{a}\int_{\mathbb{R}^d} u(y)\big\langle\big|\theta(ay)\big|\big\rangle_{\mathbb{Q}} dy,$$

$$II_a := \int_{\mathbb{H}} \big|\phi(t,x)\big| v_a(t,x)\,dt\,dx$$

and

$$u(y) := \int_{\mathbb{H}} q_{-t}(z-y)\big|\phi(t,z)\big|\,dt\,dz,$$

$$v_a(t,x) := a\bigg\langle\bigg|\int_{C_t} q_{s-t}(x-y)\big[\tilde{V}\big(a^2 s, ay\big)\cdot\tilde{\mathbf{f}}\big(a^2 s, ay\big) + \tilde{V}_p\big(a^2 s, ay\big)\big]\,ds\,dy\bigg|\bigg\rangle_{\mathbb{Q}}.$$

From Lemma 11.16 we easily conclude that $\lim_{a\to+\infty} I_a = 0$. On the other hand, using formula (11.21), we can write $\tilde{V}(t,x) = \nabla_x \cdot \tilde{b}(t,x)$, where $\tilde{b}(t,x)$ is the square integrable, stationary stream matrix. Using integration by parts, see the proof below, we obtain

$$v_a(t,x) = \bigg\langle\bigg|\int_{C_t} \tilde{\mathbf{g}}_p\big(a^2 s, ay\big)\cdot\nabla_y q_{s-t}(x-y)\,ds\,dy\bigg|\bigg\rangle_{\mathbb{Q}}. \quad (11.65)$$

Here $\tilde{\mathbf{g}}_p(s,y) = \mathbf{g}_p(\tau_{s,y}\omega)$ with $\mathbf{g}_p := b\mathbf{f} + b_p$ and $b_p := (b_{p1}, \ldots, b_{pd})$. Applying Theorem 11.18 we infer that

$$\lim_{a\to+\infty} v_a(t,x) = \bigg|\int_{C_t} \partial_\ell q_{s-t}(x-y)\,ds\,dy\bigg| = 0,$$

where $\ell := \langle \mathbf{g}_p\rangle_{\mathbb{Q}}$. Since functions $v_a(t,x)$ are uniformly bounded on compact sets in $\mathbb{H}$, as $a \to +\infty$, we can use the Lebesgue dominated convergence theorem and conclude that $\lim_{a\to+\infty} II_a = 0$. Thus (11.57) follows.

Proof of (11.65) It suffices to show the formula for $a = 1$. Let $\{\psi^{(n)}, n \geq 1\}$ be a sequence of functions from $C_b^\infty(\Omega)$ such that $\lim_{n\to+\infty} \nabla\psi^{(n)} = \mathbf{f}$ in $L^2(\mathbb{Q})$. Let $\tilde{\mathbf{g}}_p^{(n)}(s, y) = \mathbf{g}_p^{(n)}(\tau_{s,y}\omega)$ with $\mathbf{g}_p^{(n)} := b \cdot \nabla\psi^{(n)} + b_p$. Define also $\tilde{h}_p^{(n)}(s, y) = h_p^{(n)}(\tau_{s,y}\omega)$, where $h_p^{(n)} := V \cdot \nabla\psi^{(n)} + V_p$. Due to anti-symmetry of b

$$h_p^{(n)} = \nabla \cdot \mathbf{g}_p^{(n)}. \tag{11.66}$$

We are left to prove that

$$\left\langle \left| \int_{C_t} \tilde{\mathbf{g}}_p^{(n)}(s, y) \cdot \nabla_y q_{s-t}(x - y) ds dy \right| \right\rangle_{\mathbb{Q}} = \left\langle \left| \int_{C_t} \tilde{h}_p^{(n)}(s, y) q_{s-t}(x - y) ds dy \right| \right\rangle_{\mathbb{Q}}. \tag{11.67}$$

Formula (11.65) follows then from (11.67) upon taking the limit, as $n \to +\infty$.

To show (11.67) we use (11.66) and integration by parts. We conclude in this way that for any $R > 0$

$$\begin{aligned}\int_{C_{t,R}} \tilde{\mathbf{g}}_p^{(n)}(s, y) \cdot \nabla_y q_{s-t}(x - y) ds dy &= \int_{C_{t,R}} \tilde{h}_p^{(n)}(s, y) q_{s-t}(x - y) ds dy \\ &\quad + \int_{S_{t,R}} \tilde{\mathbf{g}}_p^{(n)}(s, y) \cdot \mathbf{n}(y) q_{s-t}(x - y) ds S(dy).\end{aligned} \tag{11.68}$$

Here $\mathbf{n}(y)$ is the unit outer normal to the boundary of B_R at y. Thanks to stationarity of $\tilde{\mathbf{g}}_p^{(n)}(s, y)$ we conclude easily that

$$\left\langle \left| \int_{S_{t,R}} \tilde{\mathbf{g}}_p^{(n)}(s, y) \cdot \mathbf{n}(y) q_{s-t}(x - y) ds S(dy) \right| \right\rangle_{\mathbb{Q}} \leq C \int_{S_{t,R}} q_{s-t}(x - y) ds S(dy) \to 0,$$

as $R \to +\infty$. Formula (11.67) then follows from (11.68). □

11.6.6 Proof of the Energy Identity

In this subsection we maintain the convention of suppressing the superscript indicating the direction in the notation of the corrector. Recall that χ is an $\mathcal{H}_1$-weak limit point of χ_λ, as $\lambda \to 0$. Performing the scalar product of both sides of the resolvent equation (11.22) with χ_λ and letting $\lambda \to 0+$ we obtain, by the weak lower semicontinuity of the norm $\|\cdot\|_1$, that

$$\|\chi\|_1^2 \le -\langle b_p, \nabla\chi\rangle_{\mathbb{Q}}. \tag{11.69}$$

Our purpose is to prove that the set of $\mathscr{H}_1$-weak limit points of χ_λ, as $\lambda \to 0+$, is a singleton and in fact (11.69) becomes an equality. Suppose therefore that χ_{λ_n}, $\chi_{\lambda_n'}$ converge weakly in $\mathscr{H}_1$ respectively to χ, χ_* for some $\lambda_n, \lambda_n' \to 0+$. Let $\gamma \in [0,1]$ and $\chi^{(\gamma)} := \gamma\chi + (1-\gamma)\chi_*$.

Assume that $\varphi \in C_c^\infty(\mathbb{R}^d)$ and $\psi \in C_c^\infty(-\infty, 0)$ are two probability densities. For an arbitrary $a > 0$ we set $\varphi_a(x) := a^{-d}\varphi(x/a)$ and $\psi_a(t) := a^{-2}\psi(t/a^2)$. Let $h_a(t,x) := \varphi_a(x)\psi_a(t)$ and $f_a(r) := (-a) \vee (r \wedge a)$ for $r \in \mathbb{R}$. Denote by F_a the primitive of f_a satisfying $F_a(0) = 0$ and by $h(t,x) := h_1(t,x)$. From the weak formulation (11.54), with $v = f_a(\chi^{(\gamma)})h_a$ (slightly generalized here by considering $W^{1,1}_{2,loc}(\mathbb{H})$ as the space of test functions) and integration by parts formula (11.56) we obtain

$$\begin{aligned} -\int_{\mathbb{H}} \langle \partial_t\chi^{(\gamma)} f_a(\chi^{(\gamma)})h_a\rangle_{\mathbb{Q}} dt dx &+ \frac{1}{2}\int_{\mathbb{H}} \langle \nabla_x\chi^{(\gamma)} \cdot \nabla_x[f_a(\chi^{(\gamma)})h_a]\rangle_{\mathbb{Q}} dt dx \\ &+ \int_{\mathbb{H}} \langle \tilde{b}\nabla_x\chi^{(\gamma)} \cdot \nabla_x[f_a(\chi^{(\gamma)})h_a]\rangle_{\mathbb{Q}} dt dx \\ &= \int_{\mathbb{H}} \langle \tilde{b}_p \cdot \nabla_x[f_a(\chi^{(\gamma)})h_a]\rangle_{\mathbb{Q}} dt dx. \end{aligned} \tag{11.70}$$

Denote by $L_1(a)$, $L_2(a)$ and $L_3(a)$ the respective terms appearing on the left-hand side of (11.70).

Since $f_a(\chi^{(\gamma)})\partial_t\chi^{(\gamma)} = \partial_t F_a(\chi^{(\gamma)})$ the absolute value of $L_1(a)$ equals

$$\begin{aligned} a^{-4-d}&\left|\int_{\mathbb{H}} \langle F_a(\chi^{(\gamma)}(t,x))\varphi(x/a)\psi'(t/a^2)\rangle_{\mathbb{Q}} dt dx\right| \\ &\le a^{-3-d}\int_{\mathbb{H}} \|\chi^{(\gamma)}(t,x)\|_{L^1(\mathbb{Q})}\varphi(x/a)|\psi'(t/a^2)| dt dx. \end{aligned} \tag{11.71}$$

The inequality follows from an elementary estimate $F_a(r) \le a|r|$ valid for all $r \in \mathbb{R}$ and $a > 0$. Changing variables $t := t/a^2$ and $x := x/a$ we conclude, using (11.57), that $L_1(a) \to 0$, as $a \to +\infty$.

The second term on the left-hand side of (11.70) can be written as being equal to

$$\begin{aligned} L_2(a) &= \frac{1}{2}\int_{\mathbb{H}} \langle |\nabla_x\chi^{(\gamma)}(t,x)|^2 h_a(t,x)\rangle_{\mathbb{Q}} dt dx \\ &+ \frac{1}{2}\int_{\mathbb{H}} \langle |\nabla_x\chi^{(\gamma)}(t,x)|^2[f_a'(\chi^{(\gamma)}(t,x)) - 1]h_a(t,x)\rangle_{\mathbb{Q}} dt dx \\ &+ \frac{1}{2a^{3+d}}\int_{\mathbb{H}} \langle \nabla_x[F_a(\chi^{(\gamma)}(t,x))] \cdot \nabla_x\varphi(x/a)\psi(t/a^2)\rangle_{\mathbb{Q}} dt dx. \end{aligned} \tag{11.72}$$

The first term on the right-hand side equals $(1/2)\|\nabla\chi^{(\gamma)}\|_{\mathbb{Q}}^2$. Denote the second and third terms appearing on the right-hand side of (11.72) by $L_{21}(a)$ and $L_{22}(a)$ respectively. Let $K > 0$ be arbitrary. Since $|f_a'(r) - 1| = 1[|r| > a]$ we can estimate $|L_{21}(a)|$ by

$$\frac{1}{2}\int_{\mathbb{H}}\langle|\nabla_x\chi^{(\gamma)}(t,x)|^2 h_a(t,x), |\chi^{(\gamma)}(t,x)| \geq a\rangle_{\mathbb{Q}} dtdx$$

$$\leq \frac{K^2}{2}\int_{\mathbb{H}}\langle h_a(t,x), |\nabla_x\chi^{(\gamma)}(t,x)| \leq K, |\chi^{(\gamma)}(t,x)| \geq a\rangle_{\mathbb{Q}} dtdx$$

$$+ \frac{1}{2}\langle|\nabla\chi^{(\gamma)}|^2, |\nabla\chi^{(\gamma)}| \geq K\rangle_{\mathbb{Q}}. \tag{11.73}$$

We have used the convention of writing $\langle F, A\rangle_{\mathbb{Q}}$ instead of $\langle F 1_A\rangle_{\mathbb{Q}}$ for any random variable F and event A. The first term on the right-hand side of (11.73) can be estimated by

$$\frac{K^2}{2a}\int_{\mathbb{H}}\|\chi^{(\gamma)}(t,x)\|_{L^1(\mathbb{Q})} h_a(t,x) dtdx$$

$$= \frac{K^2}{2a}\int_{\mathbb{H}}\|\chi^{(\gamma)}(a^2t, ax)\|_{L^1(\mathbb{Q})} h(t,x) dtdx$$

which vanishes, as $a \to +\infty$, by virtue of (11.57). We obtain therefore

$$\limsup_{a\to+\infty}|L_{21}(a)| \leq \frac{1}{2}\langle|\nabla\chi^{(\gamma)}|^2, |\nabla\chi^{(\gamma)}| \geq K\rangle_{\mathbb{Q}}$$

for an arbitrary $K > 0$. This leads to the conclusion that

$$\lim_{a\to+\infty} L_{21}(a) = 0. \tag{11.74}$$

As for $|L_{22}(a)|$, integrating by parts we conclude that it equals

$$\frac{1}{2a^{4+d}}\left|\int_{\mathbb{H}}\langle F_a(\chi^{(\gamma)}(t,x))(\Delta_x\varphi)(x/a)\psi(t/a^2)\rangle_{\mathbb{Q}} dtdx\right|$$

$$\leq \frac{1}{2a^{3+d}}\int_{\mathbb{H}}\|\chi^{(\gamma)}(t,x)\|_{L^1(\mathbb{Q})}|(\Delta_x\varphi)(x/a)|\psi(t/a^2) dtdx$$

$$= \frac{1}{2a}\int_{\mathbb{H}}\|\chi^{(\gamma)}(a^2t, ax)\|_{L^1(\mathbb{Q})}|\Delta_x\varphi(x)|\psi(t) dtdx$$

and

$$\lim_{a\to+\infty} L_{22}(a) = 0, \tag{11.75}$$

by virtue of (11.57).

We consider now $L_3(a)$. For any $K>0$ denote $B:=b\nabla\chi^{(\gamma)}$, $B^K:=B1[|B|\ge K]$ and $B_K:=B1[|B|<K]$. Thanks to anti-symmetry of b the absolute value of $L_3(a)$ equals

$$\begin{aligned}
&\frac{1}{a^{3+d}}\left|\int_{\mathbb{H}}\langle B\circ\tau_{t,x}\cdot\nabla_x\varphi(x/a)f_a\big(\chi^{(\gamma)}(t,x)\big)\psi\big(t/a^2\big)\rangle_{\mathbb{Q}}dtdx\right|\\
&=\frac{1}{a^{3+d}}\left|\int_{\mathbb{H}}\langle B^K\circ\tau_{t,x}\cdot\nabla_x\varphi(x/a)f_a\big(\chi^{(\gamma)}(t,x)\big)\psi\big(t/a^2\big)\rangle_{\mathbb{Q}}dtdx\right|\\
&+\frac{1}{a^{3+d}}\left|\int_{\mathbb{H}}\langle B_K\circ\tau_{t,x}\cdot\nabla_x\varphi(x/a)f_a\big(\chi^{(\gamma)}(t,x)\big)\psi\big(t/a^2\big)\rangle_{\mathbb{Q}}dtdx\right|.
\end{aligned}$$

Denote by $L_{31}(a)$ and $L_{32}(a)$ the first and the second terms appearing on the right-hand side, respectively. We have

$$L_{32}(a)\le\frac{K}{a}\int_{\mathbb{H}}\big\|\chi^{(\gamma)}\big(a^2t,ax\big)\big\|_{L^1(\mathbb{Q})}\big|\nabla_x\varphi(x)\big|\psi(t)dtdx.$$

Applying (11.57) we deduce that the right-hand side of the above estimate vanishes, as $a\to+\infty$. On the other hand, since $a^{-1}|f_a(r)|\le 1$ we have

$$L_{31}(a)\le\frac{1}{a^{2+d}}\int_{\mathbb{H}}\langle\big|B^K\circ\tau_{t,x}\big|\rangle_{\mathbb{Q}}\big|\nabla_x\varphi(x/a)\big|\psi\big(t/a^2\big)dtdx$$

and, by the ergodic theorem,

$$\limsup_{a\to+\infty}L_3(a)\le\big\|B^K\big\|_{L^1(\mathbb{Q})}\|\nabla_x\varphi\|_{L^1(\mathbb{R}^d)}.$$

This expression can be made as small as we wish, provided that K is chosen sufficiently large. We have shown therefore that

$$\lim_{a\to+\infty}L_3(a)=0. \tag{11.76}$$

An analogous argument to the one used to deal with $L_2(a)$ can be applied to show that the limit of the right-hand side of (11.70) equals $-\langle b_p,\nabla\chi^{(\gamma)}\rangle_{\mathbb{Q}}$. Summarizing, we have shown that

$$\frac{1}{2}\|\nabla\chi^{(\gamma)}\|_{\mathbb{Q}}^2=-\langle b_p,\nabla\chi^{(\gamma)}\rangle_{\mathbb{Q}} \tag{11.77}$$

for all $\gamma\in[0,1]$. This, of course, implies that $\chi=\chi_*$. The set of $\mathscr{H}_1$-weak limit points of χ_λ is therefore a singleton and in consequence $\chi=\lim_{\lambda\to0+}\chi_\lambda$ exists weakly in $\mathscr{H}_1$. In addition, using (11.39) we deduce the reverse inequality to (11.69). Thus we have shown the energy identity and therefore concluded the proof of Theorem 11.4.

11.7 Proofs of the Technical Results

11.7.1 Proof of Proposition 11.6

Let b be the stream matrix corresponding to V given by (11.10). For an arbitrary positive integer n and probability density $\phi \in C_c^\infty(\mathbb{R}^{d+1})$ define

$$b_0^{(n)}(\omega) := n^{d+1} \int_{\mathbb{R}^{d+1}} \phi(nt, nx) b(\tau_{t,x}\omega) dt dx.$$

Let $m > d/2 + 1$ be as in the statement of the proposition. One can easily verify that the entries of $b_0^{(n)}$ belong to $C^\infty(\Omega) \cap H^{0,m+1}(\Omega)$ and sequence $\{V_0^{(n)} := \nabla \cdot b_0^{(n)}, n \geq 1\}$ satisfies the conclusion of part (i) of the proposition. As far as (ii) is concerned, we can deduce that

$$\lim_{n \to +\infty} \left\langle \int_0^T \| \tilde{V}_0^{(n)}(t, \cdot) - \tilde{V}(t, \cdot) \|^2_{W_2^m(B_R)} dt \right\rangle_{\mathbb{Q}} = 0. \tag{11.78}$$

The entries of $b_0^{(n)}$ need not have deterministically bounded derivatives. However, we can approximate them further by anti-symmetric, random matrices $b^{(n)}$ whose entries belong to $C_b^\infty(\Omega)$. Suppose that $\lim_{n \to +\infty} \|b_0^{(n)} - b^{(n)}\|_{H^{0,m+1}} = 0$. In particular, $V^{(n)} = \nabla \cdot b^{(n)}$ satisfies then part (i). Note that for any $T, R > 0$

$$\begin{aligned} &\left\langle \int_0^T \| \tilde{V}_0^{(n)}(t, \cdot) - \tilde{V}^{(n)}(t, \cdot) \|^2_{W_2^m(B_R)} dt \right\rangle_{\mathbb{Q}} \\ &\quad \leq \left\| b_0^{(n)} - b^{(n)} \right\|^2_{H^{0,m+1}} |B_R| T \to 0+, \end{aligned}$$

as $n \to +\infty$. Since $m > d/2 + 1$ by the Sobolev embedding we conclude the convergence in (11.25) in the $L^1(\mathbb{Q})$ sense. Choosing an appropriate subsequence we can also deduce the almost sure convergence part of the statement (ii). □

11.7.2 Proof of Proposition 11.7

For a given $(\omega, \zeta) \in \Omega \times \Sigma$ and $R > 0$ let $\tau_{n,R}^\omega(\zeta)$, $\tau_R^\omega(\zeta)$ be the respective exit times of $X_{t,n}^\omega(\zeta)$ and $X_t^\omega(\zeta)$ from the ball $\bar{B}_R$. We shall suppress writing the arguments corresponding to the environment and path of the underlying Brownian motion, i.e. ω and ζ, when they are obvious from the context. Let $\sigma_{n,R} := \tau_R \wedge \tau_{n,R}$. The claim of the proposition follows if we show that

$$\lim_{R \to +\infty} \tau_R^\omega(\zeta) = +\infty, \quad \mathbb{Q} \otimes \mathbb{P} \text{ a.s. in } (\omega, \zeta). \tag{11.79}$$

Since $V^{(n)} = \nabla \cdot b^{(n)}$, for some anti-symmetric, C^2 smooth and bounded random matrix $b^{(n)}$, the process $\tilde{V}^{(n)}(t, X_{t,n}^\omega(\zeta); \omega) = V^{(n)}(\tau_{t, X_{t,n}^\omega(\zeta)}\omega)$ is stationary over the probability space corresponding to $\mathbb{Q}\otimes\mathbb{P}$, see Proposition 9.25. From the stochastic differential equation defining $X_{t,n}$ we deduce that

$$\Big\langle E_{\mathbb{P}} \sup_{t\in[0,T]} |X_{t,n}|^2\Big\rangle_{\mathbb{Q}} \le 2\Big(T^2\|V^{(n)}\|_{\mathbb{Q}}^2 + E_{\mathbb{P}} \sup_{t\in[0,T]} |w_t|^2\Big), \quad \forall T>0.$$

Hence, there exists $C_*>0$ such that

$$\big\langle \mathbb{P}[\tau_{n,R}<T]\big\rangle_{\mathbb{Q}} \le \frac{C_*}{R^2}, \quad \forall R>0. \tag{11.80}$$

Let $\tilde{\sigma}_{n,R}(t):=t\wedge\sigma_{n,R}$. From (11.18) we obtain that for $t\in[0,T]$

$$\begin{aligned}
|X_{\tilde{\sigma}_{n,R}(t),n}-X_{\tilde{\sigma}_{n,R}(t)}| &\le \int_0^{\tilde{\sigma}_{n,R}(t)} \big|\tilde{V}^{(n)}(s,X_{s,n})-\tilde{V}(s,X_s)\big|ds\\
&\le \int_0^{\tilde{\sigma}_{n,R}(t)} \big|\tilde{V}^{(n)}(s,X_{s,n})-\tilde{V}(s,X_{s,n})\big|ds\\
&\quad+\int_0^{\tilde{\sigma}_{n,R}(t)} \big|\tilde{V}(s,X_{s,n})-\tilde{V}(s,X_s)\big|ds\\
&\le \int_0^T \big\|\tilde{V}^{(n)}(s,\cdot)-\tilde{V}(s,\cdot)\big\|_{W_2^1(B_R)}ds\\
&\quad+L(T,R)\int_0^{\tilde{\sigma}_{n,R}(t)} |X_{s,n}-X_s|ds,
\end{aligned}$$

where $L(T,R;\omega)$ is the Lipschitz constant in the x variable of $\tilde{V}(t,x;\omega)$ on $[0,T]\times B_R$. A simple application of Gronwall's inequality and part (ii) of Proposition 11.6 yields that

$$\lim_{n\to+\infty}\sup_{t\in[0,T]} |X_{\tilde{\sigma}_n(t),n}-X_{\tilde{\sigma}_n(t)}|=0, \quad \forall T>0, \mathbb{Q}\otimes\mathbb{P}\text{ a.s.} \tag{11.81}$$

From (11.81) we conclude in particular that for (ω,ζ) such that $\tau_R^{\omega}(\zeta)<T$ there exists $n_0:=n_0(\omega,\zeta)$ for which $\tau_{n,R-1}^{\omega}(\zeta)<T$ for all $n\ge n_0$. Hence,

$$1_{[0,T)}(\tau_R)\le \liminf_{n\to+\infty} 1_{[0,T)}(\tau_{n,R-1}), \quad \mathbb{Q}\otimes\mathbb{P}\text{ a.s.}$$

By Fatou's lemma we get that

$$\mathbb{Q}\otimes\mathbb{P}[\tau_R<T]\le \liminf_{n\to+\infty}\mathbb{Q}\otimes\mathbb{P}[\tau_{n,R-1}<T] \overset{(11.80)}{\le} \frac{C_*}{(R-1)^2} \tag{11.82}$$

for all $R>1$. Since τ_R is monotonically increasing in R from (11.82) we conclude (11.79). □

11.7.3 Proof of Proposition 11.8

Denote by D the event consisting of those (ω, ζ) for which (11.26) does not hold. Choose an arbitrary $\varepsilon > 0$. According to (11.82) for a given $T > 0$ one can find a sufficiently large $R > 0$ such that $\mathbb{Q} \otimes \mathbb{P}[\tau_R < T] < \varepsilon$. From (11.81) we conclude that there exists a $\mathbb{Q} \otimes \mathbb{P}$ null measure event N such that $\liminf_{n \to +\infty} \tau_{n,R+1} \geq T$ on $[\tau_R \geq T] \setminus N$. Using again (11.81) we deduce that the latter event is included in D^c. This in turn implies $\mathbb{Q} \otimes \mathbb{P}[D] < \varepsilon$. Since ε has been arbitrarily chosen the conclusion of the proposition follows. □

11.7.4 Proof of Proposition 11.9

Let A be a set of Lebesgue measure 0. Suppose that $\phi_k : \mathbb{R}^d \to [0, 1]$ is a C^∞ smooth function equal 1 on the ball B_k and 0 outside of B_{k+1}. Let $Y^\omega_{t,k}$ be the solution of (11.18) with the drift replaced by $\phi_k(x)\tilde{V}(t, x; \omega)$ and satisfying $Y^\omega_{s,k} = x$. We have $Y^\omega_{t,k} = X^{s,x,\omega}_t$ for $t \in [0, \tau^{(\omega)}_k]$. Since for a given ω the drift of $Y^\omega_{t,k}$ is bounded and of C^1 class we conclude that $\mathbb{P}[Y^\omega_{t,k} \in A] = 0$ for each k, see Theorem 5.4 of Friedman (1975). We can write therefore

$$\mathbb{P}\big[X^{s,x,\omega}_t \in A\big] \leq \sum_{k=1}^{+\infty} \mathbb{P}\big[X^{s,x,\omega}_t \in A, t \leq \tau^{(\omega)}_k\big] \leq \sum_{k=1}^{+\infty} \mathbb{P}\big[Y^\omega_{t,k} \in A\big] = 0.$$

Suppose now that $A \subset \mathbb{R}^d$ is of positive Lebesgue measure $m_d[A]$. Let k be so large that $m_d[A \cap B_k] > 0$ and

$$\mathbb{P}\Big[\sup_{u \in [s,t]} |x + w_u| > k\Big] < 1/2\mathbb{P}[x + w_t \in A \cap B_k].$$

We write then

$$P^\omega_{s,t}(x, A) \geq \mathbb{P}\Big[Y_{t,k} \in A \cap B_k, \sup_{u \in [s,t]} |Y_{u,k}| \leq k\Big]. \tag{11.83}$$

By the Girsanov theorem (see e.g. Theorem 8.6.4 of Øksendal 2003) to show strict positivity of the right-hand side it suffices only to prove that

$$\mathbb{P}\Big[x + w_t \in A \cap B_k, \sup_{u \in [s,t]} |x + w_u| \leq k\Big] > 0.$$

The latter expression can be estimated from below by

$$\mathbb{P}[x + w_t \in A \cap B_k] - \mathbb{P}\Big[\sup_{u \in [s,t]} |x + w_u| > k\Big]$$

$$\geq \frac{1}{2}\mathbb{P}[x + w_t \in A \cap B_k] > 0.$$

□

11.7.5 Ergodic Theorem

We assume that the group $\{\tau_x, x \in \mathbb{R}^d\}$ acts on $(\Omega, \mathcal{F}, \mathbb{Q})$ and satisfies the assumptions made in Sect. 9.3.1. For an arbitrary function $\phi : \mathbb{R}^d \to \mathbb{R}$ and $\varepsilon > 0$ we let $\phi_\varepsilon(x) := \varepsilon^d \phi(\varepsilon x)$. When $\phi \in L^1(\mathbb{R}^d)$ and $f \in L^1(\mathbb{Q})$ the quantity

$$A_\varepsilon(f; \phi) := \int_{\mathbb{R}^d} f(\tau_x \omega)\phi_\varepsilon(x)dx$$

is called an *ergodic average*.

Theorem 11.18 *Under the assumptions made above we have*

$$\lim_{\varepsilon \to 0+} A_\varepsilon(f; \phi) = \langle f \rangle_{\mathbb{Q}} \int_{\mathbb{R}^d} \phi(x)dx \tag{11.84}$$

in the $L^1(\mathbb{Q})$ sense. If ϕ is compactly supported and belongs to $L^\infty(\mathbb{R}^d)$ then the convergence in (11.84) *holds also $\mathbb{Q}$ a.s.*

Proof Suppose first that $A_1, \ldots, A_d > 0$ and $\phi(x)$ is the indicator function of the box $[0, A_1] \times \cdots \times [0, A_d]$. The result is then a consequence of the multidimensional version of the ergodic theorem, see e.g. Krengel (1985), Theorem 3.5, p. 215. In fact the theorem asserts only a.s. convergence of $A_\varepsilon(f; \phi)$ but, since one can check easily that the family $A_\varepsilon(f; \phi)$ is uniformly integrable for $\varepsilon \in (0, 1]$, the L^1 convergence can be straightforwardly deduced. This can be generalized to ϕ being an indicator function of an arbitrary box $[A_1, B_1] \times \cdots \times [A_d, B_d]$, where $A_i < B_i$, $i = 1, \ldots, d$ by shifting the box $[0, A_1] \times \cdots \times [0, A_d]$. Equality (11.84) can then be generalized to ϕ that is a finite linear combination of indicator functions of disjoint intervals. We call such a ϕ a step function. An arbitrary $\phi \in L^1(\mathbb{R}^d)$ can be approximated in the L^1 sense by step functions and the assertion of the lemma can be extended to such functions as well. To show the a.s. convergence statement it suffices only to observe that any function satisfying the assumptions of the theorem can be uniformly approximated by a finite linear combination of the indicators of (disjoint) boxes. □

Theorem 11.84 can be extended to space-time, ergodic groups $\{\tau_{t,x}, (t, x) \in \mathbb{R}^{1+d}\}$, see Sect. 9.9.1. For an arbitrary function $\phi \in L^1(\mathbb{R}^{1+d})$ and $\varepsilon > 0$ we let $\phi_\varepsilon(t, x) := \varepsilon^{d+2}\phi(\varepsilon^2 t, \varepsilon x)$ and consider the averages defined by

$$\tilde{A}_\varepsilon(f; \phi) := \int_{\mathbb{R}^{1+d}} f(\tau_{t,x}\omega)\tilde{\phi}_\varepsilon(t, x)dtdx.$$

Theorem 11.19 *Under the assumptions made above we have*

$$\lim_{\varepsilon \to 0+} \tilde{A}_\varepsilon(f; \phi) = \langle f \rangle_{\mathbb{Q}} \int_{\mathbb{R}^{1+d}} \phi(t, x)dtdx \tag{11.85}$$

in the $L^1(\mathbb{Q})$ sense.

Proof We only need to consider the case when $\phi(x)$ is the indicator function of the box $\Box := [0, A_0] \times \cdots \times [0, A_d]$ with $A_0, \ldots, A_d > 0$. The remaining part of the argument then follows the proof of Theorem 11.18. Denote by $\Box^{(\varepsilon)} := [0, A_0/\varepsilon^2] \times [0, A_1/\varepsilon] \times \cdots \times [0, A_d/\varepsilon]$ and by $\varepsilon^{-1}\Box := [0, A_0/\varepsilon] \times \cdots \times [0, A_d/\varepsilon]$. We have

$$\Box^{(\varepsilon)} = \bigcup_{k=1}^{[\varepsilon^{-1}]} \Box_\varepsilon^k \cup \mathscr{R}_\varepsilon,$$

where $\Box_\varepsilon^k := kA_0/\varepsilon + \varepsilon^{-1}\Box$ and $\mathscr{R}_\varepsilon := [[\varepsilon^{-1}]A_0/\varepsilon, A_0/\varepsilon^2] \times [0, A_1/\varepsilon] \times \cdots \times [0, A_d/\varepsilon]$. Using Theorem 11.18 for any $\rho > 0$ we can find $\varepsilon_0 > 0$ sufficiently small so that for all $\varepsilon \in (0, \varepsilon_0)$ we have

$$\left\langle \left| \varepsilon^{d+1} \int_{\varepsilon^{-1}\Box} \tilde{f}(t,x)dtdx - \mathrm{vol}(\Box)\langle f \rangle_{\mathbb{Q}} \right| \right\rangle_{\mathbb{Q}} < \rho.$$

Here $\mathrm{vol}(\Box)$ is the volume of the box. Due to the shift invariance of $\mathbb{Q}$ the above inequality stays in force when the integration over $\varepsilon^{-1}\Box$ is replaced by the one over $\Box_\varepsilon^k$. This, in turn implies that

$$\left\langle \left| \varepsilon^{2+d} \int_{\Box^{(\varepsilon)}} \tilde{f}(t,x)dtdx - \mathrm{vol}(\Box)\langle f \rangle_{\mathbb{Q}} \right| \right\rangle_{\mathbb{Q}} \\ < \rho + \varepsilon^{2+d}\langle |f| \rangle_{\mathbb{Q}} \,\mathrm{vol}(\mathscr{R}_\varepsilon) + \left(1 - \varepsilon\left[\varepsilon^{-1}\right]\right) \mathrm{vol}(\Box).$$

Letting $\varepsilon \to 0+$ we conclude therefore that (11.85) holds for the indicator function of the box. □

11.8 Comments and References

The *passive tracer model*, described by (11.1), is quite often used e.g. in hydrology, meteorological sciences and oceanography to describe a turbulent transport phenomenon. A good source of information on the physical motivations for this model can be found in Csanady (1973), see also Frisch (1995); Monin and Yaglom (2007).

The proof of homogenization for diffusions with divergence free drifts that possess a deterministically bounded stream matrix has been given in Osada (1983). The result holds almost surely in the realization of the coefficients, the so-called *quenched* central limit theorem. The approach used there differs somewhat from the one taken in Sect. 11.2. Osada has applied Aronson–Moser–Nash bounds for transition probability densities to prove that the supremum norms of the corrector field grow sublinearly. This fact permits to show that the term $t^{-1/2}R_t$, appearing in the martingale decomposition (9.52), converges to 0 almost surely in the environment, which in turn allows for the *quenched formulation* of the central limit theorem, i.e. almost surely with respect to the realization of the drift.

A central limit theorem for diffusions with divergence free drift, in the case when the stream matrix is unbounded but square integrable, has been shown in Oelschläger (1988). A weak version of the homogenization result for the initial-boundary value problem for the related Kolmogorov equation has been obtained, by an application of the compensated compactness method, in Avellaneda and Majda (1991) for drifts having L^d integrable stream matrix when $d \geq 3$. When $d = 2$ the requirement is that the stream matrix is $L^{2+\epsilon}$ integrable for some $\epsilon > 0$. It has been shown in Fannjiang and Komorowski (1997) that the method of Osada (1983) can be extended to the case of drifts whose stream matrices are L^p integrable for some $p > d$. Using Moser iteration technique it is possible to prove the sublinear growth condition for the corrector field in the supremum norm. As a result one can establish the quenched central limit theorem. In the case of isotropic, divergence free drifts that do not have a square integrable stream matrix one can show that a central limit theorem fails. A counterexample has been shown in Komorowski and Olla (2002). We return to this issue in Chap. 12.

Time dependent drifts have been considered in Landim et al. (1998b). Theorem 11.4 comes from Komorowski and Olla (2001) and its proof adapts the argument of Oelschläger (1988) to the situation when the drift coefficients depend on time. A quenched version of the central limit theorem has been shown in Fannjiang and Komorowski (1999a, 2002). It requires the L^p integrability of the stream matrix for some $p > d + 2$.

The problem of convection enhanced diffusions is discussed in Fannjiang and Papanicolaou (1994). Variational principles are used there to obtain asymptotics of the effective diffusivity of the passive tracer in the regime of vanishing molecular diffusivity for two-dimensional cellular flows. Koralov (2004) uses Freidlin–Ventzel theory of averaging for randomly perturbed two-dimensional Hamiltonian systems to extend those results to a broader class of flows.

References

Avellaneda M, Majda AJ (1991) An integral representation and bounds on the effective diffusivity in passive advection by laminar and turbulent flows. Commun Math Phys 138(2):339–391

Csanady GT (1973) Turbulent diffusion in the environment. Geophysics and astrophysics monographs, vol 3. Kluwer Academic, Dordrecht, Boston

Ethier SN, Kurtz TG (1986) Markov processes: characterization and convergence. Wiley series in probability and mathematical statistics: probability and mathematical statistics. Wiley, New York

Fannjiang A, Komorowski T (1997) A martingale approach to homogenization of unbounded random flows. Ann Probab 25(4):1872–1894

Fannjiang A, Komorowski T (1999a) An invariance principle for diffusion in turbulence. Ann Probab 27(2):751–781

Fannjiang AC, Komorowski T (2002b) Correction: "An invariance principle for diffusion in turbulence" [Ann Probab **27**(2):751–781 (1999); MR1698963 (2001e:60069)]. Ann Probab 30(1):480–482

Fannjiang A, Papanicolaou G (1994) Convection enhanced diffusion for periodic flows. SIAM J Appl Math 54(2):333–408

Friedman A (1964) Partial differential equations of parabolic type. Prentice-Hall, Englewood Cliffs

Friedman A (1975) Stochastic differential equations and applications, vol 1. Probability and mathematical statistics, vol 28. Academic Press [Harcourt Brace Jovanovich Publishers], New York

Frisch U (1995) Turbulence: the legacy of AN Kolmogorov. Cambridge University Press, Cambridge

Gilbarg D, Trudinger NS (1983) Elliptic partial differential equations of second order. Grundlehren der Mathematischen Wissenschaften [Fundamental principles of mathematical sciences], vol 224, 2nd edn. Springer, Berlin

Ikeda N, Watanabe S (1989) Stochastic differential equations and diffusion processes. North-Holland mathematical library, vol 24, 2nd edn. North-Holland, Amsterdam

Il'in AM, Kalashnikov AS, Oleĭnik OA (2002) Second-order linear equations of parabolic type. J Math Sci 108:435–452

Komorowski T, Olla S (2001) On homogenization of time-dependent random flows. Probab Theory Relat Fields 121(1):98–116

Komorowski T, Olla S (2002) On the superdiffusive behavior of passive tracer with a Gaussian drift. J Stat Phys 108(3–4):647–668

Koralov L (2004) Random perturbations of 2-dimensional Hamiltonian flows. Probab Theory Relat Fields 129(1):37–62

Krengel U (1985) Ergodic theorems: with a supplement by Antoine Brunel. de Gruyter studies in mathematics, vol 6. de Gruyter, Berlin

Ladyženskaja OA, Solonnikov VA, Ural'ceva NN (1968) Lineinye i kvazilineinye uravneniya parabolicheskogo tipa. Nauka, Moscow

Landim C, Olla S, Yau HT (1998b) Convection-diffusion equation with space-time ergodic random flow. Probab Theory Relat Fields 112(2):203–220

Monin AS, Yaglom AM (2007) Statistical fluid mechanics: mechanics of turbulence, vol II: Edited and with a preface by John L Lumley. Dover, Mineola. Translated from the 1965 Russian original, English edition updated, augmented and revised by the authors, reprinted from the 1975 edition

Oelschläger K (1988) Homogenization of a diffusion process in a divergence-free random field. Ann Probab 16(3):1084–1126

Øksendal B (2003) Stochastic differential equations: an introduction with applications. Universitext, 6th edn. Springer, Berlin

Osada H (1983) Homogenization of diffusion processes with random stationary coefficients. In: Probability theory and mathematical statistics, Tbilisi, 1982. Lecture notes in math, vol 1021. Springer, Berlin, pp 507–517

Chapter 12
Diffusions with Gaussian Drifts

We continue our discussion of the passive tracer model introduced in Chap. 11 and consider Eq. (11.1) under an additional assumption that the drift is a Gaussian random field, i.e. all its finite dimensional marginals have normal distributions. Due to the fact that the invariant measure of the associated environment process is also Gaussian, we are able to obtain more detailed information on the behavior of the tracer. A key observation is that the generator of the environment process satisfies the graded sector condition introduced in Sect. 2.7.4 with respect to the decomposition of the corresponding L^2 space into the orthogonal sum of subspaces of Hermite polynomials of a given degree. Using this approach we first re-prove (see Sect. 12.4) the central limit theorem for diffusions whose drift has a finite Péclet number. Then, in Sect. 12.6 we apply the variational principles derived in Chap. 4 to prove that the result is in some sense optimal for steady (i.e. time independent) fields. Namely, we show that the motion of a tracer in an isotropic flow with an infinite Péclet number is superdiffusive.

In the second part of the chapter (Sects. 12.7–12.15) we turn our attention to the case when the drift is a time dependent, Markovian field. We have already seen that the tracer satisfies the central limit theorem, provided that the Péclet number of the flow is finite, see Theorem 11.4. On the other hand, when the decorrelation in time of the field is uniform on all spatial scales the generator of the environment process satisfies the spectral gap estimate and the central limit theorem holds regardless of whether the Péclet number is finite or not. In fact, this result can be significantly strengthened by allowing the rate of temporal decorrelation to depend also on spatial scales. In Theorem 12.13 we formulate a sufficient condition for the validity of the central limit theorem. The result remains valid also in the situation when the molecular diffusion is absent in (11.1) (the case of motions in a random flow). Furthermore, we show that the hypotheses for the central limit theorem are optimal in case of isotropic flows, see Theorem 12.26.

T. Komorowski et al., *Fluctuations in Markov Processes*,
Grundlehren der mathematischen Wissenschaften 345,
DOI 10.1007/978-3-642-29880-6_12,

12.1 Stationary Gaussian Fields

For our purposes it will be convenient to work with Ω defined as the space of all C^2 continuous, vector fields equipped with the standard Fréchet metric. Let $\mathscr{F}$ be its Borel σ-algebra and $\mathbb{Q}$ be a Borel probability measure that is invariant under the group of shifts defined by $\tau_x\omega(\cdot) := \omega(x+\cdot)$ for all $(x,\omega) \in \mathbb{R}^d \times \Omega$.

Define $\tilde{V}(x;\omega) := V(\tau_x\omega)$, where a random vector $V = (V_1, \ldots, V_d) : \Omega \to \mathbb{R}^d$ is given by

$$V(\omega) := \omega(0) \quad \text{for } \omega \in \Omega. \tag{12.1}$$

We assume that for any $N \geq 1$ and $x_1, \ldots, x_N \in \mathbb{R}^d$ the Nd-dimensional random vector $(\tilde{V}(x_1), \ldots, \tilde{V}(x_N))$ is a centered Gaussian over $(\Omega, \mathscr{F}, \mathbb{Q})$. The entries of its covariance matrix $R(x) = [R_{ij}(x)]$, given by

$$R_{ij}(x) = \langle \tilde{V}_i(x)\tilde{V}_j(0)\rangle_{\mathbb{Q}}, \quad i,j = 1, \ldots, d,$$

belong then to $C_b^4(\mathbb{R}^d)$ and the measure $\mathbb{Q}$ is ergodic with respect to the group of spatial shifts. A sufficient and necessary condition for the latter to hold is that the corresponding spectral measure has no atoms, see Example 6.2, p. 163 of Rozanov (1967). Since the field is stationary and Gaussian with sufficiently regular realizations, according to the results of Sect. 12.5.1 there exists a random variable $K(\omega)$ possessing all moments, such that

$$\left|\tilde{V}(x;\omega)\right| + \left|\nabla_x \tilde{V}(x;\omega)\right| \leq K(\omega)\big(1+|x|\big), \quad \forall x \in \mathbb{R}^d, \quad \mathbb{Q} \text{ a.s. in } \omega \in \Omega. \tag{12.2}$$

12.2 Hermite Polynomials and Graded Structure of $L^2(\mathbb{Q})$

Denote by $\mathscr{V}$ the linear space consisting of all random variables of the form

$$\langle \tilde{V}, \varphi\rangle := \sum_{i=1}^{d} \int_{\mathbb{R}^d} \tilde{V}_i(x)\varphi_i(x)dx,$$

where $\varphi = (\varphi_1, \ldots, \varphi_d)$ and $\varphi_i \in C_c^\infty(\mathbb{R}^d)$, $i = 1, \ldots, d$. For any $n \geq 0$ let $\mathscr{G}_n$ be the space of *n-th degree polynomials* defined as the L^2-closure of the set $\mathscr{G}_n^{(0)}$ consisting of random variables of the form $f = p(g_1, \ldots, g_m)$, where $g_i \in \mathscr{V}$, $i = 1, \ldots, m$ and $p(x_1, \ldots, x_m)$ is a polynomial in m variables of degree at most n. Let $\mathscr{A}_0 := \mathscr{G}_0$, and $\mathscr{A}_n = \mathscr{G}_n \ominus \mathscr{G}_{n-1}$ for an arbitrary $n \geq 1$, i.e. $\mathscr{A}_n$ is the orthogonal complement of $\mathscr{G}_{n-1}$ to $\mathscr{G}_n$. Elements of $\mathscr{A}_n$ are called *n-th degree Hermite polynomials*. By $\mathscr{G} := \bigcup_{n\geq 0}\mathscr{G}_n$ we denote the space of *elements of a finite degree*. According to Theorem 2.2.6, p. 18 of Janson (1997) we have

$$L^2(\mathbb{Q}) = \bigoplus_{n\geq 0} \mathscr{A}_n.$$

Let Π_n be the orthogonal projection of $L^2(\mathbb{Q})$ onto $\mathscr{A}_n$ and

$$T_x f := f \circ \tau_x, \quad \forall f \in L^2(\mathbb{Q}),\ x \in \mathbb{R}^d.$$

Proposition 12.1 *Spatial shifts commute with projections, i.e. for any $n \ge 0$*

$$\Pi_n T_x = T_x \Pi_n, \quad \forall x \in \mathbb{R}^d. \tag{12.3}$$

Proof Since $T_x(\mathscr{G}_n) = \mathscr{G}_n$ and T_x is an unitary isomorphism we have

$$T_x(\mathscr{A}_n) = T_x(\mathscr{G}_n) \ominus T_x(\mathscr{G}_{n-1}) = \mathscr{G}_n \ominus \mathscr{G}_{n-1} = \mathscr{A}_n.$$

Equality (12.3) then follows. □

As a direct corollary from the above proposition we deduce.

Corollary 12.2 *For any $n \ge 0$ we have $\Pi_n(H^1(\Omega)) \subset H^1(\Omega)$ and*

$$\Pi_n \nabla = \nabla \Pi_n, \quad \text{on } H^1(\Omega).$$

12.3 Environment Process and Its Properties

Consider the solution of the stochastic differential equation

$$\begin{aligned} dX_t^{x,\omega} &= \tilde{V}\big(X_t^{x,\omega}; \omega\big)dt + dw_t, \\ X_0^{x,\omega} &= x, \end{aligned} \tag{12.4}$$

where $\tilde{V} : \mathbb{R}^d \times \Omega \to \mathbb{R}^d$ is a random vector field defined in Sect. 12.1 and $\{w_t, t \ge 0\}$ is a standard d-dimensional Brownian motion given over another probability space $(\Sigma, \mathscr{A}, \mathbb{P})$. Since the coefficients of (12.4) grow at most linearly the stochastic differential equation has a unique global solution satisfying $X_{0,x}^{\omega} = x$, see e.g. Theorem 1.1, p. 98 of Friedman (1975). A Markovian environment process $\{\eta_t, t \ge 0\}$ can be defined therefore in the same way as in Sect. 11.6.3. Measure $\mathbb{Q}$ is invariant and ergodic under its transition probability semigroup $\{P_t, t \ge 0\}$, which can be extended to a strongly continuous semigroup of contractions on $L^2(\mathbb{Q})$. Because the coefficients of the diffusion are unbounded Proposition 9.8 cannot be directly applied to identify a core of its generator. To find a core we use Theorem 12.3, proved below. Before stating the result define the Banach space $W_p^m(\Omega)$ for a given $p \in [1, +\infty)$ and an integer $m \ge 0$. It consists of those $g \in L^p(\mathbb{Q})$, for which

$$\|g\|_{W_p^m}^p := \sum_{|\alpha| \le m} \|D^\alpha g\|_{L^p(\mathbb{Q})}^p < +\infty.$$

Here $\alpha = (\alpha_1, \dots, \alpha_d)$ is a non-negative integer valued multi-index, $|\alpha| = \sum_{i=1}^d \alpha_i$ and $D^\alpha g := D^{\alpha_1} \dots D^{\alpha_d} g$. In case $m = 0$ we identify $W_p^0(\Omega)$ with $L^p(\mathbb{Q})$.

Let

$$\mathscr{C}_0 := \bigcap_{p\geq 1} W_p^2(\Omega). \tag{12.5}$$

Elements of $\mathscr{A}_n^{reg} := \mathscr{A}_n \cap \mathscr{C}_0$ are called regular n-th degree Hermite polynomials. Likewise we introduce $\mathscr{G}_n^{reg}$ and $\mathscr{G}^{reg}$ the spaces of at most n-th degree and finite degree regular polynomials respectively.

Theorem 12.3 *The domains of both the generator L and its adjoint L^* contain $\mathscr{C}_0$. In addition, for any $g \in \mathscr{C}_0$ we have*

$$Lg = Sg + Ag, \tag{12.6}$$

and

$$L^* g = Sg - Ag, \tag{12.7}$$

where S and A are the symmetric and anti-symmetric parts of L given by

$$Sg := \frac{1}{2}\Delta g \quad \text{and} \quad Ag := V \cdot \nabla g. \tag{12.8}$$

The set $\mathscr{C} := \mathscr{G}^{reg}$ is a common core of both L and L^*.

We postpone the presentation of the proof until Sect. 12.5.2, where we also demonstrate the following.

Proposition 12.4 *Operator S is essentially self-adjoint and $\mathscr{C}$ is its core.*

Having introduced the symmetric part of the generator we can define the spaces $\mathscr{H}_1$ and $\mathscr{H}_{-1}$ as in Chap. 2.

Below, we formulate some of the properties of Hermite polynomials that are crucial for the application of the results of Sect. 2.7.4.

Proposition 12.5

(i)

$$\Pi_n(\mathscr{C}_0) = \mathscr{A}_n^{reg}, \quad \forall n \geq 0,$$

(ii) *$\mathscr{H}_1$ coincides with $H^1(\Omega)$ and*

$$\|f\|_1^2 = \frac{1}{2}\|f\|_{H^1(\Omega)}^2, \quad \forall f \in \mathscr{H}_1, \tag{12.9}$$

(iii)

$$L\big(\mathscr{A}_n^{reg}\big) \subset \mathscr{A}_{n-1} \oplus \mathscr{A}_n \oplus \mathscr{A}_{n+1}, \tag{12.10}$$

and

$$L : D(L) \cap \mathscr{A}_n \to \mathscr{A}_{n-1} \oplus \mathscr{A}_n \oplus \mathscr{A}_{n+1}, \quad \forall n \geq 1. \tag{12.11}$$

An analogous result holds also for L^*,

(iv) *there exists* $C > 0$ *such that*

$$\|L\Pi_n f - \Pi_n L f\|_{\mathbb{Q}} \leq C\sqrt{n}\|f\|_1, \quad \forall f \in \mathscr{C}, n \geq 0, \tag{12.12}$$

(v) *for each* $n \geq 0$ *we have* $\Pi_n(D(L)) = D(L) \cap \mathscr{A}_n$, *and*

(vi) $\Pi_n A \Pi_n = 0$ *on* $\mathscr{C}$.

Proof (i) Thanks to (12.3) and the fact that Π_n is bounded on $L^p(\mathbb{Q})$ for any $p \in [1, +\infty)$, see Theorem 2.5.14, p. 63 of Janson (1997), we conclude that $\Pi_n(\mathscr{C}_0) \subset \mathscr{C}_0$. This in particular implies that $\Pi_n(\mathscr{C}_0) \subset \mathscr{A}_n^{reg}$. The opposite inclusion is obvious, due to the trivial equality $\Pi_n(\mathscr{A}_n^{reg}) = \mathscr{A}_n^{reg}$.

(ii) Note that $\mathscr{C}_0$ is contained in both $\mathscr{H}_1$ and $H^1(\Omega)$. Equality (12.9) holds by a simple integration by parts and the first formula of (12.8). The equality of $\mathscr{H}_1$ and $H^1(\Omega)$ follows from the fact that $\mathscr{C}_0$ is dense both in $\mathscr{H}_1$ (as a core of L) and $H^1(\Omega)$, see Proposition 9.3.

(iii) Suppose that $f \in \mathscr{A}_n^{reg}$ and g is orthogonal to $\mathscr{A}_{n-1} \oplus \mathscr{A}_n \oplus \mathscr{A}_{n+1}$. From (12.8) we have $\langle Sf, g\rangle_{\mathbb{Q}} = 0$. We show that also

$$\langle Af, g\rangle_{\mathbb{Q}} = 0. \tag{12.13}$$

This would conclude the proof of (12.10).

Note that for $i = 1, \ldots, d$ and $m \leq n - 2$, where $n \geq 2$, we have $V_i \Pi_m g \in \mathscr{G}_{n-1}$ so $\langle f, V_i \Pi_m g\rangle_{\mathbb{Q}} = 0$. On the other hand, $V_i f \in \mathscr{G}_{n+1}$ when $m \geq n + 2$ hence also

$$\langle f, V_i \Pi_m g\rangle_{\mathbb{Q}} = \langle V_i f, \Pi_m g\rangle_{\mathbb{Q}} = 0.$$

In this way we have shown that $V_i g \perp \mathscr{A}_n$. Thus, (12.13) follows from the second formula of (12.8). The result for the adjoint operator can be proved in the same way.

To demonstrate (12.11) suppose that $f \in D(L) \cap \mathscr{A}_n$. It suffices only to prove that for any $g \in \mathscr{C}$ orthogonal to $\mathscr{A}_{n-1} \oplus \mathscr{A}_n \oplus \mathscr{A}_{n+1}$ we have

$$\langle Lf, g\rangle_{\mathbb{Q}} = 0. \tag{12.14}$$

From the already proven formula (12.10) for L^* we conclude that $L^* g \perp \mathscr{A}_n$, which in turn implies (12.14).

(iv) The argument from the proof of part (iii) shows that

$$\Pi_n A \Pi_m g = 0, \quad \forall g \in \mathscr{C},$$

when $|m - n| \geq 2$. From Corollary 12.2 we conclude also that $\Pi_n S = S\Pi_n$ on $\mathscr{C}$. As a result we obtain

$$\begin{aligned} L\Pi_n f - \Pi_n L f &= A\Pi_n f - \Pi_n A f \\ &= A\Pi_n f - \Pi_n A\Pi_{n-1} f - \Pi_n A\Pi_n f - \Pi_n A\Pi_{n+1} f. \end{aligned}$$

Hence the left-hand side of (12.12) can be estimated by

$$\|A\Pi_{n-1} f\|_{\mathbb{Q}} + 2\|A\Pi_n f\|_{\mathbb{Q}} + \|A\Pi_{n+1} f\|_{\mathbb{Q}}. \tag{12.15}$$

Each term in (12.15) can be estimated in the same fashion, so we deal only with the first one. Using (12.8) and Hölder inequality we can write that

$$\|A\Pi_{n-1} f\|_{\mathbb{Q}} \le \|V\|_{L^{2n}(\mathbb{Q})} \|\nabla \Pi_{n-1} f\|_{L^{2n/(n-1)}(\mathbb{Q})}. \tag{12.16}$$

Using Theorem 5.10, p. 62 of Janson (1997) we obtain that

$$\|g\|_{L^{2n/(n-1)}(\mathbb{Q})} \le \left(\frac{n+1}{n-1}\right)^{n/2} \|g\|_{\mathbb{Q}}, \quad \forall g \in \mathcal{G}_n.$$

From Stirling's formula we get $\|V\|_{L^{2n}(\mathbb{Q})} \sim \sqrt{n}$. Thus, the right-hand side of (12.16) can be estimated by

$$Cn^{1/2}\left(\frac{n+1}{n-1}\right)^{n/2} \|\nabla \Pi_{n-1} f\|_{\mathbb{Q}}$$

for some constant C independent of n. Thus estimate (12.12) follows.

(v) It suffices only to prove that $\Pi_n f \in D(L)$ for any $f \in D(L)$. Suppose that $\{g_m, m \ge 1\} \subset \mathscr{C}$ is such that $(g_m, L g_m)$ tends to (f, Lf), as $m \to +\infty$, in the epigraph norm. Therefore g_m converges to f in $\mathscr{H}_1$. Using (12.12) we conclude that $L\Pi_n g_m$, as $m \to +\infty$, is convergent in $L^2(\mathbb{Q})$ for each n. Since L is closed and $\Pi_n g_m \to \Pi_n f$ in $L^2(\mathbb{Q})$ we deduce that $\Pi_n f \in D(L)$ and $L\Pi_n g_m \to L\Pi_n f$ in $L^2(\mathbb{Q})$, as $m \to +\infty$.

(vi) It suffices only to show that

$$V_i f \perp \mathscr{A}_n \tag{12.17}$$

for each $i = 1, \ldots, d$, $n \ge 0$ and $f \in \mathscr{A}_n$. Let us introduce some terminology. Suppose that $g_j \in \mathscr{V}$, where $j = 1, \ldots, N$. A *monomial of degree* N is a random variable of the form $\prod_{i=1}^N g_i$. Since V is of zero mean we have

$$\left\langle \prod_{i=1}^N g_i \right\rangle_{\mathbb{Q}} = 0 \tag{12.18}$$

when N is odd. Suppose that $n = 2m$ for some integer $m \ge 0$. According to Theorem 3.4, p. 24 of Janson (1997) for any $f \in \mathcal{G}_n^{(0)}$ the projection $\Pi_n f$ is a sum of monomials of degree $2k$, where $k = 0, \ldots, m$. Likewise, when $n = 2m + 1$ for some integer $m \geq 0$ polynomial $\Pi_n f$ is a sum of monomials of degree $2k+1$, where $k = 0, \ldots, m$. This, together with (12.18) implies that

$$\langle V_i \Pi_n f, \Pi_n g \rangle_{\mathbb{Q}} = 0, \quad \forall f, g \in \mathscr{G}_n^{(0)}. \tag{12.19}$$

By a standard density argument we can extend (12.19) to all $f, g \in \mathscr{G}_n$. Thus (12.17) follows. □

Following the notation of Sect. 2.7.4 we introduce

$$L_0 = \sum_{n \geq 0} \Pi_n L \Pi_n, \qquad L_{\pm} = \sum_{n \geq 0} \Pi_{n\pm 1} L \Pi_n,$$

defined on $D(L)$. In fact $\mathscr{C}$ is a common core for the above operators and on this set we can write $L_0 = S_0 + B_0$, where $S_0 = S$. It is obvious that S_0 satisfies condition (2.40). According to part (vi) of Proposition 12.5, we have

$$B_0 = \sum_{n \geq 0} \Pi_n A \Pi_n = 0.$$

12.4 Central Limit Theorem

The main result of this section is contained in the following.

Theorem 12.6 *Suppose that condition* (11.9) *holds. Then,* $\{X_t^x/\sqrt{t}, t \geq 0\}$ *satisfies a central limit theorem, in probability with respect to the environment.*

Proof Using estimate (11.41) we conclude that $V_p \in \mathscr{H}_{-1}$ for each $p = 1, \ldots, d$. Furthermore we apply the decomposition (9.46) to represent the trajectory as a martingale (with c the $d \times d$ identity matrix) and an additive functional of the environment process. Since in this case $S = S_0$ the proof of the central limit theorem reduces to the verification of conditions of Theorem 2.23.

Part (i) of Proposition 12.5 implies condition (2.39), while part (vi) implies that (2.40) and (2.50) hold trivially. Since the components of V belong to $\mathscr{A}_1 \cap \mathscr{H}_{-1}$ it is also trivial to verify that $|||V_j|||_{k,-1} < \infty$ for each $j = 1, \ldots, d$ and an arbitrary $k \geq 1$. Finally, we show that (2.45) is satisfied with $\beta = 1/2$. Indeed, a simple calculation shows that

$$Lf = \sum_{k,l=1}^{d} D_k \left[\left(\frac{1}{2} \delta_{k,l} + b_{kl} \right) D_l f \right], \quad f \in \mathscr{C}, \tag{12.20}$$

with b_{kl} given by (11.10). Suppose that $f \in \mathscr{A}_n^{reg}$. Note that

$$\|L_+ f\|_{-1} = \sup\left[\left| \langle Lf, g \rangle_{\mathbb{Q}} \right| : g \in \mathscr{A}_{n+1}^{reg}, \|g\|_1 = 1 \right]. \tag{12.21}$$

Using (12.20) we can write that the right-hand side of (12.21) equals

$$\sup\big[\big|\langle b\nabla f, \nabla g\rangle_{\mathbb{Q}}\big| : g \in \mathscr{A}_{n+1}^{reg}, \|g\|_1 = 1\big]. \tag{12.22}$$

To estimate the expression in (12.22) we apply the same argument as in the proof of part (v) of Proposition 12.5 and obtain that it is less than or equal to

$$\begin{aligned}\|b\nabla f\|_{\mathbb{Q}}\|g\|_{H^1(\Omega)} &\le \|b\|_{L^{2n}(\mathbb{Q})}\|\nabla f\|_{L^{2n/(n-1)}(\mathbb{Q})}\|g\|_{H^1(\Omega)}\\ &\le \left(\frac{n+1}{n-1}\right)^{n/2}\|b\|_{L^{2n}(\mathbb{Q})}\|f\|_{H^1(\Omega)}\|g\|_{H^1(\Omega)}.\end{aligned}$$

To obtain the last inequality we have used the fact that

$$\|h\|_{L^{2n/(n-1)}(\mathbb{Q})} \le \left(\frac{n+1}{n-1}\right)^{n/2}\|h\|_{\mathbb{Q}}$$

for any $h \in \mathscr{A}_n$, see Theorem 2.5.10, p. 62 of Janson (1997). Since $\|b\|_{L^{2n}(\mathbb{Q})} \sim \sqrt{n}$, we conclude that

$$\|L_+ f\|_{-1} \le C\sqrt{n}\|f\|_1$$

for some constant $C > 0$ independent of n. The proof of an analogous estimate for L_- uses the same argument. Hence (2.45) follows. Since $\nabla \cdot V = 0$ the limiting covariance matrix $\bar{a} = [\bar{a}_{pq}]$ is given by formula (11.13), i.e. for any $e \in \mathbb{R}^d$

$$\bar{a}(e) = |e|^2 + 2\lim_{\lambda\to 0+}\big\langle(\lambda - L)^{-1}V_e, V_e\big\rangle_{\mathbb{Q}}, \tag{12.23}$$

where $V_e := V \cdot e$. □

Finally, we formulate variational principles for the limiting covariance matrix given by (12.23). Using Theorem 4.7 we infer the following.

Theorem 12.7 *Under the assumptions of Theorem* 12.6 *formula* (11.15) *holds, provided that* $\mathscr{C}$ *is chosen as the space of test functions.*

12.5 Proofs of Technical Results

12.5.1 Proof of Estimate (12.2)

Suppose that

$$\big\{\widetilde{W}(x) = \big(\widetilde{W}_1(x), \dots, \widetilde{W}_d(x)\big), x \in \mathbb{R}^d\big\}$$

is an $\mathbb{R}^d$-valued, centered Gaussian field over the probability space $(\Omega, \mathscr{F}, \mathbb{Q})$. By a d-*ball* centered at x_0 and of radius $\rho > 0$ we mean the respective ball in the pseudometric

$$\mathrm{d}(x_1, x_2) := \langle |\widetilde{W}(x_1) - \widetilde{W}(x_2)|^2 \rangle_{\mathbb{Q}}^{1/2}, \quad x_1, x_2 \in \mathbb{R}^d.$$

For a given $\varepsilon > 0$ let $N(\varepsilon)$ be the minimal number of d-balls with radius $\varepsilon > 0$ needed to cover $\mathbb{R}^d$. Let also $\Sigma(\widetilde{W}) := \sup_{x \in \mathbb{R}^d} \operatorname{tr} R(x, x)$, where $R(x, y) = [R_{ij}(x, y)]$, $i, j = 1, \dots, d$—the covariance matrix of the field—is given by

$$R_{ij}(x, y) = \langle \widetilde{W}_i(x) \widetilde{W}_j(y) \rangle_{\mathbb{Q}}.$$

The following estimate of the tail probabilities for the supremum of a Gaussian field can be found in Adler (1990), see Theorem 5.4, p. 121.

Theorem 12.8 *Suppose that*

$$N(\varepsilon) \le \frac{1}{C_1} \exp\{-C_1 \varepsilon^{-1/(1+C_2)}\}, \quad \forall \varepsilon \in (0, 1] \tag{12.24}$$

for some positive constants C_1, C_2. Then, there exist constants $C, \Lambda > 0$ depending only on C_1, C_2 such that

$$\mathbb{Q}\Big(\sup_{x \in \mathbb{R}^d} |\widetilde{W}(x)| \ge \lambda\Big) \le C \exp\left\{-\frac{\lambda^2}{8\Sigma(\widetilde{W})}\right\}, \quad \forall \lambda \ge \Lambda.$$

As a consequence of the above theorem we obtain an estimate on the growth of a stationary Gaussian field. Let $\tilde{V}(x)$ be the field as considered in Sect. 12.1 and let

$$\widetilde{W}^{(n)}(x; \omega) := \left(\frac{\tilde{V}(x; \omega)}{|x| + n}, \frac{\nabla_x \tilde{V}(x; \omega)}{|x| + n}\right).$$

Define a sequence of events as follows: $K_0 := \emptyset$ and

$$K_n := \Big[\omega : \sup_{x \in \mathbb{R}^d} |\widetilde{W}^{(n)}(x; \omega)| \le 1\Big], \quad \forall n \ge 1. \tag{12.25}$$

A straightforward calculation implies that the entropy function $N_n(\varepsilon)$ of $\widetilde{W}^{(n)}(x)$ can be estimated by $C\varepsilon^{-d}$ for some C independent of both n and ε. Both here and in what follows C stands for a generic positive constant independent of n. Since $\Sigma(\widetilde{W}^{(n)}) \le Cn^{-2}$ Theorem 12.8 implies that for some $C > 0$ we have

$$\mathbb{Q}(K_n^c) \le C \exp\{-n^2/C\}, \quad \forall n \ge 1. \tag{12.26}$$

For any $n \ge 1$ we let

$$K(\omega) := n, \quad \text{on } K_n \Big\backslash \left(\bigcup_{m=0}^{n-1} K_m\right).$$

This random variable clearly satisfies (12.2).

12.5.2 Proofs of Theorem 12.3 and Proposition 12.4

We start with the following.

Lemma 12.9 *$\mathscr{C}_0 \subset D(L)$ is a core of L. In addition, formula* (12.6) *holds on $\mathscr{C}_0$.*

Proof Obviously $\mathscr{C}_0$ is dense in $L^2(\mathbb{Q})$ so it remains only to show that

$$\mathscr{C}_0 \subset D(L) \tag{12.27}$$

and

$$P_t(\mathscr{C}_0) \subset \mathscr{C}_0 \quad \text{for all } t > 0. \tag{12.28}$$

The argument from Sect. 9.4 can be applied to show that $C_b^2(\Omega)$ is contained in $D(L)$ and that formula (12.6) holds for any g in $C_b^2(\Omega)$.

Suppose that $g \in \mathscr{C}$ and $p > 2$. One can find a sequence $\{g_n, n \ge 1\} \subset C_b^2(\Omega)$ such that $\lim_{n \to +\infty} \|g - g_n\|_{W_p^2} = 0$. This in particular implies that $(g_n, Lg_n) \to (g, h)$ in the epigraph norm, with h given by the right-hand side of (12.6). Since L is closed, from the above we conclude that g belongs to $D(L)$ and $h = Lg$. This ends the proof of (12.27).

To verify (12.28) we prove that for every fixed $p > p' > 2$ and $t \ge 0$

$$P_t\big(W_p^k(\Omega)\big) \subset W_{p'}^k(\Omega), \quad \forall k \ge 1. \tag{12.29}$$

For $f \in W_p^1(\Omega)$ and $h \in \mathbb{R}^d$ we can write

$$P_t f(\tau_h \omega) = E_{\mathbb{P}} f(\tau_{X_t^{h,\omega}} \omega). \tag{12.30}$$

Here, as we recall, the expectation $E_{\mathbb{P}}$ corresponds to the probability measure corresponding to the Brownian motion appearing in (12.4).

Differentiating both sides of (12.30) with respect to h at 0 in the $L^{p'}$ norm we obtain

$$\nabla P_t f(\omega) = E_{\mathbb{P}}\big[J_t \nabla f(\eta_t)\big], \tag{12.31}$$

where $J_t = [J_t^{ij}]$, $i, j = 1, \ldots, d$, is the Jacobian matrix given by

$$J_t^{ij} := \partial_{i|h=0} X_{t,j}^h.$$

The partial above is understood in the L^q sense, with $1/p' = 1/p + 1/q$. It satisfies the equation

$$J_t = I + \int_0^t J_s \nabla V(\eta_s)\, ds, \tag{12.32}$$

where I is the $d \times d$ identity matrix and $\nabla V = [D_i V_j]$, $i, j = 1, \dots, d$. To justify the equality in (12.31) it suffices only to prove that

$$\big\langle E_{\mathbb{P}} |J_t|^q \big\rangle_{\mathbb{Q}} < +\infty \quad \text{for all } q > 1. \tag{12.33}$$

Formula (12.29) would follow then for $k = 1$. Its generalization to an arbitrary $k \geq 1$ can be easily done by induction.

For a fixed T, $\lambda > 0$ and $n \geq 1$ let

$$L_n := \Big[\sup_{0 \leq s \leq T} |w_s| \leq n \Big].$$

By the standard estimate of the first passage time for a Brownian motion we have $\mathbb{P}(L_n^c) \leq C \exp\{-n^2/C\}$ for some $C > 0$. We maintain the convention of denoting by C a generic, deterministic, positive constant that is independent of n. From (12.26) there exists C such that

$$\mathbb{Q}\big(K_n^c\big) \leq C \exp\big\{-n^2/C\big\}, \quad \forall n \geq 1.$$

Let $x_t := \sup_{0 \leq s \leq t} |X_s|$. From (12.4) we obtain that

$$x_t \leq C_1 \bigg(\int_0^t x_s \, ds + n \bigg), \quad \text{on } L_n \times K_n.$$

Hence, $x_T \leq C_1 n e^{C_1 T}$ on $L_n \times K_n$ and in consequence then

$$\sup_{t \in [0,T]} \big| \nabla V(\eta_t) \big| \leq x_T + n \leq C_1 n. \tag{12.34}$$

From (12.32), (12.34) and Gronwall's inequality we obtain that we have

$$|J_t| \leq e^{C_1 n t}, \quad \forall t \in [0, T], \quad \text{on } L_n \times K_n.$$

We have shown therefore that

$$\mathbb{P} \otimes \mathbb{Q} \Big[\sup_{t \in [0,T]} |J_t| > e^{C_1 n T} \Big] \leq C e^{-n^2/C}, \quad \forall n \geq 1$$

for some $C, C_1 > 0$. Hence, in particular (12.33) follows. □

Suppose that $\{\hat{X}_t^{x,\omega}, t \geq 0\}$ is the solution of (12.4) with $\tilde{V}(x;\omega)$ replaced by $-\tilde{V}(x;\omega)$. We denote by $\hat{\eta}_t$, $\hat{P}_t$ the environment process and its respective transition semigroup that correspond to $\hat{X}_t^x$. Note that $\hat{P}_t = P_t^*$. This follows from Proposition 9.9 in case the corresponding stream matrix is bounded (cf. (11.12)). Such a result can be generalized to the situation when the stream function is Gaussian via an approximation argument. Namely, V can be approximated by $V^{(n)}$s that are incompressible and whose corresponding stream matrices belong to $C_b^\infty(\Omega)$ similarly as in the proof of Proposition 11.6. Minor modification in the argument consists in replacing the Sobolev embedding of $W_2^m(B_R)$ into $C_b^1(B_R)$ for $m > d/2+1$ by the respective embedding of $W_p^2(B_R)$ into $C_b^1(B_R)$ for $p > d$.

Summarizing, we have shown so far that $\mathscr{C}_0$ is a common core for both L and L^* and formula (12.6) holds. We demonstrate now that S, defined on $\mathscr{C}_0$ by (12.8), is essentially self-adjoint. Note that the operator coincides on this set with the generator of the heat semigroup "lifted" to $L^p(\mathbb{Q})$, i.e. a strongly continuous semigroup given by

$$R_t f(\omega) := \int_{\mathbb{R}^d} q_t(y) f(\tau_y \omega) dy, \quad f \in L^p(\mathbb{Q}),\ t > 0$$

where $q_t(y)$ is the heat kernel, see (11.38). For each $k \geq 1$ the set $W_p^k(\Omega)$ is invariant under the action of the semigroup, thus it is a core of its generator. For the same reason the above also holds $\mathscr{C}_0$ and $\mathscr{C}$ (density of the latter set in $L^p(\mathbb{Q})$ follows from Theorem 3.51, p. 40 of Janson 1997). Using this observation in the particular case when $p = 2$ we conclude that the closure of S is self-adjoint (as the generator of a semigroup of self-adjoint operators) and $\mathscr{C}$ is its core. This ends the proof of Proposition 12.4.

To finish the proof of Theorem 12.3 we show the following.

Proposition 12.10 *For any $f \in \mathscr{C}_0$ there exists a sequence $\{f_n, n \geq 1\}$ contained in $\mathscr{C}$ converging to f in the graph norm of the operator L. An analogous statement holds for its adjoint L^*.*

Proof Let $p > 2$. From the definition of $\mathscr{C}_0$ we conclude that $f \in W_p^2(\Omega)$. Since $\mathscr{C}$ is a core of S in any $L^p(\mathbb{Q})$ there exists a sequence $\{f_n, n \geq 1\}$ contained in $\mathscr{C}$ such that $f_n \to f$ and $Sf_n \to Sf$ in $L^p(\mathbb{Q})$, as $n \to +\infty$. We claim, see (12.36) below, that the above implies that

$$f_n \to f \quad \text{in } W_p^1(\Omega), \quad \text{as } n \to +\infty. \tag{12.35}$$

Having established this fact we obtain, using already proven formula (12.6), that in particular $Lf_n \to Lf$ in $L^2(\mathbb{Q})$, as $n \to +\infty$, which ends the proof of the proposition.

What remains to be seen is the proof of (12.35). Let $\tilde{f}_n(x;\omega) := f_n(\tau_x\omega)$ and $\tilde{f}(x;\omega) := f(\tau_x\omega)$. Invoking, from the theory of the elliptic partial differential equations, classical interior a priori L^p-estimates, see e.g. (9.36), p. 235 of Gilbarg and Trudinger (1983), we can write, using the notation introduced in (11.24),

$$\begin{aligned}
&\|\tilde{f}_n(\cdot;\omega) - \tilde{f}(\cdot;\omega)\|^p_{W_p^2(B_{1/2})} \\
&\quad \leq C_p\big[\|\tilde{f}_n(\cdot;\omega) - \tilde{f}(\cdot;\omega)\|^p_{L^p(B_1)} + \|\Delta\tilde{f}_n(\cdot;\omega) - \Delta\tilde{f}(\cdot;\omega)\|^p_{L^p(B_1)}\big]
\end{aligned}$$

for some constant C_p independent of n and ω. Averaging over ω we conclude that

$$\|f_n - f\|^p_{W^2_p(\Omega)} \le C_p\big[\|f_n - f\|^p_{L^p(\mathbb{Q})} + \|Sf_n - Sf\|^p_{L^p(\Omega)}\big], \tag{12.36}$$

which clearly implies (12.35). □

12.6 Superdiffusive Transport in a Flow with Infinite Péclet Number

In this section we consider the passive tracer model with a stationary, isotropic Gaussian drift whose energy spectrum density behaves as a power function in some neighborhood of 0. Theorem 12.6 implies then that for the flows whose Péclet number is finite the tracer satisfies the central limit theorem. On the other hand, Theorem 12.11 shows that the result is no longer true for a flow with the infinite Péclet number and the motion of a tracer in this case is superdiffusive.

12.6.1 Homogeneous, Isotropic Gaussian Flows

Let $\tilde{V}(x)$ be a stationary Gaussian field introduced in Sect. 12.1. Its spectral resolution can be written in the form

$$\tilde{V}(x;\omega) = \int_{\mathbb{R}^d} e^{ix\cdot\xi}\,\hat{V}(d\xi;\omega), \tag{12.37}$$

where $\hat{V} = (\hat{V}_1, \ldots, \hat{V}_d)$ is a $\mathbb{C}^d$-valued, stochastic Borel measure over $\mathbb{R}^d$. Since the field is real vector valued we have $\hat{V}(-A) = \hat{V}^*(A)$ for all Borel sets A. The *spectral measure* of the field $\hat{R}(d\xi) = [\hat{R}_{ij}(d\xi)]$ is defined by the relation

$$\big\langle \hat{V}_i(d\xi)\hat{V}_j(d\xi')\big\rangle_{\mathbb{Q}} = \delta(\xi + \xi')\hat{R}_{ij}(d\xi)d\xi' \tag{12.38}$$

and its Fourier transform is the covariance matrix of the field $R(x) = [R_{ij}(x)]$, where

$$R_{ij}(x) = \int_{\mathbb{R}^d} e^{ix\cdot\xi}\,\hat{R}_{ij}(d\xi), \quad i, j = 1, \ldots, d.$$

We assume furthermore that the spectral measure has a density $\hat{R}_{ij}(d\xi) = \hat{R}_{ij}(\xi)\,d\xi$, where

$$\hat{R}_{ij}(\xi) = \rho\big(|\xi|\big)\Gamma_{ij}(\xi), \tag{12.39}$$

with $\rho(\cdot)$ non-negative and

$$\Gamma_{ij}(\xi) := \delta_{ij} - \xi_i\xi_j|\xi|^{-2}.$$

The presence of this factor guarantees that the realizations of the field are almost surely incompressible. We shall use a shorthand notation

$$\hat{R}_{a,b}(\xi) := \hat{R}(\xi)a \cdot b, \qquad \hat{R}_a(\xi) := \hat{R}_{a,a}(\xi), \quad \forall a, b \in \mathbb{C}^d. \tag{12.40}$$

The product above is understood as a scalar product between complex vectors, i.e.

$$a \cdot b = \sum_{p=1}^{d} a_p b_p^*,$$

for $a = (a_1, \ldots, a_d)$ and $b = (b_1, \ldots, b_d)$. Here * denotes the complex conjugation. To simplify further the notation we shall write $\hat{R}_{p,b}(\xi)$ when $a = e_p$, $p = 1, \ldots, d$.

Recall that the group $SO(d)$ (the special orthogonal group) is made of $d \times d$ matrices g satisfying

$$gg^T = g^T g = I \quad \text{and} \quad \det g = 1,$$

where I is the $d \times d$ identity matrix. A random field whose covariance matrix satisfies

$$R(gx) = gR(x)g^T, \quad \forall x \in \mathbb{R}^d, g \in SO(d) \tag{12.41}$$

is called *statistically isotropic*. For a Gaussian field condition (12.41) is equivalent to the invariance of its law under rotations, i.e. for any $g \in SO(d)$ the laws of $\{\tilde{V}(x), x \in \mathbb{R}^d\}$ and that of $\{g^{-1}\tilde{V}(gx), x \in \mathbb{R}^d\}$ are identical. This statement can be easily verified by comparing the covariance matrices of the respective fields.

A direct computation shows that any field whose spectral measure is of the form (12.39) is statistically isotropic. It can also be proved, see e.g. Sect. 3.4 of Batchelor (1982), that the density of the spectral measure associated with an incompressible, isotropic random field has to be of the form (12.39).

We assume that the energy spectrum satisfies the *power law* in a neighborhood of 0, i.e.

$$\rho(u) := a(u)u^{\alpha+1-d}, \quad \forall u > 0 \tag{12.42}$$

for some $\alpha \in \mathbb{R}$ and $a(\cdot)$ is a bounded, measurable function, continuous at 0 with $a(0) > 0$. Its decay at infinity should guarantee a sufficient regularity of the drift, i.e.

$$\sup_{u>0}(1+u)^{2+\alpha}a(u) < +\infty. \tag{12.43}$$

12.6.2 Flows with Infinite Péclet Numbers

Integrability of the spectral measure at 0 requires that $\alpha > -1$. The Péclet number of the flow, defined in (11.9), equals then

$$\mathrm{Pe} = (d-1)s_d \int_0^{+\infty} u^{\alpha-2} a(u)\, du.$$

Here s_d is the surface area of $\mathbb{S}^{d-1}$—the unit sphere in $\mathbb{R}^d$. According to Theorem 12.6 the trajectory of the tracer satisfies the central limit theorem, provided that $\mathrm{Pe} < +\infty$. The latter is equivalent to the assumption that $\alpha > 1$. We show that the behavior of the tracer is *superdiffusive* for $\alpha \in (-1, 1)$. More precisely, for $\{X_t, t \ge 0\}$, given by (12.4), define

$$\gamma_* := \sup\left[\gamma \ge 0 : \lim_{t\uparrow+\infty} t^{-2-\gamma} \int_0^t \langle E_{\mathbb{P}}|X_s|^2\rangle_{\mathbb{Q}}\, ds = +\infty\right]. \tag{12.44}$$

If $\langle E_{\mathbb{P}}|X_t|^2\rangle \sim t$ for $t \gg 1$, i.e. the behavior of the tracer is diffusive, we should have $\gamma_* = 0$. Superdiffusivity means that $\langle E_{\mathbb{P}}|X_t|^2\rangle \gg t$ for $t \gg 1$. This situation occurs (at least in the Cesaro sense) when $\gamma_* > 0$.

Using Proposition 2.17 it is also possible to estimate $\langle E_{\mathbb{P}}|X_t|^2\rangle_{\mathbb{Q}}$ from above. Let

$$\gamma^* := \sup\left[\gamma \ge 0 : \liminf_{t\uparrow+\infty} t^{-1-\gamma} \langle E_{\mathbb{P}}|X_t|^2\rangle_{\mathbb{Q}} > 0\right]. \tag{12.45}$$

Directly from the above definition and (12.44) we conclude that $\gamma^* \ge \gamma_*$.

Theorem 12.11 *Suppose that the density of the spectral measure of a Gaussian field satisfies* (12.39), (12.42) *and* (12.43) *with* $|\alpha| < 1$. *Then,*

$$\frac{1}{4}(1-\alpha)^2 \le \gamma_* \le \gamma^* \le \frac{1}{2}(1-\alpha). \tag{12.46}$$

Proof Let $Y_t = (Y_{1,t}, \dots, Y_{d,t})$ be given by

$$Y_{p,t} := \int_0^t V_p(\eta_s) ds, \quad p = 1, \dots, d.$$

It is obvious that $\langle E_{\mathbb{P}}|X_t|^2\rangle$ can be replaced by $\langle E_{\mathbb{P}}|Y_t|^2\rangle$ in the definition (12.44). Therefore, to show (12.46) it suffices to prove that

$$\lim_{t\to+\infty} \frac{1}{t^{2+\gamma}} \int_0^t \langle E_{\mathbb{P}}|Y_s|^2\rangle_{\mathbb{Q}} ds = +\infty \tag{12.47}$$

for any $\gamma < (1-\alpha)^2/4$ and

$$\lim_{t\to+\infty} \frac{1}{t^{1+\gamma}} \langle E_{\mathbb{P}}|X_t|^2\rangle_{\mathbb{Q}} = 0 \tag{12.48}$$

for any $\gamma > (1-\alpha)/2$.

According to Proposition 4.9 in order to prove (12.47) it suffices only to show that

$$\langle V_p, \chi_\lambda^{(p)}\rangle_{\mathbb{Q}} \ge C\lambda^{-\gamma} \tag{12.49}$$

for some $C > 0$, $p = 1, \dots, d$ and all $\lambda \in (0, 1]$. Here, as we recall, $\chi_\lambda^{(p)}$ is the solution of the resolvent equation (9.48).

To demonstrate (12.49) we use the variational principle formulated in the first equality stated in Theorem 2.27. Test functions shall be chosen from an appropriate space, which we define below. Let $\mathscr{K}_0$ be the space consisting of $\mathbb{R}^d$-valued functions φ whose Fourier transform $\hat{\varphi}$ is compactly supported and bounded. On this space we define a pseudo-norm

$$\|\varphi\|_{\mathscr{K}}^2 := \int_{\mathbb{R}^d} \hat{R}_{\hat{\varphi}(\xi)}(\xi)d\xi. \tag{12.50}$$

It becomes a pre-Hilbert norm after a usual identification of functions whose difference vanishes in the pseudo-norm. Denote by $\mathscr{K}$ the completion of $\mathscr{K}_0$ under the above norm. For any $\varphi \in \mathscr{K}_0$ we let

$$f_\varphi := \int_{\mathbb{R}^d} \varphi(x) \cdot \tilde{V}(x)dx = \int_{\mathbb{R}^d} \hat{\varphi}(\xi) \cdot \hat{V}(d\xi). \tag{12.51}$$

Using the notation introduced in (12.51) we define

$$d^*(\lambda) := \sup\big[2\langle V_p, f_\varphi\rangle_{\mathbb{Q}} - \|f_\varphi\|_{1,\lambda}^2 : \varphi \in \mathscr{K}_0\big],$$

where

$$\|f\|_{1,\lambda}^2 := \lambda\|f\|_{\mathbb{Q}}^2 + \|f\|_1^2.$$

Directly from the definition of $d^*(\lambda)$ and the variational principle stated in Theorem 2.27 we have

$$\langle V_p, \chi_\lambda^{(p)}\rangle_{\mathbb{Q}} \ge d^*(\lambda). \tag{12.52}$$

□

Lemma 12.12 *For an arbitrary $c \in \mathbb{S}^{d-1}$ we have*

$$d^*(\lambda) \ge \int_{\mathbb{R}^d} \mathscr{H}^{-1}(\lambda, \xi)\hat{R}_c(\xi)\, d\xi, \tag{12.53}$$

where

$$\mathscr{H}(\lambda, \xi) := b_\lambda(\xi) + 2\int_{\mathbb{R}^d} b_\lambda^{-1}(\xi + \xi')\hat{R}_\xi(\xi')d\xi' \tag{12.54}$$

and

$$b_\lambda(\xi) := \lambda + \frac{1}{2}|\xi|^2.$$

Proof Since the right-hand side of (12.53) does not depend on c, with no loss of generality we may assume that $c = e_p$. For f_φ given by (12.51) we have

$$\langle V_p, f_\varphi \rangle_{\mathbb{Q}} = \int_{\mathbb{R}^d} \hat{R}_{\hat{\varphi}(\xi),p}(\xi)\, d\xi. \tag{12.55}$$

Using (12.8) we obtain

$$Sf_\varphi = -\frac{1}{2}\int_{\mathbb{R}^d} |\xi|^2 \hat{\varphi}(\xi)\cdot \hat{V}(d\xi),$$

$$Af_\varphi = i\int_{\mathbb{R}^{2d}} \hat{\varphi}(\xi)\cdot \hat{V}(d\xi)\xi \cdot \hat{V}(d\xi').$$

A multiple stochastic integral appearing above is defined in Sect. 12.16.1 below. Hence,

$$\|f_\varphi\|_{1,\lambda}^2 = \int_{\mathbb{R}^d} b_\lambda(\xi)\hat{R}_{\hat{\varphi}(\xi)}(\xi)d\xi \tag{12.56}$$

and

$$\|Af_\varphi\|_{-1,\lambda}^2 = \langle Af_\varphi, g\rangle_{\mathbb{Q}}, \tag{12.57}$$

where g is the unique solution of the equation

$$(\lambda - S)g = Af_\varphi.$$

We look for the solutions of this equation among the elements of $\mathscr{G}_2$ of the form

$$g = \sum_{i,j=1}^{d} \int_{\mathbb{R}^{2d}} \alpha_{ij}(\xi,\xi')\hat{V}_i(d\xi)\hat{V}_j(d\xi'), \tag{12.58}$$

where α_{ij} are functions from $C^\infty(\mathbb{R}^{2d})$ for which the double stochastic integral in (12.58) makes sense. After a straightforward calculation we obtain

$$\alpha_{ij}(\xi,\xi') = ib_\lambda^{-1}(\xi+\xi')\hat{\varphi}_i(\xi)\xi_j.$$

Using (12.57) we conclude from (12.58) that

$$\begin{aligned}\|Af_\varphi\|_{-1,\lambda}^2 = \int_{\mathbb{R}^{2d}} \big[&\hat{R}_{\xi,\hat{\varphi}(\xi)}(\xi)\hat{R}_{\xi',\hat{\varphi}(\xi')}(\xi')b_\lambda^{-1}(\xi'-\xi') \\ &+ \big(\hat{R}_{\hat{\varphi}(\xi)}(\xi)\hat{R}_\xi(\xi') + \hat{R}_{\xi',\hat{\varphi}(\xi)}(\xi)\hat{R}_{\hat{\varphi}(\xi'),\xi}(\xi')\big)b_\lambda^{-1}(\xi+\xi')\big]\, d\xi\, d\xi'.\end{aligned} \tag{12.59}$$

In the above calculation we have used the equality

$$\left\langle \prod_{k=1}^{4} \hat{V}_{i_k}(d\xi_k)\right\rangle_{\mathbb{Q}} = \sum \hat{R}_{i_k,i_l}(\xi_{i_k})\hat{R}_{i_m,i_n}(\xi_{i_m})\delta(\xi_{i_k}+\xi_{i_l})\delta(\xi_{i_m}+\xi_{i_n})\prod_{k=1}^{4} d\xi_{i_k},$$

see Sect. 12.16.1, where the summation extends over all pairings formed over the elements of the set $\{1,2,3,4\}$. Since $\hat{R}_{\xi,\hat{\varphi}(\xi)}(\xi)=0$ the first term on the right-hand side of (12.59) (corresponding to the pairing $(1,2)$, $(3,4)$) vanishes. By Cauchy–Schwartz inequality,

$$\hat{R}_{\hat{\varphi}(\xi'),\xi}(\xi') \le \hat{R}_{\xi}^{1/2}(\xi')\rho^{1/2}(|\xi'|)|\hat{\varphi}(\xi')|.$$

Hence,

$$|\hat{R}_{\xi',\hat{\varphi}(\xi)}(\xi)\hat{R}_{\hat{\varphi}(\xi'),\xi}(\xi')| \le \frac{1}{2}[\hat{R}_{\xi}(\xi')|\hat{\varphi}(\xi)|^2\rho(|\xi|)+\hat{R}_{\xi'}(\xi)|\hat{\varphi}(\xi')|^2\rho(|\xi'|)]$$

and in consequence

$$\|Af_\varphi\|_{-1,\lambda}^2 \le 2\int_{\mathbb{R}^{2d}}|\hat{\varphi}(\xi)|^2\hat{R}_{\xi}(\xi')b_\lambda^{-1}(\xi+\xi')\rho(|\xi|)\,d\xi'\,d\xi. \tag{12.60}$$

Combining (12.55), (12.56) and (12.60) we conclude that

$$d^*(\lambda) \ge \sup_{\varphi\in\mathscr{H}}\mathscr{G}(\varphi), \tag{12.61}$$

where

$$\mathscr{G}(\varphi):=2\int_{\mathbb{R}^d}\hat{R}_{\hat{\varphi}(\xi),p}(\xi)\,d\xi-\int_{\mathbb{R}^d}K(\xi)|\hat{\varphi}(\xi)|^2\,d\xi$$

and $K(\xi):=r(|\xi|)\mathscr{H}(\lambda,\xi)$. Since $\mathscr{G}(\varphi)\to-\infty$, as $\|\varphi\|_{\mathscr{H}}\to+\infty$, we conclude that the supremum appearing on the right-hand side of (12.61) is attained. The maximizer can be calculated explicitly from the corresponding Euler–Lagrange equation and it equals

$$\hat{\varphi}_*(\xi):=\mathscr{H}^{-1}(\lambda,\xi)\Gamma(\xi)e_p.$$

The respective maximum is equal to the expression on the right-hand side of (12.53) with $c=e_p$. The conclusion of the lemma then follows a simple observation that the integral on the right-hand side of (12.53) does not depend on c. □

Substitute $\tilde{\xi}:=\lambda^{-1/2}\xi$, $\tilde{\xi}':=\lambda^{-1/2}\xi'$ in the integrals appearing in (12.53) and (12.54) respectively. We conclude that

$$d^*(\lambda)\ge\lambda^{(\alpha-1)/2}\int_{\mathbb{R}^d}[\mathscr{H}'(t,\xi)]^{-1}a(\lambda^{1/2}|\xi|)|\xi|^{\alpha+1-d}\Gamma_{pp}(\xi)\,d\xi$$

and

$$\mathscr{H}'(t,\xi):=1+|\xi|^2+2\lambda^{(\alpha-1)/2}|\xi|^2\|a\|_\infty\mathscr{I}(|\xi|) \tag{12.62}$$

and

$$\mathscr{I}(v):=\int_0^{+\infty}u^\alpha[1+(u-v)^2]^{-1}\,du.$$

Consider first the case $\alpha \in (-1, 0]$. We can estimate then the integral appearing on the right-hand side of the above equality by the sum of integrals from 0 to 1 and from 1 to $+\infty$ and conclude easily that $\mathscr{I}(v) \le C$ for some constant $C > 0$. Hence,

$$d^*(\lambda) \ge C_1 \lambda^{(\alpha-1)/2} \int_{\mathbb{R}^d} \frac{a(\lambda^{1/2}|\xi|)|\xi|^{\alpha+1-d}}{1+\lambda^{(\alpha-1)/2}|\xi|^2} \Gamma_{pp}(\xi)\, d\xi$$

for some constant $C_1 > 0$. After substitution $\tilde{\xi} := \lambda^{(\alpha-1)/4}\xi$ in the integral on the right-hand side we obtain

$$d^*(\lambda) \ge C\lambda^{-(1-\alpha)^2/4} \tag{12.63}$$

for some positive constant $C > 0$ and (12.49) follows from (12.52).

In case when $\alpha \in (0, 1)$ we can write

$$\mathscr{I}(v) \le v^\alpha \int_0^{2|\xi|} \frac{du}{1+(v-u)^2} + \int_{2v}^{+\infty} \frac{u^\alpha du}{1+u^2} \le C\left(v^\alpha + 1\right)$$

and as a result

$$d^*(\lambda) \ge C\lambda^{(\alpha-1)/2} \int_{\mathbb{R}^d} \frac{|\xi|^{\alpha+1-d} a(\lambda^{1/2}|\xi|)\Gamma_{pp}(\xi) d\xi}{1+\lambda^{(\alpha-1)/2}(|\xi|^2+|\xi|^{\alpha+2})}$$

for some $C > 0$. The substitution $\tilde{\xi} := \lambda^{(\alpha-1)/4}\xi$ yields again (12.63). Thus, (12.49) follows.

To obtain (12.48) we use the variational principle formulated in the second equality of Theorem 2.27. As a test function we choose $g \equiv 0$ and obtain that

$$\left\langle (1/t - L)^{-1} V_p, V_p \right\rangle_{\mathbb{Q}} \le 2\| V_p \|_{-1,1/t}^2. \tag{12.64}$$

On the other hand, from (12.4) we get

$$\left\langle E_{\mathbb{P}} |X_{p,t}|^2 \right\rangle_{\mathbb{Q}} \le 2\left[t + \left\langle E_{\mathbb{P}} Y_{p,t}^2 \right\rangle\right] \le t\left(1 + 12\| V_p \|_{-1,1/t}^2\right). \tag{12.65}$$

The last inequality follows from Lemma 4.8. A direct computation of the norm $\| V_p \|_{-1,1/t}^2$ yields

$$\int_{\mathbb{R}^d} \hat{R}_{pp}(\xi)\left(\frac{1}{t} + \frac{1}{2}|\xi|^2\right)^{-1} d\xi = s_d \int_0^{+\infty} a(u) u^\alpha \left(\frac{1}{t} + \frac{u^2}{2}\right)^{-1} du. \tag{12.66}$$

Substituting $\tilde{u} := t^{1/2} u$ we obtain that $\langle E_{\mathbb{P}} |X_t|^2 \rangle_{\mathbb{Q}} \le ct(1+t)^{(1-\alpha)/2}$ for some $c > 0$. This ends the proof of (12.48). □

12.7 Central Limit Theorem for Diffusions in Gaussian and Markovian Flows

We return to the question of the central limit theorem for a tracer in a time dependent flow already considered in Sect. 11.5, see Corollary 11.5. Recall that such a result has been shown when the Péclet number of the flow is finite. However, the argument used in the proof has not taken into account mixing properties of the environment in the temporal variable. In this section we formulate a sufficient condition for the validity of the central limit theorem, provided that the drift is a Gaussian and Markovian flow, see Theorem 12.13 below. It turns out, see Sect. 12.12, that this condition is sharp for isotropic flows.

We assume that the particle trajectory $X_t^{s,x}(\omega)$ satisfies an Itô stochastic differential equation

$$\begin{aligned} dX_t^{s,x}(\omega) &= \tilde{V}\big(t, X_t^{s,x}(\omega); \omega\big)dt + \kappa\, dw_{t,s}, \\ X_s^{s,x}(\omega) &= x, \end{aligned} \tag{12.67}$$

where $w_{t,s} := w_{t-s}$ for $t \ge s$ and $\{w_t, t \ge 0\}$ is a standard, d-dimensional Brownian motion over probability space $(\Sigma, \mathscr{A}, \mathbb{P})$. The parameter $\kappa \ge 0$, called a *molecular diffusivity*, set to be equal 1 in (11.18), is introduced here because we wish to consider also motions in a random field, i.e. the case when $\kappa = 0$. We maintain the convention of omitting the random argument ω, when its value is obvious from context, and the superscripts when $(s, x) = (0, 0)$.

Before formulation of our main assumptions concerning the drift we present an informal argument that gives some motivation behind these hypotheses. Since for each t fixed the field $\{\tilde{V}(t, x), x \in \mathbb{R}^d\}$ is spatially stationary there exists a spectral measure $\hat{V}(t, d\xi)$ such that

$$\tilde{V}(t, x) = \int_{\mathbb{R}^d} e^{i\xi\cdot x}\hat{V}(t, d\xi).$$

Gaussianity of the field implies that its spectral measure is also (real) Gaussian in the following sense: for any $N \ge 1$ the times $t_1, \ldots, t_N$, Borel sets $A_1, \ldots, A_N \in \mathscr{B}(\mathbb{R}^d)$ the $2Nd$-dimensional, real component vector

$$\big(\mathrm{Re}\,\hat{V}(t_1, A_1), \ldots, \mathrm{Re}\,\hat{V}(t_N, A_N), \mathrm{Im}\,\hat{V}(t_1, A_1), \ldots, \mathrm{Im}\,\hat{V}(t_N, A_N)\big)$$

is Gaussian. We assume also that the temporal dynamics of the field is Markovian. Informally speaking the above means that each $\hat{V}(t, d\xi)$ is an "infinitesimal" time stationary, Ornstein–Uhlenbeck process and any two of such infinitesimal processes are mutually independent. The notion of an "infinitesimal process" is not precisely formulated at the moment but using it as if it were a "true" Ornstein–Uhlenbeck process allows us to predict the form of the covariance matrix of $\tilde{V}(t, x)$. Namely, we should have

$$\big\langle \hat{V}_p(t, d\xi)\hat{V}_q\big(s, d\xi'\big)\big\rangle_{\mathbb{Q}} = e^{-\gamma(\xi)|t-s|}\delta\big(\xi + \xi'\big)\hat{R}_{pq}(d\xi)d\xi', \tag{12.68}$$

where $\gamma(\cdot)$ is a non-negative, Borel measurable function and $\hat{R}(d\xi) = [\hat{R}_{pq}(d\xi)]$, $p, q = 1, \ldots, d$, is a non-negative Hermitian, $d \times d$ matrix valued, Borel measure. Its trace $r(\cdot) := \operatorname{tr} \hat{R}(\cdot)$ (the spatial energy spectrum) is a finite Borel measure, i.e. $r(\mathbb{R}^d) < +\infty$.

Motivated by the above intuition we consider stationary, zero mean Gaussian fields whose covariance matrix $R(t,x) = [R_{pq}(t,x)]$, $p, q = 1, \ldots, d$ satisfies

$$R_{pq}(t,x) := \langle \tilde{V}_p(h+t, z+x)\tilde{V}_q(h,z)\rangle_{\mathbb{Q}} = \int_{\mathbb{R}^d} e^{ix\cdot\xi - \gamma(\xi)|t|}\hat{R}_{pq}(d\xi) \tag{12.69}$$

for all $(t,x), (h,z) \in \mathbb{R}^{1+d}$. In Sect. 12.8 we show that the process $\{V(t,\cdot), t \geq 0\}$, representing the evolution of the spatial realization of the field, is Markovian.

The fact that the field is real valued forces us to assume also that

$$\hat{R}(-d\xi) = \hat{R}^*(d\xi) \quad \text{and} \quad \gamma(-\xi) = \gamma(\xi). \tag{12.70}$$

Furthermore, incompressibility implies that

$$\hat{R}(d\xi)\xi = 0 \quad \text{for all } \xi \in \mathbb{R}^d. \tag{12.71}$$

Denote by $\mathscr{F}_{=0}$ the σ-algebra generated by $\{\tilde{V}(0,x), x \in \mathbb{R}^d\}$.

Theorem 12.13 *In addition to assumptions* (12.69)–(12.71) *made above we suppose that* $\gamma(\cdot)$ *is continuous and for some* $\delta > 0$ *we have*

$$\int_{\mathbb{R}^d} [1 + |\xi|^4 + \gamma^\delta(\xi)] r(d\xi) < +\infty. \tag{12.72}$$

Assume also that either of the following two hypotheses hold:

(1) $\kappa > 0$ *and*

$$\int_{\mathbb{R}^d} \frac{r(d\xi)}{\gamma(\xi)} < +\infty, \tag{12.73}$$

or

(2) $\kappa = 0$, *condition* (12.73) *is in force and the function* $\beta(\xi) := |\xi|^2/\gamma(\xi)$ *satisfies*

$$\|\beta\|_{L^\infty(r)} < +\infty. \tag{12.74}$$

Then, the laws of random vectors $X_t/\sqrt{t}$ *conditioned on* $\mathscr{F}_{=0}$ *satisfy the central limit theorem in* $\mathbb{Q}$*-probability. The limiting covariance matrix* $\bar{a} = [\bar{a}_{p,q}]$ *satisfies*:

$$\bar{a}_{pq} = \lim_{t\to+\infty} \frac{1}{t} \langle E_{\mathbb{P}}(X_{p,t}X_{q,t})\rangle_{\mathbb{Q}}, \quad p, q = 1, \ldots, d. \tag{12.75}$$

12.8 Markovian Dynamics of the Environment

We present some properties of the dynamics of the Gaussian field introduced in the previous section. An important feature is that the flow can be viewed as a Markov process taking values in an appropriately defined phase space. In order to focus our attention on the proof of Theorem 12.13 and not to be distracted too much by technical details we postpone the demonstrations of the results announced here until Sect. 12.14.

12.8.1 Hermite Polynomials

Let $\rho > d/2$ and

$$\vartheta_\rho(u) := \left(1+u^2\right)^{-\rho}, \quad u \in \mathbb{R}.$$

Denote by E the Hilbert space that is the completion of the space of functions $v = (v_1, \ldots, v_d) : \mathbb{R}^d \to \mathbb{R}^d$ with components in $C_c^\infty(\mathbb{R}^d)$, under the norm

$$\|v\|_E^2 := \int_{\mathbb{R}^d} \left(\left|v(x)\right|^2 + \left|\nabla_x v(x)\right|^2\right) \vartheta_\rho\left(|x|\right) dx.$$

Here

$$\left|v(x)\right|^2 := \sum_{i=1}^d v_i^2(x) \quad \text{and} \quad \left|\nabla v(x)\right|^2 := \sum_{i,j=1}^d (\partial_j v_i)^2(x).$$

The group of spatial shifts $\tau_x : E \to E$ is given by $\tau_x v(\cdot) := v(x + \cdot)$, where $v \in E$. Define also

$$V(v) = \left(V_1(v), \ldots, V_d(v)\right) := v(0) \tag{12.76}$$

and the field $\tilde{V}(x; v) := V(\tau_x v)$.

Condition (12.72) guarantees that measure π, defined as the law of $\tilde{V}(0, \cdot)$, is supported in E. The measure is Gaussian, i.e. all finite dimensional marginals of $\{\tilde{V}(x), x \in \mathbb{R}^d\}$ are Gaussian, and spatially homogeneous, i.e. $\pi\tau_x = \pi$, for all $x \in \mathbb{R}^d$. According to Theorem 6.5.3, p. 145 of Adler (1990), the group action is ergodic if and only if the energy spectrum $r(\cdot)$ of the field $\tilde{V}(0, x)$ has no atoms, i.e. $r(\{\xi\}) = 0$ for each $\xi \in \mathbb{R}^d$. We shall always assume this condition.

Define a strongly continuous group of unitary mappings on $L^2(\pi)$ by letting $T_x f(v) := f(\tau_x v)$. Its generators shall be denoted by D_q, $q = 1, \ldots, d$. For an integer $m \geq 0$ we can introduce the space $H^m(E)$ as in Sect. 9.3.2.

Condition (12.72) implies that π almost every element of E belongs to $C^1(\mathbb{R}^d)$. This can be seen as follows. Gaussianity of π implies that

$$\sum_{i=1}^d \langle \|v_i\|^p_{W_p^2(B_R)} \rangle_\pi < +\infty, \quad \forall R > 0$$

for any $p \in [1, +\infty)$. Choosing $p > d$ we conclude, with the help of the Sobolev embedding theorem (see e.g. (7.30), p. 158 of Gilbarg and Trudinger 1983), that

$$\sum_{i=1}^{d} \langle \| v_i \|_{C^1(B_R)}^p \rangle_\pi < +\infty.$$

Hence, the components of V belong to $H^1(E)$ and (12.71) implies that

$$\sum_{p=1}^{d} D_p V_p = 0.$$

We recall that $\mathcal{K}_0$ (cf. Sect. 12.6.1) is the space of all $\mathbb{R}^d$-valued functions whose Fourier transforms are compactly supported and continuous. By $\mathcal{K}$ we denote its completion under the norm introduced in (12.50). We define the spaces of polynomials as in Sect. 12.2, i.e. the space of n-th degree polynomials $\mathcal{G}_n$ is the closure of $\mathcal{G}_n^{reg}$—the linear space spanned by monomials $\prod_{i=1}^{m} f_{\varphi_i}$, where $\varphi_1, \ldots, \varphi_m \in \mathcal{K}_0$ and $n \geq m \geq 0$. By convention the product corresponding to an empty set (i.e. when $m = 0$) equals 1. The space of all finite degree regular elements is given by $\mathcal{G}^{reg} := \bigcup_{n \geq 0} \mathcal{G}_n^{reg}$. As before the spaces of finite degree elements (polynomials) and Hermite polynomials of n-th degree are defined as $\mathcal{G} := \bigcup_{n \geq 0} \mathcal{G}_n$ and $\mathcal{A}_n := \mathcal{G}_n \ominus \mathcal{G}_{n-1}$, respectively. By Π_n we denote the corresponding orthogonal projection onto $\mathcal{A}_n$. We have

$$L^2(\pi) = \bigoplus_{n \geq 0} \mathcal{A}_n.$$

Suppose that Z is a Borel subset of $\mathbb{R}^d$. We can define in an obvious way the spaces $\mathcal{G}_n(Z)$, as the space of n-th degree polynomials generated by f_φ for which $\hat{\varphi}$ is bounded and supported in Z. This definition can be extended in an obvious way and allows us to define the space of all polynomials $\mathcal{G}(Z)$, or all Hermite polynomials of a given degree $\mathcal{A}_n(Z)$, etc. that are generated by linear functionals f_φ, for which $\hat{\varphi}$ is supported in Z. By $L_Z^2(\pi)$ we denote the closure of $\mathcal{G}(Z)$. We have

$$L_Z^2(\pi) = \bigoplus_{n=0}^{+\infty} \mathcal{A}_n(Z).$$

12.8.2 Definition of the Transition Semigroup

This section is devoted to a rigorous formulation of the Markov property of the flow. Define an E-valued stochastic process $\{V_t := \tilde{V}(t, \cdot), t \in \mathbb{R}\}$ and denote by $\{\mathcal{F}_t, t \in \mathbb{R}\}$ its natural filtration. In the following result, proved in Sect. 12.14.1, we assert the existence of the transition probability semigroup that corresponds to the process.

Proposition 12.14 *There exists a family of operators* $\{R_t, t \geq 0\}$ *such that*:

(i) *for any* $p \in [1, +\infty]$ *and* $t, h \geq 0$ *we have*

$$R_h f(V_t) = \big\langle f(V_{t+h}) \big| \mathscr{F}_t \big\rangle_{\mathbb{Q}}, \quad \forall f \in L^p(\pi), \tag{12.77}$$

(ii) *it is a strongly continuous semigroup of Markov operators on* $L^2(\pi)$ *satisfying*

$$\langle R_h f, g \rangle_\pi = \langle f, R_h g \rangle_\pi, \quad \forall f, g \in L^2(\pi), h \geq 0, \tag{12.78}$$

(iii) $R_t T_x = T_x R_t$ *for all* $t \geq 0$, $x \in \mathbb{R}^d$,

(iv) *for any Borel subset* Z *of* $\mathbb{R}^d$, *non-negative integer* n *and* $t \geq 0$ *each of the spaces* $\mathscr{G}_n^{reg}(Z)$ *is invariant under* R_t,

(v) *for any* $f \in L^2(\pi; Z)$ *such that* $\langle f \rangle_\pi = 0$ *we have*

$$\|R_t f\|_\pi \leq e^{-\gamma_*(Z)t} \|f\|_\pi, \quad \forall t \geq 0. \tag{12.79}$$

Here $\gamma_*(Z) := \inf[\gamma(\xi) : \xi \in Z]$.

12.8.3 Properties of the Generator

From part (ii) of the above proposition we conclude that π is an invariant, reversible probability measure for the transition semigroup. Its generator $L_0 : D(L_0) \to L^2(\pi)$ is therefore self-adjoint. Another direct consequence of Proposition 12.14 is:

Corollary 12.15 (Spectral gap estimate) *Suppose that* $f \in D(L_0) \cap L^2_Z(\pi)$. *Then*,

$$\big\langle (-L_0) f, f \big\rangle_\pi \geq \gamma_*(Z) \big(\|f\|_\pi^2 - \langle f \rangle_\pi^2 \big).$$

In particular, if $\gamma(\xi) \geq \gamma_0 > 0$ *for all* $\xi \in \mathbb{R}^d$ *then the generator satisfies the spectral gap estimate*:

$$\big\langle (-L_0) f, f \big\rangle_\pi \geq \gamma_0 \big(\|f\|_\pi^2 - \langle f \rangle_\pi^2 \big), \quad \forall f \in D(L_0).$$

On $\mathscr{K}$ we define an operator $A : D(A) \to \mathscr{K}$ by the formula $\widehat{A\varphi}(\xi) := \gamma(\xi)\hat{\varphi}(\xi)$. Its domain $D(A)$ consists of those functions $\varphi \in \mathscr{K}$ for which $A\varphi \in \mathscr{K}$. Observe that clearly $\mathscr{K}_0 \subset D(A)$.

Proposition 12.16

(i) *Suppose that* $\gamma_*(K) > 0$ *for any compact subset* $K \subset \mathbb{R}^d \setminus \{0\}$. *Then, measure* π *is ergodic under* $\{R_t, t \geq 0\}$.

(ii) *The set* $\mathscr{G}^{reg}$ *is a core of* L_0. *For any* $f = \prod_{i=1}^n f_{\varphi_i}$, *such that* $\varphi_1, \ldots, \varphi_n \in \mathscr{K}_0$, *we have*

$$L_0 f = \sum_{k=1}^n f_{A\varphi_k} f_k + \frac{1}{2} \sum_{k \neq l} \mathscr{R}_{k,l} f_{k,l}. \tag{12.80}$$

Here

$$f_k := \begin{cases} \prod_{i\neq k} f_{\varphi_i}, & \text{when } n\ge 2,\\ 1, & \text{when } n=1,\end{cases}$$

$$f_{k,l} := \begin{cases} \prod_{i\neq k,l} f_{\varphi_i}, & \text{when } n\ge 3,\\ 1, & \text{when } n=2,\\ 0, & \text{when } n=1\end{cases}$$

and

$$\mathscr{R}_{k,l} := \int_{\mathbb{R}^d} \gamma(\xi)\hat{R}_{\hat{\varphi}_k,\hat{\varphi}_l}(d\xi).$$

The notation for $\hat{R}_{\hat{\varphi}_k,\hat{\varphi}_l}(d\xi)$ *is the same as in* (12.40).

The proof of the proposition is presented in Sect. 12.14.2.

12.8.4 More General Formulation of the Markov Property of the Environment Process

Denote by $\mathscr{F}_{\ge s}$ the σ-algebra generated by $\{V_t, t\ge s\}$. Let $s=t_0<t_1<t_2<\cdots<t_n$ and $f_1,\ldots,f_n\in B_b(E)$. For

$$F(\omega)=\prod_{i=1}^n f_i\big(V_{t_i}(\omega)\big) \quad \text{and} \quad k=1,\ldots,n$$

we define $F_k\in B_b(E)$ by recursion as follows: let $F_n := R_{t_n-t_{n-1}} f_n$ and assuming that F_k for some $2\le k\le n$ is already defined we let

$$F_{k-1} := R_{t_{k-1}-t_{k-2}}(f_{k-1}F_k).$$

Finally, set $\mathfrak{E}F := F_1$.

One can easily verify that $\langle \mathfrak{E}F^2\rangle_\pi \le \langle F^2\rangle_{\mathbb{Q}}$. Therefore $\mathfrak{E}$ extends continuously to a linear operator from $L^2(\mathbb{Q})$ to $L^2(\pi)$. We can think of $\mathfrak{E}$ as the conditional expectation of F with respect to the path measure corresponding to $\{V_t, t\ge 0\}$, given the initial configuration $V(0)=v$. Due to this fact sometimes we shall use the notation $\mathfrak{E}_v F$. Some caution is advised, since the object in question is defined only for π almost every initial configuration.

The transition probabilities of the Markov semigroup are spatially homogeneous, see part (iii) of Proposition 12.14. This fact implies spatial homogeneity of $\mathfrak{E}_v$ with respect to the initial configuration. To formulate rigorously this property we identify Ω with the space of d-dimensional vector fields defined on $\mathbb{R}^{1+d}$ that are jointly continuous in all variables and continuously differentiable in the spatial variables, equipped with the standard Fréchet metric. The probability measure $\mathbb{Q}$ is assumed to be the law of the field $\tilde{V}(t,x)$. The stationarity of the field implies that $\mathbb{Q}\tau_{t,x} = \mathbb{Q}$, where $\tau_{t,x}\omega(\cdot,\cdot) := \omega(t+\cdot, x+\cdot)$ for all $(t,x) \in \mathbb{R}^{1+d}$ and $\omega \in \Omega$. From (12.77) and part (iii) of Proposition 12.14 we conclude that

$$\langle F \circ \tau_{t,x} | \mathscr{F}_t \rangle_{\mathbb{Q}} = \mathfrak{E}_{\tau_x V_t} F, \quad \forall (t,x) \in [0,+\infty) \times \mathbb{R}^d \tag{12.81}$$

for any F that is $\mathscr{F}_{\geq 0}$-measurable and bounded. We can generalize (12.81), replacing deterministic argument x by an appropriate random vector, provided F is sufficiently regular. Assume that ξ is a random vector. Since $f(x) := \mathfrak{E}_{\tau_x V_t} F$ is only defined $\mathbb{Q}$ a.s. for each $x \in \mathbb{R}^d$ it is not immediately clear how to make sense of $f(\xi)$. The following result allows us to define this object when $\tilde{F}(t,x;\omega) := F \circ \tau_{t,x}(\omega)$ is continuous in the spatial variable. For a fixed $n \geq 1$ and $k \in \mathbb{Z}^d$ we let

$$A_{k,n} := \left[\omega : k/2^n \leq \xi(\omega) < (k+1)/2^n\right]$$

and

$$X_n := \sum_k f\left(k2^{-n}\right) 1_{A_{k,n}}. \tag{12.82}$$

Proposition 12.17 *Suppose that ξ is a random, $\mathbb{R}^d$-valued vector that is $\mathscr{F}_t$ measurable for some $t \geq 0$ and $F \in B_b(\Omega)$ is such that $\{\tilde{F}(t,x), x \in \mathbb{R}^d\}$ possesses continuous modification. Under the above assumptions the sequence $\{X_n, n \geq 1\}$, defined in* (12.82), *is convergent in $L^1(\mathbb{Q})$. Its limit, denoted by $\mathfrak{E}_{\tau_\xi V_t} F$, satisfies the following equality*

$$\left\langle \tilde{F}(t,\xi) \middle| \mathscr{F}_t \right\rangle_{\mathbb{Q}} = \mathfrak{E}_{\tau_\xi V_t} F. \tag{12.83}$$

The proof of the proposition is postponed until Sect. 12.14.3.

12.9 Periodic Approximation of the Flow

We denote $|\mathbf{j}|_\infty := \max\{|j_1|, \ldots, |j_d|\}$ for any $\mathbf{j} = (j_1, \ldots, j_d) \in \mathbb{Z}^d$. For a given integer $N \geq 1$ we let $\Lambda_N := [\mathbf{j} \in \mathbb{Z}^d : 0 < |\mathbf{j}|_\infty \leq N2^N]$ and Λ_N^+ be its subset consisting of those $\mathbf{j} \in \Lambda_N$ for which either $j_d > 0$, or for some k we have $j_k > 0$ and $j_{k+1} = \cdots = j_d = 0$. We define $\Lambda_N^- := -\Lambda_N^+$. Of course

$$\Lambda_N = \Lambda_N^+ \cup \Lambda_N^- \quad \text{and} \quad \Lambda_N^+ \cap \Lambda_N^- = \emptyset.$$

For $\mathbf{j} \in \Lambda_N^+$ we define $\Box_{\mathbf{j},N} := \prod_{k=1}^d [2^{-N} j_k, 2^{-N}(j_k+1))$. When $\mathbf{j} \in \Lambda_N^-$ we let $\Box_{\mathbf{j},N} := -\Box_{-\mathbf{j},N}$. We write $\hat{R}^{(s)}_{\mathbf{j},N}$, $\hat{R}^{(a)}_{\mathbf{j},N}$ for the real entry $d \times d$ matrices that are the real and imaginary parts of $\hat{R}_{\mathbf{j},N} := \hat{R}(\Box_{\mathbf{j},N})$, i.e.

$$\hat{R}_{\mathbf{j},N} = \hat{R}^{(s)}_{\mathbf{j},N} + i\hat{R}^{(a)}_{\mathbf{j},N}, \quad \mathbf{j} \in \Lambda_N. \tag{12.84}$$

Since $\hat{R}_{\mathbf{j},N}$ is complex Hermitian these matrices are symmetric and anti-symmetric respectively. We define by $\hat{R}_{\mathbf{j},N}^{-1}$ the inverse matrix in case $\hat{R}_{\mathbf{j},N}$ is non-singular. If otherwise the symbol denotes the pseudo-inverse understood as the unique symmetric matrix with the same null space as $\hat{R}_{\mathbf{j},N}$ and such that $\hat{R}_{\mathbf{j},N}^{-1}\hat{R}_{\mathbf{j},N} = \hat{R}_{\mathbf{j},N}\hat{R}_{\mathbf{j},N}^{-1}$ is the projection onto the range of $\hat{R}_{\mathbf{j},N}$. Note that $\hat{R}_{-\mathbf{j},N} := \hat{R}_{\mathbf{j},N}^{*}$ and $\hat{R}_{-\mathbf{j},N}^{-1} := (\hat{R}_{\mathbf{j},N}^{-1})^{*}$ for all $\mathbf{j} \in \Lambda_N$.

Suppose that $\xi_{\mathbf{j},N} \in \square_{\mathbf{j},N}$ are such that $\xi_{-\mathbf{j},N} = -\xi_{\mathbf{j},N}$. We denote $\gamma_{\mathbf{j},N} := \gamma(\xi_{\mathbf{j},N})$. Points $\xi_{\mathbf{j},N}$ are selected in such a way that they belong to the support of $r(d\xi)$ and

$$\lim_{N\to+\infty}\left|\sum_{\mathbf{j}\in\Lambda_N}\frac{\hat{R}_{\mathbf{j},N}}{\gamma_{\mathbf{j},N}} - \int_{\mathbb{R}^d}\frac{\hat{R}(d\xi)}{\gamma(\xi)}\right| = 0.$$

Assume furthermore that $\pi_{\mathbf{j},N}$ is a zero mean Gaussian measure on $\mathbb{R}^{2d}$ whose covariance matrix equals

$$\hat{C}_{\mathbf{j},N} := \begin{bmatrix} \hat{R}_{\mathbf{j},N}^{(s)} & \hat{R}_{\mathbf{j},N}^{(a)} \\ -\hat{R}_{\mathbf{j},N}^{(a)} & \hat{R}_{\mathbf{j},N}^{(s)} \end{bmatrix}. \tag{12.85}$$

Let π_N be the product probability measure $\prod_{\mathbf{j}\in\Lambda_N^+}\pi_{\mathbf{j},N}$ defined on $E_N := \mathbb{R}^{2dK_N}$, where K_N is the cardinality of Λ_N^+.

Denote by $\mathscr{G}_n^{(N)}$ (resp. $\mathscr{A}_n^{(N)}$) the spaces of the n-th degree (resp. Hermite) polynomials that correspond to measure π_N with $\Pi_n^{(N)}$ the orthogonal projection onto $\mathscr{A}_n^{(N)}$. The space of all polynomials is defined as $\mathscr{G}^{(N)} := \bigcup_{n\ge 0}\mathscr{G}_n^{(N)}$.

Let $\{\tau_x^{(N)}, x \in \mathbb{R}^d\}$ be a group of shifts, i.e. π_N-measure preserving transformations on E_N defined by $\tau_x^{(N)}(\mathbf{v}) := \{(a_{\mathbf{j}}', b_{\mathbf{j}}'), \mathbf{j} \in \Lambda_N^+\}$, where $a_{\mathbf{j}}'$, $b_{\mathbf{j}}'$ are the real and imaginary parts of $(a_{\mathbf{j}} + ib_{\mathbf{j}})\exp\{i\xi_{\mathbf{j}}\cdot x\}$ respectively. The gradient $\nabla^{(N)} = (D_1^{(N)}, \dots, D_d^{(N)})$ is the generator corresponding to the representation of the group over $L^2(\pi_N)$, i.e.

$$\nabla^{(N)} f := \nabla_{|x=0} f \circ \tau_x^{(N)}.$$

The differentiation is understood in the L^2-sense. The invariance of $\mathscr{G}^{(N)}$ under shift transformations implies that it is a core of $\nabla^{(N)}$. In addition, on $\mathscr{G}^{(N)}$

$$\nabla^{(N)} = \sum_{\mathbf{j}\in\Lambda_N^+}\xi_{\mathbf{j}}(a_{\mathbf{j}}\cdot\nabla_{b_{\mathbf{j}}} - b_{\mathbf{j}}\cdot\nabla_{a_{\mathbf{j}}}). \tag{12.86}$$

Let $\hat{S}_{\mathbf{j},s}^{(N)}$ and $\hat{S}_{\mathbf{j},a}^{(N)}$ be the real and imaginary parts of $\hat{R}_{\mathbf{j},N}^{1/2}$. Define an E_N-valued Ornstein–Uhlenbeck process

$$\mathbf{c}_t(\mathbf{v}) := \left\{\left(a_{t,\mathbf{j}}(\mathbf{v}), b_{t,\mathbf{j}}(\mathbf{v})\right), \mathbf{j} \in \Lambda_N^+\right\}$$

whose components are the solutions of

$$\begin{cases} da_{t,\mathbf{j}}(\mathbf{v}) = -\gamma_{\mathbf{j}} a_{t,\mathbf{j}}(\mathbf{v}) dt + (2\gamma_{\mathbf{j}})^{1/2} \big[\hat{S}^{(N)}_{\mathbf{j},s} dw_{t,\mathbf{j}} + \hat{S}^{(N)}_{\mathbf{j},a} dw'_{t,\mathbf{j}} \big], \\ db_{t,\mathbf{j}}(\mathbf{v}) = -\gamma_{\mathbf{j}} b_{t,\mathbf{j}}(\mathbf{v}) dt + (2\gamma_{\mathbf{j}})^{1/2} \big[-\hat{S}^{(N)}_{\mathbf{j},a} dw_{t,\mathbf{j}} + \hat{S}^{(N)}_{\mathbf{j},s} dw'_{t,\mathbf{j}} \big], \\ a_{0,\mathbf{j}}(\mathbf{v}) = a_{\mathbf{j}}, \quad b_{0,\mathbf{j}}(\mathbf{v}) = b_{\mathbf{j}}, \end{cases} \tag{12.87}$$

where $\{w_{t,\mathbf{j}}, t \geq 0\}$, $\{w'_{t,\mathbf{j}'}, t \geq 0\}$ are mutually independent, d-dimensional, standard Brownian motions for $\mathbf{j}, \mathbf{j}' \in \Lambda_N^+$. The Dirichlet form of the process equals

$$\mathscr{E}_N(f) := \big\langle \big(-L_0^{(N)}\big) f, f \big\rangle_{\pi_N} = \sum_{\mathbf{j} \in \Lambda_N^+} \gamma_{\mathbf{j}} \mathscr{E}_{\mathbf{j}}^{(N)}(f), \quad f \in \mathscr{G}^{(N)}, \tag{12.88}$$

where $L_0^{(N)}$ is the generator of the process. Here

$$\mathscr{E}_{\mathbf{j}}^{(N)}(f) := \big\langle \delta_{\mathbf{j}} f \hat{C}^{(s)}_{\mathbf{j},N} (\delta_{\mathbf{j}} f)^T \big\rangle_{\pi_N},$$

where $\delta_{\mathbf{j}} f := (\nabla_{a_{\mathbf{j}}} f, \nabla_{b_{\mathbf{j}}} f)$. Since the spaces $\mathscr{A}_n^{(N)}$ are invariant under $L_0^{(N)}$ we have

$$\mathscr{E}_N(f) = \sum_{n=1}^{+\infty} \mathscr{E}_N\big(\Pi_n^{(N)} f\big).$$

Suppose that $\hat{U}_{\mathbf{j},N}$, $\hat{V}_{\mathbf{j},N}$ are real random vectors over $(E, \mathscr{B}(E), \pi)$ such that

$$\hat{U}_{\mathbf{j},N} + i \hat{V}_{\mathbf{j},N} = \hat{V}(\square_{\mathbf{j},N}), \quad \forall \mathbf{j} \in \Lambda_N^+,$$

where $\hat{V}(d\xi)$ is the stochastic spectral measure corresponding to $\tilde{V}(x) := V \circ \tau_x$. Let $p_N : E \to E_N$ be defined by

$$p_N(v) := \big\{ \big(\hat{U}_{\mathbf{j},N}(v), \hat{V}_{\mathbf{j},N}(v) \big), \mathbf{j} \in \Lambda_N^+ \big\}$$

and let $P_N f := f \circ p_N$ be the respective mapping between $L^2(\pi_N)$ and $L^2(\pi)$.

For any $\varphi \in \mathscr{K}$ we set

$$\hat{\varphi}_{\mathbf{j},N} := \hat{R}_{\mathbf{j},N}^{-1} \int_{\square_{\mathbf{j},N}} \hat{R}(d\xi) \hat{\varphi}(\xi).$$

The mapping $J_N : \mathscr{G}_1 \to \mathscr{G}_1^{(N)}$ is given by

$$J_N f_\varphi(\mathbf{v}) := \sum_{\mathbf{j} \in \Lambda_N} (a_{\mathbf{j}} + i b_{\mathbf{j}}) \cdot \hat{\varphi}_{\mathbf{j},N}$$

and the scalar product is taken in $\mathbb{C}^d$. Here for $\mathbf{v} = \{(a_{\mathbf{j}}, b_{\mathbf{j}}), \mathbf{j} \in \Lambda_N^+\}$ we define $(a_{\mathbf{j}}, b_{\mathbf{j}}) = (a_{\mathbf{j}}, -b_{\mathbf{j}})$, if $\mathbf{j} \in \Lambda_N^-$. A straightforward calculation shows that J_N is a contraction. It can be extended therefore, via the relation

$$J_N\left(\prod_{i=1}^{m} f_{\varphi_i}\right) := \prod_{i=1}^{m} J_N f_{\varphi_i},$$

to a contraction from the entire $L^2(\pi)$ to $L^2(\pi_N)$, see Theorem 4.5, p. 45 of Janson (1997). Denote also $V^{(N)}(\mathbf{v}) := \sum_{\mathbf{j}\in\Lambda_N} a_{\mathbf{j}}$. Our main approximation result can now be formulated as follows.

Proposition 12.18 *For any* $f \in \mathcal{G}^{reg}$ *we have*

$$\lim_{N\uparrow+\infty} \mathcal{E}_N(J_N f) = \langle(-L_0)f, f\rangle_\pi \tag{12.89}$$

and

$$\lim_{N\uparrow+\infty} \left\{\|P_N J_N f - f\|_\pi + \left\|P_N \nabla^{(N)} J_N f - \nabla f\right\|_\pi\right\} = 0. \tag{12.90}$$

Moreover

$$\lim_{N\to+\infty} \left\|P_N V^{(N)} - V\right\|_{L^p(\pi)} = 0, \quad \forall p \in [1, +\infty). \tag{12.91}$$

Proof It suffices to prove (12.89) and (12.90) for $f = f_\varphi$, where $\varphi \in \mathcal{K}_0$. From that we can easily extend the conclusion of the proposition to polynomials f belonging to $\mathcal{G}_n^{reg}$, and thus to the entire $\mathcal{G}^{reg}$. The computation of the limits for $f = f_\varphi$ uses explicit formulas for the Dirichlet form and the relevant norms involving the structure measure for the respective Gaussian fields. Verification of (12.91) follows directly from the definition of P_N. □

12.10 Environment Process

The environment process is an E-valued, stochastic process, over the product probability space $(\Omega \times \Sigma, \mathcal{F} \otimes \mathcal{A}, \mathbb{Q} \otimes \mathbb{P})$ given by formula

$$\eta_t := \tau_{X_t} V_t, \quad t \ge 0. \tag{12.92}$$

From (12.67) we can write

$$X_t = \int_0^t V(\eta_s)ds + \kappa w_t. \tag{12.93}$$

Below, we formulate several properties of the process. In order not to distract the reader we postpone their proofs until Sect. 12.15.

Proposition 12.19 *Suppose that* $\kappa \geq 0$. *Then,*

(i) $\{\eta_t, t \geq 0\}$ *is Markovian with the transition probability semigroup*

$$P_t f(v) := \mathfrak{E}_v\big\{E_{\mathbb{P}} f(\eta_t)\big\}, \quad f \in B_b(\pi), \tag{12.94}$$

(ii) *measure* π *is invariant under* $\{P_t, t \geq 0\}$,
(iii) *the semigroup extends to a strongly continuous semigroup of contractions on* $L^2(\pi)$.

Denote by $L : D(L) \to L^2(\pi)$ the generator of the semigroup. The following result provides closer description on the operator. Let $\widehat{\mathscr{C}} := D(L_0) \cap \mathscr{C}_0$, where $\mathscr{C}_0$, L_0 are defined in (12.5) and in Sect. 12.8.3, respectively.

Proposition 12.20 *Under the assumptions of Proposition* 12.19 *the following are true*:

(i) $\mathscr{G}^{reg} \subset \widehat{\mathscr{C}} \subset D(L) \cap D(L^*)$,
(ii) *on* $\widehat{\mathscr{C}}$ *the symmetric and anti-symmetric parts of L respectively equal*

$$S = L_0 + \frac{\kappa^2}{2}\Delta, \quad A = V \cdot \nabla, \tag{12.95}$$

(iii) $\mathscr{G}^{reg}$ *is a common core of L and* L^*.

Corollary 12.21 *Under the assumptions of part* (i) *of Proposition* 12.16 *measure* π *is ergodic for the semigroup* $\{P_t, t \geq 0\}$.

Proof As a consequence of Proposition 12.20 we have

$$\big\langle (-L)f, f\big\rangle_\pi \geq \big\langle (-L_0)f, f\big\rangle_\pi, \quad \forall f \in \mathscr{C}. \tag{12.96}$$

Suppose that $f \in D(L)$ is such that $Lf = 0$. From (12.96) we conclude that then also $\langle (-L_0)f, f\rangle_\pi = 0$. From Corollary 12.15 and the fact that $\mathscr{C}$ is a core of L_0 we conclude that f has to be constant π a.s. □

We introduce the spaces $\mathscr{H}_1$ and $\mathscr{H}_{-1}$ that correspond to the generator L (cf. Sect. 2.2). A direct consequence from the above corollary and Corollary 12.15 is the following.

Corollary 12.22 *Suppose that* $\kappa \geq 0$ *and* $\gamma(\xi) \geq \gamma_0 > 0$ *for all* $\xi \in \mathbb{R}^d$. *Then, L satisfies the spectral gap condition, i.e.*

$$\big\langle (-L)f, f\big\rangle_\pi \geq \gamma_0\big(\|f\|_\pi^2 - \langle f\rangle_\pi^2\big) \quad \forall f \in D(L). \tag{12.97}$$

As a simple corollary of (12.97) and Theorem 2.18 we conclude that:

Corollary 12.23 *Under the assumptions of Corollary* 12.22 *the laws of* $X_t/\sqrt{t}$ *converge, as* $t \to +\infty$, *to a normal distribution in the same sense as in Theorem* 12.13.

12.11 Proof of Part (1) of Theorem 12.13

The conclusion of part (1) of Theorem 12.13 is a consequence of the fact that the generator of the environment process satisfies the sector condition. As a result we may apply then apply the results of Sect. 2.7.3. In fact a slightly stronger estimate is available.

Theorem 12.24 *Under the assumptions of part* (1) *of Theorem* 12.13 *there exists a constant* $C>0$, *independent of* $\kappa>0$, *such that*

$$\big|\langle V\cdot\nabla f,g\rangle_\pi\big|\le C\|\nabla f\|_\pi\big\langle(-L_0)g,g\big\rangle_\pi^{1/2},\quad \forall f,g\in\mathscr{G}^{reg}. \tag{12.98}$$

Proof Estimate (12.98) implies the sector condition for generator L. Indeed, using formulas claimed in Proposition 12.20, we obtain that the anti-symmetric part of the generator satisfies

$$\begin{aligned}\big|\langle Af,g\rangle_\pi\big|&\le C\|\nabla f\|_\pi\big\langle(-L_0)g,g\big\rangle_\pi^{1/2}\\&\le\sqrt{2}C\kappa^{-1}\big\langle\big(-(\kappa^2/2)\Delta\big)f,f\big\rangle_\pi^{1/2}\big\langle(-L_0)g,g\big\rangle_\pi^{1/2}\\&\le\sqrt{2}C\kappa^{-1}\big\langle(-S)f,f\big\rangle_\pi^{1/2}\big\langle(-S)g,g\big\rangle_\pi^{1/2},\quad\forall f,g\in\mathscr{G}^{reg}.\end{aligned}$$

The proof of the theorem relies on the following auxiliary result.

Lemma 12.25 *There exists a constant* $C>0$, *independent of* κ, n, N *such that*

$$\big\|\Pi_n^{(N)}\big(V_j^{(N)}g\big)\big\|_{\pi_N}\le C\mathscr{E}_N^{1/2}(g),\quad\forall g\in\mathscr{A}_{n+1}^{(N)},\ j=1,\dots,d. \tag{12.99}$$

The quadratic form $\mathscr{E}_N(\cdot)$ *is defined in* (12.88).

Before presenting the proof we use the above result to show estimate (12.98). Assume that $f\in\mathscr{A}_n^{(N)}$ and $g\in\mathscr{G}^{(N)}$. Let also

$$B_{k,n}^jf:=\Pi_k^{(N)}\big(V_j^{(N)}\Pi_n^{(N)}g\big),\quad k\ge0.$$

The operators vanish when $|k-n|\ne1$. We can write

$$\big\langle V^{(N)}\cdot\nabla^{(N)}f,g\big\rangle_{\pi_N}=\big\langle\Pi_n^{(N)}\nabla^{(N)}f,B_{n,n+1}g\big\rangle_{\pi_N}+\big\langle\Pi_n^{(N)}\nabla^{(N)}f,B_{n,n-1}g\big\rangle_{\pi_N}.$$

Hence, for any $f,g\in\mathscr{G}^{(N)}$ we have

$$\begin{aligned}\big|\big\langle V^{(N)}\cdot\nabla^{(N)}f,g\big\rangle_{\pi_N}\big|\le{}&\sum_{n=1}^{+\infty}\big|\big\langle\Pi_n^{(N)}\nabla^{(N)}f,B_{n,n+1}g\big\rangle_{\pi_N}\big|\\&+\sum_{n=1}^{+\infty}\big|\big\langle\Pi_n^{(N)}\nabla^{(N)}f,B_{n,n-1}g\big\rangle_{\pi_N}\big|.\end{aligned} \tag{12.100}$$

From (12.99) the right-hand side of (12.100) can be estimated, using Cauchy–Schwartz inequality, by

$$C\sum_{n=1}^{+\infty}\|\Pi_n^{(N)}\nabla^{(N)}f\|_{\pi_N}\left[\mathscr{E}_N^{1/2}(\Pi_{n-1}^{(N)}g)+\mathscr{E}_N^{1/2}(\Pi_{n+1}^{(N)}g)\right]$$

$$\le 2C\|\nabla^{(N)}f\|_{\pi_N}\mathscr{E}_N^{1/2}(g)$$

where constant C does not depend on N. Estimate (12.98) can then be obtained by letting $N\to+\infty$, thanks to Proposition 12.18.

Proof of Lemma 12.25 Since each matrix $\hat{R}_{\mathbf{j},N}$, given by (12.84), is Hermitian and non-negative definite it can be written in the form $\hat{R}_{\mathbf{j},N}=T_{\mathbf{j}}^{-1}D_{\mathbf{j}}T_{\mathbf{j}}$, where $T_{\mathbf{j}}$ is unitary and $D_{\mathbf{j}}$ is a diagonal matrix $\mathrm{diag}(\lambda_{\mathbf{j},1},\dots,\lambda_{\mathbf{j},d})$ and the eigenvalues are put in descending order. Denote by $d_{\mathbf{j}}$ the rank of the matrix and by $T_{\mathbf{j}}^{(s)}$ and $T_{\mathbf{j}}^{(a)}$ the real and imaginary parts of $T_{\mathbf{j}}$, respectively. Then,

$$T_{\mathbf{j}}^{(r)}:=\begin{bmatrix}T_{\mathbf{j}}^{(s)} & T_{\mathbf{j}}^{(a)}\\ -T_{\mathbf{j}}^{(a)} & T_{\mathbf{j}}^{(s)}\end{bmatrix}$$

is a $2d\times 2d$ orthogonal matrix such that

$$T_{\mathbf{j}}^{(r)}\hat{C}_{\mathbf{j},N}\left(T_{\mathbf{j}}^{(r)}\right)^{-1}=D_{\mathbf{j}}^{(r)}:=\begin{bmatrix}D_{\mathbf{j}} & 0\\ 0 & D_{\mathbf{j}}\end{bmatrix}.$$

Here $\hat{C}_{\mathbf{j},N}$ is a real entry matrix given by (12.85). On E_N we introduce the change of coordinates

$$\Phi:\left\{(a_{\mathbf{j}},b_{\mathbf{j}}),\mathbf{j}\in\Lambda_N^+\right\}\mapsto\left\{(a'_{\mathbf{j}},b'_{\mathbf{j}}),\mathbf{j}\in\Lambda_N^+\right\}$$

letting

$$\begin{bmatrix}a'_{\mathbf{j}}\\ b'_{\mathbf{j}}\end{bmatrix}:=T_{\mathbf{j}}^{(r)}\begin{bmatrix}a_{\mathbf{j}}\\ b_{\mathbf{j}}\end{bmatrix}\quad\text{for each } \mathbf{j}\in\Lambda_N^+.$$

It induces a unitary mapping $U:L^2(\pi_{N,*})\to L^2(\pi_N)$, where $\pi_{N,*}$ is the pullback of π_N by Φ. Measure $\pi_{N,*}$ is a product of $2d$-dimensional, zero mean, Gaussian measures $\pi_{\mathbf{j}}$ whose covariances are given by diagonal matrices $D_{\mathbf{j}}^{(r)}$. The mapping preserves the respective spaces of Hermite polynomials and diagonalizes Dirichlet form $\mathscr{E}_N$, i.e.

$$\mathscr{E}_U^{(N)}(f):=\left\langle(-L_0^{(N)})Uf,Uf\right\rangle_{\pi_{N,*}}=\sum_{\mathbf{j}\in\Lambda_N^+}\gamma_{\mathbf{j}}\mathscr{E}_{\mathbf{j},U}^{(N)}(f),\tag{12.101}$$

where

$$\mathscr{E}^{(N)}_{\mathbf{j},U}(f) := \sum_{p=1}^{d} \int_{\mathbb{R}^{2d}} \lambda_{\mathbf{j},p}\big(|\nabla_{a_{\mathbf{j}}} f|^2 + |\nabla_{b_{\mathbf{j}}} f|^2\big)\pi_{\mathbf{j}}(da_{\mathbf{j}}, db_{\mathbf{j}}). \tag{12.102}$$

Formula (12.102) can be further transformed using the representation of f in the base of Hermite polynomials. The definition and some basic facts concerning these polynomials are gathered in an appendix in Sect. 12.16.

For any $\mathbf{n} = \{(n_{\mathbf{j}}, m_{\mathbf{j}}), \mathbf{j} \in \Lambda_N^+\}$, where $n_{\mathbf{j}}, m_{\mathbf{j}} \in \mathbb{Z}_+^{d_{\mathbf{j}}}$ and $\mathbf{v} = \{(a_{\mathbf{j}}, b_{\mathbf{j}}), \mathbf{j} \in \Lambda_N^+\}$ we let

$$h_{\mathbf{n}}(\mathbf{v}) := \prod_{(\mathbf{j},p)} h_{n_{\mathbf{j},p}}\big(\lambda_{\mathbf{j},p}^{-1/2} a_{\mathbf{j},p}\big) h_{m_{\mathbf{j},p}}\big(\lambda_{\mathbf{j},p}^{-1/2} b_{\mathbf{j},p}\big). \tag{12.103}$$

The product extends over $\mathbf{j} \in \Lambda_N^+$ and $p = 1, \dots, d_{\mathbf{j}}$. In case $d_{\mathbf{j}} = 0$ the respective factor equals 1 by convention. Multidimensional Hermite polynomials $h_{m_{\mathbf{j},p}}(\cdot)$ are defined by formula (12.154).

Suppose that n is a non-negative integer. The polynomials that satisfy $|\mathbf{n}| = n$ form an orthonormal base of $\tilde{\mathscr{A}}_n^{(N)}$—the space of the n-th degree Hermite polynomials in $L^2(\pi_{N,*})$. Each $h_{\mathbf{n}}$ is an eigenvector of $-U^{-1}L_0^{(N)}U$ that corresponds to the eigenvalue $\sum_{\mathbf{j}\in\Lambda_N^+} \gamma_{\mathbf{j}}(|n_{\mathbf{j}}| + |m_{\mathbf{j}}|)$. Here $|n_{\mathbf{j}}| := \sum_{p=1}^{d_{\mathbf{j}}} n_{\mathbf{j},p}$ and $|n_{\mathbf{j}}| := 0$ when $d_{\mathbf{j}} = 0$. Any element f belonging to $\tilde{\mathscr{A}}_{n+1}^{(N)}$ can be written as

$$f = \sum_{|\mathbf{n}|=n+1} \alpha(\mathbf{n}) h_{\mathbf{n}}. \tag{12.104}$$

Using (12.104) we can rewrite the form (12.101) as follows

$$\mathscr{E}_U^{(N)}(f) = \sum_{|\mathbf{n}|=n+1} \alpha^2(\mathbf{n}) \sum_{\mathbf{j}\in\Lambda_N^+} \gamma_{\mathbf{j}}\big(|\mathbf{n}_{\mathbf{j}}| + |\mathbf{m}_{\mathbf{j}}|\big). \tag{12.105}$$

Let e_p, $p = 1, \dots, 2d$ be the canonical base in $\mathbb{R}^{2d}$. For $p = 1, \dots, d$ let $f_{\mathbf{j},p}, g_{\mathbf{j},p} \in \mathbb{R}^d$ be the projections of $(T_{\mathbf{j}}^{(r)})^{-1} e_p$ and $(T_{\mathbf{j}}^{(r)})^{-1} e_{p+d}$ onto the first d coordinates. Let also $V_U^{(N)} = U^{-1} V^{(N)}$. We can write

$$V_U^{(N)}(\mathbf{v}) := 2 \sum_{(\mathbf{j},p)} (a_{\mathbf{j},p} f_{\mathbf{j},p} + b_{\mathbf{j},p} g_{\mathbf{j},p}).$$

The sum extends over $\mathbf{j} \in \Lambda_N^+$ and $p = 1, \dots, d_{\mathbf{j}}$. We wish to express the product $V_U^{(N)} h_{\mathbf{n}}$ using the orthonormal base $\{h_{\mathbf{n}}, \mathbf{n}\}$. To do so we introduce some notation. For given $\mathbf{i}$, p and a collection of integer multi-indices $\mathbf{n} := \{(n_{\mathbf{j}}, m_{\mathbf{j}}), \mathbf{j} \in \Lambda_N^+\}$ we let $\mathbf{n}^{\pm,a}_{\mathbf{i},p}$ to be a collection of multi-indices $\{(n'_{\mathbf{j}}, m'_{\mathbf{j}}), \mathbf{j} \in \Lambda^+_N\}$ defined as follows:

$$\begin{aligned} &m'_{\mathbf{j}} := m_{\mathbf{j}} \quad \text{for all } \mathbf{j}, \\ &n'_{\mathbf{j},q} := n_{\mathbf{j},q} \quad \text{when } \mathbf{j} \neq \mathbf{i}, \text{ or } q \neq p \text{ and} \\ &n'_{\mathbf{i},p} := n_{\mathbf{i},p} \pm 1. \end{aligned}$$

The multi-index $\mathbf{n}^{\pm,b}_{\mathbf{j},p}$ is defined analogously, the only difference is that in this case the modification concerns the components $m_{\mathbf{j}}$. Using formula (12.155) we obtain

$$V_U^{(N)}(\mathbf{v}) h_{\mathbf{n}}(\mathbf{v}) = 2F_{\mathbf{n}}(\mathbf{v}) + 2G_{\mathbf{n}}(\mathbf{v}),$$

where

$$F_{\mathbf{n}}(\mathbf{v}) := \sum_{(\mathbf{j},p)} \lambda^{1/2}_{\mathbf{j},p} \big[(n_{\mathbf{j},p}+1)^{1/2} h_{\mathbf{n}^{+,a}_{\mathbf{j},p}}(\mathbf{v}) + n^{1/2}_{\mathbf{j},p} h_{\mathbf{n}^{-,a}_{\mathbf{j},p}}(\mathbf{v})\big] f_{\mathbf{j},p},$$

$$G_{\mathbf{n}}(\mathbf{v}) := \sum_{(\mathbf{j},p)} \lambda^{1/2}_{\mathbf{j},p} \big[(m_{\mathbf{j},p}+1)^{1/2} h_{\mathbf{n}^{+,b}_{\mathbf{j},p}}(\mathbf{v}) + m^{1/2}_{\mathbf{j},p} h_{\mathbf{n}^{-,b}_{\mathbf{j},p}}(\mathbf{v})\big] g_{\mathbf{j},p}.$$

Therefore, we can write

$$V_U^{(N)} f = F + G,$$

where

$$F := \sum_{|\mathbf{n}|=n+1} \alpha(\mathbf{n}) F_{\mathbf{n}} \quad \text{and} \quad G := \sum_{|\mathbf{n}|=n+1} \alpha(\mathbf{n}) G_{\mathbf{n}}.$$

Hence

$$\big\| \Pi_n^{(N)} \big(V_U^{(N)} f\big) \big\|^2_{\pi_{N,*}} \le 2\big(\big\| \Pi_n^{(N)} F \big\|^2_{\pi_{N,*}} + \big\| \Pi_n^{(N)} G \big\|^2_{\pi_{N,*}} \big). \tag{12.106}$$

Note that

$$\Pi_n^{(N)} F_{\mathbf{n}} = \sum_{(\mathbf{j},p)} (\lambda_{\mathbf{j},p} n_{\mathbf{j},p})^{1/2} h_{\mathbf{n}^{-,a}_{\mathbf{j},p}} f_{\mathbf{j},p}$$

and

$$\big\| \Pi_n^{(N)} F \big\|^2_{\pi_{N,*}} = \sum \alpha(\mathbf{n}) \alpha(\mathbf{n}') \big(\lambda_{\mathbf{j},p} n_{\mathbf{j},p} \lambda_{\mathbf{j}',p'} n'_{\mathbf{j}',p'}\big)^{1/2} \delta^{\mathbf{n}',\mathbf{j}',p'}_{\mathbf{n},\mathbf{j},p} f_{\mathbf{j},p} \cdot f_{\mathbf{j}',p'}.$$

Here

$$\delta^{\mathbf{n}',\mathbf{j}',p'}_{\mathbf{n},\mathbf{j},p} := \langle h_{\mathbf{n}^{-,a}_{\mathbf{j},p}}, h_{(\mathbf{n}')^{-,a}_{\mathbf{j}',p'}} \rangle_{\pi_{N,*}}.$$

The summation extends over all appropriate indices $\mathbf{n}, \mathbf{n}', \mathbf{j}, \mathbf{j}', p, p'$. The right-hand side of the above equality can be estimated by

$$\frac{1}{2} \sum \big[\gamma_{\mathbf{j}'}^{-1} \gamma_{\mathbf{j}} \alpha^2(\mathbf{n}) \lambda_{\mathbf{j}',p'} n_{\mathbf{j},p} + \gamma_{\mathbf{j}}^{-1} \gamma_{\mathbf{j}'} \alpha^2(\mathbf{n}') \lambda_{\mathbf{j},p} n'_{\mathbf{j}',p'}\big] \delta^{\mathbf{n}',\mathbf{j}',p'}_{\mathbf{n},\mathbf{j},p}.$$

The expression corresponding to the first term inside the brackets can be further estimated by

$$\sum_{\mathbf{n},\mathbf{j},p} \alpha^2(\mathbf{n})\gamma_{\mathbf{j}} n_{\mathbf{j},p} \sum_{\mathbf{n}',\mathbf{j}',p'} \gamma_{\mathbf{j}'}^{-1} \operatorname{tr} \hat{R}_{\mathbf{j}',N} \delta_{\mathbf{n},\mathbf{j},p}^{\mathbf{n}',\mathbf{j}',p'} \le d\Big(\sum_{\mathbf{n},\mathbf{j},p} \alpha^2(\mathbf{n})\gamma_{\mathbf{j}} n_{\mathbf{j},p}\Big)\mathscr{R}^{(N)},$$

where

$$\mathscr{R}^{(N)} := \sum_{\mathbf{j}'} \gamma_{\mathbf{j}'}^{-1} \operatorname{tr} \hat{R}_{\mathbf{j}',N}.$$

Using formula (12.105) we estimate the right-hand side by $d\mathscr{E}_U^{(N)}(g)\mathscr{R}^{(N)}$. The quantity $\|\Pi_n^{(N)} G\|_{\pi_{N,*}}^2$ can be dealt with in an identical manner. Choosing $g = Uf$ we obtain (12.99). This ends the proof of Lemma 12.25. □

Finally, to finish the proof of this part of the theorem we apply (12.93) to verify that $V_p \in \mathscr{H}_{-1}$. Suppose that $\varphi \in \mathscr{K}_0$, see Sect. 12.6.1. Then, according to (12.80),

$$\langle (-L_0) f_\varphi, f_\varphi \rangle_\pi = \int_{\mathbb{R}^d} \gamma(\xi) \hat{R}_{\hat{\varphi}(\xi)}(d\xi). \tag{12.107}$$

Since V_p belongs to $\mathscr{A}_1$ it suffices only to verify that

$$\sup\big[|\langle V_p, f_\varphi \rangle_\pi| : \mathscr{E}_{L_0}(f_\varphi) = 1, \varphi \in \mathscr{K}_0 \big] < +\infty. \tag{12.108}$$

Note however that by Cauchy–Schwartz inequality

$$\begin{aligned} |\langle V_p, f_\varphi \rangle_\pi| &= \left| \int_{\mathbb{R}^d} \hat{R}_{p,\hat{\varphi}(\xi)}(d\xi) \right| \\ &\le \mathscr{R}^{1/2} \left(\int_{\mathbb{R}^d} \gamma(\xi) \hat{R}_{\hat{\varphi}(\xi)}(d\xi) \right)^{1/2} \stackrel{(12.107)}{=} \mathscr{R}^{1/2} \mathscr{E}_{L_0}^{1/2}(f_\varphi) \end{aligned}$$

where

$$\mathscr{R} := \int_{\mathbb{R}^d} \gamma^{-1}(\xi) r(d\xi).$$

Hence, (12.108) follows. □

12.12 On Superdiffusive Behavior of a Tracer in an Isotropic Flow

In this section we show that the result of part (i) of Theorem 12.13 is optimal for a class of Markovian, Gaussian isotropic drifts. The computations are quite similar to those made in Sect. 12.6 and rely on the use of variational principle. To simplify the notation we assume also that $\kappa = 1$ in (12.67).

Suppose that the covariance matrix of the field is given by (12.69) with

$$\gamma(\xi) = |\xi|^{\mu} \quad \text{for some } \mu \geq 0$$

and

$$\hat{R}_{pq}(d\xi) = \rho\big(|\xi|\big)\Gamma_{pq}(\xi)d\xi,$$

where $\rho(u) := a(u)u^{\alpha+1-d}$ for some $\alpha > -1$. Here $a(\cdot)$ is bounded, measurable, continuous at 0 with $a(0) > 0$ and satisfies growth condition (12.43). When $\alpha > 1$ one can easily verify that the Péclet number Pe, defined in (11.9), is finite and the central limit theorem holds for all $\mu \geq 0$, see Corollary 11.5. On the other hand, condition (12.73) holds when $\mu \in [0, 1+\alpha)$. Then, the tracer satisfies the central limit theorem by virtue of part (i) of Theorem 12.13. In what follows we demonstrate that for $\mu > 1+\alpha$ the behavior of the tracer is superdiffusive in the same sense as explained in Sect. 12.6. Namely, the following result holds.

Theorem 12.26 *Suppose that* $\alpha \in (-1, 1)$. *With* γ_* *and* γ^* *defined as in Sect.* 12.6, *see* (12.44) *and* (12.45), *we have*:

(i) *for* $(3+\alpha)/2 > \mu > 1+\alpha$

$$\gamma_* = \gamma^* = (\mu - 1 - \alpha)/\mu,$$

(ii) *for* $2 > \mu > (3+\alpha)/2$

$$(2-\mu)/(3+\alpha-\mu) \leq \gamma_* \leq \gamma^* \leq (\mu - 1 - \alpha)/\mu,$$

(iii) *for* $\mu > 2$

$$(1-\alpha)^2/4 \leq \gamma_* \leq \gamma^* \leq (1-\alpha)/2.$$

Proof Throughout the proof we preserve the notation from Sect. 12.6. For any $\varphi \in \mathscr{K}_0$ we have

$$\|f_\varphi\|_{1,\lambda} = \big\langle(\lambda - L)f_\varphi, f_\varphi\big\rangle_\pi = \int_{\mathbb{R}^d} c_{\lambda,\mu}(\xi)\hat{R}_{\hat{\varphi}(\xi)}(\xi)d\xi,$$

where $c_{\lambda,\mu}(\xi) = c_{\lambda,\mu}(\xi, 1)$ and

$$c_{\lambda,\mu}(\xi, z) := \lambda + |\xi|^{\mu} + \frac{z}{2}|\xi|^2.$$

To compute $\|Af_\varphi\|_{-1,\lambda}$ note that the solution of $(\lambda - S)g = f_\varphi$ is given by

$$g = i\sum_{p,q=1}^{d}\iint_{\mathbb{R}^{2d}} b_{\lambda,\mu}^{-1}(\xi,\xi')\hat{\varphi}_p(\xi)\xi_q\hat{V}_p(d\xi)\hat{V}_q(d\xi'),$$

where $b_{\lambda,\mu}(\xi,\xi') := b_{\lambda,\mu}(\xi,\xi',1)$ and

$$b_{\lambda,\mu}(\xi,\xi',z) := \lambda + |\xi|^\mu + |\xi'|^\mu + \frac{z}{2}|\xi+\xi'|^2.$$

Then, repeating the relevant part of calculations from Sect. 12.6, we get

$$\begin{aligned}\|Af_\varphi\|^2_{-1,\lambda} = &\int\int_{\mathbb{R}^{2d}} \hat{R}_{\hat{\varphi}(\xi)}(\xi)\hat{R}_\xi(\xi')b^{-1}_{\lambda,\mu}(\xi,\xi')\,d\xi\,d\xi' \\ &+ \int\int_{\mathbb{R}^{2d}} \hat{R}_{\hat{\varphi}(\xi),\xi'}(\xi)\hat{R}_{\xi,\varphi(\xi')}(\xi')b^{-1}_{\lambda,\mu}(\xi,\xi')\,d\xi\,d\xi'.\end{aligned}$$

Therefore

$$\langle V_p, \chi_\lambda^{(p)}\rangle_\pi \ge d^*(\lambda) := \sup_{\varphi\in\mathscr{K}} \mathscr{G}(\varphi), \tag{12.109}$$

where

$$\mathscr{G}(\varphi) := 2\int_{\mathbb{R}^d} \hat{R}_{\hat{\varphi}(\xi),p}(\xi)\,d\xi - \int_{\mathbb{R}^d} K(\xi)\left|\hat{\varphi}(\xi)\right|^2 d\xi,$$

with $K(\xi) = r(|\xi|)\mathscr{H}(\lambda,\xi)$ and

$$\mathscr{H}(\lambda,\xi) := c_{\lambda,\mu}(\xi') + 2\int_{\mathbb{R}^d} b^{-1}_{\lambda,\mu}(\xi,\xi')\hat{R}_\xi(\xi')d\xi'. \tag{12.110}$$

The supremum on the right-hand side of (12.109) is attained and can be explicitly calculated by solving the respective Euler–Lagrange equation. It equals

$$\int_{\mathbb{R}^d} \mathscr{H}^{-1}(\lambda,\xi)\hat{R}_{pp}(\xi)d\xi. \tag{12.111}$$

In case (i) (then necessarily $\mu < 2$), the change of variables

$$\tilde{\xi} := \lambda^{-1/\mu}\xi \quad \text{and} \quad \tilde{\xi}' := \lambda^{-1/\mu}\xi'$$

gives (tilde sign is omitted)

$$d^*(\lambda) := \lambda^{(\alpha+1-\mu)/\mu}\int_{\mathbb{R}^d} \widetilde{\mathscr{H}}^{-1}(\lambda,\xi)\Gamma_{pp}(\xi)|\xi|^{\alpha+1-d}a\left(|\xi|\lambda^{1/\mu}\right)d\xi, \tag{12.112}$$

where

$$\begin{aligned}\widetilde{\mathscr{H}}(\lambda,\xi) := &\, c_{1,\mu}\left(\xi,\lambda^{2/\mu-1}\right) + \lambda^{(\alpha+3-2\mu)/\mu} \\ &\times \int_{\mathbb{R}^d} b^{-1}_{1,\mu}\left(\xi,\xi',\lambda^{2/\mu-1}\right)a\left(|\xi'|\lambda^{1/\mu}\right)|\xi'|^{\alpha+1-d}\Gamma_\xi(\xi')d\xi'.\end{aligned} \tag{12.113}$$

We have

$$\lim_{\lambda\to 0+} \widetilde{\mathscr{H}}(\lambda,\xi) := 1+|\xi|^{\mu}$$

thus,

$$d^*(\lambda) \sim \lambda^{(\alpha+1-\mu)/\mu}, \quad \text{as } \lambda \to 0+.$$

In case (ii) we substitute $\tilde{\xi}' := |\xi|^{-1}\xi'$ in the integral appearing on the right-hand side of (12.113) and conclude that there exists a constant C such that

$$\widetilde{\mathscr{H}}(\lambda,\xi) \le C\big[c_{1,\mu}\big(\xi,\lambda^{2/\mu-1}\big) + \lambda^{(\alpha+3-2\mu)/\mu}|\xi|^{3+\alpha-\mu}\big].$$

As a result we obtain the lower bound

$$d^*(\lambda) \ge C\lambda^{(\alpha+1-\mu)/\mu} \times \int_0^{+\infty} \big(1+u^{\mu}+\lambda^{2/\mu-1}u^2+\lambda^{(\alpha+3-2\mu)/\mu}u^{3+\alpha-\mu}\big)^{-1}u^{\alpha}a\big(u\lambda^{1/\mu}\big)du \tag{12.114}$$

for some $C>0$. The right-hand side of (12.114) can be estimated from below by

$$C\lambda^{(\alpha+1-\mu)/\mu}\int_0^1 \big(1+\lambda^{(\alpha+3-2\mu)/\mu}u^{3+\alpha-\mu}\big)^{-1}u^{\alpha}a\big(u\lambda^{1/\mu}\big)\,du.$$

Substitute $\tilde{u} := \lambda^{(\alpha+3-2\mu)/[\mu(\alpha+3-\mu)]}u$ to conclude that this expression is bounded from below by $C\lambda^{(\mu-2)/(\alpha+3-\mu)}$ for some $C>0$.

Finally, in case (iii) we use the substitution

$$\tilde{\xi} := \lambda^{-1/2}\xi \quad \text{and} \quad \tilde{\xi}' := \lambda^{-1/2}\xi'$$

in the integrals appearing in (12.110) and (12.111). We obtain then

$$d_*(\lambda) \ge C\lambda^{(\alpha-1)/2}\int_{\mathbb{R}^d} \widetilde{\mathscr{H}}_1^{-1}(t,\xi)a\big(\lambda^{1/2}|\xi|\big)|\xi|^{\alpha+1-d}\Gamma_{pp}(\xi)\,d\xi,$$

where

$$\widetilde{\mathscr{H}}_1(t,\xi) := c_{1,\mu}\big(\xi,\lambda^{2/\mu-1}\big) + 2\lambda^{(\alpha-1)/2}|\xi|^2\|a\|_{\infty}\int_0^{+\infty}\frac{u^{\alpha}\,du}{1+(u-|\xi|)^2}.$$

From here on the estimates are similar to those made after (12.62) in Sect. 12.6 and they lead to the lower bound $d^*(\lambda) \ge C\lambda^{(\alpha-1)/2}$ for some $C>0$. The lower bounds for γ_* follows then from the bounds on $d^*(\lambda)$, estimate (12.109) and an application of Proposition 4.9.

To obtain an upper bound in all considered cases we use (12.65) and conclude that $\langle E_{\mathbb{P}}X_t^2\rangle_{\mathbb{Q}}$ can be estimated by

$$Ct\int_{\mathbb{R}^d} c_{1/t,\mu}^{-1}(\xi)\rho\big(|\xi|\big)d\xi.$$

This expression is of order $t^{1+(\mu-1-\alpha)/\mu}$, as $t \to +\infty$, when $\mu < 2$ and of order $t^{1+(1-\alpha)/2}$, when $\mu > 2$ (see (12.66)). □

12.13 Proof of Part (2) of Theorem 12.13

Throughout this section it is assumed that molecular diffusivity κ vanishes. Space $\mathcal{K}_0$ is defined in the same way as in Sect. 12.8.1. Recall that $\mathscr{C} := \mathscr{G}^{reg}$ (see Sect. 12.10) is a core of the generator L of the environment process. Part (2) of Theorem 12.13 can be concluded from Theorem 2.23.

First note that (2.39) is satisfied. Indeed, suppose that $f = \prod_{i=1}^{m} f_{\varphi_i}$, where $\varphi_i \in \mathcal{K}_0$. Thanks to Theorem 3.4, p. 24 of Janson (1997) we have $\Pi_n f \in \mathscr{C}$. Using an induction argument we conclude that $\Pi_n f \in \mathscr{C}$ for an arbitrary element $f \in \mathscr{C}$.

Conditions (2.39), (2.40) and (2.50) are obvious consequences of Proposition 12.20. It remains yet to conclude the graded sector condition. It is a consequence of the following.

Theorem 12.27 *Suppose that both* (12.73) *and* (12.74) *hold. Then, generator L satisfies graded sector condition* (2.45) *with* $\beta = 1/2$.

Proof The conclusion of the theorem is a consequence of Lemma 12.25 and the following estimate: there exists a constant $C > 0$ such that

$$\|\nabla f\|_\pi \le C n^{1/2} \langle (-L_0) f, f \rangle_\pi^{1/2} \quad \forall n \ge 1,\ f \in \mathscr{A}_n \cap \mathscr{C}. \tag{12.115}$$

Indeed, using (12.98) (the estimate holds even for $\kappa = 0$) we obtain for $f, g \in \mathscr{A}_n \cap \mathscr{C}$

$$\begin{aligned} \langle (-A) f, g \rangle_\pi &\le C \|\nabla f\|_\pi \langle (-L_0) g, g \rangle_\pi^{1/2} \\ &\stackrel{(12.115)}{\le} C n^{1/2} \langle (-L_0) f, f \rangle_\pi^{1/2} \langle (-L_0) g, g \rangle_\pi^{1/2} \\ &= C n^{1/2} \langle (-S) f, f \rangle_\pi^{1/2} \langle (-S) g, g \rangle_\pi^{1/2}. \end{aligned}$$

The last equality follows from part (ii) of Proposition 12.20 applied for $\kappa = 0$.

Therefore, what remains to be shown is (12.115). It is enough to prove that, cf. (12.88),

$$\left\| \nabla^{(N)} g \right\|_{\pi_N} \le C n^{1/2} \mathscr{E}_N^{1/2}(g), \quad \forall g \in \mathscr{A}_n^{(N)}, \tag{12.116}$$

where the constant $C > 0$ is independent of n, N.

Let $f \in \tilde{\mathscr{A}}_n^{(N)}$ be given by (12.104). Using Hermite polynomial identities (12.155) and (12.156) we conclude that

$$\nabla^{(N)} f = \sum_{|\mathbf{n}|=n} \alpha(\mathbf{n}) \sum_{(\mathbf{j},p)} \xi_{\mathbf{j}} \left(\varepsilon^{(1)}_{\mathbf{n},\mathbf{j},p} h^{(1)}_{\mathbf{n},\mathbf{j},p} - \varepsilon^{(2)}_{\mathbf{n},\mathbf{j},p} h^{(2)}_{\mathbf{n},\mathbf{j},p} \right), \tag{12.117}$$

with

$$\varepsilon^{(i)}_{\mathbf{n},\mathbf{j},p} := \begin{cases} \sqrt{(n_{\mathbf{j},p}+1)m_{\mathbf{j},p}}, & \text{when } i=1, \\ \sqrt{n_{\mathbf{j},p}(m_{\mathbf{j},p}+1)}, & \text{when } i=2 \end{cases}$$

and

$$h^{(i)}_{\mathbf{n},\mathbf{j},p}(\mathbf{a}) := \begin{cases} h_{\mathbf{n}^{+,-}_{\mathbf{j},p}}(\mathbf{a}), & \text{when } i=1, \\ h_{\mathbf{n}^{-,+}_{\mathbf{j},p}}(\mathbf{a}), & \text{when } i=2. \end{cases}$$

Here, given $\mathbf{n} = \{(n_{\mathbf{j}}, m_{\mathbf{j}}), \mathbf{j} \in \Lambda_N^+\}$, a multi-index $\mathbf{i}$ and $p = 1, \dots, n_{\mathbf{i}}$, we define $\mathbf{n}^{\pm,\mp}_{\mathbf{i},p} = \{(n'_{\mathbf{j}}, m'_{\mathbf{j}}), \mathbf{j} \in \Lambda_N^+\}$ as follows: $(n'_{\mathbf{j}}, m'_{\mathbf{j}}) = (n_{\mathbf{j}}, m_{\mathbf{j}})$ for $\mathbf{j} \neq \mathbf{i}$. When, on the other hand, $\mathbf{j} = \mathbf{i}$ we let $(n'_{\mathbf{i},q}, m'_{\mathbf{i},q}) = (n_{\mathbf{i},q}, m_{\mathbf{i},q})$ when $q \neq p$ and $(n'_{\mathbf{i},p}, m'_{\mathbf{i},p}) = (n_{\mathbf{i},p} \pm 1, m_{\mathbf{i},p} \mp 1)$. We have

$$\|\nabla^{(N)} f\|^2_{\pi_{N,*}} = \sum_{i,i'=1,2} \sum (-1)^{i+i'} \xi_{\mathbf{j}} \cdot \xi_{\mathbf{j}'} \alpha(\mathbf{n}) \alpha(\mathbf{n}') \varepsilon^{(i)}_{\mathbf{n},\mathbf{j},p} \varepsilon^{(i')}_{\mathbf{n}',\mathbf{j}',p'} \delta^{i',\mathbf{n}',\mathbf{j}',p'}_{i,\mathbf{n},\mathbf{j},p}.$$

Here,

$$\delta^{i',\mathbf{n}',\mathbf{j}',p'}_{i,\mathbf{n},\mathbf{j},p} := \langle h^{(i)}_{\mathbf{n},\mathbf{j},p}, h^{(i')}_{\mathbf{n}',\mathbf{j}',p'} \rangle_{\pi_{N,*}}.$$

The second summation spans over all possible $|\mathbf{n}| = |\mathbf{n}'| = n$, $\mathbf{j}, \mathbf{j}' \in \Lambda_N^+$, $p = 1, \dots, d_{\mathbf{j}}$ and $p' = 1, \dots, d_{\mathbf{j}'}$. Each term appearing under the summation over i, i' can be dealt with in the same manner so we show only how to estimate the one corresponding to $i, i' = 1$. We have

$$\begin{aligned} &\sum \xi_{\mathbf{j}} \cdot \xi_{\mathbf{j}'} \alpha(\mathbf{n}) \alpha(\mathbf{n}') \varepsilon^{(1)}_{\mathbf{n},\mathbf{j},p} \varepsilon^{(1)}_{\mathbf{n}',\mathbf{j}',p'} \delta^{1,\mathbf{n}',\mathbf{j}',p'}_{1,\mathbf{n},\mathbf{j},p} \\ &\quad \le \frac{1}{2} \sum \big\{ \big[\alpha(\mathbf{n}) |\xi_{\mathbf{j}}| \varepsilon^{(1)}_{\mathbf{n},\mathbf{j},p}\big]^2 + \big[\alpha(\mathbf{n}') |\xi_{\mathbf{j}'}| \varepsilon^{(1)}_{\mathbf{n}',\mathbf{j}',p'}\big]^2 \big\} \delta^{1,\mathbf{n}',\mathbf{j}',p'}_{1,\mathbf{n},\mathbf{j},p}. \end{aligned}$$

From the definition of $\varepsilon^{(1)}_{\mathbf{n},\mathbf{j},p}$ we can estimate the sum corresponding to the first term on the right-hand side by

$$\begin{aligned} &(n+1) \sum_{|\mathbf{n}|=n} \alpha^2(\mathbf{n}) \sum_{(\mathbf{j},p)} |m_{\mathbf{j}}| |\xi_{\mathbf{j}}|^2 \sum_{\mathbf{j}',p',\mathbf{n}'} \delta^{1,\mathbf{n}',\mathbf{j}',p'}_{1,\mathbf{n},\mathbf{j},p} \\ &\quad \le d(n+1) \sum_{|\mathbf{n}|=n} \alpha^2(\mathbf{n}) \sum_{\mathbf{j} \in \Lambda_N^+} |m_{\mathbf{j}}| |\xi_{\mathbf{j}}|^2. \end{aligned} \tag{12.118}$$

When $\mathbf{j}$ is such that $\square_{\mathbf{j},N}$ intersects the support of $r(d\xi)$ then, thanks to (12.74), we know that $|\xi_{\mathbf{j}}| \le C\gamma_{\mathbf{j}}$ for some constant $C > 0$, independent of $\mathbf{j}$. If, on the other hand, $\square_{\mathbf{j},N}$ lies outside the support of $r(d\xi)$ then $m_{\mathbf{j}} = 0$. In both cases we can estimate the right-hand side of (12.118) by

$$C(n+1)\sum_{|\mathbf{n}|=n}\alpha^2(\mathbf{n})\sum_{\mathbf{j}\in\Lambda_N^+}|m_{\mathbf{j}}|\gamma_{\mathbf{j}}\le C(n+1)\mathscr{E}_U^{(N)}(f) \tag{12.119}$$

for some constant $C>0$. The last inequality follows from formula (12.105). Estimate (12.116) follows then upon the choice $g:=U^{-1}f$. □

12.14 Proofs of the Results from Sect. 12.8

12.14.1 Construction of the Semigroup

To construct the semigroup we use the approach taken in Janson (1997), see pp. 50–53. For any $t\ge 0$ we define

$$\widehat{S_t\varphi}(\xi)=e^{-\gamma(\xi)t}\hat{\varphi}(\xi),\quad \varphi\in\mathscr{K}_0.$$

Proposition 12.28 $\{S_t,t\ge 0\}$ *extends to a strongly continuous, semigroup of contractions on* $\mathscr{K}$.

Proof Note that

$$\|S_t\varphi\|_{\mathscr{K}}^2=\int_{\mathbb{R}^d}e^{-2\gamma(\xi)t}\hat{R}_{\hat{\varphi}(\xi)}(d\xi)\le\int_{\mathbb{R}^d}\hat{R}_{\hat{\varphi}(\xi)}(d\xi)=\|\varphi\|_{\mathscr{K}}^2.$$

Verification of continuity of the semigroup is straightforward. □

Let $\mathbb{R}_+^{1+d}:=[0,+\infty)\times\mathbb{R}^d$. Suppose that

$$W(t,x)=\big(W_1(t,x),\dots,W_d(t,x)\big),\quad (t,x)\in\mathbb{R}_+^{1+d}$$

is a stationary in x, zero mean, Gaussian, random field over $(\Omega,\mathscr{F},\mathbb{Q})$ whose covariance matrix is given by $R^{(W)}(t_1,t_2,x):=[R_{pq}^{(W)}(t_1,t_2,x)]$, $p,q=1,\dots,d$ with

$$\begin{aligned}R_{pq}^{(W)}(t_1,t_2,x_1-x_2)&:=\big\langle W_p(t_1,x_1)W_q(t_2,x_2)\big\rangle_{\mathbb{Q}}\\&=\int_{\mathbb{R}^d}e^{i\xi\cdot(x_1-x_2)}\big[e^{-\gamma(\xi)|t_1-t_2|}-e^{-\gamma(\xi)(t_1+t_2)}\big]\hat{R}_{pq}(d\xi),\end{aligned}$$

for any $(t_k,x_k)\in\mathbb{R}_+^{1+d}$, $k=1,2$. Let φ be C^∞ smooth and compactly supported. Then, $W_\varphi(t):=\langle\varphi,W(t)\rangle$. An elementary calculation shows that

$$\|W_\varphi(t)\|_{\mathbb{Q}}^2=\int_{\mathbb{R}^d}\big[1-e^{-2\gamma(\xi)t}\big]\hat{R}_{\hat{\varphi}(\xi)}(d\xi)\le\|\varphi\|_{\mathscr{K}}^2. \tag{12.120}$$

Lemma 12.29

(i) *For any* $t \geq 0$ *and* $p \in [1, +\infty)$ *the mapping* $\varphi \mapsto W_\varphi(t)$ *extends to a bounded mapping from* $\mathscr{K}$ *to* $L^p(\mathbb{Q})$.

(ii) *For any* $t, s \geq 0$, $n \geq 1$ *and* $\varphi_1, \dots, \varphi_n \in \mathscr{K}$ *we have the following formula*

$$\left\langle \prod_{i=1}^n W_{\varphi_i}(t+s) \right\rangle_{\mathbb{Q}} = \sum \left\langle \prod_{i \in \Gamma} W_{\varphi_i}(t) \right\rangle_{\mathbb{Q}} \left\langle \prod_{i \in \Gamma^c} W_{S_s \varphi_i}(t) \right\rangle_{\mathbb{Q}}. \tag{12.121}$$

The summation extends over all partitions of $\{1, \dots, n\}$ *into two disjoint subsets* Γ, Γ^c. *By convention the product corresponding to an empty set is equal to* 1.

Proof Part (i) follows from (12.120) by a standard density argument and elementary properties of Gaussian random variables. We prove now (12.121). Computation of covariance allows us to conclude that the random vectors

$$\left(W_{\varphi_1}(t), \dots, W_{\varphi_n}(t) \right)$$

and

$$\left(W_{\psi_1}(t+s) - W_{S_s \psi_1}(t), \dots, W_{\psi_n}(t+s) - W_{S_s \psi_n}(t) \right)$$

are independent for any $t, s \geq 0$ and $\varphi_p, \psi_q \in \mathscr{K}$ for $p, q = 1, \dots, d$. Writing

$$W_{\varphi_p}(t+s) = U_p + V_p,$$

where

$$U_p = W_{\varphi_p}(t+s) - W_{S_s \varphi_p}(t) \quad \text{and} \quad V_p = W_{S_s \varphi_p}(t)$$

and using independence of the relevant random variables we can rewrite the left-hand side of (12.121) as being equal to

$$\sum \left\langle \prod_{p \in \Gamma} U_p \right\rangle_{\mathbb{Q}} \left\langle \prod_{p \in \Gamma^c} V_p \right\rangle_{\mathbb{Q}}$$

where the summation range is identical with the one on the right-hand side of (12.121). The formula follows then from the fact that the processes

$$\left\{ \left(W_{\psi_1}(t+s) - W_{S_s \psi_1}(t), \dots, W_{\psi_n}(t+s) - W_{S_s \psi_n}(t) \right), t \geq 0 \right\}$$

and

$$\left\{ \left(W_{S_s \psi_1}(t), \dots, W_{S_s \psi_n}(t) \right), t \geq 0 \right\}$$

have identical laws for any $s \geq 0$. Since they are both Gaussians this can be seen by comparing their respective covariances. □

Step 1. Construction for monomials.

Denote by $\mathscr{M}$ the class of all monomials, i.e. elements of the form

$$f = \prod_{i=1}^{n} f_{\varphi_i}, \quad \forall \varphi_1, \ldots, \varphi_n \in \mathscr{K}. \tag{12.122}$$

For any such f and $t \geq 0$ we let

$$f_{t,\varphi} := f_{S_t\varphi} + W_\varphi(t), \quad \forall \varphi \in \mathscr{K}, \tag{12.123}$$

and

$$R_t f := \left\langle \prod_{i=1}^{n} f_{t,\varphi_i} \right\rangle_{\mathbb{Q}}. \tag{12.124}$$

Since $W_\varphi(0) = 0$ note that R_0 coincides with the identity mapping.

Proposition 12.30 *For any $f, g \in \mathscr{M}$ and $t, h \geq 0$ formulas* (12.77) *and* (12.78) *hold.*

Proof Assume that $\varphi_1, \ldots, \varphi_{n+m} \in \mathscr{K}_0$ and

$$f = \prod_{j=1}^{m} f_{\varphi_{j+n}}.$$

We suppose also that the field $\{W(t,x), (t,x) \in \mathbb{R}_+^{1+d}\}$ is independent of $\{V_t, t \in \mathbb{R}\}$.

For $0 \leq s_1 \leq \cdots \leq s_n \leq t$ we let

$$X_i = Y_i := f_{\varphi_i}(V_{s_i}) \quad \forall i = 1, \ldots, n$$

and

$$X_{j+n} := f_{\varphi_{j+n}}(V_{t+h}), \qquad Y_{j+n} := f_{h,\varphi_{j+n}}(V_t) \quad \forall j = 1, \ldots, m.$$

Formula (12.77) can be easily concluded, provided we show the following equality

$$\left\langle \prod_{i=1}^{n+m} X_i \right\rangle_{\mathbb{Q}} = \left\langle \prod_{i=1}^{n+m} Y_i \right\rangle_{\mathbb{Q}}. \tag{12.125}$$

Indeed, assume that the above equality holds. Note that its left-hand side equals

$$\left\langle \prod_{i=1}^{n} X_i f(V_{t+h}) \right\rangle_{\mathbb{Q}}.$$

On the other hand, the right-hand side is equal to

$$\left\langle \prod_{i=1}^{n} X_i \prod_{i=1}^{m} [f_{S_h \varphi_{i+n}}(V_t) + W_{\varphi_{i+n}}(h)] \right\rangle_{\mathbb{Q}}.$$

Opening the square brackets and using independence of the relevant processes we can write that this expression equals

$$\sum_{\Gamma} \left\langle \prod_{i=1}^{n} X_i \prod_{i\in\Gamma} f_{S_h \varphi_{i+n}}(V_t) \right\rangle_{\mathbb{Q}} \left\langle \prod_{i\in\Gamma^c} W_{\varphi_{i+n}}(h) \right\rangle_{\mathbb{Q}},$$

where the summation extends over all subsets Γ of $\{1, \dots, n\}$. Combining the terms entering into the summation we conclude that the above expression coincides with

$$\left\langle \prod_{i=1}^{n} X_i R_h f(V_t) \right\rangle_{\mathbb{Q}}.$$

To show formula (12.125) we calculate the expectations appearing on the left- and right-hand sides of (12.77) using the rules for computation of moments of Gaussian variables. These involve taking all possible pairings formed over the indices $1, \dots, m+n$, see Janson (1997), Theorem 1.36, p. 16. The proof can be reduced therefore to checking that

$$\langle X_i X_j \rangle_{\mathbb{Q}} = \langle Y_i Y_j \rangle_{\mathbb{Q}} \quad \forall i, j = 1, \dots, n+m.$$

These equalities can be seen by a direct calculation. Likewise, we can verify formula (12.78). □

Step 2. Extension of the definition of the semigroup to $L^p(\pi)$. Proofs of parts (i)–(iv) of Proposition 12.14.

By linearity each R_t can be extended to span($\mathcal{M}$) consisting of all linear combinations of monomials by letting

$$R_t f := \sum_{i=1}^{n} R_t f_i \quad \text{for } f = \sum_{i=1}^{n} f_i,$$

where $f_i \in \mathcal{M}$. The extension is well defined. Indeed, suppose that $g, f_1, \dots, f_n$ are monomials and

$$0 = \sum_{i=1}^{n} f_i.$$

Then, by linearity and the already proved formula (12.78) for monomials we obtain

$$\left\langle g, R_t\left(\sum_{i=1}^n f_i\right)\right\rangle_\pi = \sum_{i=1}^n \langle g, R_t f_i\rangle_\pi$$
$$= \sum_{i=1}^n \langle R_t g, f_i\rangle_\pi = \langle R_t g, 0\rangle_\pi = 0.$$

This of course implies

$$R_t\left(\sum_{i=1}^n f_i\right) = 0.$$

The statement of Proposition 12.30 can be easily extended to span($\mathscr{M}$). Furthermore, using Jensen's inequality for conditional expectations, we obtain that for any $p \in [1, +\infty)$

$$\|R_t f\|^p_{L^p(\pi)} \le \left\langle |f(V_t)|^p\right\rangle_{\mathbb{Q}} = \left\langle |f|^p\right\rangle_\pi, \quad \forall f \in \operatorname{span}(\mathscr{M}).$$

By density each R_t extends to the entire $L^p(\pi)$. The conclusion of part (i) of Proposition 12.14 follows easily from (12.77) and a standard density argument.

To prove (ii) we verify that $R_{t+s} = R_s R_t$, on $\mathscr{M}$ for all $t, s \ge 0$. This equality can be checked directly using definition of the semigroup and formula (12.121). Indeed, from (12.124) and (12.123) we conclude that for f of the form (12.122)

$$R_t f = \sum \left\langle \prod_{i\in\Gamma} W_{\varphi_i}(t)\right\rangle_{\mathbb{Q}} \prod_{i\in\Gamma^c} f_{S_t\varphi_i}. \tag{12.126}$$

The summation extends over all partitions of $\{1,\dots,n\}$ into two disjoint subsets Γ, Γ^c. By convention a product over an empty set of indices equals 1. From here

$$R_s R_t f = \sum \left\langle \prod_{i\in\Gamma} W_{\varphi_i}(t)\right\rangle_{\mathbb{Q}} R_s\left(\prod_{i\in\Gamma^c} f_{S_t\varphi_i}\right)$$
$$= \sum \left\langle \prod_{i\in\Gamma_1} W_{\varphi_i}(t)\right\rangle_{\mathbb{Q}} \left\langle \prod_{i\in\Gamma_2} W_{S_t\varphi_i}(s)\right\rangle_{\mathbb{Q}} \prod_{i\in\Gamma_3} f_{S_{t+s}\varphi_i},$$

where the summation extends over all partitions of $\{1,\dots,n\}$ into three disjoint subsets $\Gamma_1, \Gamma_2, \Gamma_3$. Using (12.121) we conclude that the utmost right-hand side of the above equality equals

$$\sum \left\langle \prod_{i\in\Gamma} W_{\varphi_i}(t+s)\right\rangle_{\mathbb{Q}} \prod_{i\in\Gamma^c} f_{S_{t+s}\varphi_i} \tag{12.127}$$

and the range of summation is as in (12.126). This expression equals $R_{t+s} f$ as can be easily seen from (12.124) and (12.123). The assertion on the strong continuity can be concluded easily on $\mathscr{M}$ directly from the definition of R_t. Generalization to $L^2(\pi)$ follows by density.

Part (iii) is a consequence of the fact that both measure π and random field $W(t,\cdot)$ are spatially homogeneous. Part (iv) follows straightforwardly from the definition of the semigroup, see (12.126).

We prove part (v). Fix $t>0$. Denote by H and K the closed linear subspaces of $L^2(\mathbb{Q})$ generated respectively by $\langle V_t,\varphi\rangle$ and $\{\langle V_s,\varphi\rangle, s\le 0\}$, where supp $\hat{\varphi}\subset Z$ and $\varphi\in\mathscr{K}$. Note that the mapping

$$R_t f_\varphi(V_0)=\big\langle\langle V_t,\varphi\rangle\big|\mathscr{F}_0\big\rangle_{\mathbb{Q}}$$

is the orthogonal projection of H onto K. We have

$$\|R_t f_\varphi\|_\pi^2=\int e^{-2\gamma(\xi)t}\hat{R}_{\hat{\varphi}(\xi)}(d\xi)\le e^{-2\gamma_*(Z)t}\|\varphi\|_{\mathscr{K}}^2=e^{-2\gamma_*(Z)t}\|f_\varphi\|_\pi^2.$$

As a consequence

$$\|R_t\|_{\mathscr{A}_1(Z)}\le e^{-\gamma_*(Z)t},\qquad \forall t\ge 0. \tag{12.128}$$

Here, for a given subspace $\mathscr{L}$ and an operator T, we denote $\|T\|_{\mathscr{L}}:=\sup_{f\in\mathscr{L}}\|Tf\|_\pi$. Estimate (12.128) generalizes to the spaces of Hermite polynomials of higher degree thanks to Theorems 4.4.5 and 4.4.8 of Janson (1997) and we obtain that

$$\|R_t\|_{\mathscr{A}_n(Z)}=\|R_t\|^n_{\mathscr{A}_1(Z)}\le e^{-n\gamma_*(Z)t},\qquad \forall t\ge 0.$$

Hence, (12.79) follows. □

12.14.2 Proof of Proposition 12.16

To show part (i) assume that there exists f such that $\langle f\rangle_\pi=0$ and $R_t f=f$ for any $t\ge 0$. Let $K\subset\mathbb{R}^d\setminus\{0\}$ be compact and $\gamma_*(K)>0$. Denote by Π_K the orthogonal projection of $L^2(\pi)$ onto $L^2_K(\pi)$. Since the space is invariant $R_t\Pi_K f=\Pi_K f$. We also have

$$\langle\Pi_K f\rangle_\pi=\langle\Pi_K 1,f\rangle_\pi=\langle 1,f\rangle_\pi=0.$$

Thus, by (12.79), we conclude that $\Pi_K f=0$ for any K compact. This in turn implies that f has to be constant, in fact equal to 0, and ergodicity follows.

Formula (12.80) can be shown by a direct calculation differentiating (12.124) with respect to t at 0. From the definition of R_t one can also easily conclude the invariance of $\mathscr{G}^{reg}$. Since the set is dense in $L^2(\pi)$ it is a core of the generator of the semigroup, see Ethier and Kurtz (1986), Proposition 1.3.3, p. 17. □

12.14.3 Proof of Proposition 12.17

For any $N\ge 1$ we let

$$\bar{B}_N:=\big[x:|x|\le N\big]\quad\text{and}\quad C_N:=\big[\omega:\big|\xi(\omega)\big|\le N\big].$$

Define $\rho_N^\omega(h)$ as the modulus of continuity of $x \mapsto \tilde{F}(t, x; \omega)$ on $\bar{B}_N$ for fixed $\omega \in \Omega$ and $h > 0$. Let $\mathfrak{Q}^d$ be the set of points from $\mathbb{R}^d$ with rational coordinates and

$$r_N^\omega(h) := \operatorname{ess\,sup}\big[|\mathfrak{E}_{\tau_x V_t} F - \mathfrak{E}_{\tau_y V_t} F| : |x - y| \le h, x, y \in \bar{B}_N \cap \mathfrak{Q}^d\big].$$

Since

$$r_N(h) \le \langle \rho_N(h) | \mathscr{F}_t \rangle_{\mathbb{Q}}, \quad \mathbb{Q} \text{ a.s.}$$

and $\tilde{F}(t, x; \omega)$ is uniformly continuous on $\bar{B}_N$ we conclude easily that

$$\lim_{h \to 0+} \langle r_N(h) \rangle_{\mathbb{Q}} = 0, \quad \forall N \ge 1. \tag{12.129}$$

For any $\varepsilon > 0$, $m \ge n$ and a fixed $N \ge 1$ we can write

$$\begin{aligned} \|X_n - X_m\|_{L^1(\mathbb{Q})} &\le \sum_k \sum_{k'} \left\langle \left| f\left(\frac{k}{2^n}\right) - f\left(\frac{k'}{2^m}\right) \right|, A_{k,n} \cap A_{k',m} \right\rangle_{\mathbb{Q}} \\ &\le \left\langle r_N\left(\frac{1}{2^n}\right) \right\rangle_{\mathbb{Q}} \mathbb{Q}(C_N) + 2\|F\|_\infty \mathbb{Q}(C_N^c). \end{aligned} \tag{12.130}$$

Choosing N sufficiently large and then adjusting suitably n we see that $\|X_n - X_m\|_{L^1(\mathbb{Q})} < \varepsilon$ for all $m \ge n$. The first part of the conclusion of the proposition follows from completeness of $L^1(\mathbb{Q})$.

Furthermore, we have

$$\lim_{n \to +\infty} \sum_k \tilde{F}(t, k/2^n) 1_{A_{k,n}} = \tilde{F}(t, \xi), \quad \mathbb{Q} \text{ a.s.}$$

Therefore, for any bounded and $\mathscr{F}_t$-measurable G we can write that

$$\langle \tilde{F}(t, \xi) G \rangle_{\mathbb{Q}} = \lim_{n \to +\infty} \left\langle \sum_k \tilde{F}(t, k/2^n) G, A_{k,n} \right\rangle_{\mathbb{Q}}. \tag{12.131}$$

By Fubini theorem and (12.81) we conclude therefore that the utmost right-hand side of (12.131) equals

$$\lim_{n \to +\infty} \sum_k \langle G \mathfrak{E}_{\tau_{k/2^n} V_t} F, A_{k,n} \rangle_{\mathbb{Q}} = \lim_{n \to +\infty} \left\langle \sum_k G \mathfrak{E}_{\tau_{k/2^n} V_t} F, A_{k,n} \right\rangle_{\mathbb{Q}}. \tag{12.132}$$

The expression on the utmost right-hand side of (12.132) can be written as

$$\langle \mathfrak{E}_{\tau_\xi V_t} F G \rangle_{\mathbb{Q}} + \lim_{n \to +\infty} I_n,$$

where

$$I_n := \left\langle \sum_k (\mathfrak{E}_{\tau_{k/2^n} V_t} F - \mathfrak{E}_{\tau_\xi V_t} F) G, A_{k,n} \right\rangle_{\mathbb{Q}}.$$

Estimating as in (12.130) we obtain

$$|I_n| \le \left\langle r_N\left(\frac{1}{2^n}\right)\right\rangle_{\mathbb{Q}} \|G\|_\infty \mathbb{Q}(C_N) + \|F\|_\infty \|G\|_\infty \mathbb{Q}\big(C_N^c\big).$$

Letting $n \to +\infty$ first and then subsequently $N \to +\infty$ we conclude that $\lim_{n\to+\infty} I_n = 0$. Thus (12.83) follows.

12.15 Proofs of the Results from Sect. 12.10

12.15.1 Proof of Proposition 12.19

For fixed $f, g_1, \ldots, g_n \in C_b(E)$ and times $0 \le t_1 \le \cdots \le t_n \le t$ we let

$$G := \prod_{i=1}^{n} g_i(\eta_{t_i}).$$

Denote by $S_t(\omega) := \tau_{t,X_t(\omega)}\omega$. A simple calculation shows that

$$X_{t+h}(\omega) = X_t(\omega) + \tilde{X}_h\big(S_t(\omega)\big), \quad \forall t, h \ge 0,$$

where $\{\tilde{X}_h, h \ge 0\}$ is the solution of (12.67) based on $\{w_{t,h}, t \ge 0\}$ and corresponding to the initial condition $(s, x) = (0, 0)$. We can write

$$\big\langle E_{\mathbb{P}}\big[f(\eta_{t+h})G\big]\big\rangle_{\mathbb{Q}} = \big\langle E_{\mathbb{P}}\big\{f\big(\tau_{\tilde{X}_h(S_t(\omega))} V_h\big(S_t(\omega)\big)\big)G\big\}\big\rangle_{\mathbb{Q}}. \tag{12.133}$$

We use Proposition 12.17 with

$$F(\omega) := f\big(\tau_{\tilde{X}_h(\omega)} V_h(\omega)\big), \quad \xi := X_t$$

and a "frozen" Brownian path $\{w_t, t \ge 0\}$. Since

$$x + \tilde{X}_h(\tau_{t,x}\omega) = X_{t+h}^{t,x}(\omega)$$

we have

$$F(\tau_{t,x}\omega) = f(\tau_{t+h,X_{t+h}^{t,x}(\omega)}\omega).$$

Standard regularity results for solutions of stochastic differential equations (ordinary differential equations in case $\kappa = 0$) imply that the mapping $x \mapsto F(\tau_{t,x}\omega)$ is continuous for each t. The right-hand side of (12.133) equals therefore

$$\big\langle E_{\mathbb{P}}\big\{\mathfrak{E}_{\eta_t}\big[f(\tau_{\tilde{X}_h} V_h)\big]G\big\}\big\rangle_{\mathbb{Q}} = \big\langle E_{\mathbb{P}}\big[P_h f(\eta_t)G\big]\big\rangle_{\mathbb{Q}}$$

and (12.94) follows.

To prove the invariance of π we assume first that $\kappa > 0$. Note that

$$\langle P_t f\rangle_\pi = \left\langle \int_{\mathbb{R}^d} F(\tau_{t,x}\omega) p^{\omega}_{0,0}(t,x)dx \right\rangle_{\mathbb{Q}}, \quad \forall f \in B_b(\Omega),\ t \geq 0, \tag{12.134}$$

where $F(\omega) := f(V_0(\omega))$. Stationarity and incompressibility of $\tilde{V}(t,x)$ implies then, see (11.28), that the right-hand side of (12.134) equals $\langle F\rangle_{\mathbb{Q}} = \langle f\rangle_\pi$. The proof in case $\kappa = 0$ can be obtained by approximation. Let $X_t^{(\kappa)}$ be the solution of (12.67) corresponding to molecular diffusivity $\kappa > 0$. Under the assumptions made about regularity of the drift it can easily be shown that

$$\lim_{\kappa \to 0+} \sup_{t\in[0,T]} \left|X_t^{(\kappa)} - X_t^{(0)}\right| = 0 \quad \text{a.s.}$$

From this we conclude that

$$\langle P_t f\rangle_\pi = \lim_{\kappa \to 0+} \left\langle F(\tau_{t,X_t^{(\kappa)}}\omega)\right\rangle_{\mathbb{Q}} = \langle f\rangle_\pi, \quad \forall f \in C_b(E). \tag{12.135}$$

Hence, by a density argument we can extend (12.135) to the entire $B_b(E)$ and invariance of π follows. This in turn implies the possibility of extension of the semigroup to $L^2(\pi)$. Strong continuity is clear on $C_b(\Omega)$. By density we conclude the property on the entire $L^2(\pi)$. □

12.15.2 *Proof of Proposition 12.20*

Proofs of parts (i) and (ii) Suppose first that $\kappa > 0$ and $f \in C_b^2(E)$. From Itô's formula we obtain that

$$\tilde{f}(X_h; V_h) = f(V_h) + \int_0^h \mathscr{L}_s^{\omega} \tilde{f}(X_s; V_h)ds + \kappa \int_0^h \nabla_x \tilde{f}(X_s; V_h)\cdot dw_s, \tag{12.136}$$

where $\mathscr{L}^{\omega}$ is the generator of the diffusion given by (12.67) and

$$\tilde{f}(x; v) := f(\tau_x v).$$

Applying first expectation $E_{\mathbb{P}}$ and then $\mathfrak{E}_v$ to both sides of (12.136) we obtain

$$P_h f(v) = R_h f(v) + \int_0^h \mathfrak{E}_v E_{\mathbb{P}}\big[\mathscr{L}_s^{\omega} \tilde{f}(X_s; V_h)\big]ds. \tag{12.137}$$

The above equality can be extended to any $f \in \widehat{\mathscr{C}}$ by approximation. From (12.137) we conclude easily, by a direct computation of $\lim_{h\to 0+} h^{-1}(P_h - I)f$, that f belongs to $D(L)$ and $Lf = (S + A)f$ with S and A given by (12.95). We have shown therefore that $\widehat{\mathscr{C}} \subset D(L)$.

An elementary calculation shows that $D^m f_\varphi = (-1)^{|m|} f_{D^m\varphi}$ in the $L^p(\mathbb{Q})$ sense for an arbitrary multi-index of non-negative integers m and $\varphi \in \mathscr{K}_0$. Hence $\mathscr{G}_1^{reg} \subset \widehat{\mathscr{C}}$. The fact that $\mathscr{G}_n^{reg} \subset \widehat{\mathscr{C}}$ for each n, thus also $\mathscr{C} \subset \widehat{\mathscr{C}}$, can be easily concluded by an application of standard rules of differentiation.

In the next step we prove that $\widehat{\mathscr{C}}$ is in fact a core of L. The density in $L^2(\pi)$ is standard therefore it suffices only to prove its invariance under $\{P_h, h \ge 0\}$. This fact holds for $\mathscr{C}_0$ and can be shown in the same way as in the proof of Lemma 12.9. It remains to prove that

$$P_t(\widehat{\mathscr{C}}) \subset D(L_0), \quad \forall t > 0$$

and

$$L_0 P_t = P_t L_0 + [P_t, L_1], \quad \text{on } \widehat{\mathscr{C}}. \tag{12.138}$$

Here

$$L_1 = \left(\kappa^2/2\right)\Delta + V \cdot \nabla \tag{12.139}$$

and the commutator of A and B is defined as $[A, B] := AB - BA$. □

Claim There exists a sequence of random vectors

$$\left\{V^{(n)} = \left(V_1^{(n)}, \dots, V_d^{(n)}\right), n \ge 1\right\},$$

whose components belong to $C_b^\infty(E)$, approximating V in any $L^p(\mathbb{Q})$ for $p \in [1, +\infty)$ and such that

$$\sum_{q=1}^{d} D_q V_q^{(n)} = 0.$$

In addition, the solutions $X_{t,n}^{s,x}$ of (12.67), corresponding to the drifts $\tilde{V}^{(n)}(t, x) := V^{(n)}(\tau_{t,x}\omega)$, converge to $X_t^{s,x}$ almost surely, uniformly on compact intervals.

Admitting this claim, its proof will be presented below, we show how to demonstrate (12.138). Suppose that $\{\eta_t^{(n)}, t \ge 0\}$ is the environment process corresponding to $V^{(n)}$ and $\{P_t^{(n)}, t \ge 0\}$ is the respective semigroup. We shall prove that $P_t^{(n)}(\widehat{\mathscr{C}}) \subset D(L_0)$ and

$$L_0 P_t^{(n)} = P_t^{(n)} L_0 + \left[P_t^{(n)}, L_1^{(n)}\right], \quad \text{on } \widehat{\mathscr{C}}, \tag{12.140}$$

with $L_1^{(n)}$ defined by a modification of formula (12.139) with V replaced by $V^{(n)}$. This in turn implies (12.138). Indeed, elementary regularity properties of solutions of stochastic differential equations allow us to conclude that $P_t^{(n)} f \to P_t f$ in $W_p^2(E)$. Hence,

$$\lim_{n\to+\infty} L_1^{(n)} P_t^{(n)} f = L_1 P_t f \quad \text{in } L^2(\mathbb{Q}), \quad \forall t \ge 0.$$

In consequence, from formula (12.140) we obtain that $\{L_0 P_t^{(n)} f, n \ge 1\}$ converges in $L^2(\mathbb{Q})$. Since L_0 is closed we conclude that $P_t f \in D(L_0)$ and (12.138) holds.

To show (12.140) assume first that $\kappa > 0$. Then,

$$R_h P_t^{(n)} f(v) = \mathfrak{E}_v\big[\tilde{f}\big(X^{h,0}_{t+h,n}; V_{t+h}\big)\big] = \mathfrak{E}_v\bigg[\int_{\mathbb{R}^d} f(\tau_z V_{t+h}) p^{\omega,n}_{h,t+h}(0,z)dz\bigg].$$

Here $p^{\omega,n}_{s,t}(x,z)$ is the probability density of $X^{s,x}_{t,n}$. Using Kolmogorov equations, both forward and backward, it can be expanded in h as follows

$$p^{\omega,n}_{h,t+h}(0,z) = p^{\omega,n}_{0,t}(0,z) + \int_0^h C_s(z;\omega)ds, \tag{12.141}$$

where

$$C_s(z;\omega) := -\mathscr{L}^{\omega,n}_s p^{\omega,n}_{s,t+h}(0,z) + \big(\mathscr{L}^{\omega,n}_{t+s}\big)^* p^{\omega,n}_{0,t+s}(0,z).$$

Here $\mathscr{L}^{\omega,n}_t$ is the generator of $X^{s,x}_{t,n}(\omega)$ and $(\mathscr{L}^{\omega,n}_t)^*$ is its formal adjoint. These operators act on the first and second spatial variables of the transition probability kernel, respectively. Using Gaussian bounds and continuity of the transition probability densities and their partials, see Theorem 4.5, p. 141 of Friedman (1975), together with (12.141) we conclude that

$$\begin{aligned} R_h P_t f(v) &= \mathfrak{E}_v\bigg[\int_{\mathbb{R}^d} f(\tau_z V_{t+h}) p^{\omega,n}_{0,t}(0,z)dz\bigg] \\ &\quad + h\mathfrak{E}_v\bigg[\int_{\mathbb{R}^d} C_0(z;\omega) f(\tau_z V_{t+h})dz\bigg] + o(h), \end{aligned} \tag{12.142}$$

where an expression denoted by $o(h)$ satisfies

$$\lim_{h\to 0+} \frac{1}{h}\big\|o(h)\big\|_\pi = 0.$$

Since $f \in D(L_0)$ the first term on the right-hand side of (12.142) equals

$$\mathfrak{E}_v\bigg[\int_{\mathbb{R}^d} L_0 f(\tau_z V_t) p^{\omega,n}_{0,t}(0,z)dz\bigg]h + o(h) = hP_t L_0 f + o(h).$$

On the other hand, the second term on the right-hand side of (12.142) equals

$$\begin{aligned} &h\bigg\{-L_1^{(n)} P_t^{(n)} f(v) + \mathfrak{E}_v\bigg[\int_{\mathbb{R}^d} \big(L_1^{(n)} f\big)(\tau_z V_t) p^{\omega,n}_{0,t}(0,z)dz\bigg]\bigg\} + o(h) \\ &\quad = h\big[P_t^{(n)}, L_1^{(n)}\big] f(v) + o(h) \end{aligned}$$

and formula (12.138) follows. To extend it to the case $\kappa = 0$ one can use an approximation argument and consider a sequence $\kappa_n \to 0+$, as $n \to +\infty$. Formula (12.138) follows then from the already established formulas for each $\kappa_n > 0$ and standard regularity results for solutions of stochastic differential equations. We leave it to the reader to fill in the details of this argument.

The proof that $L^* = S - A$ on $\widehat{\mathscr{C}}$ is analogous. It suffices only to show that for a given $T > 0$ the semigroup of adjoint operators $\{P_t^*, t \in [0, T]\}$ corresponds to the tracer dynamics governed by the process $\{-V_{T-t}, t \in [0, T]\}$. The ensuing calculations are very similar to the ones presented above.

Proof of the claim The argument is quite similar to the one used in the proof of Proposition 11.6. Define random vectors

$$U^{(n)} := \int_{[1/n \le |\xi| \le n]} \hat{V}(d\xi).$$

Note that $U^{(n)} \in C^\infty(E)$. Gaussianity of V and condition (12.72) guarantee that

$$\lim_{n \to +\infty} \left\| U^{(n)} - V \right\|_{L^p(\pi)} = 0.$$

In consequence

$$\lim_{n \to +\infty} \int_0^T \left\| \tilde{U}^{(n)}(t, \cdot) - \tilde{V}(t, \cdot) \right\|^p_{C^1(B_R)} dt = 0$$

for any $p > 1$ and $R, T > 1$. Here

$$\tilde{U}^{(n)}(t, x) := U^{(n)}(\tau_x V_t) \quad \text{and} \quad \tilde{V}(t, x) := V \circ \tau_{t,x}.$$

However, vectors $U^{(n)}$ need not be deterministically bounded. To make a necessary modification we construct stream matrices $h^{(n)} = [h_{lm}^{(n)}]$, $l, m = 1, \ldots, d$ that correspond to $U^{(n)}$ via formula (11.5). Their entries belong to $\widehat{\mathscr{C}}$. There exists therefore a sequence of random anti-symmetric random matrices $b^{(n)} = [b_{lm}^{(n)}]$, $l, m = 1, \ldots, d$ with entries in $C_b^\infty(E)$ such that $\|h_{lm}^{(n)} - b_{lm}^{(n)}\|_{W_p^k} \to 0$ for any $p \in [1, +\infty)$ and a non-negative integer k. Define $V^{(n)} = \nabla \cdot b^{(n)}$. Putting together these two approximations we conclude that the sequence $\{V^{(n)}, n \ge 1\}$ satisfies the assertions made in the claim. The statement about the convergence of the respective solutions of (12.67) follows from regularity properties of solutions of stochastic differential equations, cf. estimates leading to (11.81). □

Proof of part (iii) We show that for any $f \in \widehat{\mathscr{C}}$ there exists a sequence $\{f_n, n \ge 1\} \subset \mathscr{C}$ such that

$$(f_n, Lf_n) \to (f, Lf) \quad \text{and} \quad \big(f_n, L^* f_n\big) \to \big(f, L^* f\big), \quad \text{as } n \to +\infty, \tag{12.143}$$

in the epigraph norm. This implies the result claimed in this part of the proposition.

Thanks to part (iii) of Proposition 12.14 set $\widehat{\mathscr{C}}$ is invariant under each R_t, hence it is a core of its generator in any $L^p(\pi)$ space. From part (iv) of the proposition we conclude that the same also holds for $\mathscr{G}^{reg}$. The above proves that for any $f \in \widehat{\mathscr{C}}$ we can find a sequence $\{g_n, n \geq 1\} \subset \mathscr{G}^{reg}$ such that

$$\lim_{n\to+\infty}\left(\|g_n - f\|_{L^p(\pi)} + \|L_0 g_n - L_0 f\|_{L^p(\pi)}\right) = 0$$

for any $p \in [1, +\infty)$. We recall the "smearing" of a random field introduced in Sect. 9.3.2, i.e. for any random variable $h(v)$ and $\delta > 0$ let

$$h_\delta(v) := \int_{\mathbb{R}^d} \varphi(x) h(\tau_x v) dx.$$

Here $\varphi \in C_c^\infty(\mathbb{R}^d)$ is a non-negative function such that $\int_{\mathbb{R}^d} \varphi(x)dx = 1$. Since the transition operators corresponding to V_t commute with the spacial shifts, we have $L_0 h_\delta = (L_0 h)_\delta$ for any $h \in D(L_0)$. In addition,

$$\lim_{\delta\to 0+} \|h - h_\delta\|_{W_p^2} = 0, \quad \forall h \in W^{2,p}(E),$$

provided $p \in [1, +\infty)$. Setting $f_n := (g_n)_{\delta_n}$, where $\delta_n \to 0$, we obtain

$$\lim_{n\to+\infty}\left[\left\|(g_n)_{\delta_n} - f\right\|_{W_p^2} + \left\|(L_0 g_n)_{\delta_n} - L_0 f\right\|_{L^p(\pi)}\right] = 0.$$

We have shown therefore that sequence $\{f_n, n \geq 1\}$ satisfies (12.143). □

12.16 Appendix: Some Auxiliary Results About Gaussian Random Fields

12.16.1 Multiple Stochastic Integrals

Suppose that $\{V(\tau_x\omega), x \in \mathbb{R}^d\}$ is a Gaussian, homogeneous, random, real vector field over $(\Omega, \mathscr{F}, \mathbb{Q})$. Denote by

$$\hat{V}(d\xi) = \left(\hat{V}_1(d\xi), \ldots, \hat{V}_d(d\xi)\right)$$

the corresponding Gaussian vector valued spectral measure on $(\mathbb{R}^d, \mathscr{B}(\mathbb{R}^d))$ that satisfies $\hat{V}_i^*(d\xi) = \hat{V}_i(-d\xi)$. Its covariance tensor is given by a non-negative definite matrix valued function $R(x) = [R_{pq}(x)]$. By Bochner theorem there exists a Hermitian matrix valued, Borel measure $\hat{R}(d\xi) = [\hat{R}_{pq}(d\xi)]$, called *the structure matrix* of the field such that

$$R(x) = \int_{\mathbb{R}^d} e^{ix\cdot\xi} \hat{R}(d\xi).$$

Since the field is real valued the measure is complex even, i.e. $\hat{R}(-d\xi) = \hat{R}^*(d\xi)$.

In what follows we suppose that the structure measure $\hat{R}(d\xi)$ has density $\hat{R}(\xi) = [\hat{R}_{pq}(\xi)]$ with respect to the Lebesgue measure, i.e.

$$\hat{R}_{pq}(\xi)\delta(\xi+\xi')d\xi d\xi' = \langle \hat{V}_p(d\xi)\hat{V}_q(d\xi')\rangle_{\mathbb{Q}}.$$

For a given $n \geq 1$ define by $\mathfrak{F}_n$ the family of all possible pairings made of $\{1,\dots,2n\}$, i.e. partitions of the set into two-element subsets. Let $r(\xi) := \operatorname{tr}\hat{R}(\xi) \vee 1$. For a given integer $n \geq 1$ define a Borel measure on $(\mathbb{R}^d)^{2n}$

$$M_{2n}(d\xi_1,\dots,d\xi_{2n}) := \sum_{\mathscr{F}\in\mathfrak{F}_n}\prod_{\{p,q\}\in\mathscr{F}}\delta(\xi_p+\xi_q)\prod_{j=1}^{2n} r(\xi_j)d\xi_j.$$

By $\mathscr{L}_n^2$ we denote the completion of the space of all complex valued, bounded Borel measurable functions $\psi : (\mathbb{R}^d)^n \to (\mathbb{C}^d)^n$ in the norm

$$\|\psi\|_{\mathscr{L}_n^2}^2 := \int\cdots\int_{(\mathbb{R}^d)^{2n}} \left(\left|\psi(\xi_1,\dots,\xi_n)\right|^2 + \left|\psi(\xi_{n+1},\dots,\xi_{2n})\right|^2\right) M_{2n}(d\xi_1,\dots,d\xi_{2n}).$$

Suppose that $i \in \mathbb{Z}^d$ and $N \geq 1$. We define

$$\square_N^{(i)} := \left[\xi \in \mathbb{R}^d : 2^{-N} i_j \leq \xi_j < 2^{-N}(i_j+1), \forall j = 1,\dots,d\right]. \tag{12.144}$$

For $\mathbf{i} := (i_1,\dots,i_n) \in (\mathbb{Z}^d)^n$ let

$$\square_N^{(\mathbf{i})} := \square_N^{(i_1)} \times \cdots \times \square_N^{(i_n)}. \tag{12.145}$$

Denote by $\mathscr{D}$ the family of all such boxes. Its subfamily Π is called *an admissible dyadic partition* of $\mathbb{R}^{2nd}$ if

(P1) $\bigcup_{\square\in\Pi}\square = \mathbb{R}^{2nd}$,
(P2) for any two boxes $\square \neq \square' \in \Pi$ we have $\square \cap \square' = \emptyset$,
(P3) there exists $d_0 > 0$ such that $|\square| \geq d_0$ for all $\square \in \Pi$.

Here $|\square|$ stands for the volume of $\square$. A set function $c : \Pi \to \mathbb{C}^{nd}$ is called *admissible* if it vanishes for all but finitely many boxes. We denote by $\mathscr{A}$ the family of all admissible set functions.

For any multi-index $\mathbf{j} := (j_1,\dots,j_n)$ and $\square_N^{(\mathbf{i})}$ given by (12.145) we let

$$\hat{V}_{\mathbf{j}}[\square_N^{(\mathbf{i})}] = \prod_{p=1}^n \hat{V}_{j_p}[\square_N^{(i_p)}]. \tag{12.146}$$

Suppose that $c(\cdot)$ is admissible. We define then $\psi(\xi) := c(\square)$ for all $\xi \in \square$. With some abuse of terminology we call such a function admissible and denote by $\mathscr{E}_n \subset \mathscr{L}_n^2$ the space of all such functions. For any $\psi \in \mathscr{E}_n$ we define the n-tuple stochastic integral letting

$$\mathscr{I}(\psi) := \sum_{\square\in\Pi}\sum_{\mathbf{j}} c_{\mathbf{j}}(\square)\hat{V}_{\mathbf{j}}[\square]. \tag{12.147}$$

Here $c_{\mathbf{j}}(\square)$ are the components of $c(\square)$. We shall also write

$$\int\int_{\mathbb{R}^{nd}} \psi(\xi_1,\ldots,\xi_n)\hat{V}(d\xi_1)\otimes\cdots\otimes\hat{V}(d\xi_n)$$

as an alternative notation for $\mathscr{I}(\psi)$. Below, we list some of the properties of the multiple stochastic integral. They are elementary and obtained straightforwardly from the definition so we leave their verification to the reader.

Proposition 12.31

(i) *$\mathscr{E}_n$ is dense in $\mathscr{L}_n^2$ in the norm $\|\cdot\|_{\mathscr{L}_n^2}$.*
(ii) *The stochastic integral given by (12.147) is well defined, i.e. if there exist two admissible set functions c_1, c_2 corresponding to a given ψ then the respective definitions of the stochastic integrals are identical.*
(iii) *We have $\mathscr{I}(a_1\psi_1 + a_2\psi_2) = a_1\mathscr{I}(\psi_1) + a_2\mathscr{I}(\psi_2)$.*
(iv) *Suppose that $\psi_1,\ldots,\psi_n \in \mathscr{E}_1$. Then $\psi_1\otimes\cdots\otimes\psi_n \in \mathscr{E}_n$ and*

$$\mathscr{I}(\psi_1\otimes\cdots\otimes\psi_n) = \prod_{j=1}^n \mathscr{I}(\psi_j).$$

(v) *We have*

$$\begin{aligned}\langle\mathscr{I}(\psi^{(1)})\mathscr{I}^*(\psi^{(2)})\rangle_{\mathbb{Q}} = \sum_{\mathscr{F}\in\mathfrak{F}_n}\sum_{j_1,\ldots,j_{2n}=1}^{d}\int\int_{\mathbb{R}^{2nd}} \psi^{(1)}\otimes\left(\psi^{(2)}\right)^*(\xi_1,\ldots,\xi_{2n}) \\ \times \prod_{\{p,q\}\in\mathscr{F}} \hat{R}_{pq}(\xi_p)\delta\left(\xi_p + (-1)^{\bar{a}(p,q)}\xi_q\right)\prod_{j=1}^{2n} d\xi_j.\end{aligned} \tag{12.148}$$

The component of $\psi^{(1)}\otimes(\psi^{(2)})^* : \mathbb{R}^{2nd}\to\mathbb{C}^{2nd}$ corresponding to a multi-index $\mathbf{j} = (j_1,\ldots,j_{2n})$ is given by

$$\begin{aligned}&\left(\psi^{(1)}\otimes\left(\psi^{(2)}\right)^*\right)_{\mathbf{j}}(\xi_1,\ldots,\xi_{2n}) \\ &:= \psi^{(1)}_{j_1,\ldots,j_n}(\xi_1,\ldots,\xi_n)\left(\psi^{(2)}_{j_{n+1},\ldots,j_{2n}}\right)^*(\xi_{n+1},\ldots,\xi_{2n}).\end{aligned}$$

In addition, $\bar{a}(p,q) = 1$ if (p,q) intersects both $\{1,\ldots,n\}$ and $\{n+1,\ldots,2n\}$, and equals 0 otherwise.

As a direct consequence of property (v) and the definition of the norm on $\mathscr{L}_n$ we obtain.

Corollary 12.32 *The mapping* $\psi \mapsto \mathscr{I}(\psi)$ *is a continuous linear functional on* $\mathscr{E}_n$. *It extends, by continuity, to the entire* $\mathscr{L}_n^2$. *The extension shall be called an* n*-tuple stochastic integral with respect to the spectral measure* $\hat{V}(d\xi)$. *It inherits properties* (iii)–(v) *from Proposition* 12.31.

Proposition 12.33 *For any* $\psi \in \mathscr{L}_n^2$ *and* $x \in \mathbb{R}^d$ *we let*

$$U^x \psi(\xi_1, \dots, \xi_n) := \exp\left\{ i \left(\sum_{p=1}^n \xi_p \right) \cdot x \right\} \psi(\xi_1, \dots, \xi_n).$$

Then, for any $x \in \mathbb{R}^d$ *we have* $U^x \psi \in \mathscr{L}_n^2$ *and*

$$\mathscr{I}(\psi)(\tau_x \omega) = \mathscr{I}\big(U^x \psi\big)(\omega), \quad \mathbb{Q} \text{ a.s.} \tag{12.149}$$

Proof We have

$$V(\tau_{x+y}\omega) = \int_{\mathbb{R}^d} e^{i\xi\cdot(x+y)} \hat{V}(d\xi;\omega) = V\big(\tau_x(\tau_y\omega)\big) = \int_{\mathbb{R}^d} e^{i\xi\cdot x} \hat{V}(d\xi;\tau_y\omega)$$

for all $x, y \in \mathbb{R}^d$. Hence $\hat{V}(d\xi;\tau_y\omega) = e^{i\xi\cdot y}\hat{V}(d\xi;\omega)$ and (12.149) follows. □

12.16.2 Some Properties of Hermite Polynomials

Suppose that ν_* is a zero mean, unit variance, Gaussian measure on $\mathbb{R}$. We can define then, see Example 3.18, p. 28 of Janson (1997), a unique orthonormal base $\{h_n(x), n \ge 0\}$ such that:

(1) $h_n(x)$ is an n-th degree polynomial,
(2) $\langle h_n, h_m \rangle_{\nu_*} = \delta_{mn}$ for all $n, m \ge 0$ and
(3) the coefficient by the leading order term is positive.

The elements of the base are called *Hermite polynomials*. We also use the convention that $h_{-1}(x) \equiv 0$.

It is easy to observe that a function

$$H_n(x) := (-1)^n e^{x^2/2} \frac{d^n}{dx^n}\big(e^{-x^2/2}\big), \quad n \ge 1 \tag{12.150}$$

is an n-th degree polynomial, with the leading order coefficient equal to 1. For $n > m$ we have

$$\langle H_n, H_m \rangle_{\nu_*} = \frac{(-1)^n}{\sqrt{2\pi}} \int_{\mathbb{R}} \frac{d^n}{dx^n}\big(e^{-x^2/2}\big) H_m(x) dx = 0.$$

On the other hand, for $n = m$ we have $\langle H_n, H_n \rangle_{\nu_*} = n!$. Sometimes, in the literature, it is these functions that are called Hermite polynomials. Using (12.150) one gets $H_0(x) \equiv 1$, $H_1(x) = x$, $H_2(x) = x^2 - 1$ and more generally

$$H_n(x) = \sum_{r=0}^{[n/2]} (-1)^r \binom{n}{2r} (2r-1)!!x^{n-2r}, \quad n \geq 2.$$

Here by convention $(-1)!! := 1$. One can show (see (3.19), p. 33 of Janson, 1997), that

$$e^{tx-t^2/2} = \sum_{n=0}^{+\infty} \frac{H_n(x)t^n}{n!}, \quad \forall t, x \in \mathbb{R}. \tag{12.151}$$

Differentiating both sides of (12.151) with respect to x and comparing the expressions obtained in that way we can conclude the recursion relation

$$H_{n+1}(x) = xH_n(x) - nH_{n-1}(x). \tag{12.152}$$

Suppose that $n \geq 2$. It is easy to see that $\langle H_n', H_m \rangle_{\nu_*} = 0$ for all $m \leq n-2$. Since H_n' is a polynomial of degree $n-1$ with the leading coefficient n we have

$$H_n'(x) = nH_{n-1}(x) \tag{12.153}$$

for $n \geq 2$. A direct calculation shows that this identity also holds for $n = 1$.

The uniqueness of the base of orthonormal polynomials satisfying conditions (1), (2) and (3) implies that

$$h_n(x) = \frac{1}{\sqrt{n!}} H_n(x). \tag{12.154}$$

Therefore, from (12.152) we get

$$xh_n(x) = (n+1)^{1/2} h_{n+1}(x) + n^{1/2} h_{n-1}(x) \tag{12.155}$$

and from (12.153) we obtain

$$h_n'(x) = n^{1/2} h_{n-1}(x) \tag{12.156}$$

for $n \geq 1$.

Finally, if $n = (n_1, \dots, n_d)$ is a non-negative integer valued multi-index we define a corresponding Hermite polynomial by $h_n(x_1, \dots, x_d) := \prod_{j=1}^d h_{n_j}(x_j)$.

12.17 Comments and References

The proof of Theorem 12.6 presented here appeared in Komorowski and Olla (2003b), where Theorem 12.13 has also been shown. This result has been obtained with a different technique in Fannjiang and Komorowski (2002), see also Fannjiang and Komorowski (2001/2002). As a further application of the graded sector condition one can show the regularity of the effective diffusivity tensor with respect to a perturbation of the drift, see Theorem 5.1 in Komorowski and Olla (2003b).

The central limit theorem for random motions, as well as random diffusions, in a flow that is Markovian and whose dynamics has the spectral gap, stated in Corollary 12.23, has been shown for Gaussian fields with finitely many spatial modes in Carmona and Xu (1997). This fact has been generalized to Gaussians with infinitely many modes in dimension two in Koralov (1999) and to a general Markovian flow with the spectral gap in Fannjiang and Komorowski (1999b).

An example of a shear layer flow that corresponds to a superdiffusive behavior of a tracer has been given earlier in Matheron and DeMarsily (1980). A problem of finding the scaling limits of solutions to an advection diffusion equation in a shear layer Gaussian flow in two dimensions has been investigated in Avellaneda and Majda (1990). This model can be described as follows. Suppose that $T_\delta(t,x,y)$ satisfies

$$\begin{aligned} &\partial_t T_\delta(t,x,y) + v_\delta(t,x)\partial_y T_\delta(t,x,y) = \nu \Delta T_\delta(t,x,y),\\ &T_\delta(0,x,y) = T_0(\delta x, \delta y), \end{aligned} \tag{12.157}$$

where $\nu > 0$, $T_0(x,y)$ is a deterministic initial condition and $v_\delta(t,x)$ is a time space stationary Gaussian random field whose covariance function is given by

$$R(t,x) = \langle v_\delta(t,x) v_\delta(0,0)\rangle_{\mathbb{Q}} = \int_{\mathbb{R}} e^{-|k|^z t + ikx} \frac{a(k)b(|k|/\delta)}{|k|^{\varepsilon-1}} dk. \tag{12.158}$$

Functions $a(k)$ and $b(k)$ are non-negative and compactly supported. The first one is continuous at 0 and such that $a(0) > 0$, while $b(k) \equiv 0$ in a neighborhood of 0. Time independent drifts can be also admitted and then the necessary adjustment in the formula for the covariance function should be made. One should simply drop the exponential involving time argument. Using explicit calculations it is possible to find an exact value of the exponent $\gamma = \gamma(\varepsilon, z)$, for which the limit

$$\lim_{\delta\to 0+} \left\langle T_\delta\left(\frac{t}{\delta^\gamma}, \frac{x}{\delta}, \frac{y}{\delta}\right)\right\rangle_{\mathbb{Q}} = \bar{T}(t,x,y)$$

exists. The complete diagram illustrating the dependence of γ on the parameters of the spectrum together with exact description of the limit has been given. The dependence of the diagram on the infrared cutoff function, represented in (12.158) by $b(\cdot)$, has been considered in Zhang and Glimm (1992). Further development on the scaling limit for the shear layer model can be found in Avellaneda and Majda (1992a,b); Fannjiang and Komorowski (2000a), see also the extensive review (Majda and Kramer, 1999).

Theorem 12.11, together with Theorem 12.26, have been shown in Komorowski and Olla (2002). The conclusions of the latter theorem stay in agreement with the results of Fannjiang (2000), obtained by a heuristic argument. They match in regions *I* and *II*, given in diagram Fig. 1 on p. 148 of that paper. In the remaining region *III* it is predicted that $\gamma_* = \gamma^* = (1-\alpha)/(3+\alpha)$, using our terminology. Note that the latter value belongs to the interval determined by the lower and upper bounds given in parts (ii) and (iii) of Theorem 12.26. Superdiffusivity in the border case when $\alpha = 1$, see Theorem 12.11, has been shown in Tóth and Valko (2010). Moreover for $\alpha \in (-1, 0)$ it is possible to show that $\gamma_* > 0$ even when Brownian motion is dropped in (12.4) and it becomes an ordinary differential

$$\frac{dX_t^{x,\omega}}{dt} = \tilde{V}\big(X_t^{x,\omega};\omega\big), \quad X_0^{x,\omega} = x \tag{12.159}$$

with the right-hand side described in Sect. 12.6.1, see Komorowski and Nieznaj (2008). Variational methods can also be used to describe the motion of a tracer whose trajectory satisfies (12.159) in a field whose mean dominates its fluctuation (one says then that the field satisfies Taylor hypothesis). Results of that type have been obtained in Komorowski and Ryzhik (2007a,b). A heuristic argument is presented that in case the behavior of the tracer is not diffusive its scaling limit is described by a superdiffusive fractional Brownian motion. A rigorous result of that type can be obtained for a time dependent field, see Fannjiang and Komorowski (2000b).

A Tauberian type theorem has been applied in Sethuraman (2000) to connect the long time asymptotics of a trajectory with the resolvent of the generator of the respective environment process in the context of a tagged particle in an exclusion process. Variational principles have been used in Bernardin (2004) to prove the lower and upper bounds on the norm of the resolvent of the respective generator. A similar technique has been applied in Landim et al. (2004b) to prove that the diffusion coefficient for the asymmetric simple exclusion process is unbounded in the spatial dimensions $d = 1, 2$.

Another application of variational principles can be found in Owhadi (2004), where the motion of a tracer in a multiscale periodic, divergence free flow is shown to be superdiffusive. This type of motion is also proved for a tracer in a two-dimensional, multiscale periodic, shear layer flow, see Ben Arous and Owhadi (2002). The argument is based on a careful analysis of the quadratic variation of the martingale appearing in the proof of the central limit theorem.

References

Adler RJ (1990) An introduction to continuity, extrema, and related topics for general Gaussian processes. IMS lect notes monogr ser, vol 12. Inst Math Stat, Hayward

Avellaneda M, Majda AJ (1990) Mathematical models with exact renormalization for turbulent transport. Commun Math Phys 131(2):381–429

Avellaneda M, Majda A (1992a) Mathematical models with exact renormalization for turbulent transport. II. Fractal interfaces, non-Gaussian statistics and the sweeping effect. Commun Math Phys 146(1):139–204

Avellaneda M, Majda AJ (1992b) Superdiffusion in nearly stratified flows. J Stat Phys 69(3–4):689–729

Batchelor GK (1982) The theory of homogeneous turbulence. Cambridge University Press, Cambridge. Reprint, Cambridge Science Classics.

Ben Arous G, Owhadi H (2002) Super-diffusivity in a shear flow model from perpetual homogenization. Commun Math Phys 227(2):281–302

Bernardin C (2004) Fluctuations in the occupation time of a site in the asymmetric simple exclusion process. Ann Probab 32(1B):855–879

Carmona RA, Xu L (1997) Homogenization for time-dependent two-dimensional incompressible Gaussian flows. Ann Appl Probab 7(1):265–279

Ethier SN, Kurtz TG (1986) Markov processes: characterization and convergence. Wiley series in probability and mathematical statistics: probability and mathematical statistics. Wiley, New York

Fannjiang AC (2000) Phase diagram for turbulent transport: sampling drift, eddy diffusivity and variational principles. Physica D 136(1–2):145–174

Fannjiang A, Komorowski T (1999b) Turbulent diffusion in Markovian flows. Ann Appl Probab 9(3):591–610

Fannjiang A, Komorowski T (2000a) Fractional Brownian motions and enhanced diffusion in a unidirectional wave-like turbulence. J Stat Phys 100(5–6):1071–1095

Fannjiang A, Komorowski T (2000b) Fractional Brownian motions in a limit of turbulent transport. Ann Appl Probab 10(4):1100–1120

Fannjiang A, Komorowski T (2001/2002) Diffusive and nondiffusive limits of transport in non-mixing flows. SIAM J Appl Math 62(3):909–923 (electronic)

Fannjiang A, Komorowski T (2002a) Diffusion in long-range correlated Ornstein-Uhlenbeck flows. Electron J Probab 7(20):22 (electronic)

Friedman A (1975) Stochastic differential equations and applications, vol 1. Probability and mathematical statistics, vol 28. Academic Press [Harcourt Brace Jovanovich Publishers], New York

Gilbarg D, Trudinger NS (1983) Elliptic partial differential equations of second order. Grundlehren der Mathematischen Wissenschaften [Fundamental principles of mathematical sciences], vol 224, 2nd edn. Springer, Berlin

Janson S (1997) Gaussian Hilbert spaces. Cambridge tracts in mathematics, vol 129. Cambridge University Press, Cambridge

Komorowski T, Nieznaj E (2008) On superdiffusive behavior of a passive tracer in a random flow. Physica D 237(24):3377–3381

Komorowski T, Olla S (2002) On the superdiffusive behavior of passive tracer with a Gaussian drift. J Stat Phys 108(3–4):647–668

Komorowski T, Olla S (2003b) On the sector condition and homogenization of diffusions with a Gaussian drift. J Funct Anal 197(1):179–211

Komorowski T, Ryzhik L (2007a) On asymptotics of a tracer advected in a locally self-similar, correlated flow. Asymptot Anal 53(3):159–187

Komorowski T, Ryzhik L (2007b) Passive tracer in a slowly decorrelating random flow with a large mean. Nonlinearity 20(5):1215–1239

Koralov L (1999) Transport by time dependent stationary flow. Commun Math Phys 199(3):649–681

Landim C, Quastel J, Salmhofer M, Yau HT (2004b) Superdiffusivity of asymmetric exclusion process in dimensions one and two. Commun Math Phys 244(3):455–481

Majda AJ, Kramer PR (1999) Simplified models for turbulent diffusion: theory, numerical modelling, and physical phenomena. Phys Rep 314(4–5):237–574

Matheron G, DeMarsily G (1980) Is transport in porous media always diffusive? A counterexample. Water Resou Bull 16:901–917

Owhadi H (2004) Averaging versus chaos in turbulent transport? Commun Math Phys 247(3):553–599

Rozanov YA (1967) Stationary random processes. Holden-Day, San Francisco. Translated from the Russian by A Feinstein

Sethuraman S (2000) Central limit theorems for additive functionals of the simple exclusion process. Ann Probab 28(1):277–302

Tóth B, Valko B (2010) Superdiffusive bounds on self-repellent Brownian polymers and diffusion in the curl of the Gaussian free field in $d = 2$. arXiv:1012.5698

Zhang Q, Glimm J (1992) Inertial range scaling of laminar shear flow as a model of turbulent transport. Commun Math Phys 146(2):217–229

Chapter 13
Ornstein–Uhlenbeck Process with a Random Potential

13.1 Random Diffusion of a Particle with Inertia

In the previous chapters we have considered particles moving in a random environment in an *overdamped* regime, where viscous effects dominate over the inertia. In a regime where damping and inertia forces are of the same order, we should model the movements as a solution of the following system of stochastic differential equations

$$\begin{aligned} dv_t &= -\big[v_t + \nabla\tilde{U}(X_t;\omega)\big]dt + \sqrt{2}\,dw_t, \\ dX_t &= v_t\,dt, \quad t \ge 0, \qquad (X_0, v_0) = (x, v). \end{aligned} \tag{13.1}$$

They describe respectively the evolution of the d-dimensional momentum v_t (assuming that the mass the particle equals one) and the position X_t of the tracer. Here, $\{U(z;\omega), z \in \mathbb{R}^d\}$ is a stationary random field that describes the potential by which the particle interacts with the environment. It is assumed to be of the form $\tilde{U}(z;\omega) := U(\tau_z\omega)$, where $U : \Omega \to \mathbb{R}^d$ is a random variable, called *the potential*, defined over a probability space $(\Omega, \mathcal{F}, \mathbb{Q})$, equipped with the group of transformations $\{\tau_z, z \in \mathbb{R}^d\}$ that satisfies the assumptions made in Sect. 9.3.1. A standard, d-dimensional, Brownian motion $\{w_t, t \ge 0\}$ is defined over another probability space $(\Sigma, \mathcal{A}, \mathbb{P})$. We suppose that the potential U belongs to $C_b^2(\Omega)$ so the existence and uniqueness hold for the solutions of (13.1). The main result of this chapter is the following.

Theorem 13.1 *Under the above assumptions about potential, random variables $X_t/\sqrt{t}$, considered over the product probability space, satisfy the central limit theorem in probability with respect to the environment.*

T. Komorowski et al., *Fluctuations in Markov Processes*,
Grundlehren der mathematischen Wissenschaften 345,
DOI 10.1007/978-3-642-29880-6_13,

13.2 Proof of the Central Limit Theorem

Define the energy of the configuration (v, ω) as

$$\mathscr{E}(v, \omega) = \frac{|v|^2}{2} + U(\omega), \tag{13.2}$$

and the respective stationary field $\tilde{\mathscr{E}}(z, v; \omega) := \mathcal{E}(v, \tau_z \omega)$. The generator of the Feller semigroup on $C_b(\mathbb{R}^{2d})$ corresponding to the diffusion $\{(X_t, v_t), t \geq 0\}$ is given by

$$\mathscr{L}_\omega f(x, v) = e^{\tilde{\mathscr{E}}(z,v)} \nabla_v \cdot \left\{ e^{-\tilde{\mathscr{E}}(z,v)} \nabla_v f(z, v) \right\} + \{\tilde{\mathscr{E}}, f\}(z, v), \tag{13.3}$$

for $f \in C_c^{1,2}(\mathbb{R}^{2d})$, where

$$\{\tilde{\mathscr{E}}, f\}(z, v) := \left\{ v \cdot \nabla_z - \nabla_z \tilde{U}(z; \omega) \cdot \nabla_v \right\} f(z, v)$$

and $C_c^{k,m}(\mathbb{R}^{2d})$ is the space of compactly supported functions that are differentiable k times in the x variable and m times in v. It follows from Theorem 8.2.5, p. 373 of Ethier and Kurtz (1986) that this set forms a core of $\mathscr{L}_\omega$.

It can easily be verified that

$$m_\omega(dz, dv) := e^{-\tilde{\mathscr{E}}(z,v;\omega)} dz dv$$

is an invariant (non-probabilistic) measure for the diffusion, i.e.

$$\int_{\mathbb{R}^{2d}} \mathscr{L}_\omega f \, dm_\omega = 0, \quad \forall f \in C_c^{1,2}(\mathbb{R}^{2d}).$$

We define an $\mathbb{R}^d \times \Omega$-valued, environment process over $(\Sigma, \mathscr{A}, \mathbb{P})$ by

$$\eta_t := (v_t, \tau_{X_t} \omega), \quad t \geq 0 \tag{13.4}$$

and functional $V = (V_1, \ldots, V_d) : \mathbb{R}^d \times \Omega \to \mathbb{R}^d$ by $V(v, \omega) := v$. The position of the particle is an additive functional of the process, namely

$$X_t = \int_0^t V(\eta_s) \, ds.$$

From the results of Sect. 13.4 we conclude the existence of (random) strictly positive, transition of probability densities $p_t^\omega(x, v; y, u)$ corresponding to (13.1), defined on $(0, +\infty) \times \mathbb{R}^{4d}$. In fact, see Theorem 13.7 below, these densities are continuously differentiable, once in t, x and y and twice in v and u. From formulas (13.46) and (13.54) we obtain

$$p_t^{\tau_z \omega}(x, v; y, u) = p_t^\omega(x + z, v; y + z, u), \quad \forall x, v, y, u, z \in \mathbb{R}^d, \ t > 0$$

for all ω. The proof of the Markov property of $\{\eta_t, t \geq 0\}$ can be conducted in the same way as in Sect. 9.4. Its transition of probability semigroup is given by

$$P_t f(v, \omega) = \int_{\mathbb{R}} \int_{\mathbb{R}} p_t^{\omega}(0, v; y, u) f(u, \tau_y \omega)\, dy du, \qquad \text{for all } f \in B_b\big(\mathbb{R}^d \times \Omega\big). \tag{13.5}$$

Arguing as in the proof of Proposition 9.8 we obtain that

$$\pi(dv, d\omega) := Z^{-1} e^{-\mathcal{E}(v,\omega)} dv \mathbb{Q}(d\omega),$$

where Z is an appropriate normalizing factor, and an ergodic, invariant measure for $\{P_t, t \geq 0\}$. As usual we will denote by $\langle \cdot \rangle_\pi$ the expectation with respect to π, and by $\| \cdot \|_\pi$ the corresponding norm in $L^2(\pi)$. The semigroup can be extended therefore to a strongly continuous semigroup of Markov contractions on $L^2(\pi)$. In addition, we have $\langle V \rangle_\pi = 0$. By virtue of the ergodic theorem

$$\frac{X_t}{t} = \frac{1}{t} \int_0^t V(\eta_s) ds \to 0, \qquad \text{as } t \to +\infty,$$

both π-a.s. and in the L^1 sense.

Let $\mathscr{C} := C_b^{2,1}(\mathbb{R}^d \times \Omega)$ be the Banach space consisting of functions $f(v, \omega)$ boundedly differentiable: twice in the v variable and once with respect to spatial shifts in the ω variable, equipped with the standard supremum norm of the appropriate derivatives. We have the following.

Proposition 13.2 *Set $\mathscr{C}$ is a common core of the generators L and L^* of the semigroup $\{P^t, t \geq 0\}$ and its adjoint. For any $f \in \mathscr{C}$ we have*

$$Lf = (S + A) f, \qquad L^* f = (S - A) f, \tag{13.6}$$

where S—the symmetric part of L—that is essentially self-adjoint, is defined by

$$Sf = e^{\mathscr{E}} \nabla_v \cdot \big(e^{-\mathscr{E}} \nabla_v f\big) = (\Delta_v - V \cdot \nabla_v) f \tag{13.7}$$

and A—the anti-symmetric part—is given by

$$Af = \{\mathscr{E}, f\} := (V \cdot \nabla - \nabla U \cdot \nabla_v) f. \tag{13.8}$$

We postpone the presentation of the proof of this result untill Sect. 13.3.

A simple integration by parts gives

$$\int_{\mathbb{R}^d} f(v)(-Sf)(v) g_*(v) dv = \int_{\mathbb{R}^d} \big|\nabla_v f(v)\big|^2 g_*(v) dv, \quad \forall f \in C_b^2\big(\mathbb{R}^d\big), \tag{13.9}$$

where

$$g_*(v) := (2\pi)^{-d/2} e^{-|v|^2/2}.$$

Operator S has the spectral gap property that holds on $L^2(g_*)$, i.e. we have

$$\int_{\mathbb{R}^d} f^2(v)g_*(v)dv - \left(\int_{\mathbb{R}^d} f(v)g_*(v)dv\right)^2 \leq \int_{\mathbb{R}^d} \left|\nabla_v f(v)\right|^2 g_*(v)dv \tag{13.10}$$

which holds for all $f \in H^1_{loc}(\mathbb{R}^d)$ such that the right-hand side is finite, see e.g. part (i) of Theorem 5.41 of Janson (1997).

The spaces $\mathscr{H}_1$ and $\mathscr{H}_{-1}$ are introduced as in Chap. 2. In particular, from (13.7) it follows that the space $\mathscr{H}_1$ consists of all functions $f(v, \omega)$ that possess a generalized derivative in v satisfying

$$\|f\|_1^2 = \left\langle |\nabla_v f|^2\right\rangle_\pi < +\infty.$$

For a given $\lambda > 0$ let $\chi_\lambda^{(p)}$ be the solution of the resolvent equation

$$(\lambda - L)\chi_\lambda^{(p)} = V_p. \tag{13.11}$$

Note that

$$\left|\langle V_p, g\rangle_\pi\right| = \left|\langle \partial_{v_p} g\rangle_\pi\right| \leq \|g\|_1, \quad g \in \mathscr{C}, \tag{13.12}$$

so $\|V_p\|_{-1} \leq 1$. As in Sect. 2.8, we obtain easy bounds

$$\left\|\chi_\lambda^{(p)}\right\|_1 \leq \|V_p\|_{-1} = 1 \quad \text{and} \quad \lambda\left\|\chi_\lambda^{(p)}\right\|^2 \leq \|V_p\|_{-1}^2 = 1. \tag{13.13}$$

According to Theorem 2.14 in order to claim the central limit theorem we need to prove that

$$\lim_{\lambda\to 0+} \lambda\left\|\chi_\lambda^{(p)}\right\|^2 = 0 \quad \text{and} \quad \lim_{\lambda\to 0+} \chi_\lambda^{(p)} = \chi^{(p)},$$

strongly in $\mathscr{H}_1$. The subtle point is that the symmetric part of the generator L is quite degenerate in this case so we cannot directly use any of the sufficient conditions for the central limit theorem stated in Sect. 2.7. In particular, note that sector condition (2.36) obviously fails. We have

$$\left\langle(-L)f, f\right\rangle_\pi = \left\langle|\nabla_v f|^2\right\rangle_\pi,$$

hence it vanishes for any f depending only on ω.

Define $\widehat{\mathscr{H}_1}$ as the closure of $\mathscr{C}$ in the norm

$$\|f\|_{\widehat{\mathscr{H}_1}}^2 := \|f\|_\pi^2 + \|f\|_1^2.$$

Operator S defined in (13.7) gives rise to a mapping between $\mathscr{C}$ and $\widehat{\mathscr{H}_1^*}$—the dual to $\widehat{\mathscr{H}_1}$—defined as follows: for any $f \in \widehat{\mathscr{H}_1}$ let $Sf \in \widehat{\mathscr{H}_1^*}$ be the continuous extension of the linear functional

$$Sf(g) := -\langle \nabla_v f, \nabla_v g\rangle_\pi, \quad g \in \mathscr{C}. \tag{13.14}$$

Observe that

$$\left|{}_{\widehat{\mathscr{H}_1^*}}\langle Sf, g\rangle_{\widehat{\mathscr{H}_1}}\right| \le \|f\|_{\widehat{\mathscr{H}_1}}\|g\|_{\widehat{\mathscr{H}_1}}.$$

Here ${}_{\widehat{\mathscr{H}_1^*}}\langle\cdot,\cdot\rangle_{\widehat{\mathscr{H}_1}}$ is the duality pairing between $\widehat{\mathscr{H}_1}$ and $\widehat{\mathscr{H}_1^*}$. We conclude easily the following.

Lemma 13.3 *Operator S defined in* (13.14) *extends to a contraction $S: \widehat{\mathscr{H}_1} \to \widehat{\mathscr{H}_1^*}$.*

For any $f \in L^2(\pi)$ we introduce its projections onto the spaces of even and odd functions in the v-variable, respectively by

$$f_e(v,\omega) := \frac{1}{2}\big[f(v,\omega) + f(-v,\omega)\big]$$

and

$$f_o(v,\omega) := \frac{1}{2}\big[f(v,\omega) - f(-v,\omega)\big].$$

Since $\mathscr{C}$ is invariant under both of these operators we infer that for any $f \in D(L)$ we have $f_e, f_o \in D(L)$. In fact,

$$(Sf)_e = Sf_e \quad \text{and} \quad (Sf)_o = Sf_o \quad \text{for any } f \in \widehat{\mathscr{H}_1}. \tag{13.15}$$

Note also that the definitions of the even (or odd) part of a function can be extended by duality to any element $f \in \widehat{\mathscr{H}_1^*}$. Namely, we let

$${}_{\widehat{\mathscr{H}_1^*}}\langle f_e, g\rangle_{\widehat{\mathscr{H}_1}} := {}_{\widehat{\mathscr{H}_1^*}}\langle f, g_e\rangle_{\widehat{\mathscr{H}_1}}.$$

Likewise, we can define f_o.

It is clear that $\chi_\lambda^{(p)} \in \widehat{\mathscr{H}_1}$ so that $S\chi_\lambda^{(p)}$ is well defined and belongs to $\widehat{\mathscr{H}_1^*}$, by virtue of Lemma 13.3. To simplify the notation, in the ensuing calculation we omit writing the superscript p of the corrector. Since

$$\|\chi_\lambda\|_1^2 = \|\nabla_v\chi_\lambda\|_\pi^2 = \|\nabla_v\chi_{\lambda,e}\|_\pi^2 + \|\nabla_v\chi_{\lambda,o}\|_\pi^2 \tag{13.16}$$

both $S\chi_{\lambda,e}$ and $S\chi_{\lambda,o}$ are also well defined as elements of the dual space $\widehat{\mathscr{H}_1^*}$. We cannot guarantee that A can be extended to a bounded operator from $\widehat{\mathscr{H}_1}$ to $\widehat{\mathscr{H}_1^*}$ (that would mean a sector condition for L) but we can prove the following.

Lemma 13.4 *For any $\lambda > 0$ the linear functional*

$$A\chi_\lambda(g) := -\langle\chi_\lambda, Ag\rangle_\pi, \quad g \in \mathscr{C} \tag{13.17}$$

can be uniquely extended to a continuous functional on $\widehat{\mathscr{H}_1}$ (it will be denoted by the same symbol). For any $\lambda, \lambda' > 0$ we have

$${}_{\widehat{\mathscr{H}_1^*}}\langle A\chi_\lambda, \chi_{\lambda'}\rangle_{\widehat{\mathscr{H}_1}} = -{}_{\widehat{\mathscr{H}_1^*}}\langle A\chi_{\lambda'}, \chi_\lambda\rangle_{\widehat{\mathscr{H}_1}}. \tag{13.18}$$

Analogous results hold also for the even and odd parts of χ_λ. *Namely, the linear functionals*

$$A\chi_{\lambda,i}(g) := -\langle \chi_{\lambda,i}, Ag\rangle_\pi, \quad g \in \mathscr{C}, i \in \{o, e\}$$

extend continuously to $\widehat{\mathscr{H}_1}$ *and satisfy*

$$_{\widehat{\mathscr{H}_1^*}}\langle A\chi_{\lambda,i}, \chi_{\lambda',j}\rangle_{\widehat{\mathscr{H}_1}} = -_{\widehat{\mathscr{H}_1^*}}\langle A\chi_{\lambda',j}, \chi_{\lambda,i}\rangle_{\widehat{\mathscr{H}_1}}, \tag{13.19}$$

for $i, j \in \{o, e\}$. *In addition,*

$$(A\chi_{\lambda,e})_e = (A\chi_{\lambda,o})_o = 0. \tag{13.20}$$

Proof From (13.12) we know that $V_p \in \widehat{\mathscr{H}_1^*}$ and $\|V_p\|_{\widehat{\mathscr{H}_1^*}} \le 1$. Equality (13.11) implies that

$$A\chi_\lambda(g) = -\lambda\langle \chi_\lambda, g\rangle_\pi - \langle \nabla_v\chi_\lambda, \nabla_v g\rangle_\pi + \langle V_p, g\rangle_\pi, \quad \forall g \in \mathscr{C},$$

which leads to the estimate

$$\left|A\chi_\lambda(g)\right| \le C\|g\|_{\widehat{\mathscr{H}_1}}, \quad \forall g \in \mathscr{C} \tag{13.21}$$

for some constant $C > 0$ possibly depending on λ and $\|\chi_\lambda\|_{\widehat{\mathscr{H}_1}}$. The first part of the lemma follows then from the fact that $\mathscr{C}$ is dense in $\widehat{\mathscr{H}_1}$.

We prove (13.18). Since $\mathscr{C}$ is a common core of L and L^* one can find sequences $\{h_n, n \ge 0\}$ and $\{g_n, n \ge 0\}$ of elements of $\mathscr{C}$ such that

$$Lh_n \to L\chi_\lambda \quad \text{and} \quad Lg_n \to L\chi_{\lambda'}, \quad \text{in } L^2(\pi)$$

and

$$h_n \to \chi_\lambda \quad \text{and} \quad g_n \to \chi_{\lambda'}, \quad \text{in } \widehat{\mathscr{H}_1}$$

as $n \to +\infty$. Using definition (13.17) and the density argument we can write that the left-hand side of (13.18) equals $-\lim_{n\to+\infty}\langle \chi_\lambda, Ag_n\rangle_\pi$. Thanks to the equality $Lg_n = Sg_n + Ag_n$ and (13.14) we conclude that this expression further equals

$$\begin{aligned}
&-\lim_{n\to+\infty}\left[\langle \chi_\lambda, Lg_n\rangle_\pi + \langle \nabla_v\chi_\lambda, \nabla_v g_n\rangle_\pi\right] \\
&\quad = -\left[\langle \chi_\lambda, L\chi_{\lambda'}\rangle_\pi + \langle \nabla_v\chi_\lambda, \nabla_v\chi_{\lambda'}\rangle_\pi\right] \\
&\quad = -\lim_{m\to+\infty}\left[\langle h_m, L\chi_{\lambda'}\rangle_\pi + \langle \nabla_v h_m, \nabla_v\chi_{\lambda'}\rangle_\pi\right].
\end{aligned}$$

Since $L^*h_m = Sh_m - Ah_m$ we obtain that the utmost right-hand side equals

$$-\lim_{m\to+\infty}\left[\langle L^*h_m, \chi_{\lambda'}\rangle_\pi + \langle \nabla_v h_m, \nabla_v\chi_{\lambda'}\rangle_\pi\right] = \lim_{m\to+\infty}\langle Ah_m, \chi_{\lambda'}\rangle_\pi.$$

The last expression, by definition (13.17), equals

$$-\lim_{m\to+\infty}{}_{\widehat{\mathscr{H}_1^*}}\langle A\chi_{\lambda'}, h_m\rangle_{\widehat{\mathscr{H}_1}} = -_{\widehat{\mathscr{H}_1^*}}\langle A\chi_{\lambda'}, \chi_\lambda\rangle_{\widehat{\mathscr{H}_1}}.$$

To prove the corresponding result for $\chi_{\lambda,e}$ note that

$$A\chi_{\lambda,e}(g) = -\langle \chi_{\lambda,e}, Ag_o\rangle_\pi = -\langle \chi_\lambda, Ag_o\rangle_\pi, \quad \forall g \in \mathscr{C}. \tag{13.22}$$

To get the second equality above we used the fact Ag_o is even, see (13.8). Since $Ag_o = -L^* g_o + Sg_o$ the utmost right-hand side can be rewritten in the form

$$\langle L\chi_\lambda, g_o\rangle_\pi + \langle \nabla_v \chi_\lambda, \nabla_v g_o\rangle_\pi.$$

Using the resolvent equation (13.11) we can further transform the expression into

$$\lambda\langle \chi_{\lambda,o}, g_o\rangle_\pi - \langle V_p, g_o\rangle_\pi + \langle \nabla_v \chi_{\lambda,e}, \nabla_v g_o\rangle_\pi,$$

which, as in (13.21), leads to the definition of the extension of the functional to the entire $\widehat{\mathscr{H}_1}$. A similar argument can be also made for $\chi_{\lambda,o}$. The proof of (13.19) is analogous to the proof of (13.18).

Finally, note that for any $g \in \mathscr{C}$

$$_{\widehat{\mathscr{H}_1^*}}\langle A\chi_{\lambda,e}, g_e\rangle_{\widehat{\mathscr{H}_1}} = -\langle \chi_{\lambda,e}, Ag_e\rangle_\pi = 0$$

because $(Ag_e)_e = 0$, see (13.8). Likewise, we show that

$$_{\widehat{\mathscr{H}_1^*}}\langle A\chi_{\lambda,o}, g_o\rangle_{\widehat{\mathscr{H}_1}} = 0.$$

This proves (13.20). □

The resolvent equation (13.11) can be now transformed into the following system of equations

$$\lambda\chi_{\lambda,e} - S\chi_{\lambda,e} - A\chi_{\lambda,o} = 0 \tag{13.23}$$

and

$$\lambda\chi_{\lambda,o} - S\chi_{\lambda,o} - A\chi_{\lambda,e} = V_p. \tag{13.24}$$

Here elements $A\chi_{\lambda,i}$, $S\chi_{\lambda,i}$ for $i \in \{e, o\}$ are elements of $\widehat{\mathscr{H}_1^*}$. We also identify $\chi_{\lambda,i}$ with its embedding in the space, which is the continuous extension in $\widehat{\mathscr{H}_1}$ of the linear functional

$$\chi_{\lambda,i}(g) := \langle \chi_{\lambda,i}, g\rangle_\pi, \quad g \in \mathscr{C}.$$

Lemma 13.5 *For any $\lambda > 0$ we have*

$$\|\chi_{\lambda,o}\|_\pi \le 1. \tag{13.25}$$

Proof Combining (13.13) and (13.16) we conclude that

$$\|\nabla_v \chi_{\lambda,o}\|_\pi \le 1. \tag{13.26}$$

Since

$$\int_{\mathbb{R}^d} \chi_{\lambda,o}(v,\omega) g_*(v) dv = 0$$

from the spectral gap estimate for S, see (13.10), we obtain

$$\int_{\mathbb{R}^d} \left[\chi_{\lambda,o}(v,\omega)\right]^2 g_*(v) dv \le \int_{\mathbb{R}^d} \left|\nabla_v \chi_{\lambda,o}(v,\omega)\right|^2 g_*(v) dv, \quad \mathbb{Q}\text{-a.s. in } \omega. \quad (13.27)$$

The desired estimate follows from (13.26) after integrating out the ω variable. □

From (13.24) we obtain that for any $\lambda' > 0$

$$\langle V_p, \chi_{\lambda',o}\rangle_\pi = \lambda\langle \chi_{\lambda,o}, \chi_{\lambda',o}\rangle_\pi + \langle \nabla_v \chi_{\lambda,o}, \nabla_v \chi_{\lambda',o}\rangle_\pi - {}_{\widehat{\mathscr{H}}_1^*}\langle A\chi_{\lambda,e}, \chi_{\lambda',o}\rangle_{\widehat{\mathscr{H}}_1}. \quad (13.28)$$

On the other hand, from (13.19)

$$\begin{aligned} -{}_{\widehat{\mathscr{H}}_1^*}\langle A\chi_{\lambda,e}, \chi_{\lambda',o}\rangle_{\widehat{\mathscr{H}}_1} &= {}_{\widehat{\mathscr{H}}_1^*}\langle A\chi_{\lambda',o}, \chi_{\lambda,e}\rangle_{\widehat{\mathscr{H}}_1} \\ &\overset{(13.23)}{=} \lambda'\langle \chi_{\lambda',e}, \chi_{\lambda,e}\rangle_\pi + \langle \nabla_v \chi_{\lambda',e}, \nabla_v \chi_{\lambda,e}\rangle_\pi. \quad (13.29) \end{aligned}$$

Combining (13.28) with (13.29) we obtain

$$\begin{aligned} \langle V_p, \chi_{\lambda'}\rangle_\pi &= \lambda\langle \chi_{\lambda,o}, \chi_{\lambda',o}\rangle_\pi + \langle \nabla_v \chi_{\lambda,o}, \nabla_v \chi_{\lambda',o}\rangle_\pi \\ &\quad + \lambda'\langle \chi_{\lambda,e}, \chi_{\lambda',e}\rangle_\pi + \langle \nabla_v \chi_{\lambda,e}, \nabla_v \chi_{\lambda',e}\rangle_\pi \\ &= \lambda\langle \chi_{\lambda,o}, \chi_{\lambda',o}\rangle_\pi + \lambda'\langle \chi_{\lambda,e}, \chi_{\lambda',e}\rangle_\pi + \langle \nabla_v \chi_\lambda, \nabla_v \chi_{\lambda'}\rangle_\pi. \quad (13.30) \end{aligned}$$

Since $\|V_p\|_{-1} \le 1$, by virtue of estimate (13.13), we conclude that the set $\{\chi_\lambda, \lambda > 0\}$ is weakly pre-compact in $\mathscr{H}_1$. Let $\lambda_n \to 0$, as $n \to +\infty$, be such that

$$\lim_{n\to+\infty} \chi_{\lambda_n} = f_* \quad (13.31)$$

weakly in $\mathscr{H}_1$. In (13.30) take $\lambda = \lambda_n$ and $\lambda' = \lambda_m$ and let subsequently $m \to +\infty$ and then $n \to +\infty$. Using Lemma 13.5 we obtain

$$\|f_*\|_1^2 = \|f_{*,e}\|_1^2 + \|f_{*,o}\|_1^2 = {}_{\mathscr{H}_{-1}}\langle V_p, f_*\rangle_{\mathscr{H}_1}. \quad (13.32)$$

Arguing as in Claim 3 of the proof of Theorem 2.14 we conclude from (13.32) that $\lim_{n\to+\infty} \lambda_n \|\chi_{\lambda_n}\|_\pi^2 = 0$ and therefore $\lim_{n\to+\infty} \chi_{\lambda_n} = f_*$ strongly in $\mathscr{H}_1$. Uniqueness of the $\mathscr{H}_1$-limit can be argued in exactly the same way as in Claim 4 of the aforementioned theorem. We are in the position to use Theorem 2.17 and therefore conclude the central limit theorem for $X_t/\sqrt{t}$, as $t \to +\infty$.

13.3 Proof of Proposition 13.2

The fact that $\mathscr{C}$ is invariant under the action of the semigroup P_t follows easily from Theorem 13.7 formulated below. By a direct calculation one can establish also that this set is contained in the domain of L and the first formula of (13.6) holds. The fact that $\mathscr{C}$ is a core of the generator is a consequence of Proposition 1.3.3 of Ethier and Kurtz (1986).

The following fact will help us to identify the adjoint semigroup.

Lemma 13.6 *We have*

$$e^{\tilde{\mathscr{E}}(y,u;\omega)} p_t^\omega(x, v; y, u) = e^{\tilde{\mathscr{E}}(x,v;\omega)} p_t^\omega(y, -u; x, -v) \tag{13.33}$$

for all $t > 0$, $x, y, u, v \in \mathbb{R}^d$ *and* $\mathbb{Q}$ *a.s.* ω.

Proof Let $\tilde{p}(t, x, v) := p_t(y, -u; x, v)$ and $\tilde{q}(t, x, v) := e^{\tilde{\mathscr{E}}(x,v)} \tilde{p}(t, x, -v)$. A direct computation shows that, cf. Theorem 13.7 formulated below,

$$(\partial_t - \mathscr{L})\tilde{q}(t, x, v) = e^{\tilde{\mathscr{E}}(x,v)} (\partial_t - \mathscr{L}^*) \tilde{p}_t(t, x, v) = 0.$$

Here $\mathscr{L}^*$ is the formal adjoint to the generator $\mathscr{L}$.

Suppose that $f \in C_c^\infty(\mathbb{R}^{2d})$ and let

$$w(t, x, v) := \int_{\mathbb{R}^{2d}} q_t(x, v; y, u) f(y, u) dy du,$$

where

$$q_t(x, v; y, u) := e^{-\tilde{\mathscr{E}}(y,u)} \tilde{q}(t, x, v).$$

Using estimate (13.48) one can verify that $w(t, x, v)$ is boundedly and continuously differentiable once, in t and x, and twice in v at least in the slab $(0, \delta) \times \mathbb{R}^{2d}$, for a sufficiently small $\delta > 0$. In addition it satisfies

$$(\partial_t - \mathscr{L}) w(t, x, v) = 0$$

with the initial condition $w(0, x, v) = f(x, v)$. The uniqueness of bounded classical solutions of the equation (following e.g. from an application of the Itô formula) implies that

$$\int_{\mathbb{R}^{2d}} f(y, u) q_t(x, v; y, u) dy\, du = \int_{\mathbb{R}^{2d}} f(y, u) p_t(x, v; y, u) dy\, du.$$

Thus $q_t(x, v; y, u) = p_t(x, v; y, u)$ and formula (13.33) follows. □

A simple calculation yields that for any $f \in L^2(\pi)$

$$P_t^* f(u, \omega) = \int_{\mathbb{R}^{2d}} e^{\mathcal{E}(u,\omega)} p_t^\omega(y, v; 0, u) e^{-\tilde{\mathscr{E}}(y,v;\omega)} f(v, \tau_y \omega)\, dy dv$$

$$\stackrel{(13.33)}{=} \int_{\mathbb{R}^{2d}} p_t^{\omega}(0,-u;y,-v) f(v,\tau_y\omega)\,dydv.$$

From this formula we obtain the second equality in (13.6). Invoking again the results of Sect. 13.4 we conclude that $\mathscr{C}$ is invariant under the adjoint semigroup, which proves that it is a common core of both L and L^*.

Note that S agrees on $\mathscr{C}$ with the generator of the semigroup

$$R_t f(v,\omega) := \int_{\mathbb{R}^{2d}} r_t(0,v;y,u) f(u,\tau_y\omega) dydu, \qquad t \ge 0,$$

where $r_t(x,v;y,u)$ are the transition probability densities corresponding to the process, defined by the solution of (13.1) with $\tilde{U} \equiv 0$, see formula (13.39) below. Since $\mathscr{C}$ is a core for the generator of this semigroup we conclude that S is essentially self-adjoint.

13.4 Gaussian Bounds on Transition Probability Densities

We consider an equation that is slightly more general than (13.1). Suppose that $(s,x) \in [0,+\infty) \times \mathbb{R}^N$. Let $\{Y_t^{s,x}, t \ge s\}$ be an $\mathbb{R}^N$-valued solution of the equation

$$\begin{cases} dY_t^{s,x} = [AY_t^{s,x} + H(t,Y_t^{s,x})]dt + Qdw_t, \\ Y_s^{s,x} = x, \end{cases} \tag{13.34}$$

where $\{w_t, t \ge 0\}$ is a standard N-dimensional Brownian motion over a probability space $(\Sigma, \mathscr{A}, \mathbb{P})$ and $A = [a_{ij}]$, $Q = [q_{ij}]$ are constant $N \times N$ matrices. Here

$$H(t,x) = \big(h_1(t,x), \ldots, h_d(t,x)\big) = QG(t,x), \tag{13.35}$$

and $G : \mathbb{R} \times \mathbb{R}^N \to \mathbb{R}^N$ is a continuous function such that $\|G\|_\infty < +\infty$. Matrix Q is assumed to be symmetric. We suppress writing the superscripts in the notation of the process if any one of them (or both) equals 0.

Let $P_{s,x}$ be the law of the solution to (13.34) on $C([s,+\infty), \mathbb{R}^N)$ and let $P_{s,t}(x;\cdot)$ be the corresponding transition of the probability function. The generator of the process is given by

$$\mathscr{L} f(x) = \frac{1}{2} \sum_{i,j=1}^N c_{ij} \partial^2_{x_i x_j} f(x) + \sum_{i=1}^N \left(h_i(t,x) + \frac{1}{2} \sum_{j=1}^N a_{ij} x_j \right) \partial_{x_i} f(x),$$
$$f \in C^2(\mathbb{R}^N).$$

Here $C = Q^2 = [c_{ij}]$. By $\mathscr{L}^*$ we denote the formal adjoint of $\mathscr{L}$.

When $H \equiv 0$ the solution of (13.34) is a Gaussian, Markovian process given by

$$Z_t^{s,x} = e^{A(t-s)} x + \int_s^t e^{A(t-u)} Q\,dw_u. \tag{13.36}$$

Its mean and the covariance matrix are respectively equal to $e^{A(t-s)}x$ and

$$\mathrm{Cov}\big(Z_t^{s,x}, Z_u^{s,x}\big) = e^{A(t-u)}C_{u-s}, \quad \text{for } t > u > s,$$

where

$$C_t := \int_0^t e^{Ar} Q^2 e^{A^T r}\, dr. \tag{13.37}$$

We assume that

$$\det C_t > 0 \quad \text{for all } t > 0. \tag{13.38}$$

Let $Q_{s,x}$ denote the law of this process and $Q_{s,t}(x;\cdot)$, $q_{s,t}(x,y)$ be the respective transition probability function and its density. Then,

$$q_{t-s}(x,y) = q_{t-s}(x,y;1),$$

where

$$q_{t-s}(x,y;m) := r_{t-s}\big(y - e^{A(t-s)}x; m\big), \tag{13.39}$$

and

$$r_u(z;m) := \frac{1}{(2\pi)^{d/2}\sqrt{\det C_u}} \exp\left\{-\frac{1}{2m} z^T C_u^{-1} z\right\}.$$

Since both matrices C_u' and C_u^{-1} are non-negative definite $\mathrm{tr}(C_u' C_u^{-\gamma}) \geq 0$ for any $\gamma > 0$. Our assumption is that for some $\gamma \in (0,1)$ we have

$$\mathcal{I}_{\gamma,T} := \int_0^T \mathrm{tr}\big(C_u' C_u^{-\gamma}\big) du < +\infty \tag{13.40}$$

and

$$\mathcal{G}_{\gamma,T} := \int_0^T \mathrm{tr}\big(C_u^{-(1-\gamma)}\big) du < +\infty, \quad \forall T > 0. \tag{13.41}$$

Remark 13.1 Note that the system (13.1) satisfies the assumptions made in the foregoing. In this case $N = 2d$, the solution can be identified with an N-dimensional column vector valued process whose components are X_t and V_t. The respective matrices and a non-linear perturbation of the drift are given by

$$A = \begin{bmatrix} 0 & I \\ 0 & -I \end{bmatrix}, \qquad Q = \begin{bmatrix} 0 & 0 \\ 0 & I \end{bmatrix}, \qquad G(t,x) = \begin{bmatrix} 0 \\ -\nabla_x \tilde{U}(x;\omega) \end{bmatrix}. \tag{13.42}$$

Here I denotes the $d \times d$ identity matrix. We have

$$e^{At} = \begin{bmatrix} I & (1-e^{-t})I \\ 0 & e^{-t} I \end{bmatrix}$$

and

$$C_t = \begin{pmatrix} f_{11}(t)I & f_{12}(t)I \\ f_{12}(t)I & f_{22}(t)I \end{pmatrix}.$$

Here

$$f_{11}(t) := \int_0^t \big(1-e^{-u}\big)^2 du, \qquad f_{12}(t) := \int_0^t e^{-u}\big(1-e^{-u}\big) du$$

and

$$f_{22}(t) := \int_0^t e^{-2u} du.$$

Let

$$\Delta(t) := \int_0^t \big(1-e^{-u}\big)^2 du \int_0^t e^{-2u} du - \left(\int_0^t e^{-u}\big(1-e^{-u}\big) du\right)^2.$$

Strict positivity of $\Delta(t)$ for $t>0$ is a consequence of the Cauchy–Schwartz inequality. Hence

$$\det C_t = \Delta^d(t) > 0$$

and hypothesis (13.38) is fulfilled. The expression appearing in (13.40) in this case equals

$$\mathscr{I}_{\gamma,T} = \int_0^T \mathrm{tr}\big(C_t' C_t^{-\gamma}\big) dt \le \int_0^T \mathrm{tr}\big(C_t' C_t^{-1}\big) |C_t|^{1-\gamma} dt.$$

Here $|A| := (\sum_{ij=1}^N a_{ij}^2)^{1/2}$ is the norm of the given matrix. A simple calculation shows that

$$\mathrm{tr}\big(C_t' C_t^{-1}\big) = d\Delta'(t)\Delta^{-1}(t) \sim t^{-1}, \quad \text{for } t \ll 1.$$

Since $|C_t| \sim t$, as $t \ll 1$, we conclude that $\mathscr{I}_{\gamma,T} < +\infty$ for any $\gamma > 0$. On the other hand, $|C_t^{-1}| \sim t^{-4}$, as $t \ll 1$, so $\mathscr{G}_{\gamma,T} < +\infty$, provided γ is sufficiently close to 1.

For any $t>0$ we let

$$V_u^t := Q e^{A^T(t-u)} C_{t-u}^{-1} \int_u^t e^{A(t-r)} Q\, dw_r, \quad u \in [0,t] \tag{13.43}$$

and let $G(t,x)$ be as in (13.35). Define

$$\rho_{s,t}\big(w, Z^{s,x}; G\big) := \int_s^t G\big(u, Z_u^{s,x}\big) \cdot dw_u - \frac{1}{2}\int_s^t \big|G\big(u, Z_u^{s,x}\big)\big|^2 du. \tag{13.44}$$

We shall omit writing parameter G if it is obvious from the context. The Girsanov theorem for degenerate diffusions, see Corollary 2.1 of Stroock and Varadhan

(1972), implies that $h_{t,s,x} : C([s,t];\mathbb{R}^N) \to \mathbb{R}_+$—the Radon–Nikodym derivative of $P_{s,x}$ with respect to $Q_{s,x}$, restricted to $\mathscr{M}_{s,t}$ (with $s < t$), satisfies

$$h_{t,s,x} \circ Z^{s,x} = \exp\{\rho_{s,t}(w, Z^{s,x})\}. \tag{13.45}$$

Formula (13.45) in particular implies that $P_{s,t}(x,\cdot)$ possesses a density $p_{s,t}(x,y)$ with respect to the Lebesgue measure given by

$$p_{s,t}(x,y) = g_{s,t}(x,y) q_{t-s}(x,y), \tag{13.46}$$

where

$$g_{s,t}(x,y) := E_{\mathbb{P}}\big[h_{t,s,x} \circ Z^{s,x} | Z_t^{s,x} = y\big]. \tag{13.47}$$

Theorem 13.7 *Under the assumptions made above functions $p_{s,t}(x,y)$ are jointly continuous and strictly positive in the region $\mathscr{R} := [(s,x,t,y) : 0 \le s < t]$. There exists a function $\mathscr{C} : [(t,s) : t > s > 0] \to (0,+\infty)$ such that*

$$p_{s,t}(x,y) \le \mathscr{C}(t,s) q_{t-s}(x,y;2). \tag{13.48}$$

If $G(t,x)$ is m times differentiable in x, then so is $p_{s,t}(x,y)$, jointly in x and y variables. In addition, estimate (13.48) *holds for $|\nabla_x^{m_1} \nabla_y^{m_2} p_{s,t}(x,y)|$ for non-negative integer valued multiindices m_1, m_2 such that $|m_1| + |m_2| = m$.*

When $m = 2$ functions $x \mapsto p_{t,s}(x,y)$, $y \mapsto p_{t,s}(x,y)$ solve the Kolmogorov backward and forward equations:

$$(\partial_s + \mathscr{L}_x) p_{s,t}(x,y) = 0 \quad \textit{and resp.} \quad (\partial_t - \mathscr{L}_y^*) p_{s,t}(x,y) = 0, \tag{13.49}$$

with the final condition (*resp. initial condition*) *given by*

$$\lim_{s \to t-} \int_{\mathbb{R}^N} p_{s,t}(x,y) f(y) dy = f(x) \tag{13.50}$$

and

$$\lim_{t \to s+} \int_{\mathbb{R}^N} p_{s,t}(x,y) f(x) dx = f(y), \quad \forall f \in C_b(\mathbb{R}^N).$$

Proof Consider a Gaussian process given by

$$\hat{Z}_{u,t,y}^{s,x} := Z_u^{s,x} - C_{u-s} e^{A^T(t-u)} C_{t-s}^{-1} (Z_t^{s,x} - y), \quad u \in [s,t]. \tag{13.51}$$

By comparing the respective mean and covariance functions it can easily be shown that the law of this process coincides with that of $\{Z_u^{s,x}, u \in [s,t]\}$ conditioned on the event $Z_t^{s,x} = y$, i.e. *the Ornstein–Uhlenbeck bridge* starting at time s at x and reaching y at time t. A direct calculation shows that the bridge process defined by (13.51) is uncorrelated with (thus independent of) the random vector $Z_t^{s,x}$. Denote by $\hat{Z}_{u,t}^s$ the bridge corresponding to $x = y = 0$.

Let

$$\tilde{w}_u := w_u - \int_s^u V_r^t dr, \quad s \le u \le t. \tag{13.52}$$

The following lemma can be verified by a direct calculation of the respective covariance matrices.

Lemma 13.8 *The process* $\{\tilde{w}_u, s \le u \le t\}$ *is a standard* N*-dimensional Brownian motion over* $(\Sigma, \mathscr{A}, \mathbb{P})$ *that is non-anti-cipative with respect to the natural filtration corresponding to* $\{(\tilde{w}_u, \hat{Z}^{s,x}_{u,t,y}), s \le u \le t\}$ *and independent of the vector* $Z_t^{s,x}$.

Our next task is to express function $g_{s,t}(x, y)$ appearing on the right-hand side of (13.46) in terms of the unconditional expectation of the Ornstein–Uhlenbeck bridge. For this purpose observe first that

$$\begin{aligned} e^{A(t-u)}\hat{Z}^s_{u,t} &= -\int_u^t e^{A(t-r)}Q dw_r + \left[C_{t-s} - e^{A(t-u)}C_{u-s}e^{A^T(t-u)}\right]C_{t-s}^{-1}Z_t^s \\ &= -\int_u^t e^{A(t-r)}Q dw_r + C_{t-u}C_{t-s}^{-1}Z_t^s. \end{aligned}$$

This equality implies that

$$b_u^{(1)}\hat{Z}^s_{u,t} = -V_u^t + b_u^{(2)}Z_t^s, \tag{13.53}$$

where

$$b_u^{(1)} := Qe^{A^T(t-u)}C_{t-u}^{-1}e^{A(t-u)} \quad \text{and} \quad b_u^{(2)} := Qe^{A^T(t-u)}C_{t-s}^{-1}.$$

Lemma 13.9 *We have*

$$g_{s,t}(x, y) = E_{\mathbb{P}} \exp\{e_{t,s}^{x,y}\}, \tag{13.54}$$

where

$$e_{t,s}^{x,y} := \rho_{s,t}\big(\tilde{w}, \hat{Z}^{s,x}_{\cdot,t,y}\big) - \int_s^t G\big(u, \hat{Z}^{s,x}_{u,t,y}\big) \cdot \big\{b_u^{(1)}\hat{Z}^s_{u,t} + b_u^{(2)}\big[e^{A(t-s)}x - y\big]\big\}\, du.$$

Proof For an integer $k \ge 1$ and $i = 0, \ldots, k$ we define

$$t_i^k := s + i(t-s)/k, \qquad G_i^k := G\big(t_i^k, Z^{s,x}_{t_i^k}\big), \qquad \Delta w_{t_i^k} := w_{t_{i+1}^k} - w_{t_i^k},$$

$\Delta \tilde{w}_{t_i^k} := \tilde{w}_{t_{i+1}^k} - \tilde{w}_{t_i^k}$ and

$$\rho_{s,t}^k\big(w, Z^{s,x}\big) := \sum_{i=0}^{k-1} G_i^k \cdot \Delta w_{t_i^k} - \frac{1}{2}\int_s^t \big|G\big(r, Z_r^{s,x}\big)\big|^2 dr. \tag{13.55}$$

Using (13.52) we can rewrite the left-hand side of (13.55) as being equal to

$$\sum_{i=0}^{k-1}\left[G_i^k \cdot \Delta\tilde{w}_{t_i^k} + \int_{t_i^k}^{t_{i+1}^k} G_i^k \cdot V_u^t\, du\right] - \frac{1}{2}\int_s^t \left|G\left(r, Z_r^{s,x}\right)\right|^2 dr$$

$$\stackrel{(13.53)}{=} \sum_{i=0}^{k-1} G_i^k \cdot \Delta\tilde{w}_{t_i^k} - \sum_{i=0}^{k-1}\int_{t_i^k}^{t_{i+1}^k} G_i^k \cdot \left\{b_u^{(1)}\hat{Z}_{u,t}^s + b_u^{(2)}\left[e^{A(t-s)}x - y\right]\right\} du$$

$$+ \sum_{i=0}^{k-1}\int_{t_i^k}^{t_{i+1}^k} G_i^k \cdot b_u^{(2)}\left(Z_t^{s,x} - y\right) du - \frac{1}{2}\int_s^t \left|G\left(r, Z_r^{s,x}\right)\right|^2 dr.$$

Using Lemma 13.8 we obtain that

$$E_{\mathbb{P}}\left[\exp\left\{\rho_{s,t}^k\left(w, Z^{s,x}\right)\right\} | Z_t^{s,x} = y\right] = E_{\mathbb{P}} \exp\left\{\tilde{\rho}_u^k\left(\tilde{w}, \hat{Z}_{\cdot;t,y}^{s,x}\right)\right\}, \tag{13.56}$$

where

$$\tilde{\rho}_{s,t}^k\left(\tilde{w}, \hat{Z}_{\cdot,t,y}^{s,x}\right) := \sum_{i=0}^{k-1} \hat{G}_i^k \cdot \Delta\tilde{w}_{t_i^k}$$

$$- \sum_{i=0}^{k-1}\int_{t_i^k}^{t_{i+1}^k} \hat{G}_i^k \cdot \left\{b_u^{(1)}\hat{Z}_{u,t}^s + b_u^{(2)}\left[e^{A(t-s)}x - y\right]\right\} du$$

$$- \frac{1}{2}\int_s^t \left|G\left(r, \hat{Z}_{r,t,y}^{s,x}\right)\right|^2 dr$$

and

$$\hat{G}_i^k := G\left(t_i^k, \hat{Z}_{t_i^k,t,y}^{s,x}\right).$$

The right-hand of (13.56) converges, as $k \to +\infty$, to the expression on the right-hand side (13.54). Moreover, under our assumptions, for any $\phi \in C_b(\mathbb{R}^N)$ we have

$$\lim_{k\to+\infty} E_{\mathbb{P}}\left[\exp\left\{\rho_{s,t}^k\left(w, Z^{s,x}\right)\right\}\phi\left(Z_t^{s,x}\right)\right] = E_{\mathbb{P}}\left[g_{s,t}\left(x, Z_t^{s,x}\right)\phi\left(Z_t^{s,x}\right)\right].$$

This allows us to identify the limit of the expressions on the left-hand side of (13.56) as being equal to $g_{s,t}(x, y)$ which ends the proof of the lemma. □

Using Cauchy–Schwartz inequality we obtain from (13.54) the following upper bound on $g_{s,t}(x, y)$

$$g_{s,t}(x, y) \le \left\{E_{\mathbb{P}} \exp\left\{2\rho_{s,t}\left(\tilde{w}, \hat{Z}_{\cdot,t,y}^{s,x}\right)\right\}\right\}^{1/2}$$

$$\times \left\{E_{\mathbb{P}} \exp\left\{2\|G\|_\infty \int_s^t \left|b_u^{(1)}\hat{Z}_{u,t}^s + b_u^{(2)}\left[e^{A(t-s)}x - y\right]\right| du\right\}\right\}^{1/2} \tag{13.57}$$

for all $x, y \in \mathbb{R}^d$, $t, s \in [0, T]$ and $s < t$. The first factor on the right-hand side of (13.57) can be rewritten in the form

$$\left\{E_{\mathbb{P}} \exp\left\{\rho_{s,t}\left(\tilde{w}, \hat{Z}^{s,x}_{\cdot,t,y}; 2G\right) + \int_s^t \left|G\left(u, Z^{s,x}_u\right)\right|^2 du\right\}\right\}^{1/2} \le e^{\|G\|^2_\infty (t-s)/2}.$$

Using (13.53) and Hölder inequality we can estimate the second factor on the right-hand side by

$$\begin{aligned}&\exp\left\{\|G\|_\infty \|b^{(2)}\|_{L^1[s,t]}\left(\left|e^{A(t-s)}x - y\right|\right)\right\}\left\{E_{\mathbb{P}} \exp\left\{6\|G\|_\infty \|b^{(2)}\|_{L^1[s,t]}\left|Z^t_s\right|\right\}\right\}^{1/6}\\ &\times \left\{E_{\mathbb{P}} \exp\left\{6\|V^t\|_{L^1[s,t]}\|G\|_\infty\right\}\right\}^{1/6},\end{aligned} \tag{13.58}$$

where

$$\|b^{(2)}\|_{L^1[s,t]} := \int_s^t \left|b^{(2)}_u\right| du \quad \text{and} \quad \|V^t\|_{L^1[s,t]} := \int_s^t \left|V^t_u\right| du.$$

An application of Cauchy–Schwartz inequality yields

$$\|b^{(2)}\|_{L^1[s,t]} \le \left(\operatorname{tr} C^{-1}_{t-s}\right)^{1/2} (t-s)^{1/2}.$$

Since C_{t-s} is the covariance matrix of a Gaussian random vector Z^s_t the second factor in (13.58) can be estimated by

$$\exp\left\{C^* \|G\|^2_\infty \|b^{(2)}\|^2_{L^1[s,t]} (\operatorname{tr} C_{t-s})^2\right\} \le \exp\left\{C^* \|G\|^2_\infty (t-s)\left(\operatorname{tr} C^{-1}_{t-s}\right)(\operatorname{tr} C_{t-s})^2\right\}$$

for some absolute constant C^*. To estimate the third factor in (13.58) we rewrite (13.43) in the form

$$V^t_u = Q e^{A^T (t-u)} C^{-\gamma/2}_{t-u} W^t_u,$$

where

$$W^t_u := C^{-(1-\gamma/2)}_{t-u} \int_u^t e^{A(t-r)} Q \, dw_r.$$

Using Cauchy–Schwartz inequality we can estimate $\|V^t\|_{L^1[s,t]}$ by $\mathscr{I}_{\gamma,t-s} \|W^t\|_{L^2[s,t]}$. Suppose now that μ is a centered Gaussian measure on $(H, \|\cdot\|)$—a separable Hilbert space. Then, using Corollary 3.2, pp. 59–60 of Ledoux and Talagrand (1991), we conclude that there exists an absolute constant C^* such that

$$\int_H e^{\xi \|x\|} \mu(dx) \le C^* e^{C^* \xi^2 \operatorname{tr} S}, \quad \forall \xi \in \mathbb{R}.$$

Here S is the covariance operator of the measure. Using this inequality we estimate the term in question by

$$C^* \exp\{C^* \mathscr{I}^2_{\gamma,t-s}\mathscr{G}^2_{\gamma,t-s}\}.$$

Hence (13.48) follows. The proof of the estimates for the derivatives of $p_{s,t}(x, y)$ in x and y can be done analogously. □

Since, it is clear from (13.46) and (13.54) that under the assumptions of twice differentiability of $G(t, x)$ in the x variable the transition probabilities are also twice continuously differentiable in the respective variables we conclude that the Kolmogorov backward equation holds, see Theorem 7.6, p. 366 of Karatzas and Shreve (1991). The fact that the Kolmogorov forward equation also holds is a classical result. Its proof is contained in e.g. Friedman (1975), see the proof of Theorem 4.7, p. 143. The uniform ellipticity assumption made there is only needed for the existence of transition probability densities and appropriate upper bounds. These facts have already been established by us in the foregoing.

13.5 Comments and References

The model described by (13.1) has been considered in Papanicolaou and Varadhan (1985), where the central limit theorem has been shown using a somewhat different argument from the one presented in this chapter. The method of the proof has required the potential to be deterministically bounded together with its two derivatives. The argument presented here has been adapted, with some modifications, from Benabou (2006), where the central limit theorem has been shown for a tagged particle in a system of Ornstein–Uhlenbeck particles interacting with each other via a certain two-body potential. In this situation the resulting potential is unbounded and the method from Papanicolaou and Varadhan (1985) cannot be applied. In the statement of Theorem 13.1 we maintained, for clarity's sake, the assumption that the potential is deterministically bounded together with its gradient. We also mention here an article Tóth (1986) that considers a related discrete model of a persistent random walk in a random environment.

The central limit theorem in case of periodic potentials has been shown in Rodenhausen (1989), where also the Einstein relation between diffusivity and mobility of the particle has been shown. When the dynamics is described by an equation analogous to (13.1), but with forcing of the potential type replaced by a general periodic vector field, the central limit theorem has been shown in Hairer and Pavliotis (2004). Using strong ergodic properties of the hypoelliptic diffusion described by that system it is possible to prove the existence of a smooth solution of the Poisson equation $-\mathscr{L}\chi^{(p)} = V_p$ (using our terminology), when V_p is of zero mean with respect to the invariant measure for the diffusion. From that point on the argument follows the method described in Sect. 2.6.

The estimates on the transition probability densities of Sect. 13.4 are obtained by a technique adapted from Goldys and Maslowski (2006). This paper also contains lower bounds of the transition probability densities corresponding to the solutions of (13.34) by the densities of the Ornstein–Uhlenbeck process corresponding to $H(t,x) \equiv 0$. One can use such bounds and a coupling argument to prove the existence of the asymptotic covariance matrix for the diffusion given by (13.1), when the gradient field $\nabla \tilde{U}(x)$ is replaced by a forcing term that is time dependent, zero mean, stationary and of finite dependence range in the temporal variable, see Komorowski and Krupa (2006).

References

Benabou G (2006) Homogenization of Ornstein-Uhlenbeck process in random environment. Commun Math Phys 266(3):699–714

Ethier SN, Kurtz TG (1986) Markov processes: characterization and convergence. Wiley series in probability and mathematical statistics: probability and mathematical statistics. Wiley, New York

Friedman A (1975) Stochastic differential equations and applications, vol 1. Probability and mathematical statistics, vol 28. Academic Press [Harcourt Brace Jovanovich Publishers], New York

Goldys B, Maslowski B (2006) Lower estimates of transition densities and bounds on exponential ergodicity for stochastic PDE's. Ann Probab 34(4):1451–1496

Hairer M, Pavliotis GA (2004) Periodic homogenization for hypoelliptic diffusions. J Stat Phys 117(1–2):261–279

Janson S (1997) Gaussian Hilbert spaces. Cambridge tracts in mathematics, vol 129. Cambridge University Press, Cambridge

Karatzas I, Shreve SE (1991) Brownian motion and stochastic calculus. Graduate texts in mathematics, vol 113, 2nd edn. Springer, New York

Komorowski T, Krupa G (2006) On the asymptotic behavior of an Ornstein-Uhlenbeck process with random forcing. Commun Math Phys 261(2):517–543

Ledoux M, Talagrand M (1991) Probability in Banach spaces: isoperimetry and processes. Ergebnisse der Mathematik und ihrer Grenzgebiete (3) [Results in mathematics and related areas (3)], vol 23. Springer, Berlin

Papanicolaou G, Varadhan SRS (1985) Ornstein-Uhlenbeck process in a random potential. Commun Pure Appl Math 38(6):819–834

Rodenhausen H (1989) Einstein's relation between diffusion constant and mobility for a diffusion model. J Stat Phys 55(5–6):1065–1088

Stroock DW, Varadhan SRS (1972) On the support of diffusion processes with applications to the strong maximum principle. In: Proceedings of the sixth Berkeley symposium on mathematical statistics and probability, vol III: Probability theory. Univ California, Berkeley, California, 1970/1971, University of California Press, Berkeley, pp 333–359

Tóth B (1986) Persistent random walks in random environment. Probab Theory Relat Fields 71(4):615–625

Chapter 14
Analytic Methods in Homogenization Theory

In this chapter we present another, more analytic in flavor, point of view on the problem of homogenization of diffusions in random environments. We have already indicated in Sect. 9.6 that this topic is closely related to the problem of convergence of solutions of partial differential equations with random coefficients. To recall we suppose that $\{X_t^{x,\omega}, t \geq 0\}$ is a diffusion given by (9.30). For a bounded domain U with a sufficiently regular boundary ∂U, parameter $\varepsilon > 0$ and $u_0 \in L^2(U)$ consider a solution of the initial boundary value problem

$$\begin{aligned} &\partial_t u^{(\varepsilon)}(t,x) = \frac{1}{2}\sum_{k,l=1}^{d} \partial_{x_k}\left[\tilde{a}_{k,l}\left(\frac{x}{\varepsilon};\omega\right)\partial_{x_l} u^{(\varepsilon)}(t,x)\right], \quad (t,x) \in (0,+\infty)\times U, \\ &u^{(\varepsilon)}(t,x) = 0, \quad (t,x) \in (0,+\infty)\times \partial U, \\ &u^{(\varepsilon)}(0,x) = u_0(x). \end{aligned} \tag{14.1}$$

The solution can be represented using the diffusion process:

$$u^{(\varepsilon)}(t,x;\omega) = E_{\mathbb{P}}\big[u_0\big(X_{t,\varepsilon}^{x,\omega}\big), \tau_{x,U}^{(\varepsilon)} > t\big],$$

where

$$X_{t,\varepsilon}^{x,\omega} := \varepsilon X_{t/\varepsilon^2}^{x/\varepsilon,\omega} \tag{14.2}$$

and $\tau_{x,U}^{(\varepsilon)}$ is the exit time of the diffusion from U. It turns out, see Theorem 14.14 below, that if the coefficients $\tilde{a}_{k,l}(x;\omega)$ satisfy assumptions made in Sect. 9.3.3 then

$$\lim_{\varepsilon\to 0+} \big\langle \big[u^{(\varepsilon)}(t,x) - \bar{u}(t,x)\big]^2 \big\rangle_{\mathbb{Q}} = 0,$$

T. Komorowski et al., *Fluctuations in Markov Processes*,
Grundlehren der mathematischen Wissenschaften 345,
DOI 10.1007/978-3-642-29880-6_14, © Springer-Verlag Berlin Heidelberg 2012

where $\bar{u}(t,x)$ is a solution of the heat equation with constant coefficients

$$\begin{aligned} \partial_t \bar{u}(t,x) &= \frac{1}{2}\sum_{k,l=1}^{d} \bar{a}_{k,l}\partial^2_{x_k,x_l}\bar{u}(t,x), \\ \bar{u}(0,x) &= u_0(x), \qquad \bar{u}(t,x)=0, x\in\partial U, \end{aligned} \tag{14.3}$$

with $\bar{a}_{k,l}$, given by (9.55). The limiting procedure described above (called homogenization, cf. Sect. 9.6) in particular implies the central limit theorem for $X_t^{x,\omega}/\sqrt{t}$, see Theorem 14.16 below. The reverse implication has already been discussed in Sect. 9.6.

Concerning the organization of the material in this chapter: in the first part, contained in Sects. 14.1 and 14.2, we introduce the notions of G-convergence of operators and Γ-convergence of quadratic forms that play important roles in the analytic theory of homogenization. In Sect. 14.3 we present the compensated compactness argument to prove the G-convergence of a family of second order elliptic differential operators. This material is used in Sect. 14.4 to derive the convergence result for the solutions of (14.1) with stationary and ergodic coefficients and prove again the central limit theorem for corresponding diffusions. The appendix contained in Sect. 14.5 is devoted to the proof of uniform ellipticity for the second order divergence form partial differential operators whose corresponding Dirichlet form is coercive.

14.1 *G*-Convergence of Operators

Suppose that H is a separable Hilbert space and H^* its dual. For fixed constants $M\ge m>0$ denote by $\mathscr{C}(m,M)$ the class of linear operators $L:H\to H^*$ that satisfy:

$$m\|u\|_H^2 \le {}_{H^*}\langle Lu,u\rangle_H, \qquad \|Lu\|_{H^*}\le M\|u\|_H, \qquad \forall u\in H. \tag{14.4}$$

Here ${}_{H^*}\langle\cdot,\cdot\rangle_H$ is the duality pairing between H and H^*.

Note that the above estimates imply in particular that each $L\in\mathscr{C}(m,M)$ is invertible and $\|L^{-1}\|\le Mm^{-1}$.

Definition 14.1 We say that a sequence of operators $\{L_n, n\ge 1\}\subset\mathscr{C}(m,M)$ is G-convergent to $L_0:H\to H^*$, and write $L_0=G\text{-}\lim_{n\to+\infty}L_n$, if L_0 is invertible and

$$L_n^{-1}f \rightharpoonup L_0^{-1}f \quad \text{as } n\to+\infty, \tag{14.5}$$

weakly in H for any $f\in H^*$.

Below we list some of the properties of G-convergence.

Proposition 14.2 *Suppose that* $L_0=G\text{-}\lim_{n\to+\infty}L_n$, *where* $\{L_n,n\ge 1\}\subset\mathscr{C}(m,M)$. *Then,* $L_0\in\mathscr{C}(m,M^2/m)$.

Proof For any $n \ge 0$ let $u_n := L_n^{-1} f$. Since $L_n \in \mathscr{C}(m, M)$ we conclude that

$$m\|u_n\|_H^2 \le {}_{H^*}\langle L_n u_n, u_n\rangle_H = {}_{H^*}\langle f, u_n\rangle_H. \tag{14.6}$$

Passing to the limit as $n \to +\infty$ we get

$$m\|u_0\|_H^2 \le {}_{H^*}\langle f, u_0\rangle_H = {}_{H^*}\langle L_0 u_0, u_0\rangle_H.$$

On the other hand, from (14.4), we have

$$\|f\|_{H^*}^2 \le M^2\|u_n\|_H^2 \le M^2 m^{-1}{}_{H^*}\langle f, u_n\rangle_H.$$

We obtain, upon letting $n \to +\infty$,

$$\|f\|_{H^*}^2 \le M^2 m^{-1}{}_{H^*}\langle f, u_0\rangle_H \le M^2 m^{-1}\|f\|_{H^*}\|u_0\|_H \tag{14.7}$$

and the proposition follows. □

Proposition 14.3 *Suppose that $\mathscr{I}$ is an arbitrary set of indices. Any family of operators $\{L_\varepsilon, \varepsilon \in \mathscr{I}\} \subset \mathscr{C}(m, M)$ contains a G-convergent sequence.*

Proof Suppose that $\{f_k, k \ge 1\}$ is a linearly dense subset of H^*. Since the set $\{\|L_\varepsilon^{-1}\|, \varepsilon \in \mathscr{I}\}$ is bounded we can use the diagonal process to extract a sequence $\{L_{\varepsilon_n}, n \ge 1\}$ such that

$$\lim_{n\to+\infty} {}_{H^*}\big\langle f_k, L_{\varepsilon_n}^{-1} f_j\big\rangle_H \quad \text{exists for each } j, k \ge 1.$$

By a simple density argument we conclude that $\lim_{n\to+\infty} {}_{H^*}\langle g, L_{\varepsilon_n}^{-1} f\rangle_H$ exists for each $g, f \in H^*$. Denote by Bf the weak limit in H of $L_{\varepsilon_n}^{-1} f$, as $n \to +\infty$. Since the norms of $L_{\varepsilon_n}^{-1}$ are uniformly bounded by Mm^{-1} we conclude that $B : H^* \to H$ is bounded and $\|B\| \le Mm^{-1}$. Combining (14.6) and (14.7) we can write

$$\|f\|_{H^*}^2 \le M^2 m^{-1}{}_{H^*}\langle f, Bf\rangle_H.$$

Hence B is invertible and $L_0 := B^{-1}$ is the G-limit of L_{ε_n}, as $n \to +\infty$. □

Define $L^* : H \to H^*$, the adjoint of L, by

$${}_{H^*}\big\langle L^* u, v\big\rangle_H := {}_{H^*}\langle Lv, u\rangle_H, \quad \forall u, v \in H.$$

It is elementary to verify that $L^* \in \mathscr{C}(m, M)$, if $L \in \mathscr{C}(m, M)$.

Proposition 14.4 *Suppose that $\{L_n, n \ge 1\}$ is G-convergent to L_0. Then, $\{L_n^*, n \ge 1\}$ is G-convergent to L_0^*.*

Proof It suffices to show that an arbitrary subsequence of $\{L_n^*, n \geq 1\}$ contains a subsequence G-converging to L_0^*. For a convenience sake we shall also denote such a subsequence by $\{L_n^*, n \geq 1\}$. From Proposition 14.3 there exists a G-converging subsequence $\{L_{n_k}^*, k \geq 1\}$. We show that it converges to L_0^*. For $f, g \in H_*$ we let

$$u_k := \left(L_{n_k}^*\right)^{-1} f, \qquad v_k := L_{n_k}^{-1} g, \qquad u := \left(L_0^*\right)^{-1} f$$

and $v := L_0^{-1} g$. We have

$$\lim_{k\to+\infty} {}_{H^*}\langle g, u_k\rangle_H = \lim_{k\to+\infty} {}_{H^*}\langle L_{n_k} v_k, u_k\rangle_H = \lim_{k\to+\infty} {}_{H^*}\langle f, v_k\rangle_H. \tag{14.8}$$

Since $\{L_{n_k}, k \geq 1\}$ is G-convergent to L_0 the utmost right-hand side of (14.8) equals

$${}_{H^*}\langle f, v\rangle_H = {}_{H^*}\langle g, u\rangle_H$$

and the conclusion of the proposition follows. □

14.2 Γ-Convergence of Quadratic Forms

Suppose that $L \in \mathscr{C}(m, M)$. We call it symmetric if the bilinear form

$$\Gamma(u, v) := {}_{H^*}\langle Lu, v\rangle_H, \qquad u, v \in H \tag{14.9}$$

is symmetric, i.e. $\Gamma(u, v) = \Gamma(v, u)$. To abbreviate we shall write $\Gamma(u) := \Gamma(u, u)$.

For a given $f \in H^*$ define *the energy functional* $\mathscr{E}_f : H \to \mathbb{R}$ by

$$\mathscr{E}_f(u) := \frac{1}{2}\Gamma(u) - {}_{H^*}\langle f, u\rangle_H, \qquad u \in H. \tag{14.10}$$

There exists a unique minimizer u_* of this functional that satisfies Euler–Lagrange equation

$$Lu_* = f. \tag{14.11}$$

Thanks to coercivity of L Eq. (14.11) has a unique solution and

$$\bar{\mathscr{E}}(f) := \min_u \mathscr{E}_f(u) = -\frac{1}{2}{}_{H^*}\langle f, L^{-1} f\rangle_H. \tag{14.12}$$

As a direct corollary of Proposition 14.3 we obtain the following.

Proposition 14.5 *The G-limit of a sequence of symmetric operators $\{L_n, n \geq 1\} \subset \mathscr{C}(m, M)$ is symmetric.*

The following result gives a criterion of the G-convergence of operators in terms of the convergence of their corresponding energy functionals.

Theorem 14.6 *The sequence of symmetric operators* $\{L_n, n \geq 1\} \subset \mathscr{C}(m, M)$ *is G convergent to* L_0 *if and only if*

$$\lim_{n\to+\infty} \bar{\mathscr{E}}^{(n)}(f) = \bar{\mathscr{E}}^{(0)}(f), \quad \forall f \in H^*. \tag{14.13}$$

Here $\bar{\mathscr{E}}^{(n)}(f)$ *is the minimum of the energy functional* $\mathscr{E}^{(n)}(f)$ *corresponding to* f *and* L_n.

Proof From (14.11) we know that the minimizers of the energy functional corresponding to L_n is given by $u_n := L_n^{-1} f$ for $n \geq 0$. The asserted equivalence follows from the definition of the G-convergence and formula (14.12). □

Definition 14.7 We say that a sequence of functionals $F_n : H \to \mathbb{R}$, $n \geq 1$ is Γ-convergent to a functional $F_0 : H \to \mathbb{R}$ if

(i) for any $u \in H$ and any sequence $u_n \rightharpoonup u$ weakly convergent in H, as $n \to +\infty$, we have

$$\liminf_{n\to+\infty} F_n(u_n) \geq F_0(u),$$

(ii) there exists a sequence $v_n \rightharpoonup u$, weakly in H, as $n \to +\infty$, such that

$$\liminf_{n\to+\infty} F_n(v_n) = F_0(u).$$

The notion of G-convergence for symmetric operators is equivalent to the notion of Γ-convergence of their corresponding energy functionals as it is asserted in the next result.

Theorem 14.8 *Suppose that* $\{L_n, n \geq 1\}$ *is a sequence of symmetric operators belonging to* $\mathscr{C}(m, M)$. *Denote by* $\{\Gamma_n,\ n \geq 0\}$ *the sequence of corresponding bilinear forms, see* (14.9). *Then,* $\{L_n, n \geq 1\}$ *is G-convergent to* L_0 *if and only if* $\{\Gamma_n, n \geq 1\}$ *is* Γ*-convergent to* Γ_0.

Proof The "if" part. Suppose that sequence $\{\Gamma_n, n \geq 1\}$ converges to Γ_0. Let $u \in H$ be arbitrary and let $u_n \rightharpoonup u$, as $n \to +\infty$, weakly in H be such that $\lim_{n\to+\infty} \Gamma_n(u_n) = \Gamma_0(u)$. Then,

$$\limsup_{n\to+\infty} \bar{\mathscr{E}}^{(n)}(f) \leq \lim_{n\to+\infty} \mathscr{E}_f^{(n)}(u_n) = \mathscr{E}_f^{(0)}(u).$$

In consequence, we obtain

$$\limsup_{n\to+\infty} \bar{\mathscr{E}}^{(n)}(f) \leq \bar{\mathscr{E}}^{(0)}(f). \tag{14.14}$$

On the other hand, suppose that u_n is such that

$$\mathscr{E}_f^{(n)}(u_n) \leq \bar{\mathscr{E}}^{(n)}(f) + \frac{1}{n}. \tag{14.15}$$

The uniform coercivity of $\{L_n, n \geq 1\}$ implies that $\{\|u_n\|_H, n \geq 1\}$ is bounded. One can choose therefore a subsequence of $\{u_n, n \geq 1\}$, denoted for simplicity in the same way, that is G-weakly convergent, say to u_0. Using property (i) from the definition of Γ-convergence we obtain

$$\liminf_{n\to+\infty} \mathscr{E}_f^{(n)}(u_n) \geq \mathscr{E}_f^{(0)}(u_0) \geq \bar{\mathscr{E}}^{(0)}(f).$$

Combining this inequality with (14.15) we get

$$\liminf_{n\to+\infty} \bar{\mathscr{E}}^{(n)}(f) \geq \bar{\mathscr{E}}^{(0)}(f).$$

Thanks to Theorem 14.6, this together with (14.14) allows us to conclude that

$$G\text{-}\lim_{n\to+\infty} L_n = L_0. \tag{14.16}$$

The "only if" part. Suppose that (14.16) holds and that $u_n \rightharpoonup u_0$, weakly in H. Let $f := L_0 u_0$. By virtue of Theorem 14.6 we have then

$$\liminf_{n\to+\infty} \mathscr{E}_f^{(n)}(u_n) \geq \lim_{n\to+\infty} \bar{\mathscr{E}}^{(n)}(f) = \bar{\mathscr{E}}^{(0)}(f) = \mathscr{E}_f^{(0)}(u_0).$$

Hence

$$\liminf_{n\to+\infty} \Gamma_n(u_n) \geq \Gamma_0(u_0)$$

thus, condition (i) from the definition of the Γ-convergence has been verified.

To verify part (ii) of the definition, choose $u_n := L_n^{-1} f$. From the definition of G-convergence we obtain $u_n \rightharpoonup u_0 = L_0^{-1} f$. In consequence,

$$\lim_{n\to+\infty} \Gamma_n(u_n) = \lim_{n\to+\infty} {}_{H^*}\langle f, u_n\rangle_H = {}_{H^*}\langle f, u_0\rangle_H = \Gamma_0(u_0).$$

Hence, $\{\Gamma_n, n \geq 1\}$ is Γ-convergent to Γ_0. □

14.3 *G*-Convergence of Matrix Valued Functions

Suppose that $U \subset \mathbb{R}^d$ is a bounded domain. For fixed $M \geq m > 0$ denote by $\mathbb{M}(m, M)$ the class of measurable, $d \times d$ matrix valued functions $a = [a_{ij}] : U \to \mathbb{M}(d)$ such that:

(i) $|a_{ij}(x)| \leq M$ for all $i, j = 1, \ldots, d$ and a.e. $x \in U$,
(ii) $\sum_{ij=1}^d a_{ij}(x)\xi_i\xi_j \geq m|\xi|^2$, $\forall \xi \in \mathbb{R}^d$ and a.e. $x \in U$.

Let $H := W_{2,o}^1(U)$ be the completion of $C_c^\infty(U)$ in the norm

$$\|u\|_{W_{2,o}^1(U)} := \|\nabla_x u\|_{L_d^2(U)}, \quad u \in C_c^\infty(U).$$

Its dual will be denoted by $W_2^{-1}(U)$. For each $a \in \mathbb{M}(m, M)$ we can define an operator $L : W_{2,o}^1(U) \to W_2^{-1}(U)$ by

$$Lu(v) := \int_U a\nabla_x u \cdot \nabla_x v dx, \quad \forall u, v \in W_{2,o}^1(U). \tag{14.17}$$

L is the divergence form elliptic operator corresponding to a given matrix valued function $a(\cdot)$ with the Dirichlet boundary condition on U. Sometimes, we shall also write $L = -\nabla_x \cdot (a\nabla_x)$. Using the notation of Sect. 14.1 we can write $L \in \mathscr{C}(m, M)$.

In this section we assume that a family of matrix valued functions $\{a^{(\varepsilon)}, \varepsilon \in (0, 1]\}$ is contained in $\mathbb{M}(m, M)$ for some fixed $0 < m < M$. Suppose also that $f \in W_2^{-1}(U)$ and u_ε is the unique solution of the equation $L_\varepsilon u_\varepsilon = f$, where L_ε is the operator corresponding to $a^{(\varepsilon)}$ via (14.17). It is the unique weak solution in $W_{2,o}^1(U)$ of the Dirichlet problem

$$\begin{aligned} &\nabla_x \cdot \left(a^{(\varepsilon)}\nabla_x u_\varepsilon\right) = f \quad \text{in } U, \\ &u_\varepsilon = 0 \quad \text{on } \partial U. \end{aligned} \tag{14.18}$$

Definition 14.9 We say that $a^{(\varepsilon)}$ is G-convergent, as $\varepsilon \to 0+$, to a matrix valued function $\bar{a}$ and write $\bar{a} = G\text{-}\lim_{\varepsilon\to 0+} a^{(\varepsilon)}$, if for each $f \in W_2^{-1}(U)$ there exists $u_0 \in W_{2,o}^1(U)$ such that

$$u_\varepsilon \rightharpoonup u_0, \quad \textit{weakly in } W_{2,o}^1(U) \tag{14.19}$$

and

$$a^{(\varepsilon)} \cdot \nabla_x u_\varepsilon \rightharpoonup \bar{a} \cdot \nabla_x u_0, \quad \textit{weakly in } L_d^2(U), \quad \textit{as } \varepsilon \to 0+. \tag{14.20}$$

Note that the above definition implies the uniqueness of a possible limit. Indeed, suppose that there exist two matrices $\bar{a}$ and $\hat{a}$ that are G-limits of $a^{(\varepsilon)}$, as $\varepsilon \to 0+$. Choose an arbitrary $\varepsilon_n \to 0+$. According to Proposition 14.3 the corresponding sequence of operators $\{L_n, n \geq 1\}$ contains a subsequence, which for convenience sake will be denoted by the same symbol, G-converging towards some L_0.

Choose an arbitrary $u_0 \in W_{2,o}^1(U)$ and set $f := L_0 u_0$. From the definition of the G-convergence of operators we obtain that $u_{\varepsilon_n} = L_n^{-1} f \rightharpoonup u_0$, weakly in $W_{2,o}^1(U)$. On the other hand, from (14.19) and (14.20) we conclude that

$$\bar{a} \cdot \nabla_x u_0 = \lim_{n\to+\infty} a^{(\varepsilon_n)} \cdot \nabla_x u_{\varepsilon_n} = \hat{a} \cdot \nabla_x u_0.$$

Since u_0 can be arbitrary from $W_{2,o}^1(U)$ this implies that $\bar{a} = \hat{a}$.

In what follows we formulate some basic properties of G-convergence of matrix valued functions.

Proposition 14.10 *Suppose that* $\bar{a} = G\text{-}\lim_{\varepsilon\to 0+} a^{(\varepsilon)}$. *Then,* $\bar{a} \in \mathbb{M}(m, M_1)$ *for some* M_1 *depending only on* m, M.

Proof Let us choose an arbitrary sequence $\varepsilon_n \to 0+$. Denote by L_n the operator corresponding to $a^{(\varepsilon_n)}$ via (14.17) and by L_0 its G-limit. Then, L_0 and $\bar{a}$ are related via (14.17). Indeed, let $u_{\varepsilon_n} := L_n^{-1} L_0 u_0$, where u_0 belongs to $W^1_{2,o}(U)$. For any $v \in W^1_{2,o}(U)$ we can write

$$L_0 u_0(v) = L_n u_{\varepsilon_n}(v) = \int_U a^{(\varepsilon_n)} \nabla_x u_{\varepsilon_n} \cdot \nabla_x v dx. \tag{14.21}$$

From (14.20) the utmost right-hand side converges to

$$\int_U \bar{a} \nabla_x u_0 \cdot \nabla_x v dx.$$

According to Proposition 14.2 we have $L_0 \in \mathscr{C}(m, M^2/m)$. This implies that $\bar{a} \in \mathbb{M}(m, M_1)$ for some M_1. Indeed, by (14.20) we have

$$\|\bar{a}\nabla_x u_0\|_{L^2_d(U)} \le \liminf_{n\to+\infty} \left\|a^{(\varepsilon_n)} \nabla_x u_{\varepsilon_n}\right\|_{L^2_d(U)} \le M \liminf_{n\to+\infty} \|u_{\varepsilon_n}\|_{W^1_{2,o}(U)}. \tag{14.22}$$

Since $\{L_n, n \ge 1\} \subset \mathscr{C}(m, M)$ and $L_0 \in \mathscr{C}(m, M^2/m)$ the utmost right-hand side is bounded from above by

$$Mm^{-1} \liminf_{n\to+\infty} \|L_n u_{\varepsilon_n}\|_{W_2^{-1}(U)},$$

which, thanks to (14.21), equals

$$Mm^{-1} \|L_0 u_0\|_{W_2^{-1}(U)} \le M^3 m^{-2} \|\nabla_x u_0\|_{L^2_d(U)}. \tag{14.23}$$

Fix $p \in \{1, \dots, d\}$. For any $z \in U$ and $R > 0$ such that $B_R(z) \subset U$ we choose

$$u_0(x) := \rho\big((x - z)R^{-1}\big)(x_p - z_p),$$

where $0 \le \rho(x) \le 1$ is a C^∞ function such that $\rho(x) \equiv 1$, $|x| \le 1/2$ and $\rho(x) \equiv 0$, when $|x| \ge 1$. With this choice of u_0 we conclude from (14.22) and (14.23) that

$$\sum_{q=1}^d \int_{B_{R/2}(z)} \big[\bar{a}_{p,q}(x)\big]^2 dx \le C_1 R^d, \quad p = 1, \dots, d$$

for some constant $C_1 > 0$ depending only on m, M. This implies the upper bound on the entries of $\bar{a}$.

To obtain the lower bound recall that $L_0 \in \mathscr{C}(m, M)$, thus

$$\int_U \bar{a} \nabla_x u_0 \cdot \nabla_x u_0 dx \ge m \|\nabla_x u_0\|^2_{L^2_d(U)}, \quad \forall u_0 \in W^1_{2,o}(U). \tag{14.24}$$

The coercivity condition implies the desired lower bound on the matrix $\bar{a}(x)$, see Theorem 14.17 below. □

The following compactness result holds.

Theorem 14.11 *Suppose that $\{a^{(\varepsilon)}, \varepsilon \in (0,1]\}$ is contained in $\mathbb{M}(m, M)$. Then, any sequence $\{a^{(\varepsilon_n)}, n \ge 1\}$ contains a G-converging subsequence.*

Proof Denote by L_n the elliptic operator associated with $a^{(\varepsilon_n)}$. According to Proposition 14.3 one can choose a subsequence, which for convenience we denote the same way as the original sequence, $\{a^{(\varepsilon_n)}, n \ge 1\}$ such that the corresponding operators G-converge, as $n \to +\infty$, to some L_0. Then, for any $f \in W_2^{-1}(U)$ we have

$$u_n := L_n^{-1} f \rightharpoonup u_0 := L_0^{-1} f, \quad \text{weakly in } W^1_{2,o}(U).$$

In fact, since matrices $a^{(\varepsilon_n)}$ are bounded, by further refinement of the subsequence we can also guarantee that $a^{(\varepsilon_n)} \nabla_x u_n$ converges weakly in $L^2_d(U)$ to some $p_0 \in L^2_d(U)$.

To finish the proof it suffices to show that

$$p_0(x) = \bar{a}(x) \nabla_x u_0(x), \quad \text{a.e. on } U \tag{14.25}$$

for a certain matrix valued function $\bar{a}$ belonging to $\mathbb{M}(m_1, M_1)$ and some positive constants m_1, M_1. Fix an arbitrary $i = 1, \ldots, d$ and define $f_n^{(i)} \in W_2^{-1}(U)$ by

$$f_n^{(i)}(u) := \int_U a^{(\varepsilon_n)} e_i \cdot \nabla_x u dx, \quad u \in W^1_{2,o}(U).$$

Here e_i is the i-th vector of the canonical base in $\mathbb{R}^d$. Let $v_n^{(i)} \in W^1_{2,o}(U)$ be the unique solution of $L_n(v_n^{(i)}) = f_n^{(i)}$. Since $\{v_n^{(i)}, n \ge 1\}$ is bounded in $W^1_{2,o}(U)$ we can find a subsequence (denoted by the same symbol) such that $v_n^{(i)} \rightharpoonup v_0^{(i)}$, weakly in $W^1_{2,o}(U)$. Define

$$\chi_n^{(i)} := L_n^{-1}\big(L_0 v_0^{(i)} - f_n^{(i)}\big) = L_n^{-1} L_0 v_0^{(i)} - v_n^{(i)}. \tag{14.26}$$

Since L_n is G-convergent to L_0 we conclude that

$$\lim_{n \to +\infty} \chi_n^{(i)} = 0,$$

weakly in $W^1_{2,o}(U)$.

Define a sequence of matrices $A^{(n)}(x) := [A_{ij}^{(n)}(x)]$, where

$$A_{ij}^{(n)}(x) := a_{ij}^{(\varepsilon_n)}(x) + \sum_{k=1}^{d} a_{ik}^{(\varepsilon_n)}(x) \partial_{x_k} \chi_n^{(j)}(x), \quad i, j = 1, \ldots, d.$$

Functions $A_{ij}^{(n)}$ are bounded in $L^2(U)$, uniformly in n, so one can find a weakly convergent subsequence. Denote the limiting matrix by $\bar{a}(x) := [\bar{a}_{ij}(x)]$. For any $\xi \in \mathbb{R}^d$ define

$$h_n(x;\xi) = \big(h_{n1}(x), \ldots, h_{nd}(x)\big) := \nabla_x u_n(x) - \xi - \sum_{i=1}^d \nabla_x \chi_n^{(i)}(x)\xi_i.$$

Note that $\lim_{n\to+\infty} h_n(x;\xi) = \nabla_x u_0(x) - \xi$ and (from the definition of $A_{ij}^{(n)}(x)$)

$$\lim_{n\to+\infty} a^{(\varepsilon_n)}(x) h_n(x;\xi) = p_0 - \bar{a}(x)\xi,$$

weakly in $L_d^2(U)$. Let ϕ be an arbitrary, non-negative function from $C_c^\infty(U)$. We have

$$0 \le r_n(\phi) := \int_U a^{(\varepsilon_n)}(x) h_n(x;\xi)\cdot h_n(x;\xi)\phi(x)dx. \tag{14.27}$$

In addition, from (14.26) we get

$$\begin{aligned} \nabla_x \cdot \big[a^{(\varepsilon_n)} h_n(\cdot;\xi)\big] &= -f - \nabla_x \cdot \big(\tilde{a}^{(n)}\xi\big) \\ &= -f + \sum_{j=1}^d \big(f_n^{(j)} + L_n \chi_n^{(j)}\big)\xi_j \\ &= -f + \sum_{j=1}^d v_0^{(j)}\xi_j. \end{aligned} \tag{14.28}$$

In particular the sequence appearing on the utmost left-hand side of (14.28) is compact in $W_2^{-1}(U)$. Since also

$$\operatorname{curl} h_n(x) = \big[\partial_{x_i} h_{nj}(x) - \partial_{x_j} h_{ni}(x)\big] \equiv 0,$$

by the div-curl lemma, see e.g. Theorem 4, p. 54 of Evans (1990), we conclude from (14.27) that

$$0 \le \lim_{n\to+\infty} r_n(\phi) = \int_U \big[p_0(x) - \bar{a}(x)\xi\big]\cdot\big[\nabla_x u_0(x) - \xi\big]\phi(x)dx,$$

for all $\xi \in \mathbb{R}^d$ and non-negative $\phi \in C_c^\infty(U)$. Choose representatives $\bar{a}(x)$, $u_0(x)$, $\nabla_x u_0(x)$, $p_0(x)$ and a set N of Lebesgue measure 0, for which

$$\big[p_0(x) - \bar{a}(x)\xi\big]\cdot\big[\nabla_x u_0(x) - \xi\big] \ge 0, \quad \forall \xi \in \mathbb{R}^d,\ x \in U \setminus N. \tag{14.29}$$

Suppose that $x \in U \setminus N$ and $(t,\eta) \in \mathbb{R}^{1+d}$ is arbitrary. Set $\xi := \nabla_x u_0(x) + t\eta$. Inequality (14.29) implies that

$$t\big[p_0(x) - \bar{a}(x)\nabla_x u_0(x)\big]\cdot\eta + t^2\bar{a}(x)\eta\cdot\eta \geq 0, \quad \forall(t,\eta)\in\mathbb{R}^{1+d},\ x\in U\setminus N.$$

This is possible only if $p_0(x) = \bar{a}(x)\nabla_x u_0(x)$ on $U\setminus N$ and (14.25) follows. □

Recall that $W_2^1(U)$ is the completion of the space (cf. (11.24))

$$\big[u\in C^\infty(U) : \|u\|^2_{W_2^1(U)} := \|u\|^2_{L^2(U)} + \|\nabla_x u\|^2_{L^2_d(U)} < +\infty\big]$$

under the norm $\|\cdot\|_{W_2^1(U)}$. The argument presented in the proof of the above theorem permits the following slight generalization of the result.

Theorem 14.12 *Suppose that $0 < m < M$ are fixed. A sequence of matrix valued functions $\{a^{(n)}, n\geq 1\}$ contained in $\mathbb{M}(m,M)$ is G-convergent to $\bar{a} := [\bar{a}_{ij}]$, provided one can construct a sequence $\{\chi_n^{(i)}, n\geq 1\}$ of elements from $W_2^1(U)$ such that*

$$\big\{\nabla_x\cdot\big[a^{(n)}\big(e_i + \nabla_x\chi_n^{(i)}\big)\big], n\geq 1\big\} \quad \textit{is compact in } W_2^{-1}(U), \tag{14.30}$$

$$\lim_{n\to+\infty}\partial_{x_j}\chi_n^{(i)} = 0, \tag{14.31}$$

and

$$\lim_{n\to+\infty} a^{(n)}\big(e_i + \nabla_x\chi_n^{(i)}\big)\cdot e_j = \bar{a}_{ij}, \quad \forall i,j = 1,\dots,d. \tag{14.32}$$

The last two limits are understood weakly in $L^2(U)$.

It is clear that the G-convergence of functions $a^{(\varepsilon)}(x)$ implies the G-convergence of the corresponding operators L_ε. In fact, the converse also holds. Namely, we have the following.

Corollary 14.13 *Suppose that $\{L_\varepsilon, \varepsilon\in(0,1]\}$ is a family of divergence form elliptic operators corresponding to matrix valued functions $\{a^{(\varepsilon)}, \varepsilon\in(0,1]\}$, contained in $\mathbb{M}(m,M)$ for some $0<m<M$. Assume also that $L_0 = G\text{-}\lim_{\varepsilon\to 0+} L_\varepsilon$. Then, functions $a^{(\varepsilon)}(\cdot)$ G-converge to some $\bar{a}(\cdot)$. In addition, L_0 corresponds to $\bar{a}(\cdot)$ via* (14.17).

Proof From Theorem 14.11 we know that for any sequence $\varepsilon_n\to 0$ there exists a G-converging subsequence, which we also denote by $\{a^{(\varepsilon_n)}, n\geq 1\}$. Denote its G-limit by $\bar{a}(\cdot)$. From the assumption made above the corresponding sequence of operators G-converges to L_0. Thus, the operator and the limiting matrix have to be related to each other via (14.17). This argument also shows that the set of the limit points of $\{a^{(\varepsilon)}, \varepsilon\in(0,1]\}$, as $\varepsilon\to 0+$, has to be a singleton and the conclusion of the corollary follows. □

14.4 Application to Homogenization of Diffusions in Random Media

Suppose that a probability space $(\Omega, \mathscr{F}, \mathbb{Q})$, an additive group $\{\tau_x, x \in \mathbb{R}^d\}$ and a random matrix $a : \Omega \to \mathbb{M}(d)$ are as described in Sect. 9.3. We assume also that $\tilde{a}(x;\omega) = a(\tau_x\omega)$ belongs to $\mathbb{M}(m, M)$ for $\mathbb{Q}$-a.s. ω and U is a bounded region in $\mathbb{R}^d$ with a sufficiently smooth boundary, e.g. of C^2 class. For any $\varepsilon > 0$ denote $\tilde{a}^{(\varepsilon)}(x;\omega) := \tilde{a}(x/\varepsilon;\omega)$. Let

$$L_\varepsilon^\omega \varphi(x) := \frac{1}{2}\sum_{k,l=1}^{d} \partial_{x_k}\big(\tilde{a}_{k,l}^{(\varepsilon)}(x;\omega)\partial_{x_l}\varphi(x)\big), \quad \varphi \in C_c^2(U). \tag{14.33}$$

Operator L_ε^ω maps $W_{2,o}^1(U)$ to $W_2^{-1}(U)$, see (14.17). Our first result deals with the question of G-convergence of these operators, as $\varepsilon \to 0+$.

Theorem 14.14 *Under the assumptions made above, for $\mathbb{Q}$ a.s. ω there exists the G-limit of $\tilde{a}^{(\varepsilon)}(\cdot;\omega)$, as $\varepsilon \to 0+$. The limit is a deterministic, constant, positive definite matrix. The corresponding operator L_0 is the G-limit of L_ε^ω, as $\varepsilon \to 0+$, for $\mathbb{Q}$ a.s. ω.*

Proof Let $\chi_\lambda^{(p)}$ be the λ-correctors defined by (9.48) and let $\chi^{(p)}$ be any $\mathscr{H}_1$ weak limiting point obtained when λ converges to 0. Then, $\nabla\chi^{(p)} \in L^2(\mathbb{Q})$ and we can define a (non-stationary) random field

$$\chi^{(p)}(x;\omega) := \int_0^1 \nabla\chi^{(p)}(\tau_{sx}\omega)\cdot x\,ds, \quad x \in \mathbb{R}^d. \tag{14.34}$$

Let $\chi^{(p,\varepsilon)}(x;\omega) := \varepsilon\chi^{(p)}(x/\varepsilon;\omega)$. It can easily be checked that for $\mathbb{Q}$-a.s. ω we have $\chi^{(p)}(\cdot;\omega) \in W_2^1(U)$ and

$$\nabla_x\chi^{(p,\varepsilon)}(x;\omega) = \nabla\chi^{(p)}(\tau_{x/\varepsilon}\omega).$$

The family $\chi^{(p,\varepsilon)}(x)$ satisfies conditions (14.30)–(14.32) formulated in Theorem 14.12 (with the obvious replacement of n by parameter ε). Indeed, from (9.47) and (9.48) we obtain that for any $\phi \in C_c^\infty(U)$

$$\lambda\int_U \tilde{\chi}_\lambda^{(p,\varepsilon)}(x)\phi(x)dx + \int_U \big\{\tilde{a}^{(\varepsilon)}(x)\big[e_i + \nabla_x\tilde{\chi}_\lambda^{(p,\varepsilon)}(x)\big]\big\}\cdot\nabla_x\phi(x)dx = 0,$$

where $\tilde{\chi}_\lambda^{(p,\varepsilon)}(x;\omega) := \varepsilon\chi_\lambda^{(p)}(\tau_{x/\varepsilon}\omega)$. Letting $\lambda \to 0+$ and using (9.49) (in this case $\pi = \mathbb{Q}$) we conclude that

$$\int_U \big\{\tilde{a}^{(\varepsilon)}(x)\big[e_i + \nabla_x\tilde{\chi}^{(p,\varepsilon)}(x)\big]\big\}\cdot\nabla_x\phi(x)dx = 0, \quad \forall\phi \in C_c^\infty(U).$$

From Theorem 11.18 we conclude that for any smooth and compactly supported vector field $F:\mathbb{R}^d\to\mathbb{R}^d$ and $f\in C_c^\infty(U)$

$$\lim_{\varepsilon\to 0+}\int_U \nabla_x\chi^{(p,\varepsilon)}(x;\omega)\cdot F(x)dx=\langle\nabla\chi^{(p)}\rangle_{\mathbb{Q}}\cdot\int_{\mathbb{R}^d}F(x)dx=0 \quad (14.35)$$

and

$$\lim_{\varepsilon\to 0+}\int_U \tilde{a}^{(\varepsilon)}(x;\omega)\big[e_i+\nabla_x\chi^{(p,\varepsilon)}(x;\omega)\big]\cdot e_j f(x)dx=\bar{a}_{ij}\int_{\mathbb{R}^d}f(x)dx,\qquad \mathbb{Q}\text{-a.s.},$$

where coefficients $\bar{a}_{ij}$ are given by formula (9.55). By Theorem 11.18 we also know that

$$\lim_{\varepsilon\to 0+}\big\|\nabla_x\chi^{(p,\varepsilon)}(\cdot;\omega)\big\|^2_{L^2_d(U)}=|U|\,\big\|\nabla\chi^{(p)}\big\|^2_{\mathbb{Q}}\quad \mathbb{Q}\text{-a.s.}$$

Putting the above facts together we conclude that the hypotheses of Theorem 14.12 are met for $\mathbb{Q}$ a.s. ω. Therefore, Theorem 14.14 is a consequence of Theorem 14.12. □

We can also formulate the corresponding result for the semigroups associated with the operators L_ε^ω. Recall that $\{X_{t,\varepsilon}^{x,\omega},t\ge 0\}$, given by (14.2), is the diffusion corresponding to generator L_ε^ω that starts at x and is defined over a probability space $(\Sigma,\mathscr{A},\mathbb{P})$. We let $H_{x,U}^{(\varepsilon)}:=\min[t: X_{t,\varepsilon}^{x,\omega}\notin U]$ be the exit time of the diffusion from U. Here $H_{x,U}^{(\varepsilon)}:=\infty$ if the respective set is empty. Each operator L_ε^ω is the generator of a contraction semigroup on $L^2(U)$ given by a formula

$$P_{t,U}^{\omega,\varepsilon}f=E_{\mathbb{P}}\big[f\big(Z_{t,\varepsilon}^{x,\omega}\big),H_{x,U}^{(\varepsilon)}>t\big],\qquad f\in L^2(U),t\ge 0.$$

For any u belonging to the domain of L_0 we let $u_\varepsilon:=(L_\varepsilon^\omega)^{-1}f$, where $f=L_0u$. From G-convergence of L_ε^ω to L_0 we conclude that

$$u_\varepsilon\rightharpoonup u,\quad \text{weakly in } W^1_{2,o}(U).$$

Since $W^1_{2,o}(U)$ is compactly embedded in $L^2(U)$, see e.g. Theorem 7.22 of Gilbarg and Trudinger (1983), the above convergence holds in fact in the strong sense in $L^2(U)$. We can use then Theorem 1.6.1, p. 28 of Ethier and Kurtz (1986) to conclude the convergence of the respective semigroups and obtain the following:

Theorem 14.15 *For any $f\in L^2(U)$ we have*

$$\lim_{\varepsilon\to 0+}\int_U\big\{E_{\mathbb{P}}\big[f\big(X_{t,\varepsilon}^{x,\omega}\big),H_{x,U}^{(\varepsilon)}>t\big]-E_{\mathbb{P}}\big[f(x+w_t),H_{x,U}>t\big]\big\}^2dx=0$$

$\mathbb{Q}$ a.s. in ω. Here $\{w_t,t\ge 0\}$ is a d-dimensional Brownian motion with zero drift and covariance matrix $\bar{a}$, while $H_{x,U}$ denotes the exit time of $\{x+w_t,t\ge 0\}$ from U.

We conclude this section with the functional central limit theorem for diffusions $\{X_{t,\varepsilon}^{x,\omega}, t \ge 0\}$. Denote by $Q_\varepsilon^{x,\omega}$ the law of the diffusion over $C([0,+\infty);\mathbb{R}^d)$ and by $\bar{Q}^x$ the respective law of $\{x + w_t, t \ge 0\}$.

Theorem 14.16 *For any $x \in \mathbb{R}^d$ measures $Q_\varepsilon^{x,\omega}$ converge weakly to $\bar{Q}^x$, as $\varepsilon \to 0+$, $\mathbb{Q}$ a.s. in ω.*

Proof The proof is conducted in two steps. First, we prove that there exists a set N for which $\mathbb{Q}(N) = 0$ and such that for any $\omega \notin N$ the family $\{Q_\varepsilon^{x,\omega}, \varepsilon \in (0,1]\}$ is tight, as $\varepsilon \to 0+$. The latter means that for any sequence of $\varepsilon_n \to 0+$ the respective family of measures is tight. Then, we show that for any limiting measure $\bar{Q}$, integer $m \ge 1$, times $0 \le t_1 \le \cdots \le t_m < s \le t$ and functions $f, g_1, \ldots, g_m \in C_c^\infty(\mathbb{R}^d)$ we have

$$\int \left\{ \left[\mathscr{M}_t(f) - \mathscr{M}_s(f) \right] \prod_{i=1}^m g_i(\Pi_{t_i}) \right\} \bar{Q}(d\sigma) = 0. \tag{14.36}$$

Here $\Pi_t(\sigma) := \sigma(t)$ for any $\sigma \in C([0,+\infty);\mathbb{R}^d)$ and $t \ge 0$, and

$$\mathscr{M}_t(f) := f(\Pi_t) - f(\Pi_0) - \int_0^t L_0 f(\Pi_\rho) d\rho.$$

The operator L_0 is the G-limit of L_ε^ω, cf. Theorem 14.14. The above property identifies the limiting measure as the unique solution of the corresponding martingale problem, see Corollary 7.1.7 of Stroock and Varadhan (1979).

To show tightness we recall that, according to assumptions made about random matrix $a(\cdot)$, outside a set N of null $\mathbb{Q}$ measure the respective matrix valued function $\tilde{a}(x;\omega)$ belongs to $\mathbb{M}(m, M)$. Hence, the transition probability densities of $X_{t,\varepsilon}^{x,\omega}$ satisfy the following Gaussian upper bounds

$$p_{t,\varepsilon}^\omega(x,y) \le \frac{C_1}{t^{d/2}} \exp\left\{ -\frac{C_2|x-y|^2}{t} \right\}, \quad \forall t > 0, x, y \in \mathbb{R}^d, \tag{14.37}$$

see Theorem 1, p. 891 and Remark 5, p. 895 of Aronson (1967). The constants $C_1, C_2 > 0$ are independent of $\varepsilon \in (0,1]$ and $\omega \notin N$. From (14.37) we immediately conclude that

$$E_{\mathbb{P}} \left| X_{t,\varepsilon}^{x,\omega} - X_{s,\varepsilon}^{x,\omega} \right|^4 \le C_3(t-s)^2, \quad \forall t > s$$

for some constant $C_3 > 0$, independent of ε and ω. This, according to Theorem 12.3, p. 95 of Billingsley (1999), implies tightness of $\{Q_\varepsilon^{x,\omega}, \varepsilon \in (0,1]\}$. Suppose that $\bar{Q}$ is a limiting point of $Q_{\varepsilon_n}^{x,\omega}$ for some sequence $\varepsilon_n \to 0+$. For a given $R > 0$ and $|x| \le R$ we let $H_R := \min[t : |\Pi_t| \ge R]$, or equal $+\infty$ if the respective set is empty. Choose arbitrary ρ and $T > 0$. Using tightness we can find $R > 0$ large enough so that $Q_{\varepsilon_n}^{x,\omega}[H_R < T] < \rho/2$. From weak convergence we conclude that

$\bar{Q}[H_{R+1} < T] \le \rho/2$. Therefore for any $0 < s < t < T$:

$$\begin{aligned}
&\left|\int \left\{ [\mathscr{M}_t(f) - \mathscr{M}_s(f)] \prod_{i=1}^{n} g_i(\Pi_{t_i}) \right\} dQ_{\varepsilon_n}^{x,\omega} \right| \\
&\quad \le 3\|f\|_{C_b^2(\mathbb{R}^d)} Q_{\varepsilon_n}^{x,\omega}[H_{R+1} < T] \\
&\quad\quad + \left| \int_{[H_{R+1} \ge T]} \left\{ [\mathscr{M}_t(f) - \mathscr{M}_s(f)] \prod_{i=1}^{n} g_i(\Pi_{t_i}) \right\} dQ_{\varepsilon_n}^{x,\omega} \right|. \qquad (14.38)
\end{aligned}$$

The first term on the right-hand side is estimated by $(3/2)\|f\|_{C_b^2(\mathbb{R}^d)}\rho$ while the second equals

$$\left| \int_{[H_{R+1} \ge T]} \left\{ [N_n^{\omega}(t - t_m, \Pi_{t_m}) - N_n^{\omega}(s - t_m, \Pi_{t_m})] \prod_{i=1}^{m} g_i(\Pi_{t_i}) \right\} dQ_{\varepsilon_n}^{x,\omega} \right|. \qquad (14.39)$$

Here

$$N_n^{\omega}(t, x) := u_n(t, x; \omega) - f(x) - \int_0^t (L_0 u_n)(r, x; \omega) dr, \qquad (14.40)$$

and

$$u_n(t, x; \omega) := E_{\mathbb{P}}\big[f\big(X_{t,\varepsilon_n}^{x,\omega}\big), H_{x, B_{R+2}}^{(\varepsilon)} > t\big]. \qquad (14.41)$$

Function $u_n(\cdot)$ solves the parabolic initial-boundary value problem

$$\begin{aligned}
&\partial_t u_n = \frac{1}{2} \sum_{k,l=1}^{d} \partial_{x_k}\big(\tilde{a}_{k,l}^{(\varepsilon_n)} \partial_{x_l} u_n\big), \quad \text{in } (0, +\infty) \times B_{R+2}, \\
&u_n = 0, \quad \text{on } (0, +\infty) \times \partial B_{R+2}, \qquad (14.42) \\
&u_n(0, \cdot) = f(\cdot).
\end{aligned}$$

From (14.41) we conclude that

$$\|u_n\|_\infty \le \|f\|_\infty. \qquad (14.43)$$

Using Harnack estimates for solutions of parabolic partial differential equations, see Moser (1964), we conclude that there exists a constants $\alpha \in (0, 1)$ and C_* independent of n and ω, such that

$$\big|u_n(t, x) - u_n\big(t', x'\big)\big| \le C_*\big(|x - y|^{\alpha} + |t - s|^{\alpha/2}\big), \quad \forall (t, x), \big(t', x'\big) \in \mathscr{C}, \qquad (14.44)$$

where $\mathscr{C} := [s - t_m, t - t_m] \times \bar{B}_{R+1}$. Estimates (14.43) and (14.44) together imply that the family $\{u_n(\cdot, \cdot; \omega), n \ge 1\}$ is relatively compact in the uniform topology on the space of continuous functions on $\mathscr{C}$. Thanks to Theorem 14.15 we also know

$$\lim_{n \to +\infty} \big\|u_n(t) - \bar{u}(t)\big\|_{L^2(B_{R+2})} = 0, \quad \forall t > 0,$$

where

$$\bar{u}(t,x) := E_{\mathbb{P}}\big[f(x+w_t),\, H_{x,B_{R+2}} > t\big].$$

These two facts imply that

$$\lim_{n\to+\infty} u_n(t,x) = \bar{u}(t,x), \qquad \text{uniformly on } \mathscr{C}.$$

Hence, the upper limit of the expression (14.39), as $n \to +\infty$, can be estimated by

$$\lim_{n\to+\infty} \int_{[H_{R+1}\ge T]} \left| \big[\bar{N}(t-t_m, \Pi_{t_m}) - \bar{N}(s-t_m, \Pi_{t_m})\big] \prod_{i=1}^{m} g_i(\Pi_{t_i}) \right| d Q^{x,\omega}_{\varepsilon_n} = 0.$$

Here $\bar{N}(t,x)$ is given by an analogue of (14.40), where u_n is replaced by $\bar{u}$. Thanks to Theorem 14.15 we know that $\bar{N}(t,x) \equiv 0$. Thus, from estimate (14.38) we conclude that

$$\int \left\{ \big[\mathscr{M}_t(f) - \mathscr{M}_s(f)\big] \prod_{i=1}^{n} g_i(\Pi_{t_i}) \right\} d\bar{Q} = 0, \qquad \forall f \in C_c^\infty(\mathbb{R}^d).$$

This identifies the measure $\bar{Q}$ as the law of the Brownian motion $\{x + w_t, t \ge 0\}$ and ends the proof of the theorem. □

14.5 Appendix: Ellipticity of the Coefficient Matrix of a Coercive Form

Theorem 14.17 *Suppose that U is a region in $\mathbb{R}^d$ and $a : U \to \mathbb{M}(d)$ is a matrix valued function with measurable entries that satisfies*

$$\int_U a\nabla_x v \cdot \nabla_x v dx \ge m\|\nabla_x v\|^2_{L^2_d(U)}, \qquad \forall v \in W^1_{2,o}(U) \tag{14.45}$$

for some $m > 0$. Then

$$a(x)\xi \cdot \xi \ge m|\xi|^2 \quad \forall \xi \in \mathbb{R}^d,\ a.e.\ x \in U. \tag{14.46}$$

Proof Suppose first that $a(x) \equiv a$ is a constant matrix and $U = \mathbb{R}^d$. Then for any $v \in W^1_2(\mathbb{R}^d)$ the left-hand side of (14.45) equals

$$\sum_{p,q=1}^{d} a_{pq} \int k_p k_q \big|\hat{v}(k)\big|^2 dk \ge m \int |k|^2 \big|\hat{v}(k)\big|^2 dk, \qquad \forall v \in W^1_2\big(\mathbb{R}^d\big).$$

Here $\hat{v}(k)$ is the Fourier transform of $v(x)$. This clearly implies that

$$\sum_{p,q=1}^{d} a_{pq} k_p k_q \ge m|k|^2, \qquad \forall k \in \mathbb{R}^d.$$

It suffices only to notice that the set $[|\hat{v}(k)|^2, v \in W_2^1(\mathbb{R}^d)]$ contains all non-negative, compactly supported, continuous, even functions.

Next, we assume that U is an arbitrary region. With no loss of generality we may suppose that $0 \in U$. Note that if (14.45) holds for this region then it also remains true for any nU, where $n \geq 1$. The conclusion then follows from the already proved result for $U = \mathbb{R}^d$ and the fact that $\bigcup_{n\geq 1} W_{2,o}^1(nU)$ is dense in $W_2^1(\mathbb{R}^d)$.

Assume now that the entries of $a(x)$ are continuous. Let us fix $z \in U$ and suppose that $\varepsilon > 0$ is arbitrary. We can choose $\delta > 0$ sufficiently small so that $B_\delta(z) \subset U$ and

$$\begin{aligned}\int_{B_\delta(z)} & a(z)\nabla_x v \cdot \nabla_x v dx \\ & \geq m\|\nabla_x v\|^2_{L^2_d(B_\delta(z))} - \int_{B_\delta(z)} \left|a(x) - a(z)\right|\left|\nabla_x v(x)\right|^2 dx \\ & \geq (m - \varepsilon)\|\nabla_x v\|^2_{L^2_d(B_\delta(z))}, \quad \forall v \in W^1_{2,o}\big(B_\delta(z)\big).\end{aligned}$$

Using the conclusion of the theorem obtained for constant coefficients we conclude that

$$a(z)\xi \cdot \xi \geq m|\xi|^2, \quad \forall \xi \in \mathbb{R}^d.$$

To finish the proof we extend the result to a matrix valued function with measurable entries. Let

$$a_R(x) := |B_R|^{-1} \int_{B_R} a(x+y)dy$$

on

$$U_R := \big[x \in U : \mathrm{dist}(x, \partial U) \geq 2R\big].$$

Then (14.45) holds for $a_R(x)$ and v from $W^1_{2,o}(U_R)$. Since the entries of the matrix are continuous we can use the result proved so far and obtain that $a_R(x)\xi \cdot \xi \geq m|\xi|^2$ for all $\xi \in \mathbb{R}^d$ and $x \in U_R$. Letting $R \to 0+$ we conclude that $a(x)$ satisfies lower bound (14.46). □

14.6 Comments and References

The notion of G-convergence (from the convergence of Green's functions) was introduced in Spagnolo (1967) and Spagnolo (1968) to study compactness properties of solutions of linear parabolic and elliptic second order differential equations in divergence form whose matrix of coefficients is symmetric, uniformly bounded and elliptic.

Γ-convergence of quadratic functionals considered in Sect. 14.2 is the special case of an abstract notion of Γ-convergence introduced in De Giorgi and Franzoni (1975), see also De Giorgi (1975). The connection between the G-convergence of symmetric elliptic operators in a divergence form and Γ-convergence of their quadratic forms has been established in Sbordone (1975). The fact that the G-convergence of generators implies the convergence in law of the respective processes has been observed in Papanicolaou and Varadhan (1981). In that paper it is assumed that the generators are self-adjoint but the argument can be used in the general case.

An application of the compensated compactness method and extension of convergence results to non-symmetric, second order, elliptic operators in divergence form is given in Murat (1977) (see also Murat and Tartar, 1997), Simon (1979); Tartar (1977, 1978). In the first three articles the term H-convergence (from homogenization) is used in case of the convergence of matrix valued functions, instead of the G-convergence used here. The div-curl lemma used in Sect. 14.3 comes from Murat (1978).

An extensive discussion concerning the G-convergence for solutions of elliptic and parabolic equations of an arbitrary order can be found in a series of papers Ngoan (1977a,b); Zhikov (1983a,b); Zhikov et al. (1979, 1981). Our proof of Theorem 14.11 follows closely the argument contained in Chap. 5 of the monograph Zhikov et al. (1994).

The choice of the material presented in this chapter is not intended to be a review of techniques used in modern analytic homogenization theory. Its main purpose is to inform the reader, with some probabilistic background, about the relationship between the results that can be obtained with probabilistic tools and their counterparts that can be shown by analytic methods. Likewise, the account of literature, concerning analytic aspects of homogenization theory, presented above is far from being complete. We direct the reader to the monographs (Bensoussan et al., 1978; Cioranescu and Donato, 1999; Dal Maso, 1993), and the already mentioned (Zhikov et al., 1994) for more information concerning this subject.

References

Aronson DG (1967) Bounds for the fundamental solution of a parabolic equation. Bull Am Math Soc 73:890–896

Bensoussan A, Lions JL, Papanicolaou G (1978) Asymptotic analysis for periodic structures. Studies in mathematics and its applications, vol 5. North-Holland, Amsterdam

Billingsley P (1999) Convergence of probability measures. Wiley series in probability and mathematical statistics: probability and mathematical statistics, 2nd edn. Wiley, New York. A Wiley-Interscience Publication.

Cioranescu D, Donato P (1999) An introduction to homogenization. Oxford lecture series in mathematics and its applications, vol 17. Clarendon Press, Oxford University Press, New York

Dal Maso G (1993) An introduction to Γ-convergence. Progress in nonlinear differential equations and their applications, vol 8. Birkhäuser Boston, Boston

De Giorgi E (1975) Sulla convergenza di alcune successioni d'integrali del tipo dell'area. Rend Mat (6) 8:277–294. Collection of articles dedicated to Mauro Picone on the occasion of his ninetieth birthday

De Giorgi E, Franzoni T (1975) Su un tipo di convergenza variazionale. Atti Accad Naz Lincei, Rend Cl Sci Fis Mat Nat (8) 58(6):842–850

Ethier SN, Kurtz TG (1986) Markov processes: characterization and convergence. Wiley series in probability and mathematical statistics: probability and mathematical statistics. Wiley, New York

Evans LC (1990) Weak convergence methods for nonlinear partial differential equations. CBMS regional conference series in mathematics, vol 74. Published for the conference board of the mathematical sciences, Washington

Gilbarg D, Trudinger NS (1983) Elliptic partial differential equations of second order. Grundlehren der Mathematischen Wissenschaften [Fundamental principles of mathematical sciences], vol 224, 2nd edn. Springer, Berlin

Moser J (1964) A Harnack inequality for parabolic differential equations. Commun Pure Appl Math 17:101–134

Murat F (1977) H-convergence. Seminaire d'analyse fonctionnelle et numerique de l'universite d'Alger

Murat F (1978) Compacité par compensation. Ann Sc Norm Super Pisa, Cl Sci (4) 5(3):489–507

Murat F, Tartar L (1997) H-convergence. In: Topics in the mathematical modelling of composite materials. Progr nonlinear differential equations appl, vol 31. Birkhäuser Boston, Boston, pp 21–43

Ngoan HT (1977a) Convergence of the solutions of boundary value problems for a sequence of elliptic equations. Usp Mat Nauk 32(3(195)):183–184

Ngoan HT (1977b) Convergence of the solutions of boundary value problems for a sequence of elliptic systems. Vestn Mosk Univ Ser I Mat Meh 5:83–92

Papanicolaou GC, Varadhan SRS (1981) Boundary value problems with rapidly oscillating random coefficients. In: Random fields, vols I, II, Esztergom, 1979. Colloq math soc János Bolyai, vol 27. North-Holland, Amsterdam, pp 835–873

Sbordone C (1975) Su alcune applicazioni di un tipo di convergenza variazionale. Ann Sc Norm Super Pisa, Cl Sci (4) 2(4):617–638

Simon L (1979) On G-convergence of elliptic operators. Indiana Univ Math J 28(4):587–594

Spagnolo S (1967) Sul limite delle soluzioni di problemi di Cauchy relativi all'equazione del calore. Ann Sc Norm Super Pisa, Cl Sci (3) 21:657–699

Spagnolo S (1968) Sulla convergenza di soluzioni di equazioni paraboliche ed ellittiche. Ann Sc Norm Super Pisa, Cl Sci (3) 22(1968):571–597. Errata, Ann Sc Norm Super Pisa, Cl Sci (3) **22**:673

Stroock DW, Varadhan SRS (1979) Multidimensional diffusion processes. Grundlehren der Mathematischen Wissenschaften [Fundamental principles of mathematical sciences], vol 233. Springer, Berlin

Tartar L (1977) Cours Peccot au collège de France. Paris

Tartar L (1978) Quelques remarques sur homogeneisation. In: Proc of the Japan-France seminar 1976, functional analysis and numerical analysis. Japan Society for the Promotion of Science, Tokyo

Zhikov VV (1983a) Asymptotic behavior and stabilization of solutions of a second-order parabolic equation with lowest terms. Tr Mosk Mat Obŝ 46:69–98

Zhikov VV (1983b) G-convergence of elliptic operators. Mat Zametki 33(3):345–356

Zhikov VV, Kozlov SM, Oleĭnik OA, Ngoan HT (1979) Averaging and G-convergence of differential operators. Usp Mat Nauk 34(5(209)):65–133, 256

Zhikov VV, Kozlov SM, Oleĭnik OA (1981) G-convergence of parabolic operators. Usp Mat Nauk 36(1(217)):11–58, 248

Zhikov VV, Kozlov SM, Oleĭnik OA (1994) Homogenization of differential operators and integral functionals. Springer, Berlin. Translated from the Russian by GA Yosifian [GA Iosif'yan]

References

Adler RJ (1990) An introduction to continuity, extrema, and related topics for general Gaussian processes. IMS lect notes monogr ser, vol 12. Inst Math Stat, Hayward

Alexander FJ, Lebowitz JL (1990) Driven diffusive systems with a moving obstacle: a variation on the Brazil nuts problem. J Phys A, Math Gen 23:L375–L381

Alexander FJ, Lebowitz JL (1994) On the drift and diffusion of a rod in a lattice fluid. J Phys A, Math Gen 27:683–696

Anshelevich VV, Khanin KM, Sinaĭ YG (1982) Symmetric random walks in random environments. Commun Math Phys 85(3):449–470

Aronson DG (1967) Bounds for the fundamental solution of a parabolic equation. Bull Am Math Soc 73:890–896

Arratia R (1983) The motion of a tagged particle in the simple symmetric exclusion system on $\mathbf{Z}$. Ann Probab 11(2):362–373

Aubin JP (1979) Applied functional analysis: with exercises by Bernard Cornet and Jean-Michel Lasry. Wiley, New York/Chichester/Brisbane. Translated from the French by Carole Labrousse

Avellaneda M, Majda AJ (1990) Mathematical models with exact renormalization for turbulent transport. Commun Math Phys 131(2):381–429

Avellaneda M, Majda AJ (1991) An integral representation and bounds on the effective diffusivity in passive advection by laminar and turbulent flows. Commun Math Phys 138(2):339–391

Avellaneda M, Majda A (1992a) Mathematical models with exact renormalization for turbulent transport. II. Fractal interfaces, non-Gaussian statistics and the sweeping effect. Commun Math Phys 146(1):139–204

Avellaneda M, Majda AJ (1992b) Superdiffusion in nearly stratified flows. J Stat Phys 69(3–4):689–729

Barlow MT (2004) Random walks on supercritical percolation clusters. Ann Probab 32(4):3024–3084

Barlow MT, Deuschel JD (2010) Invariance principle for the random conductance model with unbounded conductances. Ann Probab 38(1):234–276

Batchelor GK (1982) The theory of homogeneous turbulence. Cambridge University Press, Cambridge. Reprint, Cambridge Science Classics.

van Beijeren H, Kutner R, Spohn H (1985) Excess noise for driven diffusive systems. Phys Rev Lett 54:2026–2029

Beltrán J (2005) Regularity of diffusion coefficient for nearest neighbor asymmetric simple exclusion on $\mathbb{Z}$. Stoch Process Appl 115(9):1451–1474

Ben Arous G, Owhadi H (2002) Super-diffusivity in a shear flow model from perpetual homogenization. Commun Math Phys 227(2):281–302

Benabou G (2006) Homogenization of Ornstein-Uhlenbeck process in random environment. Commun Math Phys 266(3):699–714

T. Komorowski et al., *Fluctuations in Markov Processes*, Grundlehren der mathematischen Wissenschaften 345,

DOI 10.1007/978-3-642-29880-6,

Benois O, Esposito R, Marra R (2003) Equilibrium fluctuations for lattice gases. Ann Inst Henri Poincaré Probab Stat 39(5):743–777

Bensoussan A, Lions JL, Papanicolaou G (1978) Asymptotic analysis for periodic structures. Studies in mathematics and its applications, vol 5. North-Holland, Amsterdam

Berger N, Biskup M (2007) Quenched invariance principle for simple random walk on percolation clusters. Probab Theory Relat Fields 137(1–2):83–120

Bernardin C (2002) Regularity of the diffusion coefficient for lattice gas reversible under Bernoulli measures. Stoch Process Appl 101(1):43–68

Bernardin C (2004) Fluctuations in the occupation time of a site in the asymmetric simple exclusion process. Ann Probab 32(1B):855–879

Bertini L, Toninelli C (2004) Exclusion processes with degenerate rates: convergence to equilibrium and tagged particle. J Stat Phys 117(3–4):549–580

Bezuidenhout C, Grimmett G (1999) A central limit theorem for random walks in random labyrinths. Ann Inst Henri Poincaré Probab Stat 35(5):631–683

Billingsley P (1995) Probability and measure. Wiley series in probability and mathematical statistics: probability and mathematical statistics, 3rd edn. Wiley, New York. A Wiley-Interscience Publication

Billingsley P (1999) Convergence of probability measures. Wiley series in probability and mathematical statistics: probability and mathematical statistics, 2nd edn. Wiley, New York. A Wiley-Interscience Publication.

Biskup M, Prescott TM (2007) Functional CLT for random walk among bounded random conductances. Electron J Probab 12(49):1323–1348 (electronic)

Blumenthal RM, Getoor RK (1968) Markov processes and potential theory. Pure and applied mathematics, vol 29. Academic Press, New York

Bolthausen E (1980) The Berry-Esseen theorem for functionals of discrete Markov chains. Z Wahrscheinlichkeitstheor Verw Geb 54(1):59–73

Bolthausen E (1982) The Berry-Esseén theorem for strongly mixing Harris recurrent Markov chains. Z Wahrscheinlichkeitstheor Verw Geb 60(3):283–289

Breiman L (1968) Probability. Addison-Wesley, Reading

Brox T, Rost H (1984) Equilibrium fluctuations of stochastic particle systems: the role of conserved quantities. Ann Probab 12(3):742–759

Campillo F, Kleptsyna M, Piatnitski A (2001) Homogenization of random parabolic operator with large potential. Stoch Process Appl 93(1):57–85

Caputo P, Ioffe D (2003) Finite volume approximation of the effective diffusion matrix: the case of independent bond disorder. Ann Inst Henri Poincaré Probab Stat 39(3):505–525

Carlson JM, Grannan ER, Swindle GH (1993a) A limit theorem for tagged particles in a class of self-organizing particle systems. Stoch Process Appl 47(1):1–16

Carlson JM, Grannan ER, Swindle GH, Tour J (1993b) Singular diffusion limits of a class of reversible self-organizing particle systems. Ann Probab 21(3):1372–1393

Carmona RA, Xu L (1997) Homogenization for time-dependent two-dimensional incompressible Gaussian flows. Ann Appl Probab 7(1):265–279

Chang CC (1994) Equilibrium fluctuations of gradient reversible particle systems. Probab Theory Relat Fields 100(3):269–283

Chang CC, Yau HT (1992) Fluctuations of one-dimensional Ginzburg-Landau models in nonequilibrium. Commun Math Phys 145(2):209–234

Chang CC, Landim C, Olla S (2001) Equilibrium fluctuations of asymmetric simple exclusion processes in dimension $d \geq 3$. Probab Theory Relat Fields 119(3):381–409

Chatterjee S, Diaconis P, Meckes E (2005) Exchangeable pairs and Poisson approximation. Probab Surv 2:64–106 (electronic)

Cheeger J (1970) A lower bound for the smallest eigenvalue of the Laplacian. In: Problems in analysis (Papers dedicated to Salomon Bochner, 1969). Princeton University Press, Princeton, pp 195–199

Cherkaev AV, Gibiansky LV (1994) Variational principles for complex conductivity, viscoelasticity, and similar problems in media with complex moduli. J Math Phys 35(1):127–145

Chung KL, Walsh JB (2005) Markov processes, Brownian motion, and time symmetry. Grundlehren der Mathematischen Wissenschaften [Fundamental principles of mathematical sciences], vol 249, 2nd edn. Springer, New York

Cioranescu D, Donato P (1999) An introduction to homogenization. Oxford lecture series in mathematics and its applications, vol 17. Clarendon Press, Oxford University Press, New York

Csanady GT (1973) Turbulent diffusion in the environment. Geophysics and astrophysics monographs, vol 3. Kluwer Academic, Dordrecht, Boston

Dacunha-Castelle D, Duflo M (1986) Probability and statistics, vol II. Springer, New York. Translated from the French by David McHale

Dal Maso G (1993) An introduction to Γ-convergence. Progress in nonlinear differential equations and their applications, vol 8. Birkhäuser Boston, Boston

De Giorgi E (1975) Sulla convergenza di alcune successioni d'integrali del tipo dell'area. Rend Mat (6) 8:277–294. Collection of articles dedicated to Mauro Picone on the occasion of his ninetieth birthday

De Giorgi E, Franzoni T (1975) Su un tipo di convergenza variazionale. Atti Accad Naz Lincei, Rend Cl Sci Fis Mat Nat (8) 58(6):842–850

De Masi A, Presutti E, Spohn H, Wick WD (1986) Asymptotic equivalence of fluctuation fields for reversible exclusion processes with speed change. Ann Probab 14(2):409–423

De Masi A, Ferrari PA, Goldstein S, Wick WD (1989) An invariance principle for reversible Markov processes. Applications to random motions in random environments. J Stat Phys 55(3–4):787–855

Derriennic Y, Lin M (1996) Sur le théorème limite central de Kipnis et Varadhan pour les chaînes réversibles ou normales. C R Acad Sci Paris Sér I Math 323(9):1053–1057

Derriennic Y, Lin M (2001a) The central limit theorem for Markov chains with normal transition operators, started at a point. Probab Theory Relat Fields 119(4):508–528

Derriennic Y, Lin M (2001b) Fractional Poisson equations and ergodic theorems for fractional coboundaries. Isr J Math 123:93–130

Derriennic Y, Lin M (2003) The central limit theorem for Markov chains started at a point. Probab Theory Relat Fields 125(1):73–76

Deuschel JD, Kösters H (2008) The quenched invariance principle for random walks in random environments admitting a bounded cycle representation. Ann Inst Henri Poincaré Probab Stat 44(3):574–591

Diaconis P, Holmes S (eds) (2004) Stein's method: expository lectures and applications. IMS lect notes monogr ser, vol 46. Inst Math Stat, Beachwood. Papers from the workshop on Stein's method held at Stanford University, Stanford, CA, 1998

Doeblin W (1938) Sur deux problèmes de M Kolmogoroff concernant les chaînes dénombrables. Bull Soc Math Fr 66:210–220

Dolgopyat D, Keller G, Liverani C (2008) Random walk in Markovian environment. Ann Probab 36(5):1676–1710

Doukhan P (1994) Mixing: properties and examples. Lecture notes in statistics, vol 85. Springer, New York

Durrett R (1996) Probability: theory and examples, 2nd edn. Duxbury Press, Belmont

Esposito R, Marra R, Yau HT (1994) Diffusive limit of asymmetric simple exclusion. Rev Math Phys 6(5A):1233–1267. Special issue dedicated to Elliott H Lieb

Esposito R, Marra R, Yau HT (1996) Navier-Stokes equations for stochastic particle systems on the lattice. Commun Math Phys 182(2):395–456

Ethier SN, Kurtz TG (1986) Markov processes: characterization and convergence. Wiley series in probability and mathematical statistics: probability and mathematical statistics. Wiley, New York

Evans LC (1990) Weak convergence methods for nonlinear partial differential equations. CBMS regional conference series in mathematics, vol 74. Published for the conference board of the mathematical sciences, Washington

Faggionato A, Martinelli F (2003) Hydrodynamic limit of a disordered lattice gas. Probab Theory Relat Fields 127(4):535–608

Fannjiang A (1998) Boundary value problem with unbounded, fast oscillatory random flows. J Differ Equ 143(2):327–359

Fannjiang AC (2000) Phase diagram for turbulent transport: sampling drift, eddy diffusivity and variational principles. Physica D 136(1–2):145–174

Fannjiang A, Komorowski T (1997) A martingale approach to homogenization of unbounded random flows. Ann Probab 25(4):1872–1894

Fannjiang A, Komorowski T (1999a) An invariance principle for diffusion in turbulence. Ann Probab 27(2):751–781

Fannjiang A, Komorowski T (1999b) Turbulent diffusion in Markovian flows. Ann Appl Probab 9(3):591–610

Fannjiang A, Komorowski T (2000a) Fractional Brownian motions and enhanced diffusion in a unidirectional wave-like turbulence. J Stat Phys 100(5–6):1071–1095

Fannjiang A, Komorowski T (2000b) Fractional Brownian motions in a limit of turbulent transport. Ann Appl Probab 10(4):1100–1120

Fannjiang A, Komorowski T (2001/2002) Diffusive and nondiffusive limits of transport in nonmixing flows. SIAM J Appl Math 62(3):909–923 (electronic)

Fannjiang A, Komorowski T (2002a) Diffusion in long-range correlated Ornstein-Uhlenbeck flows. Electron J Probab 7(20):22 (electronic)

Fannjiang AC, Komorowski T (2002b) Correction: "An invariance principle for diffusion in turbulence" [Ann Probab **27**(2):751–781 (1999); MR1698963 (2001e:60069)]. Ann Probab 30(1):480–482

Fannjiang A, Papanicolaou G (1994) Convection enhanced diffusion for periodic flows. SIAM J Appl Math 54(2):333–408

Fannjiang A, Papanicolaou G (1996) Diffusion in turbulence. Probab Theory Relat Fields 105(3):279–334

Feller W (1971) An introduction to probability theory and its applications, vol II, 2nd edn. Wiley, New York

Ferrari PL, Spohn H (2006) Scaling limit for the space-time covariance of the stationary totally asymmetric simple exclusion process. Commun Math Phys 265(1):1–44

Franke B (2007) A functional non-central limit theorem for jump-diffusions with periodic coefficients driven by stable Lévy-noise. J Theor Probab 20(4):1087–1100

Friedman A (1964) Partial differential equations of parabolic type. Prentice-Hall, Englewood Cliffs

Friedman A (1975) Stochastic differential equations and applications, vol 1. Probability and mathematical statistics, vol 28. Academic Press [Harcourt Brace Jovanovich Publishers], New York

Frisch U (1995) Turbulence: the legacy of AN Kolmogorov. Cambridge University Press, Cambridge

Gilbarg D, Trudinger NS (1983) Elliptic partial differential equations of second order. Grundlehren der Mathematischen Wissenschaften [Fundamental principles of mathematical sciences], vol 224, 2nd edn. Springer, Berlin

Goldstein S (1995) Antisymmetric functionals of reversible Markov processes. Ann Inst Henri Poincaré Probab Stat 31(1):177–190

Goldys B, Maslowski B (2006) Lower estimates of transition densities and bounds on exponential ergodicity for stochastic PDE's. Ann Probab 34(4):1451–1496

Gonçalves P, Jara M (2008) Scaling limits of a tagged particle in the exclusion process with variable diffusion coefficient. J Stat Phys 132(6):1135–1143

Gordin MI (1969) The central limit theorem for stationary processes. Dokl Akad Nauk SSSR 188:739–741

Gordin MI, Lifšic BA (1978) Central limit theorem for stationary Markov processes. Dokl Akad Nauk SSSR 239(4):766–767

Häggström O, Rosenthal JS (2007) On variance conditions for Markov chain CLTs. Electron Commun Probab 12:454–464 (electronic)

Hairer M, Pavliotis GA (2004) Periodic homogenization for hypoelliptic diffusions. J Stat Phys 117(1–2):261–279

Helland IS (1982) Central limit theorems for martingales with discrete or continuous time. Scand J Stat 9(2):79–94

Horváth I, Tóth B, Vető B (2010) Diffusive limits for "true" (or myopic) self-avoiding random walks and self-repellent Brownian polymers in $d \geq 3$. arXiv:1009.0401

Ibragimov IA, Linnik YV (1971) Independent and stationary sequences of random variables: with a supplementary chapter by IA Ibragimov and VV Petrov. Wolters-Noordhoff Publishing, Groningen. Translation from the Russian edited by JFC Kingman

Iftimie B, Pardoux É, Piatnitski A (2008) Homogenization of a singular random one-dimensional PDE. Ann Inst Henri Poincaré Probab Stat 44(3):519–543

Ikeda N, Watanabe S (1989) Stochastic differential equations and diffusion processes. North-Holland mathematical library, vol 24, 2nd edn. North-Holland, Amsterdam

Il'in AM, Kalashnikov AS, Oleĭnik OA (2002) Second-order linear equations of parabolic type. J Math Sci 108:435–452

Jacod J, Shiryaev AN (1987) Limit theorems for stochastic processes. Grundlehren der Mathematischen Wissenschaften [Fundamental principles of mathematical sciences], vol 288. Springer, Berlin

Janson S (1997) Gaussian Hilbert spaces. Cambridge tracts in mathematics, vol 129. Cambridge University Press, Cambridge

Jara M (2006) Finite-dimensional approximation for the diffusion coefficient in the simple exclusion process. Ann Probab 34(6):2365–2381

Jara M (2009) Nonequilibrium scaling limit for a tagged particle in the simple exclusion process with long jumps. Commun Pure Appl Math 62(2):198–214

Jara MD, Landim C (2006) Nonequilibrium central limit theorem for a tagged particle in symmetric simple exclusion. Ann Inst Henri Poincaré Probab Stat 42(5):567–577

Jara MD, Landim C (2008) Quenched non-equilibrium central limit theorem for a tagged particle in the exclusion process with bond disorder. Ann Inst Henri Poincaré Probab Stat 44(2):341–361

Karatzas I, Shreve SE (1991) Brownian motion and stochastic calculus. Graduate texts in mathematics, vol 113, 2nd edn. Springer, New York

Karr AF (1983) Extreme points of certain sets of probability measures, with applications. Math Oper Res 8(1):74–85

Kesavan S (1989) Topics in functional analysis and applications. Wiley, New York

Kipnis C (1986) Central limit theorems for infinite series of queues and applications to simple exclusion. Ann Probab 14(2):397–408

Kipnis C (1987) Fluctuations des temps d'occupation d'un site dans l'exclusion simple symétrique. Ann Inst Henri Poincaré Probab Stat 23(1):21–35

Kipnis C, Landim C (1999) Scaling limits of interacting particle systems. Grundlehren der Mathematischen Wissenschaften [Fundamental principles of mathematical sciences], vol 320. Springer, Berlin

Kipnis C, Varadhan SRS (1986) Central limit theorem for additive functionals of reversible Markov processes and applications to simple exclusions. Commun Math Phys 104(1):1–19

Komoriya K (1998) Hydrodynamic limit for asymmetric mean zero exclusion processes with speed change. Ann Inst Henri Poincaré Probab Stat 34(6):767–797

Komorowski T, Krupa G (2004) On stationarity of Lagrangian observations of passive tracer velocity in a compressible environment. Ann Appl Probab 14(4):1666–1697

Komorowski T, Krupa G (2006) On the asymptotic behavior of an Ornstein-Uhlenbeck process with random forcing. Commun Math Phys 261(2):517–543

Komorowski T, Nieznaj E (2008) On superdiffusive behavior of a passive tracer in a random flow. Physica D 237(24):3377–3381

Komorowski T, Olla S (2001) On homogenization of time-dependent random flows. Probab Theory Relat Fields 121(1):98–116

Komorowski T, Olla S (2002) On the superdiffusive behavior of passive tracer with a Gaussian drift. J Stat Phys 108(3–4):647–668

Komorowski T, Olla S (2003a) A note on the central limit theorem for two-fold stochastic random walks in a random environment. Bull Pol Acad Sci, Math 51(2):217–232

Komorowski T, Olla S (2003b) On the sector condition and homogenization of diffusions with a Gaussian drift. J Funct Anal 197(1):179–211

Komorowski T, Ryzhik L (2007a) On asymptotics of a tracer advected in a locally self-similar, correlated flow. Asymptot Anal 53(3):159–187

Komorowski T, Ryzhik L (2007b) Passive tracer in a slowly decorrelating random flow with a large mean. Nonlinearity 20(5):1215–1239

Kontoyiannis I, Meyn SP (2003) Spectral theory and limit theorems for geometrically ergodic Markov processes. Ann Appl Probab 13(1):304–362

Koralov L (1999) Transport by time dependent stationary flow. Commun Math Phys 199(3):649–681

Koralov L (2004) Random perturbations of 2-dimensional Hamiltonian flows. Probab Theory Relat Fields 129(1):37–62

Kozlov SM (1979) The averaging of random operators. Mat Sb (NS) 109 (151):188–202, 327

Kozlov SM (1985) The averaging method and walks in inhomogeneous environments. Usp Mat Nauk 40(2(242)):61–120, 238

Krengel U (1985) Ergodic theorems: with a supplement by Antoine Brunel. de Gruyter studies in mathematics, vol 6. de Gruyter, Berlin

Künnemann R (1983) The diffusion limit for reversible jump processes on $\mathbf{Z}^d$ with ergodic random bond conductivities. Commun Math Phys 90(1):27–68

Ladyženskaja OA, Solonnikov VA, Ural'ceva NN (1968) Lineinye i kvazilineinye uravneniya parabolicheskogo tipa. Nauka, Moscow

Landim C (1998) Decay to equilibrium in L^∞ of finite interacting particle systems in infinite volume. Markov Process Relat Fields 4(4):517–534. I Brazilian School in Probability, Rio de Janeiro, 1997

Landim C, Yau HT (1997) Fluctuation-dissipation equation of asymmetric simple exclusion processes. Probab Theory Relat Fields 108(3):321–356

Landim C, Olla S, Yau HT (1996) Some properties of the diffusion coefficient for asymmetric simple exclusion processes. Ann Probab 24(4):1779–1808

Landim C, Olla S, Yau HT (1997) First-order correction for the hydrodynamic limit of asymmetric simple exclusion processes in dimension $d \geq 3$. Commun Pure Appl Math 50(2):149–203

Landim C, Olla S, Volchan SB (1998a) Driven tracer particle in one-dimensional symmetric simple exclusion. Commun Math Phys 192(2):287–307

Landim C, Olla S, Yau HT (1998b) Convection-diffusion equation with space-time ergodic random flow. Probab Theory Relat Fields 112(2):203–220

Landim C, Olla S, Varadhan SRS (2001) Symmetric simple exclusion process: regularity of the self-diffusion coefficient. Commun Math Phys 224(1):307–321. Dedicated to Joel L Lebowitz

Landim C, Olla S, Varadhan SRS (2002) Finite-dimensional approximation of the self-diffusion coefficient for the exclusion process. Ann Probab 30(2):483–508

Landim C, Olla S, Varadhan SRS (2004a) On viscosity and fluctuation-dissipation in exclusion processes. J Stat Phys 115(1–2):323–363

Landim C, Quastel J, Salmhofer M, Yau HT (2004b) Superdiffusivity of asymmetric exclusion process in dimensions one and two. Commun Math Phys 244(3):455–481

Landim C, Milanés A, Olla S (2008) Stationary and nonequilibrium fluctuations in boundary driven exclusion processes. Markov Process Relat Fields 14(2):165–184

Lawler GF (1982) Weak convergence of a random walk in a random environment. Commun Math Phys 87(1):81–87

Lawler GF, Sokal AD (1988) Bounds on the L^2 spectrum for Markov chains and Markov processes: a generalization of Cheeger's inequality. Trans Am Math Soc 309(2):557–580

Lebowitz JL, Rost H (1994) The Einstein relation for the displacement of a test particle in a random environment. Stoch Process Appl 54(2):183–196

Ledoux M, Talagrand M (1991) Probability in Banach spaces: isoperimetry and processes. Ergebnisse der Mathematik und ihrer Grenzgebiete (3) [Results in mathematics and related areas (3)], vol 23. Springer, Berlin

Lejay A (2001) Homogenization of divergence-form operators with lower-order terms in random media. Probab Theory Relat Fields 120(2):255–276
Leoni G (2009) A first course in Sobolev spaces. Graduate studies in mathematics, vol 105. Am Math Soc, Providence
Levin DA, Peres Y, Wilmer EL (2009) Markov chains and mixing times: with a chapter by James G Propp and David B Wilson. Am Math Soc, Providence
Lezaud P (2001) Chernoff and Berry-Esséen inequalities for Markov processes. ESAIM Probab Stat 5:183–201 (electronic)
Li CM, Mao MC (2008) An upper bound on the occupation time for a simple exclusion process. Teor Mat Fiz 156(1):147–158
Liggett TM (1985) Interacting particle systems. Grundlehren der Mathematischen Wissenschaften [Fundamental principles of mathematical sciences], vol 276. Springer, New York
Loulakis M (2002) Einstein relation for a tagged particle in simple exclusion processes. Commun Math Phys 229(2):347–367
Loulakis M (2005) Mobility and Einstein relation for a tagged particle in asymmetric mean zero random walk with simple exclusion. Ann Inst Henri Poincaré Probab Stat 41(2):237–254
Lu SL (1994) Equilibrium fluctuations of a one-dimensional nongradient Ginzburg-Landau model. Ann Probab 22(3):1252–1272
Majda AJ, Kramer PR (1999) Simplified models for turbulent diffusion: theory, numerical modelling, and physical phenomena. Phys Rep 314(4–5):237–574
Matheron G, DeMarsily G (1980) Is transport in porous media always diffusive? A counterexample. Water Resou Bull 16:901–917
Mathieu P (2006) Carne-Varopoulos bounds for centered random walks. Ann Probab 34(3):987–1011
Mathieu P (2008) Quenched invariance principles for random walks with random conductances. J Stat Phys 130(5):1025–1046
Mathieu P, Piatnitski A (2007) Quenched invariance principles for random walks on percolation clusters. Proc R Soc Lond, Ser A, Math Phys Eng Sci 463(2085):2287–2307
Maxwell M, Woodroofe M (2000) Central limit theorems for additive functionals of Markov chains. Ann Probab 28(2):713–724
Monin AS, Yaglom AM (2007) Statistical fluid mechanics: mechanics of turbulence, vol II: Edited and with a preface by John L Lumley. Dover, Mineola. Translated from the 1965 Russian original, English edition updated, augmented and revised by the authors, reprinted from the 1975 edition
Moser J (1964) A Harnack inequality for parabolic differential equations. Commun Pure Appl Math 17:101–134
Murat F (1977) H-convergence. Seminaire d'analyse fonctionnelle et numerique de l'universite d'Alger
Murat F (1978) Compacité par compensation. Ann Sc Norm Super Pisa, Cl Sci (4) 5(3):489–507
Murat F, Tartar L (1997) H-convergence. In: Topics in the mathematical modelling of composite materials. Progr nonlinear differential equations appl, vol 31. Birkhäuser Boston, Boston, pp 21–43
Nagaev SV (1957) Some limit theorems for stationary Markov chains. Teor Veroâtn Ee Primen 2:389–416
Nagahata Y (2005) Regularity of the diffusion coefficient matrix for the lattice gas with energy. Ann Inst Henri Poincaré Probab Stat 41(1):45–67
Nagahata Y (2006) Regularity of the diffusion coefficient matrix for generalized exclusion process. Stoch Process Appl 116(6):957–982
Nagahata Y (2007) Regularity of the diffusion coefficient matrix for lattice gas reversible under Gibbs measures with mixing condition. Commun Math Phys 273(3):637–650
Newman CM (1980) Normal fluctuations and the FKG inequalities. Commun Math Phys 74(2):119–128
Newman CM (1983) A general central limit theorem for FKG systems. Commun Math Phys 91(1):75–80

Newman CM (1984) Asymptotic independence and limit theorems for positively and negatively dependent random variables. In: Inequalities in statistics and probability, Lincoln, NE, 1982. IMS lect notes monogr ser, vol 5. Inst Math Stat, Hayward, pp 127–140

Newman CM, Wright AL (1981) An invariance principle for certain dependent sequences. Ann Probab 9(4):671–675

Newman CM, Wright AL (1982) Associated random variables and martingale inequalities. Z Wahrscheinlichkeitstheor Verw Geb 59(3):361–371

Ngoan HT (1977a) Convergence of the solutions of boundary value problems for a sequence of elliptic equations. Usp Mat Nauk 32(3(195)):183–184

Ngoan HT (1977b) Convergence of the solutions of boundary value problems for a sequence of elliptic systems. Vestn Mosk Univ Ser I Mat Meh 5:83–92

Oelschläger K (1988) Homogenization of a diffusion process in a divergence-free random field. Ann Probab 16(3):1084–1126

Øksendal B (2003) Stochastic differential equations: an introduction with applications. Universitext, 6th edn. Springer, Berlin

Olla S (1994a) Homogenization of diffusion processes in random fields. Centre de mathématiques appliquées, ecole polytechnique. Available at http://www.ceremade.dauphine.fr/~olla/lho.ps

Olla S (1994b) Lectures on homogenization of diffusion processes in random fields. Publications de l'école doctorale de l'école polytechnique

Olla S, Tremoulet C (2003) Equilibrium fluctuations for interacting Ornstein-Uhlenbeck particles. Commun Math Phys 233(3):463–491

Osada H (1983) Homogenization of diffusion processes with random stationary coefficients. In: Probability theory and mathematical statistics, Tbilisi, 1982. Lecture notes in math, vol 1021. Springer, Berlin, pp 507–517

Osada H (1998a) An invariance principle for Markov processes and Brownian particles with singular interaction. Ann Inst Henri Poincaré Probab Stat 34(2):217–248

Osada H (1998b) Positivity of the self-diffusion matrix of interacting Brownian particles with hard core. Probab Theory Relat Fields 112(1):53–90

Osada H, Saitoh T (1995) An invariance principle for non-symmetric Markov processes and reflecting diffusions in random domains. Probab Theory Relat Fields 101(1):45–63

Ouchti L, Volný D (2008) A conditional CLT which fails for ergodic components. J Theor Probab 21(3):687–703

Owhadi H (2003) Approximation of the effective conductivity of ergodic media by periodization. Probab Theory Relat Fields 125(2):225–258

Owhadi H (2004) Averaging versus chaos in turbulent transport? Commun Math Phys 247(3):553–599

Papanicolaou G, Varadhan SRS (1985) Ornstein-Uhlenbeck process in a random potential. Commun Pure Appl Math 38(6):819–834

Papanicolaou GC, Varadhan SRS (1981) Boundary value problems with rapidly oscillating random coefficients. In: Random fields, vols I, II, Esztergom, 1979. Colloq math soc János Bolyai, vol 27. North-Holland, Amsterdam, pp 835–873

Papanicolaou GC, Varadhan SRS (1982) Diffusions with random coefficients. In: Statistics and probability: essays in honor of CR Rao. North-Holland, Amsterdam, pp 547–552

Pardoux E, Piatnitski A (2006) Homogenization of a singular random one dimensional PDE. In: Multi scale problems and asymptotic analysis. GAKUTO internat ser math sci appl, vol 24. Gakkōtosho, Tokyo, pp 291–303

Peligrad M, Sethuraman S (2008) On fractional Brownian motion limits in one dimensional nearest-neighbor symmetric simple exclusion. ALEA Lat Am J Probab Math Stat 4:245–255

Peligrad M, Utev S (2005) A new maximal inequality and invariance principle for stationary sequences. Ann Probab 33(2):798–815

Quastel J (1992) Diffusion of color in the simple exclusion process. Commun Pure Appl Math 45(6):623–679

Quastel J (2006) Bulk diffusion in a system with site disorder. Ann Probab 34(5):1990–2036

Quastel J, Valko B (2007) $t^{1/3}$ superdiffusivity of finite-range asymmetric exclusion processes on $\mathbb{Z}$. Commun Math Phys 273(2):379–394

Quastel J, Jankowski H, Sheriff J (2002) Central limit theorem for zero-range processes. Methods Appl Anal 9(3):393–406. Special issue dedicated to Daniel W Stroock and Srinivasa SR Varadhan on the occasion of their 60th birthday

Rassoul-Agha F, Seppäläinen T (2005) An almost sure invariance principle for random walks in a space-time random environment. Probab Theory Relat Fields 133(3):299–314

Rassoul-Agha F, Seppäläinen T (2008) An almost sure invariance principle for additive functionals of Markov chains. Stat Probab Lett 78(7):854–860

Ravishankar K (1992) Fluctuations from the hydrodynamical limit for the symmetric simple exclusion in $\mathbf{Z}^d$. Stoch Process Appl 42(1):31–37

Rebolledo R (1980) Central limit theorems for local martingales. Z Wahrscheinlichkeitstheor Verw Geb 51(3):269–286

Rhodes R (2006) Homogénéisation en milieu aléatoire. These de l'universite de provence, ecole doctorale de mathématiques et informatique de Marseille

Rhodes R (2007) On homogenization of space-time dependent and degenerate random flows. Stoch Process Appl 117(10):1561–1585

Rhodes R (2008) On homogenization of space-time dependent and degenerate random flows. II. Ann Inst Henri Poincaré Probab Stat 44(4):673–692

Rhodes R, Vargas V (2009) Scaling limits for symmetric Itô-Lévy processes in random medium. Stoch Process Appl 119(12):4004–4033

Riesz F, Sz.-Nagy B (1990) Functional analysis. Dover books on advanced mathematics. Dover, New York. Translated from the second French edition by Leo F Boron, reprint of the 1955 original

Rodenhausen H (1989) Einstein's relation between diffusion constant and mobility for a diffusion model. J Stat Phys 55(5–6):1065–1088

Rogers LCG, Williams D (2000) Diffusions, Markov processes, and martingales, vol 1: foundations. Cambridge mathematical library. Cambridge University Press, Cambridge. Reprint of the second 1994 edition

Rost H, Vares ME (1985) Hydrodynamics of a one-dimensional nearest neighbor model. In: Particle systems, random media and large deviations, Brunswick, Maine, 1984. Contemp math, vol 41. Am Math Soc, Providence, pp 329–342

Rozanov YA (1967) Stationary random processes. Holden-Day, San Francisco. Translated from the Russian by A Feinstein

Saada E (1987) A limit theorem for the position of a tagged particle in a simple exclusion process. Ann Probab 15(1):375–381

Sasada M (2010) Hydrodynamic limit for two-species exclusion processes. Stoch Process Appl 120(4):494–521

Sbordone C (1975) Su alcune applicazioni di un tipo di convergenza variazionale. Ann Sc Norm Super Pisa, Cl Sci (4) 2(4):617–638

Schmitz T (2006) Diffusions in random environment and ballistic behavior. Ann Inst Henri Poincaré Probab Stat 42(6):683–714

Schmitz T (2009) On the equivalence of the static and dynamic points of view for diffusions in a random environment. Stoch Process Appl 119(8):2501–2522

Sellami S (1999) Equilibrium density fluctuations of a one-dimensional non-gradient reversible model: the generalized exclusion process. Markov Process Relat Fields 5(1):21–51

Seppäläinen T, Sethuraman S (2003) Transience of second-class particles and diffusive bounds for additive functionals in one-dimensional asymmetric exclusion processes. Ann Probab 31(1):148–169

Sethuraman S (2000) Central limit theorems for additive functionals of the simple exclusion process. Ann Probab 28(1):277–302

Sethuraman S (2003) An equivalence of H_{-1} norms for the simple exclusion process. Ann Probab 31(1):35–62

Sethuraman S (2006a) Correction: "Central limit theorems for additive functionals of the simple exclusion process" [Ann Probab **28**(1):277–302 (2000); mr1756006]. Ann Probab 34(1):427–428

Sethuraman S (2006b) Superdiffusivity of occupation-time variance in 2-dimensional asymmetric exclusion processes with density $\rho = 1/2$. J Stat Phys 123(4):787–802

Sethuraman S (2007) On diffusivity of a tagged particle in asymmetric zero-range dynamics. Ann Inst Henri Poincaré Probab Stat 43(2):215–232

Sethuraman S, Xu L (1996) A central limit theorem for reversible exclusion and zero-range particle systems. Ann Probab 24(4):1842–1870

Sethuraman S, Varadhan SRS, Yau HT (2000) Diffusive limit of a tagged particle in asymmetric simple exclusion processes. Commun Pure Appl Math 53(8):972–1006

Shen L (2002) Asymptotic properties of certain anisotropic walks in random media. Ann Appl Probab 12(2):477–510

Shen L (2003) On ballistic diffusions in random environment. Ann Inst Henri Poincaré Probab Stat 39(5):839–876

Shiga T (1988) Tagged particle motion in a clustered random walk system. Stoch Process Appl 30(2):225–252

Sidoravicius V, Sznitman AS (2004) Quenched invariance principles for walks on clusters of percolation or among random conductances. Probab Theory Relat Fields 129(2):219–244

Simon L (1979) On G-convergence of elliptic operators. Indiana Univ Math J 28(4):587–594

Soardi PM (1994) Potential theory on infinite networks. Lecture notes in math, vol 1590. Springer, Berlin

Spagnolo S (1967) Sul limite delle soluzioni di problemi di Cauchy relativi all'equazione del calore. Ann Sc Norm Super Pisa, Cl Sci (3) 21:657–699

Spagnolo S (1968) Sulla convergenza di soluzioni di equazioni paraboliche ed ellittiche. Ann Sc Norm Super Pisa, Cl Sci (3) 22(1968):571–597. Errata, Ann Sc Norm Super Pisa, Cl Sci (3) **22**:673

Spitzer F (1970) Interaction of Markov processes. Adv Math 5:246–290

Spohn H (1986) Equilibrium fluctuations for interacting Brownian particles. Commun Math Phys 103(1):1–33

Stroock DW, Varadhan SRS (1972) On the support of diffusion processes with applications to the strong maximum principle. In: Proceedings of the sixth Berkeley symposium on mathematical statistics and probability, vol III: Probability theory. Univ California, Berkeley, California, 1970/1971, University of California Press, Berkeley, pp 333–359

Stroock DW, Varadhan SRS (1979) Multidimensional diffusion processes. Grundlehren der Mathematischen Wissenschaften [Fundamental principles of mathematical sciences], vol 233. Springer, Berlin

Sued M (2003) PhD dissertation, IMPA. http://www.preprint.impa.br/Shadows/SERIE_C/2003/22.html

Sued M (2005) Regularity properties of the diffusion coefficient for a mean zero exclusion process. Ann Inst Henri Poincaré Probab Stat 41(1):1–33

Szász D, Tóth B (1987a) A dynamical theory of Brownian motion for the Rayleigh gas. In: Proceedings of the symposium on statistical mechanics of phase transitions—mathematical and physical aspects, Trebon, 1986, vol 47, pp 681–693

Szász D, Tóth B (1987b) Towards a unified dynamical theory of the Brownian particle in an ideal gas. Commun Math Phys 111(1):41–62

Sznitman AS (2004) Topics in random walks in random environment. In: School and conference on probability theory. ICTP lect notes, vol XVII. Abdus Salam Int Cent Theoret Phys, Trieste, pp 203–266 (electronic)

Sznitman AS, Zeitouni O (2006) An invariance principle for isotropic diffusions in random environment. Invent Math 164(3):455–567

Tanemura H (1989) Ergodicity for an infinite particle system in $\mathbf{R}^d$ of jump type with hard core interaction. J Math Soc Jpn 41(4):681–697

Tanemura H (1993) Central limit theorem for a random walk with random obstacles in $\mathbb{R}^d$. Ann Probab 21(2):936–960

Tartar L (1977) Cours Peccot au collège de France. Paris

Tartar L (1978) Quelques remarques sur homogeneisation. In: Proc of the Japan-France seminar 1976, functional analysis and numerical analysis. Japan Society for the Promotion of Science, Tokyo

Toninelli C, Biroli G (2004) Dynamical arrest, tracer diffusion and kinetically constrained lattice gases. J Stat Phys 117(1–2):27–54

Tóth B (1986) Persistent random walks in random environment. Probab Theory Relat Fields 71(4):615–625

Tóth B, Valko B (2010) Superdiffusive bounds on self-repellent Brownian polymers and diffusion in the curl of the Gaussian free field in $d = 2$. arXiv:1012.5698

Varadhan SRS (1994a) Nonlinear diffusion limit for a system with nearest neighbor interactions II. In: Asymptotic problems in probability theory, stochastic models and diffusions on fractals. Pitman res notes math ser, vol 283. Wiley, New York, pp 75–128

Varadhan SRS (1994b) Regularity of self-diffusion coefficient. In: The Dynkin festschrift. Progr probab, vol 34. Birkhäuser Boston, Boston, pp 387–397

Varadhan SRS (1995) Self-diffusion of a tagged particle in equilibrium for asymmetric mean zero random walk with simple exclusion. Ann Inst Henri Poincaré Probab Stat 31(1):273–285

Varadhan SRS (2001) Probability theory. Courant lecture notes in mathematics, vol 7. New York University, Courant Institute of Mathematical Sciences, New York

von Weizsäcker H, Winkler G (1979/1980) Integral representation in the set of solutions of a generalized moment problem. Math Ann 246(1):23–32

von Weizsäcker H, Winkler G (1980) Noncompact extremal integral representations: some probabilistic aspects. In: Functional analysis: surveys and recent results, II. Proc second conf functional anal, Univ. Paderborn, Paderborn, 1979. Notas mat, vol 68. North-Holland, Amsterdam, pp 115–148

Whitt W (2007) Proofs of the martingale FCLT. Probab Surv 4:268–302

Winkler G (1988) Extreme points of moment sets. Math Oper Res 13(4):581–587

Wu L (1999) Forward-backward martingale decomposition and compactness results for additive functionals of stationary ergodic Markov processes. Ann Inst Henri Poincaré Probab Stat 35(2):121–141

Wu WB, Woodroofe M (2004) Martingale approximations for sums of stationary processes. Ann Probab 32(2):1674–1690

Xu L (1993) PhD dissertation, Courant Institute, New York University

Yaglom AM (1962) An introduction to the theory of stationary random functions. Prentice-Hall, Englewood Cliffs. Revised English edition. Translated and edited by Richard A Silverman

Yau HT (2004) $(\log t)^{2/3}$ law of the two dimensional asymmetric simple exclusion process. Ann Math (2) 159(1):377–405

Yosida K (1995) Functional analysis. Classics in mathematics. Springer, Berlin. Reprint of the sixth 1980 edition

Zeidler E (1995) Applied functional analysis: main principles and their applications. Applied mathematical sciences, vol 109. Springer, New York

Zeitouni O (2004) Random walks in random environment. In: Lectures on probability theory and statistics. Lecture notes in math, vol 1837. Springer, Berlin, pp 189–312

Zhang Q, Glimm J (1992) Inertial range scaling of laminar shear flow as a model of turbulent transport. Commun Math Phys 146(2):217–229

Zhao O, Woodroofe M (2008a) Law of the iterated logarithm for stationary processes. Ann Probab 36(1):127–142

Zhao O, Woodroofe M (2008b) On martingale approximations. Ann Appl Probab 18(5):1831–1847

Zhikov VV (1983a) Asymptotic behavior and stabilization of solutions of a second-order parabolic equation with lowest terms. Tr Mosk Mat Obŝ 46:69–98

Zhikov VV (1983b) G-convergence of elliptic operators. Mat Zametki 33(3):345–356

Zhikov VV, Kozlov SM, Oleĭnik OA, Ngoan HT (1979) Averaging and G-convergence of differential operators. Usp Mat Nauk 34(5(209)):65–133, 256

Zhikov VV, Kozlov SM, Oleĭnik OA (1981) G-convergence of parabolic operators. Usp Mat Nauk 36(1(217)):11–58, 248

Zhikov VV, Kozlov SM, Oleĭnik OA (1994) Homogenization of differential operators and integral functionals. Springer, Berlin. Translated from the Russian by GA Yosifian [GA Iosif'yan]

Notation

T. Komorowski et al., *Fluctuations in Markov Processes*,
Grundlehren der mathematischen Wissenschaften 345,
DOI 10.1007/978-3-642-29880-6,

Subject Index

T. Komorowski et al., *Fluctuations in Markov Processes*,
Grundlehren der mathematischen Wissenschaften 345,
DOI 10.1007/978-3-642-29880-6, © Springer-Verlag Berlin Heidelberg 2012

CPSIA information can be obtained at www.ICGtesting.com
Printed in the USA
LVOW01*1621130715

446008LV00013B/389/P

9 783642 298790